THE LIBRARY
ST. MARY'S COLLEGE OF MARYLAND
ST. MARY'S CITY, MARYLAND 20686

D1651661

TRANSFER RNA:
Structure,
Properties,
and Recognition

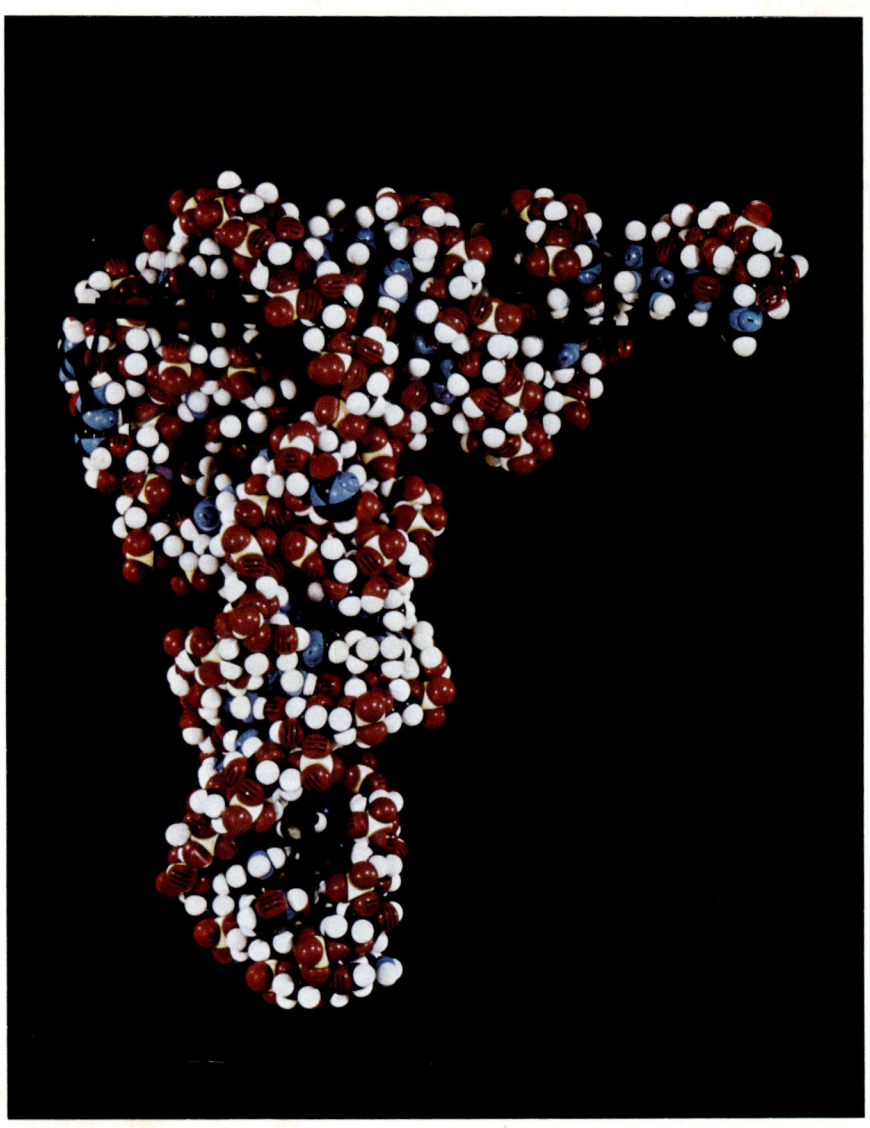
Space-filling model of tRNA^{Phe}. (Photo courtesy of S.-H. Kim, this volume.)

TRANSFER RNA: Structure, Properties, and Recognition

Edited by

Paul R. Schimmel
Massachusetts Institute of Technology

Dieter Söll
Yale University

John N. Abelson
University of California, San Diego

Cold Spring Harbor Laboratory
1979

**COLD SPRING HARBOR
MONOGRAPH SERIES**

The Lactose Operon [1]
The Bacteriophage Lambda [2]
The Molecular Biology of Tumour Viruses [3]
Ribosomes [4]
RNA Phages [5]
RNA Polymerase [6]
The Operon [7]
The Single-Stranded DNA Phages [8]
Transfer RNA (*in 2 parts*):
  Structure, Properties, and Recognition, 9A
  Biological Aspects, 9B

TRANSFER RNA: Structure, Properties, and Recognition
© 1979 by Cold Spring Harbor Laboratory
All rights reserved
Printed in the United States of America
Book design by Emily Harste

Library of Congress Cataloging in Publication Data
Main entry under title:

Transfer RNA: structure, properties, and recognition

  (Cold Spring Harbor monograph series ; 9A)
  Based on lectures presented at a meeting held in Aug. 1978 at Cold Spring Harbor, N. Y.
  Includes indexes.
  1. Ribonucleic acid, Transfer--Congresses.
I. Schimmel, Paul R. II. Söll, Dieter III. Abelson, John N. IV. Series.
QP623.T72   574.8'734   79-17143
ISBN 0-87969-128-X

Contents

Preface, xi

Primary Structures and Sequencing of tRNAs

Recent Developments in tRNA Sequencing Methods as Applied to Analyses of Mitochondrial tRNAs, 3
U. L. RajBhandary, J. E. Heckman, S. Yin, B. Alzner-DeWeerd, and E. Ackerman

The Primary Structure of tRNAs and Their Rare Nucleosides, 19
G. Dirheimer, G. Keith, A.-P. Sibler, and R. P. Martin

[^{3}H]Borohydride: A Versatile Reagent for the Analysis of tRNA—Methods and Applications, 43
K. Randerath, R. C. Gupta, and E. Randerath

Modified Nucleosides in tRNA, 59
S. Nishimura

Crystal Structure Analysis of tRNAs

Crystal Structure of Yeast tRNA^{Phe} and General Structural Features of Other tRNAs, 83
S.-H. Kim

Recent Progress in tRNA Structural Analysis, 101
A. Rich, G. J. Quigley, M. M. Teeter, A. Decruix, and N. Woo

The Structure, Conformation, and Interaction of tRNA, 115
M. Sundaralingam

A Crystallographic Analysis of Yeast Initiator tRNA, 133
R. W. Schevitz, A. D. Podjarny, N. Krishnamachari, J. J. Hughes, and P. B. Sigler

The Structure of Baker's Yeast tRNA^{Gly}: A Second tRNA Conformation, 145
H. T. Wright, P. C. Manor, K. Beurling, R. L. Karpel, and J. R. Fresco

tRNA Structure and Dynamics in Solution

Physical Studies of tRNA in Solution, 163
D. M. Crothers

High-resolution NMR Studies on tRNA Structure in Solution, 177
B. R. Reid and R. E. Hurd

Proton FT NMR Studies of tRNA Structure and Dynamics, 191
P. D. Johnston and A. G. Redfield

Effects of Aminoacylation and Solution Conditions on the Structure of tRNA, 207
R. O. Potts, C.-C. Wang, D. C. Fritzinger, N. C. Ford, Jr., and M. J. Fournier

Aminoacyl-tRNA Synthetases

Structural Studies of Aminoacyl-tRNA Synthetases, 223
B. S. Hartley

Molecular Enzymology of Beef Pancreas Tryptophanyl-tRNA Synthetase, 235
L. L. Kisselev, O. O. Favorova, and G. K. Kovaleva

Editing Mechanisms in the Aminoacylation of tRNA, 247
A. R. Fersht

The Tryptophanyl- and Tyrosyl-tRNA Synthetases from *Bacillus stearothermophilus*, 255
G. Winter, G. L. E. Koch, A. Dell, and B. S. Hartley

Mechanism of Aminoacyl-tRNA Synthetases: Recognition and Proofreading Processes, 267
F. Cramer, F. von der Haar, and G. L. Igloi

Methionyl-tRNA Synthetase from *Escherichia coli*: Structure-Function Relationships of a Dimeric Enzyme with Repeated Sequences, 281
S. Blanquet, P. Dessen, and G. Fayat

Recognition of tRNAs by Proteins

Similarities in the Structural Organization of Complexes of tRNAs with Aminoacyl-tRNA Synthetases and the Mechanism of Recognition, 297
P. R. Schimmel

Chemical Approaches to the Study of Protein-tRNA Recognition, 311
L. H. Schulman

Interaction between tRNA and Aminoacyl-tRNA Synthetase in the Valine and Phenylalanine Systems from Yeast, 325
J.-P. Ebel, M. Renaud, A. Dietrich, F. Fasiolo, G. Keith, O. O. Favorova, S. Vassilenko, M. Baltzinger, R. Ehrlich, P. Remy, J. Bonnet, and R. Giegé

2'-OH vs 3'-OH Specificity in tRNA Aminoacylation, 345
S. M. Hecht

tRNA Interactions with Ribosomes

tRNA-Ribosome Interactions, 363
C. R. Cantor

Ribosome Structure and tRNA Binding Sites, 393
J. A. Lake

Affinity Labeling of tRNA Binding Sites on Ribosomes, 413
E. Kuechler and J. Ofengand

Studies on tRNA Conformation and Ribosome Interaction with Fluorescent tRNA Derivatives, 445
W. Wintermeyer, J. M. Robertson, H. Weidner, and H. G. Zachau

Codon-induced Structural Transitions in tRNA, 459
A. Möller, U. Manderschied, R. Lipecky, S. Bertram, M. Schmitt, and H. G. Gassen

Role of the 2',3'-Isomerization of Aminoacyl-tRNA during Ribosomal Protein Synthesis, 473
M. Sprinzl and T. Wagner

Analogs of Lysyl-tRNA as Probes of Ribosome and Elongation Factor Tu Structure and Function, 487
A. E. Johnson

The Relationship of the Accuracy of Aminoacyl-tRNA Synthesis to That of Translation, 501
M. Yarus

Appendices

Appendix I Proposed Numbering System of Nucleotides in tRNAs Based on Yeast tRNA^{Phe}, 518

   A. Compilation of tRNA Sequences, 520
      D. H. Gauss, F. Grüter, and M. Sprinzl

      Table 1 Names of Some Rare Nucleosides and Citations Regarding Their Identification, 521
      Table 2 Compilation of Known tRNA Sequences, 522

   B. Collection of Mutant tRNA Sequences, 539
      J. E. Celis

      Table 3 Mutant Suppressor tRNAs^{Tyr}, 540
      Table 4 Mutant T4 tRNA^{Gln} (psu^+2, 0531) and tRNA^{Ser} (psu^+1, 1631), 543
      Table 5 Mutant T4 tRNA^{Ser}, tRNA^{Gln}, and tRNA^{Leu} Sequenced at the Precursor Level, 544

Appendix II Structures of Modified Nucleosides Found in tRNA, 547
S. Nishimura

Appendix III Chromatographic Mobilities of Modified Nucleotides, 551
S. Nishimura

Appendix IV Characteristics of Aminoacyl-tRNA Synthetases, 553
D. Söll and P. R. Schimmel

  Table 1 Molecular Weight, Quarternary Structure, and Substrate Binding Sites of Highly Purified Aminoacyl-tRNA Synthetases, 554
  Table 2 Aminoacyl-tRNA Synthetase Mutants, 560

Author Index, 565

Subject Index, 567

Preface

During recent years, tRNA has been the subject of intense investigation for several reasons. First, it plays a central role in protein synthesis and it is a major challenge to sort out the complexities and diversity of interactions it undergoes in this process. Second, it is a relatively small, single-stranded RNA (approximate m.w. of 25,000) that can be analyzed by structural methods, including sequencing and X-ray diffraction, that are often not feasible with larger, more complex RNA molecules. In this regard, it is hoped that analysis of the structural features of tRNA will be helpful in gaining insight into the structural organization of other RNA molecules, such as rRNAs and mRNAs. Third, there are a number of reactions in which tRNAs are specifically recognized by proteins so that, with a well-characterized nucleic acid structure available, these systems are excellent for exploring the molecular basis of specific protein–nucleic acid interactions. Fourth, it is now well established that tRNAs play a role in the regulation of gene expression and in a number of other cellular processes, an observation that points to the striking versatility of the molecule. And finally, because they are relatively well characterized and because of their obvious biological importance, tRNAs have become popular for studies of gene organization, cloning, and biosynthesis.

In this volume, we treat tRNA structure, properties, and recognition. There are six sections that cover these subjects: primary structure and sequencing of tRNAs, crystal structure analysis of tRNAs, tRNA structure and dynamics in solution, aminoacyl-tRNA synthetases, recognition of tRNAs by proteins, and tRNA interactions with ribosomes. These areas are the foundations of the tRNA field. For example, our knowledge of primary structures of tRNAs, and of the crystal structure of a particular tRNA, has provided a concrete framework within which we may rationalize experimental data and pose new questions. This structural information has also enabled us to understand in considerable depth, particularly through the use of nuclear magnetic resonance, the behavior of tRNA in solution and to design experiments that explore subtle conformational rearrangements that may have physiological consequences. These studies of isolated tRNA molecules are, of course, a prerequisite for approaching the more difficult questions of how tRNAs react with proteins and participate in protein synthesis. Here we consider the aminoacyl-tRNA synthetases, which are the most prominent class of enzymes that utilize tRNA as a substrate; the recognition of tRNAs by specific, well-defined proteins; and the interactions of tRNAs with ribosomes.

The purpose of the volume is to summarize major facts and concepts, rather than to communicate the latest preliminary, unpublished observations. With this in mind, a large number of summary figures and tables are included in some of the articles. Also, many of the articles have included extensive bibliographies that are invaluable reference sources. In addition, a special feature of this volume are the four appendices that give the primary structures of over 100 tRNAs, structural information on modified nucleotides found in tRNAs, chromatographic mobilities of modified nucleotides, and a tabulation of the characteristics of aminoacyl-tRNA synthetases. These appendices, and many of the articles, are intended to be used as a primary reference or handbook for the fundamental aspects of tRNA.

Transfer RNA: Biological Aspects, the second portion of the Transfer RNA set, summarizes current knowledge in the areas of tRNA biosynthesis, organization of tRNA genes, suppression and coding, and involvement of tRNA in regulatory processes. The volume concisely summarizes and organizes this material to give an overall perspective of ongoing research and to set the stage for future investigations.

The chapters for both texts were solicited by the editors from individuals who delivered invited lectures at the August 1978 Cold Spring Harbor Laboratory meeting on tRNA. Many of the chapters have been updated to early 1979. We wish to thank our colleagues for their help and suggestions.

In the organization of the meeting, we were helped enormously by Gladys Kist and Winifred Modzeleski of the Cold Spring Harbor Laboratory Meetings Office. This volume was made possible only through the conscientious and skillful editorial help of Mary-Teresa Halpin, Annette Zaninovic, and Nancy Ford, Director of Publications. We are also indebted to Jim Watson for his enthusiastic advice and counsel.

Paul R. Schimmel
Dieter Söll
John N. Abelson

Primary Structures
and Sequencing of tRNAs

Recent Developments in tRNA Sequencing Methods as Applied to Analyses of Mitochondrial tRNAs

Uttam L. RajBhandary, Joyce E. Heckman, Samuel Yin, and Birgit Alzner-DeWeerd
Department of Biology
Massachusetts Institute of Technology
Cambridge, Massachusetts 02139

Eric Ackerman
Department of Biochemistry and Biophysics
University of Chicago
Chicago, Illinois 60637

With the extensive literature that has already accumulated on tRNAs, it may not be evident that this is a relatively new area of research. Discovered only about 22 years ago (Hoagland et al. 1957), following the prediction of the existence of molecules with similar properties by Crick, the first sequence of a tRNA and indeed of any nucleic acid was published only in 1965 (Holley et al. 1965). Since then the sequence of at least 120 different tRNAs from a variety of biological sources have been established (Sprinzl et al. 1978). The two factors that have contributed most to this rapid progress in tRNA sequencing have been the development (1) of column chromatographic and gel electrophoretic methods (Gillam et al. 1967; Cherayil and Bock 1965; Pearson et al. 1971; Ikemura and Dahlberg 1973) suitable for purification of tRNAs and (2) of rapid sequencing methods requiring only very small amounts of tRNAs. Thus, whereas the first sequence analysis of a tRNA, which utilized spectrophotometric procedures for identification of nucleotides (Holley 1968), required several hundred milligrams to a gram of purified tRNA and several years of effort, methods currently available enable one to sequence a tRNA with just a few micrograms (2–10 μg) within a few weeks.

The first few tRNAs to be sequenced were all from yeast (Holley et al. 1965; Zachau et al. 1966; Madison et al. 1966; RajBhandary et al. 1967b; Baev et al. 1967). To a large extent this was a reflection of yeast being a relatively inexpensive and convenient source for the large amounts of tRNAs necessary then for sequencing. With the development of the two-dimensional electrophoretic method for the rapid separation of oligonucleotides and a rapid method for sequencing them (Sanger et al. 1965), tRNA sequence analysis could be carried out on material uniformly labeled with ^{32}P and most of the early applications of this method were in fact on *Escherichia coli* tRNAs that could be easily labeled in vivo with ^{32}P (Goodman et al. 1968). More recent developments have focused on methods involving the use of in vitro ^{32}P labeling for sequencing tRNAs that are available only in small quantity and cannot be labeled efficiently in vivo (Silberklang et al. 1979). This paper describes the basic principles behind such a method (details are available in a review to be published soon by Silberklang et al. [1979] and are not described here) and discusses some surprising features in the four *Neurospora crassa* mitochondrial tRNAs (mttRNAs) that have been sequenced using this method. Randerath et al. (this volume) describe a method utilizing ^{3}H labeling for the analysis and identification of modified nucleosides and for tRNA sequencing. This latter method is based on the specific labeling of 2′,3′-diol end groups in nucleosides, oligonucleotides, and RNAs (RajBhandary et al. 1967a; RajBhandary 1968) by treatment of the RNA with sodium periodate followed by reduction with [^{3}H]borohydride.

With the availability of so many tRNA sequences it is now possible to compare these sequences and to look for the overall structural features and sequences that have been conserved in tRNAs. This had led to the finding that the cloverleaf structure originally proposed by Holley et al. (1965) as one of the possible secondary structures for tRNA is a feature common to all tRNAs sequenced to date (Rich and RajBhandary 1976). In addition, tRNAs contain several "invariant" and "semiinvariant" residues located in the same relative position in all tRNAs. Largely through the work of Nishimura and coworkers (see Nishimura 1978), a substantial number of modified nucleosides present in tRNA have now been characterized. A summary of the overall general features of tRNA structures, including a list of the modified nucleosides, their occurrence in various tRNAs, and their location at specific sites within tRNAs, is provided by Dirheimer et al. (this volume). These workers also describe their recent work on yeast mttRNAs. Nishimura (this volume) describes recent results on some modified nucleosides, their characterization, and their biosynthesis.

USE OF IN VITRO ^{32}P LABELING FOR SEQUENCING tRNAs

The basic procedure used in our laboratory for the sequence analysis of a nonradioactive tRNA using in vitro ^{32}P labeling is summarized in Table 1 and consists of four steps. These are:

1. Analysis of modified nucleotide content of the tRNA.
2. Analysis of oligonucleotides present in complete T1 and pancreatic RNase digests.

Table 1 Basic steps in the sequence analysis of nonradioactive tRNA using in vitro ^{32}P labeling

---

Analysis of modified nucleotide content
1. Complete digestion of tRNA with T2 RNase to Np.
2. 5'-^{32}P labeling of Np to yield ^{32}pNp.
3. Conversion of ^{32}pNp to ^{32}pN using the 3'-phosphatase activity present in nuclease P1.
4. Identification of ^{32}pN by two-dimensional thin-layer chromatography.

Analysis of oligonucleotides present in T1 or pancreatic RNase digests
1. Complete digestion of tRNA with T1 or pancreatic RNase.
2. Removal of 3'-terminal phosphomonoester group by treatment with alkaline phosphatase.
3. 5'-^{32}P labeling of oligonucleotides.
4. Separation of 5'-^{32}P-labeled oligonucleotides by two-dimensional electrophoresis.
5. Sequencing of 5'-^{32}P-labeled oligonucleotides:
   a. 5'-end group analysis.
   b. Partial digestion with snake venom phosphodiesterase and/or nuclease P1 and analysis by two-dimensional homochromatography.

Alignment of these oligonucleotides (above) by isolation and sequencing of large oligonucleotide fragments
1. Controlled digestion with T1 or pancreatic RNase or specific chemical cleavage.
2. 5'-^{32}P labeling of the large fragments.
3. Separation of the 5'-^{32}P-labeled large fragments by two-dimensional gel electrophoresis and their sequencing as below.

Direct sequence analysis of ^{32}P end-labeled tRNAs or large fragments
1. End group labeling of tRNAs:
   a. 5' end using polynucleotide kinase and [γ-^{32}P]ATP.
   b. 3' end using tRNA nucleotidyl transferase and [α-^{32}P]ATP.
2. Partial digestion of end-labeled tRNAs with nuclease P1 and analysis by two-dimensional homochromatography.
3. Partial digestion with specific enzymes and analysis by polyacrylamide gel electrophoresis.

---

3. Alignment of oligonucleotides present in these complete digests into a unique sequence by isolation and sequencing of large oligonucleotide fragments.
4. Direct sequence analysis of 5'- or 3'-end-labeled [^{32}P]tRNAs.

In most instances, steps 1, 2, and 4 provide the information necessary for the derivation of the total sequence. In such cases, step 3, which requires relatively larger amounts (50 μg or so) of tRNAs, is dispensed with.

Szekely and Sanger (1969) demonstrated that oligonucleotides with a free 5'-OH end that are present in digests of nonradioactive nucleic acids could be labeled to high specific activity using T4 polynucleotide kinase and [γ-^{32}P]ATP. With certain modifications and improvements, we have used this approach for labeling with ^{32}P the 5' ends of both mononucleotides present in complete tRNA hydrolysates (Silberklang et al. 1977b) and oligonucleotides present in T1 or pancreatic RNase digests of tRNA (Simsek et al. 1973). The 5'-^{32}P-labeled mononucleotides (see Table 1) are identified by two-dimensional thin-layer chromatography and can, therefore, be used for analysis of the modified nucleotide content of a tRNA or of an oligonucleotide (Silberklang et al. 1979). The 5'-^{32}P-labeled oligonucleotides are separated by fingerprinting (Fig. 1) (Sanger et al. 1965). The labeled oligonucleotides are then recovered and partially digested with snake venom phosphodiesterase and/or nuclease P1, a relatively nonspecific endonuclease from *Penicillium citrinum*. These partial digestion products are separated by two-dimensional homochromatography (Sanger et al. 1973) and the sequence of the oligonucleotide in question is deduced (Fig. 2) from the characteristic mobility shifts resulting from the successive removal of nucleotides from the 3' end (Gillum et al. 1975; Silberklang et al. 1977b). All the oligonucleotides present in complete T1 RNase and pancreatic RNase digests of an RNA can, therefore, be sequenced. Using oligonucleotides present in tRNAs of known sequence as

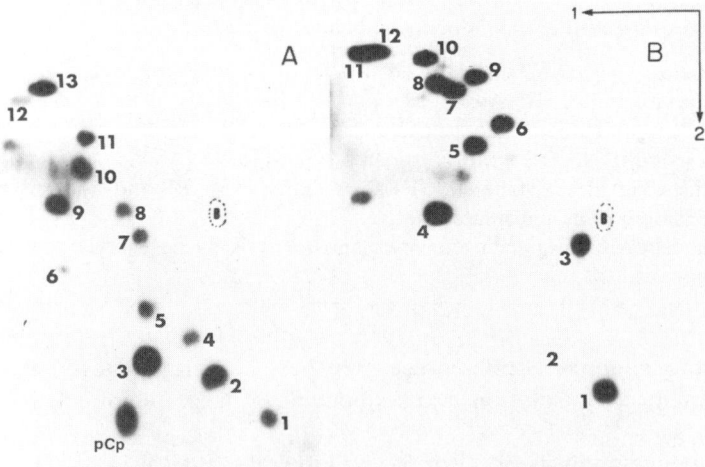

Figure 1 Fingerprints of 5'-^{32}P-labeled oligonucleotides obtained from pancreatic RNase digestion (*A*) and T1 RNase digestion (*B*) of *N. crassa* mttRNA^{fMet}. B. circled is blue dye marker. (Reprinted, with permission, from Heckman et al. 1978.)

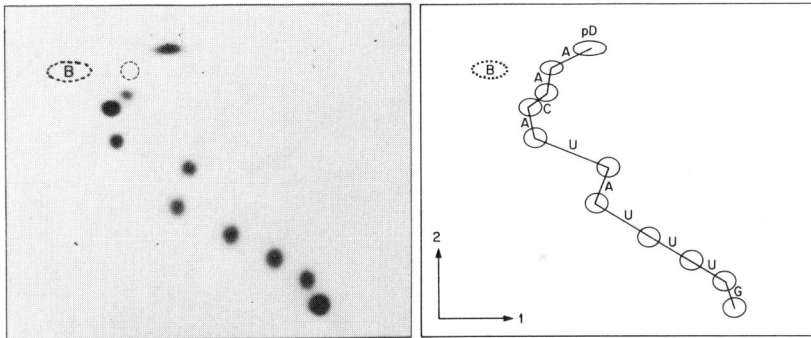

Figure 2 Autoradiograph of partial snake venom phosphodiesterase digest on spot 12 of Fig. 1B as analyzed by two-dimensional homochromatography. First dimension, electrophoresis on cellulose acetate at pH 3.5; second dimension, homochromatography. B circled is blue dye marker.

standards, we have also characterized the mobility shifts of many commonly occurring modified nucleotides, which has proven useful in recognizing these modified nucleotides within unknown sequences (Silberklang et al. 1979).

To order the shorter oligonucleotides present in complete T1 or pancreatic RNase digests into a total tRNA sequence, larger (overlapping) oligonucleotide fragments may be obtained by specific chemical cleavage of the tRNA or by limited digestion with base-specific nucleases. The use of in vitro ^{32}P labeling allows sequence analysis of such fragments on a picomole scale. In addition, using [5'-^{32}P]tRNA or [3'-^{32}P]tRNA (Silberklang et al. 1977a), we have used two methods for direct sequence analysis of large nucleotide stretches at either end of the molecule. One method involves partial digestion of [5'-^{32}P]tRNA or [3'-^{32}P]tRNA with nuclease P1, followed by two-dimensional homochromatographic separation and mobility shift analysis of the digestion products (Fig. 3) (Silberklang et al. 1977a). This method allows one to derive the sequence of 15–25 nucleotides directly from either end of a tRNA. Figure 3 shows an example of this method as applied to the 5'-terminal sequence analysis of *N. crassa* initiator mitochondrial formylmethionine tRNA (mttRNA^{fMet}).

The other method, similar in principle to the Maxam and Gilbert (1977) method for DNA sequencing, involves partial digestion of the end-labeled tRNA with base-specific nucleases followed by mapping of the base-specific cleavage sites by polyacrylamide gel electrophoresis (Donis-Keller et al. 1977; Simoncsits et al. 1977; Lockard et al. 1978). Figures 4 and 5 provide examples of this procedure as applied to 5'-^{32}P-labeled *N. crassa* mttRNA^{fMet} and mttRNA^{Tyr}, respectively. These last two methods for direct

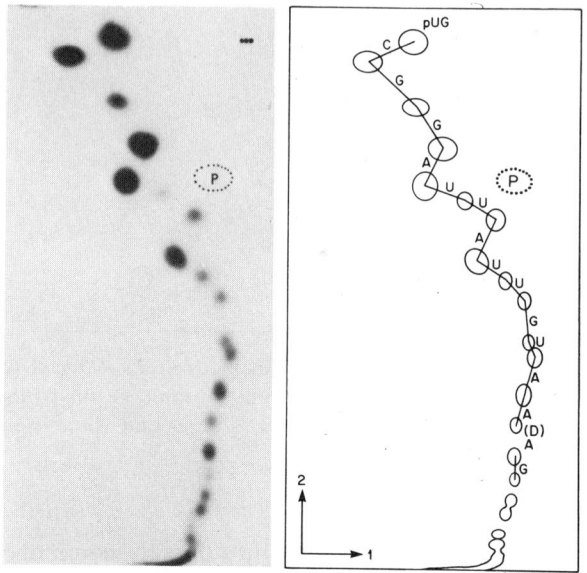

Figure 3 Autoradiograph of partial nuclease P1 digest on 5'-^{32}P-labeled *N. crassa* mttRNA^{fMet} as analyzed by two-dimensional homochromatography. P circled is pink dye marker. (Reprinted, with permission, from Heckman et al. 1978.)

sequence analysis usually provide all the information necessary for aligning the oligonucleotides present in complete T1 RNase and pancreatic RNase digests into a unique sequence.

Although the direct sequencing method utilizing gel electrophoretic analysis is applicable to sequencing mRNAs and most other RNAs, an application of this method alone is not sufficient for establishing the total sequence of tRNAs, since tRNAs contain many modified nucleotides—some of which are quite resistant to partial digestion with nucleases and others, such as 2'-O-methylated nucleotides, are resistant toward digestion by either nucleases or alkali. In addition, because of their stable secondary structure, stem regions of certain tRNAs are often not accessible to partial digestions with enzymes even in 7 M urea and at 50°C. Consequently, it is still necessary to have in hand a knowledge of oligonucleotide sequences present in complete RNase digests of the tRNA. The usefulness of the direct sequencing methods for sequencing tRNA lies in the fact that these methods provide most of the overlap information necessary for ordering these oligonucleotides into a unique sequence within a relatively short period and require less than a microgram of the tRNA.

An alternative approach for direct sequencing of tRNAs has been

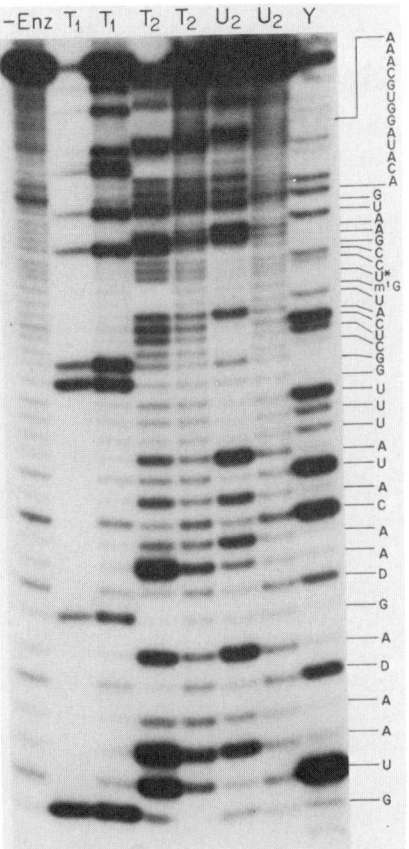

Figure 4 Autoradiograph of partial enzymatic digests on 5'-^{32}P-labeled *N. crassa* mttRNA^{fMet} as analyzed by electrophoresis on 20% polyacrylamide gel. (-Enz) Incubated in the absence of enzyme; (T1, T2, U2) incubation of RNA at two different levels of these enzymes (T1 is G specific; T2 is nonspecific, and U2 is A specific); (Y) incubation with a pyrimidine-specific RNase from *Bacillus cereus*. (Details are in Lockard et al. 1978.) (Reprinted, with permission, from Heckman et al. 1978.)

recently proposed by Stanley and Vassilenko (1978), although no applications using this technique alone for sequencing an unknown tRNA have appeared as yet.

SEQUENCE OF *N. CRASSA* mttRNAs

Besides its application to the sequencing of chloroplast and several prokaryotic and eukaryotic tRNAs, the above method has been used by us for the sequencing of *N. crassa* mttRNA^{fMet} (Heckman et al. 1978), mttRNA^{Tyr} (Heckman et al. 1979), mttRNA^{Ala}, and mttRNA^{Val}, and by Dirheimer et al. (this volume) for yeast mttRNA^{Phe}. The sequences of the four *N. crassa* mttRNAs are shown in Figure 6. The most interesting and surprising finding from these sequences is that every one of these mttRNAs contains in its sequence novel features that differ from some of the features

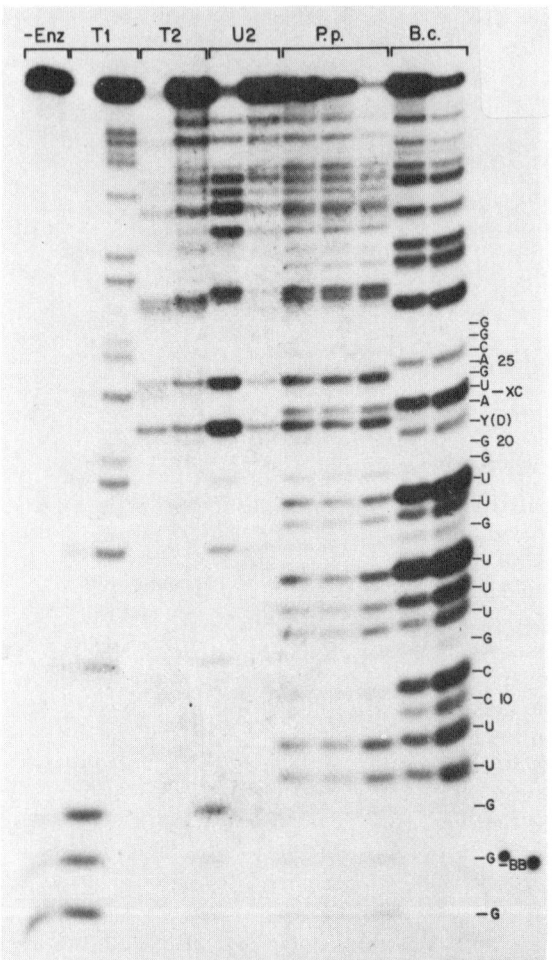

Figure 5 Autoradiograph of partial enzymatic digests on 5'-^{32}P-labeled *N. crassa* mttRNA^{Tyr} as analyzed by electrophoresis on 20% polyacrylamide gel. (BB, XC) Positions of bromphenol and xylene cyanol blue dye markers, respectively; (-Enz) incubated in the absence of enzyme; (T1, T2, U2) incubation of tRNA at two different levels of these ribonucleases (T1 is G specific, T2 is nonspecific, and U2 is A specific); (P.p) incubation of tRNA at three different levels of cytidine-negative RNase from *Physarum polycephalum*; (B.c) incubation of tRNA with two different levels of a pyrimidine-specific RNase from *B. cereus*. (Details in Lockard et al. 1978.) (Reprinted, with permission, from Heckman et al. 1979.)

of normal prokaryotic and eukaryotic cytoplasmic tRNAs. These novel features are indicated as shadowed regions in Figure 6. For instance, the *N. crassa* mttRNA^{fMet} has several interesting features. The first potential base pair in the D stem is a UU and not the more usual Watson-Crick or GU base

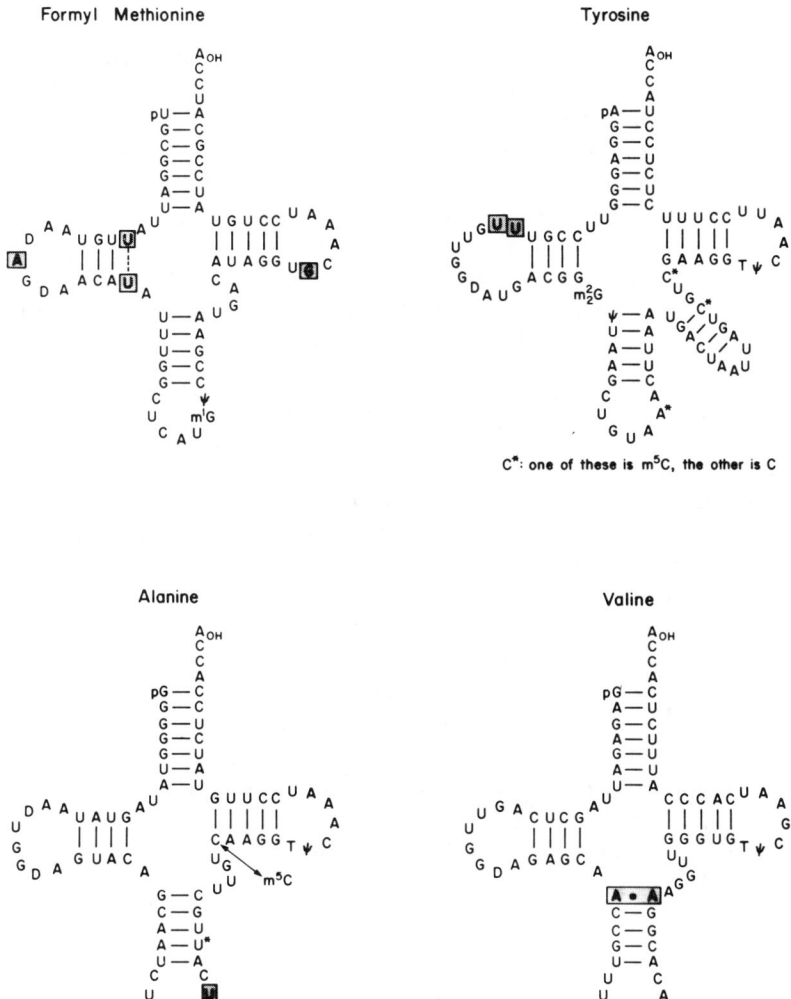

Figure 6 Nucleotide sequence in cloverleaf form of *N. crassa* mttRNA^fMet, mttRNA^Tyr, mttRNA^Ala, and mttRNA^Val. Shadowed regions indicate novel features present in these mttRNAs, which are absent in normal prokaryotic and eukaryotic cytoplasmic tRNAs.

pair as found in other tRNAs. In addition, it lacks the sequence GG in the D loop and ψC or UC in the TψC loop that have so far been found in all tRNAs that are active in protein synthesis; instead, it contains the sequence AG in the D loop and GC in the TψC loop. In the three-dimensional structure of yeast tRNA^Phe (Kim et al. 1974; Robertus et al. 1974), the two G residues in the D loop interact, respectively, with the ψC of the TψC loop to form Gψ and GC base pairs and provide the main tertiary interaction between the D loop and the TψC loop. It is, therefore, interesting to note that along with a change from G to A in the D loop of mttRNA^fMet, there is a corollary change in the TψC loop from ψ to G. Whether an AG interaction instead of a Gψ can be accommodated into the same three-dimensional framework as that of yeast tRNA^Phe with minimal perturbation of the overall three-dimensional structure or whether the tertiary interaction between the D and TψC loops in *N. crassa* mttRNA^fMet is significantly different remains to be seen.

In contrast to mttRNA^fMet, the three noninitiator mttRNAs that we have sequenced contain the more usual GG sequence in the D loop and TψC sequence in the TψC loop. This suggests that in the case of *N. crassa* mttRNAs, the unusual sequence features of mttRNA^fMet are probably unique to the initiator species and not to mttRNAs in general.

N. crassa mttRNA^fMet was the first mttRNA to be sequenced (Heckman et al. 1978). One of our main reasons for sequencing this tRNA was based on the finding that initiator tRNAs from both prokaryotes and eukaryotic cytoplasm possess in their sequence unique features that distinguish them from each other as a class and from most noninitiator tRNAs (Rich and RajBhandary 1976). The structural feature common to all of the six prokaryotic initiator tRNAs that have been studied (Dube et al. 1968; Delk and Rabinowitz 1974; Sprinzl et al. 1978) is the lack of a Watson-Crick base pair in the acceptor stem between the 5'-terminal nucleotide and the fifth nucleotide from the 3' end (Fig. 7). Eukaryotic cytoplasmic initiators do not share this feature (Fig. 7). Instead, they all possess the sequence AU(or ψ)CG in place of TψCG(or A) found at the beginning of loop 4 in

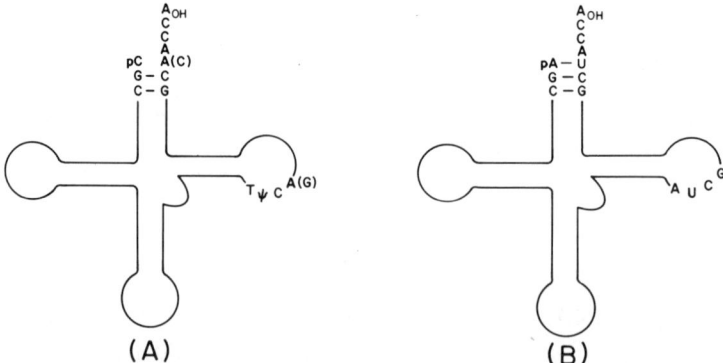

Figure 7 Unique sequence features present in initiator tRNAs from prokaryotes (*A*) and from eukaryotic cytoplasm (*B*).

the prokaryotic initiators and most other tRNAs that function in protein synthesis (Rich and RajBhandary 1976).

In view of this clear-cut distinction in the structural features between prokaryotic and eukaryotic cytoplasmic initiator tRNAs, it was of interest to sequence an initiator mttRNA to determine whether it too has a unique sequence feature different from noninitiator tRNAs and, if so, whether it resembles the prokaryotic or eukaryotic type of initiator tRNA structure. From a comparison of N. crassa mttRNA^{fMet} sequence to other initiator tRNAs, it can be seen that mttRNA^{fMet} differs from both prokaryotic and eukaryotic cytoplasmic initiator tRNAs in these respects. Unlike prokaryotic initiator tRNAs, the 5'-terminal nucleotide of mttRNA^{fMet} can form a Watson-Crick base pair to the fifth nucleotide from the 3' end (Fig. 7); however, the base pair (5'pU·3'pA) is the reverse of that found (5'pA·3'pU) in the eukaryotic initiator tRNAs. mttRNA^{fMet} lacks the sequence TψCG(or A). However, the corresponding sequence in mttRNA^{fMet} is UGCA and not AU(or ψ)CG as in the eukaryotic cytoplasmic initiator tRNAs. Whether the unique sequence features found in N. crassa mttRNA^{fMet} represents a third class of initiator tRNA structure common to all initiator mttRNAs or whether other initiator mttRNAs will turn out to have yet other distinguishing features is an interesting question. An answer to this should be forthcoming when sequence studies on other initiator mttRNAs are completed.

Besides the structural features probably unique to N. crassa mttRNA^{fMet}, the presence of novel features in the three other N. crassa mttRNAs (Fig. 6) and in a yeast mttRNA (Dirheimer et al., this volume) raises several very interesting questions.

1. What is the effect of the novel features of tRNAs shown in Figure 6 on their secondary and tertiary structure?
2. Does the presence of these novel features in the mttRNAs imply that the mitochondrial protein-synthesizing system, in general, is more flexible in what it requires of its tRNAs compared to prokaryotic and eukaryotic-cytoplasmic protein-synthesizing systems?
3. Is there any functional significance associated with these unusual features in mttRNAs?

For example, is it possible that the presence of a pyrimidine next to the anticodon in mttRNA^{Ala} instead of the usual purine — often a hypermodified purine — enables this tRNA to read all the four alanine codons? If so, this might explain the general observation (Borst and Grivell 1978) that the number of tRNA species in most mitochondria is less than the 32 needed as a minimum to read all the 61 codons of the genetic code using the rules of the wobble hypothesis (Crick 1966). A plausible explanation as to how mttRNA^{Ala} would acquire the ability to read all four alanine codons would be the destabilization of the anticodon loop structure. This destabilization would be due to the loss of stacking interactions caused by a change of the base next to the anticodon from a purine to a pyrimidine. This could result in a decrease in the energy of interaction required to trigger a conformational change of tRNA that is induced upon codon binding and that is postulated to be necessary (Kurland et al. 1975) for tRNA binding to ribosomes. Consequently, a matching of only two, instead of three, nucleotides between the codon and anticodon (Lagerkvist 1978) might suffice for such a conformational change to occur in tRNA^{Ala}. A similar explanation might also apply to tRNA^{Val}, which is lacking the base pair in the anticodon stem near the hinge region between the D stem and the anticodon stem resulting in a destabilization of its overall tertiary structure.

Leaving aside such speculations, it is clear that answers to the questions raised above can only come from further studies on sequences of other mttRNAs, a more rigorous analysis of the number of isoacceptor tRNAs present in various mitochondria, detailed studies on the secondary and tertiary structure of mttRNAs, and functional studies on mttRNAs. The amount of mttRNAs that can be isolated and purified is unfortunately very limited and not enough for such detailed studies. Consequently, as a first step toward obtaining larger amounts of mttRNAs, we have recently been involved in the mapping and cloning of *N. crassa* mttRNA genes (J. E. Heckman and U. L. RajBhandary, in prep.) in the hope that eventually it might be possible to have these tRNA genes transcribed and the transcripts processed in *E. coli*. Virtually all of the modified nucleosides present in *N. crassa* mttRNA are also present in *E. coli* (de Vries et al. 1978). It should, therefore, be possible to use such an approach to obtain mttRNAs made within *E. coli* that are similar to those isolated from mitochondria. Using the tRNAs labeled in vitro with ^{32}P that are used for sequence studies (Figs. 3–5) as hybridization probes, we have mapped the genes for all the tRNAs listed in Figure 6 on the mitochondrial DNA and have cloned them separately in *E. coli* using the plasmid pBR322 as vector.

Zachau (1978) has listed some of the highlights of the last two decades of tRNA research and predicts for the third decade not only the filling in of details but also some surprises and achievements. The novel features noted in here for the *N. crassa* mttRNAs and by Dirheimer et al. (this volume) for yeast mttRNA^{Phe} could prove to be among these surprises.

ACKNOWLEDGMENTS

Work in this laboratory was supported by grants GM-17151 from the National Institutes of Health and NP114 from the American Cancer Society. E. A.'s stay at Massachusetts Institute of Technology was supported by grant GM-07183 from the National Institutes of Health to the University of Chicago.

REFERENCES

Baev, A. A., T. V. Venkstern, A. D. Mirzabekov, A. I. Krutilina, L. Li, and V. D. Aksel'rod. 1967. Primary structure of valine transfer RNA 1 of baker's yeast. *Mol. Biol. (Mosc.)* **1:**631.

Borst, P. and L. A. Grivell. 1978. The mitochondrial genome of yeast. *Cell* **15:**705.

Cherayil, J. D. and R. M. Bock. 1965. A column chromatographic procedure for the fractionation of s-RNA. *Biochemistry* **4:**1174.

Crick, F. H. C. 1966. Codon-anticodon pairing: The wobble hypothesis. *J. Mol. Biol.* **19:**548.

Delk, A. S. and J. C. Rabinowitz. 1974. Partial nucleotide sequence of a prokaryote initiator tRNA that functions in its non-formylated form. *Nature* **252:**106.

DeVries, H., J. C. DeJonge, J.-M. Schneller, R. P. Martin, G. Dirheimer, and A. J. C. Stahl. 1978. *Neurospora crassa* mitochondrial transfer RNAs. *Biochim. Biophys. Acta* **520:**419.

Donis-Keller, H., A. M. Maxam, and W. Gilbert. 1977. Mapping adenines, guanines and pyrimidines in RNA. *Nucleic Acids Res.* **4:**2527.

Dube, S. K., K. A. Marcker, B. F. C. Clark, and S. Cory. 1968. Nucleotide sequence of N-formyl-methionyl transfer RNA. *Nature* **218:**232.

Gillam, I., S. Millward, D. Blew, M. von Tigerstrom, E. Wimmer, and G. M. Tener. 1967. The separation of soluble ribonucleic acids on benzoylated diethylaminoethyl cellulose. *Biochemistry* **6:**3043.

Gillum, A. M., N. Urquhart, M. Smith, and U. L. RajBhandary. 1975. Nucleotide sequence of salmon testes and salmon liver cytoplasmic initiator tRNA. *Cell* **6:**395.

Goodman, H. M., J. Abelson, A. Landy, S. Brenner, and J. D. Smith. 1968. Amber suppression: A nucleotide change in the anticodon of a tyrosine transfer RNA. *Nature* **217:**1019.

Heckman, J. E., B. Alzner-DeWeerd, and U. L. RajBhandary. 1979. Interesting and unusual features in the sequence of *N. crassa* mitochondrial tyrosine tRNA. *Proc. Natl. Acad. Sci.* **76:**717.

Heckman, J. E., L. I. Hecker, S. D. Schwartzbach, W. Edgar Barnett, B. Baumstark, and U. L. RajBhandary. 1978. Structure and function of initiator methionine tRNA from the mitochondria of *Neurospora crassa. Cell* **13:**83.

Hoagland, M. B., P. C. Zamecnik, and M. L. Stephenson. 1957. Intermediate reactions in protein biosynthesis. *Biochim. Biophys. Acta* **24:**215.

Holley, R. W. 1968. Experimental approaches to the determination of the nucleotide sequences of large oligonucleotides and small nucleic acids. *Prog. Nucleic Acid Res. Mol. Biol.* **8:**37.

Holley, R. W., J. Apgar, G. A. Everett, J. T. Madison, M. Marquisee, S. H. Merrill, J. R. Penswick, and A. Zamir. 1965. Structure of a ribonucleic acid. *Science* **147:**1462.

Ikemura, T. and J. E. Dahlberg. 1973. Small ribonucleic acids of *Escherichia coli*: Characterization by polyacrylamide gel electrophoresis and fingerprint analysis. *J. Biol. Chem.* **248:**5024.

Kim, S.-H., F. L. Suddath, G. J. Quigley, A. McPherson, J. S. Sussman, A. H. J. Wang, N. C. Seeman, and A. Rich. 1974. Three-dimensional tertiary structure of yeast phenylalanine transfer RNA. *Science* **185:**435.

Kurland, C. G., R. Rigler, M. Ehrenberg, and C. Blomberg. 1975. Allosteric mechanism for codon-dependent tRNA selection on ribosomes. *Proc. Natl. Acad. Sci.* **72**:4248.

Lagerkvist, U. 1978. "Two out of three": An alternative method for codon reading. *Proc. Natl. Acad. Sci.* **75**:1759.

Lockard, R. E., B. Alzner-DeWeerd, J. E. Heckman, J. MacGee, M. W. Tabor, and U. L. RajBhandary. 1978. Sequence analysis of 5'-[^{32}P]-labeled mRNA and tRNA using polyacrylamide gel electrophoresis. *Nucleic Acids Res.* **5**:37.

Madison, J. T., G. A. Everett, and H. K. Kung. 1966. Nucleotide sequence of a yeast tyrosine transfer RNA. *Science* **153**:531.

Maxam, A. M. and W. Gilbert. 1977. A new method for sequencing DNA. *Proc. Natl. Acad. Sci.* **74**:560.

Nishimura, S. 1978. Modified nucleosides and isoaccepting tRNA. In *Transfer RNA* (ed. S. Altman), p. 168. MIT Press, Cambridge, Massachusetts.

Pearson, R. L., J. F. Weiss, and A. D. Kelmers. 1971. Improved separation of transfer RNAs on polychlorotrifluoroethylene-supported reversed-phase chromatography columns. *Biochim. Biophys. Acta* **228**:70.

RajBhandary, U. L. 1968. The specific labeling of 2',3'-diol end groups in RNA. *J. Biol. Chem.* **243**:556.

RajBhandary, U. L., A. Stuart, R. D. Faulkner, S. H. Chang, and H. G. Khorana. 1967a. Nucleotide sequence studies on yeast phenylalanine sRNA. *Cold Spring Harbor Symp. Quant. Biol.* **31**:425.

RajBhandary, U. L., S. H. Chang, A. Stuart, R. D. Faulkner, R. M. Hoskinson, and H. G. Khorana. 1967b. The primary structure of yeast phenylalanine transfer RNA. *Proc. Natl. Acad. Sci.* **57**:751.

Rich, A. and U. L. RajBhandary. 1976. Transfer RNA: Molecular structure, sequence and properties. *Annu. Rev. Biochem.* **45**:805.

Robertus, J. D., J. E. Ladner, J. T. Finch, D. Rhodes, R. S. Brown, B. F. C. Clark, and A. Klug. 1974. Structure of yeast phenylalanine transfer RNA at 3Å resolution. *Nature* **250**:546.

Sanger, F., G. G. Brownlee, and B. G. Barrell. 1965. A two-dimensional fractionation procedure for radioactive nucleotides. *J. Mol. Biol.* **13**:373.

Sanger, F., J. E. Donelson, A. R. Carlson, H. Kössel, and D. Fischer. 1973. Use of DNA polymerase I primed by a synthetic oligonucleotide to determine a nucleotide sequence in phage fl DNA. *Proc. Natl. Acad. Sci.* **70**:1209.

Silberklang, M., A. M. Gillum, and U. L. RajBhandary. 1977a. The use of nuclease P_1 in sequence analysis of end group labeled RNA. *Nucleic Acids Res.* **4**:4091.

———. 1979. Use of *in vitro* ^{32}P-labeling in the sequence analysis of non-radioactive tRNAs. *Methods Enzymol.* **59**:58.

Silberklang, M., A. Prochiantz, A. L. Haenni, and U. L. RajBhandary. 1977b. Studies on the sequence of the 3'-terminal region of turnip-yellow mosaic-virus RNA. *Eur. J. Biochem.* **72**:465.

Simoncsits, S., G. G. Brownlee, R. S. Brown, J. R. Rubin, and H. Guilley. 1977. New rapid gel sequencing method for RNA. *Nature* **269**:833.

Simsek, M., J. Ziegenmeyer, J. E. Heckman, and U. L. RajBhandary. 1973. Absence of the sequence G-T-ψ-C-G(A)- in several eukaryotic cytoplasmic initiator transfer RNAs. *Proc. Natl. Acad. Sci.* **70**:1041.

Sprinzl, M., F. Grüter, and D. H. Gauss. 1978. Compilation of tRNA sequences. *Nucleic Acids Res.* (special suppl.) **5**:r15.

Stanley, J. and S. Vassilenko. 1978. A different approach to RNA sequencing. *Nature* **274**:87.

Szekely, M. and F. Sanger. 1969. Use of polynucleotide kinase in fingerprinting non-radioactive nucleic acids. *J. Mol. Biol.* **43**:607.

Zachau, H. G. 1978. Transfer RNA coming of age. In *Transfer RNA* (ed. S. Altman), p. 1. MIT Press, Cambridge, Massachusetts.

Zachau, H. G., D. Dutting, and H. Feldman. 1966. The structures of two serine transfer ribonucleic acids. *Hoppe Seyler's Z. Physiol. Chem.* **347**:212.

The Primary Structure of tRNAs and Their Rare Nucleosides

Guy Dirheimer, Gerard Keith, Annie-Paule Sibler, and Robert Pierre Martin
Institut de Biologie Moléculaire et Cellulaire du CNRS
67084 Strasbourg, France
and Faculté de Pharmacie, Université Louis Pasteur
67083 Strasbourg, France

Since the first determination of a primary structure of a tRNA by Holley et al. (1965), 113 structures have been published of which 108 are different. This number is sometimes difficult to calculate. Two structures are defined as different if they differ by one basic nucleotide, for example, if an A replaces a G, but not if they differ by posttranscriptional modifications. Some structures have been determined on mixtures of isoacceptors that were impossible to separate, like the one of beef liver tRNA^{Trp} (Fournier et al. 1978). Figure 1 shows that this tRNA has several incompletely modified nucleotides, such as m^{2}G in the amino acid stem, Cm in the anticodon loop, and m^{7}G in the variable loop. But in two positions, C is partially replaced by D. If these Ds arise from a posttranscriptional modification of C, then we would have only two tRNAs^{Trp}, one having a G in position 56 and the other having an A in this position. In fact, so far, only U → D heterogeneities have been found in tRNAs. There is no evidence that a C can be modified to a D. Therefore, we conclude that two more basic nucleotides are different in the tRNA^{Trp}. There are several combinations possible between tRNAs differing in three different positions; we have at least three tRNAs^{Trp}. In fact, by two-dimensional gel electrophoresis, five spots have now been found for this tRNA (M. Fournier, pers. comm.). It will be interesting to test if each functions as primer for reverse transcriptase as does the mixture of them.

Recently all published tRNA sequences have been compiled by Gauss et al. (1979) (see Appendix I, this volume.) These tRNAs and 28 sequenced but still unpublished tRNA species have been arranged in Table 1 indicating the source of the tRNA. As shown, *Escherichia coli* and yeast tRNAs primary structures have been determined, and several bacteriophage T4-coded, *Bacillus subtilis*, and mammalian tRNAs have also been sequenced. The families of initiator and tRNAs^{Phe} are the best known. Recently, six tRNAs from organelles, the tRNA^{Phe} from *Euglena* (Chang et al. 1976) and bean (Guillemaut and Keith 1977) chloroplasts, the initiator tRNA^{fMet} (Heckman et al. 1978), tRNA^{Ala}, and tRNA^{Tyr} (U. L.

Table 1 tRNA structures that have been determined as of September 1978

	Ala	Arg	Asp	Asn	Cys	Gln	Glu	Gly
T4		+				++		+
Mycoplasma								
E. coli	+	++	+	+	+	++	++	+++ ↕
S. typhimurium								+
S. aureus and epidermidis								6+
B. stearothermophilus								
B. subtilis	×							××
T. thermophilus								
Blue green algae								
A. nidulans								
Euglena gracilis								
Scenedesmus obliquus (green algae)								
Yeast (Saccharomyces cerevisae)	+	+++	+		+		+	+
Schizosaccharomyces pombe							×	
Yeast (Torula)	+							
N. crassa								
Plants								+
B. mori		++						++
Drosophila								
Xenopus laevis (oocytes)								
Fish								
Chicken cell								
Mammals		××		+				××
Chloroplast Euglena								
Chloroplast bean								
Mitochondria Neurospora	×							
Mitochondria yeast								

Double arrow indicates that the structures are the same; (+) published structures; (×) not yet published structures; (P) partial structures; (i) initiator; (m) noninitiator. The numerous mutants obtained by Smith, McClain, Carbon, Hill, and Singer et al. (pers. comm.) are not counted in this table.

The following 28 sequences have been determined but not yet published: Ala 3 from *E. coli* (J. E. Dahlberg, pers. comm.) and Ala from *B. subtilis* (K. Murao and I. Ishikura, pers. comm.) and *N. crassa* mitochondria (U. L. RajBhandary, pers. comm.); Arg 1 and 2 from mouse L1210 tissue culture (F. Harada, pers. comm.); Glu from *S. pombe* (D. Söll, pers. comm.); Gly 1 and 2 from *B. subtilis* (K. Murao and I. Ishikura, pers. comm.) and human placenta (R. C. Gupta et al., pers. comm.); Leu 5 from *E. coli* (Z. Yamaizumi et al., pers. comm.) and Leu from *X.*

RajBhandary, pers. comm.) from *Neurospora crassa* mitochondria, and the tRNA^{Phe} from yeast mitochondria (Martin et al. 1978), have been sequenced.

YEAST MITOCHONDRIAL tRNAs

For some years our laboratory has been engaged, in collaboration with J. M. Schneller and A. Stahl (Faculty of Pharmacy, Strasbourg), in the study of yeast mitochondrial tRNAs (mttRNAs). There are several interesting

His	Ile	Leu	Lys	iMet	mMet	Phe	Pro	Ser	Thr	Trp	Tyr	Val	Total +
		+					+	+	+				8
			+		+								2
+	+	++×	+	+	+	+		++	+	++	+	+++	30
↕+	↕+												3
													6
						+					+	+	3
		+	+	×	+			+			×	×	4
			+										1
					×								
			+										1
					×								1
				×									
		++	++	+	+	+		++	++	+	+	+++	24
		×					+	×					1
	+										+	+	4
				+									1
				×		+							2
								PP					4
		×	×										
	×		××										
			↕+										1
			↑					××		+			1
		PPP	↓+++	+	+	+	+	+P+		×××		++	13
					+								1
					P		+						1
			+								×		1
						×							

laevis (S. Clarkson et al., pers. comm.); Lys 2 from *Drosophila* (D. Söll, pers. comm.); iMet from the green algae *S. obliquus* (P. O. Olins and D. S. Jones, pers. comm.), wheat germ (H. P. Ghosh et al., pers. comm.) and *Drosophila* (D. Söll, pers comm.), and iMet 1 and 2 from *X. laevis* (S. Clarkson et al., pers. comm.); Phe from blue green algae (S. H. Chang et al., pers. comm.) and yeast mitochondria (R. P. Martin et al., in prep.); Pro 1 and 2 from chicken (F. Harada et al., pers. comm.); Ser sup 3-e from *S. pombe* (D. Söll, pers. comm.); Thr from mouse (F. Harada, pers. comm.); Trp from beef liver (Fournier et al. 1978); Tyr from *B. subtilis* (G. Keith et al., pers. comm.) and *N. crassa* mitochondria (U. L. RajBhandary, pers. comm.); and Val 1 from *B. subtilis* (K. Murao and I. Ishikura, pers. comm.).

questions concerning mttRNAs. Do they contain only mitochondrial DNA (mtDNA)-coded tRNAs, and how many, or are there tRNAs imported from the cytoplasm?

It has been proposed (Borst 1972; Suyama and Hamada 1976) that cytoplasmic tRNAs are necessary for mitochondrial protein synthesis. Chromatographic studies of mttRNA from different origins have shown the presence of cytoplasmic tRNAs in mttRNA preparations. We have improved the purification of yeast mitochondria so that cytoplasmic tRNA contamination is no longer observed by RPC-5 column chroma-

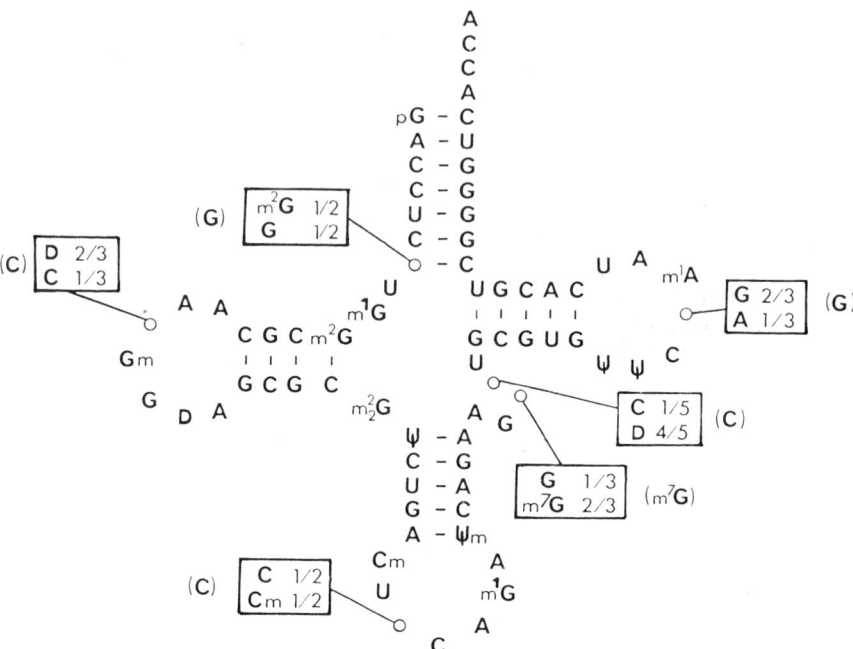

Figure 1 Structure of beef liver tRNA^{Trp}. (Reprinted, with permission, from Fournier et al. 1978.)

tography for several aminoacyl tRNAs: tRNA^{Leu} and tRNA^{Tyr} (Schneller et al. 1975a); tRNA^{Phe} (Schneller et al. 1975b); tRNA^{Met} (Martin et al. 1976a); and tRNA^{Arg}, tRNA^{Asn}, tRNA^{Ser}, and tRNA^{Val} (R. P. Martin et al., unpubl.).

Identification of Yeast mttRNA Isoacceptors

The isoacceptor tRNAs for 19 amino acids were characterized by preparative RPC-5 column chromatography (Fig. 2). The results obtained independently by Martin and Rabinowitz (1978) are consistent with ours, except in the case of tRNA^{Asn}, which was not detected by these authors.

We could not aminoacylate mttRNA with glutamine but we detected two peaks that could be charged with glutamic acid. N. C. Martin et al. (1977) had shown that one of these two tRNA species recognizes the glutamine codon CAA. We could show that in the mitosol, one of these tRNAs is first charged by glutamic acid that is later modified by a mitochondrial transamidase. In contrast we showed that yeast mttRNA contains a tRNA that can be directly aminoacylated with asparagine.

The two-dimensional polyacrylamide gel electrophoresis, developed by Fradin et al. (1975), proved to be a suitable technique for complete resolution of all mttRNA species. About 30 major spots were observed

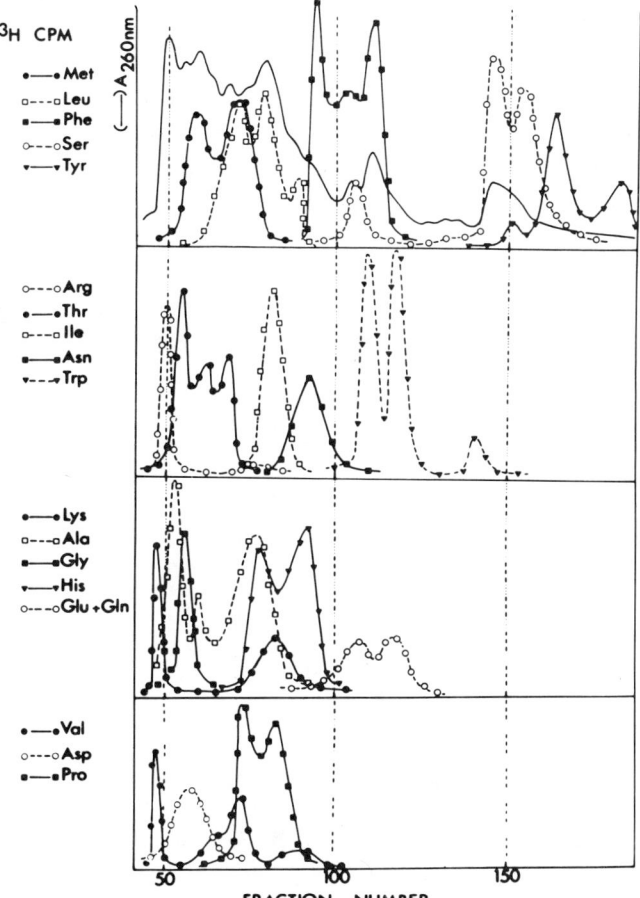

Figure 2 RPC-5 chromatography at pH 4.5 for uncharged total yeast mttRNA.

(R. P. Martin et al. 1977). By cutting out the spots and extracting the tRNAs with a mixture of buffer and phenol, followed by ethanol precipitation in the presence of rRNA, we could test their aminoacylation. Table 2 shows the number of isoacceptors found by two-dimensional electrophoresis and by RPC-5 column chromatography.

Coding Origin of mttRNAs; Presence of a Single Nuclear DNA-coded tRNA in Mitochondrial Preparations

All but six mttRNAs migrated distinctly from cytoplasmic tRNA in two-dimensional electrophoresis (R. P. Martin et al. 1977). To determine the coding origin of the mttRNA spots, especially of those just men-

Table 2 Number of isoacceptors compared to the number of codons

Amino acid	Number of isoacceptors fractionated by		Number of codons (genetic code)
	two-dimensional electrophoresis	RPC-5 chromatography	
Ala	2	3	4
Arg	2	2	6
Asn	1	1	2
Asp	1	1	2
Glu + Gln	2	1 + 1	2 + 2
Gly	2	2	4
His	1	2	2
Ile	1	1	3
Leu	2	3	6
Lys	2	2	2
Met	2	3	1
Phe	1 (2)	2–3	2
Pro	2	2	4
Ser	3	3	6
Thr	2	3	4
Trp	1	3	1
Tyr	1	3	2
Val	2–3	3	4
Total	30–32	42	

tioned, [^{32}P]mttRNA was hybridized to mtDNA (tRNA in excess). The tRNAs were then eluted from the hybrids and analyzed by two-dimensional electrophoresis. The autoradiograph obtained was compared to the control, which consisted of nonhybridized mttRNA. Only one spot is lacking in the hybridized mttRNA pattern. It corresponds to one of the two lysine isoacceptors found in yeast mitochondria. Sequence analysis showed that this tRNA corresponds to tRNA$_1^{Lys}$ (anticodon CUU) from the yeast cytoplasm, which was sequenced by Smith et al. (1971). tRNA$_1^{Lys}$ was shown to be present in RNase-treated mitochondria. In addition to all the mtDNA-coded tRNAs, it was also present in mitochondrial innermembrane and mitosol preparations. No tRNA could be detected in mitochondrial outer-membrane preparations. Therefore, tRNA$_1^{Lys}$ can be considered as an imported tRNA species. It is not aminoacylated by mitochondrial aminoacyl-tRNA synthetase preparations. The second tRNA^{Lys} present in yeast mitochondria (tRNA$_2^{Lys}$) is coded by the mitochondrial genome, aminoacylated by the mitochondrial enzyme preparations, and recognizes both AAA and AAG lysine codons; tRNA$_1^{Lys}$, however, recognizes only the AAG codon. The imported tRNA$_1^{Lys}$ seems, therefore, not to be essential for the transfer of lysine into mtDNA-coded polypeptide chains and may have a regulatory role.

Primary Structure of mtDNA-coded tRNA^{Phe}

Since mttRNA^{Phe} is clearly separated from the other mttRNAs on the two-dimensional gel (R. P. Martin et al. 1977), purification of either in vivo ^{32}P-labeled or unlabeled mttRNA^{Phe} could be done in one step. The isolation of unlabeled mttRNA^{Phe} gave a recovery of 10.5 μg pure tRNA^{Phe} from 250 μg of total mttRNA put on the gel. This amount is sufficient for subsequent sequencing by postlabeling, and also for assignment of the gene location on mtDNA restriction fragments. The nucleotide sequence is shown in Figure 3 (R. P. Martin et al., in prep.).

The basic feature that distinguishes this tRNA from all other sequenced tRNAs, whatever their origin, is its very low G + C content: 25 out of 75 nucleotides. When the sequence is arranged in the cloverleaf form the amino acid stem and the D stem contain only 2 GC base pairs each, and the anticodon stem consists exclusively of AU (or ψ) base pairs. It was difficult to deduce the ψC stem because when written conventionally (as shown in Fig. 3b) it has only 2 base pairs, 1 AU and 1 GC, which is insufficient for stability. Moreover, this places at position 47

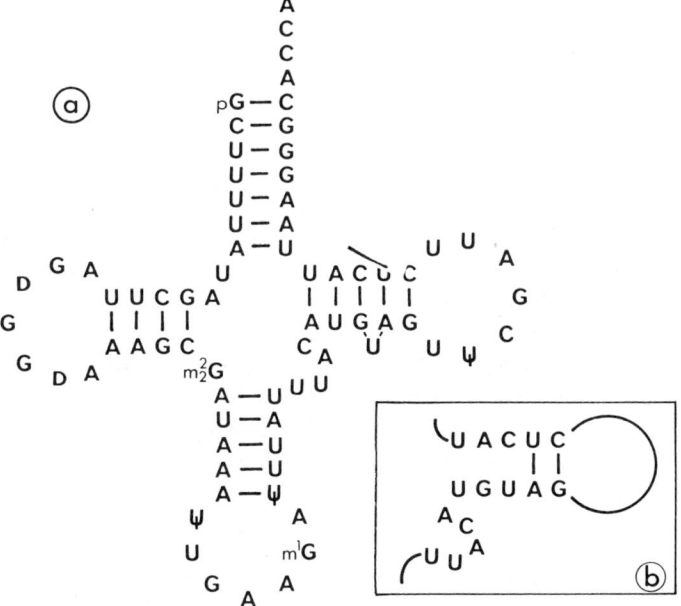

Figure 3 Structure of yeast mttRNA^{Phe}.

an A that then cannot pair with G15 in the tertiary structure. This pairing is essential for the stability of the tertiary structure. Therefore, we prefer the representation in Figure 3a, where the variable loop has 4 nucleotides. This type of loop has already been found in other tRNAs. Second, C46 can pair with G15 and, third, the ψC stem has 5 base pairs (3 AU and 2 GC). This representation permits the good positioning of all variant and semiinvariant nucleotides, but it excludes U50 from base pairing and gives a bulge in the ψC stem. Such a bulge has never been shown in a tRNA cloverleaf model, but is frequent in both ribosomal and viral secondary-structure models.

Yeast mttRNA^{Phe} has only 7 rare nucleotides: m^1G, m_2^2G, 3 ψ, and 2 D. This number is low, as in prokaryotic tRNAs, but like eukaryotic tRNAs^{Phe}, yeast mttRNA^{Phe} has an m_2^2G between the D stem and the anticodon stem. There is an m^1G adjacent in 3' to the anticodon and T is lacking in the ψC loop as in *Mycoplasma* tRNA^{Phe} (Kimball et al. 1974). Finally, it lacks m^7G present in the variable loop of all sequenced tRNAs^{Phe} (Keith and Dirheimer 1978).

Yeast mttRNA^{Phe} has from 56.6% to 63.1% sequence homology with other sequenced tRNAs^{Phe}, whatever their origin. This percentage is relatively low, since eukaryotic tRNAs^{Phe} have 64.5–84.2% sequence homology and prokaryotic tRNAs^{Phe} have 72.4–96% homology (R. P. Martin et al., in prep.). Therefore, no conclusion can be drawn concerning the endosymbiotic theory of mitochondria evolution (Margulis 1970).

The comparison of *N. crassa*-sequenced mttRNAs with yeast mttRNA^{Phe} does not show any structural feature particular to mttRNAs.

GENERAL FEATURES OF tRNAs

Invariant and Semiinvariant Nucleotides in tRNAs

By comparing the nucleotides of the sequenced tRNA species one finds that some of them are always on the same position. There are 14 such common or invariant nucleotides (Dirheimer et al. 1972). In addition, some positions are generally occupied by purines or pyrimidines. They are called semiinvariant nucleotides. Their number is 9. The invariant positions are shown in Figure 4 by nucleoside letters, whereas semiinvariant positions are shown by R (for a purine nucleoside) or Y (for a pyrimidine). Some authors also consider that in position 21^1 there is an invariant A, although two *E. coli* tRNAs^{Leu} have a G in this position. Therefore, we

---

[1]The numbering of nucleotides we have adopted is that of yeast cytoplasmic tRNA^{Phe}, which contains 76 nucleotides (8 nucleotides in the D loop and 5 nucleotides in the variable loop). The 3'-terminal end always carries number 76, even in tRNAs having less nucleotides. The numbering of the nucleotides in the D and variable loops is as follows: the nucleotides between the invariable A-Pu15 and G18-G19 sequences are numbered 16, 17, 17:1, 17:2, etc.; those nucleotides between the invariable G18-G19 sequence and the semivariable Pu22 (Fig. 7) are numbered 20, 20:1, 20:2, 21, etc.

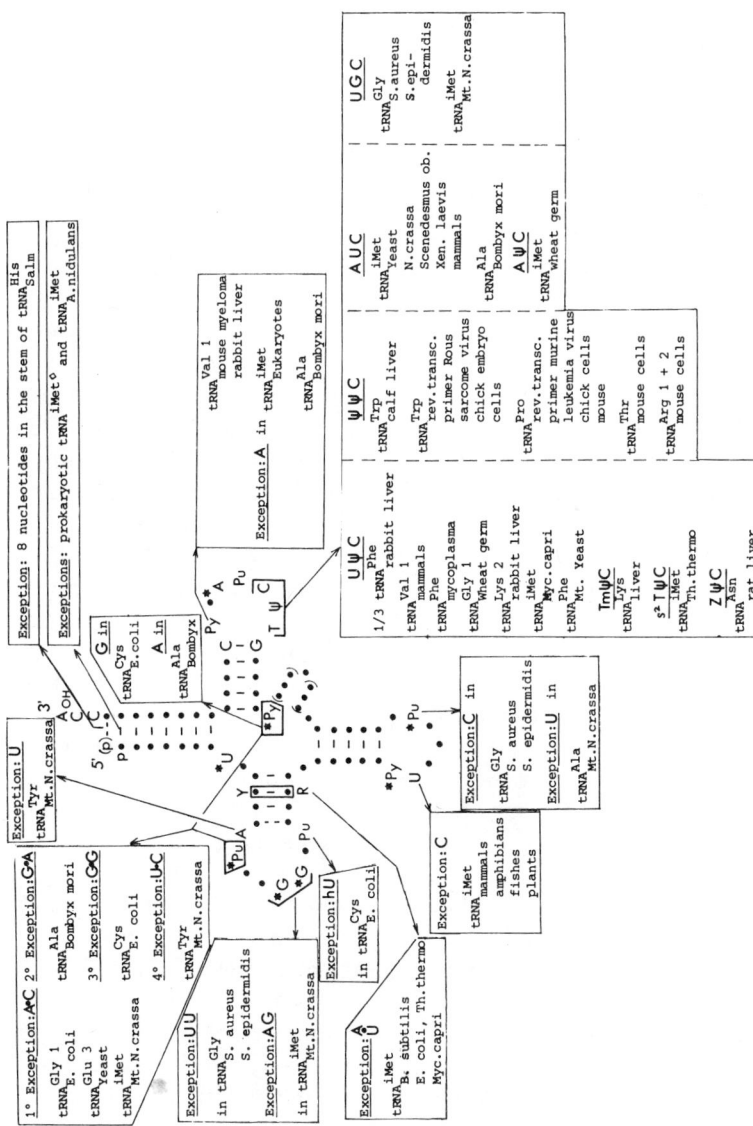

Figure 4 Exceptions in the invariant and semiinvariant nucleotides in the sequenced tRNAs. (★) Often modified; (◇) with exception of tRNA$_{Halobacter\ cutirubrum}^{fMet}$. When not specified, all eukaryotic tRNAs are of cytoplasmic origin. (Reprinted, with permission, from Martin et al. 1978.)

prefer to consider position 21 as being a semiinvariant R position. (For general reviews on tRNA, see Dirheimer et al. [1972, 1976]; Barrel and Clark [1974]; Clark and Klug [1975]; Sodd [1976]; Rich and RajBhandary [1976]; and Goddard [1977].)

There are exceptions in the invariant and semiinvariant nucleotides in sequenced tRNAs (Fig. 4).

There is always a U or an s^4U in position 8 from the 5' end. In a lot of tRNAs there is a pyrimidine-11–purine pair. This feature does not exist in prokaryotic initiator tRNAs, which have a purine-11–pyrimidine ($A11 \cdot U$) pair in position.

There is always a sequence A-Pu in positions 14 and 15 except in the *N. crassa* mttRNA^{Tyr} (U. L. RajBhandary, pers. comm.) (Fig. 4). According to Lewitt (1969), the tRNAs that contain G in position 15 have C in the position located before the ψC stem, whereas those that contain A in position 15 have U before the ψC stem. Several exceptions to this general rule have been observed in GGG-specific tRNA$_1^{Gly}$ from *E. coli* and *Salmonella typhimurium* (Hill et al. 1973; Riddle and Carbon 1973), in tRNA$_3^{Glu}$ from yeast (Kobayashi et al. 1974), and in *N. crassa* mitochondrial initiator tRNA^{Met} (Heckman et al. 1978). These three tRNAs have an AC pair in this position. Therefore, the base pairing between these positions, although possible in almost all tRNA sequences reported so far, is not a universal feature of tRNA structure, but according to A. Klug (pers. comm.), an AC pairing could take place by a distortion of the molecule. Finally, a GA pair has recently been reported for tRNA^{Ala} from *Bombyx mori* (Sprague et al. 1977) and a GG pair for tRNA^{Cys} from *E. coli* (Mazzara and McClain 1977).

Almost all tRNAs that participate in the protein synthesis have a GG or a GmG sequence in the D loop. This sequence was not found in the unusual species of tRNAs^{Gly} in *Staphylococcus aureus* and *Staphylococcus epidermidis*, which do not function in protein synthesis but are able to participate in cell-wall peptidoglycan synthesis (Roberts 1972, 1974). The tertiary structure of these tRNAs must be significantly different from those that participate in protein synthesis because these two Gs join the D loop to the ψC loop. *N. crassa* initiator tRNA has an AG sequence in this position (Heckman et al. 1978). The pairing of the A with the ψ in the ψC loop cannot take place by classical Watson-Crick hydrogen bonds because this tRNA has a G in place of the ψ in this loop.

There is always a pyrimidine in the first position of the anticodon loop and a purine after the anticodon, but this purine is not found in the tRNAs^{Gly} just mentioned that do not function in protein synthesis. It is probable that the purine after the anticodon is essential for a correct codon-anticodon pairing. This pairing is not required for the function of these tRNAs^{Gly}, which work in the peptidoglycan synthesis, which does not require ribosomes and is not required for the function of mRNAs.

Another exception is the recently sequenced *N. crassa* mitochondrial tRNA^{Ala}, which has a U in this position (U. L. RajBhandary, pers. comm.). The coding properties of this tRNA would be interesting to study.

The U before the anticodon has been found in all sequenced tRNAs except the methionine-accepting tRNAs, whose role is restricted to the initiation of protein synthesis in mammals, amphibians, and fish. But in the tRNA^{fMet} from yeast, blue green algae, and prokaryotes, the U is still present in that place. Virtually nothing is known at present about the molecular geometry of different anticodon loops and, therefore, the significance of this base substitution remains unclear.

Over the past few years most of the changes have been shown to occur in the ψC loop. Six years ago everybody thought that all tRNAs had a TψC sequence in this loop but, in fact, several other sequences have now been identified there.

First, some tRNAs have an incomplete modification or no modification at all of U into T in the TψC sequence. It has recently been found that three liver tRNAs^{Lys} have a 2'-*O*-ribose-methylated T (H. J. Gross, pers. comm.) and that initiator tRNA from *Thermus thermophilus* contains an s^{2}T (Watanabe et al. 1976).

Second, we have found the sequence $\psi\psi$C instead of TψC in tRNA^{Trp} from beef liver (Fournier et al. 1978). This sequence was also found in tRNA^{Trp} from Rous-sarcoma-virus-infected chicken cells; this tRNA^{Trp} is the primer for reverse transcriptase (Harada et al. 1975). Dahlberg has also recently shown that in murine-leukemia-virus-infected cells a tRNA^{Pro} is the primer for reverse transcriptase and has a $\psi\psi$C sequence, too (J. E. Dahlberg, pers. comm.).

Third, all the cytoplasmic initiator tRNAs^{fMet} so far sequenced have an AUC sequence instead of a TψC sequence in this loop (Gillum et al. 1975). This strong conservation of this specific feature implies an essential, functional role for this sequence. Whether it prevents initiator tRNA from binding to the A site of the ribosome remains to be proven. Let us emphasize that the prokaryotic initiator tRNAs^{fMet} from *E. coli*, *B. subtilis*, *Streptococcus faecalis*, *Mycoplasma*, blue green algae, and *T. thermophilus* have the TψC or s^{2}TψC sequence.

It is surprising that tRNA^{Ala} from *B. mori* also has the AUC sequence. It is not known if this tRNA can act as an initiator of protein synthesis.

The tRNAs^{Gly} that participate in the peptidoglycan synthesis have the sequence UGC. This is not surprising because these tRNAs do not interact with the ribosome. What is more surprising is that initiator tRNA from *Neurospora* mitochondria also has this sequence.

Finally, the last base of the TψC loop is always a pyrimidine with the exception of the eukaryotic initiator tRNAs that have an A in this position. However, this last A is also found in mammalian tRNA$^{Val}_1$ (Piper

1) N-Methylated nucleosides

m^1A, m^1G, m^7G, m^1I

$m^2G = N^2$-Methylguanosine

$m_2^2G = N^2,N^2$-Dimethylguanosine

$m^6A = N^6$-Methyladenosine

m^3C

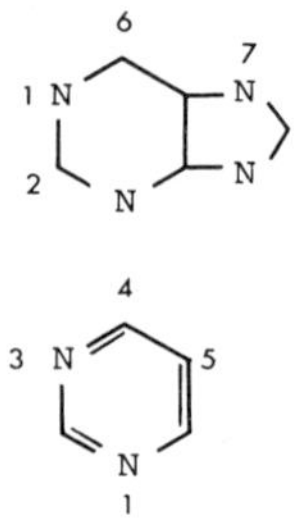

2) C-Methylated nucleosides

m^2A ; m^5C ; $T = m^5U$

3) O-2' Methylated nucleosides

Gm ; Cm ; Um ; ψm ; Tm

4) Pseudouridine   = ψ
   Inosine   = I
   5,6-Dihydrouridine = D or hU

5) Other derivatives of U in position 5

mo^5U = 5-Methoxyuridine : $-OCH_3$

mcm^5U = 5-Methoxycarbonylmethyluridine : $-CH_2-COOCH_3$

cmo^5U = V = 5-Carboxymethoxyuridine : $-O-CH_2-COOH$

$mcmo^5U$ = 5-Methoxycarbonylmethoxyuridine : $-O-CH_2-COOCH_3$

mam^5U = 5-Methylaminomethyluridine : $-CH_2-NH-CH_3$

$cmam^5U$ = 5-Carboxymethylaminomethyluridine : $-CH_2-NH-CH_2-COOH$

6) Derivative of U in position 3

acp^3U = 3-(3-Amino-3-carboxypropyl)uridine

$= X = -CH_2-CH_2-CH\begin{smallmatrix}COOH\\NH_2\end{smallmatrix}$

7) Derivative of C in position 4

ac^4C = N^4-Acetylcytidine : $-NH-CO-CH_3$

8) Thioderivatives of U and C

s^4U = 4-Thiouridine

m^5s^2U = 5-Methyl-2-thiouridine = s^2T

mcm^5s^2U = 5-(Methoxycarbonylmethyl)-2-thiouridine

mam^5s^2U = 5-Methyl(aminomethyl)-2-thiouridine

$cmam^5s^2U$ = 5-Carboxymethylaminomethyl-2-thiouridine

s^2C = 2-Thiocytidine

Figure 5 Modified nucleosides from tRNAs.

9) Isopentenyl derivatives of A

i^6A = N^6-Isopentenyladenosine

ms^2i^6A = 2-Methylthio-i^6A

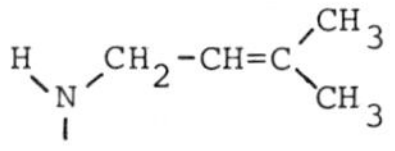

$$\text{H}\diagdown \underset{|}{\text{N}} - \text{CH}_2-\text{CH=C} \diagup_{\text{CH}_3}^{\text{CH}_3}$$

10) N^6-Threoninocarbonyl derivatives of A

t^6A = N^6-Threoninocarbonyladenosine (R_1 = H)

mt^6A = N^6-Methyl-t^6A (R_1 = CH_3)

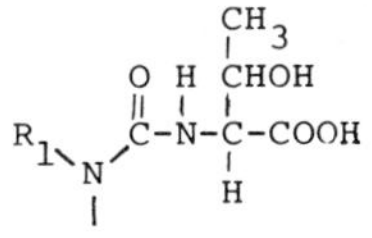

$$R_1\diagdown \underset{|}{\text{N}} \diagup \overset{\text{O}}{\underset{||}{\text{C}}}-\overset{\text{H}}{\underset{|}{\text{N}}}-\overset{\overset{\text{CH}_3}{|}}{\underset{|}{\underset{\text{H}}{\text{C}}}}-\text{COOH}$$

with CHOH on middle carbon

11) <u>Wye derivatives</u> (formerly Y derivatives)

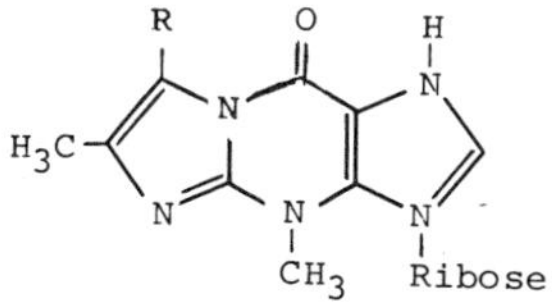

- Wyo = Wyosine (Formerly "Yt")
  R = H

- $(MeO_2)Fn$ Bto Wyo
  = Wybutosine (Formerly "Y")

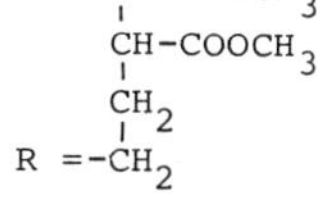

$$R = -\underset{|}{\text{CH}_2}-\underset{|}{\text{CH}_2}-\underset{|}{\text{CH}}-\text{COOCH}_3 \text{ with NH-COOCH}_3$$

- $O_2(MeO)_2Fn$ Bto Wyo
  = Peroxywybutosine (Formerly "Yr, Yw or peroxy Y")

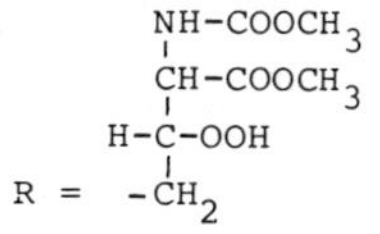

$$R = -\text{CH}_2-\text{C(OOH)H}-\text{CH(COOCH}_3)-\text{NH-COOCH}_3$$

12) <u>Q derivatives</u>

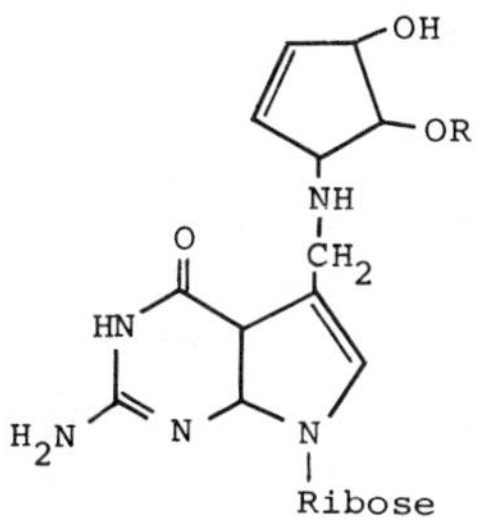

Q or Quo (Quenosine) : R = H

man Q : R = β-D-Mannosyl

gal Q : R = β-D-Galactosyl

Figure 5 (Continued)

and Clark 1974; Jank et al. 1977) and in tRNA^{Ala} from *B. mori* so it does not appear to be a specific feature of eukaryotic initiator tRNAs.

It must be mentioned that there are three exceptions to the generalized tRNA structure. The first one is tRNA$^{His}_1$ from *E. coli* and *S. typhimurium*, which contains 8 bp in the amino acid stem (Singer and Smith 1972; Singer et al. 1972). This stem is thus longer by 1 nucleotide at the 5' end and the single-stand region at the 3' end of the amino acid stem is restricted to the CCA triplet, but these tRNAs, which do act in protein synthesis, are also involved, when aminoacylated, in regulation of histidine biosynthesis.

The second is the absence of a Watson-Crick base pair between the 5'-terminal base and the fifth base from the 3' end, which is found in all, formylated prokaryotic initiator tRNAs including the blue-green algae *Anacystis nidulans* (Ecarot-Charrier and Cedergren 1976), but the non-formylated initiator tRNA from the prokaryotic *Halobacter cutirubrum* possesses a normal Watson-Crick base pair at the end of the acceptor stem (Baumstark et al., as quoted by Rich and RajBhandary 1976). Schulman and Her (1973) and Schulman and Pelka (1975) postulated that the inability of *E. coli* tRNA^{fMet} to bind to *E. coli* Tu factor and the resistance of *E. coli* formylmethionyl-tRNA^{fMet} toward peptidyl-tRNA hydrolase are due to this nonpairing.

The last exceptions, which would necessitate further investigations, are tRNA^{Val} from Torula yeast (Takemura et al. 1968; Mizutani et al., 1968) and tRNA$^{Gly}_1$ from yeast (Yoshida 1973). They have only three nucleotides in the variable loop, which, as pointed out by Rich and RajBhandary (1976), does not permit their accommodation within the three-dimensional structure of tRNA^{Phe}.

Rare Nucleosides of tRNAs

All tRNAs contain a variety of atypical (also called modified or rare) nucleosides (Nishimura, this volume). To date more than 50 of them have been characterized. In addition to dihydrouridine (D) and pseudouridine (ψ), many base or 2'-*O*-ribose-methylated nucleosides have been found (Fig. 5). Also so-called hypermodified nucleosides having very complex structures, such as wyosine or queuosine and their derivatives, have been identified in sequenced tRNAs (Fig. 5) (for recent reviews on rare nucleosides, see Hall and Dunn 1975; Sodd 1976; McCloskey and Nishimura 1977; Feldman 1977; Nishimura, this volume). Several tRNAs for which primary structures have been published contain unknown nucleosides (Fig. 6). This is often the case with tRNAs whose structures were determined by radioactive methods. These unknown nucleotides come mostly from T4-coded or *E. coli* tRNAs.

Several other rare nucleosides will probably be found when more

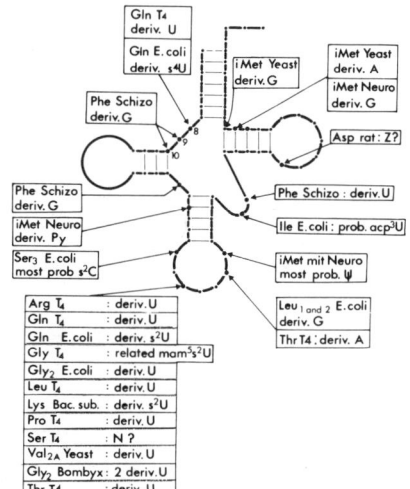

Figure 6 Previously unknown nucleosides in published tRNA structures.

mammalian or plant tRNA sequences are known, since these tRNAs are very rich in modified bases. For example, N^6-(cis-4-hydroxyisopentenyl) adenosine has been characterized in plant tRNA (Hall and Dunn 1975) and ribosyl-*cis*-zeatin has been isolated and identified from *Corynebacterium fascians* tRNA (Einset and Skoog 1977), but they are not yet described in sequenced tRNAs.

The positions of modified nucleosides in the tRNAs are very distinctive. For example (Fig. 7), in the amino acid stem, modified nucleosides are relatively rare: one finds only ψ in position 1, Um or C in position 4, m^2G in positions 6 and 7, and ψ in position 67. Only eukaryotes have rare nucleosides in these positions. The numbers of eukaryotic tRNAs, where the rare nucleosides are present, are circled in Figure 7. s^4U in position 8 is found only in prokaryotic tRNAs, 33 altogether, as is shown by the uncircled 33 in the figure. Some positions, such as positions 9 or 13, are often modified in both eukaryotes and prokaryotes. Some modified nucleosides are present only in eukaryotes, for example, m^2G in position 10 or m_2^2G in position 26. ac^4C is present in eukaryotes only in position 12, but in prokaryotes only in the anticodon loop position 34. Therefore, the modifying enzymes exist in prokaryotes as well as in eukaryotes, but have different specificities. There are some positions in tRNAs, such as positions 8, 10, and 12, where only one type of modified nucleoside is found and other positions, such as positions 4, 9, and 26, where several different types of modified nucleosides have been found.

The D loop of eukaryotic and of prokaryotic tRNAs contains mainly D

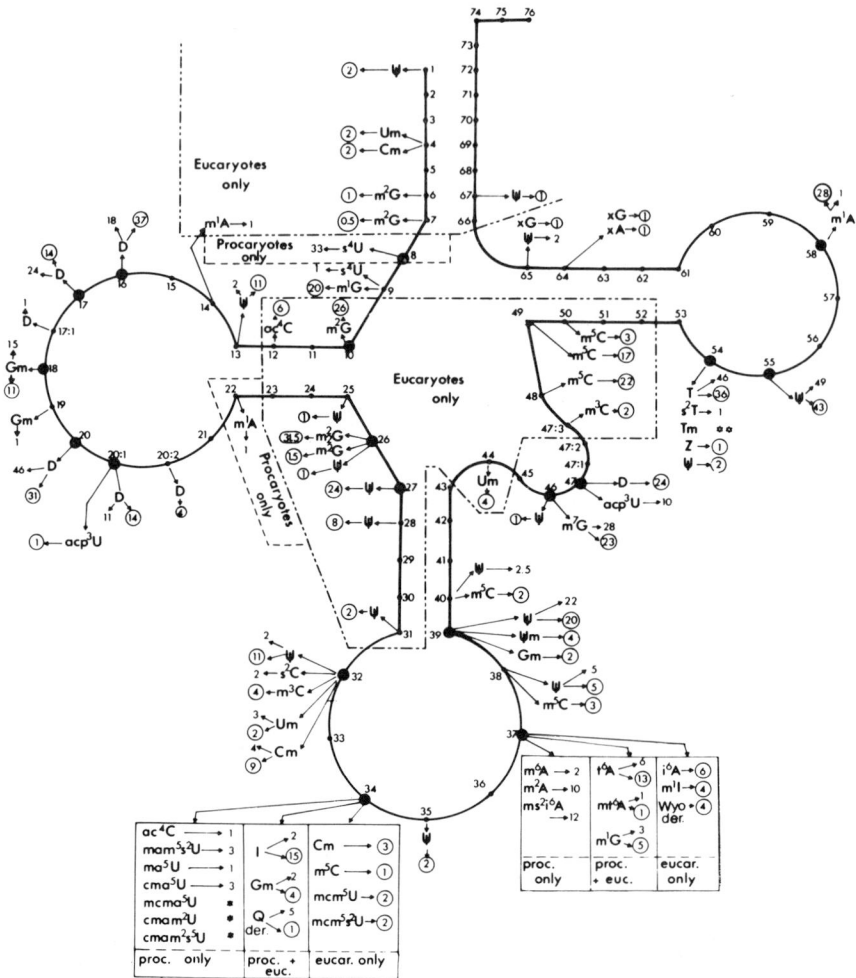

Figure 7 Localization of known nucleosides in published tRNA sequences. The uncircled numbers beside the rare nucleosides correspond to the prokaryotic tRNAs; the circled numbers correspond to the eukaryotic tRNAs where the rare nucleosides were found. (★) H. Ishikura et al. (pers. comm.). (✡✡) H. Gross (pers. comm.). ψ was also found in position 68 and m^2G in position 9 in eukaryotes. xG and xA are unknown modified G and A.

and Gm, but recently acp^3U has been found at position 20:1 of rat tRNA^{Asn} (Chen and Roe 1978). The m^1A in position 14 has been found only in eukaryotes.

The anticodon stem often has several ψs at its 5' side but only in eukaryotes. In the anticodon loop there are five positions where modifications often occur. At the wobble position, in addition to I, there are modified nucleosides not found in other positions in tRNAs, such as several derivatives of U (Fig. 7), as well as hypermodified nucleosides like mam^{5}s^{2}U in prokaryotes and mcm^{5}U in eukaryotes. Adjacent to the anticodon in 3' position there are often modified and hypermodified purines, some of which are only in prokaryotes and others only in eukaryotes (Fig. 7).

The variable arm often has m^{7}G in position 46. In position 47, there is D only in eukaryotes and acp^{3}U only in prokaryotes. Positions 48, 49, and 50 are m^{5}C only in eukaryotes. As it has already been discussed, the Tψ sequence is very common, but the T can be replaced by either an s^{2}T, a Tm, or an unknown nucleotide Z. Finally, the m^{1}A was once thought to be characteristic only of eukaryotes, but the *T. thermophilus* initiator tRNA^{Met} also has an m^{1}A at this position (Watanabe et al. 1976).

Table 3 gives the list of rare nucleosides found only in prokaryotes or only in eukaryotes, but this list, as well as all the characteristics given above, applies to the 113 structures published and will certainly be modified when new structures have been published. Supporting evidence for this prediction comes from study of specificity of tRNA-modifying enzymes. For example, the existence of an enzyme methylating A22 in *B. subtilis* was demonstrated (Raettig et al. 1977) before such modification could be found in that position in tRNA from a related bacteria: *Bacillus stearothermophilus* (Brown et al. 1978). Furthermore, plant mitochondrial or chloroplastic methyl transferases methylate an A on a cytoplasmic

Table 3 Rare nucleotides present

In prokaryotic tRNAs only	In eukaryotic tRNAs only
s^{2}C	m^{3}C
s^{4}U	m^{5}C
o^{5}U	mcm^{5}U
mo^{5}U	Tm
cmo^{5}U	m^{2}G
mam^{5}s^{2}U	m$^2_2$G
cmam^{5}U	i^{6}A
cmam^{5}s^{2}U	m^{1}I
mcmo^{5}U	Wye and derivatives
s^{2}T	
m^{2}A	
m^{6}A	
ms^{2}i^{6}A	

tRNA in position 7 in which no methylation was found up to now (Dubois et al. 1974).

mttRNAs have lower G + C contents than their cytoplasmic counterparts (for a review, see Buetow and Wood 1978) and their modified nucleoside content is distinctive.

It has been shown that mitochondria of different origin contain a variety of methylases capable of modifying specific residues in mttRNAs of yeast (Martin et al. 1976b), *N. crassa* (DeVries et al. 1978), *Locusta migratoria* (Feldmann and Kleinow 1976), rat liver (Chia et al. 1976), BHK21 cells (Davenport et al. 1976; Dubin and Friend 1974), and HeLa

Table 4 Rare base composition and G + C content of bulk mttRNA from different origins

	Fungi		Insects	Mammals		
Minor nucleoside	yeast[a] (%)	*N. crassa*[b] (%)	*L. migratoria*[c] (%)	rat liver[d] (%)	hamster cells BHK21[e] (%)	HeLa cells[f] (%)
m^1A	—	—	present	1.03	0.60	0.75
t^6A	0.35	0.20		—		
ψ	2.80	1.55	4.5	2.80		
D	2.60	2.50		1.20		
rT	0.95	1.20	1.0	0.19	0.09	0.18
s^4U	—	—	—			
acp^3U				0.17		
m^5C	—	—		1.01 ⎫	0.43	0.48
m^3C	—	—		0.14 ⎭		
m^1G	0.75	0.50		0.54	0.12	0.14
m^2G	0.20	0.25		0.98	0.75	0.62
m_2^2G	0.80	0.35	present	0.36	0.53	0.41
m^7G	—	—	present	0.28	<0.03	<0.02
I	<0.07	0.05		0.14		
Wye	—	—	—			
Ribose-methylated	—	—			<0.08	<0.08
Methylated nucleosides content	2.70	2.55		4.53	2.70	2.80
G + C content	35.1	43.2	30.7	43.4	38.5	39.2

[a]Martin et al. 1976b; Schneller et al. 1975a.
[b]DeVries et al. 1978.
[c]Feldmann and Kleinow 1976.
[d]Chia et al. 1976.
[e]Dubin and Friend 1974; Davenport et al. 1976.
[f]Davenport et al. 1976.

cells (Davenport et al. 1976). The level of methylation is about half the amount for corresponding cytoplasmic tRNAs. Yeast mttRNAs are methylated to about the same extent as *E. coli* tRNAs, but their methylation pattern is quite different (Martin et al. 1976b). The rare nucleosides found in the different mttRNAs studied so far are listed in Table 4. mttRNAs from fungi share the presence of rT and the absence of m^1A, m^7G, and m^5C (Martin et al. 1976b; DeVries et al. 1978). These nucleosides as well as m^3C are, however, present in mttRNAs from mammalian cells, where, in contrast, the rT content is very low (Dubin and Friend 1974; Chia et al. 1976; Davenport et al. 1976).

Among the hypermodified nucleosides usually adjacent to the 3' end of the anticodon was found t^6A in both yeast and *N. crassa* mttRNA. Rat liver mttRNA contains the hypermodified acp^3U nucleoside. The absence of the prokaryotic s^4U base and of the eukaryotic Wye base seems to be general in mttRNAs in both lower and higher eukaryotic cells (Schneller et al. 1975b; Fairfield and Barnett 1971; DeVries et al. 1978; Feldmann and Kleinow 1976).

Several modifying enzymes would be required to produce the number of rare bases found in mttRNAs. These enzymes, as in the case of aminoacyl-tRNA synthetases, are different from those found in the cytoplasm and are presumably of nuclear origin, at least in yeast (A. J. C. Stahl, pers. comm.).

ACKNOWLEDGMENTS

We are grateful to Mrs. J. Canaday-Blum for correcting the English of the manuscript, Mrs. M. Schneider for the preparation of the figures, and Mrs. G. Issler for her success in overcoming the chaos in the manuscript pages. We thank the numerous investigators who made unpublished data available to us.

REFERENCES

Barrell, B. G. and B. F. C. Clark, eds. 1974. *Handbook of nucleic acids sequences.* Joynson-Bruvvers Ltd, Oxford.

Borst, B. 1972. Mitochondrial nucleic acids. *Annu. Rev. Biochem.* **41**:333.

Brown, R. S., J. R. Rubin, D. Rhodes, H. Guilley, A. Simoncsits, and G. G. Brownlee. 1978. The nucleotide sequence of tyrosine tRNA from *Bacillus stearothermophilus. Nucleic Acids Res.* **5**:23.

Buetow, D. E. and W. M. Wood. 1978. The mitochondrial translation system. In *Subcellular biochemistry* (ed. D. B. Roodyn), vol. 5, p.1. Plenum Press, New York.

Chang, S. H., C. K. Brum, M. Silberklang, U. L. RajBhandary, L. I. Hecker, and W. E. Barnett. 1976. The first nucleotide sequence of an organelle tRNA: Chloroplastic tRNA^{Phe}. *Cell* **9**:717.

Chen, E. Y. and B. A. Roe. 1978. The nucleotide sequence of rat liver tRNA^{Asn}. *Biochem. Biophys. Res. Commun.* **82**:235.

Chia, L.-L. S. Y., H. P. Morris, K. Randerath, and E. Randerath. 1976. Base composition studies on mitochondrial 4S RNA from rat liver and Morris hepatomas 5123D and 7777. *Biochim. Biophys. Acta* **425**:49.

Clark, B. F. C. and A. Klug. 1975. Structure and function of tRNA with special reference to the three-dimensional structure of yeast phenylalanine tRNA. *FEBS Proc. Meet.* **10**:183.

Davenport, L. W., R. H. Taylor, and D. T. Dubin. 1976. Comparison of human and hamster mitochondrial transfer RNA. Physical properties and methylation states. *Biochim. Biophys. Acta* **447**:285.

DeVries, H., J. C. DeJonge, J.-M. Schneller, R. P. Martin, G. Dirheimer, and A. J. C. Stahl. 1978. *Neurospora crassa* mitochondrial transfer RNAs. *Biochim. Biophys. Acta* **520**:419.

Dirheimer, G., G. Keith, R. Martin, and J. Weissenbach. 1976. Primary sequences of tRNAs. In *Synthesis, structure and chemistry of tRNAs and their components*, p. 273. Polish Acad. Sci., Poznan.

Dirheimer, G., J.-P. Ebel, J. Bonnet, G. Keith, B. Krebs, B. Kuntzel, A. Roy, J. Weissenbach, and C. Werner. 1972. Structure primaire des tRNAs. *Biochimie* **54**:127.

Dubin, D. T. and D. A. Friend. 1974. Methylation properties of mitochondrion-specific transfer RNA from cultured hamster cells. *Biochim. Biophys. Acta* **340**:269.

Dubois, E. G., G. Dirheimer, and J. H. Weil. 1974. Methylation of yeast tRNA^{Asp} by enzymes from cytoplasm, chloroplasts and mitochondria of *Phaseolus vulgaris*. *Biochim. Biophys. Acta* **374**:332.

Einset, J. W. and F. K. Skoog. 1977. Isolation and identification of ribosyl-*cis*-zeatin from tRNA of *Corynebacterium fascians*. *Biochem. Biophys. Res. Commun.* **79**:1117.

Ecarot-Charrier, B. and R. J. Cedergren. 1976. The preliminary sequence of tRNA$_f^{Met}$ from *Anacystis nidulans* compared with other initiator tRNAs. *FEBS Lett.* **63**:287.

Fairfield, S. A. and W. E. Barnett. 1971. On the similarity between the tRNAs of organelles and procaryotes. *Proc. Natl. Acad. Sci.* **68**:2972.

Feldman, M. Y. 1977. Minor components in transfer RNA: The location-function relationships. *Prog. Biophys. Mol. Biol.* **32**:83.

Feldmann, H. and W. Kleinow. 1976. Base composition of mitochondrial RNA species and characterization of mitochondrial 4S RNA from *Locusta migratoria*. *FEBS Lett.* **69**:300.

Fournier, M., J. Labouesse, G. Dirheimer, C. Fix, and G. Keith. 1978. Primary structure of bovine liver tRNA^{Trp}. *Biochim. Biophys. Acta* **521**:198.

Fradin, A., H. Grühl, and H. Feldmann. 1975. Mapping of yeast tRNAs by two-dimensional electrophoresis on polyacrylamide gels. *FEBS Lett.* **50**:185.

Gauss, D. H., F. Grüter, and M. Sprinzl. 1979. Compilation of tRNA sequences. *Nucleic Acids Res.* **6**:r1.

Gillum, A. M., N. Urquhart, M. Smith, and U. L. RajBhandary. 1975. Nucleotide sequence of salmon testes and salmon liver cytoplasmic initiator tRNA. *Cell* **6**:395.

Goddard, J. P. 1977. The structures and functions of tRNA. *Prog. Biophys. Mol. Biol.* **32**:233.
Guillemaut, P. and G. Keith. 1977. Primary structure of bean chloroplastic tRNA^{Phe}. *FEBS Lett.* **84**:351.
Hall, R. H. and D. B. Dunn. 1975. Natural occurrence of modified nucleosides. In *Handbook of biochemistry and molecular biology*, 3rd ed., *Nucleic acids* (ed. G. D. Fasman), vol. 1, p. 216. CRC Press, Cleveland, Ohio.
Harada, F., R. C. Sawyer, and J. Dahlberg. 1975. A primer ribonucleic acid for initiation of *in vitro* Rous sarcoma virus deoxyribonucleic acid synthesis. Nucleotide sequence and amino acid acceptor activity. *J. Biol. Chem.* **250**:3487.
Heckman, J. E., L. I. Hecker, S. D. Schwartzbach, W. E. Barnett, B. Baumstark, and U. L. RajBhandary. 1978. Structure and function of initiator methionine tRNA from the mitochondria of *Neurospora crassa*. *Cell* **13**:83.
Hill, C. W., W. Combriato, W. Steinhart, D. L. Riddle, and J. Carbon. 1973. The nucleotide sequence of the G-G-G specific glycine tRNA of *Escherichia coli* and of *Salmonella typhimurium*. *J. Biol. Chem.* **248**:4252.
Holley, R. W., J. Apgar, G. A. Everett, J. T. Madison, M. Marquisee, S. H. Merril, J. R. Penswick, and A. Zamir. 1965. Structure of a ribonucleic acid. *Science* **147**:1462.
Jank, P., N. Shindo-Okada, S. Nishimura, and H. J. Gross. 1977. Rabbit liver tRNA$_1^{Val}$. 1. Primary structure and unusual codon recognition. *Nucleic Acids Res.* **4**:1999.
Kimball, M. E., K. S. Szeto, and D. Söll. 1974. The nucleotide sequence of phenylalanine tRNA from *Mycoplasma* sp. (Kid). *Nucleic Acids Res.* **1**:1721.
Kobayashi, T., T. Irie, M. Yoshida, K. Takeishi, and T. Ukita. 1974. The primary structure of yeast glutamic acid tRNA specific to the G-A-A codon. *Biochim. Biophys. Acta* **366**:168.
Lewitt, M. 1969. Detailed molecular model for tRNA. *Nature* **224**:759.
Margulis, L. 1970. *Origins of eukaryotic cells*. Yale University Press, New Haven, Connecticut.
Martin, N. C. and M. Rabinowitz. 1978. Mitochondrial transfer RNAs in yeast. Identification of isoaccepting tRNAs. *Biochemistry* **17**:1628.
Martin, N. C., M. Rabinowitz, and H. Fukuhara. 1977. Yeast mitochondrial DNA specifies tRNA for 19 amino acids. Deletion mapping of the tRNA genes. *Biochemistry* **16**:4672.
Martin, R. P., J. M. Schneller, A. J. C. Stahl, and G. Dirheimer. 1976a. Isoacceptor tRNA species in yeast mitochondria. Methionine and formylmethionine specific tRNAs coded by mitochondrial DNA. In *Genetics and biogenesis of chloroplasts and mitochondria* (ed. T. Bücher et al.), p. 755. Elsevier/North-Holland, Amsterdam.
―――. 1976b. Studies of odd bases in yeast mitochondrial tRNA. II. Characterization of rare nucleosides. *Biochem. Biophys. Res. Commun.* **70**:997.
―――. 1977. Study of yeast mitochondrial tRNAs by two-dimensional polyacrylamide gel electrophoresis: Characterization of isoaccepting species and search for imported cytoplasmic tRNAs. *Nucleic Acids Res.* **4**:3497.
Martin, R. P., A.-P. Sibler, J.-M. Schneller, G. Keith, A. J. C. Stahl, and G. Dirheimer. 1978. Primary structures of yeast mitochondrial DNA-coded phenylalanine-tRNA. *Nucleic Acids Res.* **5**:4579.

Mazzara, G. P. and W. H. McClain. 1977. Cysteine tRNA of *Escherichia coli*: Nucleotide sequence and unusual metabolic properties of the 3'C-C-A terminus. *J. Mol. Biol.* **117:**1061.

McCloskey, J. A. and S. Nishimura. 1977. Modified nucleosides in tRNA. *Acc. Chem. Res.* **10:**403.

Mizutani, T., T. Miyazaki, and S. Takemura. 1968. The primary structure of valine-1 tRNA from *Torulopsis utilis*. II. Partial digestion with RNase T_1 and derivation of the complete sequence. *Jpn. J. Biochem.* **64:**839.

Piper, P. W. and B. F. C. Clark. 1974. The nucleotide sequence of cytoplasmic methionine and valine tRNAs from mouse myeloma cells. *FEBS Lett.* **47:**56.

Raettig, R., H. Kersten, J. Weissenbach, and G. Dirheimer. 1977. Methylation of an adenosine in the D-loop of specific tRNAs from yeast by a procaryotic tRNA (adenine-1) methyltransferase. *Nucleic Acids Res.* **4:**1769.

Rich, A. and U. L. RajBhandary. 1976. Transfer RNA: Molecular structure, sequence and properties. *Annu. Rev. Biochem.* **45:**805.

Riddle, D. L. and J. Carbon. 1973. Frameshift suppression: A nucleotide addition in the anticodon of a glycine transfer RNA. *Nat. New Biol.* **242:**230.

Roberts, R. J. 1972. Structures of two glycyl-tRNAs from *Staphylococcus epidermidis*. *Nat. New Biol.* **237:**44.

———. 1974. Staphylococcal tRNAs. II. Sequence analysis of isoaccepting glycine tRNAs IA and IB from *Staphylococcus epidermidis* Texas 26. *J. Biol. Chem.* **249:**4787.

Schneller, J. M., A. J. C. Stahl, and H. Fukuhara. 1975a. Coding origin of isoaccepting tRNA in yeast mitochondria. *Biochimie* **57:**1051.

Schneller, J. M., R. Martin, A. Stahl, and G. Dirheimer. 1975b. Studies of odd bases in yeast mitochondrial tRNA: Absence of the fluorescent "Y" base in mitochondrial DNA coded tRNA^{Phe}, absence of 4-thiouridine. *Biochem. Biophys. Res. Commun.* **64:**1046.

Schulman, L. A. and M. O. Her. 1973. Recognition of altered *E. coli* formylmethionine tRNA by bacterial T factor. *Biochem. Biophys. Res. Commun.* **51:**275.

Schulman, L. A. and H. Pelka. 1975. The structural basis of the resistance of *E. coli* formylmethionyl tRNA to cleavage by *E. coli* peptidyl tRNA hydrolase. *J. Biol. Chem.* **250:**542.

Singer, C. E. and G. R. Smith. 1972. Regulation in *Salmonella typhimurium*. XIII. Nucleotide sequence of histidine transfer ribonucleic acid. *J. Biol. Chem.* **247:**2989.

Singer, C. E., G. R. Smith, R. Cortese, and B. N. Ames. 1972. Transfer RNA: Mutant histidine tRNA ineffective in repression and lacking two pseudouridine modifications. *Nat. New Biol.* **238:**72.

Smith, C. J., A. N. Ley, P. D'Obrenan, and S. K. Mitra. 1971. The structure and coding properties of a lysine transfer ribonucleic acid from the haploid yeast *Saccharomyces cerevisiae* αS288C. *J. Biol. Chem.* **246:**7817.

Sodd, M. A. 1976. Analysis of the primary and secondary structure of tRNA. In *Handbook of biochemistry and molecular biology*, 3rd ed., *Nucleic acids* (ed. G. D. Fasman), vol. 2, p. 423. CRC Press, Cleveland, Ohio.

Sprague, K. U., O. Hagenbüchle, and M. C. Zuniga. 1977. The nucleotide sequence of two silk gland alanine tRNAs: Implication for fibroin synthesis and for initiator tRNA structure. *Cell* **11:**561.

Suyama, Y. and J. Hamada. 1976. Imported tRNA: Its synthetase as a probable transport protein. In *Genetics and biogenesis of chloroplasts and mitochondria* (ed. T. Bücher et al.), p. 763. Elsevier/North-Holland, Amsterdam.

Takemura, S., T. Mizutani, and M. Miyazaki. 1968. The primary structure of valine 1 tRNA from *Torulopsis utilis*. I. Complete digestion with pancreatic ribonuclease and ribonuclease T_1. *Jpn. J. Biochem.* **64:**827.

Watanabe, K., T. Oshima, and S. Nishimura. 1976. C. D. spectra of 5-methyl-2-thiouridine in $tRNA_f^{Met}$ from an extreme thermophile. *Nucleic Acids Res.* **3:**1703.

Yoshida, M. 1973. The nucleotide sequence of $tRNA^{Gly}$ from yeast. *Biochem. Biophys. Res. Commun.* **50:**779.

[³H]Borohydride: A Versatile Reagent for the Analysis of tRNA— Methods and Applications

Kurt Randerath, Ramesh C. Gupta, and Erika Randerath
Department of Pharmacology, Baylor College of Medicine
Texas Medical Center
Houston, Texas 77030

This review serves two purpose: (1) to outline ³H derivative methods for base composition and sequence analysis of nonradioactive RNA developed in our laboratory over the past 10 years and (2) to summarize results obtained in our laboratory by the application of these methods.

The first experiments designed to incorporate radioactive label into the end groups of nucleic acids by chemical means were reported 15–20 years ago by Khorana and his co-workers (Khorana 1959; Ralph et al. 1963; RajBhandary et al. 1964), but the principles enunciated by these workers have only recently found wide application in structural studies on nucleic acids. Thus far, the reagent most often used for labeling RNA 3' ends has been [³H]borohydride. The labeling reaction entails the oxidation of the vicinal hydroxyl groups of the ribose moiety of nucleosides or nucleotides with sodium metaperiodate (NaIO₄), followed by the reduction of the resulting dialdehydes to diols:

$$\text{Nucleoside} \xrightarrow{\text{NaIO}_4} \text{Dialdehyde} \xrightarrow{[^3\text{H}]\text{KBH}_4 \text{ or } [^3\text{H}]\text{NaBH}_4} \text{Nucleoside trialcohol (R = H) or Nucleotide dialcohol (R = phosphate, mononucleotide, or polynucleotide residue)} \quad (1)$$

In contrast to enzymatic ^{32}P labeling by the [γ-^{32}P]ATP-polynucleotide kinase reaction (see RajBhandary, this volume), ^{3}H-label incorporation into the dialdehydes according to Scheme 1 is stoichiometric if the reduction is performed with excess borotritide (Randerath et al. 1972, 1979a). This has to be regarded as a major advantage of the ^{3}H-labeling method, because the reaction is thus suitable for accurate quantitative determinations, e.g., of the base composition of polyribonucleotides or the molar ratios of oligoribonucleotides in enzymatic digests of RNA. Other advantages of the ^{3}H-derivative methods include the availability of reproducible separation methods of high resolving power for the diol derivatives (Randerath et al. 1972, 1979a) and the possibility of identifying each position in a polynucleotide chain directly as a ^{3}H-labeled derivative of the nucleoside occupying that position. Compared with ^{32}P, ^{3}H label has the disadvantage of a lower specific radioactivity. [^{3}H]Borohydride is available at a maximum specific activity of about 50 Ci/mmole, whereas carrier-free ^{32}P label is available at a specific activity of about 5000 Ci/mmole. For physical reasons, ^{3}H-labeling procedures are therefore less well suited than is labeling with ^{32}P whenever a high specific activity of the isotope is a critical factor, as appears to be the case for the gel sequencing techniques.

BASE ANALYSIS OF RNA BY ^{3}H LABELING
Method

The method is based on the initial complete enzymatic degradation of RNA to nucleosides followed by stoichiometric ^{3}H-label incorporation into the nucleosides (Scheme 1). After separation of the ^{3}H-labeled nucleoside trialcohols by two-dimensional partition chromatography on a cellulose thin layer (Fig. 1) and treatment of the chromatogram with scintillator, the radioactive compounds are located by low-temperature fluorography on X-ray film (Randerath 1970), extracted from the chromatogram with 2 N ammonia, and assayed by liquid scintillation counting (Randerath et al. 1972, 1979a). The base composition is calculated by dividing the radioactivity of each individual compound by the radioactivity recovered from all radioactive nucleoside derivatives and expressed in mole %. The method allows one to determine trace constituents (a single nucleoside in 5000 total nucleosides) in microgram amounts of RNA. For comparative studies (see below), it is essential to free the 4S RNA preparation obtained by phenol extraction from contaminating RNAs prior to base analysis. This is best accomplished by polyacrylamide gel electrophoresis (Chia et al. 1973).

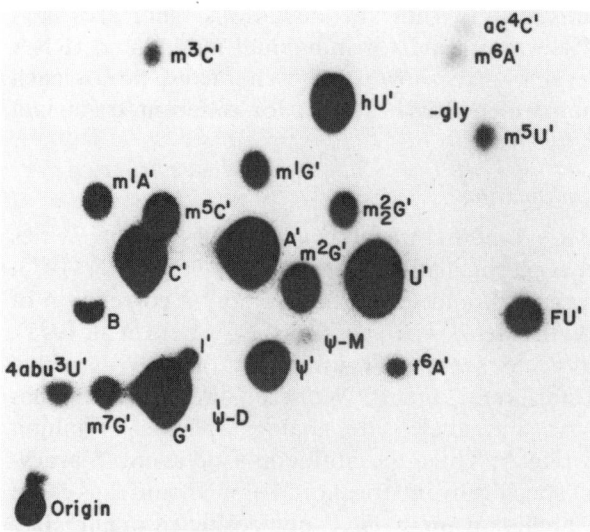

Figure 1 Fluorograph of nucleoside trialcohols on a cellulose thin-layer map. Bulk liver tRNA from 5-fluorouridine-treated mice was digested to nucleosides. The digest was treated with periodate and [^{3}H]borohydride. N' is a ^{3}H-labeled trialcohol derivative of the indicated nucleoside; FU' is 5-fluorouridine trialcohol; ψ-D and ψ-M are traces of labeled products derived from ψ; B is a background spot (not from RNA); and gly is glycerol.

Applications

Tumor tRNA

The method for base composition analysis was originally developed about 10 years ago for the purpose of measuring the state of methylation of tRNA from neoplastic cells. At that time, experiments in various laboratories had indicated increased activities and capacities of tRNA methyl transferases in many neoplastic tissues (Borek and Kerr 1972), but little was known about the methylated bases in tRNA from tumors. Using the ^{3}H derivative method, we have analyzed the base composition of tRNA from human and animal brain tumors; Morris hepatomas; chemically induced, as well as SV40- and polyoma virus-induced, hamster tumors; avian and rodent leukemias; and chemically induced mouse mammary tumors. The data were compared with those for tRNA of normal tissues (K. Randerath et al. 1971; E. Randerath et al. 1974; Randerath 1971; Randerath and Randerath 1973; Chia et al. 1976). These experiments showed that tRNA isolated from the neoplastic tissues was not generally overmethylated. In several instances, tRNA from neoplastic sources was found to be slightly less modified and methylated than its normal counterpart. Statistically significant deviations (increases and decreases) of about 2–17% were found for individual methylated tRNA constituents. These deviations were shown to be characteristic for each tumor type, as was demonstrated, for example, for different transplant generations of Morris hepatomas.

Drug Effects on tRNA Modification

Substantial elevations of tRNA methyl transferase activities and capacities have been found in preneoplastic tissues (Hancock and Forester 1973; Wainfan et al. 1978) and tumors, and for certain tumors a correlation of enzyme activities with growth rates has been established (Sheid et al. 1971). It is of interest, therefore, to search for inhibitors of tRNA methyl transferases as potential antitumor agents. Work in our laboratory has indicated that certain base and nucleoside analogs specifically inhibit modification reactions of tRNA. Thus, the antineoplastic agent, 5-azacytidine, was found to inhibit specifically the formation of m^5C, and thus cause a deficiency of this most abundant methylated nucleoside in mammalian tRNA (Lu et al. 1976a; Lu and Randerath 1979). The effect appears to be due to inhibition of tRNA (cytosine-5) methyl transferase, because the activity of this enzyme is strongly inhibited after the administration of 5-azacytidine. In addition, a delayed increase in the activities of tRNA m^1A, m^2G, and m_2^2G methyl transferases following the exposure to 5-azacytidine was observed.

Inhibitory effects on tRNA modification reactions have also been observed for other pyrimidine nucleoside analogs, such as 5-fluorocytidine (Lu et al. 1979), 5-fluorouridine (Lu et al. 1976b), and 5-fluorouracil (Tseng et al. 1978), the latter two compounds specifically inhibiting modification reactions taking place at the 5-position of uracil, but not of cytosine. tRNA (uracil-5) methyl transferase was shown to be rapidly inactivated following in vivo exposure to 5-fluoro-pyrimidines (Tseng et al. 1978; Lu et al. 1979). Our results on the effects of these compounds on tRNA and tRNA-modifying enzymes are summarized in Table 1.

When the m^5C-deficient tRNA was compared with normal tRNA in aminoacylation experiments, no differences were observed (Harris and Randerath 1978). Thus, the methylation of the 5-position of C in tRNA appears to play no role in the aminoacylation reaction.

SEQUENCE ANALYSIS OF RNA BY ^3H LABELING

Methods

The readout-sequencing methods, in which an end-labeled polynucleotide is specifically cleaved at the positions of the four major bases and the products obtained are separated by size (Maxam and Gilbert 1977; Gupta

Table 1 Summary of effects of 5-fluorocytidine, 5-fluorouracil, 5-fluorouridine, and 5-azacytidine on tRNA and tRNA modification

Analog	Incorporation	Inhibitory effects on			
		m^5C	m^5U	ψ	D
5-fluorocytidine	++++[a] ++[b]	++++	+++	++	+
5-fluorouracil and 5-fluorouridine	++	—	++++	+++	+
5-azacytidine	n.d.[c]	++++	—	—	—

[a]Incorporation of 5-fluorocytidine.
[b]Incorporation of 5-fluorouridine.
[c]Not detected (<1 in 50 C residues).

and Randerath 1977a,b; Donis-Keller et al. 1977), are not suitable for the determination of the location of most modified bases in a polynucleotide chain. The identification of modified bases by mobility shift methods is also difficult or impossible. The methods developed in our laboratory enable the direct identification of the position of each major and modified nucleoside in the polynucleotide chain as a radioactively labeled derivative. The basic features of the methods are illustrated schematically in Figure 2. The polynucleotide may be digested to a series of smaller polynucleotides having the same 5′ terminal and differing in length only at the 3′ end (Fig. 2A) or the digestion products may have the same 3′

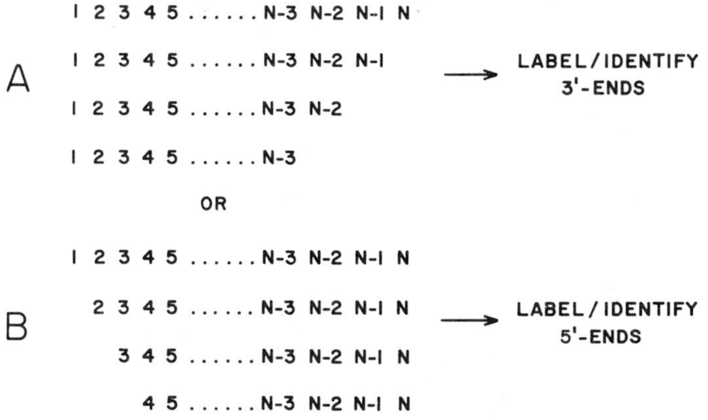

```
          1 2 3 4 5 ...... N-3 N-2 N-1 N

          1 2 3 4 5 ...... N-3 N-2 N-1
A                                              ──▶   LABEL/IDENTIFY
          1 2 3 4 5 ...... N-3 N-2                    3'-ENDS

          1 2 3 4 5 ...... N-3

                      OR

          1 2 3 4 5 ...... N-3 N-2 N-1 N

            2 3 4 5 ...... N-3 N-2 N-1 N
B                                              ──▶   LABEL/IDENTIFY
              3 4 5 ...... N-3 N-2 N-1 N              5'-ENDS

                4 5 ...... N-3 N-2 N-1 N
```

Figure 2 Sequence analysis of a polynucleotide by controlled digestion and incorporation of label into 3′- or 5′-terminal positions, respectively. The labeled products are separated by size and each radioactive terminal constituent is identified.

terminal and different 5' terminals (Fig. 2B). The products of digests A and B may then be labeled at the 3' and 5' ends, respectively. The labeled derivatives are resolved according to size by chromatography or electrophoresis and digested so as to liberate the labeled terminal constituents. Chromatographic identification of the labeled terminal groups (N, N-1, N-2, N-3, etc., in Fig. 2A; 1, 2, 3, 4, etc., in Fig. 2B) then yields the sequence of these constituents in the original intact polynucleotide chain. Various techniques are available for the initial digestion step. Treatment of the polyribonucleotide with snake venom phosphodiesterase or sodium metaperiodate and bacterial alkaline phosphatase proceeds according to Figure 2A (K. Randerath et al. 1974; Sivarajan et al. 1974). Figure 2B has been realized by nuclease S1 (Gupta et al. 1976) or formamide (Stanley and Vassilenko 1978) digestion followed by labeling of the 5' ends.

In the following discussion, three standard techniques developed in our laboratory for the sequence analysis of oligoribonucleotides and their application to the sequence analysis of tRNAs will be outlined. In the first procedure (Sivarajan et al. 1974), the oligonucleotide is partially digested with snake venom phosphodiesterase to products of decreasing chain length (cf. Fig. 2A). The 3' terminals of the partial digestion products are oxidized by periodate to oligonucleotide-3'-dialdehydes, which in turn are reduced with [^{3}H]borohydride to ^{3}H-labeled oligonucleotide 3'-dialcohols. After predevelopment with 4 M lithium formate, 7 M urea (pH 3.5) to remove radioactive contaminants present in some batches of borotritide (Randerath et al. 1979a), the products are separated by size on a polyethyleneimine (PEI)-cellulose thin layer (Fig. 3) and located by low-temperature fluorography. The labeled 3'-terminal constituents are liberated as

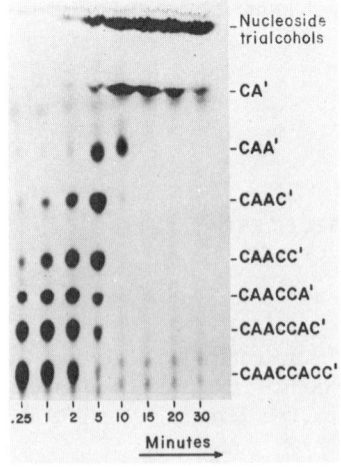

Figure 3 Resolution by size of ^{3}H-labeled oligonucleotide-3'-dialcohols on a PEI-cellulose thin layer in a Tris-urea (pH 8) system. The labeled products were obtained by controlled snake venom phosphodiesterase-phosphatase digestion of CAACCACC$_{OH}$ followed by treatments with periodate and [^{3}H]borohydride. Film detection by fluorography. CAACCACC', etc., is ^{3}H-labeled dialcohol of CAACCACC, etc.

nucleoside trialcohols by in situ treatment of the spots with RNase T2 and identified by two-dimensional partition chromatography on a silica gel thin layer (10 × 20 cm) after direct contact transfer of the nucleoside derivatives (K. Randerath et al. 1974). The separation of these compounds is illustrated schematically in Figure 4. The separation may also be performed on a cellulose thin layer, but the more rapid silica gel separation is usually preferred, particularly when the base composition of the oligonucleotide under investigation has been determined beforehand. The location of modified bases in a polynucleotide chain is readily determined by this procedure.

In the second procedure (Gupta et al. 1976a), the oligonucleotides in a ribonuclease digest of RNA are first dephosphorylated and ^{3}H labeled at the 3' terminal by periodate-borotritide treatment. The compounds are separated by two-dimensional PEI-cellulose thin-layer chromatography and located by fluorography (see Fig. 9). Individual labeled oligonucleotides are isolated (see Fig. 8) and partially digested endonucleolytically with nuclease S1. The 5'-phosphomonoester groups are removed from the digestion products by treatment with alkaline phosphatase. Nonradioactive digestion products are eliminated by periodate oxidation causing these compounds to remain bound to the origin area upon subsequent chromatography on PEI-cellulose. After the separation of the radioactive products by size (Fig. 5), their 5' terminals are determined by enzymatic incorporation of ^{32}P label, release of ^{32}P-labeled nucleoside 5'-monophosphates by digestion with nuclease S1, two-dimensional PEI-cellulose thin-layer chromatography, and autoradiography. The separation of major and modified nucleoside 5'-monophosphates has been detailed (Gupta et al. 1976b).

In the third procedure (Gupta and Randerath 1977a,b), which is not

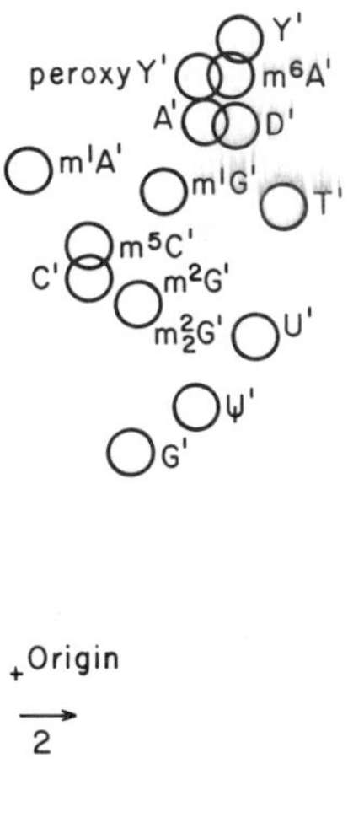

Figure 4 Two-dimensional separation of nucleoside trialcohols on silica gel, schematic. For conditions of contact transfer of 3'-terminal nucleoside trialcohols from PEI-cellulose to silica gel and chromatographic conditions, see K. Randerath et al. (1974).

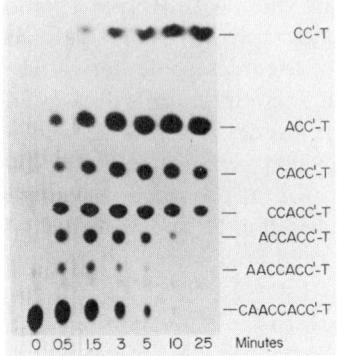

Figure 5 Resolution by size on a PEI-cellulose thin layer of ³H-labeled oligonucleotide 3′-dialcohols obtained by controlled nuclease S1 digestion of CAACCACC′-T, followed by dephosphorylation with alkaline phosphatase and periodate treatment. Film detection by fluorography. CAACCACC′-T, etc., is ³H-labeled dialcohol of CAACCACC, etc. (Reprinted, with permission, from Gupta et al. 1976a.)

suitable for the identification of modified nucleosides, the end-labeled oligoribonucleotide is partially degraded by treatment with RNases cleaving at specific positions of the chain, the products are separated by size (Fig. 6) and located on film. The sequence is then deduced by

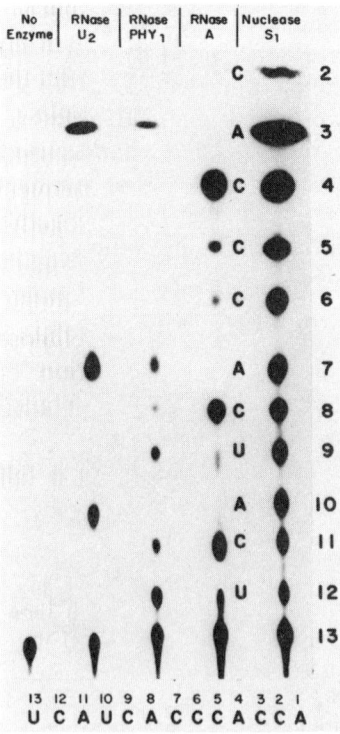

Figure 6 Resolution by size on a PEI-cellulose thin layer of ³H-labeled oligonucleotide 3′-dialcohols obtained by controlled RNase U2, RNase Phy1, RNase A, and nuclease S1 digestion of UCAUCACCCACCA′. Detection by fluorography. UCAUCACCCACCA′, etc., is ³H-labeled dialcohol of the parent compound. The numbers indicate both the chain lengths of the compounds and the positions of individual residues within the sequence. The vertical column of letters refers to individual residues identified on the basis of cleavage patterns. (Reprinted, with permission, from Gupta and Randerath 1977a.)

inspection of the cleavage patterns (readout sequencing). Both 5'-^{32}P- and 3'-^{3}H-labeled polynucleotides may be used, and separation may be by thin-layer chromatography or gel electrophoresis (Gupta and Randerath 1977a,b; Donis-Keller et al. 1977; Simoncsits et al. 1977; Lockard et al. 1978; Stanley and Vassilenko 1978). RNases A, T1, and U2 have been used to cleave at pyrimidine, G, and A residues, respectively; RNase Phy1, an extracellular RNase isolated from the culture fluid of *Physarum polycephalum* (Pilly et al. 1978), allows one to distinguish C from U residues (Gupta and Randerath 1977b; Simoncsits et al. 1977). Since the readout-sequencing procedure, in conjunction with gel electrophoresis, is capable of providing information about long nucleotide sequences, it is very useful to establish overlaps of oligonucleotides in tRNA digests.

The use of these procedures for sequence analysis of tRNAs is illustrated by specific examples in the following section.

Applications

To determine the structure of yeast tRNA$_{UAG}^{Leu}$ (Randerath et al. 1975; Randerath et al. 1979b), information from ^{3}H derivative analysis of the products in complete RNase digests was combined with data obtained by subjecting partial RNase T1 digestion products (half molecules) to further digestions with RNases A, T1, and U2 and ^{3}H derivative analysis of the digests. ^{3}H derivative analysis was carried out according to the first procedure described in the preceding section. The 3' half molecule was also analyzed by gel readout sequencing (Randerath et al. 1979b). Thus, each position in the digestion products obtained was identified in the form of a radioactive derivative and the necessary overlaps were established for the determination of the primary structure of the RNA.

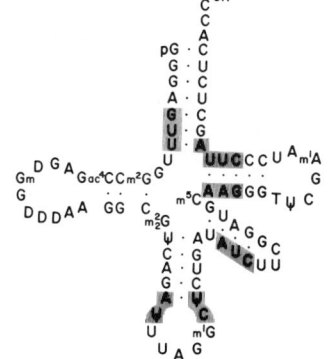

Figure 7 The sequence of tRNA$_{UAG}^{Leu}$ of baker's yeast. The shaded sequences have been found only in tRNAs^{Leu} from yeast, but not in any other sequenced tRNA.

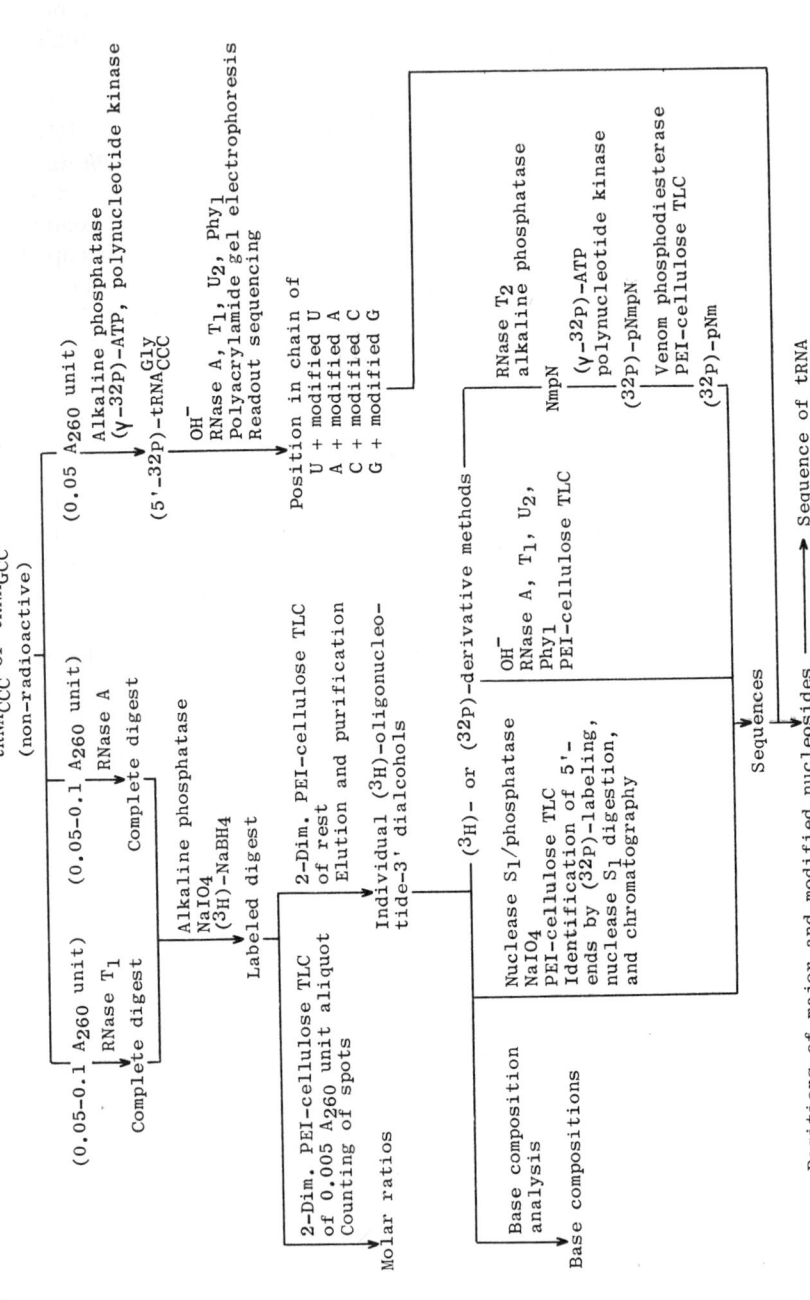

Figure 8 Radioactive derivative procedures used in the sequence analysis of nonradioactive tRNA$_{CCC}^{Gly}$ and tRNA$_{GCC}^{Gly}$ from human placenta. Oligonucleotide 3'-dialcohols were frequently sequenced by combining base composition data with readout-sequencing data. Nuclease S1 digestion and 5'-terminal analysis of oligonucleotide dialcohol intermediates were used mainly to establish the positions of modified nucleosides. TLC is thin-layer chromatography.

The sequence of yeast tRNA$^{Leu}_{UAG}$ is shown in the familiar cloverleaf form in Figure 7. The shaded areas indicate nucleotide sequences that are uniquely common to yeast tRNA$^{Leu}_{UAG}$ and yeast tRNA$^{Leu}_{m^5CAA}$, i.e., they do not occur in any other sequenced tRNAs. Although either of the shaded anticodon arm sequences (Aψ and Cψ) occurs in a few other tRNAs, only the tRNAs^{Leu} contain both sequences hydrogen bonded to each other. These sequences may conceivably be involved in interactions with cognate aminoacyl-tRNA synthetases and/or other proteins. It is interesting to note (Fig. 7) that the shaded sequences GUU and CUUA are located next to U8, a constant nucleotide, which appears to participate in synthetase binding (Schoemaker and Schimmel 1977).

Chen and Roe (1978) have recently utilized the first procedure outlined above in their work on the sequence of mammalian tRNA^{Asn}.

The structures of two human tRNAs^{Gly}, tRNA$^{Gly}_{CCC}$ and tRNA$^{Gly}_{GCC}$ (R. C. Gupta et al., in prep.), were determined recently in our laboratory by the procedures summarized in Figure 8. This approach is distinguished from the one used previously for the analysis of tRNA$^{Leu}_{UAG}$ mainly by the ^{3}H labeling of the oligonucleotides present in the initial RNase digests and the extensive use of gel readout-sequencing procedures for establishing overlaps (Fig. 8). A total of about 0.2 A$_{260}$ unit of each tRNA was required for determining the sequence. This contrasts with 6–10 A$_{260}$ units of tRNA in the older procedure. This difference in sensitivity is due in large part to the different techniques used for the detection of the

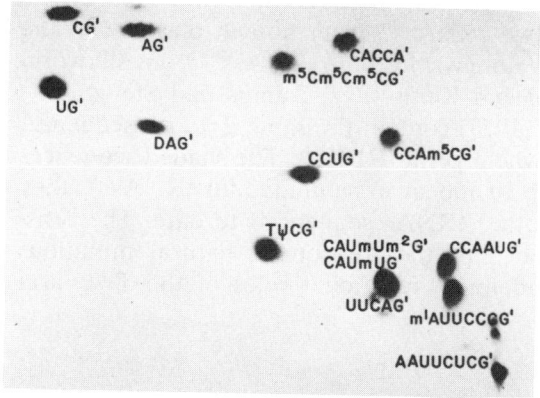

Figure 9 PEI-cellulose map of a ^{3}H-labeled RNase T1 digest of human placenta tRNA$^{Gly}_{GCC}$. First dimension (stepwise elution with lithium chloride solution), from right to left; second dimension (stepwise elution with ammonium formate [pH 2.6]), from bottom to top. Film detection by fluorography. UUCAG' was separated from CAUmUm^{2}G' and CAUmUG' by rechromatography in a Tris system (pH 8) containing 8.5 M urea (K. Randerath et al. 1974).

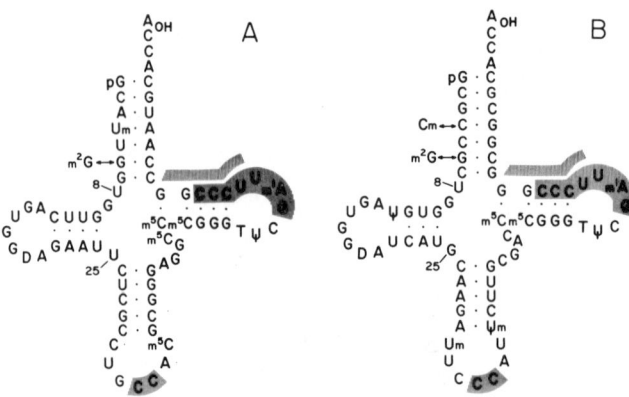

Figure 10 The sequences of tRNA$_{GCC}^{Gly}$ (*A*) and tRNA$_{CCC}^{Gly}$ (*B*) from human placenta arranged in the cloverleaf form. The possible existence of an additional Watson-Crick base pair in the anticodon stems of the tRNAs has not been indicated. Shading denotes sequences unique to tRNAs^{Gly} of higher eukaryotes (R. C. Gupta et al., in prep.).

oligonucleotides in the initial digests, i.e., examination in UV light, requiring 2–3 A_{260} units of material, versus fluorography, requiring less than 0.05 A_{260} unit of digest. In addition, determination of overlaps by gel electrophoresis requires very little material compared to partial digestions of nonradioactive RNA and chromatographic analysis.

The separation of ^{3}H-labeled oligonucleotide-3'-dialcohols in a ^{3}H-labeled RNase T1 digest of tRNA$_{CCC}^{Gly}$ is illustrated in Figure 9. Figure 10 shows the structures of the two tRNAs^{Gly} from human placenta in the familiar cloverleaf form. It is noteworthy that tRNAs^{Gly} from silkworm (Kawakami et al. 1978; Garel and Keith 1977; Zúñiga and Steitz 1977) and wheat germ (Marcu et al. 1977) contain the same Tψ arm sequence, except for U in place of T in wheat germ tRNA^{Gly}. The shaded sequences of the Tψ arm shown in Figure 10 appear to be unique for tRNAs^{Gly}; they have not been found in any other tRNAs sequenced to date. The resistance of the Tψ arm sequence to the fixation of natural mutations suggests important, as yet undefined, functional roles of the TψC arm in eukaryotic tRNAs^{Gly}.

ACKNOWLEDGMENTS

Work in our laboratory was supported by U.S. Public Health Service grants CA-13591, CA-10893-P8 (center grant), and CA-16840. K. R. was a recipient of a Faculty Research Award from the American Cancer Society (PRA-108) while part of this work was carried out.

REFERENCES

Borek, E. and S. Kerr. 1972. Atypical transfer RNAs and their origin in neoplastic cells. *Adv. Cancer Res.* **15:** 163.

Chen, E. Y. and B. A. Roe. 1978. The nucleotide sequence of rat liver tRNA^{Asn}. *Biochem. Biophys. Res. Commun.* **82:** 235.

Chia, L. S. Y., K. Randerath, and E. Randerath. 1973. Analysis of ribopolynucleotides by tritium incorporation following analytical polyacrylamide gel electrophoresis. *Anal. Biochem.* **55:** 102.

Chia, L. S. Y., H. P. Morris, K. Randerath, and E. Randerath. 1976. Base composition studies on mitochondrial 4S RNA from rat liver and Morris hepatomas 5123D and 7777. *Biochim. Biophys. Acta* **425:** 49.

Donis-Keller, H., A. M. Maxam, and W. Gilbert. 1977. Mapping adenines, guanines, and pyrimidines in RNA. *Nucleic Acids Res.* **4:** 2527.

Garel, J. P. and G. Keith. 1977. Nucleotide sequence of *Bombyx mori* L. tRNA$_1^{Gly}$. *Nature* **269:** 250.

Gupta, R. C. and K. Randerath. 1977a. Use of specific endonuclease cleavage in RNA sequencing. *Nucleic Acids Res.* **4:** 1957.

———. 1977b. Use of specific endonuclease cleavage in RNA sequencing: An enzymic method for distinguishing between cytidine and uridine residues. *Nucleic Acids Res.* **4:** 3441.

Gupta, R. C., E. Randerath, and K. Randerath. 1976a. A double-labeling procedure for sequence analysis of picomole amounts of nonradioactive RNA fragments. *Nucleic Acids Res.* **3:** 2895.

———. 1976b. An improved separation procedure for nucleoside monophosphates on polyethyleneimine (PEI)-cellulose thin layers. *Nucleic Acids Res.* **3:** 2915.

Hancock, R. L. and P. I. Forester. 1973. Increase in soluble RNA methylase activities by chemical carcinogens. *Cancer Res.* **33:** 1747.

Harris, J. S. and K. Randerath. 1978. Aminoacylation of undermethylated mammalian transfer RNA. *Biochim. Biophys. Acta* **521:** 566.

Kawakami, M., K. Nishio, and S. Takemura. 1978. Nucleotide sequence of tRNA$_2^{Gly}$ from the posterior silk glands of *Bombyx mori*. *FEBS Lett.* **87:** 288.

Khorana, H. G. 1959. Studies on polynucleotides. VII. Approaches to the marking of end groups in polynucleotide chains: The methylation of phosphomonoester groups. *J. Am. Chem. Soc.* **81:** 4657.

Lockard, R. E., B. Alzner-DeWeerd, J. E. Heckman, J. MacGee, M. W. Tabor, and U. L. RajBhandary. 1978. Sequence analysis of 5'(^{32}P)-labeled mRNA and tRNA using polyacrylamide gel electrophoresis. *Nucleic Acids Res.* **5:** 37.

Lu, L. W. and K. Randerath. 1979. Effects of 5-azacytidine on transfer RNA methyltransferases. *Cancer Res.* **39:** 940.

Lu, L. W., W.-C. Tseng, and K. Randerath. 1979. Effects of 5-fluorocytidine on mammalian transfer RNA and transfer RNA methyltransferases. *Biochem. Pharmacol* **28:** 489.

Lu, L. W., G. H. Chiang, D. Medina, and K. Randerath. 1976a. Drug effects on nucleic acid modification. I. A specific effect of 5-azacytidine on mammalian transfer RNA methylation *in vivo*. *Biochem. Biophys. Res. Commun.* **68:** 1094.

Lu, L. W., G. H. Chiang, W.-C. Tseng, and K. Randerath. 1976b. Effects of

5-fluorouridine on modified nucleosides in mouse liver transfer RNA. *Biochem. Biophys. Res. Commun.* **73:** 1075.

Marcu, K. B., R. E. Mignery, and B. S. Dudock. 1977. Complete nucleotide sequence and properties of the major species of glycine transfer RNA from wheat germ. *Biochemistry* **16:** 797.

Maxam, A. M. and W. Gilbert. 1977. A new method for sequencing DNA. *Proc. Natl. Acad. Sci.* **74:** 560.

Pilly, D., A. Niemeyer, M. Schmidt, and J. P. Bargetzi. 1978. Enzymes for RNA sequence analysis. Preparation and specificity of exoplasmodial ribonucleases I and II from *Physarum polycephalum*. *J. Biol. Chem.* **253:** 437.

RajBhandary, U. L., R. J. Young, and H. G. Khorana. 1964. Studies on polynucleotides. XXXII. The labeling of end groups in polynucleotide chains: The selective phosphorylation of phosphomonoester groups in amino acid acceptor ribonucleic acids. *J. Biol. Chem.* **239:** 3875.

Ralph, R. K., R. J. Young, and H. G. Khorana. 1963. Studies on polynucleotides. XXI. Amino acid acceptor ribonucleic acids (2). The labeling of terminal 5'-phosphomonoester groups and a preliminary investigation of adjoining nucleotide sequences. *J. Am. Chem. Soc.* **85:** 2002.

Randerath, E., R. C. Gupta, and K. Randerath. 1979a. Tritium derivative methods for base composition and sequence analysis of RNA. *Methods Enzymol.* **65:** (in press).

Randerath, E., C.-T. Yu, and K. Randerath. 1972. Base analysis of ribopolynucleotides by chemical tritium labeling: A methodological study with model nucleosides and purified tRNA species. *Anal. Biochem.* **48:** 172.

Randerath, E., L. S. Y. Chia, H. P. Morris, and K. Randerath. 1974. Transfer RNA base composition studies in Morris hepatomas and rat liver. *Cancer Res.* **34:** 643.

Randerath, E., R. C. Gupta, L. S. Y. Chia, S. H. Chang, and K. Randerath. 1979b. Yeast tRNA$^{Leu}_{UAG}$: Purification, properties, and determination of the nucleotide sequence by radioactive derivative methods. *Eur. J. Biochem.* **93:** 79.

Randerath, K. 1970. An evaluation of film detection methods for weak β-emitters, particularly tritium. *Anal. Biochem.* **34:** 188.

———. 1971. Application of a tritium derivative method to human brain and brain tumor transfer RNA analysis. *Cancer Res.* **31:** 658.

Randerath, K. and E. Randerath. 1973. Chemical characterization of unlabeled RNA and RNA derivatives by isotope derivative methods. *Methods Cancer Res.* **9:** 3.

Randerath, K., S. K. MacKinnon, and E. Randerath. 1971. An investigation of the minor base composition of transfer RNA in normal human brain and malignant brain tumors. *FEBS Lett.* **15:** 81.

Randerath, K., E. Randerath, L. S. Y. Chia, R. C. Gupta, and M. Sivarajan. 1974. Sequence analysis of nonradioactive RNA fragments by periodate-phosphatase digestion and chemical tritium labeling: Characterization of large oligonucleotides and oligonucleotides containing modified nucleosides. *Nucleic Acids Res.* **1:** 1121.

Randerath, K., L. S. Y. Chia, R. C. Gupta, E. Randerath, E. R. Hawkins, C. K. Brum, and S. H. Chang. 1975. Structural analysis of nonradioactive RNA by postlabeling: The primary structure of baker's yeast tRNA$^{Leu}_{CUA}$. *Biochem. Biophys. Res. Commun.* **63:**157.

Schoemaker, H. J. P. and P. R. Schimmel. 1977. Effect of aminoacyl transfer RNA synthetases on H-5 exchange of specific pyrimidines in transfer RNAs. *Biochemistry* **16:**5454.

Sheid, B., S. M. Wilson, and H. P. Morris. 1971. Transfer RNA methylase activity in normal rat liver and some Morris hepatomas. *Cancer Res.* **31:**774.

Simoncsits, A., G. G. Brownlee, R. S. Brown, J. R. Rubin, and H. Guilley. 1977. New rapid gel sequencing method for RNA. *Nature* **269:**833.

Sivarajan, M., R. C. Gupta, L. S. Y. Chia, E. Randerath, and K. Randerath. 1974. Tritium sequence analysis of oligoribonucleotides: A combination of postlabeling and thin-layer chromatographic techniques for the analysis of partial snake venom phosphodiesterase digests. *Nucleic Acids Res.* **1:**1329.

Stanley, J. and S. Vassilenko. 1978. A different approach to RNA sequencing. *Nature* **274:**87.

Tseng, W.-C., D. Medina, and K. Randerath. 1978. Specific inhibition of transfer RNA methylation and modification in tissues of mice treated with 5-fluorouracil. *Cancer Res.* **38:**1250.

Wainfan, E., S. S. Randle, N. M. Relya, and M. E. Balis. 1978. Effects of 2-acetylaminofluorene on transfer RNA methyltransferases of rat organs. *Cancer Res.* **38:**1852.

Zúñiga, M. C. and J. A. Steitz. 1977. The nucleotide sequence of a major glycine transfer RNA from the posterior silk gland of *Bombyx mori* L. *Nucleic Acids Res.* **4:**4175.

Modified Nucleosides in tRNA

Susumu Nishimura
Biology Division
National Cancer Center Research Institute
Chuo-ku, Tokyo, 104, Japan

One of the characteristics of tRNA is the presence of a variety of modified nucleosides, more than 50 of which have been isolated and characterized so far. The structures of the modified nucleosides thus far found in tRNA, except the 2′-O-methylated nucleosides, are shown in Appendix II. Many of these modified nucleosides result from methylation at the base or at 2′-OH of ribose, but there are also a number of hypermodified nucleosides that reflect more complex modification.

I, ψ, s^4U, and a number of methylated nucleosides were isolated and characterized over 10 years ago (Nishimura 1972; Dunn and Hall 1975). These modified nucleosides are present in most tRNA molecules. For this reason their detection and isolation in relatively large quantities are straightforward, with unfractionated tRNA being used as a source. By contrast, hypermodified nucleosides that are usually located in the anticodon region are present in only one or a few tRNA species. As a result, their detection in unfractionated tRNA is difficult. Their detection in pure tRNA species is easier, since they constitute at least 1% of the total nucleoside content. Therefore, in general, one must have a large quantity of pure tRNA species for isolation and characterization of a new modified nucleoside. In addition, knowledge of the exact location of the modified nucleoside in a specific tRNA molecule may be the only way to definitely establish that it is a legitimate component of tRNA—not an impurity carried through the isolation procedure. Using unfractionated tRNA as a source, this difficulty can be handled by isolating the corresponding mono- or oligonucleotide prior to its conversion to the nucleoside on which most of the structural work is to be carried out.

To detect modified nucleosides in a particular tRNA, the tRNA is hydrolyzed by RNase T2. The resulting nucleotide mixture is subjected to two-dimensional thin-layer or two-dimensional paper chromatography (Nishimura 1972). With this procedure, about 0.1 mg of purified tRNA is sufficient for detection of a modified component present at the level of one residue per tRNA molecule. A model chromatogram indicating the relative positions of modified nucleoside 3′-monophosphates is shown in Appendix III. This chromatogram is also useful for the identification of ^{32}P-labeled modified nucleotides. In general, 0.2 A_{260} unit of a modified nucleoside is sufficient for preliminary examination of its UV and mass spectra and measurement of its chromatographic and electrophoretic mobilities. Limitation in quantity of material poses a great problem in structural characterization of unknown modified nucleosides. Physicochemical techniques, such as Fourier transform nuclear magnetic resonance (FT NMR) and mass spectrometry, have proved to be especially valuable when sample quantities are limited. With FT NMR, spectra can be acquired with as little as 50–100 μg. Mass spectrometry generally requires 2–10 μg. Although adaptation of these advanced techniques greatly reduced the amount of material necessary for structural characterization, the isolation of these quantities of a new modified nucleoside is still usually a laborious process, requiring a large quantity of pure tRNA. Even more material is needed when the nucleoside structure is complicated. For example, 200 g of *Escherichia coli* tRNA were actually required for structural characterization of modified nucleoside Q (queuosine). In this sense, characterization of a new modified nucleoside is not easy when compared with determination of primary sequence of tRNA, which can be finished with less than 50 μg of tRNA when postlabeling techniques are adopted, as discussed by RajBhandary et al. and Randerath et al. (both this volume).

FUNCTIONS OF MODIFIED NUCLEOSIDES IN tRNA

The structural complexities of many modified nucleosides, their locations at specific sites in the tRNA molecule, and their constantly encountered presence in a large number of different organisms suggest they play an important role in tRNA function, but no real proof of this exists. The following discussion offers strong evidence that modified nucleosides perform certain specific functions; contradictory data is also presented.

Figure 1 illustrates that between the presence of particular modified nucleosides adjacent to the 5' terminal of the tRNA anticodon and recognition of the corresponding mRNA codon there exist distinct regularities. For example, hydrophobic nucleosides, such as i^6A or its derivatives, are almost always found in tRNAs that recognize codons starting with U; and most tRNAs that recognize codons starting with A contain hydrophilic modified nucleosides, such as t^6A or its derivatives. Thus it can be said that as far as tRNA is concerned, the genetic code is read as four, rather than three, letters. No similar consistency appears in tRNAs that recognize a codon starting with C or G. In these tRNAs, the rather simple methylated purine nucleosides or unmodified A are present. The AU pair is energetically weaker than the GC pair. Hypermodification may be necessary for stabilization of an AU pair and enhancement of the fidelity of the base pair. The fact that the presence of modified nucleosides next to the anticodon is essential for the amino acid transfer function of tRNA has been demonstrated by numerous researchers. *E. coli* suppressor $tRNA^{Tyr}$, which can be aminoacylated and contains A adjacent to the anticodon, is nonetheless only

Codon	Prokaryote (*E. coli*)	Eukaryote: yeast	Eukaryote: others	Codon	Prokaryote (*E. coli*)	Eukaryote: yeast	Eukaryote: others	Codon	Prokaryote (*E. coli*)	Eukaryote: yeast	Eukaryote: others	Codon	Prokaryote (*E. coli*)	Eukaryote: yeast	Eukaryote: others
UUU	(N)[1] ms^{2}i^{6}A (m^{1}G)[2]	yW, W	oyW	UCU				UAU	ms^{2}i^{6}A	i^{6}A	m^{1}G[6]	UGU	ms^{2}i^{6}A	i^{6}A	
UUC				UCC		i^{6}A	i^{6}A	UAC				UGC			
UUA	ms^{2}i^{6}A			UCA	ms^{2}i^{6}A			UAA	ms^{2}i^{6}A			UGA	ms^{2}i^{6}A		
UUG		m^{1}G		UCG			i^{6}A	UAG	ms^{2}i^{6}A			UGG	ms^{2}i^{6}A	A	m^{1}G
CUU	m^{1}G			CCU				CAU	m^{2}A (m^{1}G)[4]			CGU			
CUC				CCC			m^{1}G	CAC				CGC	m^{2}A (m^{1}G)[4]	A	m^{1}G
CUA		m^{1}G		CCA	(m^{1}G)[3]		m^{1}G	CAA	m^{2}A			CGA			
CUG	m^{1}G			CCG				CAG	m^{2}A			CGG			
AUU	t^{6}A			ACU	mt^{6}A			AAU	t^{6}A			AGU	t^{6}A		mt^{6}A
AUC		t^{6}A	t^{6}A	ACC		t^{6}A		AAC				AGC			
AUA	t^{6}A			ACA	t^{6}A			AAA	t^{6}A (N)[5]	t^{6}A	N	AGA		t^{6}A	
AUG	fMet A; Met t^{6}A(m^{2}A)[4]	t^{6}A; t^{6}A	t^{6}A; t^{6}A (m^{1}G)[5]	ACG				AAG		t^{6}A	t^{6}A	AGG			
GUU	A			GCU				GAU	m^{2}A	m^{1}G		GGU	A	t^{6}A A	
GUC		A	A	GCC		m^{1}I		GAC				GGC			
GUA	m^{6}A	A		GCA	A (m^{6}A)[4]			GAA	m^{2}A	A		GGA	A		
GUG				GCG				GAG				GGG	A		

Figure 1 Relationship between the nucleoside located adjacent to the anticodon and codon recognition of tRNA. Data were taken from the references in the text and the references in the article by Sprinzl et al. (1978). (-----) Uncertain codon recognitions; (N) unidentified modified nucleoside. Footnotes: [1]*Bacillus stearothermophilus*; [2]*Mycoplasma* sp. (kid); [3]bacteriophage T4; [4]*B. subtilis* (H. Ishikura, pers. comm.); [5]*Neurospora crassa* (mitochondrial); [6]Y. Kuchino et al. (unpubl.), earlier report by Brambilla et al. (1976) showing the presence of t^{6}A seems to be incorrect.

slightly active with respect to amino acid transfer function (Gefter and Russell 1969). Similar behavior has been observed with *E. coli* tRNA^{Ile}, which lacks t^{6}A (Miller et al. 1976). The *E. coli* mutant *trpX* exhibits an increased level of expression of all *trp* operon enzymes. This seems to be due to the reduced ability of tRNA^{Trp} lacking ms^{2}i^{6}A in amino acid transfer function (M. Yarus, pers. comm.). It was also shown that a mutant of *Saccharomyces cerevisiae* that was isolated by reduced ability to function as a tyrosine-inserting UAA nonsense suppressor contained very little i^{6}A in tRNA (Laten et al. 1978).

Several hypermodified nucleosides are present in the first position of the anticodon (Fig. 2) and are involved in codon-anticodon interaction. When located in this position, they have been described by Crick as wobbling bases (Crick 1966). I is the first such nucleoside, pairing with U, C, and A in the third position of the codon sequence. o^{5}U, found in *E. coli*, recognizes A, G, and U in the third position of the codon when assayed by binding of tRNA to ribosomes (Takemoto et al. 1973). Recently, mo^{5}U was found instead of o^{5}U in *Bacilus subtilis* tRNA (Murao et al. 1976; Albani et al. 1976). Organisms may be classified into three groups depending on whether o^{5}U, mo^{5}U, or I acts as a wobbling base in the anticodon of tRNA^{Val}, tRNA^{Ser}, or tRNA^{Ala}. Contrary to the wobble base hypothesis, s^{2}U achieves strict base pairing with A, but not with G, in the third letter of the codon sequence, as clearly shown in the case of yeast tRNA$_3^{Glu}$, which contains mcm^{5}s^{2}U (Sekiya et al. 1969). Preferential recognition of A in the first position of the anticodon has been found in other s^{2}U-containing tRNAs, such as *E. coli* tRNA^{Glu} (Ohashi et al. 1970) and tRNA^{Gln} (Yaniv and Folk 1975); *B. subtilis* tRNA^{Lys} and tRNA^{Glu} (H. Ishikura, pers. comm.); and rabbit (reticulocyte and liver) tRNA^{Lys}, tRNA^{Glu}, and tRNA^{Gln} (Hilse and Rudloff 1975; Rudloff and Hilse 1975). However, yeast tRNA$_3^{Arg}$ that contained mcm^{5}U also preferentially recognized A (Weissenbach and Dirheimer 1978). Another possible function of s^{2}U derivatives is to prevent miscoding in protein synthesis. Modification of U to the s^{2}U derivatives may prevent mispairing of U to U. s^{2}U derivatives have only been found in some of the tRNAs mentioned above. *B. subtilis* tRNA^{Lys} contained cmnm^{5}s^{2}U (H. Ishikura, pers. comm.); *B. subtilis* tRNA^{Gly} contained cmnm^{5}U and was recognized by GGG and GGA, and by GGU at 50% of efficiency (Murao and Ishikura 1978). *supK* mutation of *Salmonella* was thought to be a miscoding of UGA nonsense codon by tRNA^{Ser} that lacks formation of methyl ester in o^{5}U located in the first position of the anticodon (Pope et al. 1978). The *E. coli* tRNA$^{Ile}_{minor}$, which contains an unidentified modified nucleoside, recognizes only AUA codon (Harada and Nishimura 1974). *E. coli* tRNA$^{Leu}_{UUR}$ that contains an unknown A derivative in the first position of the anticodon presumably codon (Harada and Nishimura 1974). *E. coli* tRNA$^{Leu}_{UUR}$ that contains an unknown A

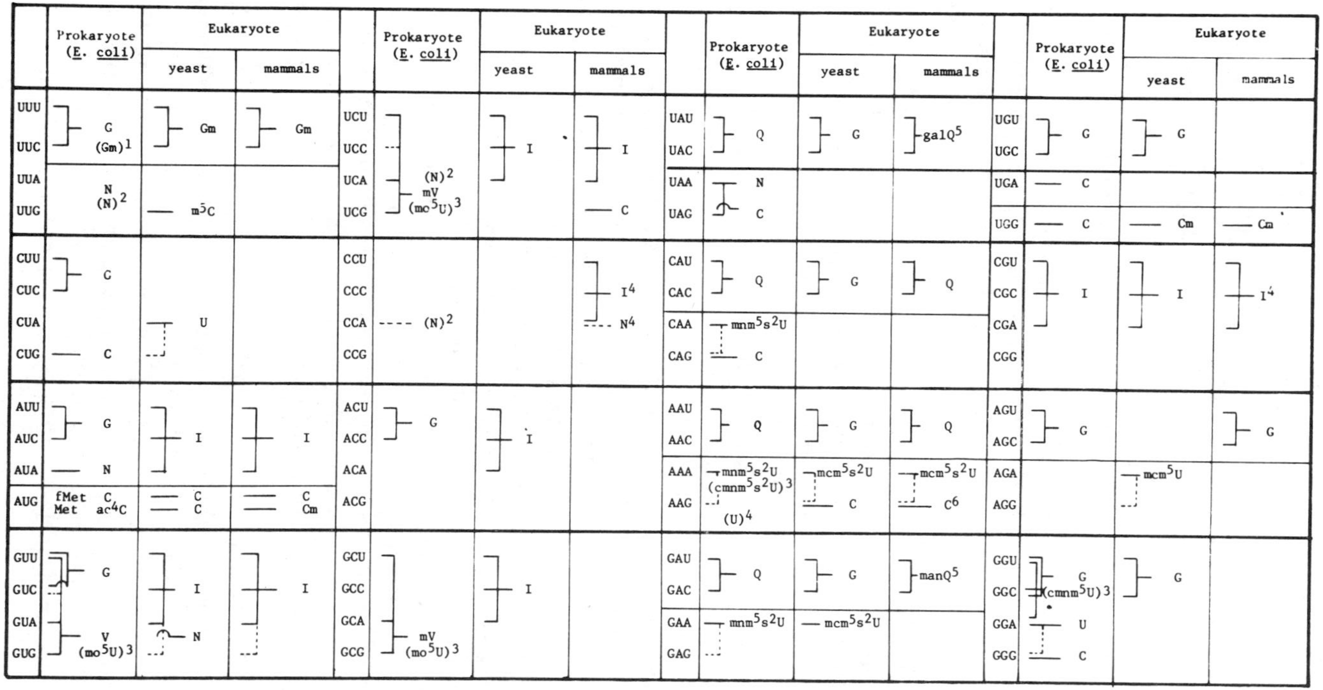

Figure 2 Relationship between the nucleoside located in the first position of the anticodon and codon recognition of tRNA. Data were taken from the references in the text and the references in the article by Sprinzl et al. (1978). (———) Uncertain codon recognitions; (N) unidentified modified nucleoside. Footnotes: [1]*B. stearothermophilus*; [2]bacteriophage T4; [3]*B. subtilis* (H. Ishikura, pers. comm.); [4]F. Harada (pers. comm.); [5]Okada et al. (1977a); [6]H. J. Gross (pers. comm.).

derivative in the first position of the anticodon presumably recognizes UUA and/or UUG, but not the codon for phenylalanine, UUU (Z. Yamaizumi et al., unpubl.). With very few exceptions, almost all tRNAs sequenced so far contain modified U and A in the first position of the anticodon. Apparently Nature does not allow the occurrence of normal U or A, which is presumably lethal to cells by causing miscoding, if present.

There are data that suggest functions for modified nucleosides located in regions other than the anticodon. They are:

1. Involvement of T and ψ in binding of tRNA to ribosomes (Sprinzl et al. 1976).
2. Stabilization of the conformation of tRNA (Watanabe et al. 1976; Igo-Kelmers and Zachau 1971; Wintermeyer and Zachau 1975).
3. Enhancement of resistance of tRNA to attack by RNase and nuclease.
4. Enhancement of specificity of the recognition of aminoacyl-tRNA synthetase (Roe et al. 1973).

The results presented above indicate positive functions of modified nucleosides in tRNA; other data, however, contradict these indications. It is known that tRNAs from *Lactobacillus acidophilus* and *Mycoplasma* sp. (kid), both of which lack i^6A, function normally in amino acid transfer reaction (Hayashi et al. 1969; Litwack and Peterkofsky 1971; Kimball and Söll 1974). Recently it was shown that the tRNAs^{Val} from yeast and *E. coli* recognize the four valine codons in MS2 coat protein mRNA, regardless of whether tRNA^{Val} contains I, o^5U, or G (Mitra et al. 1977). Mutants of *E. coli* have been isolated with tRNAs partially or completely lacking m^7G, s^4U, mnm^5s^2U, or T (Svensson et al. 1971; Marinus et al. 1975; Ramabhadran et al. 1976), yet they grow normally. Purified methyl-deficient tRNA^{fMet} of *E. coli* functions normally in aminoacylation, formylation, recognition of code, and interaction with initiation factor (Marmor et al. 1971). *E. coli* tRNA$_1^{Val}$, in which T and ψ in the GTψC loop are replaced by 5-fluoruridine, can function in aminoacylation and in in vitro protein synthesis (Horowitz et al. 1974; Ofengand et al. 1974). Moreover, several amino-acid-specific tRNAs from mammalian or plant tissues do not contain T (Sprinzl et al. 1968). Finally, some species of microorganisms almost completely lack T in their tRNA (Johnson et al. 1970; Delk et al. 1976; Vani et al. 1979).

A mutant of *E. coli* has been found that lacks T but appears to be normal with respect to growth rate and other criteria. However, when the mutant was grown together with the original wild-type strain the mutants were easily overcome by the wild type (Björk and Neidhardt 1975). The difference in growth rate in a glucose-limiting chemostat between the wild type and the mutant was only 4% (G. R. Björk, pers. comm.). It is likely that the presence of T slightly enhances tRNA function in a way difficult to detect by an in vitro biochemical assay. Nevertheless just such a marginal advantage gained by having modified nucleosides may be decisive for maintaining the organism in its natural environment.

ψ in tRNA^{His} from *Salmonella* has been found to be important in the function of tRNA in regulatory expression of the histidine operon (Singer et al. 1972), possibly by the mechanism in which tRNA^{His}, having two ψs in the anticodon stem and loop, is more efficiently utilized in reading repeating histidine codons in the leader region of the histidine operon (Nocera et al. 1978; Barnes 1978). Possibly other modified nucleosides are involved in the function of tRNA in regulation, although clear evidence to prove this is still lacking. Modified nucleosides must constitute suitable sites for specific interaction of tRNA with protein or other components related to regulation, as reflected by their differences in structure, charge, and presence of certain reactive groups, when compared with normal nucleosides. The major difficulty in studying tRNA function in regulatory process is our inability to find a good experimental system in which a marginal effect of modified nucleosides can be easily assayed. Development of such in vitro system is required for further investigation of the function of modified nucleosides.

NOVEL MECHANISM OF POSTTRANSCRIPTIONAL MODIFICATION OF tRNA

E. coli tRNA Transglycosylase

Modified nucleosides are all thought to be synthesized by modification of parental nucleosides originally present in polynucleotides (Söll 1971). Among these modified nucleosides, the modified nucleoside Q is unique in that its purine skeleton itself is modified to a 7-deaza structure (Kasai et al. 1975). We previously showed that the carbon atom at position 8 of guanine was expelled during biosynthesis of Q (Kuchino et al. 1976). It is difficult to imagine how such a drastic modification could occur if the parental G present in the polynucleotide is directly utilized for Q biosynthesis, because the N—C glycoside bond should be labile after breakage of the heterocyclic nucleus. An alternative mechanism of biosynthesis of Q in tRNA is the insertion of queuine (base of Q) into tRNA to replace guanine by a transglycosylase reaction.

Insertion of guanine into tRNA by a lysate of rabbit reticulocyte was previously reported by Farkas and Singh (1973). We recently demonstrated that this guanine insertion reaction is the specific exchange of guanine with queuine or guanine located in the first position of the anticodon of tRNA^{Tyr}, tRNA^{His}, tRNA^{Asn}, or tRNA^{Asp} (Okada et al. 1976). Thus the enzyme is a kind of tRNA transglycosylase. It was thought that the actual substrate for the guanine insertion enzyme might not be guanine but queuine or a base of

Q precursor: namely, in the biosynthesis of Q, queuine or a base of Q precursor might be synthesized without participation of tRNA and then inserted into tRNA by the guanine insertion enzyme.

To prove this hypothesis, an attempt was made to isolate the guanine insertion enzyme from *E. coli* rather than rabbit reticulocytes because the enzyme was found to be unstable and seemed to have low total activity in the cells, making it difficult to study its properties in detail. In fact, considerable activity was found in an *E. coli* extract. A guanine insertion enzyme (tRNA transglycosylase) was purified to a homogeneous state from *E. coli* B cells by standard procedures (Okada and Nishimura 1979). Its molecular weight was deduced to be 46,000 by SDS gel electrophoresis. Gel filtration of the enzyme in the absence of SDS showed that the molecular weight of the enzyme is 58,000. These results indicate that the native form of the enzyme is probably a monomer. Magnesium ion was essential for the guanine insertion reaction, and its optimal concentration was found to be 20 mM. No energy source, such as ATP, was required for the reaction, as in the case of rabbit-reticulocyte guanine insertion enzyme. *E. coli* tRNA transglycosylase is one of the few enzymes involved in tRNA modification that have been purified to homogeneity, and it should be useful in future biochemical and physicochemical studies on the interaction of tRNA with modification enzyme.

Insertion of a Base of Q Precursors into Undermodified tRNA by *E. coli* tRNA Transglycosylase

We found recently that several precursors of modified nucleoside Q were present in *E. coli* methyl-deficient tRNA and in tRNA of *E. coli* mutants selected for deficiency of modified nucleosides (Okada et al. 1978b; Noguchi et al. 1978). Two of those precursors were characterized as 7-(aminomethyl)-7-deazaguanosine ($preQ_1$) and 7-(cyano)-7-deazaguanosine ($preQ_0$) (Fig. 3). Therefore, if Q biosynthesis in tRNA is catalyzed by tRNA transglycosylase reaction, the actual substrate is either $preQ_1$ or $preQ_0$ rather than the Q base (queuine). For tests on whether bases corresponding to these modified nucleosides are actually incorporated into tRNA by tRNA transglycosylase, $preQ_1$, $preQ_0$, and queuine were synthesized chemically, since they could not be obtained from naturally occurring nucleosides (Ohgi et al. 1978).

[^{14}C]Guanine-labeled G-tRNA^{Tyr}, prepared by incubating G-tRNA^{Tyr} with tRNA transglycosylase and [^{14}C]guanine, was incubated with these bases and the enzyme and release of [^{14}C]guanine from tRNA was measured. The addition of $preQ_1$ or $preQ_0$ led to their enzymatic incorporation into G-tRNA^{Tyr}, with concomitant release of [^{14}C]guanine from the tRNA. Bases of Q precursors were good substrates in the tRNA transglycosylase reaction,

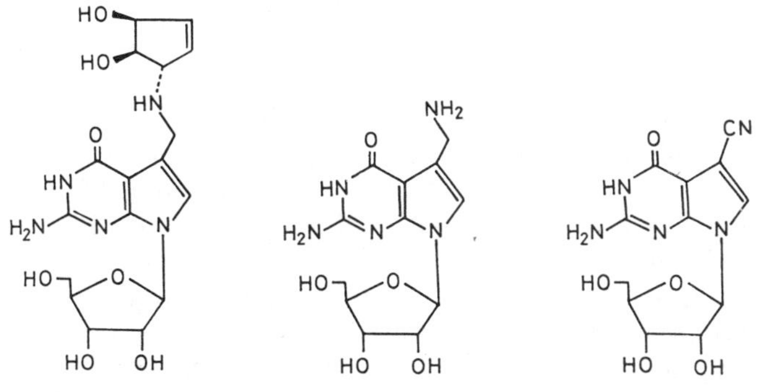

Figure 3 Structure of Q precursors isolated from tRNA of *E. coli* mutants.

but queuine or other analogs of queuine, such as 7-deazaguanine, 7-(methyl)-7-deazaguanine, 2-thio-7-(methyl)-7-deazaguanine, and 2-methylthio-7-(methyl)-7-deazaguanine, did not act as substrate. These bases were incorporated into tRNA, as shown by nucleotide analysis (postlabeling techniques) of the resulting tRNA.

Kinetic experiments with the tRNA transglycosylase show the following K_m values: for guanine, 5.3×10^{-8} M; for preQ$_1$, 1.4×10^{-8} M; and for preQ$_0$, 2.8×10^{-8} M. These data clearly indicated that the *E. coli* tRNA transglycosylase reaction with preQ$_1$ or preQ$_0$ as substrate proceeds much more efficiently than that with guanine as substrate.

When the incorporation of [^{14}C]guanine into tRNAs containing G, Q, preQ$_1$, or preQ$_0$ in the first position of the anticodon was tested, *E. coli* preQ$_0$-tRNA^{Tyr} and *E. coli* G-tRNA^{Tyr} were found to be good acceptors. On the contrary, *E. coli* preQ$_1$-tRNA^{Tyr} and normal *E. coli* tRNA^{Tyr} were not acceptors. The acceptor activity of unfractionated *E. coli* methyl-deficient tRNA was probably due to tRNA species that contain G in place of Q (Okada et al. 1976b). It is interesting to note that the tRNA transglycosylase reaction with preQ$_1$ as substrate, unlike that with guanine as substrate, is not reversible; although preQ$_1$ was efficiently incorporated into G-tRNA, preQ$_1$ in tRNA could not be replaced by guanine.

Detection of Free preQ$_1$ in *E. coli* Cells

The bases of the two Q analogs preQ$_1$ and preQ$_0$ that are present in tRNA of *E. coli* mutants were good substrates for tRNA transglycosylase as they were efficiently incorporated into G-tRNA. If Q biosynthesis in tRNA in vivo proceeds by the same mechanism as that in vitro, those bases should be found in *E. coli* cells as free molecules. This was the case, since a fraction of the acid-soluble *E. coli* extract released radioactivity from [^{14}C]guanine-labeled G-tRNA^{Tyr}. This fraction contained preQ$_1$. This was confirmed by in vitro incorporation into *E. coli* G-tRNA^{Tyr} and subsequent analyses. The amount of the base present in *E. coli* cells was calculated to be 0.3 A$_{260}$ unit per 60 grams of wet *E. coli* cells (approximately 1000 molecules of preQ$_1$ per *E. coli* cell) on the basis of its ability to release [^{14}C]guanine from G-tRNA^{Tyr}.

The present work showed that tRNA transglycosylase from *E. coli* catalyzed the incorporation into tRNA of both modified Q precursors and guanine. Although it has not been proved that the biosynthesis of Q in tRNA in vivo actually proceeds in this way, several facts strongly support this idea. First, the incorporated bases were specifically located in the first position of the anticodon of tRNAs that recognize XA$_C^U$ codons, namely, into the position where the modified bases are naturally present in vivo. Second, most other purine analogs tested so far were not incorporated into tRNA by *E. coli* tRNA transglycosylase; the only modified bases incorporated by the enzyme were precursors of queuine. Third, the enzyme was unable to catalyze the reverse reaction, namely, the conversion of queuine in tRNA to guanine. Fourth, one of the bases of Q precursors, preQ$_1$, was found in the acid-soluble fraction of *E. coli* cells. Fifth, addition of cold preQ$_1$ or preQ$_0$ in a medium markedly reduced incorporation of radioactivity into Q in tRNA when a guanine-requiring *E. coli* mutant (*guaA*) was grown in the presence of [^{14}C]guanine (H. Kasai et al., unpubl.). Since free preQ$_1$ was found in *E. coli* cells, it is likely that preQ$_1$ is the actual substrate incorporated into tRNA in vivo. The presence of preQ$_0$ in tRNA of an *E. coli* mutant may be due to the accumulation of preQ$_0$ by block of its enzymatic conversion to preQ$_1$ in the mutant, and its subsequent incorporation into tRNA. In this case, preQ$_0$ would be an intermediate in the pathway of biosynthesis of preQ$_1$.

Nucleoside transdeoxyribosidase, an enzyme that catalyzes the exchange of bases at the deoxyribonucleoside level, was isolated from various bacteria by McNutt (1955). The reaction mechanism of *E. coli* tRNA transglycosylase is similar to that of nucleoside transdeoxyribosidase. Both enzymes catalyze base exchange by cleavage of an N—C glycoside bond without requirement of an energy source. However, tRNA transglycosylase has stricter base specificity and, unlike nucleoside transdeoxyribosidase, catalyzes the reaction in a specific location of the polynucleotide chain. Modification of polynucleotides by the transglycosylase reaction without breakage of the phosphodiester bond is a novel posttranscriptional modification of polynucleotide. It will be interesting to reinvestigate whether hypermodified nucleosides other than Q are also formed by the transribosylase reaction.

ISOACCEPTING tRNA SPECIES IN TUMOR CELLS

Many new isoaccepting tRNA species have been found in particular tissues, cells at various stages of differentiation, tumor tissues, transformed cells, and cells grown under different culture conditions (for reviews, see Sueoka and Kano-Sueoka 1970; Littauer and Inoue 1973; Borek and Kerr 1972). It is thought that such new isoaccepting tRNA species may have a specific role in regulating cell function. Table 1 summarizes the data reported so far, indicating what isoaccepting species appears to be unique in tumors as derived from an analysis of the changes in the chromatographic profile of amino acid acceptor activity. It should be noted that new tRNA species were only found in the species for particular amino acids, but not for all 20 amino acids. Isoacceptor species for phenylalanine, lysine, tyrosine, histidine, asparagine, and aspartic acid were frequently found in various cancer cells. In the case of new tRNA^{Phe} species, Kuchino and Borek (1978) demonstrated that the tRNA^{Phe} that specifically appeared in Novikoff hepatoma and Ehrlich ascites tumor cells contains m^{1}G unlike tRNA^{Phe} in normal tissues. It was also reported that the new tRNA^{Phe} species that appears in some tumor tissues or tissue culture cells grown under methionine-starvation conditions is due to lack of modified base Y in the position next to the anticodon (Grunberger et al. 1975; Katze 1975b; Pergolizzi et al. 1978). Base Y was not detected in tRNA from mouse neuroblastoma by immunoreaction of tRNA against the antibody for Y base (Salomon et al. 1976). The isoaccepting tRNA^{Val} that appears in rat ascites hepatoma cells was found to be due to lack of 2'-O-methylation in the guanylate residue in the D-loop region (N. Shindo-Okada et al., unpubl.), as in the case of rat brain tRNA^{Ser} vs rat liver tRNA^{Ser} (Rogg et al. 1977). The primary sequence of tRNA$_4^{Lys}$ that specifically appears in the SV40-transformed mouse fibroblasts is identical with that of tRNA$_2^{Lys}$, which is the major isoaccepting species for normal cells; the only difference between the two tRNAs is that tRNA$_4^{Lys}$ is not modified in several positions (H. J. Gross, pers. comm.).

DETECTION OF UNIQUE UNDERMODIFIED tRNA SPECIES IN TUMOR TISSUES BY *E. COLI* tRNA TRANSGLYCOSYLASE

Abnormal chromatographic profiles of tRNA^{Tyr}, tRNA^{His}, tRNA^{Asn}, and tRNA^{Asp} from many tumors and transformed cells have been reported (Table 1). Since all these tRNAs are known to contain modified nucleoside Q or its analogs, it was thought that appearance of the isoaccepting species was due to lack of Q. In fact, White et al. (1973) showed that the appearance in *Drosophila* of two isoaccepting tRNA species for each of those four amino acids was due to the presence or absence of Q. Sequence analysis of tRNA^{Asn} from Walker carcinosarcoma

Table 1 Altered tRNA species in tumor tissues

Tumor	Altered tRNA species	Reference
P388 lymphocytic leukemia cells	His, Lys, Ser, Val	Morton and Rogers (1965)
Adenovirus-7-transformed hamster cells, SV40-transformed hamster cells, and HeLa cells	Tyr	Holland et al. (1967)
Mouse sarcoma-1	Phe	Taylor et al. (1967)
Ehrlich ascites cells	Gly, Phe, Ser, Tyr	Taylor et al. (1967)
Mouse L cells and Rous-sarcoma-virus-transformed hamster cells	Tyr	Taylor et al. (1968)
Mouse plasma cell tumors: immunoglobulin-A producer (MPC62) and immunoglobulin-G producer (MPC47)	Ser	Yang and Novelli (1968)
Novikoff hepatoma cells	Asn, His, Tyr	Baliga et al. (1969)
Novikoff ascites cells	Phe, Ser, Val	Goldman et al. (1969)
Mouse plasma cell tumor: κ-type immunoglobulin light chain producer	Leu	Mushinski and Potter (1969)
L-M cells in culture and L-M cell-induced tumor	Asp, His, Phe, Tyr	Yang et al. (1969)
Leukemic human lymphoblasts	Gln, Tyr	Gallo and Pestka (1970)
Morris hepatoma 5123 and 5123c	Asn, Gln, Phe	Gonano et al. (1971)
Morris hepatoma 5123d	His, Phe, Ser	Volkers and Taylor (1971b)
Mouse fibrosarcoma, mouse reticulum cell sarcoma, adenovirus-31-transformed hamster cells, Reuber hepatoma cells, and HTC cells	Asp, Tyr	Yang (1971)
Malignant lymphomas and chronic lymphatic and myelogenous leukemias	Phe	Mittelman (1971)
Morris hepatoma 9618A	Phe	Volkers and Taylor (1971a)
Vaccinia virus-infected HeLa cells	Phe	Clarkson and Runner (1971)
Morris hepatoma 5123c	Glu	Gonano and Pirro (1971)

Cell/Tumor	Modified tRNA(s)	Reference
Polyoma- and SV40-transformed rat and mouse embryo cells	Asp	Gallagher et al. (1972)
SV40-induced hamster tumors	Asp	Briscoe et al. (1972)
Zajdela hepatoma cells	Ala, Asp, Lys, Thr, Trp, Glu, Ser, Val	Befort et al. (1972)
Avian myeloblastosis virus	Arg, His	Gallagher and Gallo (1973)
Ehrlich ascites tumors (nitrogen mustard resistant)	Tyr, Phe	Hayashi et al. (1973)
SV40-transformed mouse fibroblast cells	Lys, Tyr, Arg	Portugal et al. (1973)
Rous sarcoma virus	Glu, Arg	Taylor et al. (1974)
Polyoma virus-transformed mouse fibroblast cells	Lys	Juarez et al. (1974)
Leukemic leukocyte	Tyr, Asp, Ler, Thr, Gly, Ser	Rainer et al. (1974)
BHK tumor cells	Asp	Briscoe et al. (1975b)
Morris hepatoma 7777	Phe	Grunberger et al. (1975)
SVT$_2$ (SV40-transformed BALB-3T3)	Asp, Asn, Tyr, His	Katze (1975a)
Human tumors	Asp	Briscoe et al. (1975a)
Murine virus-induced leukemia cells	Lys	Agris (1975)
Be-Wo cells	Phe, Tyr, Ser	Kuchino and Borek (1976)
Mice mammary tumors	Leu, Ser	Qvist et al. (1976)
Mouse neuroblastoma	Phe	Salomon et al. (1976)
Mouse tumors:		
DBAH, DBA$_3$	Lys	Mukerjee and Goldfeder (1976)
DBAH, DBAH$_2$	Phe	Mukerjee and Goldfeder (1976)
DBA$_3$	His	Mukerjee and Goldfeder (1976)
Human leukemic cells	Tyr, Asp	Griffin et al. (1976)
Novikoff hepatoma cells	Phe	Kuchino and Borek (1978)
Rat ascites hepatoma cells	Val	N. Shindo-Okada et al. (unpubl.)

A part of the data was taken from the review article by Borek and Kerr (1972). Collection of other data was done by Y. Kuchino in our laboratory.

cells showed that it lacked Q nucleoside, but its primary sequence was the same as tRNA^{Asn} from the normal cells (Roe et al. 1979). For detection of tRNA species lacking Q, use of *E. coli* tRNA transglycosylase was found to be very useful, since the enzyme catalyzed exchange of guanine with guanine, but not of queuine in tRNA with guanine. With this enzyme, much guanine was incorporated into tRNA from various tumor tissues, but scarcely any was incorporated into tRNA from normal tissues (Okada et al. 1978a). The tRNAs from all tumors tested, originating from various organs and having different growth rates, were found to be acceptors in the tRNA transglycosylase reaction, whereas none of those from normal tissues, such as liver, kidney, spleen, and testis, were acceptors. Slight incorporation of guanine was observed with tRNA isolated from rat fetal liver and rabbit reticulocyte, but the extent of guanine incorporation into these was less than 5% of that into tRNA from most tumor tissues.

Because tRNA from regenerating liver also accepts considerable amounts of guanine, the presence of undermodified tRNA containing guanine instead of queuine is not directly linked with neoplasia. However, the presence of undermodified tRNA in tumor tissues is not merely linked to the growth rate because slowly growing Morris hepatoma 7794A, a minimum deviation hepatoma that takes about 2 months to kill the host animal, had higher guanine-accepting activity than the rapidly growing ascites hepatoma AH7974 cells.

In the case of the rat ascites hepatoma, a tRNA species that accepts guanine was purified by RPC-5 column chromatography and identified as a minor species of tRNA^{Asp}. Nucleotide sequence study of this minor tRNA^{Asp} by postlabeling technique showed that its primary structure is very similar to that of major tRNA^{Asp}, but it contains guanine instead of queuine. However, guanine-accepting tRNA species differ in different cell types; for example, Ehrlich ascites tumor tRNA contained several tRNA species that accept guanine. Our method to identify unique tRNA species by using the *E. coli* tRNA transglycosylase has several advantages: (1) incorporation of guanine was observed in all tumor tissues tested; (2) the method is very simple; and (3) because the *E. coli* enzyme is very active, only a small quantity of cells is needed for the assay—less than 0.05 A$_{260}$ unit of unfractionated tRNA is enough.

Thus isoacceptor tRNA species that appeared in tumor tissues are mostly undermodified tRNAs whose primary structures are the same as those of tRNAs present in normal tissues. It is contrary to the earlier observation that unfractionated tumor tRNA contains more methylated modified nucleosides than normal tRNA (Borek and Kerr 1972). It is not known at present how such undermodification of tRNA occurs in tumor tissues. At least it is not simply due to the fast growth rate of tumor cells. It is still an open question as to whether such undermodified tRNA has some specific role in tumor cells or whether it is just less functional. It is also still uncertain whether some new tRNA transcripts also appear in tumor cells. More sequence work on mammalian tRNAs, as well as on tRNA genes, is definitely needed to provide information on the real nature of isoaccepting species and their importance in cell function.

ACKNOWLEDGMENTS

I thank my colleagues for their contributions, especially Dr. N. Okada, who participated in most of the work done for this paper.

REFERENCES

Agris, P. F. 1975. Alterations of tRNA during erythroid differentiation of murine virus-induced leukemia cells. *Arch. Biochem. Biophys.* **170**:114.

Albani, M., W. Schmidt, H. Kersten, K. Geibel, and I. Luderwald. 1976. 5-Methoxyuridine, a new modified constituent in tRNAs of *Bacillaceae*. *FEBS Lett.* **70**:37.

Baliga, B. S., E. Borek, I. B. Weinstein, and P. R. Srinivasan. 1969. Differences in the tRNAs of normal liver and Novikoff hepatoma. *Proc. Natl. Acad. Sci.* **62**:899.

Barnes, W. M. 1978. DNA sequence from the histidine operon control region: Seven histidine codons in a row. *Proc. Natl. Acad. Sci.* **75**:4281.

Befort, J. J., J. Mercier, N. Berfort, G. Beck, and J. P. Ebel. 1972. Comparative study of tRNAs extracted from rat liver and Zajdela hepatoma. II. Comparison of chromatographic fractionation patterns for aminoacyl-tRNA from rat liver and Zajdela hepatoma. *Biochimie* **54**:1327.

Björk, G. R. and P. C. Neidhardt. 1975. Physiological and biochemical studies on the function of 5-methyluridine in the transfer ribonucleic acid of *Escherichia coli*. *J. Bacteriol.* **124**:99.

Borek, E. and S. J. Kerr. 1972. Atypical transfer RNAs and their origin in neoplastic cells. *Adv. Cancer Res.* **15**:163.

Brambilla, R., H. Rogg, and M. Staehelin. 1976. Unexpected occurrence of an aminoacylated nucleoside in mammalian tRNA^{Tyr}. *Nature* **263**:167.

Briscoe, W. T., A. C. Griffin, C. McBride, and J. M. Bowan. 1975a. The distribution and properties of aspartyl-tRNA in human and animal tumors. *Cancer Res.* **35**:2586.

Briscoe, W. T., J. J. Syrewiez, M. V. Marshall, and A. C. Griffin. 1975b. Regulation of an aspartyl-tRNA species in BHK cells in culture and in solid tumor form. *Biochim. Biophys. Acta* **383**:441.

Briscoe, W. T., M. W. Taylor, A. C. Griffin, R. Duff, and F. Rapp. 1972. Aspartyl-tRNA profiles in normal and cancer cells. *Cancer Res.* **32**:1753.

Clarkson, S. G. and M. N. Runner. 1971. Transfer RNA changes in HeLa cells after vaccinia virus infection. *Biochim. Biophys. Acta* **238**:498.

Crick, F. H. C. 1966. Codon-anticodon pairing: The wobble hypothesis. *J. Mol. Biol.* **19**:548.

Delk, A. N., J. M. Romeo, D. P. Nagle, Jr., and J. C. Rabinowitz. 1976.

Biosynthesis of ribothymidine in the transfer RNA of *Streptococcus faecalis* and *Bacillus subtilis. J. Biol. Chem.* **251:**7649.
Farkas, W. R. and R. Singh. 1973. Guanylation of transfer ribonucleic acid by a cell-free lysate of rabbit reticulocytes. *J. Biol. Chem.* **248:**7780.
Gallagher, R. E. and R. C. Gallo. 1973. Chromatographic analysis of isoaccepting tRNAs from avian myeloblastosis viruses. *J. Virol.* **12:**449.
Gallagher, R. E., R. C. Ting, and R. C. Gallo. 1972. A common change of aspartyl-tRNA in polyoma- and SV40-transformed cells. *Biochim. Biophys. Acta* **272:**568.
Gallo, R. C. and S. Pestka. 1970. Transfer RNA species in normal and leukemic human lymphoblasts. *J. Mol. Biol.* **52:**195.
Gefter, M. L. and R. L. Russell. 1969. Role of modification in tyrosine transfer RNA: A modified base affecting ribosome binding. *J. Mol. Biol.* **39:**145.
Goldman, M., W. M. Johnston, and A. C. Griffin. 1969. Comparison of tRNAs and aminoacyl synthetases of liver and ascites tumor cells. *Cancer Res.* **29:**1051.
Gonano, F. and G. Pirro. 1971. Coding properties of tRNA^{Glu} of mammalian origin. Comparison between rat liver and minimal deviation hepatoma 5123c tRNAs^{Glu}. *Biochem. Biophys. Res. Commun.* **45:**984.
Gonano, F., V. P. Chiarugi, G. Pirro, and M. Marini. 1971. Transfer RNAs in rat liver and Morris 5123 minimal deviation hepatoma. *Biochemistry* **10:**900.
Griffin, G., W. K. Yang, and G. D. Novelli. 1976. Transfer RNA species in human lymphocytes stimulated by mitogens and in leukemic cells. *Arch. Biochem. Biophys.* **176:**187.
Grunberger, D., I. B. Weinstein, and J. F. Mushinski. 1975. Deficiency of the Y base in a hepatoma phenylalanine tRNA. *Nature* **253:**66.
Hall, R. H. and D. B. Dunn. 1975. Natural occurrence of the modified nucleosides. In *Handbook of biochemistry and molecular biology*, 3rd ed., *Nucleic acids* (ed. G. D. Fasman), vol. 1, p. 216. CRC Press, Cleveland, Ohio.
Harada, F. and S. Nishimura. 1974. Purification and characterization of AUA specific isoleucine transfer ribonucleic acid from *Escherichia coli* B. *Biochemistry* **13:**300.
Harada, F., R. C. Sawyer, and J. E. Dahlberg. 1975. A primer ribonucleic acid for initiation of *in vitro* Rous sarcoma virus deoxyribonucleic acid synthesis: Nucleotide sequence and amino acid acceptor activity. *J. Biol. Chem.* **250:**3487.
Hashimoto, S., M. Sakai, and M. Muramatsu. 1975. 2'-O-Methylated oligonucleotides in ribosomal 18S and 28S RNA of a mouse hepatoma, MH134. *Biochemistry* **14:**1956.
Hayashi, H., H. Fisher, and D. Söll. 1969. Transfer ribonucleic acid from *Mycoplasma. Biochemistry* **8:**3680.
Hayashi, M., A. C. Griffin, R. Duff, and F. Rapp. 1973. Chromatographic studies of tyrosyl- and phenylalanyl-tRNAs of liver and tumor cells. *Cancer Res.* **33:**902.
Hilse, K. and E. Rudloff. 1975. Glutamine cognate codons in rabbit haemoglobin mRNAs. *FEBS Lett.* **60:**380.
Holland, J. J., M. W. Taylor, and C. A. Buck. 1967. Chromatographic differences between tyrosyl tRNA from different mammalian cells. *Proc. Natl. Acad. Sci.* **58:**2437.
Horowitz, J., C.-N. Ou, M. Ishaq, J. Ofengand, and J. Bierbaum. 1974. Isolation and partial characterization of *Escherichia coli* valine transfer RNA with

uridine and uridine-derived residues replaced by 5-fluorouridine. *J. Mol. Biol.* **88:**301.

Igo-Kelmers, T. and H. G. Zachau. 1971. Involvement of 1-methyladenosine and 7-methylguanosine in the three-dimensional structure of (yeast) tRNA^{Phe}. *Eur. J. Biochem.* **18:**292.

Johnson, L., K. Hayashi, and D. Söll. 1970. Isolation and properties of a transfer ribonucleic acid deficient in ribothymidine. *Biochemistry* **9:**2823.

Juarez, H., D. Jaurez, and C. Hedgcoth. 1974. Seven lysine isoaccepting tRNAs from polyoma virus transformed cells. *Biochem. Biophys. Res. Commun.* **61:**110.

Kasai, H., K. Murao, S. Nishimura, J. G. Liehr, P. F. Crain, and J. A. McCloskey. 1976a. Structure determination of a modified nucleoside isolated from *Escherichia coli* transfer ribonucleic acid. N-[N-[(9-β-D-Ribofuranosylpurin-6-yl)-carbamoyl]threonyl]2-amido-2-hydroxymethylpropane-1,3-diol. *Eur. J. Biochem.* **69:**435.

Kasai, H., K. Nakanishi, R. D. Macfarlane, D. F. Torgerson, Z. Ohashi, J. A. McCloskey, H. J. Gross, and S. Nishimura. 1976b. The structure of Q^* nucleoside isolated from rabbit liver transfer ribonucleic acid. *J. Am. Chem. Soc.* **98:**5044.

Kasai, H., Z. Ohashi, F. Harada, S. Nishimura, N. J. Oppenheimer, P. F. Crain, J. G. Liehr, D. L. von Minden, and J. A. McCloskey. 1975. Structure of the modified nucleoside Q isolated from *Escherichia coli* transfer ribonucleic acid. 7-(4,5-*cis*-dihydroxy-1-cyclopenten-3-ylaminomethyl)-7-deazaguanosine. *Biochemistry* **14:**4198.

Katze, J. R. 1975a. Alterations in SVT2 cell tRNAs in response to cell density and serum type. *Biochim. Biophys. Acta* **383:**131.

——. 1975b. Relation of cell type and cell density to the degree of post-transcriptional modification of tRNA^{Lys} and tRNA^{Phe}. *Biochim. Biophys. Acta* **407:**392.

Kimball, M. E. and D. Söll. 1974. The phenylalanine tRNA from *Mycoplasma* sp. (kid): A tRNA lacking hypermodified nucleosides functional in protein synthesis. *Nucleic Acids Res.* **1:**1713.

Kuchino, Y. and E. Borek. 1976. Changes in tRNAs in human malignant trophoblastic cells (BeW$_o$ line). *Cancer Res.* **36:**2932.

——. 1978. Tumor-specific phenylalanine tRNA contains two supernumerary methylated bases. *Nature* **271:**126.

Kuchino, Y., H. Kasai, K. Nihei, and S. Nishimura. 1976. Biosynthesis of the modified nucleoside Q in transfer RNA. *Nucleic Acids Res.* **3:**393.

Laten, H., J. Gorman, and R. M. Bock. 1978. Isopentyladenosine deficient tRNA from an antisuppressor mutant of *Saccharomyces cerevisiae*. *Nucl. Acids Res.* **5:**4329.

Lesiewicz, J. and B. Dudock. 1977. *In vitro* methylation of a *E. coli* alanine tRNA with homologous *E. coli* methylase. *Fed. Proc.* **36:**705.

Littauer, U. Z. and H. Inoue. 1973. Regulation of tRNA. *Annu. Rev. Biochem.* **42:**439.

Litwack, M. D. and A. Peterkofsky. 1971. Transfer ribonucleic acid deficient in N^6-(Δ^2-isopentenyl)-adenosine due to mevalonic acid limitation. *Biochemistry* **10:**994.

Marinus, M. G., N. R. Morris, D. Söll, and T. C. Kwong. 1975. Isolation and partial characterization of three *Escherichia coli* mutants with altered transfer ribonucleic acid methylases. *J. Bacteriol.* **122**:257.

Marmor, J. B., H. W. Dickerman, and A. Peterkofsky. 1971. Studies on methyl-deficient methionine transfer ribonucleic acid from *Escherichia coli*. *J. Biol. Chem.* **246**:3464.

McNutt, W. S. 1955. Nucleoside transdeoxyribosidase from bacteria. *Methods Enzymol.* **2**:464.

Miller, J. P., Z. Hussoin, and M. P. Schweizer. 1976. The involvement of the anticodon adjacent modified nucleoside N-[9-(β-D-ribofuranosyl)purin-6-yl-carbamoyl]threonine in the biological function of *E. coli* tRNA^{Ile}. *Nucleic Acids Res.* **3**:1185.

Mitra, S. K., F. Lustig, B. Åkesson, U. Largerkvist, and L. Strid. 1977. Codon-anticodon recognition in the valine codon family. *J. Biol. Chem.* **252**:471.

Mittelman, A. 1971. Patterns of isoaccepting phenylalanine tRNA in human leukemia and lymphoma. *Cancer Res.* **31**:647.

Morton, M. J. and W. I. Rogers. 1965. A micromethod for the differentiation of amino-acid-specific sRNA molecules. *Anal. Biochem.* **13**:108.

Mukerjee, H. and A. Goldfeder. 1976. Transfer RNA species in tumors of different growth rates. *Cancer Res.* **36**:3330.

Murao, K. and H. Ishikura. 1978. A new uridine derivative located in the anticodon of tRNA$_1^{\text{Gly}}$ from *Bacillus subtilis*. *Nucleic Acids Res.* (special publication) **5**:s333.

Murao, K., T. Hasegawa, and H. Ishikura. 1976. 5-Methoxyuridine: A new minor constituent located in the first position of the anticodon of tRNA^{Ala}, tRNA^{Thr} and tRNA^{Val} from *Bacillus subtilis*. *Nucleic Acids Res.* **3**:2851.

Mushinski, J. F. and M. Potter. 1969. Variations in leucine tRNA in mouse plasma cell tumors producing κ-type immunoglobin light chains. *Biochemistry* **8**:1684.

Nishimura, S. 1972. Minor components in transfer RNA: Their characterization, location and function. *Prog. Nucleic Acid Res. Mol. Biol.* **12**:49.

Nocera, P. P. D., F. Blasi, R. D. Lauro, R. Frunzio, and C. B. Bruni. 1978. Nucleotide sequence of the attenuator region of the histidine operon of *Escherichia coli* K12. *Proc. Natl. Acad. Sci.* **75**:4276.

Noguchi, S., Z. Yamaizumi, T. Ohgi, T. Goto, Y. Nishimura, Y. Hirota, and S. Nishimura. 1979. Isolation of Q nucleoside precursor present in tRNA of an *E. coli* mutant and its characterization as 7-(cyano)-7-deazaguanosine. *Nucleic Acids Res.* **5**:4215.

Ofengand, J., J. Bierbaum, J. Horowitz, C.-N. Ou, and M. Ishaq. 1974. Protein synthetic ability of *Escherichia coli* valine transfer RNA with pseudouridine, ribothymidine, and other uridine-derived residues replaced by 5-fluorouridine. *J. Mol. Biol.* **88**:313.

Ohashi, Z., M. Saneyoshi, F. Harada, H. Hara, and S. Nishimura. 1970. Presumed anticodon structure of glutamic acid tRNA from *E. coli*: A possible location of a 2-thiouridine derivative in the first position of the anticodon. *Biochem. Biophys. Res. Commun.* **40**:866.

Ohgi, T., T. Kondo, and T. Goto. 1978. Synthesis of Q base and related compounds. *Nucleic Acids Res.* (Special Publication) **5**:s285.

Okada, N. and S. Nishimura. 1979. Isolation and characterization of a guanine insertion enzyme, a specific tRNA transglycosylase, from *Escherichia coli. J. Biol. Chem.* **254:**3061.

Okada, N., F. Harada, and S. Nishimura. 1976. Specific replacement of Q base in the anticodon of tRNA by guanine catalyzed by a cell-free extract of rabbit reticulocyte. *Nucleic Acids Res.* **3:**2593.

Okada, N., N. Shindo-Okada, and S. Nishimura. 1977a. Isolation of mammalian tRNA^{Asp} and tRNA^{Tyr} by lectin-Sepharose affinity column chromatography. *Nucleic Acids Res.* **4:**415.

Okada, N., T. Yasuda, and S. Nishimura. 1977b. Detection of nucleoside Q precursor in methyl-deficient *E. coli* tRNA. *Nucleic Acids Res.* **4:**4063.

Okada, N., N. Shindo-Okada, S. Sato, Y. H. Itoh, K. Oda, and S. Nishimura. 1978a. Detection of unique tRNA species in tumor tissues by *Escherichia coli* guanine insertion enzyme. *Proc. Natl. Acad. Sci.* **75:**4247.

Okada, N., S. Noguchi, S. Nishimura, T. Ohgi, T. Goto, P. F. Crain, and J. A. McCloskey. 1978b. Structure determination of a nucleoside Q precursor isolated from *E. coli* tRNA: 7-(aminomethyl)-7-deazaguanosine. *Nucleic Acids Res.* **5:**2289.

Pergolizzi, R. G., D. L. Engelhardt, and D. Grunberger. 1978. Formation of phenylalanine transfer RNA lacking the Wye base in Vero cells during methionine starvation. *J. Biol. Chem.* **253:**6341.

Pope, W. T., A. Brown, and R. H. Reeves. 1978. The identification of the tRNA substrates for the *supK* tRNA methylase. *Nucleic Acids Res.* **5:**1041.

Portugal, F. H., J. S. Simionds, D. Twardzik, and M. Oskarsson. 1973. Effect of SV40-induced transformation on isoaccepting species of tRNA from mouse fibroblasts. *J. Virol.* **12:**1616.

Qvist, R., C. Palin, and I. Heiberg. 1976. Transfer RNA and aminoacyl-tRNA synthetases in hormone dependent and independent mammary tumors of GR mice. II. Isoacceptor tRNAs in hormone dependent and independent mammary tumors of GR mice. *Cancer Biochem. Biophys.* **1:**317.

Rainer, H., P. Hoecker, A. Stacher, K. Moser, I. Streit, and E. Deutsh. 1974. Transfer RNA species in normal and leukemic leukocytes. *Neoplasma* **21:**409.

Ramabhadran, T. V., T. Fossum, and J. Jagger. 1976. *Escherichia coli* mutant lacking 4-thiouridine in its transfer ribonucleic acid. *J. Bacteriol.* **128:**671.

Roe, B., M. Michael, and B. Dudock. 1973. Function of N^2-methylguanine in phenylalanine transfer RNA. *Nat. New Biol.* **246:**135.

Roe, B. A., A. F. Stankiewicz, H. L. Rizi, C. Weiss, M. N. DiLauro, D. Pike, C. Y. Chen, and E. Y. Chen. 1979. Comparison of rat liver and Walker 256 carcinosarcoma tRNAs. *Nucl. Acids Res.* **6:**673.

Rogg, H., P. Müller, G. Keith, and M. Staehelin. 1977. Chemical basis for brain-specific serine transfer RNAs. *Proc. Natl. Acad. Sci.* **74:**4243.

Rudloff, E. and K. Hilse. 1975. Properties of isoaccepting tRNA^{Lys}, tRNA^{Glu} and tRNA^{Gln} from rabbit reticulocytes and liver multiplicity, codon recognition and inactivation by iodine. *Hoppe-Seyler's Z. Physiol. Chem.* **356:**1359.

Salomon, R., D. Giveon, Y. Kimihi, and U. Z. Littauer. 1976. Abundance of tRNA^{Phe} lacking the peroxy Y-base in mouse neuroblastoma. *Biochemistry* **15:**5258.

Sekiya, T., K. Takeishi, and T. Ukita. 1969. Specificity of yeast glutamic

acid transfer RNA for codon recognition. *Biochim. Biophys. Acta* **182:**411.
Singer, C. E., G. R. Smith, R. Cortese, and B. N. Ames. 1972. Mutant tRNA^{His} ineffective in repression and lacking two pseudouridine modifications. *Nat. New Biol.* **238:**72.
Sprinzl, M., F. Grüter, and D. H. Gauss. 1978. Collection of published tRNA sequences. *Nucleic Acids Res.* (special suppl.) **5:**r15.
Sprinzl, M., T. Wagner, D. Lorenz, and V. A. Erdmann. 1976. Regions of tRNA important for binding to ribosomal A and P sites. *Biochemistry* **15:**3031.
Söll, D. 1971. Enzymatic modification of transfer RNA: Modified nucleosides form at the polynucleotide level, but their function is not established. *Science* **173:**293.
Sueoka, N. and T. Kano-Sueoka. 1970. Transfer RNA and cell differentiation. *Prog. Nucleic Acid Res. Mol. Biol.* **10:**23.
Svensson, I., K. Isaksson, and A. Henningson. 1971. Aminoacylation and polypeptide synthesis with tRNA lacking ribothymidine. *Biochim. Biophys. Acta* **238:**331.
Takemoto, T., K. Takeishi, S. Nishimura, and T. Ukita. 1973. Transfer of valine into rabbit haemoglobin from various isoaccepting species of valyl-tRNA differing in codon recognition. *Eur. J. Biochem.* **38:**489.
Taylor, M. W., C. A. Buck, G. A. Granger, and J. F. Holland. 1968. Chromatographic alterations in tRNAs accompanying speciation, differentiation and tumor formation. *J. Mol. Biol.* **33:**809.
Taylor, M. W., G. A. Granger, C. A. Buck, and J. F. Holland. 1967. Similarities and differences among specific tRNAs in mammalian tissues. *Proc. Natl. Acad. Sci.* **57:**1712.
Taylor, M. W., S. Wang, R. M. Kothari, and P. P. Hung. 1974. Chromatographic analysis of isoaccepting tRNAs from avian tumor viruses. *J. Virol.* **14:**1092.
Vani, B. R., T. Ramakrishnan, Y. Taya, S. Noguchi, Z. Yamaizumi, and S. Nishimura. 1979. Occurrence of 1-methyladenosine and absence of ribothymidine in transfer ribonucleic acid in *Mycobacterium smegmatis*. *J. Bacteriol.* **137:**1084.
Volkers, S. A. S. and M. W. Taylor. 1971a. Alterations in the tRNA population of hepatoma 9618A as compared with normal rat liver. *Biochim. Biophys. Acta* **254:**415.
———. 1971b. Chromatographic comparison of the tRNAs of rat livers and Morris hepatomas. *Biochemistry* **10:**487.
Watanabe, K., T. Oshima, and S. Nishimura. 1976. CD spectra of 5-methyl-2-thiouridine in tRNA$_f^{Met}$ from an extreme thermophile. *Nucleic Acids Res.* **3:**1703.
Weissenbach, J. and G. Dirheimer. 1978. Pairing properties of the methyl ester of 5-carboxymethyl uridine in the wobble position of yeast tRNA$_3^{Arg}$. *Biochim. Biophys. Acta* **518:**530.
White, B. N., G. M. Tener, J. Holden, and D. T. Suzuki. 1973. Activity of a transfer RNA modifying enzyme during the development of *Drosophila* and its relationship to the su(s) locus. *J. Mol. Biol.* **74:**635.
Wintermeyer, W. and H. G. Zachau. 1975. Tertiary structure interaction of 7-methylguanosine in yeast tRNA^{Phe} as studied by borohydride reduction. *FEBS Lett.* **58:**306.

Yang, W. K. 1971. Isoaccepting tRNAs in mammalian differentiated cells and tumor tissues. *Cancer Res.* **31:**639.

Yang, W. K. and G. D. Novelli. 1968. Isoaccepting tRNAs in mouse plasma cell tumors that synthesize different myeloma protein. *Biochem. Biophys. Res. Commun.* **31:**534.

Yang, W. K., A. Hellman, D. H. Martin, K. B. Hellman, and G. D. Novelli. 1969. Isoaccepting tRNAs of L-M cells in culture and after tumor induction in C_3H mice. *Proc. Natl. Acad. Sci.* **64:**1411.

Yaniv, M. and W. R. Folk. 1975. The nucleotide sequences of the two glutaminic transfer ribonucleic acids from *Escherichia coli. J. Biol. Chem.* **250:**3243.

Crystal Structure Analysis of tRNAs

Crystal Structure of Yeast tRNA^{Phe} and General Structural Features of Other tRNAs

Sung-Hou Kim
Department of Biochemistry, Duke University
Durham, North Carolina 27710
and Department of Chemistry and Laboratory of Chemical Biodynamics
University of California
Berkeley, California 94720

Since the discovery of the "L-shaped" backbone structure of yeast tRNA^{Phe} in 1973, considerably more details are known for this tRNA structure in two different crystal forms. Examination of structural details of this tRNA allows one to make the following general conclusions. First, all free tRNAs are likely to assume the overall shape of an "L" with each arm being approximately 60 Å long and 20 Å thick. Second, most of the conserved and semiconserved bases in all tRNAs are involved in forming tertiary hydrogen bonds essential for maintaining the L-shaped three-dimensional structure.

There are several recent review articles on the three-dimensional structure of yeast tRNA^{Phe} emphasizing different aspects (Sigler 1975; Rich and RajBhandary 1976; Clark 1977; Kim 1976, 1978). In this paper, I will briefly summarize the structural features and relative confidence levels with which they have been determined by X-ray crystallographic technique from four different research groups (Ladner et al. 1975; Quigley et al. 1975; Sussman and Kim 1976; Stout et al. 1976; and subsequent works cited in Kim 1978) and point out the structural features of functional interest based on the most recent available results (Jack et al. 1976, 1977; Holbrook et al. 1977, 1978; Quigley et al. 1978; Sussman et al. 1978). To avoid lengthy and often confusing descriptions, many structural features are represented with figures.

CRYSTAL STRUCTURE OF YEAST tRNA^{Phe}

Overall Structure

The crystal structure of yeast tRNA^{Phe} has an overall L shape (Kim et al. 1973), with the polynucleotide backbone of this model folded so that the amino acid stem and the T stem form one continuous double-helical arm, and the D stem and the anticodon stem form the other long double-helical arm of the L. Each extension of the L is about 60 Å long with a diameter of about 20 Å and the distance between the two extremes is about 80 Å. Each stem is an antiparallel, right-handed double helix similar to A-RNA. The polynucleotide backbone structure of the molecule is shown in Figure 1, with corresponding cloverleaf model, and three views of a space-filling model are shown in Figure 2 to provide a more realistic and spatial appearance of this molecule. The 3' end, where peptide elongation occurs, is at one extreme of the molecule; the anticodon triplet, which recognizes the codon on the mRNA, is at the other. The T loop, which has been implicated as an rRNA recognition site (for a review, see Erdmann 1976), appears at the corner of the L. Thus, these three functionally important sites are maximally separated, which may help minimize mutual interference among them and their corresponding sites on the ribosomes. There is a hole 10 Å in diameter near the corner of the L, surrounded by the T loop and the T stem, but the possible functional role for it is not known.

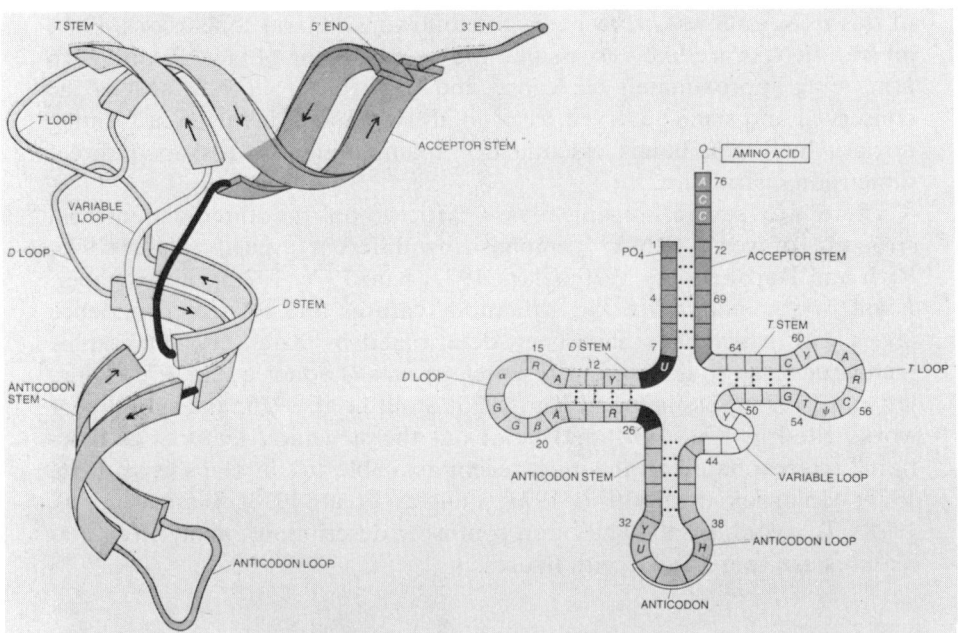

Figure 1 (*Left*) Helical segments of the tRNA molecule, corresponding to the four stems of the cloverleaf diagram, are represented by ribbons in this schematic view. (*Right*) Cloverleaf diagram of tRNA molecule. The conserved and the semiconserved bases are indicated. The number of nucleotides in the various stems and the loops is generally constant, except for two parts of the D loop designated α and β and the variable loop. (R) Purine; (Y) pyrimidine; (H) a highly modified purine.

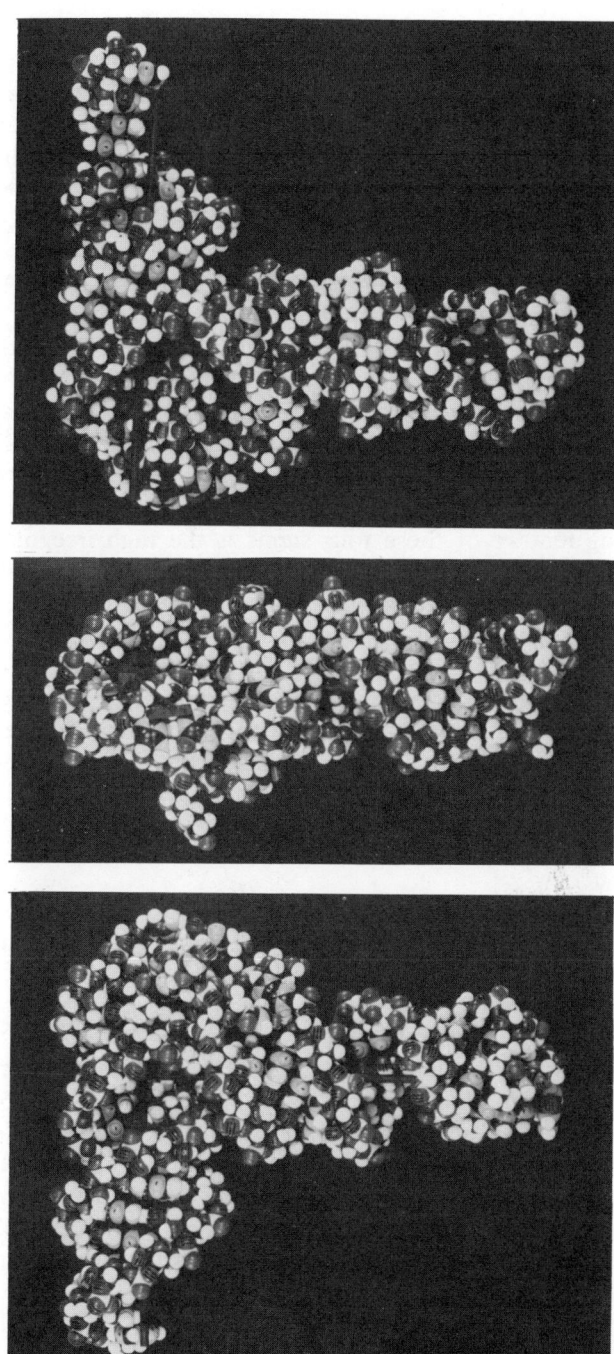

Figure 2 Three views of a space-filling model of yeast tRNA^{Phe} structure. The model is successively rotated with a 90° increment around a vertical axis.

Double-helical Stems

Although all four stems have a conformation similar to A-RNA, i.e., all have a shallow groove and a deep groove, they show considerable variations in their helical parameters, such as the number of base pairs per turn, and base-pair tilt and twist angles. Various helical parameters whose calculations are based on the highly refined atomic coordinate of this tRNA in an orthorhombic crystal form (Sussman et al. 1978; Holbrook et al. 1978) reveal that the number of residues per turn for each stem is usually less than 11 except for the anticodon stem; helix rotation is usually larger and the displacement shorter than those of A-RNA; base pairs have a wide range of tilt (~13°–21°). Among the four stems, the D stem differs most from A-RNA. This is probably due to the extensive tertiary interactions made by this stem with other parts of the molecule. The angle between the two helical axes of the acceptor and T stems is 14°, and the corresponding angle between the two helical axes of the D stem and the anticodon stem is 26°. The angle between the two arms of the L is about 92°.

One interesting feature of these four stems is the high irregularity of their helicity. The extent of this irregularity is beyond the range of error (0.1 Å) associated with the atomic coordinates, therefore, these irregularities must be considered as real in this crystal structure. It is not clear, however, whether these irregularities reflect the sequence dependence or an inherent flexibility of the backbone conformation locked into one form in this particular crystal form. An average twist angle between two bases that are base paired in each stem is considerably greater than that of A-RNA, i.e., 5° for the anticodon stem and 11° for the amino acid stem.

Base-Base Interaction

There are two classes of interaction in this category: one is base stacking and the other is base pairing by hydrogen bonds. Although only 55% of the bases in this molecule are in the stems, the three-dimensional structure reveals that all except five bases (93% of the bases) are stacked as schematically shown in the center of Figure 3. (Unstacked bases are D16, D17, G20, U47, and A76.) Such extensive base stacking, including bases in the loops, is likely to be a universal feature of all free nucleic acids.

There are 9 tertiary base pairs in the crystal structure of this tRNA as shown schematically in Figure 3. Two remarkable points are that none of these tertiary base pairs are of the Watson-Crick type except the G19·C56 pair, and that most of these involve bases that are conserved or semiconserved in all tRNAs, thus giving credence to the hypothesis that all tRNAs have the same general structural frame (Kim et al. 1974a,b; Klug et

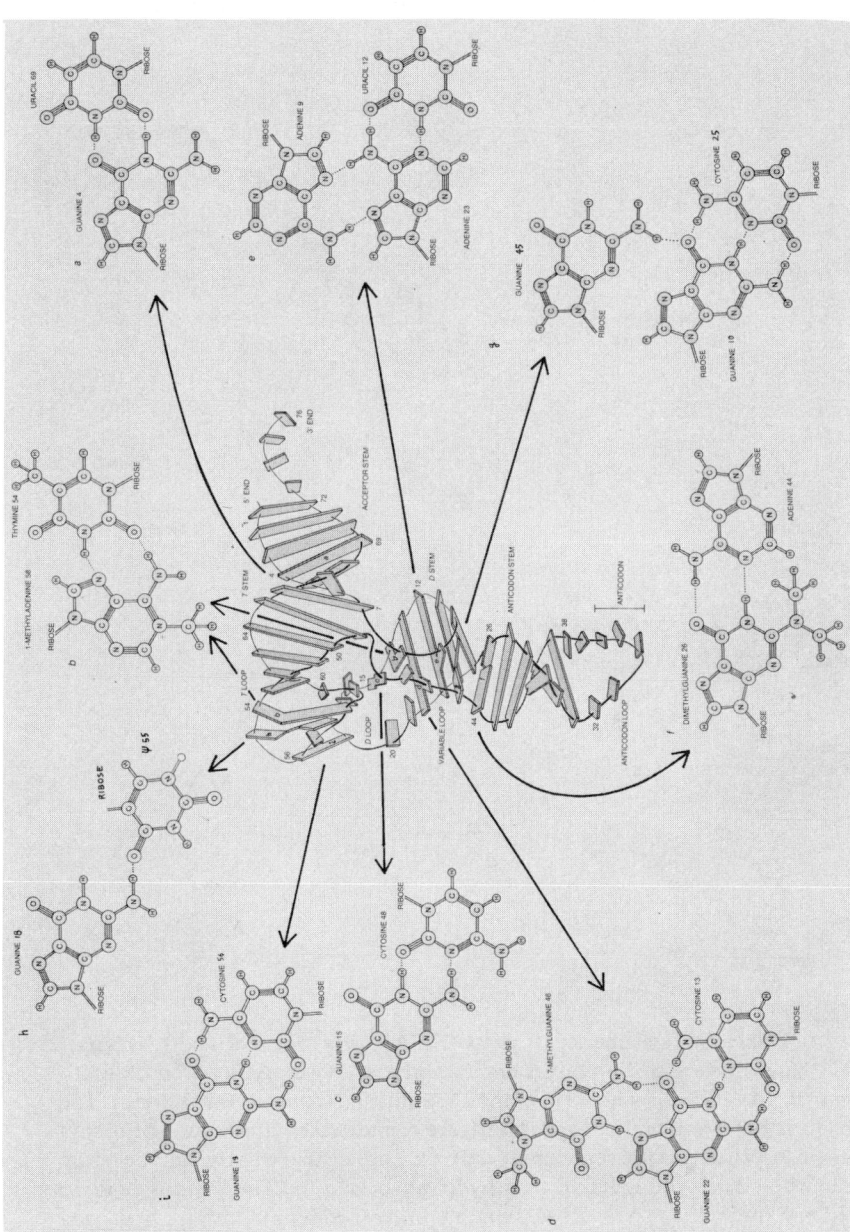

Figure 3 Nine tertiary base pairs and one GU base pair in the crystal structure of yeast tRNA[Phe]. The locations where these tertiary base pairs occur are indicated. Notice that all tertiary base pairs are different from the Watson-Crick base-pair type, and that all are located near the corner of the L.

87

al. 1974). All of these are located at the intersection of the two arms of the L and are essential for maintaining the L-shaped basic frame. Location of conserved and semiconserved bases and tertiary base pairs are indicated in the cloverleaf, L-diagram, and crystal structures in Figure 4.

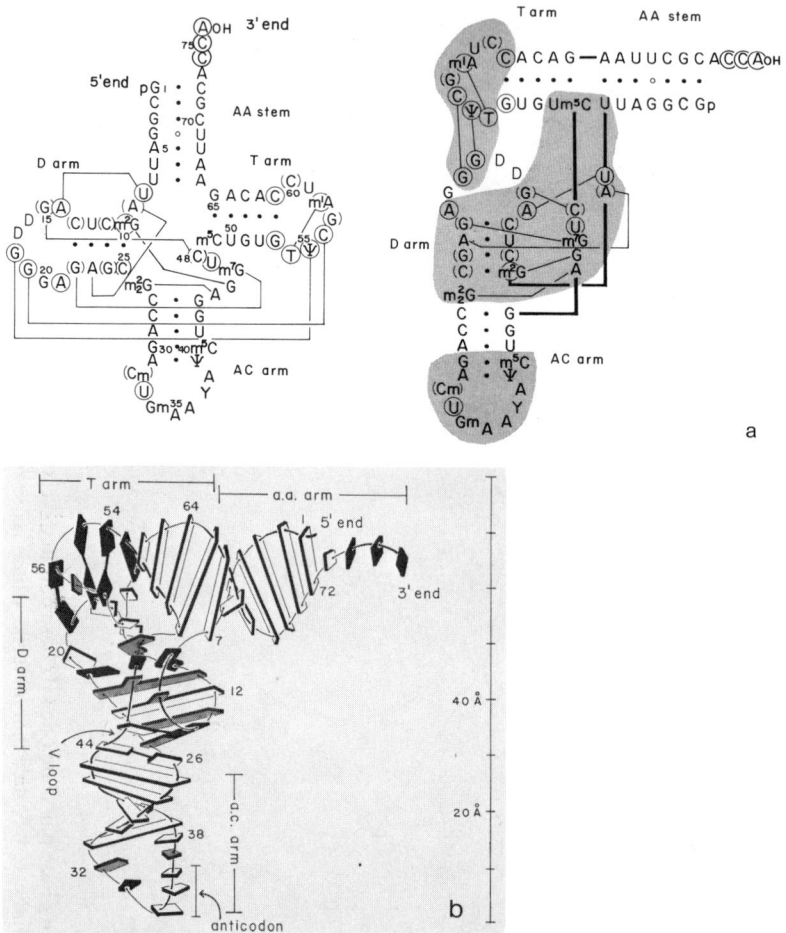

Figure 4 (*a*) Nucleotide sequence of yeast tRNA^{Phe} in cloverleaf and L arrangements. The bases that are conserved are circled and semiconserved in parentheses for D_4V_5 class tRNAs (tRNAs with 4 bp in the D stem and 5 bases in the V loop). The bases that form base-tertiary hydrogen bonds are connected by thin lines. Notice that most of the conserved and the semiconserved bases are localized to the middle of the L and are involved in forming the tertiary hydrogen bonds. (*b*) The location of these conserved and the semiconserved bases and their involvement in tertiary hydrogen bonds in the crystal structure of the yeast tRNA^{Phe}. For the details of the actual structure for the shaded regions in *a*, see Holbrook et al. (1978).

Base-Backbone Interaction

Assignments of secondary and tertiary hydrogen bonds between bases in electron density maps are relatively easy because the number of relative locations of the possible paired bases are limited and predictable. However, assignment of hydrogen bonds involving the backbone is much more difficult and arbitrary. Therefore, assignments of this class vary considerably among four research groups and should be considered as tentative at the present time. Twelve hydrogen bonds of this category have been assigned for the tRNA in an orthorhombic crystal form (Holbrook et al. 1978). Of these, five involve 2'-OH groups.

Backbone-Backbone Interaction

The assignment of hydrogen bonds of this class is perhaps the most difficult and arbitrary. Here again 2'-OH groups of riboses are involved many times.

Extensive utilization of 2'-OH groups to form tertiary hydrogen bonds may be the reason RNA rather than DNA is used for structural nucleic acids, such as tRNAs and rRNAs.

Conformational Details of Anticodon Loop

The anticodon loop is rather free from tertiary hydrogen bonds. Five bases, including the anticodon triplet, are stacked on one side of the loop leaving only two bases on the other side (see Fig. 4b). This loop conformation is further stabilized by a magnesium hydrate ion; Mg^{++} forms a direct coordination bond to a phosphate oxygen of residue 37, and the water molecules (coordinated to Mg^{++} ion) form several hydrogen bonds to nearby bases in this loop (Holbrook et al. 1977; Quigley et al. 1978). However, it is conceivable that once the magnesium hydrate ion is removed by, say, ribosomal components, the anticodon loop will become flexible and render them changeable as may be required during, for example, translocation from the aminoacyl site to the peptidyl site.

Metal-tRNA Interaction

It is known that site-specific, bound magnesium ions help to stabilize the functional conformation of tRNAs (Fresco et al. 1966). This subject has been recently reexamined (Stein and Crothers 1976). There are two magnesium-hydrate binding sites in the D loop, one in the anticodon loop, and one in the sharp turn formed by residues 8–12. The assignment and environment of these four magnesium binding sites are listed in Table 1 and shown in Figure 5. From these locations it is clear that their functional role is to stabilize the loops and sharp turns of the tRNA

Table 1 tRNA-ligands interaction

Assignment	Location	Direct coordination to:	Hydrogen bonded to:	References
$[Mg(H_2O)_5]^{++}$	D loop	phosphate oxygen of G19	phosphate oxygen (G19) bases (G20, U59, C60) bases (G20, U59, C56)	Holbrook et al. (1977) Quigley et al. (1978) Holbrook et al. (1977) Quigley et al. (1978)
$[Mg(H_2O)_4]^{++}$	D loop	phosphate oxygens of G20, A21		Holbrook et al. (1977) Quigley et al. (1978)
$[Mg(H_2O)_6]^{++}$	8–12 turn		phosphate oxygens (U8, A9, C11, U12)	Holbrook et al. (1977) Quigley et al. (1978)
$[Mg(H_2O)_5]^{++}$	anticodon loop	phosphate oxygen of Y37	bases (C32, Y37, A38, ψ39)	Holbrook et al. (1977) Quigley et al. (1978)
Spermine	wide grooves of D stem and part of anticodon stem		PO_4 (23, 24, 25, 42, 44)	Holbrook et al. (1978) Quigley et al. (1978)
Spermine	wide grooves of amino acid and T stems			Holbrook et al. (1978)
Spermine	variable loop G10		PO_4 (10, 46, 47, 45)	Quigley et al. (1978)

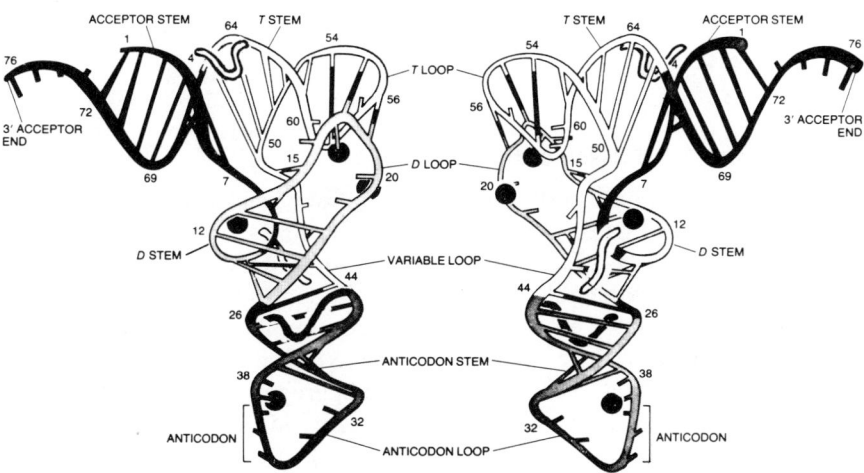

Figure 5 Two views of the folding of backbone of yeast tRNA^{Phe} are shown diagrammatically. The sugar-phosphate backbone of the molecule is represented as a coiled tube with the cross rungs standing for the nucleotide base pairs in the stem regions. The short rungs indicate bases that are not involved in base-base hydrogen bonding. Dark circles represent four magnesium hydrate ions, solid "spaghetti" represents spermine molecules found with more certainty, and open spaghetti represents the spermine positions determined with less certainty.

structure to maintain its functional conformation. Positions of randomly bound Mg^{++} are not obtainable by the X-ray crystallographic method.

Spermine-tRNA Interaction

Polyamines such as spermine are known to stabilize the functional conformation of tRNA structure. Three possible spermine molecules have been located: Holbrook et al. (1978) assigned one in the deep groove of the double helix formed by the D and the anticodon stems and the other in the deep groove of the double helix formed by the amino acid and the T stems; Quigley et al. (1978) assigned one in the same deep groove between the D and the anticodon stems, but the other between the variable loop and residue 10. Although the one in the D-anticodon stem is easily recognizable in the different electron-density maps, the assignments of the other ones are not reliable. Thus, one of the roles spermine plays is to stabilize the RNA double helix by binding to its deep (major) groove. These are also shown in Figure 5.

Water-tRNA Interaction

There are many electron-density peaks interpretable as bound water molecules mostly along the deep grooves of double-helical stems and some on shallow grooves. The positional accuracies of these peaks are quite poor.

Solvent-accessible Surface of tRNA

X-ray crystallographic studies of tRNA allow ready identification of intramolecular interactions from the proximity and orientation of potential ligands such as Mg^{++} and spermines; however, environmental influences are less amenable to such direct visualization, partly because the solvent is seldom completely ordered about the surface of the molecule and also because the solvent effects are most influential in the dynamic events leading to chain folding and aggregation. Thus, the change in water-accessible surface area of tRNA can provide useful insights on the problem of tRNA folding.

The method of calculation employed in this study is very similar to that described by Lee and Richards (1971) in their calculation of polypeptide surface areas. We imagine a spherical probe or solvent molecule of radius r_w free just to touch but not penetrate the van der Waals surface of the examined molecule. The closed surface defined by all possible loci for the center of the probe is defined as the accessible surface of the molecule.

Based on these calculations, three interesting results have been obtained (C. Alden and S.-H. Kim, in prep.): (1) the folded tRNA molecule has only about 35% of the total water-accessible surface areas of extended, unfolded tRNA, and the principal reduction in exposure surface occurs for nonpolar atoms; (2) contrary to common concept, the backbone atoms of the loop regions are more buried than those of the stems; (3) the modification groups on the bases increase the exposure areas as much as 20% of total base exposure in the molecule, suggesting that such modification may significantly increase the specificity or strength of binding during tRNA-protein recognition.

Flexibility of the Molecule

Analysis of thermal motion of tRNA crystal structure provides information about the flexibility of the molecule. Using a least-squares procedure, Sussman et al. (1978) have refined thermal parameters of bases, riboses, and phosphate groups of each residue using 8426 X-ray diffraction data. The results of this refinement are shown in Figure 6, where the radius of each circle is proportional to the corresponding group thermal motion of each nucleotide.

There are two interesting trends noticeable in the thermal motion of this molecule. The first is that both extreme ends (and three producing residues, 16, 17, and 47) of the L have very high thermal motion. An

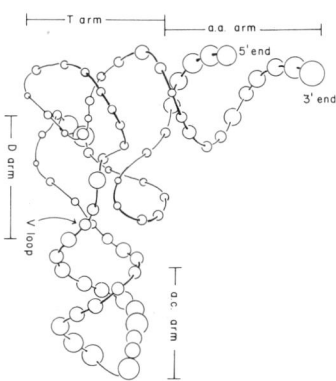

Figure 6 The thermal vibrational motions of each nucleotide of yeast tRNA^{Phe} are shown. The larger the circle is, the higher is the thermal vibration of that nucleotide in the crystal. The thermal motion pattern reflects the flexibilities of different parts of the molecule, i.e., the anticodon arm and the amino acid stem are more flexible than the rest of the molecule.

average mean squares displacement at the extreme ends is about 5.5 Å^2 compared to about 0.8 Å^2 at the corner of the L. The second is that the average thermal parameters for the bases is smaller than that for the ribose, which in turn is smaller than that for the phosphates. This trend is still more pronounced in the stem regions. This can be interpreted in one of two ways: (1) each long helical arm of the L has flexing or processing motion around the average helical axis of each long arm, or (2) each long double-helical arm of the L is partially opening up (unwinding) and closing (winding) at the extreme end.

Since the thermal parameters of residues are found to be not correlatable with the lattice contacts, we interpret that they reveal the intrinsic flexibility of the molecule at both extreme ends of the L.

Modified Nucleosides

None of the modifications appear to be essential for maintaining the integrity of the three-dimensional structure of this tRNA, suggesting that they are there to be recognized by various proteins. As mentioned earlier, the modification groups account for a 20% increase in the accessible surface area of the bases in yeast tRNA^{Phe} crystal structure.

Complete Crystal Structure of Yeast tRNA^{Phe}

Entire atomic positions of the crystal structure can be visualized in three stereo views shown in Figure 7. The only uncertainties are the positions of 3'-terminal base A76 and the side chain of the highly modified base H37. Visualization of the complete molecule is possible because it is only 20 Å thick. Although it looks complex, a few minutes of staring will reveal structural details not obvious in other diagrams.

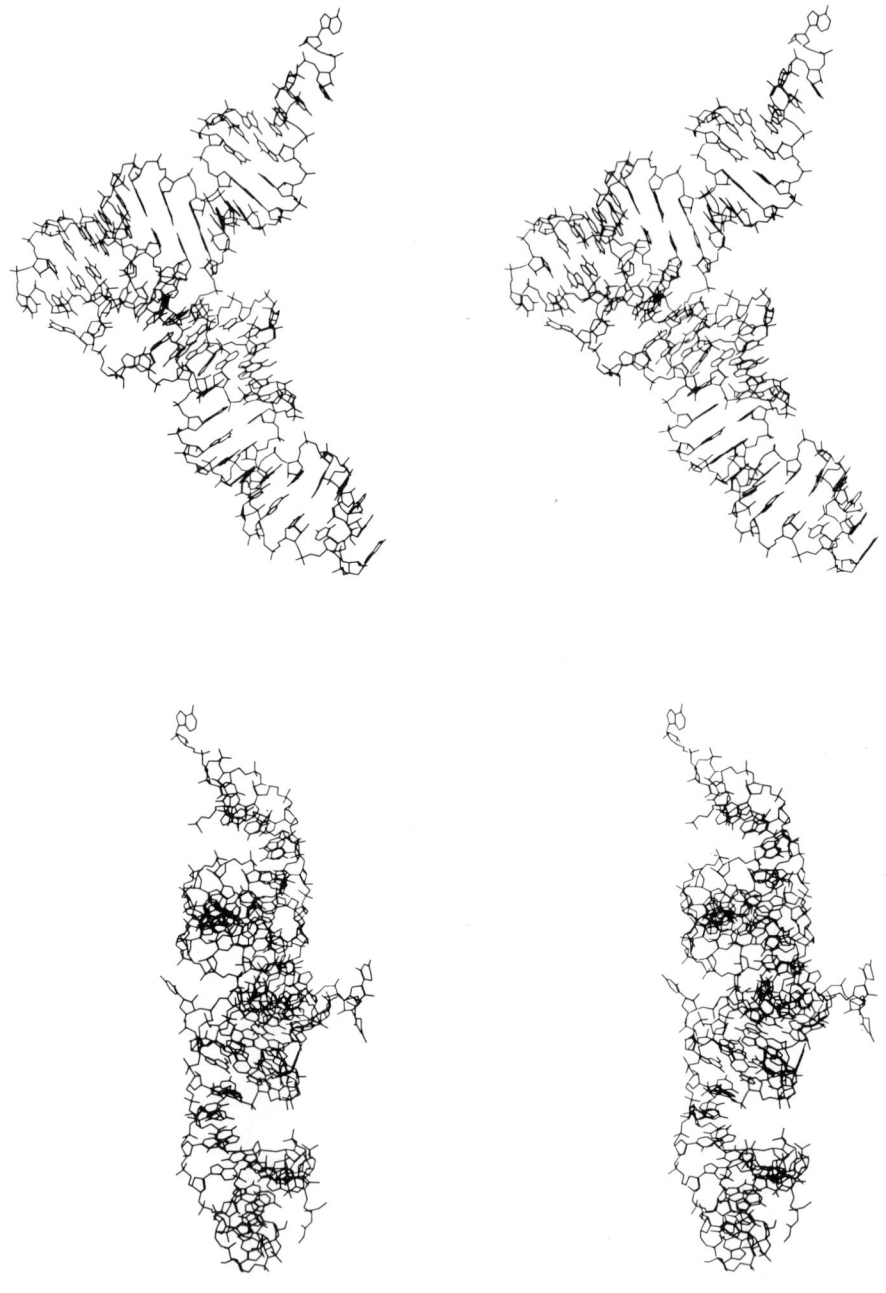

Figure 7 (*See facing page for legend.*)

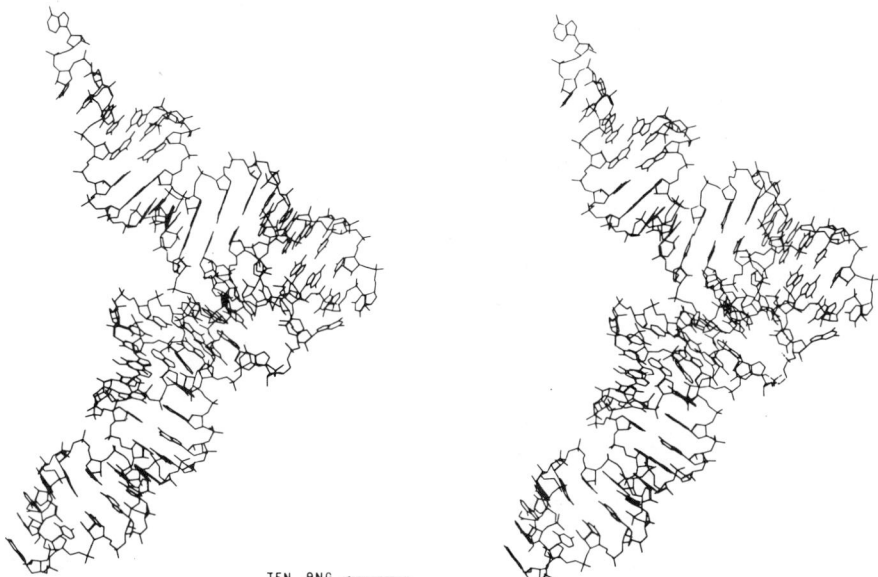

Figure 7 Three stereoviews of the complete atomic positions in the crystal structure of yeast tRNA^{Phe}. The scale at the bottom is 10 Å. Try to follow the progression from the 5' end to the 3' end by staring, for a few minutes with a pair of stereoglasses, at the top-left and the above pairs. The structural details obtainable from these stereoviews are more accurate and better than any other display methods.

GENERAL STRUCTURAL FEATURES OF ALL tRNAs

Tertiary Base Pairs Are Conserved in All tRNAs

The first indication that all tRNAs have the same overall structure comes from the observation that all tRNA sequences can be arranged into a cloverleaf model. The second convincing indication was derived from the crystal structure of yeast tRNA^{Phe}, where most of the conserved and semiconserved bases are involved in tertiary hydrogen bonds. Moreover, most of these tertiary base pairs in yeast tRNA^{Phe} can be replaced with equivalent tertiary base pairs for other tRNAs without disturbing the backbone structures significantly (Kim et al. 1974a,b; Klug et al. 1974). Besides the replaceable tertiary base pairs already suggested (see, for example, Kim 1978), additional such base pairs are listed in Table 2. Thus, one can consider the crystal structure of yeast tRNA^{Phe} as a representative structure of all tRNAs with the exception of those with long variable loops.

Table 2 Replaceable tertiary base pairs in D_4V_5 tRNAs

In yeast tRNA^{Phe} residue no. (bases)	In all D_4V_5 tRNAs (51 sequences examined)
U8·A14	51 UA
A9·A23	33 AA
	10 GC
	3 GG
	3 AC[a]
	1 AG
	1 GA
G10·G45	38 GG
	2 GU
	1 GA[b]
G15·C48	42 GC
	7 AU
	1 GA[a]
	1 AC
G18·ψ55	51 Gψ
G19·C56	51 GC
G22·G46	46 GG
	3 AA
	1 CA[a]
	1 GU
G26·A44	29 GA
	13 AG
	3 AC
	2 AU
	1 CU
	1 GG[a]
	1 AA
	1 ψG
T54·A58	51 TA

D_4V_5 tRNAs are tRNAs with 4 bp in the D stem and 5 bases in the V loop.
[a]Requires considerable twisting around the hydrogen bond.
[b]Cannot make approximately equivalent hydrogen bonds.

Stems Are Rich in G + C Content

G + C contents in the stems are considerably higher than 50%. Furthermore, one end of each stem is always very high in G + C content except the anticodon stem, where a penultimate base pair from the anticodon loop is most often a GC pair. They are G1·C72, G10·C25, G53·C61, and G30·C40 in yeast tRNA^{Phe}. At equivalent positions in all other tRNAs, the first three are GC pairs in most of the cases, but the fourth one is either a GC or a CG pair.

The G + C content of the long helix formed by the acceptor and the T stems is highest at extremes and average (55%) at the middle. Such richness in GC pairs at extremes may stabilize this long helix from unwinding at the end (fraying).

The lowest GC frequency (51%) among all stems is found at the base pair immediately next to the anticodon loop, suggesting that this base pair may be easily broken during tRNA interaction with ribosome to provide additional flexibility to the AC loop.

Conserved Presence and Conserved Absence of Bases in tRNA

As mentioned earlier in this paper, certain bases are conserved at particular locations in the nucleotide sequence and most of them are involved in forming hydrogen bonds to maintain the tertiary structure of tRNA. One can ask a converse question: Are there certain bases always absent at particular locations in nucleotide sequences and if there are, what are the functional significances? Figure 8 shows the bases and locations of such conserved presence and conserved absence.

Among the conserved bases, U8, A14, G18, G19, A21, T54, ψ55, C56, and A58 participate in forming tertiary hydrogen bondings, but A16, U33, C74, C75, and A76 are not involved in tertiary base pairing. The latter three are, of course, essential for aminoacylation. The first two, A16 and U33, probably are essential for ribosomal functions, as are the first set of conserved bases, which should be common to all tRNAs.

Among the conserved absence of certain bases, the absence of Y9,

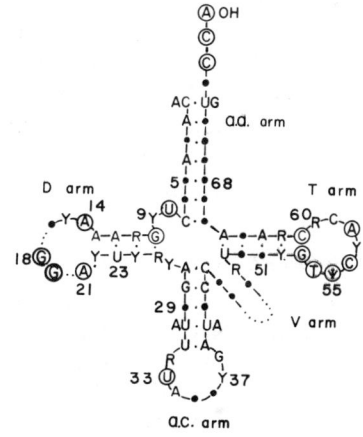

Figure 8 Conserved presence and conserved absence of bases in D_4V_5-class tRNAs. When a particular base is found at a certain position in more than 90% of the tRNA sequences (conserved presence), they are shown in circles. When a base occurs at a particular position less than 10% of the time (conserved absence), they are shown without a circle.

A12, and U23 can be explained from the necessity of forming a replaceable tertiary base triple 9·23·12. The same reasoning can be applied to the absence of Y15 and R48. The conserved absence of C59 and R60 may be due to the tight spatial requirement, i.e., the residues 59 and 60 are buried in a pocket inside the tertiary structure of tRNA, forming two hydrogen bonds by phosphate 60 with the base 61 and ribose 58. The remaining conserved absences must have other functional reasons. For example, the absence of A34 is due to the conversion of adenine to inosine presumably because the amino group of adenine is incompatible with necessary interactions with, say, the ribosome. In general, reasons for many of the conserved absences are not understood at the present time.

CONCLUSIONS

The crystal structure of yeast tRNA^{Phe} reveals a simple, overall architectural design that is maintained by a set of intricate hydrogen bondings among conserved and semiconserved bases. The sites for three important ribosomal functions are maximally separated and two of them appear to have considerable flexibility. The crystal structure is consistent with vast amounts of data on tRNA in aqueous solution, suggesting that it has the functional conformation of free tRNA in solution. Thus, this structure provides a solid starting point for studies on tRNA function, be it static or dynamic.

In addition, this structure is the first three-dimensional structure of a nucleic acid determined with the single-crystal X-ray crystallographic method. As such it provides detailed structural information that can be extended to other nucleic acids as well.

ACKNOWLEDGMENTS

Works performed in the author's laboratory have been supported by grants from the National Institutes of Health (CA-15802, K04-CA-00352) and the National Science Foundation (PCM76-04248). I thank Joel Sussman, Steve Holbrook, Wade Warrant, George Church, and Charles Alden for their critical contributions in various stages of crystallographic refinement of yeast tRNA^{Phe} structure and interpretation of the results, some of which are described here.

REFERENCES

Clark, B. F. C. 1977. Correlation of biological activities with structural features of transfer RNA. *Prog. Nucleic Acid Res. Mol. Biol.* **20**:1.

Erdmann, V. A. 1976. Structure and function of 5S and 5.8S RNA. *Prog. Nucleic Acid Res. Mol. Biol.* **18**:45.

Fresco, J. R., A. Adams, R. Ascione, D. Henley, and T. Lindahl. 1966. Tertiary structure in transfer ribonucleic acids. *Cold Spring Harbor Symp. Quant. Biol.* **31:**527.

Holbrook, S. R., J. L. Sussman, R. W. Warrant, and S.-H. Kim. 1978. Crystal structure of yeast phenylalanine transfer RNA. II. Structural features and functional implications. *J. Mol. Biol.* **123:**631.

Holbrook, S. R., J. L. Sussman, W. R. Warrant, G. M. Church, and S.-H. Kim. 1977. RNA-ligand interactions. I. Magnesium binding sites in yeast tRNA^{Phe}. *Nucleic Acids Res.* **4:**2811.

Jack, A., J. E. Ladner, and A. Klug. 1976. Crystallographic refinement of yeast phenylalanine transfer RNA at 2.5 Å resolution. *J. Mol. Biol.* **108:**619.

Jack, A., J. E. Ladner, D. Rhodes, R. S. Brown, and A. Klug. 1977. A crystallographic study of metal-binding to yeast phenylalanine transfer RNA. *J. Mol. Biol.* **111:**315.

Kim, S.-H. 1976. Three-dimensional structure of transfer RNA. *Prog. Nucleic Acid Res. Mol. Biol.* **17:**181.

———. 1978. Three-dimensional structure of transfer RNA and its functional implications. *Adv. Enzymol.* **46:**279.

Kim, S.-H., G. J. Quigley, F. L. Suddath, A. McPherson, D. Sneden, J. J. Kim, J. Weinzierl, and A. Rich. 1973. Three-dimensional structure of yeast phenylalanine transfer RNA: Folding of the polynucleotide chain. *Science* **179:**285.

Kim, S.-H., F. L. Suddath, G. J. Quigley, A. McPherson, J. L. Sussman, A. Wang, N. Seeman, and A. Rich. 1974a. Three-dimensional tertiary structure of yeast phenylalanine transfer RNA. *Science* **185:**435.

Kim, S.-H., J. L. Sussman, F. L. Suddath, G. J. Quigley, A. McPherson, A. H. J. Wang, N. C. Seeman, and A. Rich. 1974b. The general structure of transfer RNA molecules. *Proc. Natl. Acad. Sci.* **71:**4970.

Klug, A., J. Ladner, and J. D. Robertus. 1974. The structural geometry of coordinated base changes in transfer RNA. *J. Mol. Biol.* **89:**511.

Ladner, J. E., A. Jack, J. D. Robertus, R. S. Brown, D. Rhodes, B. F. C. Clark, and A. Klug. 1975. Atomic coordinates for yeast phenylalanine tRNA. *Nucleic Acids Res.* **2:**1629.

Lee, B. K. and F. M. Richards. 1971. The interpretation of protein structures: Estimation of static accessibility. *J. Mol. Biol.* **55:**379.

Quigley, G., N. C. Seeman, A. Wang, F. L. Suddath, and A. Rich. 1975. Yeast phenylalanine transfer RNA: Atomic coordinates and torsion angles. *Nucleic Acids Res.* **2:**2329.

Quigley, G. J., M. M. Teeter, and A. Rich. 1978. Structural analysis of spermine and magnesium ion binding to yeast phenylalanine transfer RNA. *Proc. Natl. Acad. Sci.* **75:**64.

Rich, A. and U. L. RajBhandary. 1976. Transfer RNA: Molecular structure, sequence and properties. *Annu. Rev. Biochem.* **45:**805.

Sigler, P. 1975. An analysis of the structure of tRNA. *Annu. Rev. Biophys. Bioeng.* **4:**477.

Stein, A. and D. Crothers. 1976. Conformational changes of tRNA. The role of Mg^{++}. *Biochemistry* **15:**157.

Stout, C. D., H. Mizuno, J. Rubin, T. Brennan, S. T. Rao, and M. Sundaralingam. 1976. Atomic coordinates and molecular conformation of yeast phenylalanine tRNA. *Nucleic Acids Res.* **3:**1111.

Sussman, J. L. and S.-H. Kim. 1976. Idealized atomic coordinates of yeast phenylalanine transfer RNA. *Biochem. Biophys. Res. Commun.* **68**:89.

Sussman, J. L., S. R. Holbrook, R. W. Warrant, G. M. Church, and S.-H. Kim. 1978. Crystal structure of yeast phenylalanine transfer RNA. I. Crystallographic refinement. *J. Mol. Biol.* **123**:607.

Recent Progress in tRNA Structural Analysis

Alexander Rich, Gary J. Quigley, Martha M. Teeter, Arnaud Ducruix, and Nancy Woo
Department of Biology
Massachusetts Institute of Technology
Cambridge, Massachusetts 02139

tRNA plays a central role in the expression of genetic information. Structural studies on tRNA have a dual goal. One is to learn about the three-dimensional structure of this molecule in sufficient detail so we can understand its solution chemistry, reactivity, and stability. And the other is to identify those structural features that are used by nature in the translation mechanism. How is tRNA recognized by enzymes and components of the ribosomal machinery so that it can carry out its functions? Structural studies eventually lead to functional considerations.

We now understand a great deal about the three-dimensional structure of one tRNA molecule, yeast tRNA^{Phe}. The first tRNA was sequenced by Holley and coworkers (1965); at the present time there are almost 100 sequences known (Gauss and Sprinzl 1978). These sequences have amply verified the universality of the cloverleaf folding of the molecule. By 1973, the three-dimensional folding of yeast tRNA^{Phe} was solved by X-ray diffraction analysis of orthorhombic crystals (Kim et al. 1973). In 1974 higher resolution analyses of both the orthorhombic and monoclinic yeast tRNA^{Phe} crystals revealed finer details of tertiary base-base interactions (Kim et al. 1974a; Robertus et al. 1974). Subsequent analysis and refinement of the molecule have produced a startlingly clear picture of the three-dimensional conformation of this one tRNA molecule. We can see how the invariant nucleotides are utilized in the three-dimensional folding of yeast tRNA^{Phe} and from this we can predict how this tRNA molecule may be used as a pattern for understanding the three-dimensional folding of all tRNA molecules (Kim et al. 1974b).

Because of an interest in function and structure, we would like to know how tRNA interacts with other molecules. What is the role of cationic molecules in the function of tRNA? In addition, we would like to know what happens to tRNA when it is aminoacylated or when it has a peptidyl group attached to it. The yeast tRNA^{Phe} molecule is typical of those involved in peptide chain elongation. However, another class of initiator tRNAs are used for starting protein synthesis. Do these differ in a significant way from chain elongating tRNA molecules?

In this paper, we describe progress in each of the following areas. First, we discuss the location and structural role of ions in the refined three-dimensional structure of yeast tRNA^{Phe}. Next, we consider our analysis of the structure of N-acetylated aminoacyl-tRNA^{Phe}. Finally, we describe progress in the study of the three-dimensional structure of *Escherichia coli* initiator, tRNA^{fMet}.

CATIONS STABILIZE THE THREE-DIMENSIONAL FOLDING OF YEAST tRNA^{Phe}

Figure 1 shows two views of yeast tRNA^{Phe}. By using a refinement program with constraints, it was possible to reduce the residual factor to a value of approximately 20% at 2.5-Å resolution (Quigley et al. 1978). The agreement between the observed and calculated model is so good that a difference map in which the calculated electron density is subtracted from the observed electron density shows many of the remaining atoms present in the lattice. In particular, this revealed a number of magnesium ions and

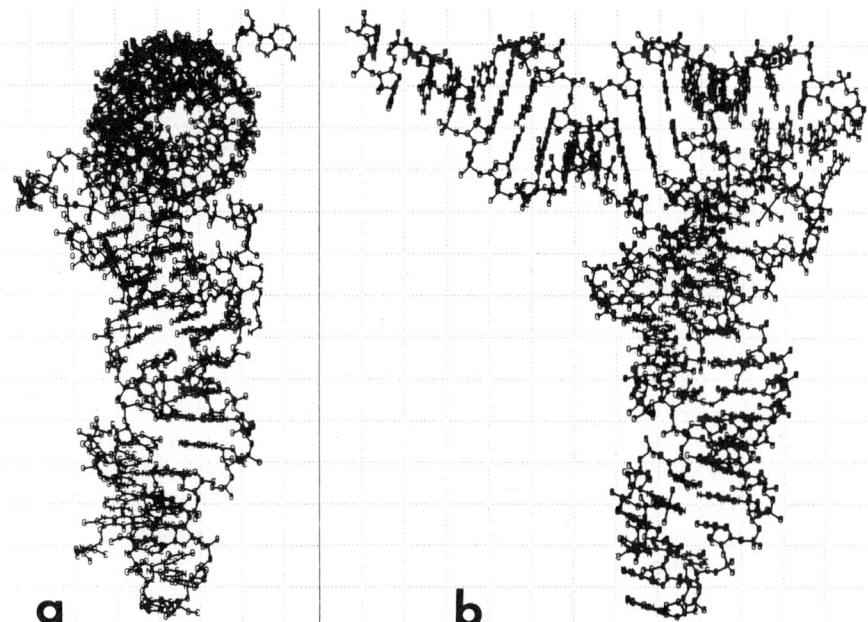

Figure 1 Skeletal diagram of the refined model of yeast tRNA^{Phe} superimposed on a 5-Å grid of lightly dotted lines. (*a*) A view down the acceptor and T stems. It can be seen at the top that the RNA double helix has a hole down its center. The 3′-terminal A is seen at upper right. Two bulges are seen in the left side of the molecule. The upper one is at residues D16 and D17, and the lower one is the large side chain of the wye base. The anticodon is at the bottom. (*b*) A side view showing the inverted "L" shape of the molecule. A bend can be seen near the bottom between the anticodon stem and the D stem. Both of these figures contain hydrated magnesium ions and spermine molecules.

spermine molecules, as well as fixed waters that are bound tightly in the lattice. Some of these are shown diagrammatically in Figure 2. Two spermine molecules are seen clearly. One is near the variable loop where it interacts with the sharp bend of the polynucleotide shown at P10. The other is in a lower position in Figure 2 where it occupies the deep groove at the top of the anticodon stem. The four positive charges of the elongated spermine molecule neutralize phosphate groups, three on one side and one on the other side of the deep groove of the RNA double helix. Because of the positively charged spermine molecule, the two ribophosphate chains on either side of the deep groove are brought closer together, and there is a 25° bend between axes of the D stem and the anticodon stem. The spermine molecule has thus produced a conformational change in the position of the anticodon stem and loop relative to the rest of the molecule. If that spermine molecule were removed and the double helix straightened out, codon bases would move a distance of 10 Å because of the long lever arm on the anticodon stem. The distance between one codon and the next in mRNA is also 10 Å. This makes one wonder whether some part of the molecular machinery of translocation may be related to the removal of the polyamines and a subsequent conformational change leading to straightening this end of the molecule.

Hydrated magnesium ions are generally found where the polynucleotide chains coils. An exception to this is the magnesium ion that is coordinated to P1 and P2 (M. Teeter et al., in prep.). This binding may be associated with the fact that P1 is a monoester and has a doubly negative charge. All the magnesium ions have an octahedral coordination with either water molecules or phosphate oxygens. An example is shown for the magnesium ion in the anticodon loop in Figure 3, which also shows that the structure of the anticodon loop is stabilized in a "u-turn" formed

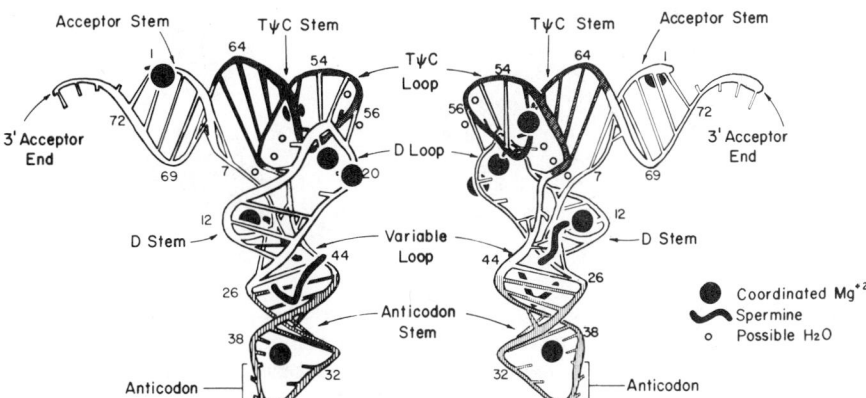

Figure 2 A schematic diagram showing two sides of yeast tRNA^{Phe} with bound ions and water molecules. The backbone is represented as a coiled tube. Cross rungs are used to represent base pairs and shorter rungs indicate bases.

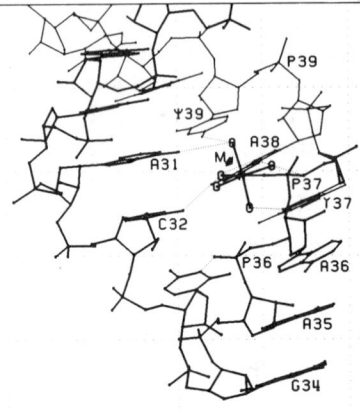

Figure 3 A skeletal diagram showing the anticodon region of yeast tRNA^{Phe}. The diagram is shown on a 5-Å grid of lightly dotted lines. Bonds in front are drawn darker than bonds behind. The hydrated magnesium ion (M) and water molecules (O) are illustrated. An oxygen from residue P37 coordinates directly to the magnesium. Hydrogen bonds, indicated by dotted lines, are formed between coordinated water molecules and surrounding bases. U33 is hydrogen bonded to an oxygen of P36, which stabilizes the anticodon conformation. The anticodon bases are G34, A35, and A36. The side chain of Y37 has been omitted for clarity.

by the invariant residue U33 that hydrogen bonds to an oxygen of P36 (Quigley and Rich 1976). Magnesium ions are found at points where the negatively charged polynucleotide chain makes close approaches. The polynucleotide chain has a sharp bend at P10, forming a pocket for a positively charged cation (Fig. 4a). We believe this is an example of a tightly bound magnesium ion (Römer and Hach 1975). Figure 4b shows the magnesium ion that is located adjacent to the T loop at G53. There are a number of phosphates near the entrance to this pocket that may

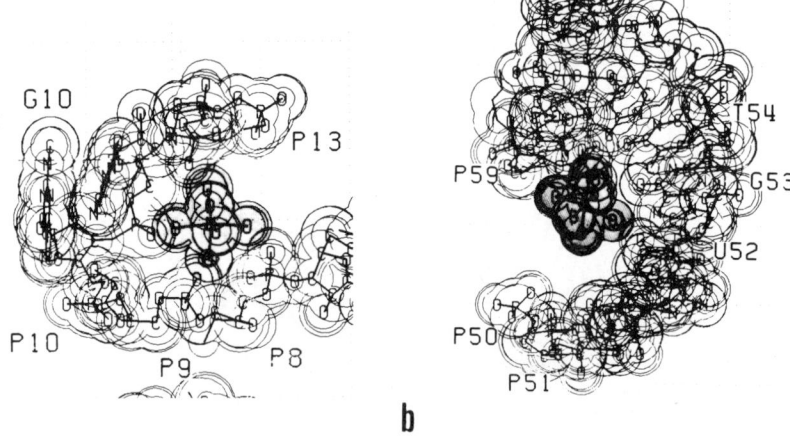

Figure 4 A van der Waals diagram showing hydrated magnesium ions (shaded) in contact with the coiled polynucleotide chain. (*a*) The magnesium octahedron found at the P10 bend of the polynucleotide chain. Water molecules in the hydration shell of magnesium are hydrogen bonded to P8, P9, P11, and P12. The fitting is rather precise. (*b*) A hydrated magnesium ion is shown interacting with bases U52, G53, and T54 adjacent to the T loop. Several phosphate groups line the entrance to this pocket. This hydrated magnesium ion fits less precisely than the magnesium ion in *a*.

stabilize the ion, but this magnesium does not fit tightly within the loop.

Analysis of the bound ions found in the yeast tRNA^{Phe} allows us to draw a few general conclusions. Magnesium and the spermine stabilize the three-dimensional folding of the molecule. In their absence it is likely that the molecule would unfold due to electrostatic repulsion. Each hydrated ion appears to be highly specific in the way in which it interacts with the polynucleotide chain. This specificity is undoubtedly used in biological systems that contain both magnesium ions and polyamines. The spermine bound in the deep groove at the top of the anticodon stem (Fig. 2) may stabilize one conformation of the molecule, and this raises the possibility that conformational changes in the molecule may be induced by modifying the cationic environment of the molecule. These changes could be brought about by tRNA interacting with positively charged side chains of proteins that could have effects similar to those seen for the spermine molecule in the deep groove of the helix.

THE STRUCTURE OF N-ACETYL AMINOACYL-tRNA

What happens to the tRNA molecule when it is aminoacylated or changed into peptidyl-tRNA? Here we describe experiments aimed at learning something about the structure of yeast tRNA^{Phe} in one of its functional states. We have aminoacylated tRNA^{Phe} molecules forming phenylalanyl-tRNA^{Phe}. To carry out crystallographic analysis, one has to overcome the intrinsic lability of the ester linkage that holds the amino acid and tRNA together. The half-life for deacylation at neutral pH is 1 hour or less at 37°C, which does not allow enough time for crystallization. The strategy was to work at lower pH, which stabilizes the OH-catalyzed cleavage of ester bonds, and to acetylate the α-amino group, which adds further stabilization. Thus, we have elected to crystallize N-acetyl phenylalanyl-tRNA^{Phe}.

Phenylalanyl-tRNA^{Phe} was made using partially purified yeast tRNA^{Phe} synthetase (Fasiolo et al. 1970). The pH was lowered to 4.5 and specific acetylation of the α-amino group was carried out by using the N-hydroxysuccinamide ester of acetic acid (Lapidot et al. 1967). The stability of the resulting compound was assayed by measuring the time-dependent, acid-precipitable radioactivity through the use of [^{14}C]phenylalanine. The ester linkage is substantially stabilized; after more than 20 days in solution under crystallizing conditions (at pH 4.5 and 4°C) there was no detectable loss of radioactivity attached to the tRNA. A complete description of the preparation and properties of the N-acetyl phenylalanyl-tRNA^{Phe} will be published elsewhere (A. Ducruix and A. Rich, in prep.).

Crystallization was carried out using a slight modification of the standard crystallization conditions that have been used previously (Kim et al. 1973). The principal modification is that the pH was maintained at 4.5 with sodium acetate. Crystals were formed of both N-acetyl phenylalanyl-tRNA^{Phe} and tRNA^{Phe} itself. The presence of radioactive [^{14}C]phenylalanine was measured in the crystals after washing them five to ten times in mother liquor to remove virtually all of the radioactive [^{14}C]phenylalanine. The washed crystals were dissolved and both tRNA and trichloroacetic acid precipitability were measured. By comparing this with the original level of aminoacylation, we could determine how many molecules in the crystals retained the radioactive amino acid. The extent of aminoacylation varied from 50% to 90% with most of the crystals containing about 70% aminoacylation. These experiments were also carried out on single crystals that yielded diffraction patterns. The radioactivity in them was consistent with the presence of at least 70% aminoacylated tRNA molecules in the crystal lattice. This number could be as high as 100% but the small size of the crystals prevents greater accuracy.

The low pH crystals of yeast tRNA^{Phe} were found to be in the monoclinic form with cell dimensions $a = 33$ Å, $b = 56$ Å, and $c = 63$ Å and $\alpha = 90°$. Examination of the N-acetylated phenylalanyl-tRNA^{Phe} revealed that they were also in the monoclinic form with the same cell dimensions. A comparison of the hko patterns of the low pH native and N-acetyl phenylalanyl-tRNA^{Phe} crystals reveals that the broad distribution of intensities is very similar in the two crystal forms. A comprehensive study of the diffraction patterns obtained by these two crystal forms is underway. However, a number of tentative conclusions can be drawn from the observation that N-acetyl phenylalanyl-tRNA^{Phe} can crystallize in a monoclinic lattice very similar to that of the native molecule at low pH and very similar to the native monoclinic form that has been analyzed at neutral pH (Robertus et al. 1974).

A general conclusion is that there has not been a substantial conformational change in yeast tRNA^{Phe} upon aminoacylation and subsequent N-acetylation. This conclusion is drawn from the similarities of both the lattice cell dimensions and the diffraction patterns. The native tRNA crystals have a high degree of hydration and there is room in the lattice for the addition of an N-acetylated phenylalanine at the 3' end of the molecule. By looking at the structure of yeast tRNA^{Phe}, one cannot make any simple statements about the most probable position for the phenylalanine residue. There is mobility at the 3'-CCA end of the molecule and this does not appear to be an important component in determining the overall packing of the molecules in the crystal lattice. A full characterization of the position of the phenylalanine residue in the crystal lattice will have to await the results of a three-dimensional analysis.

It is likely that N-acetylated phenylalanyl-tRNA^{Phe} has a conformation very similar to that of the native tRNA^{Phe}. However, we cannot rule out the possibility that there have been small conformational changes in the molecule, since these results would only become apparent upon completion of the three-dimensional analysis.

THE STRUCTURE OF *E. COLI* tRNA^{fMet}

A survey of the chemical and physical properties of tRNA molecules has led to the general conclusion that all tRNA molecules must be fairly similar (Rich and RajBhandary 1976). This conclusion is also supported by the existence of common features in the nucleotide sequence of these molecules and the manner in which these conserved components are utilized in the three-dimensional folding of the yeast tRNA^{Phe} molecule. However, it is of great interest to obtain more direct evidence and to acquire structural information about tRNA molecules that have specialized functions.

The prime example of a specialized tRNA molecule is that which initiates the synthesis of proteins by setting down the first amino acid, tRNA^{fMet}. Here we describe our progress in determining the three-dimensional structure of the *E. coli* initiator tRNA^{fMet}.

The initiator tRNA has many specialized properties. It is a substrate for the transformylase that formylates the α-amino group. It is inserted into the ribosomal peptidyl site, rather than into the aminoacyl site, the entry site for all other aminoacyl-tRNAs in the ribosome. It also acts in a manner not fully understood in modulating the activity of RNA polymerase and modifying the extent to which it is active in transcription (Pongs and Ulbrich 1976).

The *E. coli* tRNA^{fMet} molecule crystallizes readily but most of the crystals are disordered. To trace the conformation of a polynucleotide chain, a resolution of at least 2–3 Å is required. In the spermine-stabilized yeast tRNA^{Phe} crystals, the resolution is close to 2 Å. Most of the crystals produced by *E. coli* tRNA^{fMet} have a resolution of 10–15 Å and are useless in carrying out three-dimensional studies. More than 20 different crystal forms were examined and one proved to be well ordered. Figure 5 shows a diffraction pattern of an orthorhombic crystal of *E. coli* tRNA^{fMet} that has a 2.8-Å resolution. These crystals were grown in the presence of a variety of cations including spermine. The full details of the structure analysis will be published elsewhere (N. Woo et al., in prep.).

Although the initiator tRNA is functionally distinct from all other tRNAs, evidence from a variety of studies (Chang 1973; Goddard and Schulman 1972; Schulman and Pellca 1976) strongly suggests that the three-dimensional folding of this molecule is substantially similar to that observed in the yeast tRNA^{Phe}. Consequently, the initial structure determination of the initiator tRNA molecule was accomplished by using the method of rotation and translation function searches with the yeast tRNA^{Phe} as the search model. This technique assumes a general similarity in the overall conformation of the search model (yeast tRNA^{Phe}) and the unknown (*E. coli* tRNA^{fMet}). The search model is used to locate the unknown in its unit cell. The result of these searches is a location of the model in the unit cell of the unknown. This location is determined without regard to the constraints on possible locations of the model that

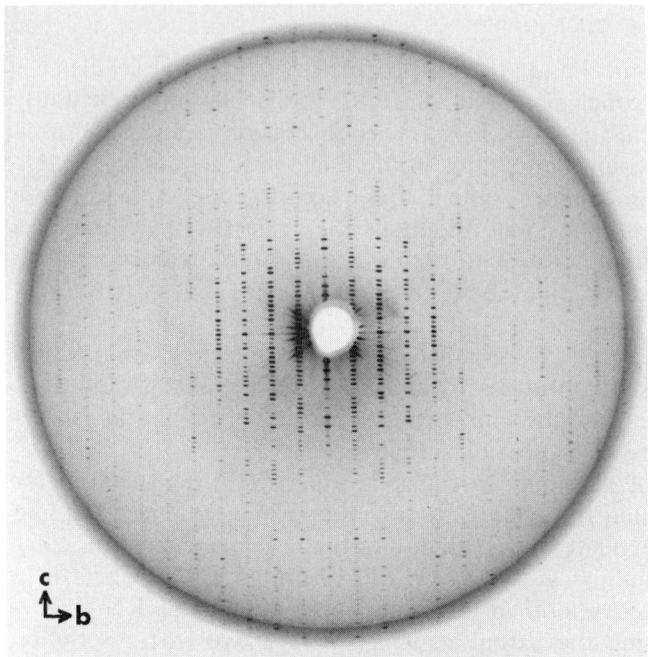

Figure 5 A crystal diffraction pattern of *E. coli* tRNA^{fMet}. A 12° precession photograph is shown of the *okl* zone and the limit of resolution in this photograph is 3.7 Å. The crystals diffract to 2.8-Å resolution. The crystal cell dimensions are $a = 32$ Å, $b = 85$ Å, and $c = 240$ Å and the space group is C222 (orthorhombic).

are imposed by the crystal symmetry. Thus, one criterion that can be used to evaluate the plausibility of a solution resulting from this method is whether the position of the model allows a crystal lattice to be formed. That is, the position must be one in which symmetrically related molecules make contact and do not interpenetrate. Another criterion is the calculation of the crystallographic residue, R, which measures the goodness-of-fit of the model positioned according to the search results and the observed data.

This calculation was carried out successfully and the correct lattice was found. Calculation of the residual function revealed a value of 44% at 6-Å resolution. This value is quite comparable to the value one would obtain from solving the structure by the more conventional method of multiple isomorphous replacement.

The *E. coli* tRNA^{fMet} lattice is shown in Figure 6, together with the yeast tRNA^{Phe} orthorhombic lattice. In the yeast tRNA^{Phe} lattice, a major packing interaction occurs at the anticodon end of the tRNA molecule where a twofold rotation axis relates one tRNA to another through

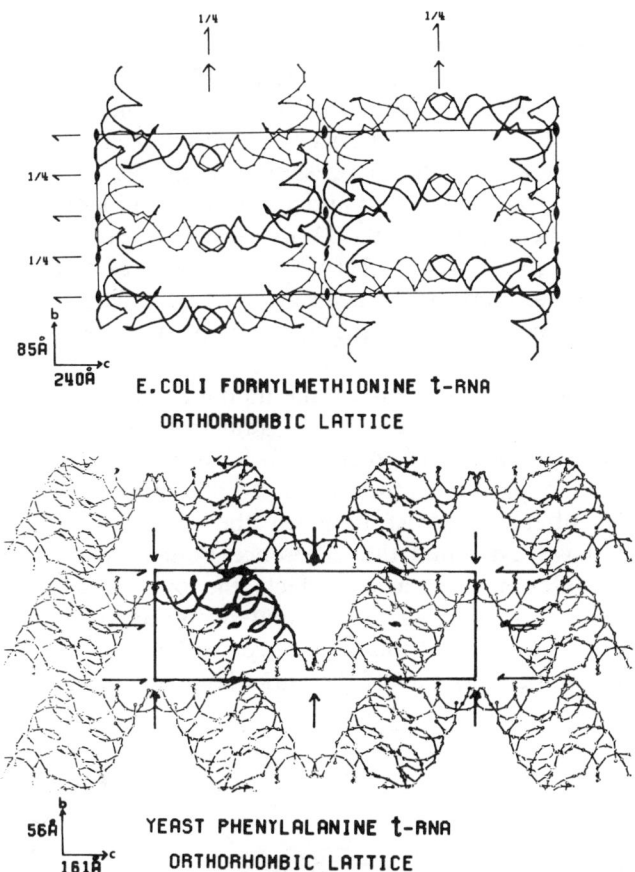

Figure 6 Lattice packing diagrams of yeast tRNA^{Phe} and *E. coli* tRNA^{fMet} in their orthorhombic lattices. It can be seen that both lattices have large channels that are filled with water running through them. Even though both lattices are orthorhombic, the tRNA interactions are quite different in the two crystal structures. The yeast tRNA^{Phe} has a twofold axis in which the anticodons are stacked upon each other. In *E. coli* tRNA^{fMet}, the anticodons are side by side rather than stacked one upon the other. (❘) Denotes twofold axes perpendicular to this plane.

stacking of the anticodons. In the *E. coli* tRNA^{fMet} lattice, the anticodons face each other, related by a twofold axis. The lattice shapes differ in a significant way, as can be seen in Figure 6.

Furthermore, the packing interactions between molecules in the two crystals are entirely distinct and there are no common packing interactions in the two lattices. Thus, the similarity in conformation of the two molecules cannot be attributed to crystal packing forces.

At the present time, the crystal structure has been solved to 3.5 Å with a residual of 48%. In Figure 7 we show a projection down the *b* axis of the present unrefined model of the structure. One molecule is shown in heavy lines with fragments of two symmetrically related molecules in lighter lines. This model has the phosphate-ribose conformation of the yeast tRNA^{Phe} search model with the base sequence of the *E. coli* tRNA^{fMet}. In this projection the D stem and anticodon stem and loop are clearly seen, as well as the bend between the D and anticodon stems. As discussed earlier, this bend in the yeast tRNA^{Phe} structure is associated with a spermine molecule that binds in the deep groove of the tRNA. As noted above, tRNA^{fMet} was crystallized in the presence of spermine, but we must await refinement to actually find it.

A residual of 48% indicates that, in general, the model agrees with the observed data, but it also indicates that the model is far from perfect and in the process of refinement, there are likely to be a number of small alterations that must be made in the conformation. To determine in what regions conformational differences exist between the model and the observed structure, we have examined the difference Fourier synthesis $(F_o - F_c)$ at 3.5 Å. The difference synthesis reflects the differences between the observed diffraction data and the calculated diffraction based on the model. In Figure 8, a portion of this difference synthesis is shown. This section is 24 Å thick and contains all of the anticodon loop and stem,

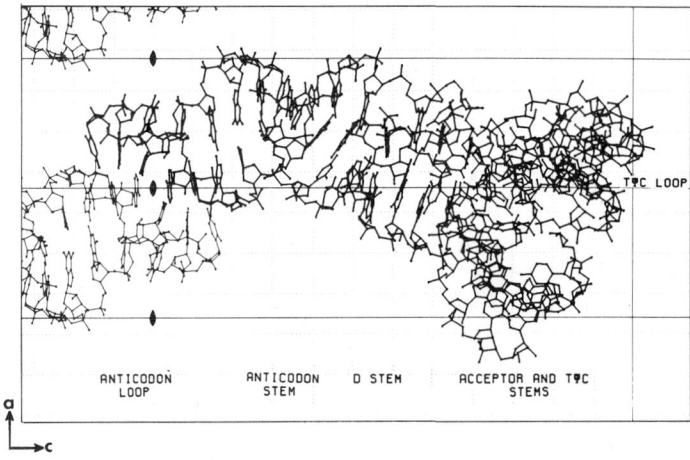

Figure 7 The molecular structure of *E. coli* tRNA^{fMet}. The molecule is shown viewed down the crystallographic *b* axis. The two lighter molecules that are partially shown are symmetrically related molecules to illustrate the side-by-side packing of the anticodon loops. (♦) Denotes twofold axes perpendicular to this plane.

as well as most of the D stem. In a difference synthesis, areas where difference density, either positive (solid lines) or negative (dashed lines) occur, are areas where the model is not in good agreement with the observed data.

There are a number of points that should be emphasized. First of all, there is reasonable agreement in the region of the two stems. There are a number of small peaks in that region that suggest some minor adjustment should be made, but the remarkable feature is the absence of any very large peaks. In contrast, there are some significant differences that are observed at the bottom of the anticodon stem and in the anticodon loop. These suggest that the conformation of the tRNA^{fMet} molecule differs from the model used in the anticodon loop. We have made some preliminary studies about the manner in which the conformation of the anticodon loop must be changed to better fit the observed diffraction data. Although it is not certain at the present time, the suggestion is very strong that the fit will require a slight opening up of the anticodon loop.

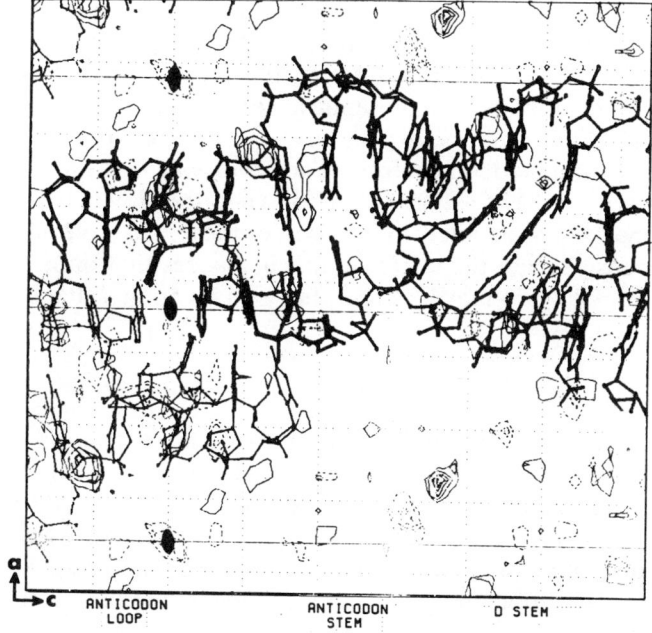

Figure 8 A portion of the difference Fourier synthesis $(F_o - F_c)$ at 3.5 Å. Solid contours indicate positive electron density, dotted contours indicate negative density. It can be seen that the difference map is fairly clean in the region of the anticodon and D stems but there are significant differences near the anticodon loop.

Specifically, there may be a movement of the constant U in the anticodon loop (analogous to U33 in yeast tRNA^{Phe}) so that the hydrogen bond between that U and the anticodon triplet would not be made (see Fig. 3). In addition to this difference in the anticodon loop, other significant differences in the structure are seen in the D loop where there are a different number of nucleotides in *E. coli* tRNA^{fMet} relative to yeast tRNA^{Phe}. There are also significant differences observed near the acceptor end of the molecule, specifically involving the first base pair in the acceptor stem that is unpaired in the prokaryotic initiator tRNA sequences. These differences will be described more fully elsewhere.

At the present time we have a good unrefined model for the structure of *E. coli* tRNA^{fMet} that, although broadly similar to the yeast tRNA^{Phe} structure, appears to differ in several regions, in the acceptor stem, the D loop, and, surprisingly, in the anticodon loop. It is clear that we will have to await the results of the fully refined structure so that an adequate comparison can be made between these two molecules, one that inserts amino acids in the middle of the polypeptide chains and one that inserts only the first amino acid. Even at this stage, however, we are tempted to wonder whether the differences in the anticodon loop conformation may somehow reflect differences that are relevant to the question as to which ribosomal site the tRNA goes. Is it possible that the yeast tRNA^{Phe} with its tightly bonded anticodon loop represents a conformation that the ribosome recognizes in the aminoacyl site, whereas the possibly more flexible anticodon loop with an open conformation represents that which is found in the ribosomal peptidyl site? It is too early to know the answer to this question but it is hoped that a detailed comparison of these two molecules may provide further insight into the conformational changes that occur in tRNA molecules, some of which may be associated with functional changes in protein synthesis.

CONCLUSIONS

Here we have described ongoing research in the structural analysis of tRNA. Increasingly, the type of question we now address relates to the biochemical function of tRNA. We wish to know how the molecule is changed by contact with other molecular species. In particular we would like to learn something about structural changes rather than structure per se as this field of investigation moves toward an understanding of the biochemical function of tRNA in protein synthesis.

ACKNOWLEDGMENTS

This research was supported by grants from the National Institutes of Health, the National Science Foundation, the American Cancer Society, and the National Aeronautics and Space Administration.

REFERENCES

Chang, S. E. 1973. Selective modification of cytidine and uridine bases in *Escherichia coli* formyl methionine transfer ribonucleic acid. *J. Mol. Biol.* **75**:533.

Fasiolo, F., N. Befort, A. Boulangery, and J. P. Ebel. 1970. Purification et quelques proprietes de la phenylalanyl tRNA synthetase de levure de Boulangerie. *Biochim. Biophys. Acta* **217**:305.

Gauss, D. and M. Sprinzl. 1978. Collection of published tRNA sequences. *Nucleic Acids Res.* **5**:r15.

Goddard, J. P. and L. H. Schulman. 1972. Conversion of exposed cytidine residues to uridine residues in *Escherichia coli* formylmethionine transfer ribonucleic acid. *J. Biol. Chem.* **247**:3864.

Holley, R. W., J. Apgar, G. A. Everett, J. T. Madison, M. Marquisee, S. H. Merrill, J. R. Perswick, and A. Zamir. 1965. Structure of a ribonucleic acid. *Science* **147**:1462.

Kim, S. H., G. J. Quigley, F. L. Suddath, A. McPherson, D. Sneden, J. J. Kim, J. Weinsierl, and A. Rich. 1973. The three-dimensional structure of yeast phenylalanine transfer RNA: Folding of the polynucleotide chain. *Science* **179**:285.

Kim, S. H., F. L. Suddath, G. J. Quigley, A. McPherson, J. L. Sussman, A. H.-J. Wang, N. C. Seeman, and A. Rich. 1974a. Three-dimensional tertiary structure of yeast phenylalanine transfer RNA. *Science* **185**:435.

Kim, S. H., J. L. Sussman, F. L. Suddath, G. J. Quigley, A. McPherson, A. H.-J. Wang, N. C. Seeman, and A. Rich. 1974b. The general structure of transfer RNA molecules. *Proc. Natl. Acad. Sci.* **71**:4970.

Lapidot, Y., N. DeGroot, and I. Fry-Shafrir. 1967. A general method for the preparation of acylaminoacyl tRNA. *Biochim. Biophys. Acta* **165**:397.

Pongs, O. and N. Ulbrich. 1976. Specific binding of formylated initiator-tRNA to *Escherichia coli* RNA polymerase. *Proc. Natl. Acad. Sci.* **73**:3064.

Quigley, G. J. and A. Rich. 1976. Structural domains of transfer RNA molecules. *Science* **194**:796.

Quigley, G. J., M. M. Teeter, and A. Rich. 1978. Structural analysis of spermine and magnesium ion binding to yeast phenylalanine transfer RNA. *Proc. Natl. Acad. Sci.* **75**:64.

Rich, A. and U. L. RajBhandary. 1976. Transfer RNA: Molecular structure, sequence, and properties. *Annu. Rev. Biochem.* **45**:805.

Robertus, J. D., J. E. Ladner, J. T. Finch, D. Rhodes, R. S. Brown, B. F. C. Clark, and A. Klug. 1974. The structure of yeast phenylalanine tRNA at 3 Å resolution. *Nature* **250**:546.

Römer, R. and R. Hach. 1975. tRNA configuration and magnesium binding. *Eur. J. Biochem.* **55**:271.

Schulman, L. H. and H. Pellca. 1976. Location of accessible bases in *Escherichia coli* formyl methionine transfer RNA as determined by chemical modification. *Biochemistry* **15**:5769.

The Structure, Conformation, and Interaction of tRNA

Muttaiya Sundaralingam
Department of Biochemistry, College of Agricultural and Life Sciences
University of Wisconsin, Madison
Madison, Wisconsin 53706

tRNA molecules are perhaps the smallest polynucleotides (75–93 residues) that possess profound biological functions. The most important function of tRNAs is their role in protein synthesis. Information on the three-dimensional structure of tRNA is of paramount importance for an understanding of the molecular basis of this and the other functions of tRNAs. The structure of yeast tRNA^{Phe} has been determined in both the orthorhombic and monoclinic forms (Kim et al. 1974; Robertus et al. 1974; Quigley and Rich 1975; Jack et al. 1976; Stout et al. 1976, 1978) using the single crystal X-ray diffraction technique. The structure of the monoclinic form was determined in our laboratory in late 1974 at low resolution and the complete structure describing the coordinates and the stereochemistry of the nucleotide and internucleotide phosphoesters were reported in 1976 (Stout et al. 1976, 1978; Sundaralingam et al. 1976; Sundaralingam 1978). Further details of the earlier structural work on tRNAs may be found in a recent review by Goddard (1978), as well as in the Proceedings of the Steenbock Symposium (Sundaralingam and Rao 1975). In this paper the salient stereochemical features of yeast tRNA^{Phe} in relation to the established principles of the stereochemistry of nucleic acids and their constituents are presented.[1]

THREE-DIMENSIONAL STRUCTURE
Backbone Fold
The folding of the polynucleotide backbone chain in tRNA^{Phe} is shown in Figure 1 and stereoviews of the complete and partial structures in Figure 2. The helical segments predicted by the Holley cloverleaf diagram are preserved in their entirety. The acceptor (A) and the pseudouridine (TψC) stems are stacked to form a quasi-continuous helix (domain 1); the D and the anticodon (AC) stems are stacked to form a second quasi-continuous helix (domain 2) and are arranged at right angles such that the AC loop and the 3'-CCA end are farthest apart (Fig. 2). The TψC and D loops are juxtaposed

---

[1]The structure of *Escherichia coli* arginine tRNA determined by the molecular replacement technique at 4-Å resolution will be published elsewhere (R. Bott et al., in prep.).

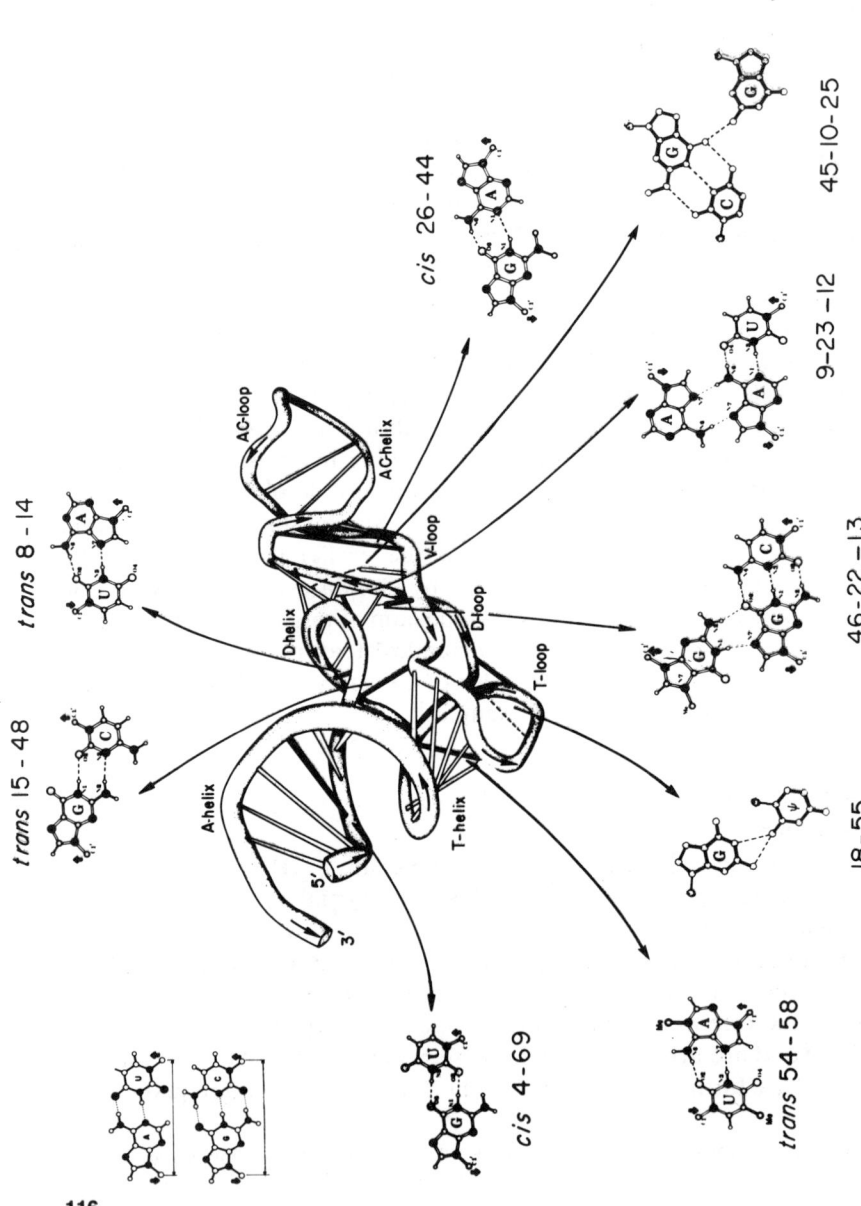

Figure 1 A drawing of the backbone fold of tRNA showing the base-paired helical stems and the non-Watson-Crick tertiary base-base interactions. The Watson-Crick base-pairing and the GU wobble-pairing schemes are shown on the left. The chain direction is indicated by arrows within the backbone trace.

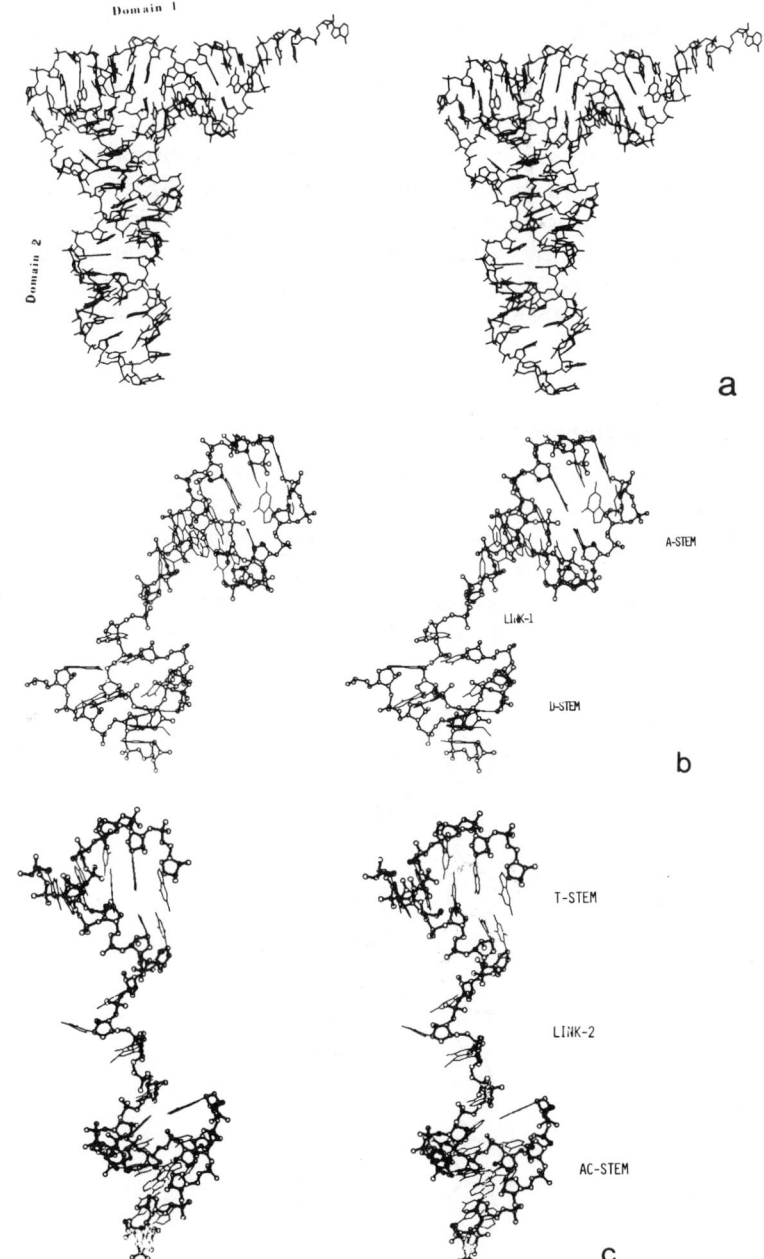

Figure 2 Stereoviews of: the complete structure of tRNA[Phe] (*a*); the A and D stems aligned roughly at right angles by link 1 (U8·A9) (*b*); and the TψC and AC stems aligned roughly at right angles by link 2 (A44·C48) (*c*).

at the corner of the molecule to engage in interloop base-pairing interactions. The two major helical domains of the tRNA lie essentially in a plane having a thickness of about 20 Å, which is close to the diameter of the RNA double helix. The domains are connected by two short stretches of polynucleotide chains: link 1 (U8-A9) connects the A and D stems and link 2 (A44-C48) connects the TψC and AC stems. Link 1 makes an abrupt turn at residue 10 to orient the D stem roughly normal to the A stem (Fig. 2b). Link 2 does not fold on itself but is extended such that the AC stem is roughly perpendicular to the D stem. This allows the AC stem to stack under the D stem and the TψC stem under the A stem (Fig. 2c). The link 1 always has two nucleotide residues in all tRNAs, whereas the link 2 varies from 4–19 residues.

Watson-Crick and Crick Wobble Pairings

The bases in the four helical stems of the cloverleaf are involved in the Watson-Crick complementary pairing scheme. Yeast tRNA^{Phe} has a GU pair in the middle of the A stem which is base paired in the same manner as the Crick GU wobble pair. It is found that when the G base is on the 5′ side of the Watson-Crick pair, strikingly greater stacking overlap with the Watson-Crick pair is observed than when it is on the 3′ side (Fig. 3).

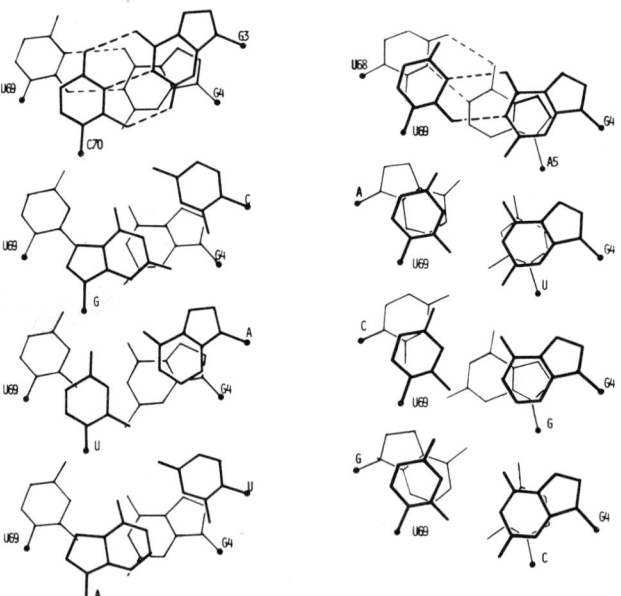

Figure 3 The base stacking between the G4·U69 wobble pair and the adjacent Watson-Crick pairs. G3·C70 (*left*) and A5·U68 (*right*). The stacking of the GU pair with the various Watson-Crick base-pairing combinations is also shown.

Thus, a GU pair at a helix terminal would stack better with the adjacent Watson-Crick pair if the G base lies on the 5' side. Indeed, an analysis of the occurrence of the terminal GU base pairs in the known tRNA sequences shows a strong preference for the G on the 5' side. It may be noted that switching the bases of the Watson-Crick pairs does not affect the stacking significantly (Fig. 3). Further implications of this rule in codon-anticodon wobble interactions are discussed elsewhere (Mizuno and Sundaralingam 1978).

Non-Watson-Crick Pairings

Several non-Watson-Crick pairing schemes occur in tRNA^{Phe} involving both Watson-Crick complementary and noncomplementary bases in different regions of the molecule (Fig. 1) (Robertus et al. 1974; Klug et al. 1974; Kim et al. 1974; Brennan and Sundaralingam 1975). The tertiary base-base interactions are correlated with the backbone chain conformation, separation, and direction. These pairs are conserved in virtually all tRNAs with appropriate coordinate base changes and are important for stabilizing the characteristic tRNA fold. A novel feature of tRNA is the occurrence of base triple interactions involving the base pairs of the D helix. The base triples are not always present in tRNAs, especially those possessing a long variable loop (Fig. 4) (Brennan and Sundaralingam 1976).

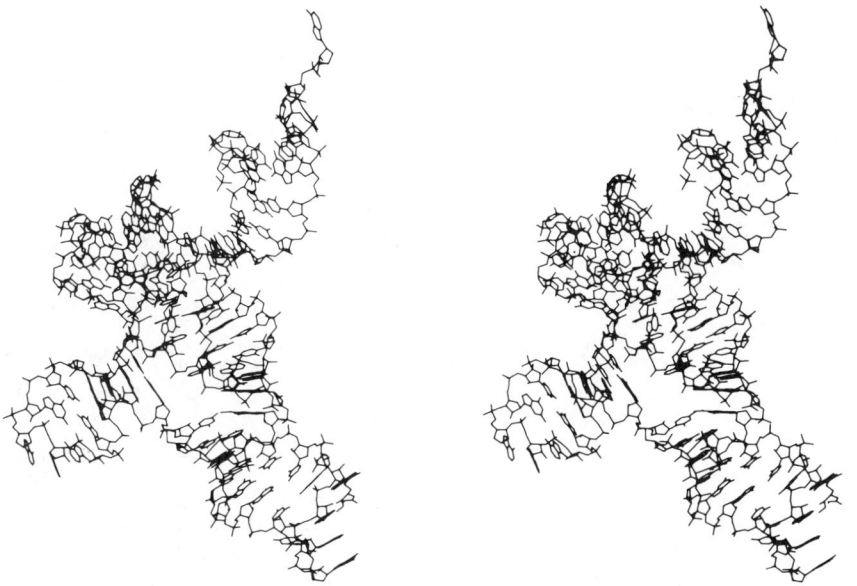

Figure 4 A stereoview of a model for *E. coli* tRNA$_1^{Ser}$ containing a long V loop.

Base Stacking

The base stacking in the helical sections of the tRNA is similar to that found in the fibrous RNA double helices. Perhaps a slight difference is in the somewhat greater base overlap in the short tRNA helices. The enhanced overlap produces a tightening of the helices so that the helical parameters evaluated from the tRNA helices show that the number of residues per turn is somewhat less than RNA-11. The base stacking patterns at the interface of the helical stems are shown in Figure 5. Although the adjacent pyrimidine bases U7 and m⁵C49 (Fig. 5a) are not on a continuous strand, there is some base-base interaction, especially involving the 5-cytosine methyl group and suggesting a possible role for the base modification. The stacking between the first D-stem base pair

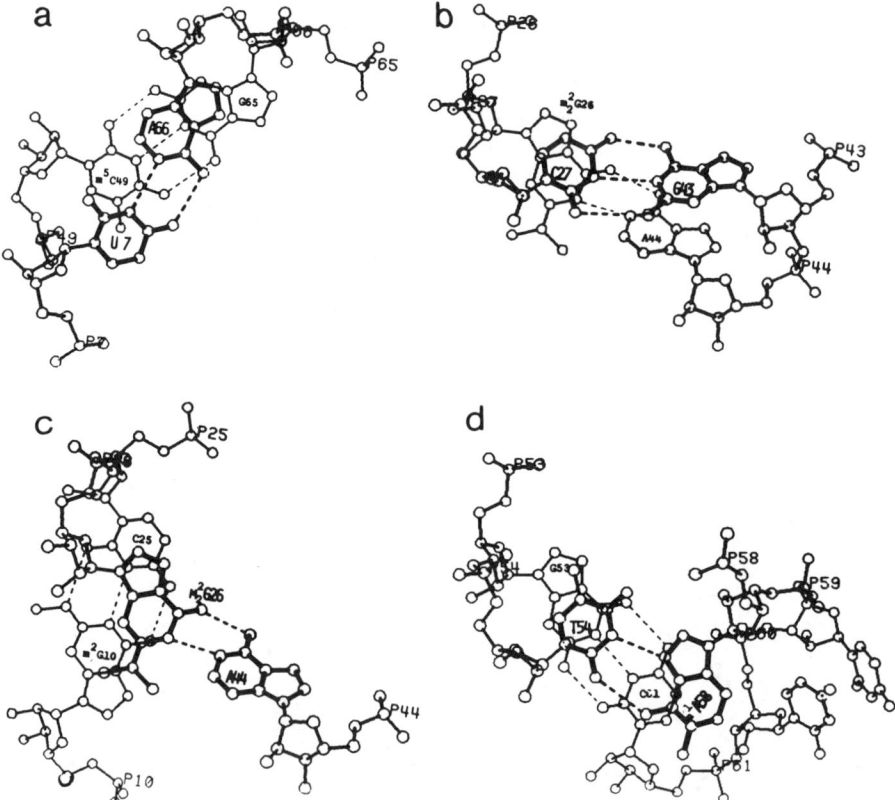

Figure 5 Base stacking. (*a*) Base pairs U7·A66 of the A stem with the m⁵C49·G65 pair of the TψC stem. (*b*) The non-Watson-Crick base pair m²₂G26·A44 with C27·G43 of the AC stem. (*c*) The m²₂G26·A44 with m²G10·C25 of the D helix. (*d*) The non-Watson-Crick intraloop base pair T54·m¹A58 with the G53·C61 base pair of the TψC stem.

$m^2G10 \cdot C25$ and the long Pu·Pu base pair $m_2^2G26 \cdot A44$ is shown in Figure 5c. The bases C25 and m_2^2G26 of the continuous strand exhibit some stacking, and there is some "cross-strand" stacking between m_2^2G26 and m^2G10. But, there is no stacking at all between m^2G10 and A44, which are not linked directly by a backbone, and the $A44 \cdot m_2^2G26$ base pair is rotated considerably from the $m^2G10 \cdot C25$ base pair resulting in a tilt (~30°) between the D helix and the AC helix. The stacking pattern between the constant reverse Hoogsteen $T54 \cdot m^1A58$ base pair in the TψC loop and the constant $G53 \cdot C61$ base pair in the TψC stem is also shown (Fig. 5d). T54 is stacked extensively with G53, but m^1A58 and A62, which are interleaved by two (pyrimidine) nucleotide residues, exhibit skewed stacking. This stacking pattern, which involves complementary bases of Watson-Crick and reverse Hoogsteen base pairs intervened by a 2-nucleotide bulge, may provide some clues into possible structures that may arise in similar situations in double-helical regions of other nucleic acids. The base stacking in single-strand segments, e.g., ACCA and AC loop, not constrained by base-pairing interactions, generally exhibits greater overlap than in the double helices. The single-strand helical stretches strive to compensate the loss in base-pairing energies by nestling the bases to achieve greater base-stacking energies, which include a significant component of base-backbone short-range interactions (Sundaralingam 1975). Generally, the bases point inward to enhance hydrophobic interactions, whereas the sugar-phosphate backbone forms the charged surface on the perimeter of the molecule.

Base Interdigitation

An interesting feature in the tRNA structure is that at the interloop sites of the P10 and V loops and the TψC and D loops the bases penetrate such that the bases of one chain are sandwiched by the bases in the other (Fig. 6). This type of stereochemical arrangement appears to be favored when loops or bends in the polynucleotide chain approach. This observation further suggests possible interdigitation of aromatic side chains of proteins, as well as the planar dyes and mutagens, to single-strand regions of nucleic acids, as was also suggested by the crystal structure of puromycin (Sundaralingam and Arora 1972).

The interdigitation brings out some interesting stereochemical features: where endo is endocyclic, alternating C3'-endo–C2'-endo puckering with concerted changes in the glycosyl torsion angle χ, as in $G57 \cdot m^1A58$ and $m^7G46 \cdot G45$, which, respectively, sandwich G18 and A9, and C2'-endo–C2'-endo sugars, as in G18 and G19. Thus, changes in the sugar pucker coupled with changes in the base glycosyl torsion angle apparently provide the principal mechanics for extension of the polynucleotide chain to permit interdigitation. In addition, the backbone C4'—C5' bond and the correlated

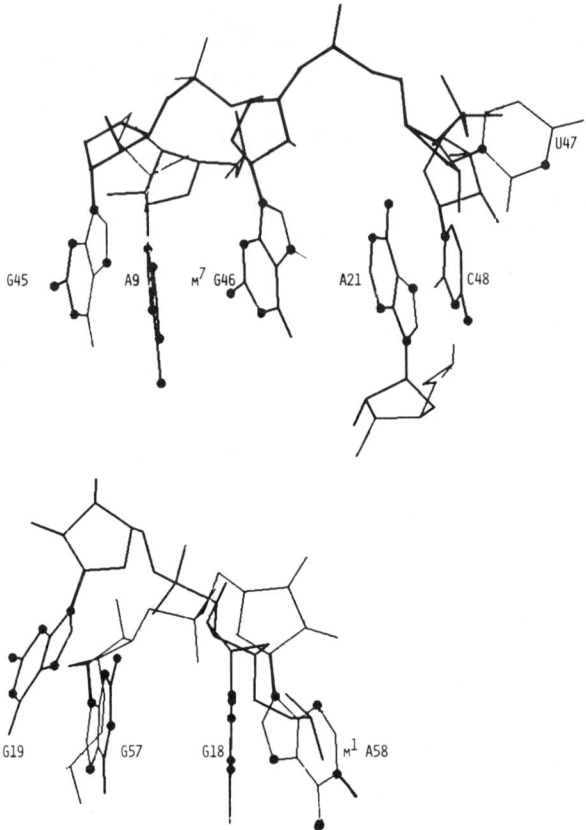

Figure 6 The base interdigitation patterns in yeast tRNA^{Phe}. (*Top*) A9 sandwiched between G45 and m^{7}G46, and A21 sandwiched between m^{7}G46 and C48. (*Bottom*) G57 sandwiched between G18 and G19, and G18 sandwiched between G57 and m^{1}A58.

P—O ester bond rotations could also be involved in the extension of the polynucleotide chain and moving the bases apart for interdigitation.

Cation and Dye Binding

The high concentration of negatively charged phosphates in the core region of the molecule forms a good trap for counter ions as well as cationic dyes (Liebman et al. 1977; Sundaralingam 1978). The counter ions and their concentration probably play an important role in the structural stability and activity of tRNAs. The polyamine, spermine, used in the crystallization experiments (Ichikawa and Sundaralingam 1972) appears to be lodged principally in the deep grooves of the tRNA helical domains and most probably engaged in interstrand cross-linking of the helices (Stout et al. 1978). The cationic dyes ethidium (Liebman et al. 1977) and proflavine (Sundaralingam 1978; Sundaralingam and Liebman 1978) are bound in a nonintercalative mode with hydrogen bonding and electrostatic interactions to the anionic oxygen atoms of the backbone phosphates dominating their interactions to the tRNA. This suggests that similar modes of interaction with nucleic acids are possible for other dyes, drugs, and mutagens.

CONFORMATIONAL ANALYSIS

The crystal structure of yeast tRNA^{Phe} has provided an excellent source of stereochemical information, offering a unique opportunity to investigate the stereochemical aspects of helical regions, tertiary structural loops and bends, and modified residues. A comparison of these in relation to the structural concepts that have already been evolved would place the principles of nucleic acid folding on a firmer basis.

The Helices: Chirality and Conformation

The four double-helical segments of the tRNA^{Phe} molecule predicted by the Holley cloverleaf diagram are confirmed and shown to be right-handed using anomalous scattering of X rays by the Sm^{+++} ions in the samarium derivative. The continuous single strands that form the common thread of each of the two major helical domains form a right-handed helix slightly greater than one turn of the helix. This observation provides a direct support for the proposed intertwined right-handed models of RNA and DNA double helices.

The Loops

Other than the three loops TψC, D, and AC predicted by the cloverleaf, an additional loop (P10 loop) around the phosphodiester P10 is manifested by the tRNA tertiary structure. The two hairpin loops, TψC and AC, and the P10 loop form tight loops and the chain reversal is accomplished by preferential rotation around the P—O5' bond (ω) of only one of the phosphodiesters (Sundaralingam et al. 1976). Such a conformation for the sugar-phosphate backbone may well be a repeating structural motif in nucleic acid folds. The folding of the D loop is more gradual and a different mechanism is adopted for the chain reversal (Sundaralingam et al. 1976).

The characteristic turn of the loops displaces the bases outward so that they can either participate in interloop base-pairing interactions, as between the TψC and D loops, or the bases are free, as is the case of the anticodon triplet, which can engage in base pairing with the corresponding codon triplets of mRNA without major conformational changes of the loop. It is of interest to note that the P10-loop bases point outward and engage in Watson-Crick base pairs with the complementary strand of the D-stem bases. Notice that this is somewhat reminiscent of the anticodon-codon triplet base interactions between tRNA and mRNA.

Conformation Wheels and Maps

The conformation wheels for the six backbone bond torsion angles, ω, ϕ, ψ, ψ', ϕ', ω' (Fig. 7) have been plotted separately for the 42 residues in the base-paired helices and the 34 residues in single-strand regions (Fig. 8) to demonstrate the nature of the nucleotide and phosphodiester conformations in helical and single-strand regions. It should be pointed out that the nucleotide residues involving the terminal base pairs of the helices having the 5'-phosphate outside the helix cannot strictly be regarded as helix nucleotides and they display, as expected, greater conformational flexibility, as seen by some of the torsion angles shown in Figure 8.

Conformations ω and ω' represent the torsion angles around the

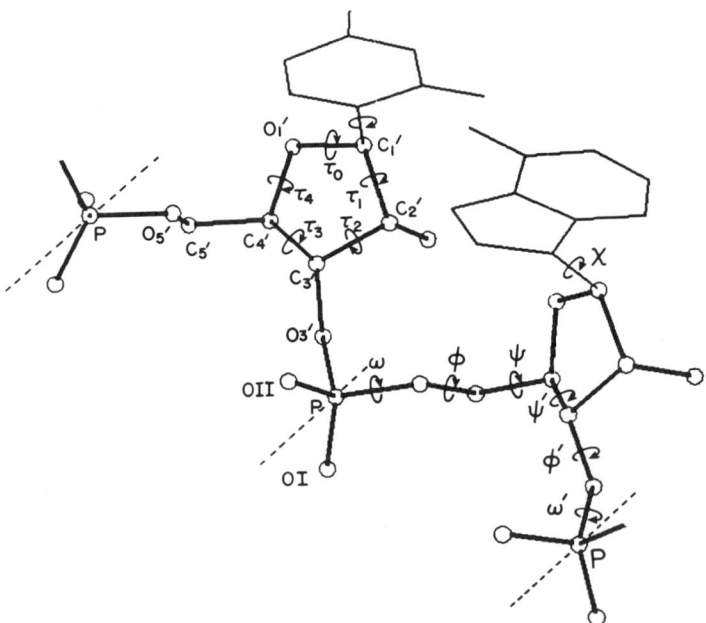

Figure 7 The notations for the torsion angles of the sugar-phosphate backbone bonds, the sugar endocyclic bonds, and the sugar-base glycosyl bond of a polynucleotide chain. Conformational abbreviations used are: *gauche*$^+$ ($g^+ \sim 60°$), *trans* ($t \sim 180°$), and *gauche*$^-$ ($g^- \sim 300°$).

internucleotide phosphodiester P—O bonds, P—O5' and P—O3'. The most favored conformations occur around the value of g^- for both bonds, and this characterizes the preferred helical state of the polynucleotide chain. A number of phosphodiester groups occur outside these angular ranges where one or both P—O bonds are in the t or g^+ sector of the wheel. Such conformations of the phosphodiester lead to either extension of the chain or reversal in the chain direction with concomitant loss of base-stacking interactions of adjacent residues. Thus, the distortion in the favored helical backbone phosphodiester conformation is necessarily correlated with a loss in adjacent base-stacking interactions.

Conformations ϕ and ϕ' represent the rotations around the two C—O bonds, C5'—O5' and C3'—O3'. They generally favor the *trans* conformation more than any of the other bonds of the backbone chain with mean values of 180° and 210°, respectively. Conformations ψ and ψ' represent the torsion angles around the two C—C bonds of the nucleotide unit, C4'—C5' and C4'—C3', respectively. The most preferred conformation for ψ is the g^+ range, which gives the nucleotide residue a compact structure with the O5' atom lying over the sugar ring and the phosphate group in close proximity to the C8—H side of the purine or the C6—H side of the pyrimidine bases when they are in their preferred anti-states. Conformation ψ' shows an overwhelming preference for the quasi g^+ domain, which corresponds to the C3'-endo pucker of the ribofuranose ring. There are only a few riboses that exhibit ψ' values outside the quasi g^+ range with values around $\psi' = 160°$ (quasi *trans*) that correspond to the C2'-endo pucker. These are associated with residues in the kinks or loops where extension of the polynucleotide chain is necessitated. Although the phosphodiester group possesses relatively greater flexibility than the other backbone bonds C—O (ϕ, ϕ') and C—C (ψ, ψ'), it is nevertheless constrained to the preferred states by steric and electrostatic interactions between interleaving phosphates of the polynucleotide chain (Sundaralingam 1975; Yathindra and Sundaralingam 1974).

The torsion angle χ that gives the disposition of the base relative to the sugar shows an overwhelming preference for the low anti-range that is a phenomenon coupled with the C3'-endo sugar. There are a few χ angles exhibiting high values and these mainly involve the C2'-endo sugars (Stout et al. 1978).

Two-dimensional correlation maps for adjacent bond rotations of the backbone in tRNA have been presented elsewhere (Stout et al. 1978) and the conformational correlations observed are in remarkable agreement with those described earlier (Sundaralingam 1969, 1972; Yathindra and Sundaralingam 1973).

The nucleotides in the double-helical regions generally show a preference for certain domains that are favored by the nucleotides themselves. The nucleotides are all essentially in the C3'-endo sugar pucker,

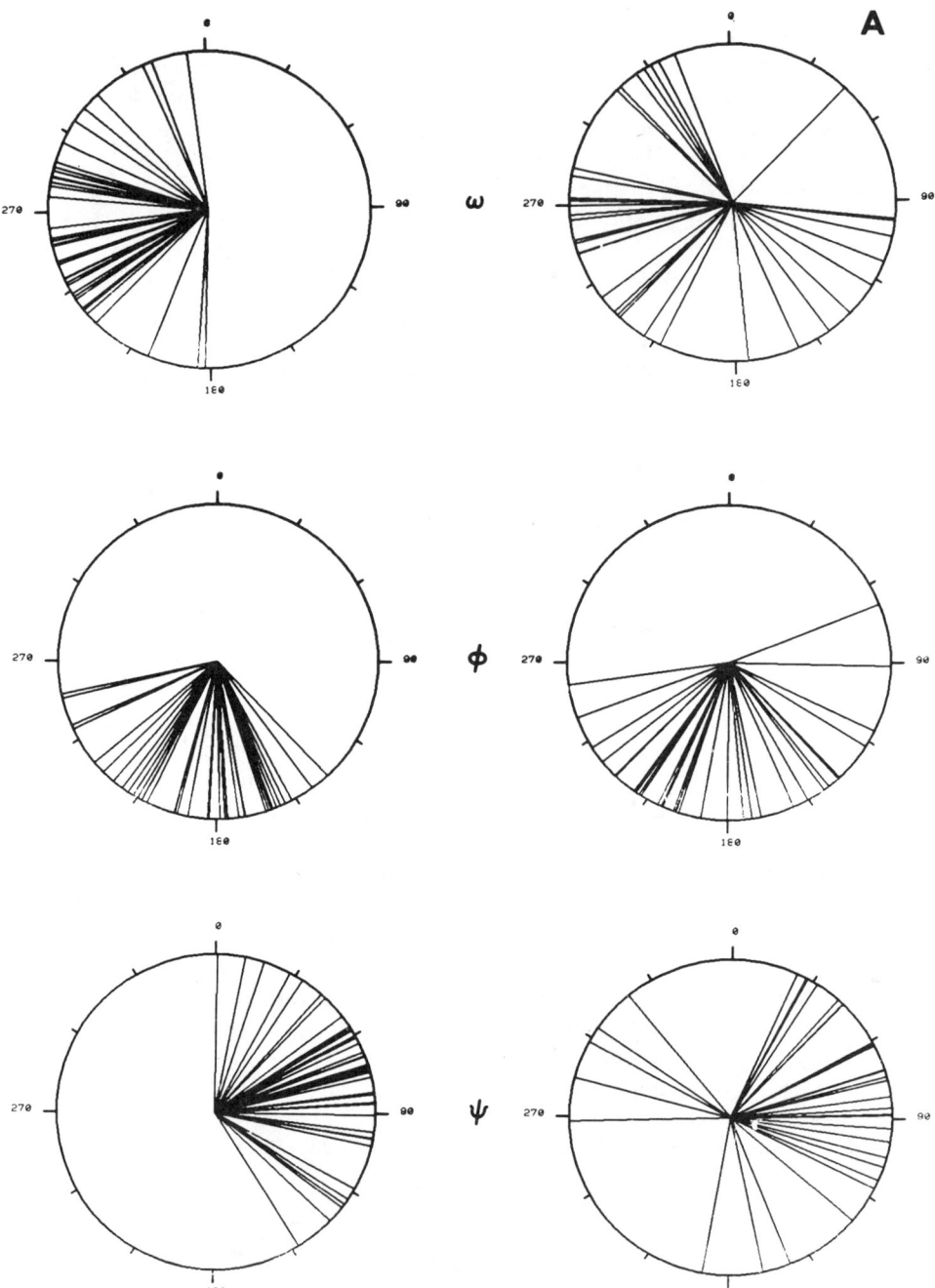

Figure 8 Conformation wheels for the backbone torsions. (*A*) ω, ϕ, and ψ wheels for helical (*left*) and single-strand (*right*) regions of yeast tRNA^{Phe}.

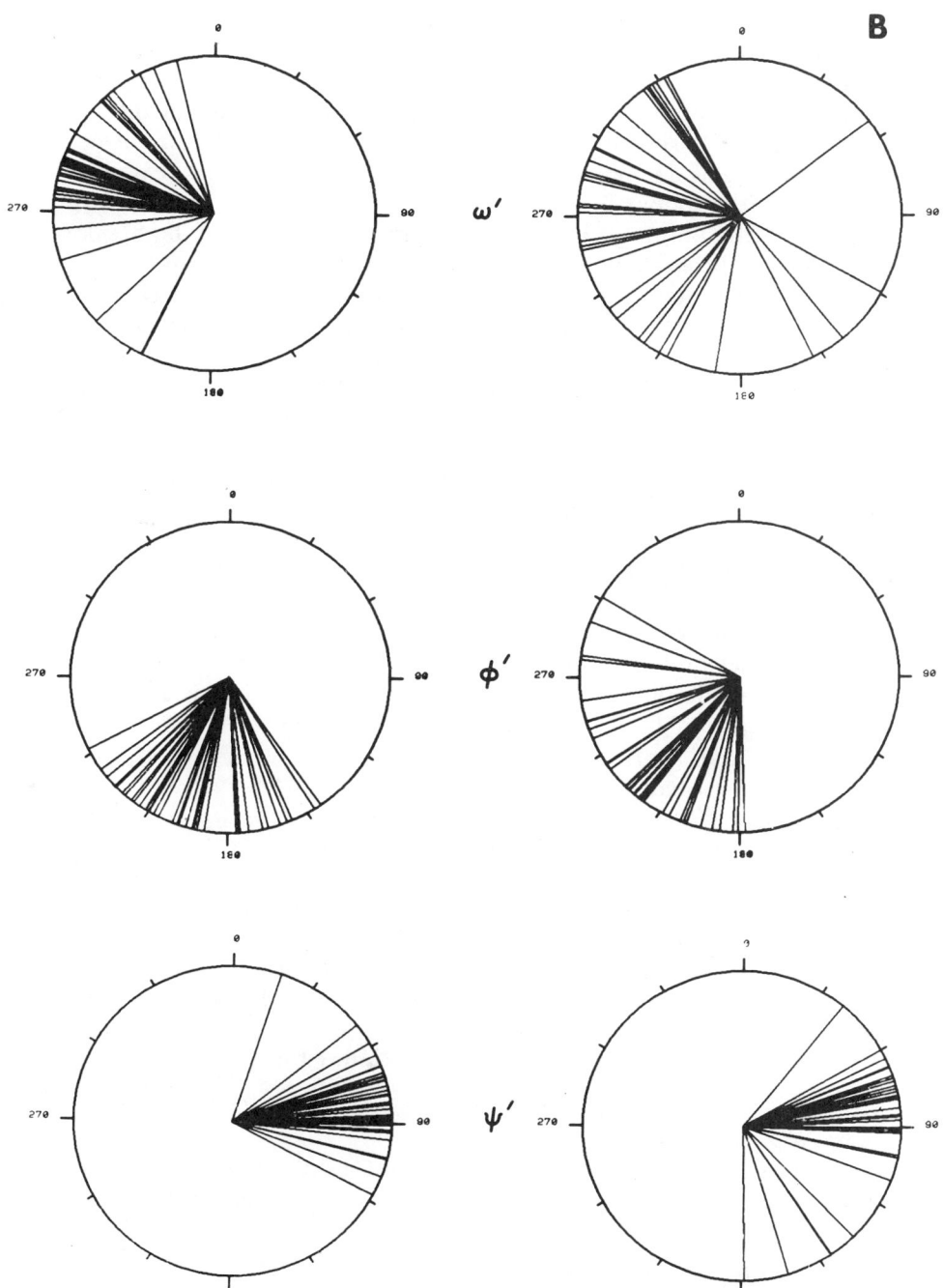

Figure 8 (continued) (B) ω', ϕ', and ψ' wheels for helical (*left*) and single-strand (*right*) regions.

the g^+ conformation for the exocyclic C4'—C5' bond (ψ) and the extended *trans* or *trans*-like conformation for the C—O bonds (ψ and ψ'). The internucleotide phosphodiester P—O bonds (ω', ω) are both in the g^- conformations. The spread in the torsion angle values in the preferred sector of the wheels demonstrates the degree of conformational flexibility within the preferred domains.

The residues in the single-strand regions, including those on the 5' side of the double helices, show greater conformational diversity. However, it should be noted that a large proportion of these also display conformations similar to the preferred helical conformations. This is probably best illustrated by the anticodon triplet in the 3' stacked AC loop and the 3'-CCA end where stacking overlap of the adjacent bases in the single strands is enhanced by minor adjustments in the backbone bond rotations. The latter nucleotides form the single-strand helical stretches emphasizing the intrinsic preference of the polynucleotide chain to assume a helical conformation. In other words, the short-range intramolecular interactions tend to enhance the stacking interactions in lieu of the loss of base-pair interactions and further stabilize the helical conformation of polynucleotides and nucleic acids.

The major deviation from the helical conformations occurs around one or both internucleotide phosphodiester P—O bonds and the intranucleotide C4'—C5' bond. Only a few sugars exhibit the C2'-endo puckers. It is seen that virtually all of these changes are associated with the backbone in the bends and turns of the tRNA molecule. The C2'-endo conformations of the sugar apparently are encountered whenever an extension of the chain is necessitated; but the P—O bond rotations, especially in ω, are crucial for abrupt turns in the chain. Variations in the C4'—C5' bond conformations (ψ) almost always result in coupled rotations in the P—O5' bond (ω) with or without changes in the sugar pucker (Sundaralingam et al. 1976; Yathindra and Sundaralingam 1976).

The P—O bonds of the polynucleotide chain exhibit relatively greater conformational flexibility than any of the other single bonds of the backbone. The folding of the polynucleotide chain into hairpin loops, bends, and kinks can be achieved essentially by three mechanisms:

1. By mere rotations around the internucleotide P—O bonds alone as in the tight AC, TψC, and P10 loops.
2. By coupled rotations around the P—O and C4'—C5' bonds.
3. The above changes accompanied by changes in the sugar pucker.

STRUCTURE-FUNCTION RELATIONSHIPS

The facts that all tRNAs can be represented by the cloverleaf pattern and that the invariant bases are involved in an intricate pattern of tertiary as well as secondary (G53–C61) base-base interactions suggest that all tRNAs can be folded into similar three-dimensional structures. The tRNAs are 75–93 nucleotides long. The extra residues are inserted in the D and V loops. The insertions of D16 and D17, for instance, in the D loop of yeast tRNA^{Phe} simply form a collapsed bulge in the loop and their deletion is not expected to affect the D-loop conformation. The V loop can be enlarged from 5 residues, as found in tRNA^{Phe}, to 19, as in a specific *E. coli* tRNA$^{Ser}_1$. The long V loop forms a helical hairpin loop similar to that formed by the TψC helix and TψC loop and protrudes from the center of the molecule (Fig. 5) so that the basic framework of the tRNA^{Phe} structure is maintained with the loss of some or all of the base triples.

The tertiary interactions of tRNAs are conserved (TψC-D region), variable (D-helix region and 26·44 base pair at AC- and D-helix interface), and hypervariable (anticodon triplet). All tertiary base interactions, conserved or semiconserved, occur in domain 2 of the tRNA. The degree of variation increases from the corner of the tRNA molecule down the D-AC helix (domain 2) to the loop.

The highly conserved tertiary interaction involving the TψC loop (T54·m^{1}A58), the TψC and D loops (G18·ψ55, G19·C56), and the D loop and links 1 and 2 (U8·A14; G15·C48), as well as regions of the sugar-phosphate backbone, will serve as a common structural domain and is of paramount importance for recognition of tRNAs by the ribosomal machinery. The other tertiary interactions seen in yeast tRNA^{Phe} that involve the D helix and the m$^2_2$G26·A44 base pair at the interface of the D and AC helices are more variable. The semiconserved tertiary base-base interaction may serve as synthetase recognition sites; in addition, the helical segments and/or the anticodon triplet may be involved. The 26·44 base pair shows the least conservation among the tertiary base interactions and exhibits a variety of pairings (with one, two, and three hydrogen bonds), including complementary Watson-Crick, wobble GU, long purine·purine, short pyrimidine·pyrimidine, and noncomplementary purine·pyrimidine base pairs. Correlations between the variable and hypervariable regions would endow each tRNA with an inherent conformational adaptability that may guarantee the fidelity of translation. In fact, a statistical analysis of the variation of the 26·44 base pair with the anticodon triplet shows that there is a possible significant correlation, particularly with the 3'-side base of the anticodon (R. Bott and M. Sundaralingam, in prep.). The structural variations of tRNAs caused by the 26·44 base pair at the critical junction of the D and AC stems and possibly even the tertiary base pairs involving the D helix located over it could be transmitted to the anticodon, thereby steering it for optimal mating with its codon. The same phenomenon could position the CCA ends of the aminoacyl- and peptidyl-tRNAs optimally for peptide-bond formation.

It is remarkable that the diversity of the interfacial 26·44 base-pairing opposition in tRNAs mentioned above is reminiscent of the variety of noncomplementary *cis* base-pairing interactions at the wobble position in anticodon-codon interaction. It then appears that, in as much as the wobble pairings are important in translation, they can also be relevant in the transcription or replication process. Thus, such anti-*cis* pairs involving noncomplementary bases on antiparallel strands (Sundaralingam 1977) may explain the origin of point mutations without having to invoke rare tautomeric base forms or *syn* nucleotides (Topal and Fresco 1976; Garduno et al. 1977).

The exposed anticodon triplet bases can engage in interactions with the mRNA codon triplets with no major conformational changes in the AC loop. It has been suggested (Kurland et al. 1975; Schwarz et al. 1976) that during anticodon-codon interaction, conformational changes occur in tRNA leading to the disruption of the TψC-D-loop tertiary base interactions, which expose the constant TψCG segment for complementary base-pairing interactions with the 5S rRNA (Erdmann 1976). Several schemes may be visualized for the transmission of conformational change from the AC to the D-TψC loops. Besides local conformational changes in the AC loop during codon-anticodon interaction that may be transmitted to the D-TψC loops, one may also visualize the flipping of the AC loop from the observed 3'-stacked to the 5'-stacked conformation in the ribosomal aminoacyl site. The latter conformation, although unfavorable, since it requires a concomitant flipping of the base-pair tilts of the AC helix to the opposite orientation (Sundaralingam 1978), could be induced by binding to the ribosome. This conformational change in the AC helix would then be transmitted via the interfacial base pair m$_2^2$G26·A44 and D-helix base pairs to the D-TψC loop, thereby disengaging the interloop base pairs. Of course, this argument can be reversed; the binding of the TψCG region to the 5S RNA in the aminoacyl site could promote conformational changes in the anticodon. It is also possible that the opening of the TψCG segment and the conformational changes in the anticodon occur independently of each other. Further experimentation is undoubtedly required to place these speculations on a firmer basis.

ACKNOWLEDGMENTS

It gives me great pleasure to acknowledge the contributions made by my colleagues, especially R. Bott, T. Brennan, T. Ichikawa, M. Liebman, H. Mizuno, S. T. Rao, J. R. Rubin, and C. D. Stout, in the X-ray structural studies of tRNA. I am also grateful to S. T. Rao, N. Yathindra, and E. Westhof for invaluable assistance in the preparation of the manuscript. This research was supported (in part) by grants from the National Institutes of Health (GM-17378 and GM-18455), the University of Wisconsin Graduate School, and the College of Agricultural and Life Sciences.

REFERENCES

Brennan, T. and M. Sundaralingam. 1975. Discussion of paper of R. G. Shulman. In *Structure and conformation of nucleic acids and protein-nucleic acid interaction* (ed. M. Sundaralingam and S. T. Rao), p. 165. University Park Press, Baltimore, Maryland.

―――. 1976. Structure of tRNA molecules containing the long variable loop. *Nucleic Acids Res.* **3**: 3235.

Erdmann, V. A. 1976. Structure and function of 5S. *Prog. Nucleic Acid Res. Mol. Biol.* **18**: 45.

Garduno, R., R. Rein, J. T. Egan, Y. Coeckelenbergh, and R. D. MacElroy. 1977. Purine-purine base pairs and the origin of transversion-type mutation. *Int. J. Quantum Chem. Sym.* **4**: 197.

Goddard, J. P. 1978. The structures and functions of transfer RNA. *Prog. Biophys. Mol. Biol.* **32**: 233.

Ichikawa, T. and M. Sundaralingam. 1972. X-ray diffraction study of a new crystal form of yeast phenylalanine tRNA. *Nat. New Biol.* **236**: 174.

Jack, A., J. E. Ladner, and A. Klug. 1976. Crystallographic refinement of yeast phenylalanine transfer RNA at 2.5 resolution. *J. Mol. Biol.* **108**: 619.

Kim, S.-H., F. L. Suddath, G. J. Quigley, A. McPherson, J. L. Sussman, A. H. J. Wang, N. C. Seeman, and A. Rich. 1974. The tertiary structure of yeast phenylalanine transfer RNA. *Science* **185**: 435.

Klug, A., J. Ladner, and J. D. Robertus. 1974. The structural change of coordinate base changes in transfer RNA. *J. Mol. Biol.* **89**: 511.

Kurland, C. G., R. H. Rigler, M. Ehrenberg, and C. Bloomberg. 1975. Allosteric mechanism for codon-dependent tRNA selection on ribosomes. *Proc. Natl. Acad. Sci.* **72**: 4248.

Liebman, M., J. Rubin, and M. Sundaralingam. 1977. Non-intercalative binding of ethidium bromide to nucleic acids. Crystal structure of an ethidium tRNA molecular complex. *Proc. Natl. Acad. Sci.* **74**: 4821.

Mizuno, H. and M. Sundaralingam. 1978. Stacking of Crick wobble pair and Watson-Crick pair: Stability rules of G-U pairs at ends of helical stems in tRNAs and the relation to codon-anticodon wobble interaction. *Nucleic Acids Res.* **5**: 4451.

Quigley, G. J. and A. Rich. 1975. Structural domains of transfer RNA molecules. *Science* **194**: 796.

Robertus, J. D., J. E. Ladner, J. T. Finch, D. Rhodes, S. R. Brown, B. F. C. Clark, and A. Klug. 1974. Structure of yeast phenylalanine tRNA at 3 Å resolution. *Nature* **250**: 564.

Schwarz, V., H. M. Menzel, and H. G. Gassen. 1976. Codon-dependent rearrangement of the three-dimensional structure of phenylalanine tRNA, exposing the T-ψ-C-G sequence for binding to the 50S ribosomal subunit. *Biochemistry* **15**: 2484.

Stout, C. D., H. Mizuno, J. Rubin, T. Brennan, S. T. Rao, and M. Sundaralingam. 1976. Atomic coordinates and molecular conformation of yeast phenylalanine tRNA. An independent investigation. *Nucleic Acids Res.* **3**: 111.

Stout, C. D., H. Mizuno, S. T. Rao, P. Swaminathan, J. Rubin, T. Brennan, and M. Sundaralingam. 1978. Crystal and molecular structure of yeast phenylalanine transfer RNA. Structure determination, difference Fourier refinement, molecular conformation, metal and solvent binding. *Acta Cryst.* **B34**: 1529.

Sundaralingam, M. 1969. Stereochemistry of nucleic acids and their constituents. IV. Allowed and preferred conformations of nucleosides, nucleotides, mono-, di-, tri-, tetraphosphates, nucleic acids and polynucleotides. *Biopolymers* **7**:821.

———. 1972. The concept of a conformationally "rigid" nucleotide and its significance in polynucleotide conformational analysis. In *Conformation of biological molecules and polymers* (ed. E. D. Bergmann and B. Pullman), vol. 5, p. 417. Academic Press, New York.

———. 1975. Principles governing nucleic acid and polynucleotide conformations. In *Structure and conformation of nucleic acid and nucleic acids-protein interactions* (ed. M. Sundaralingam and S. T. Rao), p. 487. University Park Press, Baltimore, Maryland.

———. 1977. Non-Watson-Crick base pairs in ribonucleic acids. *Int. J. Quantum-Chem.:Quantum Biol. Symp.* **4**:11.

———. 1978. Nucleic acid principles and transfer RNA. In *Aspects of biological molecules* (ed. R. Srinivasan et al.). (In press.)

Sundaralingam, M. and S. K. Arora. 1972. Crystal structure of the amino-glycosyl antibiotic puromycin dihydrochloride pentahydrate. Models for 3'-aminoacyl-adenosine moieties of transfer RNAs and protein-nucleic acid interactions. *J. Mol. Biol.* **71**:49.

Sundaralingam, M. and M. Liebman. 1978. Drug-tRNA interactions. Crystal structure of a molecular complex of proflavine and yeast tRNA^{Phe}. In *Collected abstracts XI; International Congress on Crystallography*, Warsaw, Poland: S54.

Sundaralingam, M. and S. T. Rao, eds. 1975. *Structure and conformation of nucleic acids and protein-nucleic acids interaction.* University Park Press, Baltimore, Maryland.

Sundaralingam, M., H. Mizuno, C. D. Stout, S. T. Rao, M. Liebman, and N. Yathindra. 1976. Mechanisms of chain folding in nucleic acids. The (ω', ω) plot and its correlation to the nucleotide geometry in yeast tRNA^{Phe}. *Nucleic Acids Res.* **3**:2471.

Topal, M. D. and J. R. Fresco. 1976. Complementary base pairing and the origin of substitution mutations. *Nature* **263**:285.

Yathindra, N. and M. Sundaralingam. 1973. Correlation between the backbone and side chain conformation in 5'-nucleotides. The concept of a "rigid" nucleotide conformation. *Biopolymers* **12**:297.

———. 1974. Backbone conformations in secondary and tertiary structural units of nucleic acids. Constraint in the phosphodiester conformation. *Proc. Natl. Acad. Sci.* **71**:3325.

———. 1976. Analysis of possible helical structures of nucleic acids and polynucleotides. Application of (n-h) plots. *Nucleic Acids Res.* **3**:729.

A Crystallographic Analysis of Yeast Initiator tRNA

**Richard W. Schevitz, Alberto D. Podjarny,*
Narasimhan Krishnamachari,† John J. Hughes,‡
and Paul B. Sigler**
Department of Biophysics and Theoretical Biology
The University of Chicago
Chicago, Illinois 60637

The initiation step in the complex process of protein synthesis always begins with the incorporation of a methionine residue at the amino terminal of the growing protein chain (for review, see Grunberg-Manago and Gros 1977). A specific methionine-accepting initiator tRNA that is excluded from the elongation steps on the ribosome is the unique adaptor for the "start" signals in mRNA. Together with initiation factors and GTP, the methionylated initiator tRNA is aligned in the correct frame on the ribosome so that the first peptide bond can be formed with the first aminoacylated elongator tRNA on the A site of the ribosome.

Our understanding of the molecular structure-function relationships in initiator tRNA comes from an evaluation of a diversity of experimental results. There are the characteristic sequence patterns that have emerged from a careful examination of a great many primary structures of both initiator and elongator tRNAs (Barrell and Clark 1974; Sprinzl et al. 1978). Added to this are the variety of biochemical and biophysical studies of tRNA that have described the dynamics of molecular response to environmental perturbants (for review, see Sigler 1975; Kim 1976; Rich and RajBhandary 1976). Another very important constraint on the structure-function relationships in initiator tRNA is the detailed high-resolution structure of the elongator tRNA, yeast tRNA^{Phe} (Quigley et al. 1975; Jack et al. 1976; Stout et al. 1976; Sussman and Kim 1976; Holbrook et al. 1978), which is known to bear many structural similarities and even functional similarities outside the ribosome. Reported here is a 4.5-Å resolution X-ray analysis of yeast initiator tRNA^{fMet} that has yielded its general fold and some important details. (For the X-ray analysis of the prokaryotic *Escherichia coli* initiator, see Rich et al., this volume.)

Present addresses: *Labotatorio de Rayos-X, Departmento de Fisica, Universidad de La Plata, CC67 La Plata 1900, Argentina; †Computation Center, The University of Chicago, Chicago, Illinois 60637; ‡Whitlow Corporation, Englewood Cliffs, New Jersey 07632.

THE CRYSTALS

The details of the crystal system have been presented previously (Schevitz et al. 1975) but it is worth stressing the following points. First, the crystals are grown from approximately 2 M $(NH_4)_2SO_4$, rather than organic solvents. Magnesium ion and spermine (which may play an important role in defining the conformation of crystalline yeast $tRNA^{Phe}$) are present at 5 mM and 2 mM, respectively; however, the 2 M NH_4^+ may mitigate some of these electrostatic interactions. Second, only 17.5% of the crystal volume is occupied by tRNA; treating the remaining volume as disordered solvent of uniform density can be helpful in the structure analysis. And third, the crystal is poorly ordered, as evidenced by the fact that the mean intensity at 4 Å is less than 1% of the mean intensity at 20-Å resolution. This not only prevents the collection of higher resolution intensities, but also suggests that either the packing interactions, the structure of the molecules themselves, or both vary from one molecule to the next.

MULTIPLE ISOMORPHOUS ANALYSIS

Data for the parent and two isomorphous heavy atom derivatives were collected to 4-Å resolution by oscillation photographs in which a single crystal was devoted to each of about 30 overlapping 2.5° or 2° sectors. Figure 1 shows that there is useful phasing power in the best derivatives to about

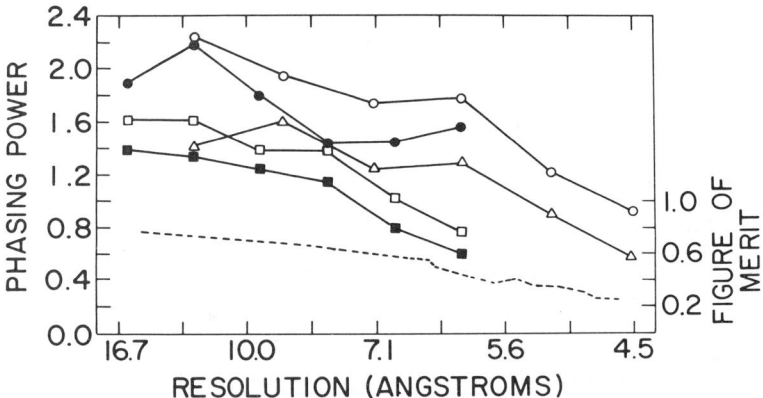

Figure 1 Phasing power $(\Sigma|F_H|^2)^{1/2}/[\Sigma(|F_{PH}^{obs}|-|F_{PH}^{calc}|)^2]^{1/2}$ of five isomorphous derivatives plotted against resolution: gadolinium, lightly substituted, precession data (■); gadolinium, heavily substituted, precession data (□); gadolinium, oscillation data (△); pyridyl mercury acetate, precession data (●); and pyridyl mercury acetate, oscillation data (○). Figure of merit (---) is determined from MIR phasing and has a mean value of 0.49 for all reflections to 4.5-Å resolution. F_H is the calculated heavy atom structure amplitude; $(|F_{PH}^{obs}|-|F_{PH}^{calc}|)$ is the lack of closure error between observed and calculated isomorphous derivative structure amplitudes.

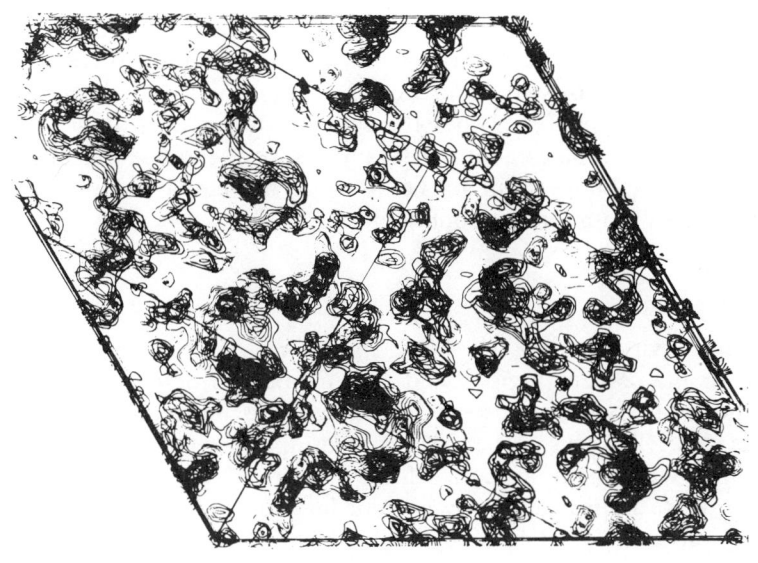

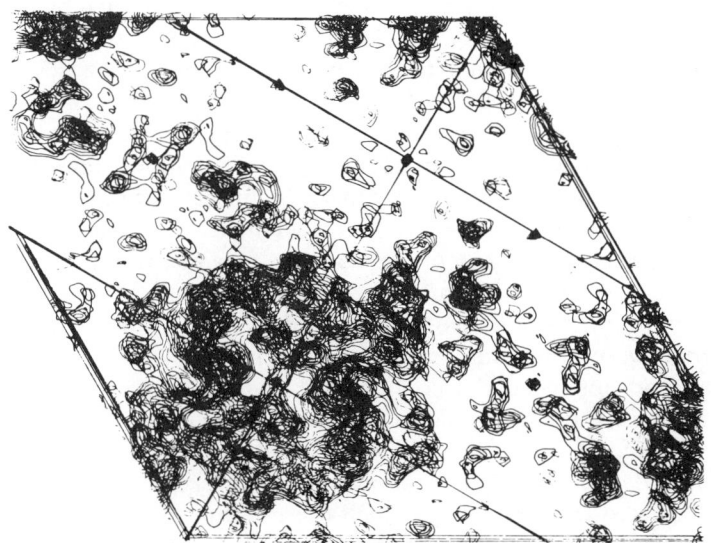

Figure 2 Comparison of ten sections of the 4.5-Å map of yeast tRNA^{fMet} before (*top*) and after (*bottom*) the inclusion of 27 intense low-resolution terms determined by matrical phase predictions. Both maps are contoured at the same intervals. Density cleared by this technique was ultimately shown to be noise; density that was enhanced improved connectivity in the helical stems. Density on the dyad axes diminished.

4.5-Å resolution. Schevitz et al. (1975) presented a 6-Å analysis of the yeast tRNA^{fMet} crystal structure. Despite two heavy atom markers at io^{5}C73 and Os^{5,6}C38, the electron density could not be convincingly interpreted in terms of the molecular backbone. Even when the data for the parent structure and multiple isomorphous replacement (MIR) phases were extended to 4.5-Å resolution and a third heavy atom marker at io^{5}U8 was employed, the map was still too "noisy" to interpret confidently.

PHASE IMPROVEMENT

A troublesome symptom of the MIR maps was the unusually high background in the solvent regions and the inability to define the exact molecular boundary with confidence, although the molecule's general position was clear. This was particularly frustrating, since over 80% of the unit cell volume was solvent and, therefore, this structure was particularly amenable to phase improvement by solvent leveling once the molecular boundary was accurately established (Bricogne 1976; Hendrickson 1978). The boundary was established by including in the Fourier synthesis 27 very intense inner reflections (d greater than 14 Å), which had previously been omitted from both parent and derivative because extreme darkening on even the weakest films prevented the accurate measurements needed for MIR phasing. Phases for these inner reflections were determined by applying matrical methods to the amplitudes and MIR phases of 107 known reflections. The technique is similar to that used to extend the phases of triclinic lysozyme and yeast tRNA^{Phe} (Podjarny et al. 1976; Podjarny and Yonath 1977) to a higher resolution than the starting phase set. Here the procedure was used to extend the phases to lower resolution than the MIR phase set. The details of this work will be published elsewhere (A. D. Podjarny et al., in prep.); however, it is evident from Figure 2 that there was substantial enhancement of the contrast between the molecule and the solvent. In particular, several dense regions highly suspected (and subsequently confirmed) to be noise disappeared, whereas other regions of the map that were ultimately shown to be part of the molecule remained intact and enhanced the connectivity of the helical stems. At the same time, density on nearby dyad axes diminished, giving strong independent evidence of phase improvement. With this improved map, we conservatively defined 50% of the unit cell volume as solvent. New phases were calculated from a map in which the noise in the solvent was leveled to a uniform average value and negative regions in the molecule attenuated (Barrett and Zwick 1971; Collins et al. 1975). These phases were merged with MIR phases and the density modification procedure was repeated for three cycles. The net result was a striking improvement in the 4.5-Å map. Figure 3 shows that density modification causes the largest phase change in those reflections having the lowest original figure of merit; that is, those reflections whose MIR phases were, in general, least accurate.

HEAVY ATOM MARKERS

Four residues have been reacted stoichiometrically with heavy atoms. Three of these were readily seen with high occupancy on difference maps (Fig. 4). The markers are nicely distributed along the molecular backbone to provide a clear guide to the interpretation. These attachment sites were established chemically.

1. io^{5}C73: The third residue from the 3′ terminal was labeled with iodine, either by enzymatically rebuilding the 3′ terminal with io^{5}CTP and nucleotidyl transferase (Pasek et al. 1973) or by directly iodinating the crystals (J. Tropp and P. B. Sigler, in prep.). The latter procedure also labels C74, which shows a small peak that cannot be refined, presumably because of disorder.
2. io^{5}U8: Produced by direct iodination of the crystal (J. Tropp and P. B. Sigler, in prep.).
3. Os^{5,6}C38: Direct reaction of the crystals with Os(VI) and pyridine (Schevitz et al. 1972), forming an osmate diester (Subbaraman et al. 1971; Neidle and Stuart 1976; Kistenmacher et al. 1976) with C38 (Rosa and Sigler 1974).

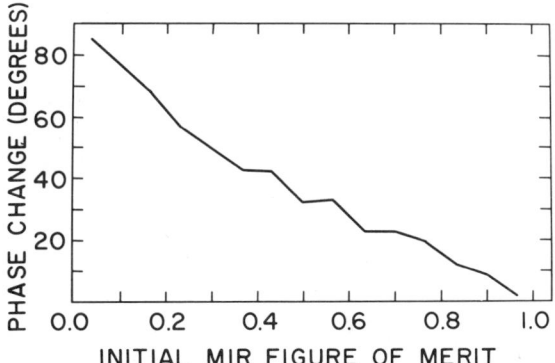

Figure 3 Final phase change in degrees resulting from direct methods (see text) as a function of the original figure of merit determined by conventional MIR phasing. Phase information from both sources was merged by multiplying together their respective phase probability curves after each cycle of combined solvent leveling and density modification. Note that those reflections predicted to have a high reliability by MIR phasing (figure of merit near 1.0) remained closely tethered to their MIR phase, whereas those with a low MIR reliability (figure of merit near 0.0) tend to be more easily shifted toward phases calculated by direct methods.

FITTING THE MOLECULAR MODEL

Although our initial efforts to interpret the map were carried out with a conventional optical comparator and skeletal models, it was more convenient to search for significantly different orientations of large molecular fragments with computerized molecular graphics. Using an adaptation (A. D. Podjarny and B. Honig, unpubl.) of a graphics package developed by Levinthal's group at Columbia (Honig et al. 1973), the density was searched in a graphics terminal linked to an IBM 370/195. Extensive use was made of stable helical substructures operating within the constraints imposed by the heavy atom markers and the crystallographic dyad axes upon which the molecular density could not intrude.

A residue-by-residue fit of the entire model to the map was attempted at the Computergraphics Laboratory at the University of North Carolina. The capacity to orient model components in real time and the continuum of views allowed one to readily rule out possible alternatives. We found the map was not sufficiently well resolved to impose a direct correspondence between phosphate, ribose, and base moieties in the model and the peaks in the electron density.

The coordinates from the graphics fitting were used as guides to position a tRNA^{Phe} model (Sussman et al. 1978) in the unit cell of yeast tRNA^{fMet} that is

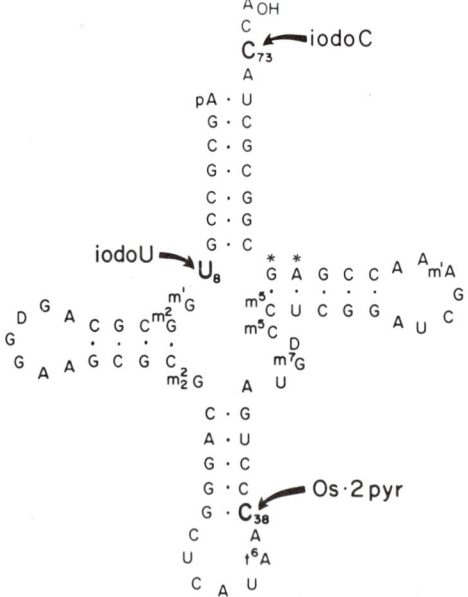

Figure 4 Nucleotide sequence of yeast tRNA^{fMet} (Simsek and RajBhandary 1972) displayed in the cloverleaf hydrogen bonding scheme. Positions of three covalent heavy atom markers are shown.

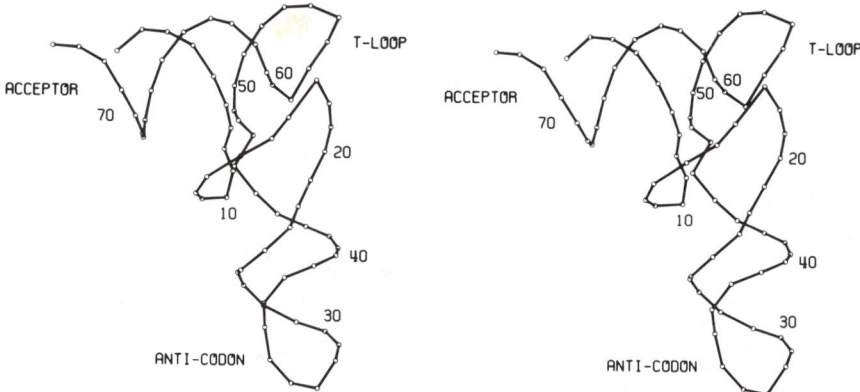

Figure 5 Tertiary structure of yeast tRNA^{fMet} phosphate backbone shown in stereo. Every tenth phosphate is numbered.

in space group P6$_4$22. The root-mean-square deviation between the guide points and the tRNA^{Phe} model coordinates was 5.4 Å. This model was divided into four rigid groups corresponding to the helical arms and refined against the observed intensities using the constrained/restrained least squares (CORELS) refinement procedure of Sussman et al. (1977) that restrained the groups to remain within reasonable bonding distance of each other. The refinement was started at 12.5 Å and carried to 6 Å (Schevitz et al. 1979). The final crystallographic R[1] factor between calculated and observed data at 6 Å was 42.3%; to 12 Å, 33%. These values indicate an essentially correct interpretation of a macromolecular crystal structure.[2]

Figure 5 shows the results of the partial structure refinement described above. In view of the improved correlation between the observed diffraction data and that calculated from the partially refined model, our current interpretation is substantially better than our best attempts to simply optimize the fit of the model to the electron density map. The refined model shows that the main helical density does not follow the sugar-phosphate backbone as we had intuitively expected but, rather, bridged the backbones in the regions of the base pairs—a phenomenon observed in the low-resolution analysis of the monoclinic crystal structure of yeast tRNA^{Phe} (A. Klug, pers. comm.).

The molecular architecture is generally similar to that of crystalline yeast tRNA^{Phe}. The acceptor and TψC stems stack nearly coaxially into a continuous A-form helix of 12 bp. The D loop appears to interact strongly

---

[1] $R = \Sigma[|F_P^{obs}| - |F_P^{calc}|]/\Sigma|F_P^{obs}|$, where $|F_P^{obs}|$ and $|F_P^{calc}|$ are the observed and calculated structure amplitudes of the parent structure.

[2] *Note added in proof*: The accuracy of the interpretation at both low and high resolution was improved by further refinement of a more flexible model that gave an R factor of 25% for all data to 4-Å resolution.

with the TψC loop much as it does in crystalline yeast tRNA^{Phe}. Although specific base interaction cannot be visualized at this stage of the analysis, electron density does bridge the two chains, and the backbone is compatible with the tertiary base interactions that stabilize this region in yeast tRNA^{Phe}.

It is the position of the anticodon arm where the crystal structure of yeast tRNA^{fMet} differs most noticeably from the crystal structure of yeast tRNA^{Phe}. Unfortunately, a detailed analysis of this difference is compromised by the fact that the anticodon arm is currently the most difficult region of the electron-density map to interpret, despite the fact that the position of C38 is indicated by an osmium marker attached through a bispyridyl osmate diester to the 5 and 6 positions of the cytosine base (Rosa and Sigler 1974), and that stringent constraints on the course of the backbone are imposed by a nearby intersection of three mutually orthogonal lattice dyads. The refinement parameters also show that the anticodon arm is the most poorly localized segment of the crystal structure. These observations, coupled with the fact that it is the orientation of the anticodon arm that deviates most noticeably from the structure of crystalline yeast tRNA^{Phe}, suggest that the union of the D and anticodon stems may indeed serve as a hinge point (Robertus et al. 1974; Holbrook et al. 1978) and allow alterations in the orientation of the anticodon during functional adjustments of the structure. Except for the possible effect of a nonstandard base pair, m^{2}G26·A44, no stabilizing tertiary interaction can be invoked to insure a particular orientation of the anticodon arm.

MOLECULAR PACKING

The packing of the molecules in the unit cell is quite unusual. All 12 of the acceptor TψC helices in the unit cell are clustered around the crystallographic 6$_4$ axis at an angle of approximately 20° to it. Pairs of neighboring molecules having opposing acceptor stems appear to stack upon one another roughly in helical register. This explains the intense, continuous, helical diffraction pattern along the reciprocal lattice direction c^* observed at approximately 3-Å resolution, i.e., well beyond the lattice transform (Schevitz et al. 1975).

Figure 6 is a projection of the cell down the 6$_4$ axis showing the crowding of helical backbones surrounding the 6$_4$ axis. The anticodon arms project out roughly perpendicular to the 6$_4$ axis and contact neighboring molecules at their anticodon loop and the base of the D stem. Thus, many of the intermolecular contacts involve the anticodon arm, which we believe to be extended on a flexible hinge. The fact that the lattice is formed through a network of flexible anticodon-arm contacts may be the reason this and possibly many other tRNA crystals are poorly ordered, although they are

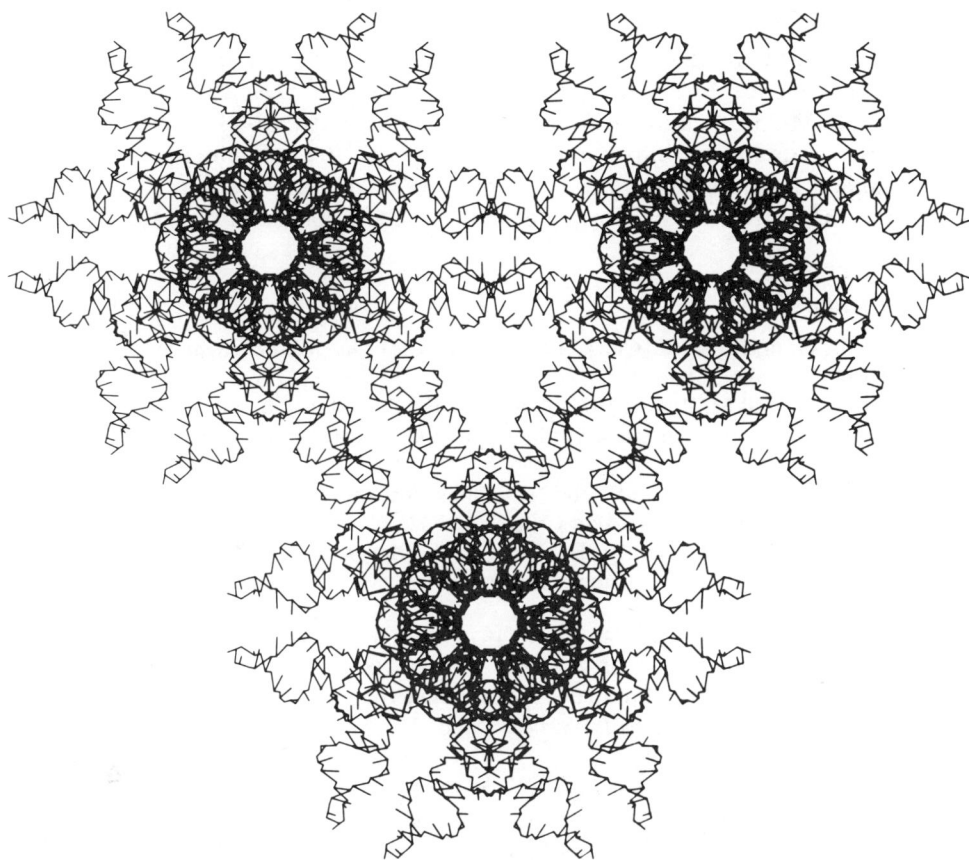

Figure 6 Axial projection of the yeast tRNA^{fMet} crystal showing the packing of molecular backbones. A phosphate backbone with lines pointing to the 2′-C ribose atoms is used. The ribose phosphate is angulated at the phosphate and 3′-O positions, and the spokes point from the 5′-phosphate atom to the 1′-C. The open network with anticodon arms projecting out roughly perpendicular from the acceptor T stems (themselves tightly clustered around the 6_4 axis) results in a unit cell over 80% solvent.

readily grown. The situation is reminiscent of the characteristically poorly ordered crystalline immunoglobulins—another macromolecule where flexibly hinged domains may be necessary for function (Silverton et al. 1977).

ACKNOWLEDGMENTS

During this work, R. W. S. was a fellow of the American Cancer Society (PF-737), the Damon Runyon/Walter Winchell Foundation (DRG 24F), and the U.S. Public Health Service (HL 16005). A. D. P. is a fellow of the Damon Runyon/Walter Winchell Foundation (231F), and N. K. was a fellow of the U.S. Public Health Service (F22-590). J.J.H. was supported by training grant USPHS GM-780. This work was supported by grants from the National Science Foundation (BMS 74-15075), the U.S. Public Health Service (GM-25234), and the American Cancer Society (NP-98A). We thank Joel Sussman for his important contributions to the refinement and the Weizmann Institute Computer Center and the Computergraphics Laboratory of the University of North Carolina for their computational support.

REFERENCES

Barrell, B. G. and B. F. C. Clark. 1974. *Handbook of nucleic acid sequences.* p. 5. Joynson-Bruvvers, Oxford.

Barrett, A. N. and M. Zwick. 1971. A method for the extension and refinement of crystallographic protein phases utilizing the fast Fourier transform. *Acta Cryst.* **27:**6.

Bricogne, G. 1976. Methods and programs for direct-space exploitation of geometrical redundancies. *Acta Cryst.* **A32:**832.

Collins, D. M., F. A. Cotton, E. E. Hazen, E. F. Meyer, Jr., and C. N. Morimoto. 1975. Protein crystal structures: Quicker, cheaper approaches. *Science* **190:**1047.

Grunberg-Manago, M. and F. Gros. 1977. Initiation mechanisms of protein synthesis. *Prog. Nucleic Acid Res. Mol. Biol.* **20:**209.

Hendrickson, W. A. 1978. Phase evaluation in macromolecular crystallography. In *Current trends in biomolecular structure* (ed. R. Srinivasan). Pergamon Press, London. (In press).

Holbrook, S. R., J. L. Sussman, R. W. Warrant, and S.-H. Kim. 1978. Crystal structure of yeast phenylalanine transfer RNA. II. Structural features and functional implications. *J. Mol. Biol.* **123:**631.

Honig, B., E. A. Kabat, L. Katz, C. Levinthal, and T. T. Wu. 1973. Model-building of neurohypophyseal hormones. *J. Mol. Biol.* **80:**277.

Jack, A., J. E. Ladner, and A. Klug. 1976. Crystallographic refinement of yeast phenylalanine transfer RNA at 2.5 Å resolution. *J. Mol. Biol.* **108:**619.

Kim, S.-H. 1976. Three-dimensional structure of transfer RNA. *Prog. Nucleic Acid Res. Mol. Biol.* **17:**181.

Kistenmacher, T. J., L. G. Marzilli, and M. Rossi. 1976. Conformational properties of the osmium tetraoxide bispyridine ester of 1-methylthymine and a comment on the linearity of trans $O = O_s = O$ group. *Bioinorg. Chem.* **6:**347.

Neidle, S. and D. T. Stuart. 1976. The crystal and molecular structure of an osmium bispyridine adduct of thymine. *Biochim. Biophys. Acta* **418:**226.

Pasek, M., M. P. Venkatappa, and P. B. Sigler. 1973. Enzymatic synthesis and crystallographic characterization of an isomorphous derivative of yeast formylatable methionine transfer ribonucleic acid containing iodocytidine. *Biochemistry* **12:**4834.

Podjarny, A. D. and A. Yonath. 1977. Use of matrix direct methods for low-resolution phase extension. *Acta Cryst.* **A33:**655.

Podjarny, A. D., A. Yonath, and W. Traub. 1976. Application of multivariate distribution theory to phase extension for a crystalline protein. *Acta Cryst.* **A32:**281.

Quigley, G. J., A. H. J. Wang, F. L. Suddath, A. Rich, J. L. Sussman, and S.-H. Kim. 1975. Hydrogen bonding in yeast phenylalanine transfer RNA. *Proc. Natl. Acad. Sci.* **72:** 4866.

Rich, A. and U. L. RajBhandary. 1976. Transfer RNA: Molecular structure, sequence, and properties. *Annu. Rev. Biochem.* **45:** 805.

Robertus, J. D., J. E. Ladner, J. T. Finch, D. Rhodes, R. S. Brown, B. F. C. Clark, and A. Klug. 1974. Structure of yeast phenylalanine tRNA at 3-Å resolution. *Nature* **250:** 546.

Rosa, J. J. and P. B. Sigler. 1974. The site of covalent attachment in the crystalline osmium—tRNA^{fMet} isomorphous derivative. *Biochemistry* **13:** 5102.

Schevitz, R. W., A. D. Podjarny, N. Krishnamachari, J. J. Hughes, P. B. Sigler, and J. L. Sussman. 1979. Crystal structure of a eukaryotic initiator tRNA. *Nature* **278:** 188.

Schevitz, R. W., M. A. Navia, D. A. Bantz, G. Cornick, J. J. Rosa, M. D. H. Rosa, and P. B. Sigler. 1972. An isomorphous heavy-atom derivative of crystalline formylmethionine transfer RNA. *Science* **177:** 429.

Schevitz, R. W., N. Krishnamachari, J. J. Hughes, J. Rosa, M. Pasek, G. Cornick, M. A. Navia, and P. B. Sigler. 1975. The crystal structure of yeast tRNA$_f^{Met}$—A map of an initiator tRNA at 6-Å resolution. In *Structure and conformation of nucleic acids and protein-nucleic acid interactions* (ed. M. Sundaralingam and S. T. Rao), p. 85. University Park Press, Baltimore, Maryland.

Sigler, P. B. 1975. An analysis of the structure of tRNA. *Ann. Rev. Biophys. Bioeng.* **4:** 477.

Silverton, E. W., M. A. Navia, and D. R. Davies. 1977. Three-dimensional structure of an intact human immunoglobulin. *Proc. Natl. Acad. Sci.* **74:** 5140.

Simsek, M. and U. L. RajBhandary. 1972. The primary structure of yeast initiator transfer ribonucleic acid. *Biochem. Biophys. Res. Commun.* **49:** 508.

Sprinzl, M., F. Grüter, and D. H. Gauss. 1978. Collection of published tRNA sequences. *Nucleic Acids Res.* (special suppl.) **6:** r15.

Stout, C. D., H. Mizuno, J. Rubin, T. Brennan, S. T. Rao, and M. Sundaralingam. 1976. Atomic coordinates and molecular conformation of yeast phenylalanyl-tRNA. An independent investigation. *Nucleic Acids Res.* **3:** 1111.

Subbaraman, L. R., J. Subbaraman, and E. J. Behrman. 1971. The reaction of osmium tetroxide-pyridine complexes with nucleic acid components. *Bioinorg. Chem.* **1:** 35.

Sussman, J. L. and S.-H. Kim. 1976. Idealized atomic coordinates of yeast phenylalanine transfer RNA. *Biochem. Biophys. Res. Commun.* **68:** 89.

Sussman, J. L., S. R. Holbrook, G. M. Church, and S.-H. Kim. 1977. A structure-factor least-squares refinement procedure for macromolecular structures using constrained and restrained parameters. *Acta Cryst.* **A33:** 243.

Sussman, J. L., S. R. Holbrook, R. W. Warrant, G. M. Church, and S.-H. Kim. 1978. Crystal structure of yeast phenylalanine transfer RNA. I. Crystallographic refinement. *J. Mol. Biol.* **123:** 607.

The Structure of Baker's Yeast tRNA^{Gly}: A Second tRNA Conformation

H. Tonie Wright, Philip C. Manor, Karin Beurling, Richard L. Karpel,* and Jacques R. Fresco
Department of Biochemical Sciences
Frick Laboratory, Princeton University
Princeton, New Jersey 08544

Yeast tRNA^{Phe} is the first nucleic acid whose structure has been determined in atomic detail (Quigley et al. 1975; Sussman and Kim 1976; Jack et al. 1976). The crystal structure of this molecule contains the double-helical regions anticipated by the general cloverleaf scheme derived from tRNA sequence information (Holley et al. 1965; Zachau et al. 1966; RajBhandary et al. 1967) and from spectroscopic studies (Fresco et al. 1964; Reid and Robillard 1975). In addition, the tRNA^{Phe} tertiary structure observed in the two crystal forms appears to depend for its integrity upon hydrogen bonds and, to a lesser extent, upon stacking interactions between bases from parts of the molecule that are remote in the sequence (Klug et al. 1974a; Kim et al. 1974). The conservation of these interacting bases in the available tRNA sequences supports the notion that all tRNAs have similar structures under physiological solvent and temperature conditions, as was first deduced from consideration of their common functions, narrow size distribution, and very similar hydrodynamic properties in the native state (Lindahl et al. 1965; Fresco et al. 1967). Furthermore, chemical modification (Robertus et al. 1974), nuclear magnetic resonance (Shulman et al. 1973; Reid and Robillard 1975), and relaxation kinetic studies (Riesner et al. 1973) on several tRNAs are consistent with the general features revealed by the crystal structure, so that it seems unlikely that crystallization has selected out a particular minor conformer of tRNA^{Phe}.

Although the weight of circumstantial evidence supports a common, general structure for all tRNA molecules, small but significant differences in structural detail, reflecting sequence and length differences, must exist among them. Moreover, if all tRNA molecules had the identical chain conformation, it is difficult to see how a simple lock-and-key recognition process could account for the great specificity of reactions such as

---

*Present address: Baltimore County Department of Chemistry, University of Maryland, Catonsville, Maryland 21228.

aminoacylation and base modification. Proposals for kinetic (Hopfield 1974) and chemical proofreading (von der Haar and Cramer 1976) have been made in which the high level of fidelity in the aminoacylation reaction is the result of either a kinetically driven discrimination against formation of the wrong product or a specific hydrolytic mechanism that aborts formation of incorrect product at an intermediate step. Both of these hypotheses can account for faithful aminoacylation, but they must depend, at least in part, upon a structurally determined specificity for the correct and against the incorrect tRNA substrate.

In addition to the problem of specificity raised by a common structure for all tRNAs is the question of whether there is a single native structure for any particular tRNA in the various reactions in which it plays a role. Proteins have a narrow distribution of conformations, with conformational differences restricted to relatively small parts of the molecule. On the other hand, during the course of many reactions, nucleic acids can undergo extensive structural changes involving the melting of individual helical regions and various condensations. Hence there may be several substantially different, biologically relevant conformational states accessible to a tRNA that depend upon local conditions and the presence of ligands, such as proteins and multivalent cations. Indeed, the occurrence of metastable, biologically inactive conformations has been demonstrated for many tRNAs (Lindahl et al. 1966; Fresco et al. 1967; Gartland and Sueoka 1966) and for other RNAs as well (Aubert et al. 1968; Traub and Nomura 1970). This, along with the recent findings that RNA-unwinding proteins can facilitate RNA renaturation (Karpel et al. 1974), lends credence to the notion of significant conformational mobility for tRNA. Such behavior is consistent with the central function of nucleic acids, to encode and read out information, which requires that their information contents in the form of nucleotide sequences exist in highly shielded and highly accessible conformations at different times and under different conditions.

The structure described here of a second tRNA, that of tRNA^{Gly} from baker's yeast, affords a comparison with that of tRNA^{Phe} that should provide some measure of the differences among tRNAs from the same species. Because the conditions under which the yeast tRNA^{Gly} crystals were grown are considerably different from those for tRNA^{Phe}, the crystal structure described here also gives information on the possible effect of solvent conditions upon tRNA structure. Although the general similarity of the tRNA^{Gly} structure to that of tRNA^{Phe} supports the idea that under the same solvent conditions all tRNAs will have the same conformation, specific differences between the structures of tRNA^{Gly} and tRNA^{Phe} suggest that other stable, biologically important conformations are accessible to the tRNA molecule. In fact, the conformation observed for tRNA^{Gly} has several potential new implications for tRNA biological function.

CRYSTAL STRUCTURE DETERMINATION

Crystals of yeast tRNA^{Gly} grow from aqueous solutions of unfractionated tRNA equilibrated with 50% dioxane-water (Fresco et al. 1968). These crystals were originally characterized as containing many tRNA species, based upon qualitative analysis of their amino acid-acceptor activities when dissolved (Blake et al. 1970). Subsequently, both quantitative amino acid acceptor analyses, as well as polyacrylamide gel electrophoretic analyses, of carefully washed crystals showed that these crystals contain at least 60% glycine-accepting tRNA (Webb et al. 1973). Reverse phase chromatography of crystals showed the presence of at least both major tRNA^{Gly} species in the relative abundance that they occur in unfractionated tRNA. Since the glycine-acceptor activity dominates that displayed by these crystals, and the measured proportion of tRNA^{Gly} in the crystals represents a minimum value, the crystals are designated as tRNA^{Gly}. For present purposes, the tRNA^{Gly} sequence determined preliminarily by Yoshida (1973), which is that of one of the two major tRNA^{Gly} species, is assumed to be that of the crystalline material. The tRNA^{Gly} sequence is shown in Figure 1. Note that it can be folded in the usual cloverleaf arrangement, and that it contains a very short extra arm. Determination of the sequences of all three tRNA^{Gly} species from baker's yeast is being undertaken in our laboratory.

It is pertinent to note that we have also succeeded in crystallizing purified samples of the two major tRNA^{Gly} species under conditions similar to those used to crystallize tRNA^{Phe}. This crystal form is less favorable for a structure determination because it has two molecules per asymmetric unit and is extremely fragile (Webb et al. 1973). It is interesting that we have not succeeded in crystallizing any of these purified tRNA^{Gly} preparations under the same conditions that are employed to successfully grow the crystals used for the present structure determination.

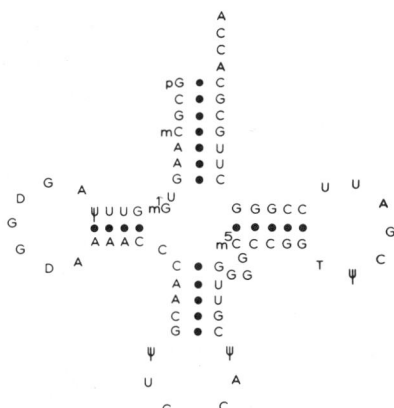

Figure 1 The sequence of baker's yeast tRNA^{Gly} folded in the cloverleaf scheme (Yoshida 1973).

These crystals were grown in the following way. A 50-μl aliquot of the tRNA (2500 A_{260} units/ml) in 0.01 M cacodylate (pH 6.1), 0.15 M NaCl, 0.5 mM EDTA, and 10 mM in $MgCl_2$ (or some other divalent cation or polyamine) is equilibrated at 32°C in a closed bottle with a reservoir of 50% dioxane-water (v/v) (Fresco et al. 1968). Crystals often take 6 months or more to grow to useable size. The space group is P222, where a is 43.9, b is 128.1, and c is 50.9 Å. Heavy atom isomorphous replacements were obtained by soaking crystals in $Pt(NH_3)_4Cl_2$, $Au(NH_2CH_2CH_2NH_2)_2Cl_3$ (Karpel et al. 1975), and K_2OsO_4·pyridine (presumably OsO_4·pyr_2) (Schevitz et al. 1972), added directly to their mother liquor.

Details of the crystal structure determination are being reported elsewhere, so they are described here only briefly. Diffraction data were collected on an Enraf Nonius CAD4 diffractometer mounted on an Elliott GX6 rotating anode X-ray source (Massey and Manor 1976) run at 38,000 V, 60 mA. Parent data were measured initially and only those reflections with nonzero intensities in this data set were measured on derivative crystals. Data were measured first to 5-Å resolution and used to calculate phases by the isomorphous replacement method. Only 397 observations of the theoretical 1483 total reciprocal lattice points in the 5-Å sphere were phased, but these were sufficient to unambiguously locate the molecular envelope in the electron-density map. Later, additional reflections were measured to a resolution of approximately 3.5 Å, and parameters were refined and phases calculated for this larger data set, which consisted of 1330 unique reciprocal lattice points.

The electron-density map based upon the expanded data set was contoured at a scale of 8 Å per inch for the unit cell from x = −0.07 to 1.07, y = −0.05 to 0.55, and z = −0.18 to 0.60 of the unit cell. These boundaries enclose one molecule. Figure 2 shows the section z = 0 to 0.45 of this electron-density map. Clear regions of electron density contrast with large volumes of interstitial solvent. Close intermolecular contacts exist at three crystallographic twofold axes, and at one other point not mediated by crystal symmetry. Figure 3 shows the density for a single molecule. The five sections of the cloverleaf structure can be discerned, though the density linking these regions, particularly at sharp turns, is often weak. It is possible, nevertheless, to trace the polynucleotide chain throughout most of the molecule with only local ambiguities at a few places. Phosphate groups are visible in many places and there is plausible density corresponding to the double-helical regions of the D and the anticodon arms. Such density for the TψC arm also exists, but it is more difficult to interpret. In contrast, the acceptor stem is clearly not helical completely, existing instead as two well-ordered single strands. Bases are evident in these separated strands of the acceptor arm, as well as in the other single-strand sections of the molecule, particularly in the

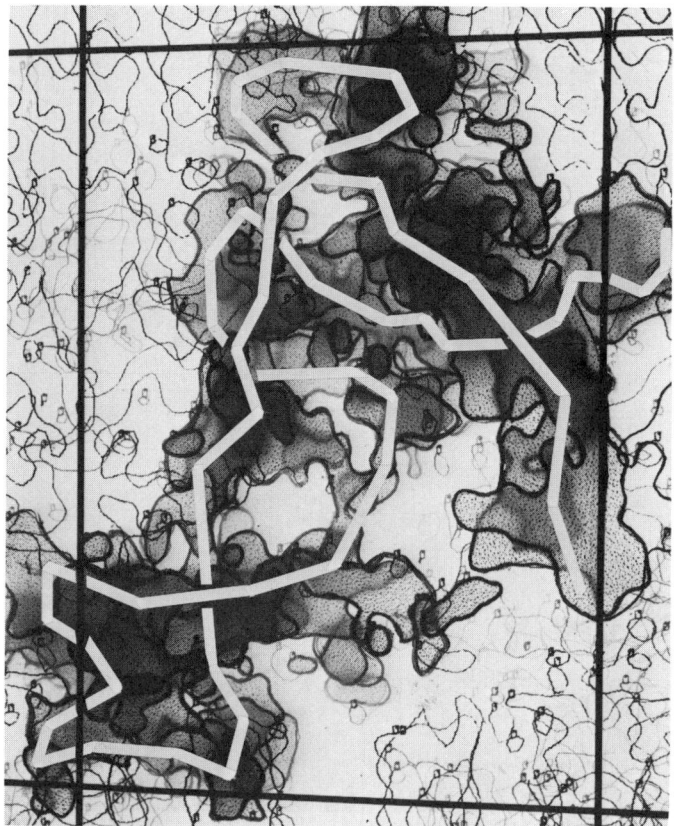

Figure 2 Electron-density map of yeast tRNA^{Gly} molecule with tracing of polynucleotide backbone. Every other section has been removed and only lowest contour level displayed for clarity. X is horizontal.

anticodon and D loops. A skeletal model of the structure is under construction.

CONFORMATION OF THE POLYNUCLEOTIDE BACKBONE

The relative disposition of the cloverleaf arms of tRNA^{Gly} is quite similar to that of tRNA^{Phe} with the exception of the acceptor stem (Fig. 4). As noted, in tRNA^{Gly} the acceptor stem is not a complete double helix, but rather exists as two well-ordered single strands diverging from each other. There may be two base pairs at the base of the stems but this interpretation is tentative. The 3' side of the acceptor arm follows an arc that brings its terminal into proximity with the 5' side of the anticodon stem. The 5' side of

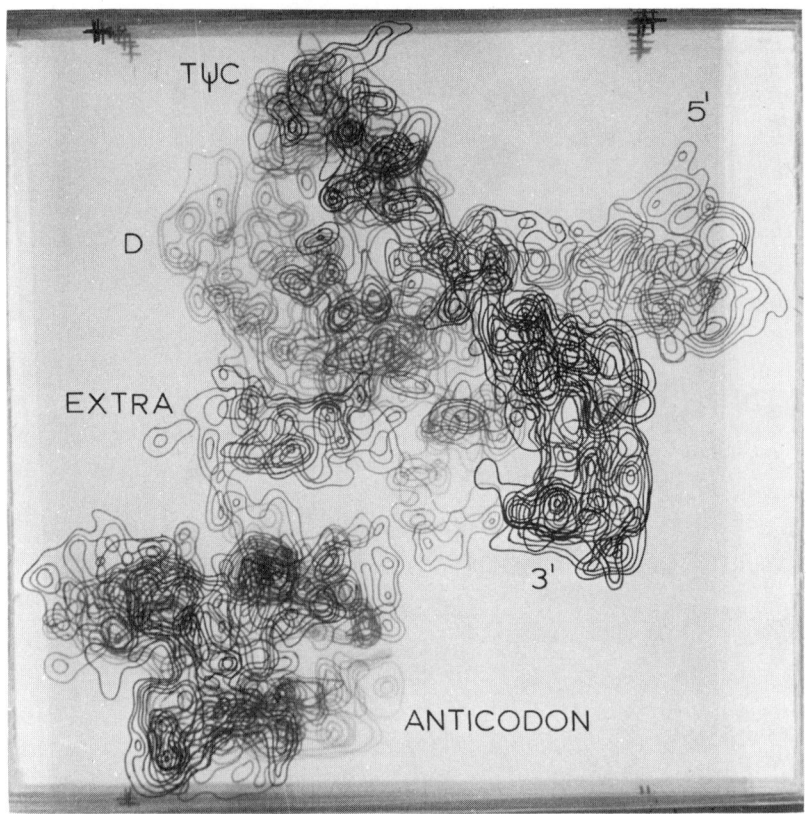

Figure 3 Electron density for a single tRNA^{Gly} molecule, with the density of adjacent molecules suppressed.

the acceptor stem extends towards an adjacent molecule, where its terminal makes a close contact with the junction of the D, TψC, and the short extra arms of that adjacent molecule. Neither single strand of the acceptor arm is in a helical, stacked conformation, but both follow comparable zigzag paths. Comparison of distances from the 5' and 3' terminals to a few other tentatively identified groups in the structure with corresponding distances in the tRNA^{Phe} structure indicates that the 5' terminal of tRNA^{Gly} is located essentially as in tRNA^{Phe}, whereas the 3' terminal of tRNA^{Gly} is displaced by more than 30 Å from its relative position in tRNA^{Phe}.

The central region where the D arm and the extra arm approach each other to form tertiary structural base pairs and triplets in tRNA^{Phe} is complicated in the tRNA^{Gly} map. These same tertiary interactions may exist in tRNA^{Gly}, but the weak density at the sharp turn of the extra arm

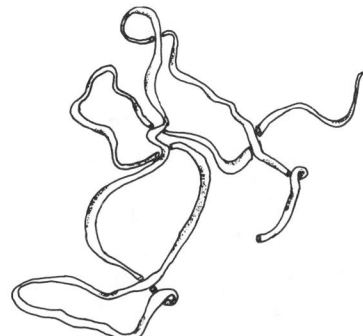

Figure 4 Backbone conformation of yeast tRNA^{Gly} in the same orientation as Figs. 2 and 3.

makes this difficult to interpret. Even though there is a close contact of residues G18 and G19 in the D loop with ψ55 and C56 in the TψC loop, only one appears close enough to make a base-pair interaction comparable to that observed in tRNA^{Phe}. The anticodon loop of tRNA^{Gly} resembles that of tRNA^{Phe}. Although precise identification of the residues has not yet been made, the most likely interpretation is that the anticodon triplet region is comparable to that of tRNA^{Phe}. The electron density corresponding to the TψC arm is confused by a close contact about a crystallographic twofold axis. The loop region is clearer than the stem for which insufficient electron density is visible to accommodate 5 bp. The precise details of secondary and tertiary structures in this region may have to await an improved electron-density map.

Little or no density exists at several points in the current tRNA^{Gly} electron-density map. The turn at residues 8 and 9 shows two conspicuous breaks in the chain density. Also, the residues on either side of the extra arm (probably 43 and 47) are not visible, nor is a piece of chain at the turn between the D and anticodon arms (residues 24, 25, and 26). These deficiencies do not change the general interpretation of the structure. It is quite likely that the further improvement in the electron-density map of tRNA^{Gly} that is underway will reveal more detailed differences in the two tRNA structures. Such differences will be of obvious interest in understanding the structure, but they will probably be less dramatic than that observed in the acceptor stem.

CRYSTAL PACKING AND INTERMOLECULAR CONTACTS

The space group of the tRNA^{Gly} crystals, P222, contains only twofold rotational symmetry operations. This simplifies interpretation of the electron density around these twofold axes by exclusion of that density related by the twofold operation to density already included in the molecule (i.e., a twofold axis cannot relate two chain segments of the same molecule). Although the central region of the tRNA^{Gly} molecule is removed from the symmetry elements, each of the cloverleaf arms extends to make a close contact with its counterpart in another molecule at a twofold axis. These contacts are quite close, and although their details have been elucidated for only the 3′ strand of the acceptor stem, there appears to be no single type of contact that dominates the interactions. Figures 5 and 6 show the relationships between molecules in which the 3′ acceptor strands and the D and TψC arms are involved, respectively.

The anticodon loop in tRNA^{Gly} is related to that of an adjacent molecule by a crystallographic twofold axis resembling that in the orthorhombic and monoclinic forms of tRNA^{Phe} (Fig. 7). This similarity is made more conspicuous by the fact that of the ten close intermolecular contacts in the tRNA^{Gly} crystal lattice, only this one appears identical to any of those in the two crystal forms of tRNA^{Phe} studied (Klug et al. 1974b).

The only close intermolecular contact in tRNA^{Gly} crystals that does not occur about a crystallographic twofold axis is that between the 5′ terminal of the acceptor arm and the D-TψC-extra arm junction of an adjacent molecule (Fig. 7). This contact and that involving the 3′ strands of the acceptor stems of adjacent molecules would not be possible if the acceptor stem were in a double-helical conformation.

FUNCTIONAL SIGNIFICANCE OF THE tRNA^{Gly} STRUCTURE

The crystal structure of tRNA^{Gly} has generally confirmed the expectation that different acceptor tRNAs have similar conformations. It seems likely that the large difference in conformation in the acceptor stems of these two molecules is a consequence of the effect of the 50% dioxane solvent from which the tRNA^{Gly} crystals are grown. The two crystal forms of tRNA^{Phe} whose structure was solved were grown from 10–20% dioxane (Ladner et al. 1972) and from 10–12% isopropanol or 2-methyl-2,4-tentanediol (Kim et al. 1971). Prinz et al. (1974) have obtained circular dichroism data that suggest that tRNA^{Phe} undergoes some transfor-

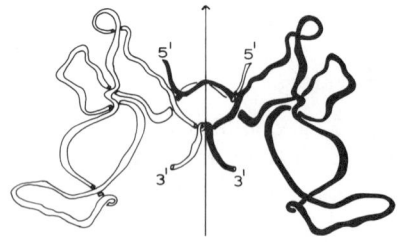

Figure 5 Contact of the 3′ strands of two tRNA^{Gly} molecules about the crystallographic twofold axis at x = 0, z = 0.5.

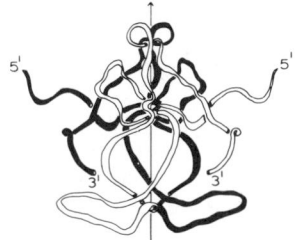

Figure 6 Orientation of two tRNA^{Gly} molecules related by the crystallographic twofold axis at x = 0.5, z = 0.

mation in solutions of 50–60% dioxane. It is noteworthy that other crystal forms of different tRNA species have been obtained by equilibration against relatively high dioxane concentrations. Yeast tRNA^{Phe} has been crystallized from solutions equilibrated with 35% dioxane (Cramer et al. 1970). The first microcrystals of tRNA, those of *Escherichia coli* tRNA^{fMet}, were also produced by equilibration against 35% dioxane (Clark et al. 1968). The lattice parameters and space-group identification from the powder pattern of these microcrystals are remarkably similar to those of the yeast tRNA^{Gly} crystals reported here. We also obtained crystals from unfractionated *E. coli* tRNA equilibrated with 50% dioxane (E. Ziff and J. R. Fresco, unpubl.).

Possibly, the most significant aspect of the tRNA^{Gly} structure is that it demonstrates, indeed confirms, that the tRNA molecule can exist in more than one stable conformation, depending upon local conditions (Fresco et al. 1967; Adams et al. 1967). It is surprising that the difference in conformation between tRNA^{Phe} and tRNA^{Gly} is greatest in a region of secondary rather than tertiary structure. The observed lability of the acceptor-stem double helix in the crystal structure stands in contrast with the melting behavior of one of the major yeast tRNA^{Gly} species (Hilbers et al. 1976). In this tRNA^{Gly}, the D arm and tertiary structure base pairs were found to melt first, followed by those of the anticodon arm, the acceptor stem, and the TψC arm. There is no reason to expect

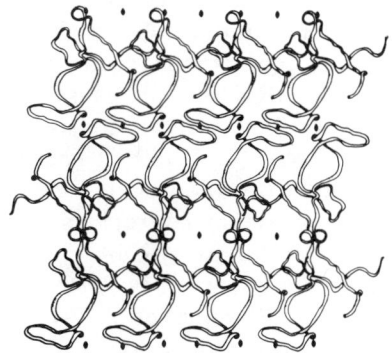

Figure 7 Contacts between tRNA^{Gly} molecules in the x–y plane. (●) Denotes twofold axes perpendicular to this plane.

equivalency between thermal melting and helix dissociation resulting from a change in solvent milieu. There are little data on this question and the most elementary viewpoint might attribute the consequences of dioxane to an influence primarily on base stacking and those of elevated temperature to an influence on both base stacking and hydrogen bonding. The observation that in dioxane-water the acceptor stem is the most extensively unwound of the four cloverleaf helices probably reflects the differential stability of the other helices conferred by the constraining loops.

The finding that tRNA secondary structure can be affected by solvent in such a way as to expose bases raises new possibilities for understanding protein-nucleic acid recognition phenomena. The current emphasis upon nucleic acid tertiary structure as a determinant of such recognition may be exaggerated. It is difficult to see how subtle differences among otherwise similar tRNA tertiary structures could occur in such a way as to guarantee the high fidelity of recognition realized in the aminoacylation reaction, for instance. On the other hand, the availability of segments of single-strand polynucleotide for recognition, whether through complementary base pairing or by direct interaction with protein, places the burden of information storage and transfer in the nucleic acid sequence. Sequence is the central feature of nucleic acid information and such a function for tRNA would make it resemble more closely the function of other nucleic acids.

It is tempting to consider ways in which the observed acceptor-stem lability of tRNA may be exploited in the enzymatic reactions in which tRNA is a substrate. The dissociation of a double helix into single strands presents the encoded sequence information of the strand for reading by a complementary strand. Conceivably, binding of tRNA to a hydrophobic specificity site of an enzyme might result in dissociation of the acceptor-stem double helix. The exposed single strands could then interact with a second identical molecule that has been similarly dissociated at its acceptor stem through binding at a second, similar site on the enzyme (Fig. 8). Matching of one or both complementary strands from separate but identical molecules might be required for productive aminoacylation. Mismatch of strands resulting from binding to the enzyme of two different acceptor tRNAs could cause the complex to abort. Such a scheme would increase fidelity and is consistent with what is known about many aminoacyl-tRNA synthetases. These enzymes, with only a few exceptions, display a twofoldness in subunit structure or pseudo-twofoldness in the form of sequence repeats (Ofengand 1977). This may reflect their specificity for two tRNA molecules in the aminoacylation reaction. Evidence that there are two tRNA binding sites in some synthetases has been reported (Blanquet et al. 1973; Bartmann et al. 1975). Such a property of twofoldness in the enzyme could impose the requirement of identity upon

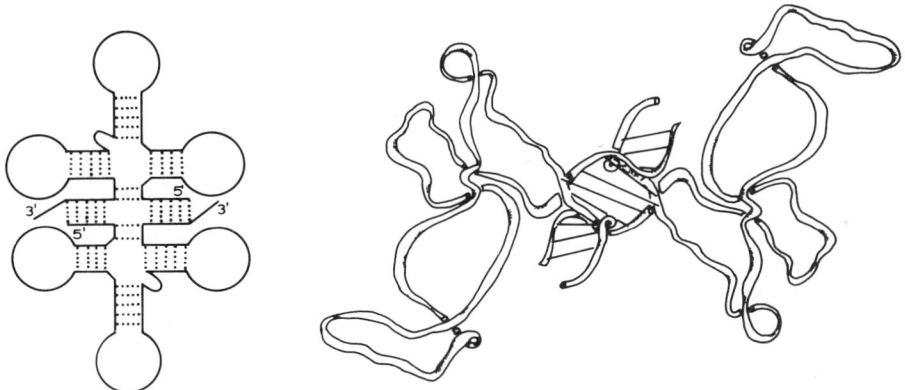

Figure 8 A hypothetical recognition complex of two tRNA molecules with identical acceptor-stem sequences as it might be bound to an aminoacyl-tRNA synthetase or other protein. The actual number of base pairs formed in this complex may not be 6 as depicted here.

the two bound tRNA substrate molecules so that product formation occurs. It may be relevant in this regard that the acceptor stem of tRNAs is one of the most variable regions in the sequences (Sprinzl et al. 1978). This is consistent with a role as a specificity determinant. There has been evidence implicating the acceptor stem in synthetase-tRNA recognition in several cases (Schulman and Chambers 1965; Mirzabekov et al. 1971; Smith and Celis 1973).

The observation of deformability in the tRNA^{Gly} structure provides a resolution to the problem raised by the relative dimensions of the tRNA^{Phe} structure and the aminoacyl-tRNA^{Tyr} synthetase molecule (Irwin et al. 1976). The work of Schimmel and his colleagues (Schoemaker and Schimmel 1974, 1977; Schimmel 1977) has shown that in several cases of cognate synthetase-tRNA complexes, the D arm, the extra arm, and the anticodon loop of the tRNA can be photo cross-linked to the enzyme. Reid (1977) pointed out that the binding of this surface of the tRNA substrate molecule to the synthetase, taken with the requirement that the CCA end must also be bound and the known tyrosyl-adenylate binding site in the structure of aminoacyl-tRNA^{Tyr} synthetase, would place the two extremes of the yeast tRNA^{Phe} molecule at the very outer edges of opposite subunits of the aminoacyl-tRNA^{Tyr} synthetase molecule. Although possible, such a model depends critically upon the distance from the CCA terminal to the anticodon loop for its validity. The crystal structure of tRNA^{Gly} relaxes the extent of synthetase molecule required to accommodate both of these parts of the tRNA substrate molecule, since the 3′ acceptor end comes within 20 to 50 Å of all parts of the anticodon arm. The CCA terminal does not come within contact distance of the anticodon triplet of its own molecule, though it does come to within approximately 10 Å of the anticodon of an adjacent molecule. This adjacent molecule is unrelated by any twofold axis to the molecule whose CCA terminal approaches its anticodon.

One function of tRNA where it is clear that the double-helical-stem region must dissociate to single strands is priming of DNA synthesis catalyzed by reverse transcriptase of certain RNA viral genomes (Waters and Mullin 1977). For example, the tRNA^{Trp} (Harada et al. 1975) was shown to bind tightly at its 3′ end to the viral 35S RNA of Rous sarcoma virus (Eiden et al. 1976) and to have the first 16 nucleotides following the terminal A complementary to the viral RNA (Cordell et al. 1976). Hybridization of the 3′ hexadecanucleotide to the viral template clearly requires an opening of the acceptor stem as well as the TψC stem of the tRNA^{Trp} cloverleaf structure. The mechanism of this complex formation is not known, but protein-mediated unwinding of the tRNA acceptor stem seems likely.

It is noteworthy that in the tRNA^{Gly} lattice all parts of the tRNA cloverleaf except the 5′ strand of the acceptor stem are related by twofold crystallographic axes to their counterparts in other molecules. This is not generally true of the monoclinic and orthorhombic lattices of tRNA^{Phe}. The closeness of these contacts in the tRNA^{Gly} lattice, in contrast to the large intermolecular solvent regions elsewhere, suggests that the different limbs of the tRNA molecule may have some self-affinity, at least under these solvent conditions. Such a self-association mediated by twofold rotational operations might be matched to corresponding symmetry elements in an enzyme or the ribosome in a way comparable to that suggested by Kim (1975) for the pseudosymmetry of a single tRNA molecule on its cognate synthetase. Such matched symmetry elements could be determinants of the specificity, cooperativity, or anticooperativity observed in some of the reactions in which tRNA is involved. The particular tRNA conformation involved in such a reaction may be determined by the solvent or macromolecular environment which potentiates the self-association.

CONCLUSIONS

In summary, the crystal structure of yeast tRNA^{Gly} has revealed a tRNA structure similar to that of yeast tRNA^{Phe}. We believe that the notable difference in the acceptor stems is the result of differences in solvent conditions in which the two crystal forms are grown. The fact that tRNA has been crystallized in two well-defined conformations suggests that both may have functional significance, since mere solvent denaturation would likely lead to a distribution of noncrystallizable structures. We have suggested one hypothesis by which this conformational variability could be exploited as a specificity determinant in a reaction in which tRNA serves as substrate. We anticipate that other functions of tRNA will depend upon its ability to assume several different conformational states depending upon its solvent environment and the other molecules with which it interacts.

ACKNOWLEDGMENTS

This work was supported by grants from the National Institutes of Health (GM-23598 and GM-07654), the National Science Foundation (GB-35595), and the American Heart Association, with funds contributed in part by the New Jersey Heart Association. H. T. W. was an Established Investigator of the American Heart Association (1973–1978), and P. C. M. and R. L. K. held NIH postdoctoral fellowships. We are pleased to acknowledge the contributions of R. D. Blake, P. K. Webb, G. Muscovites, P. Chang, J. Lee, and R. Massey to earlier phases of this investigation.

REFERENCES

Adams, A., T. Lindahl, and J. R. Fresco. 1967. Conformational differences between the biologically active and inactive forms of a transfer ribonucleic acid. *Proc. Natl. Acad. Sci.* **57:**1684.

Aubert, M., J. Scott, M. Reynier, and R. Monier. 1968. Rearrangement of the conformation of *Escherichia coli* 5S RNA. *Proc. Natl. Acad. Sci.* **61:**292.

Bartmann, P., T. Hanke, and E. Höller. 1975. Active site stoichiometry of L-phenylalanine: tRNA ligase from *Escherichia coli* K-10. *J. Biol. Chem.* **250:**7668.

Blake, R. D., J. R. Fresco, and R. Langridge. 1970. High resolution X-ray diffraction by single crystals of mixtures of transfer ribonucleic acids. *Nature* **225:**32.

Blanquet, S., M. Iwatsubo, and J. P. Waller. 1973. The mechanism of action of methionyl-tRNA synthetase from *E. coli*. *Eur. J. Biochem.* **36:**213.

Clark, B. F. C., B. P. Doctor, K. C. Holmes, A. Klug, K. A. Marcker, S. J. Morris, and H. H. Paradies. 1968. Crystallization of transfer RNA. *Nature* **219:**1222.

Cordell, B., E. Stavnezer, R. Friedrich, J. M. Bishop, and H. M. Goodman. 1976. Nucleotide sequence that binds primer for DNA synthesis to the avian sarcoma virus genome. *J. Virol.* **19:**548.

Cramer, F., F. von der Haar, K. C. Holmes, W. Saenger, E. Schlimme, and G. E. Schulz. 1970. Crystallization of yeast phenylalanine transfer ribonucleic acid. *J. Mol. Biol.* **51:**523.

Eiden, J. J., K. Quade, and J. L. Nichols. 1976. Interaction of tryptophan transfer RNA with Rous sarcoma virus 35S RNA. *Nature* **259:**245.

Fresco, J. R., R. D. Blake, and R. Langridge. 1968. Crystallization of transfer ribonucleic acids from unfractionated mixtures. *Nature* **220:**1285.

Fresco, J. R., L. C. Klotz, and E. G. Richards. 1964. A new spectroscopic approach to the determination of helical secondary structure in ribonucleic acids. *Cold Spring Harbor Symp. Quant. Biol.* **28:**83.

Fresco, J. R., A. Adams, R. Ascione, D. Henley, and T. Lindahl. 1967. Tertiary structure in transfer ribonucleic acids. *Cold Spring Harbor Symp. Quant. Biol.* **31:**527.

Gartland, N. and J. Sueoka. 1966. Two interconvertible forms of tryptophanyl sRNA in *E. coli*. *Proc. Natl. Acad. Sci.* **55:**948.

Harada, T., R. C. Sawyer, and J. E. Dahlberg. 1975. A primer ribonucleic acid for initiation of in vitro Rous sarcoma virus deoxyribonucleic acid synthesis. *J. Biol. Chem.* **250:**3487.

Hilbers, C. W., G. T. Robillard, R. G. Shulman, R. D. Blake, P. K. Webb, J. R. Fresco, and D. Riesner. 1976. Thermal unfolding of yeast glycine transfer RNA. *Biochemistry* **15:**1874.

Holley, R. W., J. Apgar, G. A. Everett, J. T. Madison, M. Marquisse, S. H. Merrill, J. R. Penswick, and Z. Zamir. 1965. Structure of a ribonucleic acid. *Science* **147:**1462.

Hopfield, J. J. 1974. Kinetic proofreading: A new mechanism for reducing errors in biosynthetic processes requiring high specificity. *Proc. Natl. Acad. Sci.* **71:**4135.

Irwin, M. J., J. Nyborg, B. R. Reid, and D. M. Blow. 1976. The crystal structure of tyrosyl-transfer RNA synthetase at 2.7 Å resolution. *J. Mol. Biol.* **105:**577.

Jack, A., J. E. Ladner, and A. Klug. 1976. Crystallographic refinement of yeast phenylalanine transfer RNA at 2.5 Å resolution. *J. Mol. Biol.* **108:**619.

Karpel, R. L., N. S. Miller, A. M. Lesk, and J. R. Fresco. 1975. Stabilization of the native tertiary structure of yeast tRNA$_3^{Leu}$ by cationic metal complexes. *J. Mol. Biol.* **97:**519.

Karpel, R. L., D. G. Swistel, N. S. Miller, M. E. Geroch, C. Lu, and J. R. Fresco. 1974. Acceleration of RNA renaturation by nucleic acid unwinding proteins. *Brookhaven Symp. Biol.* **26:**165.

Kim, S.-H. 1975. Symmetry recognition hypothesis model for tRNA binding of aminoacyl-tRNA synthetase. *Nature* **256:**679.

Kim, S-H., G. Quigley, F. L. Suddath, and A. Rich. 1971. High resolution X-ray diffraction patterns of crystalline transfer RNA that show helical regions. *Proc. Natl. Acad. Sci.* **68:**841.

Kim, S.-H., F. L. Suddath, G. J. Quigley, A. McPherson, J. L. Sussman, A. H. J. Wang, N. C. Seeman, and A. Rich. 1974. Three dimensional tertiary structure of yeast phenylalanine transfer RNA. *Science* **185:**435.

Klug, A., J. Ladner, and J. D. Robertus. 1974a. The structural geometry of coordinated base changes in transfer RNA. *J. Mol. Biol.* **89:**511.

Klug, A., J. D. Robertus, J. E. Ladner, R. S. Brown, and J. T. Finch. 1974b. Conservation of the molecular structure of yeast phenylalanine transfer RNA in two crystal forms. *Proc. Natl. Acad. Sci.* **71:**3711.

Ladner, J. E., J. T. Finch, A. Klug, and B. F. C. Clark. 1972. High resolution X-ray diffraction studies on a pure species of transfer RNA. *J. Mol. Biol.* **72:**99.

Lindahl, T., A. Adams, and J. R. Fresco. 1966. Renaturation of transfer ribonucleic acids through site binding of magnesium. *Proc. Natl. Acad. Sci.* **55:**941.

Lindahl, T., D. D. Henley, and J. R. Fresco. 1965. Molecular weight and molecular weight distribution of unfractionated yeast transfer ribonucleic acid. *J. Am. Chem. Soc.* **87:**4961.

Massey, W. R. and P. C. Manor. 1976. A four circle single crystal diffractometer with a rotating anode source. *J. Appl. Cryst.* **9:**119.

Mirzabekov, A. D., D. Lastity, E. S. Levina, and A. A. Bayev. 1971. Localization of two recognition sites in yeast valine tRNA I. *Nat. New Biol.* **229:**21.

Ofengand, J. 1977. tRNA and aminoacyl-tRNA synthetases. In *Molecular mechanisms of protein biosynthesis* (ed. H. Weissbach and S. Pestka), p. 8. Academic Press, New York.

Prinz, H., A. Maelicke, and F. Cramer. 1974. Unfolding of yeast transfer ribonucleic acid species caused by addition of organic solvents and studied by circular dichroism. *Biochemistry* **13:**1322.

Quigley, G. J., N. C. Seeman, A. H. J. Wang, and A. Rich. 1975. Yeast phenylalanine transfer RNA: Atomic coordinates and torsion angles. *Nucleic Acids Res.* **2:**2329.

RajBhandary, U. L., S. H. Chang, A. Stuart, R. D. Faulkner, R. M. Hoskinson, and H. G. Khorana. 1967. Studies on polynucleotides. LXVIII. The primary structure of yeast phenylalanine transfer RNA. *Proc. Natl. Acad. Sci.* **57:**751.

Reid, B. R. 1977. Synthetase-tRNA recognition. In *Nucleic acid-protein recognition* (ed. Henry J. Vogel), p. 375. Academic Press, New York.

Reid, B. R. and G. T. Robillard. 1975. Demonstration and origin of six tertiary base pair resonances in the NMR spectrum of *E. coli* tRNA$_1^{Val}$. *Nature* **257:**287.

Riesner, D., G. Maas, R. Thiebe, P. Philippsen, and H. G. Zachau. 1973. The conformational transitions in yeast tRNA^{Phe} as studied with tRNA^{Phe} fragments. *Eur. J. Biochem.* **36:**76.

Robertus, J. D., J. E. Ladner, J. T. Finch, D. Rhodes, R. S. Brown, B. F. C. Clark, and A. Klug. 1974. Correlation between three-dimensional structure and chemical reactivity of transfer RNA. *Nucleic Acids Res.* **1:**927.

Schevitz, R. W., M. A. Navia, D. A. Bantz, G. Cornick, J. J. Rosa, M. D. H. Rosa, and P. B. Sigler. 1972. An isomorphous heavy-atom derivative of crystalline formylmethionine transfer RNA. *Science* **177:**429.

Schimmel, P. R. 1977. Approaches to understanding the mechanism of specific protein-transfer RNA interactions. *Accts. Chem. Res.* **10:**411.

Schoemaker, H. J. P. and P. R. Schimmel. 1974. Photo-induced joining of a transfer RNA with its cognate aminoacyl-transfer RNA synthetase. *J. Mol. Biol.* **84:**503.

―――. 1977. Effect of aminoacyl-tRNA synthetases on H-5 exchange of specific pyrimidines in tRNAs. *Biochemistry* **16:**5454.

Schulman, L. H. and R. W. Chambers. 1965. Transfer RNA. II. A structural basis for alanine acceptor activity. *Proc. Natl. Acad. Sci.* **61:**308.

Shulman, R. G., C. N. Hilbers, D. R. Kearns, B. R. Reid, and Y. P. Wong. 1973. Ring current shifts in the 300 MHz nuclear magnetic resonance spectra of six purified transfer RNA molecules. *J. Mol. Biol.* **78:**57.

Smith, J. D. and J. E. Celis. 1973. Mutant tyrosine transfer RNA that can be charged with glutamine. *Nat. New Biol.* **243:**66.

Sprinzl, M., F. Grüter, and D. H. Gauss. 1978. Collection of published tRNA, 5S and 5.8S rRNA sequences. *Nucleic Acids Res.* (suppl.) **5:**r15.

Sussman, J. L. and S.-H. Kim. 1976. Idealized atomic coordinates of yeast phenylalanine transfer RNA. *Biochem. Biophys. Res. Commun.* **68:**89.

Traub, P. and M. Nomura. 1970. Studies on the assembly of ribosomes in vitro. *Cold Spring Harbor Symp. Quant. Biol.* **34:**63.

von der Haar, F. and F. Cramer. 1976. Hydrolytic action of aminoacyl-tRNA synthetase from baker's yeast: "Chemical proofreading" preventing acylation of tRNA^{Ile} with misactivated valine. *Biochemistry* **15:**4131.

Waters, L. C. and B. C. Mullin. 1977. Transfer RNA in RNA tumor viruses. *Prog. Nucleic Acid Res. Mol. Biol.* **20:**131.

Webb, P. K., H. T. Wright, G. Moscovitis, P. Chang, R. Karpel, R. D. Blake, and J. R. Fresco. 1973. A crystalline tRNA^{Gly} from yeast. *Fed. Proc.* **32:**617.

Yoshida, M. 1973. The nucleotide sequence of tRNA^{Gly} from yeast. *Biochem. Biophys. Res. Commun.* **50:**779.

Zachau, H. G., D. Dutting, and H. Feldmann. 1966. Nucleotidsequenzen zweier serinspezifischer Transfer-Ribonucleinsauren. *Angew. Chem.* **78:**392.

tRNA Structure and Dynamics in Solution

Physical Studies
of tRNA in Solution

Donald M. Crothers
Department of Chemistry
Yale University
New Haven, Connecticut 06520

RNA-RNA interactions are formed and broken during the complex process of ribosome-mediated protein synthesis. In addition to the obvious interaction between the mRNA codons and the two tRNA anticodons, proposals have included binding between (1) tRNA and 5S RNA (Erdman 1976), (2) 5S RNA and 23S RNA (Herr and Noller 1975), (3) 23S RNA and 16S RNA (van Duin et al. 1976), (4) 16S RNA and tRNA (Noller and Chaires 1972), and (5) 16S RNA and mRNA during initiation (Shine and Delgarno 1974; Steitz and Jakes 1975). Clearly, a full understanding of the detailed mechanism of protein synthesis will require study of the equilibria and dynamics of RNA conformational changes. tRNA has been the most convenient and intensively studied system of this kind.

Our knowledge of the molecular mechanisms of functional RNA conformational changes is in its infancy, usually limited to direct (Steitz and Jakes 1975) or indirect (Noller and Chaires 1972; Shine and Delgarno 1974; Herr and Noller 1975; Erdman 1976; van Duin et al. 1976) evidence that a pairing occurs. In contrast, conformational changes of isolated tRNA molecules have been intensively studied, especially between 1965 and 1975 (Crothers and Cole 1978). These early studies established the general conformational and ion-binding properties of tRNA and have now given way to more sophisticated techniques used to study specialized properties, of the kind reported by Reid and Hurd, Johnston and Redfield, and Potts et al. (all this volume). In particular, nuclear magnetic resonance (NMR) studies play a central role in present and future studies of tRNA in solution.

Even though there is presently no way of knowing whether the conformational changes observed in solution are functionally relevant, the solution studies provide an essential basis for approaching the more difficult problem of functional processes. For example, it is probable that tRNA tertiary structure is capable of transient, localized opening (Crothers et al. 1974; Rhodes 1977), which is opposed by the addition of Mg^{++} ions (Rhodes 1977). In addition, the tertiary structure can undergo a concerted disruption (Fresco et al. 1967; Cole et al. 1972; Riesner et al. 1973; Crothers et al. 1974), coupled to release of bound Mg^{++} (Römer and

Hach 1975; Stein and Crothers 1976), leaving only cloverleaf double helices as stable structural elements. The transient opening of structure could help initiate localized pairing interactions, for example, between 5S RNA and the TψC loop of tRNA. Alternatively, the general opening of tertiary structure could provide tRNA with the flexibility that has been proposed as a means of facilitating translocation (Crothers et al. 1974; Crothers 1975). At present we simply do not know which of the many interactions that stabilize tRNA structure in solution remain intact during its function.

It would be a mistake to infer that all of the solution conformational changes of tRNA are now well understood. We do have a general knowledge of the molecular mechanism of thermal unfolding, including such details as the thermodynamic (Hinz et al. 1976) and kinetic (Riesner et al. 1973; Crothers et al. 1974) characteristics of individual steps. Also, we know in specific cases the permissible limits of solution variables that leave the native structure intact, such as salt concentrations (Cole et al. 1972) and pH (Bina-Stein and Crothers 1974). However, some vital areas still need intensive investigation. For example, there is evidence that aminoacylation of tRNA causes significant conformational changes far removed from the 3' terminal (Potts et al. 1977), although other measurements reveal no important structural alteration (Liesch 1977). The nature of any change that may be induced remains a mystery, along with the mechanistic problem of how the allosteric effects of acylation could be transferred through the relatively flexible phosphodiester bonds at the single-stranded CCA terminal to other regions of the molecule at which the influence of acylation apparently is felt. In addition, current NMR methods allow a much more detailed dissection of the lability of individual parts of the tRNA structure, down to the level of individual base pairs. More than a decade of work has gone into our present understanding of conformational changes of tRNA in solution; probably at least another decade will be required before the job can be considered complete.

CONFORMATIONAL PHASE DIAGRAMS

Figure 1 (Cole et al. 1972) summarizes the general conformational equilibria of tRNA when the Na^+ ion concentration is varied in Mg^{++}-free solution. When Mg^{++} (or other multivalent) ions are absent, tRNA tertiary structure is greatly destabilized. In most cases the native structure is still intact at moderate Na^+ concentration and low temperature. As the temperature is raised, the tertiary structure usually melts in the first conformational transition that occurs. The product of this transition, called cloverleaf or close variant in Figure 1, has some cloverleaf double-helical sections intact, although the least stable of these may have melted

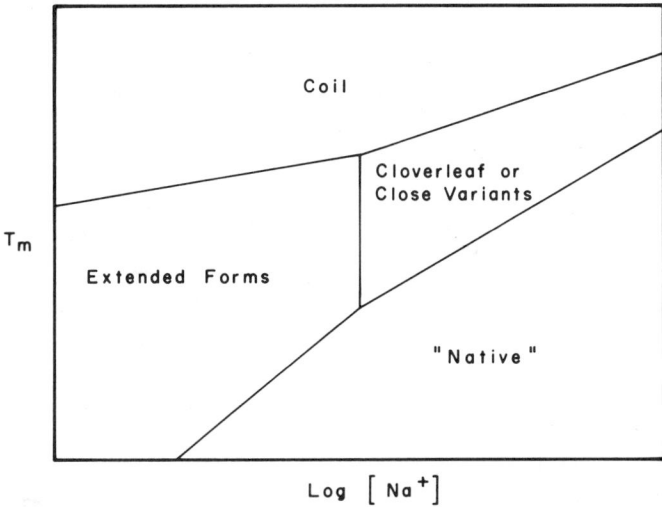

Figure 1 Simplified diagram identifying the major conformational classes of tRNA at varying salt concentrations and temperatures in salt-free solution (Cole et al. 1972).

together with the tertiary structure in the first transition. Subsequent melting steps involve the loss of these helical sections, producing the coil form at high temperature.

Most tRNAs probably adopt a nonnative conformation at low enough ion concentration in the absence of Mg^{++}, the extended-forms region of the phase diagram in Figure 1. In most cases this conformational change is not detected by biochemical assay methods, since the native structure is rapidly reformed when the tRNA is added to the Mg^{++}-containing assay mixture. However, some tRNAs can become trapped in metastable states by exposure to Mg^{++}-free conditions or other variations of solvent composition. These are the so-called denaturable tRNAs (Lindahl et al. 1966; Gartland and Sueoka 1966; Muench 1967; Adams et al. 1967; Lindahl et al. 1967a,b; Sueoka et al. 1967; Ishida and Sueoka 1967, 1968a,b, 1970; Gartland et al. 1969; Ishida et al. 1971; Bina-Stein and Crothers 1974), with recent studies focusing on $tRNA_3^{Leu}$ (Webb and Fresco 1973; Hawkins and Chang 1974; Kearns et al. 1974; Uhlenback et al. 1974; Rordorf et al. 1976; Hawkins et al. 1977), $tRNA^{Trp}$ (Greenspan and Litt 1974; Jones et al. 1976; Buckingham 1976), and $tRNA_2^{Glu}$ (Eisinger and Gross 1973; Bina-Stein and Stein 1976; Bina-Stein et al. 1976).

The physical basis for formation of extended forms is the increase of the electrostatic free energy of native tRNA as the salt concentration is lowered. These conformers almost certainly contain secondary structural elements different from the cloverleaf present in native tRNA (Cole et al. 1972). Evidence for this conclusion comes, for example, from the relatively slow conversion to the native structure when the salt or Mg^{++} concentration is increased. The large observed activation energy implies that incorrect structural elements must be dissociated before the correct structure can be formed.

UNFOLDING PATHWAY AND DYNAMICS

Considerable effort has gone into elucidation of the mechanism of thermal unfolding of tRNA, especially yeast $tRNA^{Phe}$ (Wintermeyer et al. 1969; Römer et al. 1969, 1970a; Thiebe and Zachau 1970; Robinson and Zimmerman 1971; Kearns et al. 1973; Hilbers et al. 1973; Riesner and Römer 1973; Riesner et al. 1973; Urbanke et al. 1973, 1975; Schreier and Schimmel 1974; Coutts et al. 1974; Römer and Hach 1975; Kan et al. 1974, 1975, 1977; Hinz et al. 1976; Beltchev et al. 1976; Römer and Varadi 1977), $tRNA^{Gly}$ (Hilbers et al. 1976), $tRNA^{Asp}$ (Coutts et al. 1974; Robillard et al. 1976), $tRNA^{Ala}$ (Riesner et al. 1969; 1970; Römer et al. 1970b), and $tRNA^{Ser}$ (Pilz et al. 1977), along with *Escherichia coli* $tRNA^{fMet}$ (Cole et al. 1972; Cole and Crothers 1972; Goldstein et al. 1972; Wilderauer et al. 1974; Crothers et al. 1974; Bina-Stein and Crothers 1975; Wong et al. 1975; Stein and Crothers 1976; Kyogoku et al. 1977), $tRNA^{Tyr}$ (Dourlent et al. 1971; Cole et al. 1972; Yang and Crothers 1972; Bina-Stein and Crothers 1974; Rodorf and Kearns 1976a; Leon et al. 1977), $tRNA^{Glu}$ (Hilbers and Shulman 1974; Bina-Stein et al. 1976), $tRNA^{Val}$ (Dourlent et al. 1971; Kastrup and Schmidt 1975), and $tRNA^{Leu}$ (Kearns et al. 1974). As a result, more is known about the detailed mechanism of folding these molecules than about any other class of globular macromolecules.

The two techniques most successfully used to elucidate the pathway of thermal unfolding are T-jump relaxation kinetics and NMR spectroscopy of hydrogen-bonded protons. In the T-jump method, the absorbance increase that accompanies melting is measured as a function of time. In this way a complex process can frequently be resolved into separate steps with different time dependences. It is often found that the individual helical sections that make up a tRNA melt separately, each giving rise to an exponential decay of the solution absorbance toward its new equilibrium value following the rapid temperature jump.

When the NMR method is used to measure thermal unfolding, the primary quantity observed is increased exchange of the hydrogen-bonded ring NH protons with solvent water as the structure melts. Of the NH protons, only those engaged in hydrogen bonds retain their chemical environment long enough to yield a high-resolution nuclear resonance. When the temperature is raised, the hydrogen bond becomes labile, and the proton resonance is broadened or disappears (Crothers et al. 1973). Reid and Hurd (this volume) and Johnston and Redfield (this volume) describe how magnetization transfer measurements can be used to determine the relative lability of individual base pairs in a tRNA molecule.

The detailed pathway of melting depends on the tRNA studied. If Mg^{++} is absent, one usually finds that the first transition includes loss of tertiary structure. For example, for *E. coli* tRNA^{fMet} both the tertiary structure and (probably) also the D helix melt in the first observed transition (Crothers et al. 1974). It is noteworthy that replacement of m^7G by A (tRNA$_1^{fMet}$ to tRNA$_3^{fMet}$) reduces the tertiary structure T_m by 15°C. The remaining double helices melt in the order of their predicted stability (Gralla and Crothers 1973; Tinoco et al. 1973; Borer et al. 1974): first the TψC helix, followed by the anticodon helix, and finally the acceptor-stem helix. The rates of these processes follow the rule that closure of a large loop (such as formation of only the acceptor-stem helix) is slower than closure of a small loop (such as the TψC or anticodon loop). Relaxation times vary from a few microseconds for small loops to several milliseconds for the acceptor stem, and sometimes even slower for tertiary structure melting.

Another method for resolving the melting processes in tRNA is calorimetry. For example, using the Privalov scanning calorimeter, Hinz et al. (1976) were able to show that the heat capacity of yeast tRNA^{Phe} could be resolved into components whose differential melting curves are essentially identical to those found for the same tRNA by T-jump measurements (Urbanke et al. 1973).

ION BINDING

Crystallographic analysis indicates that tRNA binds divalent ions such as Mg^{++} at specific sites (Jack et al. 1977; Holbrook et al. 1977), and it is of interest to compare this with the results of solution studies. The binding of Mg^{++} and other divalent ions by tRNA has been studied for many years (Steiner and Beers 1961; Millar and Steiner 1966; Vournakis and Scheraga 1966; Cohn et al. 1969; Reeves et al. 1970; Willick and Kay 1970; Danchin and Gueron 1970; Sander and Ts'o 1971; Danchin 1972; Rialdi et al. 1972; Willick et al. 1973; Krakauer 1974; Schreier and Schimmel 1974; Wolfson and Kearns 1974; Jones and Kearns 1974; Kayne and Cohn 1974; Lynch and Schimmel 1974; Römer and Hach 1975; Stein and Crothers 1976; Bina-Stein and Stein 1976; Rodorf and Kearns 1976; Jack et al. 1977; Holbrook et al. 1977; Leroy et al. 1977). This subject has recently become much better understood in its outline, and can readily be summarized with reference to the phase diagram in Figure 1. There are two different approaches to the study of Mg^{++} binding; the experiment begins with an Mg^{++}-free tRNA solution in which the tRNA either does, or does not, possess the native tertiary structure. Examples of the latter are tRNA samples in the cloverleaf or extended-form regions of the phase diagram.

The binding isotherms that result can be classified into two categories, anticooperative and cooperative, depending on whether the tRNA began in the native form or some other conformation. Cooperative binding results when binding of the first Mg^{++} facilitates binding of subsequent ions, whereas in anticooperative binding the subsequent ions are bound more weakly than the first ion. Binding of Mg^{++} to native tRNA is not cooperative (Wolfson and Kearns 1974; Jones and Kearns 1974; Römer and Hach 1975; Stein and Crothers 1976); there are a few (Noller and Chaires 1972; Shine and Delgarno 1974; Herr and Noller 1975; Erdman 1976; van Duin et al. 1976) sites with greater affinity than the remaining 20–25 sites. The actual number of strong sites is still in dispute and may depend on conditions. The weak sites are roughly comparable in Mg^{++} affinity to the Mg^{++} binding sites on standard double-helical nucleic acids. In contrast, binding of Mg^{++} to nonnative tRNA is cooperative (Cohn et al. 1969; Sander and Ts'o 1971; Danchin 1972; Rialdi et al. 1972; Lynch and Schimmel 1974; Schreier and Schimmel 1974) because Mg^{++} acts as an "allosteric effector" to convert tRNA from one conformation to another (Bina-Stein and Stein 1976), and more than one Mg^{++} ion is usually required for the conversion.

tRNA-tRNA INTERACTIONS

tRNAs can interact in solution by a variety of mechanisms, forming either $\alpha\beta$ or α_2 dimers, and a number of poorly characterized higher aggregates. These complexes vary greatly in stability. For example, the association between two tRNAs with complementary anticodons (Eisinger 1971) is stable only below about 30°C (Grosjean et al. 1976), whereas the metastable dimer of E. coli tRNA^{Tyr} (Yang et al. 1972) must be heated above 50°C in Mg^{++}-containing buffer before it dissociates to the thermodynamically more stable monomer.

The $\alpha\beta$ complex between two tRNAs that have complementary anticodons, discovered originally by Eisinger (1971), has been further characterized by Grosjean and his collaborators (Grosjean et al. 1976, 1978). The strength of the interaction between yeast tRNA^{Phe} (anticodon GmAA) and E. coli tRNA^{Glu} (anticodon s^{2}UUC) is about six orders of magnitude greater than expected for the association between two trinucleotides, based on the properties of model oligonucleotides. Since the formation rate constant (3×10^6 M^{-1} sec^{-1}) is approximately the same as found for combination of two oligonucleotides (Pörschke and Eigen 1971; Craig et al. 1971), the anomalously high complex stability must be due to a much slower dissociation rate than expected for complementary trinucleotides. Grosjean et al. (1976) showed that no single factor is responsible for the stabilization. Important contributions come from the modified purine adjacent to the 3' side of the anticodon, the closure of the single-strand anticodon region into a hairpin loop, and the dangling-end effect, which refers to the stabilization of double-helical oligomers by the presence at the ends of noncomplementary nucleotides (Martin et al. 1971).

In the model proposed by Grosjean et al. (1976), both tRNA anticodon loops are stacked on the two purines at the 3' side of the anticodon loop. The trinucleotide anticodon-anticodon helix thus is stacked between two almost continuous helices. The increased stacking energy of such a structure was proposed as the source of the anomalously large enthalpy of stabilization of the complex.

Further work by Grosjean et al. (1978) has shown that the stability of the tRNA-tRNA complex does not depend systematically on the $G+C$ content of the anticodons, in contrast to the rules for longer double-helical oligonucleotides. They also characterized quantitatively the effect of mismatching base pairs. Wobble pairs, including the short (pyrimidine·pyrimidine) wobble pairs originally proposed by Crick (1966), are only slightly destabilizing. In addition, it was found that the codon response rules for the modified nucleotides V and s^2U are reflected in the base-pairing stability of tRNA-tRNA complexes involving these nucleotides. U was found to be the nucleotide that is thermodynamically most susceptible to mispairing.

Other tRNA-tRNA complexes that have been studied by physical techniques include yeast $tRNA^{Ala}$ (Loehr and Keller 1968), *E. coli* $tRNA^{Tyr}$ (Rodorf and Kearns 1976b; Yang et al. 1972), along with several other α_2 dimers (Kowalski and Fresco 1971; Adams and Zachau 1968). Specific structural models have been proposed in some cases (Rodorf and Kearns 1976b; Yang et al. 1972).

FUTURE PROSPECTS

With the exception of the problem of conformational changes induced in tRNA by aminoacylation, the general conformational properties of tRNA in solution are well understood. Sophisticated physical techniques of the kind reported by Reid and Hurd, Johnston and Redfield, and Potts et al. (all this volume) promise to deepen our knowledge in this area substantially in the future. The major c' ⊥llenge that remains is to find methods to study the change in tRNA conformation while it is bound to the ribosome in the course of protein synthesis.

REFERENCES

Adams, A. and H. G. Zachau. 1968. Serine specific transfer ribonucleic acids. *Eur. J. Biochem.* **5:**556.

Adams, A., T. Lindahl, and J. Fresco. 1967. Conformational differences between the biologically active and inactive forms of a transfer ribonucleic acid. *Proc. Natl. Acad. Sci.* **57**:1684.

Beltchev, B., M. Yaneva, and D. Staynov. 1976. Thermal melting curves of tRNA^{Phe} from yeast lacking different numbers of nucleotides from the 3'-end. *Eur. J. Biochem.* **64**:507.

Bina-Stein, M. and D. M. Crothers. 1974. Conformational changes of transfer ribonucleic acid. The pH phase diagram under acidic conditions. *Biochemistry* **13**:2771.

———. 1975. Localization of the structural change induced in tRNA^{fMet} (*Escherichia coli*) by acidic pH. *Biochemistry* **14**:4185.

Bina-Stein, M. and A. Stein. 1976. Allosteric interpretation of Mg^{2+} binding to the denaturable *Escherichia coli* tRNA$_2^{Glu}$. *Biochemistry* **15**:3912.

Bina-Stein, M., D. M. Crothers, C. W. Hilbers, and R. G. Shulman. 1976. Physical studies of denatured tRNA$_2^{Glu}$ from *Escherichia coli*. *Proc. Natl. Acad. Sci.* **73**:2216.

Borer, P. N., B. Dengler, I. Tinoco, Jr., and O. C. Uhlenbeck. 1974. Stability of ribonucleic acid double-stranded helices. *J. Mol. Biol.* **86**:843.

Buckingham, R. H. 1976. Anticodon conformation and accessibility in wild-type and suppressor tryptophan tRNA from *E. coli*. *Nucleic Acids Res.* **3**:965.

Cohn, M., A. Danchin, and M. Grunberg-Manago. 1969. Proton magnetic relaxation studies of manganous complexes of transfer RNA and related compounds. *J. Mol. Biol.* **39**:199.

Cole, P. E., and D. M. Crothers. 1972. Conformational changes of transfer ribonucleic acid. Relaxation kinetics of the early melting transition of methionine transfer ribonucleic acid (*Escherichia coli*). *Biochemistry* **11**:4368.

Cole, P. E., S. K. Yang, and D. M. Crothers. 1972. Conformational changes of transfer ribonucleic acid. Equilibrium phase diagrams. *Biochemistry* **11**:4358.

Coutts, S. M., J. Gangloff, and G. Dirheimer. 1974. Conformational transitions in tRNA^{Asp} (brewer's yeast). Thermodynamic, kinetic, and enzymatic measurements on oligonucleotide fragments and the intact molecule. *Biochemistry* **13**:3938.

Craig, M. E., D. M. Crothers, and P. Doty. 1971. Relaxation kinetics of dimer formation by self-complementary oligonucleotides. *J. Mol. Biol.* **62**:383.

Crick, F. H. C. 1966. Codon-anticodon pairing: The wobble hypothesis. *J. Mol. Biol.* **19**:548.

Crothers, D. M. 1975. Some basic principles of transductive coupling. In *Functional linkage in biomolecular systems* (ed. F. D. Schmitt et al.), p. 24. Raven Press, New York.

Crothers, D. M. and P. E. Cole. 1978. Conformational changes of tRNA. In *Transfer RNA* (ed. S. Altman), p. 196. MIT Press, Cambridge, Massachusetts.

Crothers, D. M., C. W. Hilbers, and R. G. Shulman. 1973. Nuclear magnetic resonance study of hydrogen-bonded ring protons in Watson-Crick base pairs. *Proc. Natl. Acad. Sci.* **70**:2899.

Crothers, D. M., P. E. Cole, C. W. Hilbers, and R. G. Shulman. 1974. The molecular mechanism of thermal unfolding of *Escherichia coli* formylmethionine transfer RNA. *J. Mol. Biol.* **87**:63.

Danchin, A. 1972. tRNA structure and binding sites for cations. *Biopolymers* **11**:1317.

Danchin, A. and M. Gueron. 1970. Cooperative binding of manganese (II) to transfer RNA. *Eur. J. Biochem.* **16:**532.

Dourlent, M., M. Yaniv, and C. Helene. 1971. Temperature-jump relaxation studies on transfer ribonucleic acids. *Eur. J. Biochem.* **19:**108.

Eisinger, J. 1971. Complex formation between transfer RNAs with complementary anticodons. *Biochem. Biophys. Res. Commun.* **43:**854.

Eisinger, J. and N. Gross. 1973. Conformers, dimers, and anticodon complexes of tRNA^{Glu}2 (*Escherichia coli*). *Biochemistry* **14:**4031.

Erdman, V. A. 1976. Structure and function of 5S and 5.8S RNA. *Prog. Nucleic Acid Res. Mol. Biol.* **18:**45.

Fresco, J. R., A. Adams, R. Accione, D. Henley, and T. Lindahl. 1967. Tertiary structure in transfer ribonucleic acids. *Cold Spring Harbor Symp. Quant. Biol.* **31:**527.

Gartland, W. J., T. Ishida, N. Sueoka, and M. W. Nirenberg. 1969. Coding properties of two conformations of tryptophanyl-tRNA in *Escherichia coli*. *J. Mol. Biol.* **44:**403.

Gartland, W. and N. Sueoka. 1966. Two interconvertible forms of tryptophanyl sRNA in *E. coli*. *Proc. Natl. Acad. Sci.* **55:**948.

Goldstein, R. N., S. Stefanovic, and N. R. Kallenbach. 1972. On the conformation of transfer RNA in solution: Dependence of denaturation temperature and structural parameters of mixed and formylmethionyl *Escherichia coli* transfer RNA on sodium ion concentration. *J. Mol. Biol.* **69:**217.

Gralla, J. and D. M. Crothers. 1973. Free energy of imperfect nucleic acid helices. *J. Mol. Biol.* **73:**497.

Greenspan, C. M. and M. Litt. 1974. Characterization of the native and denatured forms of tRNA^{Trp} by reaction with kethoxal. *FEBS Lett.* **41:**297.

Grosjean, H., S. de Henau, and D. M. Crothers. 1978. On the physical basis for ambiguity in genetic coding interactions. *Proc. Natl. Acad. Sci.* **75:**610.

Grosjean, H., D. G. Söll, and D. M. Crothers. 1976. Studies of the complex between transfer RNAs with complementary anticondons. *J. Mol. Biol.* **103:**499.

Hawkins, E. R. and S. H. Chang. 1974. Differences in the secondary structures of native and denatured yeast leucine transfer ribonucleic acid. *Nucleic Acids Res.* **1:**1531.

Hawkins, E. R., S. H. Chang, and W. L. Mattice. 1977. Kinetics of the renaturation of yeast tRNA$_3^{Leu}$. *Biopolymers* **16:**1557.

Herr, W. and H. Noller. 1975. A fragment of 23S RNA containing a nucleotide sequence complementary to a region of 5S RNA. *FEBS Lett.* **53:**248.

Hilbers, C. W. and R. G. Shulman. 1974. Assignment of the hydrogen-bonded proton resonances in (*Escherichia coli*) tRNA^{Glu} by sequential melting. *Proc. Natl. Acad. Sci.* **71:**3239.

Hilbers, C. W., R. G. Shulman, and S. H. Kim. 1973. High resolution NMR study of the melting of tRNA$^{Phe}_{yeast}$. *Biochem. Biophys. Res. Commun.* **55:**953.

Hilbers, C. W., G. T. Robillard, R. G. Shulman, R. D. Blake, P. K. Webb, R. Fresco, and D. Riesner. 1976. Thermal unfolding of yeast glycine transfer RNA. *Biochemistry* **15:**1874.

Hinz, H.-J., V. V. Filimonov, and P. Privalov. 1976. Calorimetric studies on melting of tRNA^{Phe} (yeast). *Eur. J. Biochem.* **72:**79.

Holbrook, S. R., J. L. Sussman, R. W. Warrant, G. M. Church, and S.-H. Kim.

1977. RNA-ligand interactions: (1) Magnesium binding sites in yeast tRNA^{Phe}. *Nucleic Acids Res.* **4:**2811.

Ishida, T. and N. Sueoka. 1967. Rearrangement of the secondary structure of tryptophan sRNA in *Escherichia coli. Proc. Natl. Acad. Sci.* **58:**1080.

———. 1968a. Elimination of magnesium ions as an absolute requirement for the native conformation of tryptophan transfer ribonucleic acid. *J. Mol. Biol.* **37:**313.

———. 1968b. Effect of ambient conditions on conformations of tryptophan transfer ribonucleic acid of *Escherichia coli. J. Biol. Chem.* **243:**5329.

———. 1970. The use of conformational changes to purify tryptophan tRNA from *Escherichia coli. Biochem. Biophys. Acta* **217:**209.

Isheda, T., D. Snyder, and N. Sueoka. 1971. The interconvertibility of various bacterial transfer ribonucleic acids between an active and an inactive stable configuration. *J. Biol. Chem.* **246:**5965.

Jack, A., J. E. Ladner, D. Rhodes, R. S. Brown, and A. Klug. 1977. A crystallographic study of metal-binding to yeast phenylalanine transfer RNA. *J. Mol. Biol.* **111:**315.

Jones, C. R. and D. R. Kearns. 1974. Investigation of the structure of yeast tRNA^{Phe} by nuclear magnetic resonance: Paramagnetic rare earth ion probes of structure. *Proc. Natl. Acad. Sci.* **71:**4237.

Jones, C. R., D. R. Kearns, and K. H. Muench. 1976. Nuclear magnetic resonance of the base-pairing structure of the native and denatured conformers of *Escherichia coli* transfer RNA^{Trp}. *J. Mol. Biol.* **103:**747.

Kan, L. S., P. O. P. Ts'o, M. Sprinzl, F. von der Haar, and F. Cramer. 1977. 'H nuclear magnetic resonance studies of transfer RNA: The methyl and methylene resonances of bakers yeast phenylalanine transfer RNA and its fragments. *Biochemistry* **16:**3143.

Kan, L. S., P. O. P. Ts'o, F. von der Haar, M. Sprinzl, and F. Cramer. 1974. NMR study on the methyl and methylene proton resonances of tRNA$^{Phe}_{yeast}$. *Biochem. Biophys. Res. Commun.* **59:**22.

———. 1975. Proton magnetic resonance studies on the conformation of the hexanucleotide, GmpApApYpApψp, and related fragments from the anticodon loop of bakers yeast phenylalanine transfer ribonucleic acid. *Biochemistry* **14:**3278.

Kastrup, R. V. and P. G. Schmidt. 1975. 1H nuclear magnetic resonance of modified bases of valine transfer ribonucleic acid (*Escherichia coli*). A direct monitor of sequential thermal unfolding. *Biochemistry* **14:**3612.

Kayne, M. S. and M. Cohn. 1974. Enhancement of Tb (III) and Eu (III) fluorescence in complexes with *Escherichia coli* tRNA. *Biochemistry* **13:**4159.

Kearns, D. R., K. L. Wong, and Y. P. Wong. 1973. Effect of the removal of the Y base on the conformation of yeast tRNA^{Phe}. *Proc. Natl. Acad. Sci.* **70:**3843.

Kearns, D. R., Y. P. Wong, S. H. Chang, and E. Hawkins. 1974. Investigation of the structures of native and denatured conformations of tRNA^{Leu}3 by high-resolution nuclear magnetic resonance. *Biochemistry* **13:**4736.

Kearns, D. R., Y. P. Wong, E. Hawkins, and S. H. Chang. 1974. Model for the secondary structure of the denatured conformer of yeast tRNA$^{Leu}_3$. *Nature* **247:**541.

Kowalski. S. and J. R. Fresco. 1971. Preparation of highly labeled ^{32}P nucleic acids from yeast; isolation of "denaturable" leucine acceptor transfer RNA. *Science* **172**: 384.

Krakauer, H. 1974. A thermodynamic analysis of the influence of simple mono- and divalent cations on the conformational transitions of polynucleotide complexes. *Biochemistry* **13**: 2579.

Kyogoku, Y., T. Inubushi, I. Morishima, K. Watanabe, T. Oshima, and S. Nishimura. 1977. Proton magnetic resonance spectra of tRNA$_f^{Met}$ from *Thermus thermophilus*. *Nucleic Acids Res.* **4**: 585.

Leon, V., S. Altman, and D. M. Crothers. 1977. Influence of the A15 mutation on the conformational energy balance in *Escherichia coli* tRNA^{Tyr}. *J. Mol. Biol.* **113**: 253.

Leroy, J.-L., M. Gueron, G. Thomas, and A. Favre. 1977. Role of divalent ions in folding of tRNA. *Eur. J. Biochem.* **74**: 567.

Liesch, J. 1977. "Conformational forms of tRNA." Ph.D. thesis, University of Illinois, Urbana.

Lindahl, T., A. Adams, and J. R. Fresco. 1966. Renaturation of transfer ribonucleic acids through site binding of magnesium. *Proc. Natl. Acad. Sci.* **55**: 941.

———. 1967a. Isolation of "renaturable" transfer ribonucleic acids. *J. Biol. Chem.* **242**: 3129.

Lindahl, T., H. Adams, M. Geroch, and J. Fresco. 1967b. Selective recognition of the native conformation of transfer ribonucleic acids by enzymes. *Proc. Natl. Acad. Sci.* **57**: 178.

Loehr, J. S. and E. B. Keller. 1968. Dimers of alanine transfer RNA with acceptor activity. *Proc. Natl. Acad. Sci.* **61**: 1115.

Lynch, D. C. and P. R. Schimmel. 1974. Cooperative binding of magnesium to transfer ribonucleic acid studied by a fluorescent probe. *Biochemistry* **13**: 1841.

Martin, F. H., O. C. Uhlenbeck, and P. Doty. 1971. Self-complementary oligoribonucleotides: Adenylic-uridylic block copolymers. *J. Mol. Biol.* **57**: 201.

Millar, D. B. and R. F. Steiner. 1966. The effect of the environment on the structure and helix-coil transition of soluble ribonucleic acid. *Biochemistry* **5**: 2289.

Muench, K. H. 1967. Chloroquine-mediated conversion of transfer ribonucleic acid of *Escherichia coli* from an inactive to an active state. *Cold Spring Harbor Symp. Quant. Biol.* **31**: 539.

Noller, H. F. and J. B. Chaires. 1972. Functional modification of 16S ribosomal RNA by kethoxal. *Proc. Natl. Acad. Sci.* **69**: 3115.

Pilz, I., F. Malnig, O. Kratky, and F. von der Haar. 1977. On the conformation of serine-specific transfer RNA. *Eur. J. Biochem.* **75**: 35.

Pörschke, D. and M. Eigen. Co-operative non-enzymic base recognition. *J. Mol. Biol.* **62**: 361.

Potts, R., M. J. Fournier, and N. C. Ford, Jr. 1977. Effect of aminoacylation on the conformation of yeast phenylalanine tRNA. *Nature* **268**: 563.

Reeves, R. H., C. R. Cantor, and R. W. Chambers. 1970. Effect of magnesium ions on the conformation of two highly purified yeast alanine transfer ribonucleic acids. *Biochemistry* **9**: 3993.

Rhodes, D. 1977. Initial stages of the thermal unfolding of yeast phenylalanine transfer RNA as studied by chemical modification: The effect of magnesium. *Eur. J. Biochem.* **81**: 91.

Rialdi, G., J. Levy, and R. Biltonen. 1972. Thermodynamic studies of transfer ribonucleic acids. I. Magnesium binding to yeast phenylalanine transfer ribonucleic acid. *Biochemistry* **11**:2472.

Riesner, D. and R. Römer. 1973. Thermodynamics and kinetics of conformational transitions in oligonucleotides and tRNA. In *Physico-chemical properties of nucleic acids* (ed. J. Duchesne), vol. 2, p. 237. Academic Press, New York.

Riesner, D., R. Römer, and G. Maass. 1969. Thermodynamic properties of the three conformational transitions of alanine specific transfer RNA from yeast. *Biochem. Biophys. Res. Commun.* **35**:369.

———. 1970. Kinetic study of the three conformational transitions of alanine specific transfer RNA from yeast. *Eur. J. Biochem.* **15**:85.

Riesner, D., G. Maass, R. Thiebe, P. Philippsen, and H. G. Zachau. 1973. The conformational transitions in yeast tRNA^{Phe} as studied with tRNA^{Phe} fragments. *Eur. J. Biochem.* **36**:76.

Robillard, G. T., C. W. Hilbers, B. R. Reid, J. Gangloff, G. Dirheimer, and R. G. Shulman. 1976. A study of secondary and tertiary solution structure of yeast tRNA^{Asp} by nuclear magnetic resonance. Assignment of G·U ring NH and hydrogen-bonded base pair proton resonances. *Biochemistry* **15**:1883.

Robinson, B. and T. P. Zimmerman. 1971. A conformational study of yeast phenylalanine transfer ribonucleic acid. *J. Biol. Chem.* **246**:110.

Römer, R. and R. Hach. 1975. tRNA conformation and magnesium binding. *Eur. J. Biochem.* **55**:271.

Römer, R. and V. Varadi. 1977. Hydrogen-bonded protons in the tertiary structure of yeast tRNA^{Phe} in solution. *Proc. Natl. Acad. Sci.* **74**:1561.

Römer, R., D. Riesner, and G. Maass. 1970a. Resolution of five conformational transitions in phenylalanine-specific tRNA from yeast. *FEBS Lett.* **10**:352.

Römer, R., D. Riesner, S. M. Coutts, and G. Maass. 1970b. The coupling of conformational transitions in alanine specific transfer ribonucleic acid from yeast studied by a modified differential absorption technique. *Eur. J. Biochem.* **15**:77.

Römer, R., D. Riesner, G. Maass, W. Wintermeyer, R. Thiebe, and H. G. Zachau. 1969. Cooperative helix-coil transitions in half molecules of phenylalanine specific tRNA from yeast. *FEBS Lett.* **5**:15.

Rodorf, B. F. and D. R. Kearns. 1976a. Effect of europium (III) on the thermal denaturation and cleavage of transfer ribonucleic acids. *Biopolymers* **15**:1491.

———. 1976b. Nuclear magnetic resonance investigation of the base pairing structure of *Escherichia coli* tRNA^{Tyr} monomer and dimer conformations. *Biochemistry* **15**:3320.

Rodorf, B. F., D. R. Kearns, E. Hawkins, and S. H. Chang. 1976. High-resolution NMR study of yeast tRNA$^{Leu}_{CUA}$ and the native and denatured conformers of yeast tRNA$^{Leu}_{UUG}$. *Biopolymers* **15**:325.

Sander, C. and P. O. P. Ts'o. 1971. Interaction of nucleic acids. *J. Mol. Biol.* **55**:1.

Schreier, A. A. and P. R. Schimmel. 1974. Interaction of manganese with fragments, complementary fragment recombinations, and whole molecules of yeast phenylalanine specific transfer RNA. *J. Mol. Biol.* **86**:601.

Shine, J. and L. Delgarno. 1974. The 3'-terminal sequence of *Escherichia coli* 16S ribosomal RNA: Complementary to nonsense triplets and ribosome binding sites. *Proc. Natl. Acad. Sci.* **71:**1342.

Stein, A. and D. M. Crothers. 1976. Conformational changes of transfer RNA. The role of magnesium (II). *Biochemistry* **15:**160.

Steiner, R. F. and R. F. Beers. 1961. *Polynucleotides.* Elsevier, New York.

Steitz, J. and K. Jakes. 1975. How ribosomes select initiator regions in mRNA: Base pair formation between the 3' terminus of 16S rRNA and the mRNA during initiation of protein synthesis in *Escherichia coli. Proc. Natl. Acad. Sci.* **72:**4734.

Sueoka, N., T. Kano-Sueoka, and W. T. Gartland. 1967. Modification of sRNA and regulation of protein synthesis. *Cold Spring Harbor Symp. Quant. Biol.* **31:**571.

Thiebe, R. and H. G. Zachau. 1970. Further studies on amino acid acceptance and physical properties of tRNA$_{\text{yeast}}^{\text{Phe}}$. *Biochim. Biophys. Acta* **217:**294.

Tinoco, I., Jr., P. N. Borer, B. Dengler, M. D. Levine, O. C. Uhlenbeck, D. M. Crothers, and J. Gralla. 1973. Improved estimation of secondary structure in ribonucleic acids. *Nat. New Biol.* **246:**40.

Uhlenbeck, O. C., J. C. Chirikjian, and J. R. Fresco. 1974. Oligonucleotide binding to the native and denatured conformers of yeast transfer RNA$_3^{\text{Leu}}$. *J. Mol. Biol.* **89:**495.

Urbanke, C., R. Römer, and G. Maass. 1973. The binding of ethidium bromide to different conformations of tRNA. *Eur. J. Biochem.* **33:**511.

———. 1975. Tertiary structure of tRNA^{Phe} (yeast): Kinetics and electrostatic repulsion. *Eur. J. Biochem.* **55:**439.

van Duin, J., C. G. Kurland, J. Dondon, M. Grunberg-Managel, C. Branlant, and J. P. Ebel. 1976. New aspects of the IF3-ribosome interaction. *FEBS Lett.* **62:**111.

Vournakis, J. N. and H. A. Scheraga. 1966. Optical rotary dispersion studies of yeast alanine and tyrosine transfer ribonucleic acids. Evidence for intramolecular hydrogen bonding and discussion of conformational aspects. *Biochemistry* **5:**2997.

Webb, P. and J. R. Fresco. 1973. Tritium exchange studies of transfer RNA in native and denatured conformations. *J. Mol. Biol.* **74:**387.

Wilderauer, D., H. J. Gross, and D. Riesner. 1974. Enzymatic methylations. III. Cadavarine-induced conformational exchanges of *E. coli* tRNA^{fMet} as evidenced by the availability of a specific cytidine residue for methylation. *Nucleic Acids Res.* **1:**1165.

Willick, G. E. and C. M. Kay. 1970. Magnesium-induced conformational change in transfer ribonucleic acid as measured by circular dichroism. *Biochemistry* **10:**2216.

Willick, G., K. Oikawa, and C. M. Kay. 1973. Circular dichroism studies on the conformation of transfer ribonucleic acid in the presence of different divalent cations. *Biochemistry* **12:**899.

Wintermeyer, W., R. Thiebe, H. G. Zachau, D. Riesner, R. Römer, and G. Maass. 1969. Association and dissociation of half molecules of phenylalanine specific tRNAs from yeast and wheat germ. *FEBS Lett.* **5:**23.

Wolfson, J. M. and D. R. Kearns. 1974. Europium as a fluorescent probe of metal binding sites on transfer ribonucleic acid. I. Binding to *Escherichia coli* formylmethionine transfer ribonucleic acid. *J. Am. Chem. Soc.* **96**:3653.

Wong, K. L., Y. P. Wong, and D. R. Kearns. 1975. Investigation of the thermal unfolding of secondary and tertiary structure in *E. coli* tRNA^{fMet} by high resolution NMR. *Biopolymers* **14**:749.

Yang, S. K. and D. M. Crothers. 1972. Conformational changes of transfer ribonucleic acid. Comparison of the early melting transition of two tyrosine-specific transfer ribonucleic acids. *Biochemistry* **11**:4375.

Yang, S. K., D. G. Söll, and D. M. Crothers. 1972. Properties of a dimer of tRNA$_1^{Tyr}$ (*Escherichia coli*). *Biochemistry* **11**:2311.

High-resolution NMR Studies on tRNA Structure in Solution

Brian R. Reid and Ralph E. Hurd
Biochemistry Department
University of California
Riverside, California 92521

The advent of the yeast tRNA^{Phe} crystal structure (Kim et al. 1974; Robertus et al. 1974) stimulated a variety of biophysical studies on the conformations and dynamics of tRNA in solution. Proton nuclear magnetic resonance (NMR) spectroscopy has emerged as one of the most important approaches to this problem. The low-field (−15 to −11 ppm) region of the proton NMR spectrum of most tRNAs contains 20 secondary-structure and 7 ± 1 tertiary-structure base-pair resonances (the hydrogen-bonded ring nitrogen proton of U [UN3H] or G [GN1H] of base pairs with hydrogen-bond lifetimes of 5 msec or greater). Once assigned, these resonances provide the basis by which the conformation and dynamics of individual base pairs in the tRNA molecule can be monitored. The first priority of this type of research is reliable assignments; without these, studies on conformational changes, helix-coil dynamics, etc., are either useless or misleading. The first part of this paper deals with assignment and resolution. Poor spectral resolution limits the precision of NMR analysis and interpretation. A second type of resolution, namely, time resolution, will be discussed in the second part of this paper.

OBTAINING NMR SPECTRA

The low-field NMR spectra that we present were obtained at 360 MHz on a modified Bruker HXS360 spectrometer at the Stanford Magnetic Resonance Laboratory. Typically, a 6-mg sample of a pure tRNA was dissolved in 0.18 ml of appropriate buffer and transferred to a Wilmad 508CP NMR microtube. Spectra were obtained using either correlation spectroscopy (Dadok and Sprecher 1974) with a 2500-Hz/sec sweep rate (20- to 30-min signal averaging) or by using the Redfield 214 pulse Fourier transform (FT) method (Redfield et al. 1975) with a weak 380-μsec pulse (5- to 10-min signal averaging/spectrum).

ASSIGNMENT OF SECONDARY BASE-PAIR RESONANCES

The position of base-pair resonances that arise from the cloverleaf structure of tRNA (Fig. 1) can be predicted (assigned) using a combination of the known nearest-neighbor-sequence, ring-current shifts from these proximal bases (Arter and Schmidt 1976), and the crystallography-determined screw pitch of the tRNA secondary helices. This approach was calibrated by studying simple spectra of helical hairpin fragments derived from three tRNA species. The resulting shift values were then used to assign the secondary base-pair resonance positions in the low-field NMR spectra of several additional class 1 (D_4V_5) tRNAs. (D_4V_5 tRNAs are tRNAs with 4 bp in the D stem and 5 bases in the V loop.) The secondary base-pair assignments for *Escherichia coli* tRNA$_1^{Val}$ are shown in Figure 2.

Two things are apparent from these studies. First, AU base pairs resonate upfield of -14.35 ppm and GC base pairs resonate upfield of -13.45 ppm, the extent depending on the strength of the ring-current shift from the stacked neighboring bases ($A > G > C > U$). Second, although resonances from the interior base pairs of a helix are relatively easy to assign, terminal base pairs, which may or may not be stacked on one side, are more difficult to predict. A specific example of a secondary base pair that is particularly difficult to assign is C13·G22; it is located at the end of the D helix in many tRNAs, including yeast tRNA^{Phe} (see Fig. 1). The C13·G22 resonance has been assigned at various positions between -13.1 ppm and -11.5 ppm (Robillard et al. 1976; Kearns 1976; Geerdes and Hilbers 1977). The difficulty in predicting the position of this resonance is derived from the uncertainty of the geometry of A14 with respect to the GN1H proton of the C13·G22 base pair. As will be discussed later, the C13·G22 base-pair resonance has now been located at -12.0 ppm in native tRNA from the observed divalent paramagnetic ion relaxation effects in several tRNA spectra.

TERTIARY BASE PAIRS

In the crystal structure of yeast tRNA^{Phe} there are at least five tertiary interactions involving ring NH hydrogen bonds that contribute to the 7 ± 1 "extra" base-pair resonances that have been observed in the solution structure of several related tRNAs (Reid et al. 1977).

U8·A14 (s^4U8 in bacterial tRNA) and T54·A58 are the only AU-type tertiary interactions involving ring NH hydrogen bonds and, as such, should account for the two lowest-field tertiary resonances, since UN3H is more deshielded than GN1H. The s^4U8·A14 resonance has been unambiguously assigned at -14.8 ± 0.1 ppm by three independent methods and is the lowest-field resonance observed in tRNA spectra (Wong et al. 1975; Reid et al. 1975; Reid and Hurd 1977). The exceptionally low-field position of this proton is due to the extremely deshielded ring nitrogen

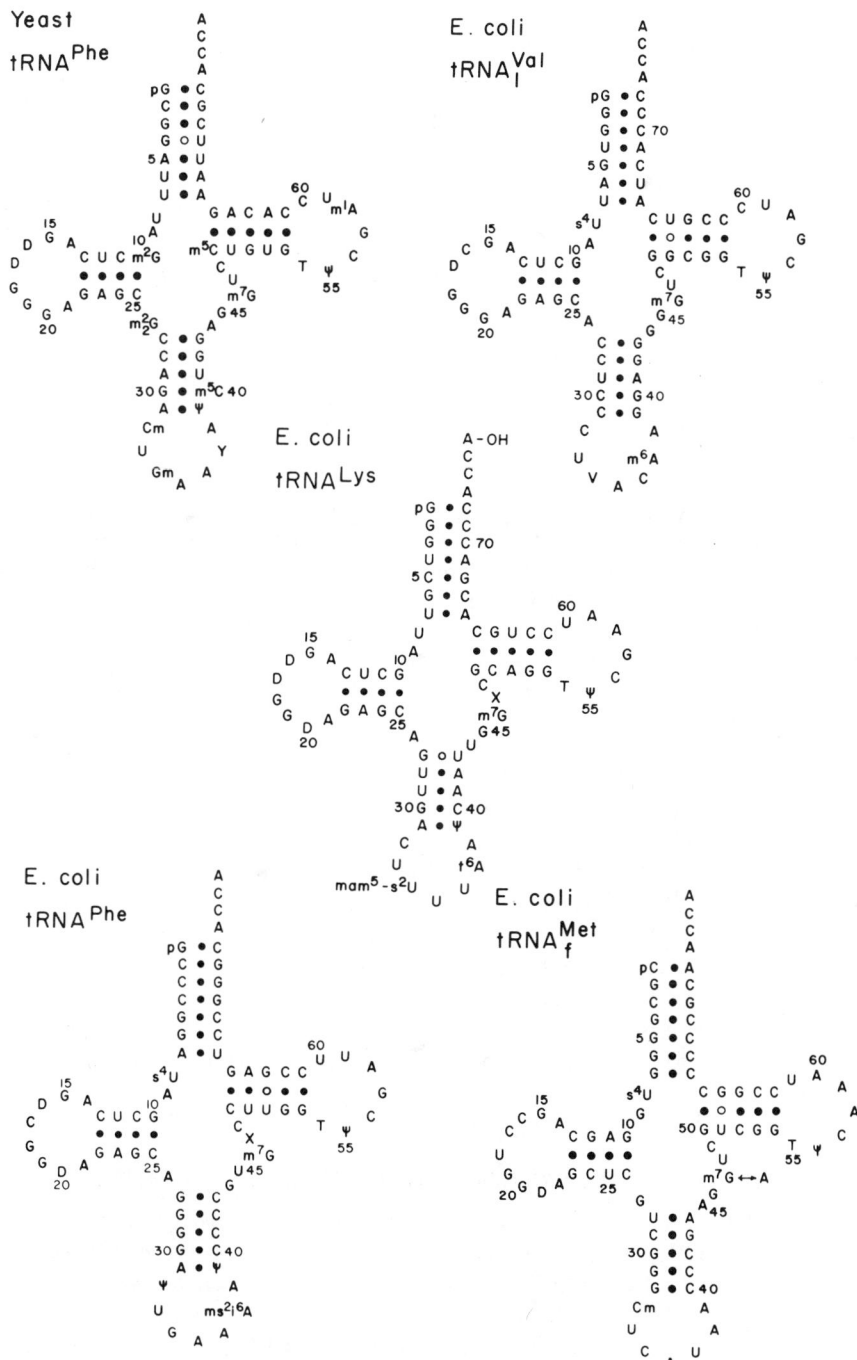

Figure 1 The cloverleaf sequences (secondary structure) of five class 1 (D_4V_5) tRNA species.

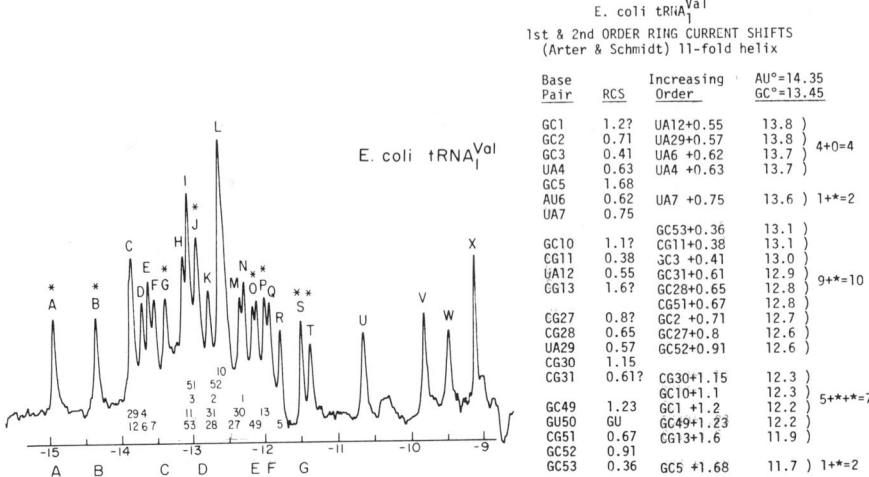

Figure 2 The assignment of secondary base-pair resonances in the 360-MHz NMR spectrum of *E. coli* tRNA$_1^{Val}$. (*Left*) Experimental spectrum with secondary resonances designated by cloverleaf base-pair numbers and with extra resonances designated by asterisks. (*Right*) Tabulation of the ring-current shifts on, and the predicted resonance positions of, the 20 secondary Watson-Crick base-pair resonances. The experimental spectrum was obtained at 37°C on a 6-mg sample of pure *E. coli* tRNA$_1^{Val}$ dissolved in 0.18 ml of 10 mM sodium cacodylate, 10 mM EDTA, 9 mM MgCl$_2$ (pH 7).

proton of s^{4}U (s^{4}UN3H); comparison of the monomer NMR spectra reveals that s^{4}UN3H is 1.5 ppm lower field than UN3H.

Assignment of the T54·A58 resonance is slightly more difficult. The best candidate for T54·A58 is the extra resonance found between −14.3 ppm and −13.9 ppm in most tRNAs (peak B at −14.3 ppm in the spectrum of *E. coli* tRNA$_1^{Val}$ shown in Fig. 2). The most compelling evidence we have for this assignment is that the TψC-stem fragment of *E. coli* tRNA$_1^{Val}$ (fragment 47–76, which contains the intact TψC loop) gives a spectrum with an extra resonance at −14.3 ppm at temperatures below 35°C.

G15·C48 ASSIGNMENT AND PARAMAGNETIC ION STUDIES

The G15·C48 tertiary base pair is a reversed Watson-Crick interaction that links the D loop with the variable loop. Unlike normal Watson-Crick GC pairs, the N1H of G15 is hydrogen bonded to the carbonyl oxygen of C48. We were able to assign this atypical GC base pair by studying the paramagnetic effects of Co^{++} and Mn^{++}. C13·G22, s^{4}U8·A14, and G15·C48 are broadened by both Co^{++} and Mn^{++}. The effect of 100 μM Co^{++} on the low-field NMR spectrum of 1.2 mM *E. coli* tRNA$_1^{Val}$ is shown in Figure 3 and the effects of low levels of Mn^{++} on the same spectrum are shown in Figure 4. Only three resonances are significantly broadened by both Co^{++} and Mn^{++} (−14.9 ppm, −12.3 ppm, and −12.0 ppm). The resonance at −14.9 ppm has already been assigned to the s^{4}U8·A14 interaction. This observation is consistent with Co^{++} binding at G15 as seen crystallographically (Jack et al. 1977) and with Mn^{++} binding to the crystallographically observed Mg^{++} site at P8 and P9 (Holbrook et al. 1977). The crystallographic Co^{++} and Mg^{++} binding sites are shown in Figure 5. C13·G22, s^{4}U8·A14, and G15·C48 are the only base-pair ring NH protons within 10 Å of both sites. Thus, we conclude that the −12.0 ppm and −12.3 ppm resonances in the *E. coli* tRNA$_1^{Val}$ spectrum belong to C13·G22 and G15·C48. To determine which resonance was from G15·C48 and which was from C13·G22, we also studied the Co^{++} and Mn^{++} effects on the low-field NMR spectrum of *E. coli* tRNA^{Lys}. This tRNA has C13·G22 as the terminal base pair of the same D helix found in *E. coli* tRNA$_1^{Val}$, but it contains A59 rather than U59 adjacent to the G15·C48 base pair (Figs. 1 and 5). Based on these comparisons, we expect the C13·G22 base-pair resonance at the same position in the spectra of *E. coli* tRNA$_1^{Val}$ and tRNA^{Lys}, whereas we expect a different resonance position for the G15·C48 proton. Only three resonances (−14.8 ppm, −12.6 ppm, and −12.0 ppm) in the low-field spectrum of *E. coli* tRNA^{Lys} are strongly affected by low levels of both Co^{++} and Mn^{++}. The −14.8 ppm resonance belongs to s^{4}U8·A14. Based on these arguments, we were able to assign the −12.0 ppm

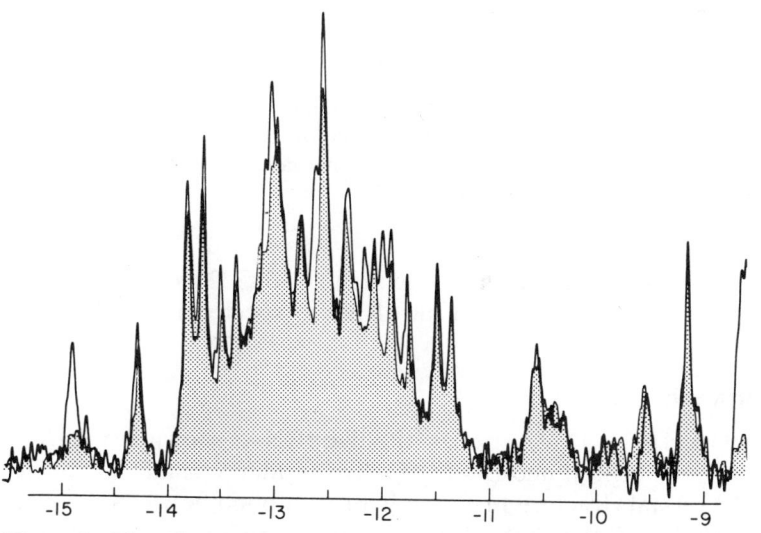

Figure 3 The effect of 0.1 mM CoCl$_2$ on the NMR spectrum of *E. coli* tRNA$_1^{Val}$ in the presence of excess magnesium ion. The Co^{++} spectrum (stippled) has been overlayered on the spectrum in the absence of Co^{++}.

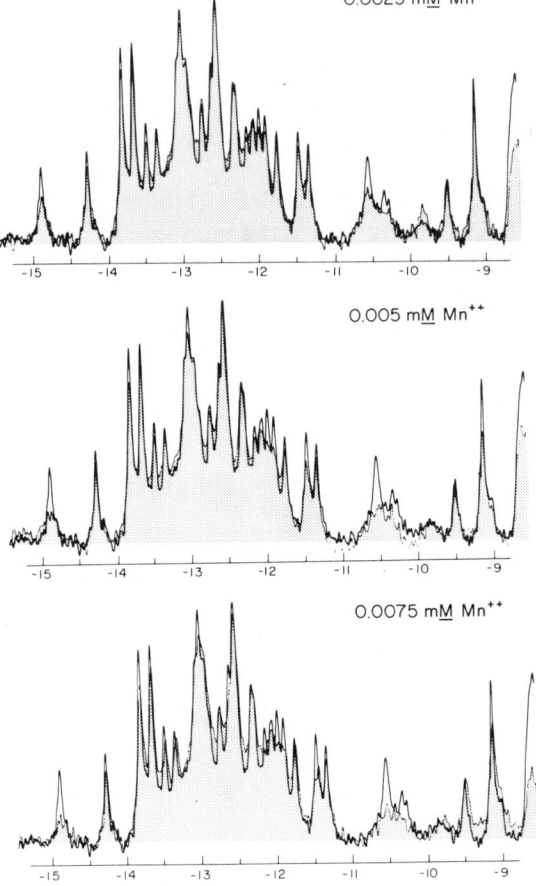

Figure 4 The effect of 0.0025 mM, 0.005 mM, and 0.0075 mM Mn^{++} on the NMR spectrum of 1.2 mM *E. coli* tRNA$_1^{Val}$ in the presence of excess magnesium ion. In each case, the Mn^{++} spectrum (stippled) has been overlayered on the spectrum in the absence of Mn^{++}.

resonance in the tRNA^{Lys} and tRNA$_1^{Val}$ spectra to C13·G22 and to assign the −12.6 ppm resonance (−12.3 ppm in the tRNA$_1^{Val}$ spectrum) to G15·C48.

m⁷G46·G22 INTERACTION

Like most other class 1 (D$_4$V$_5$) species, the five tRNAs shown in Figure 1 all have m⁷G as the central nucleoside of a five-residue variable loop. The yeast tRNA^{Phe} crystal structure reveals a hydrogen bond between the ring nitrogen proton of m⁷G46 and the N7 of G22 in the major groove of the D stem. The m⁷G46-G22-C13 base triple stabilizes the interaction between the D stem and the variable loop.

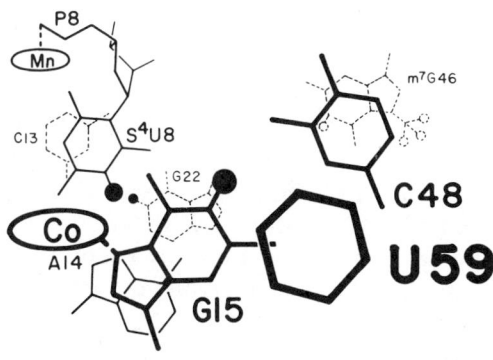

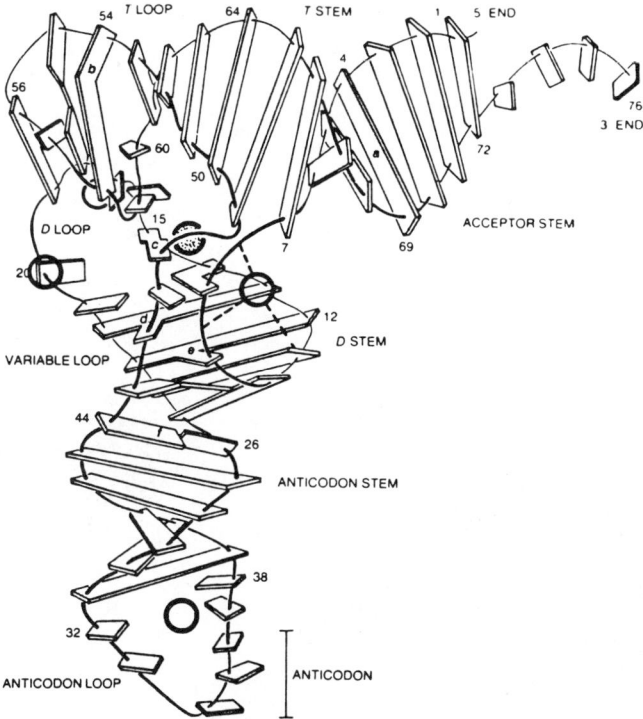

Figure 5 The Co^{++} and Mg^{++} binding sites in tRNA. (*Top*) A diagrammatic view down the tRNA D-helix axis showing the ring NH protons of 15-48, 8-14, and 13-22 (dark spheres) and the approximate location of the cobalt site and the P8-P9 Mg^{++} site described in the text (sketch by W. Reid [University of California] from a pair of stereo photographs of the tRNA^{Phe} crystal structure kindly provided by S.-H. Kim [Duke University]). (*Bottom*) The positions of the four magnesium binding sites (open circles) and the cobalt binding site (stippled circle) are drawn at their approximate positions on a schematic view of the yeast tRNA^{Phe} crystal structure (from Rich and Kim 1978).

Comparison of NMR spectra of tRNAs that contain the m⁷G46·G22 interaction with those that do not led to our first, albeit tentative, assignment of this resonance at about −13.4 ppm. This was somewhat surprising, however, in that we had never observed a base-pair GN1H resonance lower field than −13.2 ppm. Thus, to make an unambiguous assignment, we chemically removed the m⁷G46 from several class 1 (D_4V_5) tRNAs. The spectra of yeast tRNA^{Phe} and *E. coli* tRNA$_1^{Val}$ with and without m⁷G46 are shown in Figure 6. The extent of removal of m⁷G was estimated to be about 70% by hydrolysis and quantitative thin-layer chromatography analysis of the mononucleotides. In both spectra, removal of m⁷G46 results in the loss of a low-field resonance at −13.4 ppm (peak E in the yeast tRNA^{Phe} spectrum and peak F in the *E. coli* tRNA$_1^{Val}$

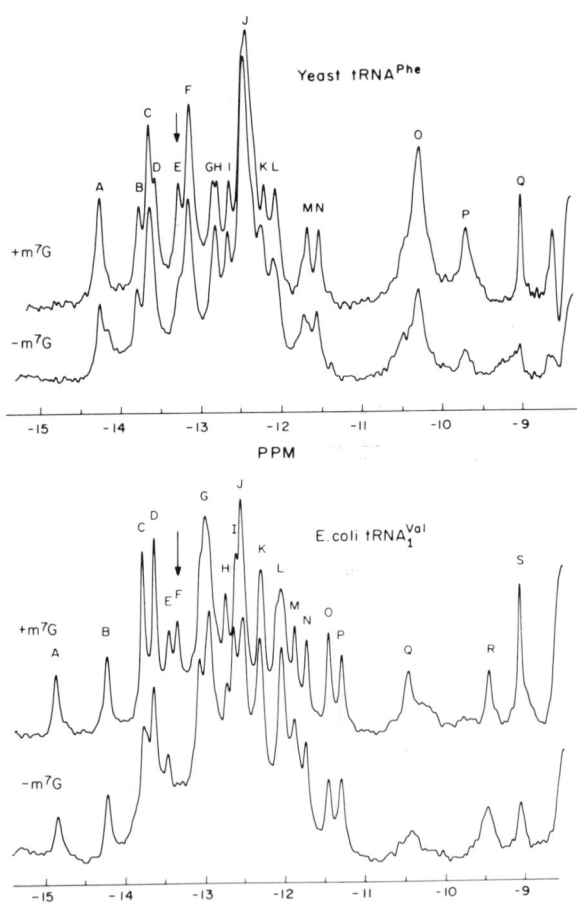

Figure 6 The NMR spectra of yeast tRNA^{Phe} and *E. coli* tRNA$_1^{Val}$, before and after m⁷G46 removal; spectra were all obtained at 45°C in 10 mM sodium cacodylate, 100 mM NaCl, and 15 mM MgCl₂ (pH 7).

spectrum [Fig. 6]). A quantitatively similar loss of intensity at −9.1 ppm in both spectra was also observed. These data led to our assignment of the ring nitrogen proton of m⁷G46 (m⁷G46N1H) at about −13.4 ppm and the proton attached to C8 of m⁷G (m⁷GC8H) at −9.1 ppm in yeast tRNA^{Phe} and *E. coli* tRNA$_1^{Val}$. A similar result was obtained for *E. coli* tRNA^{Phe}, tRNA^{fMet}, and tRNA^{Lys}. It is not surprising to find the m⁷G46N1H resonance at the same position in the spectra of these tRNAs, since in all cases their immediate environment is defined by A9, A21, A23, and U47 or modified U47 (Figs. 1 and 7). To further test our assignment we removed the m⁷G from *E. coli* tRNA^{fMet}. The m⁷G in this tRNA has a much less upfield shifting environment in which A9 is replaced by G and A23 is replaced by C; hence, the m⁷GN1H should resonate at much lower field in this environment. The spectra of *E. coli* tRNA^{fMet}, with and without 70% of its m⁷G, are shown in Figure 8. Consistent with our expectations, a lower-field resonance (70% of peak B at −14.5 ppm) is lost when m⁷G is removed from *E. coli* tRNA^{fMet}. The surprisingly low field position of the m⁷G tertiary hydrogen-bond resonance (as low as −14.5 ppm when GC resonances are above −13.4 ppm) can be explained by a comparison of the low-field NMR spectra of the monomers, m⁷G and G, in dry deuterated dimethylsulfoxide (d₆DMSO) (see Fig. 8). Methylation of G at N7 leads, via a delocalized positive charge, to a large downfield shift not only of C8H, but also of N1H and the amino protons. Qualitatively and quantitatively consistent with our assignment, N1H of m⁷G is 1.2 ppm further downfield than its counterpart in G. The m⁷GC8H is at −9.5 ppm in DMSO and appears to be minimally shifted in the tRNA molecules studied.

DYNAMICS IN SOLUTION

Once assigned, the low-field NMR spectrum of tRNA becomes a valuable probe of conformation and dynamics. Using the Redfield 214 pulse FT method (Redfield et al. 1975) we can obtain time-resolved low-field NMR spectra in H₂O solvents. As demonstrated by Johnston and Redfield (1977), under the appropriate conditions the recovery rate of a base-pair resonance that has been saturated by preirradiation is a direct measure of

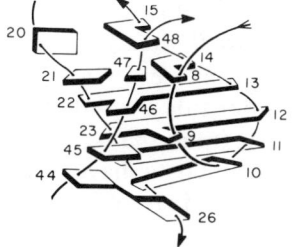

Figure 7 The base-stacking environment surrounding m⁷G46 in the crystal structure of yeast tRNA^{Phe} (redrawn from Sussman and Kim 1976).

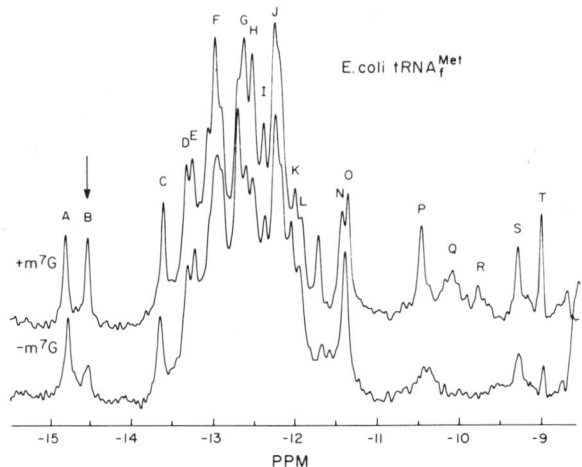

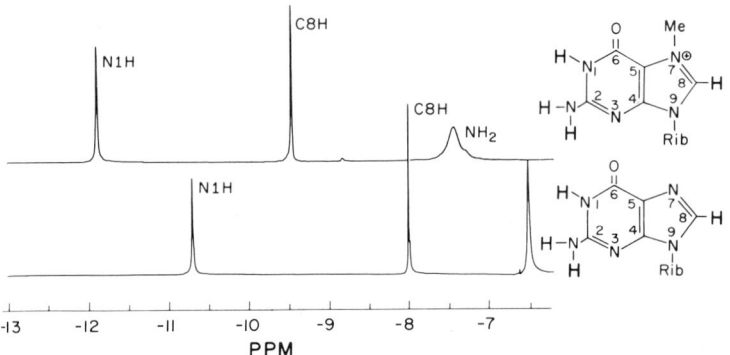

Figure 8 (*Top*) The NMR spectrum of *E. coli* tRNA^{fMet} before and after m^{7}G removal; both spectra were obtained at 45°C in 10 mM sodium cacodylate, 100 mM NaCl, 15 mM MgCl$_2$ (pH 7). (*Bottom*) Spectra of m^{7}G and G in DMSO.

the helix-coil exchange rate of that base pair; i.e., the replacement of the saturated ring NH proton with an unsaturated proton from H$_2$O dominates the recovery process.

We have used this method to study the dynamics of individual base pairs within a single helix in an intact tRNA molecule. As shown in Figure 9, the 27-resonance spectrum of *E. coli* tRNA^{Phe} can be reduced to 6 resonances by heating the sample to 68°C in a 10 mM sodium cacodylate buffer (pH 7) containing 15 mM MgCl$_2$ and 100 mM NaCl. The resonances lost from the spectrum are from base pairs that at 68°C are either disrupted or have helix lifetimes of less than 5 msec. The 6 resonances observed at 68°C belong to the 6 GC base pairs of the amino acid

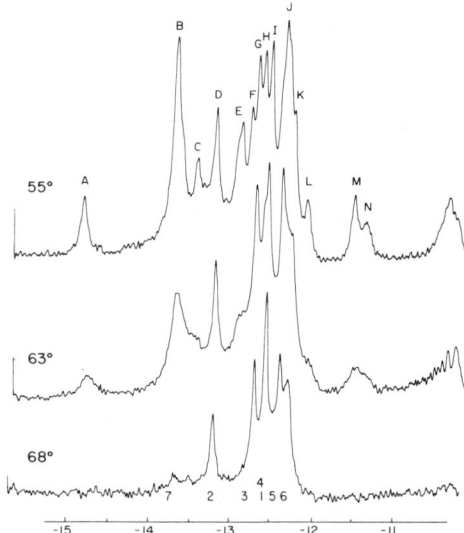

Figure 9 The NMR spectrum of *E. coli* tRNA^{Phe} at 55°C, 63°C, and 68°C in 10 mM sodium cacodylate, 100 mM NaCl, 15 mM MgCl$_2$ (pH 7).

acceptor stem (Fig. 1). Only a trace of the terminal A7·U66 base-pair resonance remains, indicating a hydrogen-bond lifetime somewhat less than 5 msec at 68°C. To lower and widen the temperature window in which only these 6 GC base-pair resonances are observed, this tRNA was studied in 10 mM sodium phosphate (pH 7) containing 100 mM NaCl but no Mg. Under these conditions, the molecule has melted to a 6-resonance spectrum by 51°C and remains as such up to 63°C.

Data obtained from a typical saturation recovery experiment is shown in Figure 10 in which the recovery of the G5·C68 base-pair resonance is measured at 58°C and 62°C. The resonance that has been preirradiated (indicated with an arrow) is allowed to recover over various delay times (1–200 msec) before the data acquisition pulse is taken. The saturation recovery rate of the G5·C68 base-pair resonance (exchange rate) at 58°C is 14 sec^{-1}; this rate increases 2.5-fold to 35 sec^{-1} at 62°C. The rate of the helix-coil exchange for all 6 GC base pairs was determined at 52°C, 58°C, and 62°C. In addition to the marked temperature dependence of these exchange rates, we find the exchange rates for the terminal base pairs are much greater than for penultimate pairs; at 58°C the rates are G1·C72 = 68 sec^{-1}; C2·G71 = 25 sec^{-1}; G5·C68 = 14 sec^{-1}; and G6·C67 = 64 sec^{-1}.

We are currently extending these studies to more complex systems that include internal AU base pairs, GU base pairs, and tertiary interactions.

Before we make the assumption that the saturation recovery rates are helix-coil exchange rates, however, we must check to see if the necessary

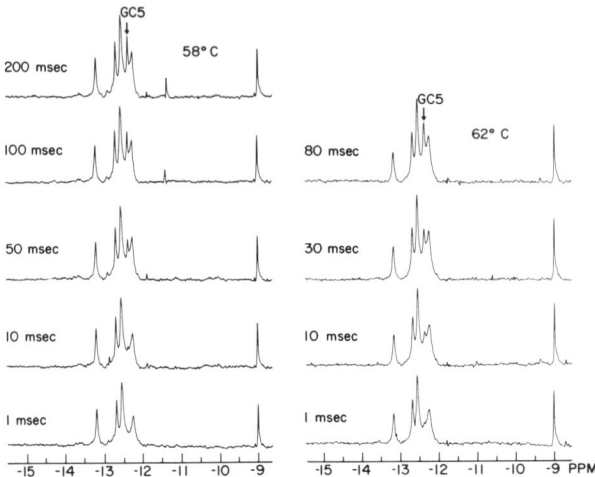

Figure 10 Saturation recovery of G5·C68 at 58°C and 62°C in 10 mM sodium phosphate, 100 mM NaCl (pH 7). The narrow resonance at −9.1 ppm belongs to m^{7}GC8H.

conditions are met. First, if saturation recovery occurs predominantly by exchange with protons from H_2O, then we should also observe loss of the base-pair resonances within 100 msec when H_2O is saturated. As expected, in the case of the *E. coli* tRNA^{Phe} acceptor system, saturation of H_2O results in the loss of over 95% of the 6 GC base-pair resonances but does not affect the nonexchangeable m^{7}GC8H aromatic proton at −9.1 ppm. This observation and the large temperature dependence of the saturation recovery rates indicate that we are directly measuring exchange. The assumption in this interpretation (justified from all available model studies) is that the coil ring NH-H_2O exchange rate is much more rapid than the observed saturation recovery rates. Experimentally, recovery rates for base-pair resonances vary from about 3 sec^{-1} to about 200 sec^{-1}; above 200 sec^{-1} resonances are not observed due to excessively short helix lifetimes, and recovery rates of about 3 sec^{-1} do not directly measure exchange, since they contain a significant component of other spin-lattice relaxation mechanisms. Thus an observed recovery rate of 3 sec^{-1} only represents an upper limit for the helix-coil exchange rate.

SUMMARY

We have shown that low-field NMR spectroscopy can be an extremely exact and informative probe of not only tRNA base pairing in solution but also of dynamics and tertiary conformation. The recent use of the Redfield 214 pulse techniques (Redfield et al. 1975) provides us with a method by which we can obtain time-resolved spectra on only a few milligrams of sample. With many reliable assignments we should now be able to deduce changes in both the conformation and dynamics of tRNA molecules.

ACKNOWLEDGMENTS

We are grateful for the use of the Stanford Magnetic Resonance Laboratory (supported by National Science Foundation grant no. GR-23633 and National Institutes of Health grant no. RR-00711) and the advice of W. W. Conover. We would especially like to thank Lillian McCollum and Susan Ribeiro for their expert technical assistance in purifying homogeneous tRNA species and Edward Azhderian for his help in obtaining high-resolution spectra.

REFERENCES

Arter, D. B. and P. G. Schmidt. 1976. Ring current shielding effects in nucleic acid double helices. *Nucleic Acids Res.* 3:1437.

Dadok, J. and R. F. Sprecher. 1974. Correlation NMR spectroscopy. *J. Magn. Resonance* 13:243.

Geerdes, H. A. M. and C. W. Hilbers. 1977. The iminoproton NMR spectrum of yeast tRNA^{Phe} predicted from crystal coordinates. *Nucleic Acids Res.* 4:207.

Holbrook, S. R., J. L. Sussman, R. W. Warrent, G. M. Church, and S.-H. Kim. 1977. RNA-ligand interactions. 1. Magnesium binding sites in yeast tRNA^{Phe}. *Nucleic Acids Res.* 4:2811.

Jack, A., J. E. Ladner, D. Rhodes, R. S. Brown, and A. Klug. 1977. A crystallographic study of metal-binding to yeast phenylalanine transfer RNA. *J. Mol. Biol.* 111:315.

Johnston, P. D. and A. G. Redfield. 1977. An NMR study of the exchange rates for protons involved in the secondary and tertiary structure of yeast tRNA^{Phe}. *Nucleic Acids Res.* 4:3599.

Kearns, D. R. 1976. High-resolution nuclear magnetic resonance investigations of the structure of tRNA in solution. *Prog. Nucleic Acid Res. Mol. Biol.* 18:91.

Kim. S.-H., J. L. Sussman, F. L. Suddath, G. J. Quigley, A. McPherson, A. H. J. Wang, N. C. Seeman, and A. Rich. 1974. Three-dimensional tertiary structure of yeast phenylalanine transfer RNA. *Science* 185:435.

Redfield, A. G., S. D. Kunz, and E. K. Ralph. 1975. Dynamic range in Fourier transform proton magnetic resonance. *J. Magn. Resonance* 19:114.

Reid, B. R. and R. E. Hurd. 1977. Application of high-resolution nuclear magnetic resonance spectroscopy in the study of base pairing and the solution structure of transfer RNA. *Accts. Chem. Res.* 10:396.

Reid, B. R., N. S. Ribeiro, L. McCollum, J. Abbate, and R. E. Hurd. 1977. High-resolution nuclear magnetic resonance determination of transfer RNA tertiary base pairs in solution. 1. Species containing a small variable loop. *Biochemistry* 16:2086.

Reid, B. R., N. S. Ribeiro, G. Gould, G. Robillard, C. W. Hilbers, and R. G. Shulman. 1975. Tertiary hydrogen bonds in the solution structure of transfer RNA. *Proc. Natl. Acad. Sci.* **72**:2049.

Rich, A. and S.-H. Kim. 1978. The three-dimensional structure of transfer RNA. *Sci. Am.* **238**:52.

Robertus, J. D., J. E. Ladner, J. T. Finch, D. Rhodes, R. S. Brown, B. F. C. Clark, and A. Klug. 1974. Structure of yeast phenylalanine tRNA at 3 Å resolution. *Nature* **250**:546.

Robillard, G. T., C. E. Tarr, F. Vosman, and H. J. C. Berendsen. 1976. Similarity of the crystal and solution structure of yeast tRNA^{Phe}. *Nature* **262**:363.

Sussman, J. C. and S.-H. Kim. 1976. Three-dimensional structure of a transfer RNA in two crystal forms. *Science* **192**:853.

Wong, K. L., P. H. Bolton, and D. R. Kearns. 1975. Tertiary structure in *E. coli* tRNA^{Arg} and tRNA^{Val}. *Biochim. Biophys. Acta* **383**:446.

Proton FT NMR Studies of tRNA Structure and Dynamics

Paul D. Johnston and Alfred G. Redfield*
Departments of Biochemistry and *Physics
and *Rosenstiel Basic Medical Sciences Research Center
Brandeis University
Waltham, Massachusetts 02154

For several years we have been studying tRNA with the primary goal of learning something about its points of flexibility. We have developed a method for measuring the out-exchange rate, over a wide temperature range, of each exchangeable proton whose nuclear magnetic resonance (NMR) line can be resolved or partially resolved. Our hope is that this will reflect the rate at which portions of the structure open, or open partially, to expose these protons to the solvent and that this information will be useful in unraveling modes of tRNA interactions with proteins and other nucleic acids.

More recently we have developed the double-resonance nuclear Overhauser effect (NOE) method for exchangeable resonances in tRNA. If there is a resonance at a frequency f_2 in the spectrum that is at least partially resolved, then, as we will explain, this method allows us to pick out resonances at f_3, f_4, etc., of those one, two, or three protons that are closest to the semiresolved proton (usually within 3.5 Å) that resonates at f_2. This multiplies the information content concerning the original resonance f_2, helping in its identification. When identification is possible for the resonance at f_2, then, generally, the linked resonances at f_3 and f_4 can be assigned and used to extract further conformation and dynamic information. The resonances at f_3 and f_4 are usually in the confused central region of the spectrum and would otherwise be completely useless and unobservable.

We explain the technology behind these experiments and then show how the combination of these measurements has allowed us to describe the kinetic and equilibrium aspects of the thermal stability for different elements of tRNA structure.

TECHNOLOGY

Observation of the Spectrum

In conventional Fourier transform (FT) NMR one excites the entire NMR spectrum with a very short ($\sim 10\ \mu$sec) pulse of frequency f_1. This scheme does not work for the study of exchangeable protons in tRNA because the tRNA must be dissolved in H_2O, not D_2O, and the molarity of water protons is 10^5 greater than that of a single tRNA proton that we would like to observe. For technical reasons, it is impossible to analyze the weak tRNA signal in the presence of the much stronger H_2O signal.

We avoid this problem by using a special observation pulse (Redfield 1976) that has a power spectrum with a complete null at the H_2O frequency. In this way we obtain the sensitivity of FT NMR but, perhaps more important, we retain its property of time resolution. That is, the line intensities of the spectrum reflect the state of the spins at the time of the pulse, permitting us to do time-resolved kinetic studies relatively easily.

Real-time Solvent Exchange

We have been able to exchange the solvent of a tRNA sample from H_2O to D_2O buffer and obtain usable NMR spectra in 2–5 minutes. A tRNA^{Phe} spectrum monitored every 5 minutes (at 15°C) after the rapid solvent replacement (at 0°C) indicated 5 ± 1 low-field imino proton resonances that disappeared in minutes and hours (Johnston et al. 1979). This NMR version of a tritium-exchange experiment (Englander et al. 1972) extends our measurements of exchange to rates below about 0.005 sec^{-1}.

Double-resonance Solvent-exchange Measurements

Exchange more rapid than about 5–10 sec^{-1} can be measured by the NMR technique of saturation recovery (Johnston and Redfield 1977). The observation pulse is preceded by a long (~ 0.1 sec), weak preirradiation pulse at frequency f_2, and there is a delay τ_1 between this pulse and the observation pulse. The frequency f_2 is set equal to the resonance frequency of some specific resonance line in the spectrum. The preirradiation pulse does not produce any signal; its function is to induce spin-flips for those particular spins that produce this resonance line and not others. These spin-flips tend to destroy the slight Boltzmann population difference between spin up and spin down that gives rise to the nuclear magnetization of those spins. Because this same magnetization gives rise to the NMR signal stimulated by the observation pulse, the result of a correctly chosen f_2 pulse, with zero delay τ_1, is that the line at f_2 is missing from the output spectrum. In NMR jargon, these spins have been "saturated."

The sequence (f_2-τ_1-f_1) is repeated for several (usually about eight) values of τ_1. At very long τ_1, the spins "forget" that they were irradiated, and the signal of the irradiated spin species evolves exponentially and is fit to the first-order recovery $A(1 - \exp[-\tau_1 R_{1_{obs}}])$ to obtain the first-order rate $R_{1_{obs}}$.

We then study $R_{1_{obs}}$ vs temperature (a lengthy measurement!) and find that at high temperatures it is strongly temperature-dependent and reflects solvent-exchange kinetics. That is, the protons that resonate at f_2 recover from the preirradiation by being replaced with solvent protons that have not been affected by the preirradiation. Thus, we are measuring chemical exchange directly, just as we would if we could label specific protons with a tracer.

At low temperatures we find that the rate $R_{1_{obs}}$ levels off in its temperature dependence and is apparently being masked by a magnetic process in which the protons resonating at f_2 are recovering by mutual spin-flips with neighboring spins (Johnston and Redfield 1979). That is, one spin flips up and a near neighbor flips down, transferring spin magnetization in a process often called "spin diffusion"; such a process occurs because there is a magnetic dipolar interaction between two protons. It is strongly analogous to the Förster energy transfer between chromophores that is electric-dipole-induced; like Förster transfer, it is strongly distance-dependent, with a rate proportional to the inverse sixth power of the distance. This means that we cannot extend this method for measurement of exchange kinetics down to rates below about 5–10 sec^{-1}.

NOE

Although this Förster-like transfer is a loss for kinetics, it is a gain for spectroscopy! It means that when we preirradiate a resonance at f_2 with zero delay τ_1, protons near the directly saturated proton will be likewise affected, and their spin populations will be partially destroyed. The intensity of their resonances will then be decreased by the preirradiation. Therefore, the differences observed away from f_2 (Fig. 1) are assigned to resonances of protons spatially near the irradiated protons.

Such a phenomenon is called "saturation transfer," and when it takes place via a magnetic transfer it is called NOE. Because the transfer rate is strongly distance-dependent and because the majority of protons around a typical proton in tRNA are more than 3.5 Å removed, the observable NOE is limited to only a few spins within about 3.5 Å of the irradiated spins, almost always on the same base pair.

INTERPRETATION OF LINE-BROADENING, DOUBLE-RESONANCE, AND DIRECT SOLVENT-EXCHANGE MEASUREMENTS

As the temperature is raised, individual lines in the spectrum usually either broaden or disappear without broadening. These changes sometimes occur before any structural change is observable by UV spectroscopy. This and the double-resonance solvent-exchange results can be discussed by means of Equation 1:

$$\text{closed} \underset{k_{cl}}{\overset{k_{op}}{\rightleftarrows}} \text{open} \xrightarrow{k_x} \text{exchanged with solvent} \qquad (1)$$
$$\text{H*} \qquad\quad \text{H*} \qquad\qquad\qquad \text{H}$$

Here the closed state represents a conformer from which exchange with solvent is not possible; the open state denotes a conformer from which exchange is possible, at a rate k_x. H* denotes a proton that has been "labeled" by the presaturating pulse as described above. If a ring nitrogen proton were exposed to solvent in the open state, k_x would be base-catalyzed with a first-order rate constant of about 10^5 sec^{-1} (Johnston and Redfield 1977). The open state does not necessarily imply a melted state as seen by UV methodology (i.e., base unstacking); however, in certain situations this state can be identified with a UV-melted conformation. The assumptions implicit in Equation 1 may not be valid, but it is pointless to consider a more detailed set of mechanisms.

Line-width Increase

If the proton resonance frequency in the open state differs from that in the closed state (this difference is represented as $\Delta\omega = 2\pi\Delta f$) by an amount greater than k_{cl} (this is likely to be true most of the time), then NMR line-shape theory predicts a simple result, namely, $k_{op} = \pi\Delta f$, where Δf (in Hz) is the additional lifetime broadening of a resonance. In practice, this yields a useful measure of k_{op} over a rather narrow range, namely, for k_{op} between about 100 and 500 sec^{-1}, because a given line becomes difficult to resolve once it is appreciably broadened. Thus, a group of lines are observed to disappear, as the temperature is raised, when the opening rate k_{op} of the part of the structure to which they belong increases to about 200 sec^{-1}.

At higher temperatures a line or lines sometimes disappear without broadening. This indicates that a relevant conformational change occurs slowly ($k_{op} \ll 100$ sec^{-1}) but that the equilibrium constant $K_{eq} = k_{op}/k_{cl}$ is not small compared to unity. In either case, we speak of such a disappearance of an NMR line as an "NMR melt." Studies describing line broadening according to Equation 1 have been discussed in more detail elsewhere (Crothers et al. 1974).

Kinetic Experiments

The rate of solvent exchange is given by the following equation:

$$R_1 = k_{op}k_x/(k_{cl} + k_x) \qquad (2)$$

At higher temperatures the quantity R_1 is the observed rate of recovery, R_{1obs}, in a double-resonance (saturation-recovery) experiment from which we have subtracted the estimated contribution due to magnetic processes. We can estimate the latter from the value of R_{1obs} at low temperatures (~20°C), where magnetic relaxation dominates. This correction is about

$5-7\,\text{sec}^{-1}$, and all rates we discuss henceforth are R_1 for which the correction for magnetic relaxation has been made.

NOE AND ASSIGNMENTS

GU Base Pairs

An NOE result that establishes conclusively the existence of the wobble GU base pair in solution is shown in Figure 1. The resonance at the frequency f_2 is marked by an arrow in each difference spectrum, and we have demonstrated for yeast tRNA^{Phe} that the two resonances (peaks L and P) are the only two related by an NOE in the -15 to -11 ppm region.

As these are the only resonances related by an NOE in this chemical-shift range, we now ask which base pairs are capable of simultaneously producing two low-field imino proton resonances. Each of these three tRNAs has only one GU base pair, and the GU wobble is the only base pair that fulfills this criterion (i.e., can produce two resonances with chemical shifts below -11 ppm [Johnston and Redfield 1979]). Thus, we assign one proton resonance at the following positions to the imino protons of the GU wobble: (1) yeast tRNA^{Phe}, -11.8 and -10.4 ppm; (2) *Escherichia coli* tRNA^{fMet}, -12.44 and -11.59 ppm; (3) *E. coli* tRNA^{Val}, -11.95 and -11.44 ppm.

These assignments agree quite well with model studies (Kallenbach et al. 1976; Early et al. 1978) and studies on tRNA (Reid et al. 1975; Robillard et al. 1976b; Kearns 1976). In addition, these results confirm the G4·U69 wobble base-pairing scheme as seen in the crystal structure (Ladner et al. 1975; Rich 1977).

Carbon and Amino Protons

We have observed many other NOEs in yeast tRNA^{Phe}, *E. coli* tRNA$_1^{\text{Val}}$, and *E. coli* tRNA^{fMet} (Johnston and Redfield 1979 and unpubl.). Most NOEs from the AU-rich imino region (-15 to -13 ppm) are to aromatic carbon protons, whereas those from the GC-rich imino region (-13 to -11 ppm) are to amino protons.

Assignments of Yeast tRNA^{Phe} Proton Resonances

The assignments required for the interpretation of the melting experiments come from several different and distinct sources. Direct experimental NOE results can assign G4·U69 (peak L). A combination of NOE and rate measurements, correlations from s^{4}U8·A14 *E. coli* tRNAs (Daniel and Cohn 1975; Reid et al. 1975) that have tertiary fold potentials similar to yeast tRNA^{Phe}, and ring-current calculations (Robillard et al. 1976a; Geerdes and Hilbers 1977) all place U8·A14 in yeast tRNA^{Phe} at peak A or B. Chemical removal of m^{7}G46 indicates that peak E is the imino proton resonance of the m^{7}G46·G22 tertiary interaction (Reid and Hurd,

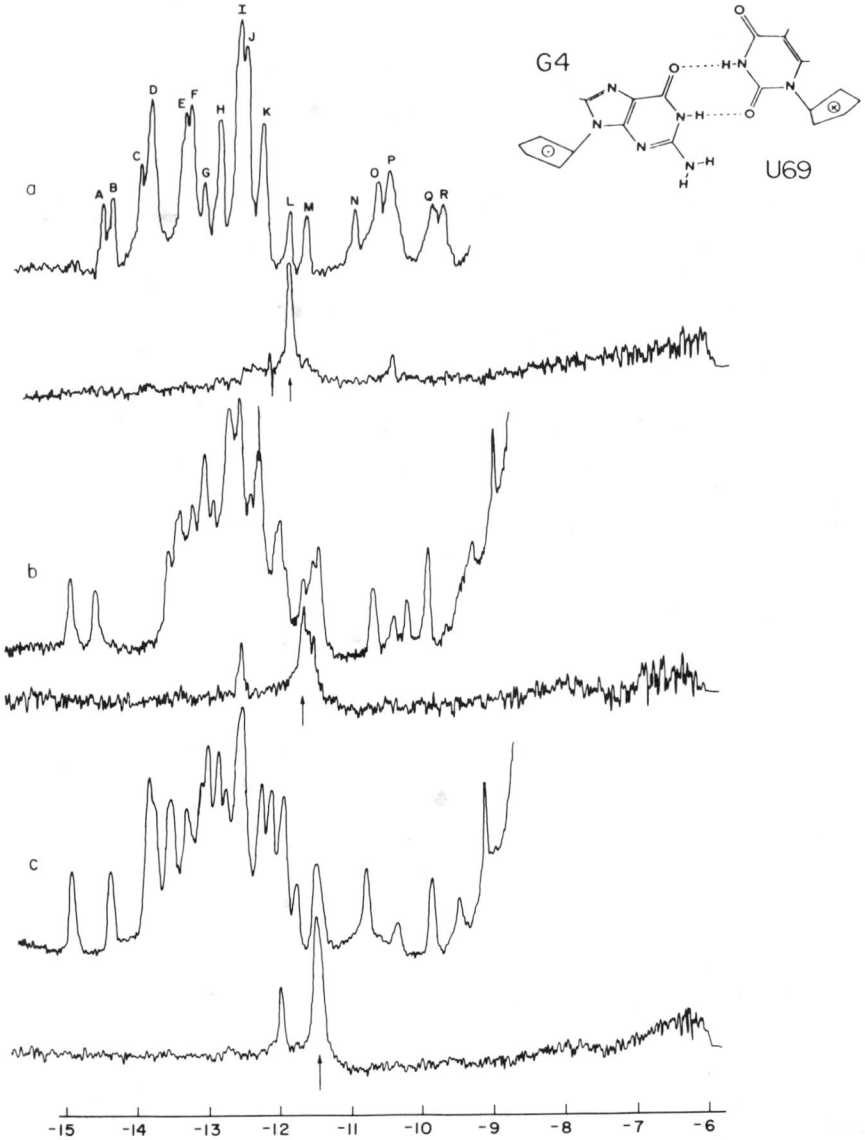

Figure 1 GU wobble NOE difference spectra. The top spectrum in each case was taken in the absence of preirradiation. The bottom difference spectrum of each tRNA is a control spectrum with f_2 set at a blank region minus a spectrum with f_2 set at a resonance of interest. In all difference spectra the arrow indicates the frequency f_2 of the preirradiated resonance. All spectra are 10 mM EDTA, 0.1 M NaCl, and 10 mM sodium cacodylate (pH 7.0) except c, which contained no added NaCl and chemical shifts are relative to the standard DSS (2,2-dimethyl-silapentane-5-sulfonate) reference: (*a*) yeast tRNA[Phe], (*b*) *E. coli* tRNA[fMet], and (*c*) *E. coli* tRNA[Val]. (The inset GU base pair is reprinted, with permission, from Rich and RajBhandary [1976].)

this volume). Paramagnetic metal studies suggest that the tertiary interaction G15·C48 produces an imino proton resonance at peak K (B. Reid, pers. comm.). Three tertiary resonances, G19·C56, T54·m¹A58 (Robillard et al. 1976a; Geerdes and Hilbers 1977), and m$_2^2$G26·A44 (Geerdes and Hilbers 1977), have been assigned to peaks G, A or B, and I–J, respectively, on the basis of ring-current calculations, and we find these assignments best explain our melting data. In addition, there is an unidentified resonance at peak M, which we believe is derived from a tertiary interaction; this conclusion is based upon melting, exchange-rate, and NOE results (P. Johnston, unpubl.). The best candidates are the ring nitrogen proton of ψ55 (ψN3H) or U33 (UN3H) that is hydrogen-bonded to the phosphate oxygens in the U turns (Quigley and Rich 1976) and the hydrogen bond between the ring nitrogen proton of G18 (GN1H) and the carbonyl of ψ55.

The G4·U69 assignment was found to be indispensable for the interpretation of our melting and exchange-rate data. The other resonances of the acceptor stem were assigned by the similarity of the melting and exchange rates to those of G4·U69. At the same time, these assignments are in agreement with ring-current calculations that take into account the first- and second-order ring-current shifts from positions calculated on 11-fold RNA geometry (Arter and Schmidt 1976).

The isolated helical fragments of yeast tRNA^{Phe} provide an estimate of the positions of the secondary imino resonances for the D, TψC, and anticodon stems (Kearns et al. 1973). These fragment spectra can be compared with our spectra taken at intermediate temperatures, where we believe that only these three stems are intact (see below); and they provide a rationale for assigning these peaks that works well for the TψC and anticodon stems but less well for the D stem.

A combination of these methods has provided a consistent set of secondary resonance assignments that fit our data: G1·C72 (I or J), C2·G71 (I or J), G3·C70 (I or J), G4·U69 (L), A5·U68 (D), U6·A67 (C), U7·A66 (F), m²G10·C25 (I), C11·G24 (H), U12·A23 (D), C13·G22 (K), C27·G43 (J), C28·G42 (K), A29·U41 (E), G30·m⁵C40 (I or J), A31·ψ39 (F), m⁵C49·G65 (J), U50·A64 (F), G51·C63 (I), U52·A62 (D), and G53·C61 (H). In the absence of MgCl$_2$ we find one less resonance at peak F, consistent with previous reports (Robillard et al. 1977). There is evidence that A31·ψ39, which resonates at this position, is particularly unstable (Hilbers et al. 1973; Kearns et al. 1973), and the interpretation of our melting data obtained in the absence of Mg^{++} assumes that A31·ψ39 is missing.

NMR MELT OF EDTA-DIALYZED YEAST tRNA^{Phe}

It is generally agreed that in the absence of Mg^{++} the thermal melting of yeast tRNA^{Phe} can be resolved into four or five transitions (Coutts et al.

1975; Privalov and Filimonov 1978). The most unstable element of the structure when Mg^{++} is absent has been assigned to the tertiary structure (Coutts et al. 1975; Privalov and Filimonov 1978). The preferential destablization of tertiary structure in the absence of Mg^{++} has been used to help identify tertiary proton resonances in the NMR spectrum of yeast tRNA^{Phe} (Römer and Varadi 1977). Similar experiments have been extended to include melting of the entire tRNA^{Phe} molecule and, correlated with UV transitions (Coutts et al. 1975), to support ring-current calculations of resonance positions (Robillard et al. 1977).

We shall describe NMR results that extend these measurements and permit identification of specific rates for structural events defined by specific, experimentally assigned proton resonances of yeast tRNA^{Phe} spectra.

A combination of rate measurements and NMR melts were performed at several MgCl$_2$ concentrations. These results have been described in part previously (Johnston and Redfield 1977) but they are enlarged and somewhat reinterpreted here. The Mg^{++}-devoid studies were performed on tRNA^{Phe} that was extensively dialyzed against EDTA. A set of EDTA-dialyzed tRNA^{Phe} spectra are presented in Figure 2 as difference spectra between successive temperatures. The exchange rates for these protons at several temperatures are reported in Table 1.

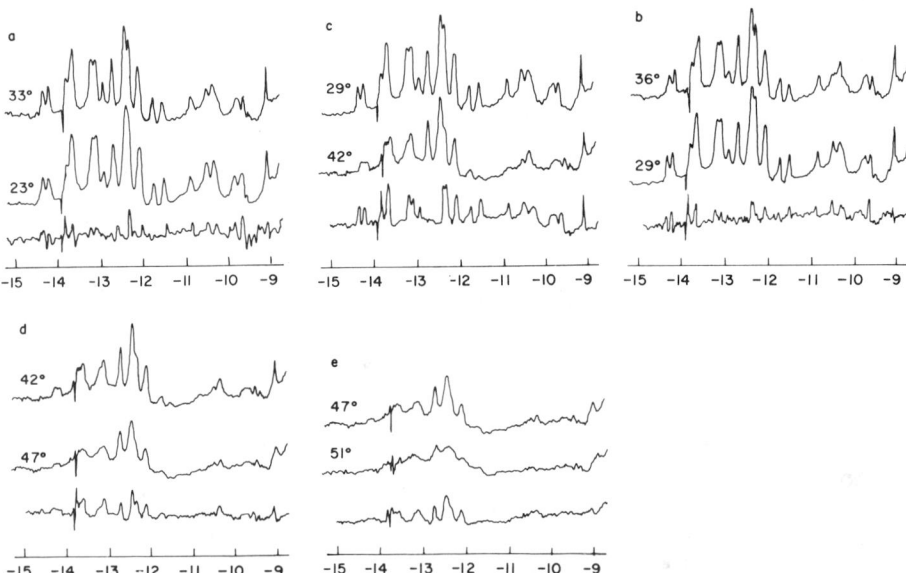

Figure 2 Temperature dependence of the resolved low-field proton resonances of yeast tRNA^{Phe} in 10 mM EDTA, 0.1 M NaCl, and 10 mM sodium cacodylate (pH 7.0). Selected difference spectra: (*a*) 23–33°C, (*b*) 29–36°C. (*c*) 29–42°C, (*d*) 42–47°C, and (*e*) 47–51°C.

Table 1 Temperature dependence of exchange rates for Mg^{++}-free yeast tRNA^{Phe} (10 mM EDTA, 0.1 M NaCl)

Peak	Exchange rates (sec^{-1})			
	30°C	36°C	38°C	42°C
A	10	14	24	120
B	11	15	21	60
C	7	14	—	39
D	17	21	—	34/gone
E	19	32	—	gone
F	14	33	—	7/120
G	+	37	—	95
H	+	7	—	10
I and J	51	5/28	—	10/45
K	21	18	—	7/gone
L	+	5	15	28
M	+	10	—	120

These data are partially those of Johnston and Redfield (1977) and have been corrected for magnetic relaxation as described in the text. The accuracy of these measurements is ±20%. Below 20°C all exchange rates are much less than 5–7 sec^{-1}. (+) Indicates that a measurement was made but that the solvent-exchange rate is too small (<5 sec^{-1}) to separate from magnetic relaxation. Biphasic behavior is indicated by a slash, which separates the two extracted values. (—) Indicates that no measurement was made.

Incipient Opening, 23°C to 33°C: Increased R_1 and No Broadening or Intensity Loss

There are two general characteristics of the melting behavior over the 23–33°C temperature range. The major one we describe as an incipient opening; the other represents a longer-lived local transition.

The latter involves the earliest detected melting (by loss of integrated intensity) as observed in the difference spectrum, peak I or J (Fig. 2a), and involves either m$_2^2$G26·A44 (tertiary) or G1·C72 (terminal fraying of the acceptor stem). Exchange is proceeding at an apparent rate of 51 sec^{-1}, at 30°C (Table 1), for a proton resonating in this region. It is likely that this proton is at a location undergoing a noncooperative local transition that is over by about 36°C.

The difference spectrum shown in Figure 2a clearly shows the lack of intensity loss for any other resonances below 30°C. However, as shown in Table 1 (30°C), there is a marked increase in exchange kinetics for other hydrogen-bonded imino protons of the base pairs T54·m^{1}A58, U8·A14, U6·A67, A5·U68, m^{7}G46·G22, U7·A66, and G15·C48. Such behavior is observed in the absence of a melting transition (Privalov and Filimonov 1978; P. D. Johnston, unpubl.) and can be best described by either of two limiting cases of Equation 2. When $k_{cl} \ll k_x$, the exchange rate R_1 is a direct measure of k_{op}; in addition, k_{op} must be much less than 100 sec^{-1} because of the lack of line-width broadening. The other limiting case exists when $k_{cl} \gg k_x$; then $R_1 = K_{eq}k_x$. For either limiting situation we know that K_{eq} must be much less than unity because of the lack of intensity loss. Specifically for the latter case, K_{eq} must be small such that $K_{eq}(k_{cl}^{-1}\Delta\omega^2 + k_x) \ll 100$ sec^{-1} to satisfy line-width requirements; but it is not possible to decide between the cases. However, it is clear that $k_{op} \ll k_{cl}$ so that the open state exists in a low concentration. Interestingly, the tertiary protons of G19·C56 and peak M are exchanging at a rate of much less than 7 sec^{-1}, which indicates that the outer corner of the molecule (interactions between D loop and TψC loop) is still stable.

Slow Kinetics and Fast Exchange, 33°C to 36°C: Increased R_1 and Intensity Loss but No Significant Broadening

By inspection of the 36°C spectrum (Fig. 2b) it can be seen that very little broadening has taken place between 23°C and 36°C; however, by 36°C, as shown in the difference spectrum (Fig. 2b), there is significant loss of intensity for the resonances of T54·m^{1}A58, U8·A14, A5·U68, U6·A67, m^{7}G46·G22, G15·C48, peak M (unidentified), and part of peak I–J (G1·C72 or m$_2^2$G26·A44). In addition, each of these protons shows increased exchange kinetics (R_1) as presented in Table 1. An intensity loss without significant broadening indicates that K_{eq} is not small compared to unity, which means that k_{op} and k_{cl} are not very different. These changes coincide with UV and calorimetric changes at roughly the same NaCl concentrations (Privalov and Filimonov 1978), and the optical relaxation time (msec range) for this transition (Riesner and Römer 1973; Coutts et al. 1975) suggests that $k_{cl} \ll k_x$, assuming k_x is not much slower than the base-catalyzed exchange rate for the free nucleic acid ($\sim 10^5$ sec^{-1}). Thus, it is likely that R_1 is essentially equal to k_{op}. Furthermore, any additional line broadening ($\pi\Delta f$) should equal k_{op} in this situation, and the value of R_1 is consistent with the lack of significant broadening (i.e., $R_1 = k_{op} <$ 100 sec^{-1}). Therefore, the open state for these base pairs exists for a length of time such that exchange takes place every time the open state occurs, and the open state can be correlated with a UV-melted state. Most of the tertiary peaks showing intensity loss (U8·A14, m$_2^2$G26·A44, and m^{7}G46·G22) involve interactions that are stabilized against phosphate electrostatic repulsions by direct coordination with Mg^{++} and/or spermine (Jack et al. 1977; Holbrook et al. 1977; Quigley et al. 1978), and here they are found to be particularly unstable in the absence of those cations. Accordingly, G15·C48 should show instability when the core tertiary interactions are weakened, as is observed.

Besides these intensity losses, there are three peaks (G, I, and L) reporting an incipient opening reaction (i.e., increased R_1 without intensity loss or broadening) for the interactions G19·C56, G1·C72, C2·G71, G3·C70, and G4·U69. Thus, we see that the outer-corner tertiary interactions and the upper GC-rich part of the acceptor stem show more rapid exchange kinetics, but only after significant weakening of the core tertiary interactions and of the AU-rich region of the acceptor stem.

Completion of Tertiary Structure and Acceptor-stem Melt, 36°C to 42°C: Increased R_1, Broadening, and Intensity Loss

By 42°C (Fig. 2c; Table 1) there is a complete loss of G15·C48, m⁷G46·G22, and A5·U68, and nearly complete loss of the rest of the tertiary structure and acceptor stem (G1·C72, C2·G71, G3·C70, G4·U69, U6·A67, U7·A66, T54·m¹A58, U8·A14, G19·C56, m²₂G26·A44, and peak M). This transition involves 12–13 resonances (including loss of a resonance in peak I–J between 23°C and 33°C), by integrated intensity, which is in agreement with the 14 expected from the tertiary structure and acceptor stem. The resonance missing from the melt can easily be accounted for by the partial resonances present in the 42°C spectrum (e.g., T54·m¹A58, U8·A14, and G4·U69; see Fig. 2, 42°C). Note that the assignment of this transition relies on the known resonance positions of G4·U69, m⁷G46·G22, and U8·A14.

Furthermore, the exchange rates at 42°C (Table 1) really begin to distinguish tertiary resonances from those in the acceptor stem. The tertiary imino protons of T54·m¹A58, U8·A14, G19·C56, and peak M are characterized by exchange rates of 60–120 sec⁻¹ (note that A5·U68, m⁷G46·G22, G15·C48, and m²₂G26·A44 or G1·C72 have melted to such an extent that rates at these positions do not reflect these resonances). The acceptor-stem imino protons, reported by G4·U69, U6·A67, C2·G71, and G3·C70, have exchange rates between 28 and 45 sec⁻¹. The exchange rate for U7·A66 of 120 sec⁻¹ is anomalous and could reflect terminal fraying as observed for the terminal AU base pair of the acceptor stem of *E. coli* tRNA^Phe (Reid and Hurd, this volume). There is also observable line-width broadening at this temperature for the resonances involved in this transition.

The rate behavior, both R_1 and $\pi\Delta f$, can best be described by the $k_{cl} \ll k_x$ limiting case of Equations 1 and 2; in other words, both R_1 and $\pi\Delta f$ are measuring the same rate, namely, the opening rate k_{op}. This conclusion is based on several facts. First, we estimate the additional line broadening ($\pi\Delta f$) to be less than 100 sec⁻¹ for resonances A, B, and L. Therefore, the values of R_1 and $\pi\Delta f$ are in good agreement. Second, this melting transition can be correlated to the first and part of the second melting transitions (Privalov and Filimonov 1978); furthermore, the optical relaxation times for these two UV transitions are in the millisecond range (Riesner and Römer 1973; Coutts et al. 1975), which suggests that $k_{cl} \ll k_x$ (assuming 10^5 sec^{-1} is a good estimate of k_x).

D, TψC, and Anticodon Stems, 42°C to 51°C

Over the temperature range 42°C to 51°C the melt becomes more difficult to interpret because we do not have many double-resonance exchange rates to help define the processes. However, at 42°C we have the D, TψC, and anticodon stems intact. By comparing our spectra at 42°C and above with spectra of the isolated fragments of these stems (Kearns et al. 1973) and using the ring-current calculation method of Arter and Schmidt (1976), we have been able to define these transitions and extract some kinetic information from the line widths and double-resonance exchange rates (at 42°C).

There are four positions at 42°C (Table 1) where R_1 is probably reporting kinetics of the D stem, namely, U12·A23, m^{2}G10·C25, C13·G22, and possibly C11·G24. Note that the largest intensity loss is for U12·A23, which also shows the fastest rate, and that the internal base pair C11·G24 shows only slight intensity loss. The optical relaxation rate for the third transition is the slowest, 2–59 sec^{-1} (Coutts et al. 1975). The NMR rates generally agree with this: U12·A23, 34 sec^{-1}; C11·G24, 9 sec^{-1}; m^{2}G10·C25, 10 sec^{-1}; and C13·G22, 7 sec^{-1}. Accordingly, R_1 is likely equal to k_{op}. Finally, these data suggest that the D-stem melt is responsible for the third melting transition, and the difference spectra suggest that the anticodon-stem melt is the fourth transition.

The difference spectrum (Fig. 2d) shows the partial melt of the D and anticodon stems. An interesting point to note is that both of these stems (in fact all stems left at this point) are GC-rich by almost 3 to 1; therefore, the intensity loss indicates a proportionally larger fraction of AU pairs melting than GC pairs. Inspection of the line widths for the 47°C spectrum (Fig. 2) confirms this observation; there is considerably more broadening in the AU-rich region of the spectrum than in the GC-rich region. The 47°C spectrum, which should be predominantly the TψC stem, is fit well by the spectrum of the TψC-stem fragment and the GC resonances of the anticodon-stem fragment (Kearns et al. 1973).

The last difference spectrum (Fig. 2e) shows the latter part of the D- and anticodon-stem melts. In all, we have a loss between 42°C and 51°C of about 10 resonances; we would expect a loss of 14 resonances if all three stems were melting over this range. Thus, about 70% of the TψC stem is still intact. At this temperature (51°C) our spectrum is fit well by the TψC-stem fragment (Kearns et al. 1973), with a partial contribution from C28·G42. This behavior is consistent with optical relaxation and equilibrium studies. In particular, the last calorimetric melting transition extends out to 80°C under approximately our salt conditions (Privalov and

Filimonov 1978), but the relaxation times for the last melting transitions are in the microsecond range (Riesner and Römer 1973). This would predict significant broadening and disappearance of an NMR line before much UV melting had taken place; this is observed at 51°C and above. Note that at this temperature AU base pairs are again much broader than the GC base pairs.

INTERMEDIATE MgCl$_2$ CONCENTRATIONS

We can correlate these conclusions with results at higher MgCl$_2$ concentrations. At 45°C and intermediate MgCl$_2$ concentrations (10–20 Mg^{++}/tRNA^{Phe}), only three protons, m^{7}G46·G22 (161 sec^{-1}), G4·U69 (16 sec^{-1}), and A31·ψ39 (51 sec^{-1}, 40°C), show relatively fast exchange. No melting, as judged by intensity loss, is observed for this sample up to 40°C; by 50°C there is a loss of nearly one proton at peak F, presumably A31·ψ39. From 50°C to 61°C there is a marked broadening and loss of intensity of G1·C72, C2·G71, G3·C70, G4·U69, A5·U68, U6·A67, and U7·A66 of the acceptor stem and the two tertiary interactions m^{7}G46·G22 and G15·C48 between the extra loop and the D stem and loop. Thus, we see that the two points of structure involved in the initiation of melting in zero MgCl$_2$ are still points of increased exchange and susceptibility to thermal unfolding in the presence of Mg^{++} (P. D. Johnston and A. G. Redfield, in prep.).

CONCLUSIONS

NOE measurements have provided direct evidence for the existence of the GU wobble base pair in solution and for the subsequent assignment of the resonances for this base pair in yeast tRNA^{Phe}, *E. coli* tRNA$_1^{Val}$, and *E. coli* tRNA^{fMet}. This assignment, the assignment of m^{7}G46·C22 by Reid and Hurd (this volume), and the measurement of the out-exchange rates of the -15 to -11 ppm proton resonances have allowed a detailed description of the transitions involved in the thermal unfolding of yeast tRNA^{Phe}. The melt, in the absence of Mg^{++}, is initiated at the AU base pairs of the acceptor stem and tertiary structure, along with the interactions between the extra loop and the D stem and loop. This is followed by a general and nearly concerted melt of all the tertiary structure and the rest of the acceptor stem. Following this is the melt of the D stem, the anticodon stem, and the TψC stem, respectively. The greater stability of GC base pairs as compared with AU base pairs seems to be a general feature during the melting process. It is possible that for stability reasons G19·C56 is a normal Watson-Crick base pair (instead of AU-type or other nonstandard base pairs). Note that G19·C56 does not start losing intensity until the latter part of the tertiary structure melts.

In the presence of intermediate Mg^{++}, the acceptor stem, A31·ψ39, and

$m^7G46 \cdot G22$ are still the most unstable elements of the structure. It is likely that the instability of the acceptor stem is, in part, reflecting the slight distortion of the acceptor helix to accommodate the $G4 \cdot U69$ base pair (Rich 1977). This also is reflected in the exchange rate of $G4 \cdot U69$ in the absence of melting.

It is interesting to note that the events described here for the early melting transition (29–42°C) for yeast $tRNA^{Phe}$ are similar to those inferred from chemical reactivity studies (Rhodes 1977). These studies suggested that $G19 \cdot C56$ and $G18 \cdot \psi 55$ are the most stable tertiary interactions. In addition, $G4 \cdot U69$, $A5 \cdot U68$, $U6 \cdot A67$, and $U7 \cdot A66$ appear more unstable than $G1 \cdot C72$, $C2 \cdot G71$, and $G3 \cdot C70$; $G4 \cdot U69$ also appears less stable than the tertiary interactions in the presence of Mg^{++} (Rhodes 1977). Thus, it appears that chemical modification is sensitive to the same early events to which NMR exchange rates are sensitive.

ACKNOWLEDGMENTS

This work was supported by U.S. Public Health Service grant GM20168, by National Institutes of Health grant 5-T01-GM00212-19, and by the Research Corporation. The studies on *E. coli* $tRNA^{fMet}$ and *E. coli* $tRNA_1^{Val}$ were done in collaboration with Professor B. R. Reid, and the real-time D_2O exchange experiments were done in collaboration with Nara Figueroa. This is publication number 1223 from the Department of Biochemistry, Brandeis University.

REFERENCES

Arter, D. B. and P. G. Schmidt. 1976. Ring current shielding effects in nucleic acid double helices. *Nucleic Acids Res.* **3**:1437.

Crothers, D. M., P. E. Cole, C. W. Hilbers, and R. G. Shulman. 1974. The molecular mechanism of thermal unfolding of *Escherichia coli* formylmethionine transfer RNA. *J. Mol. Biol.* **87**:63.

Coutts, S. M., D. Riesner, R. Römer, C. R. Rabl, and G. Maass. 1975. Kinetics of conformational changes in $tRNA^{Phe}$ (yeast) as studied by the fluorescence of the Y-base and formycin substituted for the 3′-terminal adenine. *Biophys. Chem.* **3**:275.

Daniel, W. E., Jr. and M. Cohn. 1975. Proton nuclear magnetic resonance of spin-labeled *Escherichia coli* $tRNA_f^{met}$. *Proc. Natl. Acad. Sci.* **72**:2582.

Early, T. A., J. Olmsted III, D. R. Kearns, and A. G. Lezius. 1978. Base pairing structure in the poly d(G-T) double helix: Wobble base pairs. *Nucleic Acids Res.* **5**:1955.

Englander, J. J., N. R. Kallenbach, and S. W. Englander. 1972. Hydrogen exchange study of some polynucleotides and transfer RNA. *J. Mol. Biol.* **63**:153.

Geerdes, H. A. M. and C. W. Hilbers. 1977. The amino-proton NMR spectrum of yeast tRNA^{Phe} predicted from crystal coordinates. *Nucleic Acids Res.* **4:**207.

Hilbers, C. W., R. G. Shulman, and S. H. Kim. 1973. High resolution NMR study of the melting of tRNA$^{Phe}_{yeast}$. *Biochem. Biophys. Res. Commun.* **55:**953.

Holbrook, S. R., J. L. Sussman, R. W. Warrant, G. M. Church, and S. H. Kim. 1977. RNA-ligand interactions. 1. Magnesium binding sites in yeast tRNA^{Phe}. *Nucleic Acids Res.* **4:**2811.

Jack, A., J. E. Ladner, D. Rhodes, R. S. Brown, and A. Klug. 1977. A crystallographic study of metal-binding to yeast phenylalanine transfer RNA. *J. Mol. Biol.* **111:**315.

Johnston, P. D. and A. G. Redfield. 1977. An NMR study of the exchange rates for protons involved in the secondary and tertiary structure of yeast tRNA^{Phe}. *Nucleic Acids Res.* **4:**3599.

———. 1979. Pulsed FTNMR double resonance studies of yeast tRNA^{Phe}: Specific nuclear Overhauser effects and reinterpretation of low temperature relaxation data. *Nucleic Acids Res.* **5:**3913.

Johnston, P. D., N. Figueroa, and A. G. Redfield. 1979. Real-time solvent exchange studies of the imino and amino protons of yeast phenylalanine transfer RNA by Fourier transform NMR. *Proc. Natl. Acad.* (in press).

Kallenbach, N. R., W. E. Daniel, Jr., and M. A. Kaminker. 1976. Nuclear magnetic resonance study of hydrogen-bonded ring protons in oligonucleotide helices involving classical and nonclassical base pairs. *Biochemistry* **15:**1218.

Kearns, D. R. 1976. High-resolution nuclear magnetic resonance investigations of the structure of tRNA in solution. *Prog. Nucleic Acid Res. Mol. Biol.* **18:**91.

Kearns, D. R., D. R. Lightfoot, K. L. Wong, Y. P. Wong, B. R. Reid, L. Cary, and R. G. Shulman. 1973. High-resolution NMR investigation of base pairing structure of transfer RNA. *Ann. N.Y. Acad. Sci.* **222:**324.

Ladner, J. E., A. Jack, J. D. Robertus, R. S. Brown, D. Rhodes, B. F. C. Clark, and A. Klug. 1975. Structure of yeast phenylalanine transfer RNA at 2.5 Å resolution. *Proc. Natl. Acad. Sci.* **72:**4414.

Privalov, P. L. and V. V. Filimonov. 1978. Thermodynamic analysis of transfer RNA unfolding. *J. Mol. Biol.* **122:**447.

Quigley, G. J. and A. Rich. 1976. Structural domains of transfer RNA molecules. *Science* **194:**796.

Quigley, G. J., M. M. Teeter, and A. Rich. 1978. Structural analysis of spermine and magnesium ion binding to yeast phenylalanine transfer RNA. *Proc. Natl. Acad. Sci.* **75:**64.

Redfield, A. G. 1976. How to build a Fourier transform NMR spectrometer for biochemical applications. In *NMR: Basic principles and progress* (ed. P. Diehl et al.), vol. 13, p. 137. Springer-Verlag, Berlin.

Reid, B. R., N. S. Ribeiro, G. Gould, G. Robillard, C. W. Hilbers, and R. G. Shulman. 1975. Tertiary hydrogen bonds in the solution structure of transfer RNA. *Proc. Natl. Acad. Sci.* **72:**2049.

Rhodes, D. 1977. Initial stages of the thermal unfolding of yeast phenylalanine transfer RNA as studied by chemical modification: The effect of magnesium. *Eur. J. Biochem.* **81:**91.

Rich, A. 1977. Three-dimensional structure and biological function of transfer RNA. *Accts. Chem. Res.* **10**:388.

Rich, A. and U. L. RajBhandary. 1976. Transfer RNA: Molecular structure, sequence, and properties. *Annu. Rev. Biochem.* **45**:805.

Riesner, D. and R. Römer. 1973. Thermodynamics and kinetics of conformation transitions in oligonucleotides and tRNA. In *Physico-chemical properties of nucleic acids* (ed. J. Duchesne), vol. 2, p. 237. Academic Press, London.

Robillard, G. T., C. E. Tarr, F. Vosman, and H. J. C. Berendsen. 1976a. Similarity of the crystal and solution structure of yeast tRNA^{Phe}. *Nature* **262**:363.

Robillard, G. T., C. E. Tarr, F. Vosman, and B. R. Reid. 1977. A nuclear magnetic resonance study of secondary and tertiary structure in yeast tRNA^{Phe}. *Biochemistry* **16**:5261.

Robillard, G. T., C. W. Hilbers, B. R. Reid, J. Gangloff, G. Dirheimer, and R. G. Shulman. 1976b. A study of secondary and tertiary solution structure of yeast tRNA^{asp} by nuclear magnetic resonance. Assignment of G·U ring NH and hydrogen-bonded base pair proton resonances. *Biochemistry* **15**:1883.

Römer, R. and V. Varadi. 1977. Hydrogen-bonded protons in the tertiary structure of yeast tRNA^{Phe} in solution. *Proc. Natl. Acad. Sci.* **74**:1561.

Effects of Aminoacylation and Solution Conditions on the Structure of tRNA

Russell O. Potts,*† Chun-Chen Wang,‡§
David C. Fritzinger,* Norman C. Ford, Jr.,‡
and Maurille J. Fournier*
Departments of *Biochemistry and ‡Physics
University of Massachusetts
Amherst, Massachusetts 01003

The bulk of the work performed to date on the solution structure of tRNA has quite naturally been directed toward the elucidation of the tertiary structure and the identification of functionally important structural features. With the recent elucidation of the crystal structure of tRNA (Quigley et al. 1975; Ladner et al. 1975) and the recognition that that structure is also obtained in solution (Chen et al. 1975; Robillard et al. 1976; and see results from less direct experiments reviewed by Rich and RajBhandary 1976), it is now possible to approach many other outstanding questions of tRNA structure in ways previously not possible. Of special and long-standing interest are questions related to (1) the effects of solution conditions on tertiary structure, (2) the tertiary structures of aminoacyl- and peptidyl-tRNAs and precursor tRNA, and (3) the molecular nature of the protein-nucleic acid interactions in which tRNA participates in the course of being synthesized, utilized, and degraded. Because at least some, and perhaps all, of these phenomena are likely to involve dynamic changes in tertiary structure, it will be important to study these questions with biophysical and chemical techniques that are sensitive to different features of the solution structure and that allow analyses to be conducted under physiological solution conditions.

Here we review results from our investigations into the effects of aminoacylation and solution conditions on tRNA structure, utilizing the relatively new technique of laser light scattering. This technique is especially well suited for the assessment of structural transitions of biomolecules in solution. (For reviews on the application of this technique to biochemical problems, see Ford [1972] and Schurr [1977].)

Present addresses: †Department of Chemistry, Yale University, New Haven, Connecticut 06520; §Department of Chemistry, Stanford University, Palo Alto, California 94305.

Analysis by Laser Light Scattering

Changes in the translational diffusion coefficient, $D_{20,w}$, can be explained in terms of the molecular mass, size, shape, and electrostatic properties of the scatterers (see Olson et al. 1976). The dependence of $D_{20,w}$ upon the concentration of scatters yields information about these parameters through the following equation:

$$D_{20,w} = D^0_{20,w}\left[1+\left(0.45\bar{v}_2 + \frac{10^3\bar{v}_1 Z^2}{2\mu M}\right)C\right] \quad (1)$$

$D^0_{20,w}$ is the diffusion constant in water at 20°C, extrapolated to zero concentration; $\bar{v}_1$ and $\bar{v}_2$ are the partial specific volumes of solvent and tRNA, respectively; Z is the average total macroion charge; μ is the ionic strength; M is the molecular mass; and C is the concentration of tRNA in mg/ml. Calculations have shown that excluded-volume effects for a charged molecule like tRNA are negligible when compared to the electrostatic effects. Hence, a plot of $D_{20,w}$ vs C at constant ionic strength will yield a straight line of intercept $D^0_{20,w}$ and slope proportional to the square of the average charge.

tRNA Structure and Aminoacylation

Over the past two decades there have been more than two dozen reports of biochemical and biophysical studies on the effects of aminoacylation on tRNA structure; many of these reports are cited by Potts et al. (1977). In approximately 50% of these investigations no effect was observed; in the remainder, however, significant structural changes were detected. Upon first inspection, many of the results appear to be mutually contradictory. However, close examination reveals the occurrence of experimental differences that could provide the basis for the apparent discrepancies.

First, a number of negative results were obtained with techniques that are primarily sensitive to changes in secondary structure only. Such findings do not preclude the possibility that changes may, in fact, occur only at the tertiary level. Second, quantitative differences in the solution conditions used could affect or mitigate structural transitions associated with aminoacylation. This view is supported by previous findings from our laboratories, which reveal that conformational changes can occur for non-aminoacylated tRNA under near-physiological solution conditions over rather narrow ranges of magnesium concentration and ionic strength (Olson et al. 1976; and see below). In an effort to gain additional insight into this important matter we have used the laser light-scattering technique to study the effect of aminoacylation on the structure and electrostatic charge of yeast tRNA^{Phe} in a variety of solution conditions. Some of the earliest results have already been described (Potts et al. 1977).

Solution Effects on tRNA Structure

Although the results of numerous investigations have shown striking similarities between the solution structures of various tRNAs and the crystal structure of yeast tRNA^{Phe}, it is becoming clear that tRNA can also assume different conformational states under near-physiological solution conditions. Using laser light scattering, we have been able to detect and partially describe changes in the tertiary structures of bulk *Escherichia coli* tRNA and yeast tRNA^{Phe} that result from changes in the Mg^{++} concentration and ionic strength (Olson et al. 1976). Of special interest was the observation that these tRNAs became more compact when the magnesium concentration was reduced from 10 to 1 mM at an ionic strength of 0.1 M. To better understand the molecular bases and possible functional relevance of this change, this work has been extended by analyzing a variety of single tRNAs under similar solution conditions. The results of these experiments are also presented here.

The preparation of tRNAs have been or will be described elsewhere (Olson et al. 1976; Potts et al. 1977 and in prep.).

EFFECTS OF AMINOACYLATION ON tRNA STRUCTURE

In our experiments, the structures of aminoacylated and nonacylated tRNA are compared by continuous collection of light-scattering data from a sample of aminoacyl-tRNA undergoing spontaneous deacylation. Changes in the light-scattering properties are then correlated with the extent of aminoacylation. The data in Figure 1A show a large increase in $D_{20,w}$ for tRNA in 10 mM Mg^{++}, whereas no changes occur in the translational diffusion constant for tRNA deacylating in 1 mM Mg^{++} (Fig. 1B), both at 20°C in 10 mM Tris-Cl (pH 7.2) and 0.1 M NaCl. When the data are linearly extrapolated to pure, completely aminoacylated tRNA (1800 pmoles/A$_{260}$), the 10-mM data show an 18 ± 3% increase in $D_{20,w}$; the 1-mM data, however, show no change (0.4 ± 0.6%). In each case, the final value of $D_{20,w}$ obtained upon complete deacylation was nearly identical to that obtained for nonacylated tRNA under the same conditions.

Light-scattering theory suggests several molecular explanations consistent with the results shown in Figure 1A. These hypotheses suggest that the decreased value of $D_{20,w}$ upon aminoacylation results from conformational and/or electrostatic changes. Other, less likely explanations are that $D_{20,w}$ changes result from (1) the added mass and volume of the amino acid or (2) aggregation of aminoacylated tRNA.

In the absence of any changes in $D_{20,w}$ upon deacylation in 1 mM Mg^{++}, it is very unlikely that the changes observed are due simply to the added mass and volume of the phenylalanine itself. Similarly, the possibility that

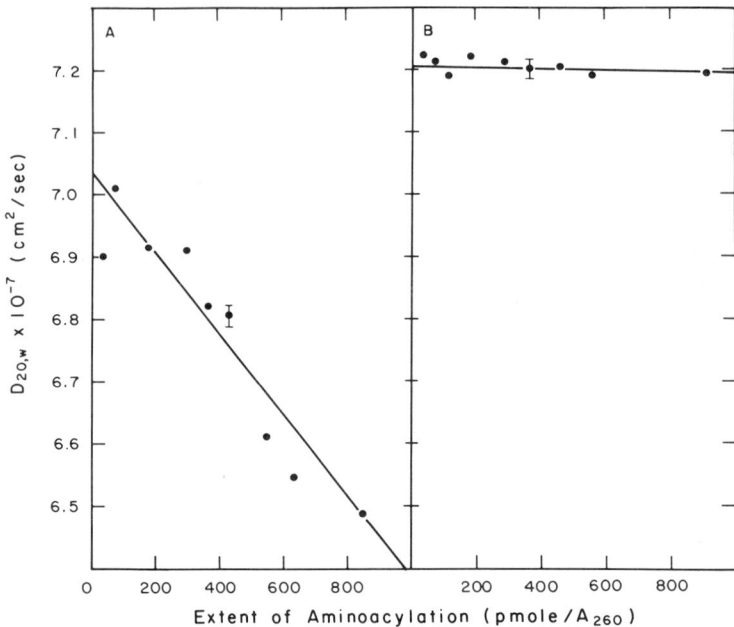

Figure 1 The $D_{20,w}$ of yeast tRNA^{Phe} as a function of the extent of aminoacylation. Changes in $D_{20,w}$ and the extent of aminoacylation were monitored as [^{3}H]phenylalanyl-tRNA^{Phe} deacylated spontaneously. Each point represents the average of at least six determinations of $D_{20,w}$ taken within a 15-min period, with the error bars representing ±1 s.d. about the point. The data were fit to straight lines by a least-squares technique. Measurements were made at 20°C for samples in 0.1 M Tris-Cl (pH 7.2), 0.1 M NaCl, and MgCl$_2$ at the concentrations indicated. The magnesium ion concentration, tRNA concentration, and halftime of deacylation were 10 mM, 1.2 mg/ml, and 75 min (*A*) or 1 mM, 1.5 mg/ml, and 55 min (*B*). The tRNA used had an acceptor activity in excess of 1100 pmoles/A$_{260}$.

the changes in diffusivity that accompany aminoacylation result from aggregation can be discounted. Measurements made of the changes in the intensity of the scattered light, a property very sensitive to aggregate formation, showed only slight variance over the course of the experiment.

To determine the relative contribution of conformational and electrostatic effects to the changes in $D_{20,w}$, the tRNA concentration dependence of the effect was determined. When plotted as relative change in $D_{20,w}$ (extrapolated to 1800 pmoles/A$_{260}$) vs tRNA concentration, the results shown in Figure 2 are obtained. The data with 1 mM Mg^{++} show no dependence on tRNA concentration, indicating that no electrostatic or conformational changes occur for aminoacyl-tRNA in that solution condition. For experiments done in 10 mM Mg^{++}, however, the $D_{20,w}$ values show a distinct tRNA concentration dependence, with an extrapolated

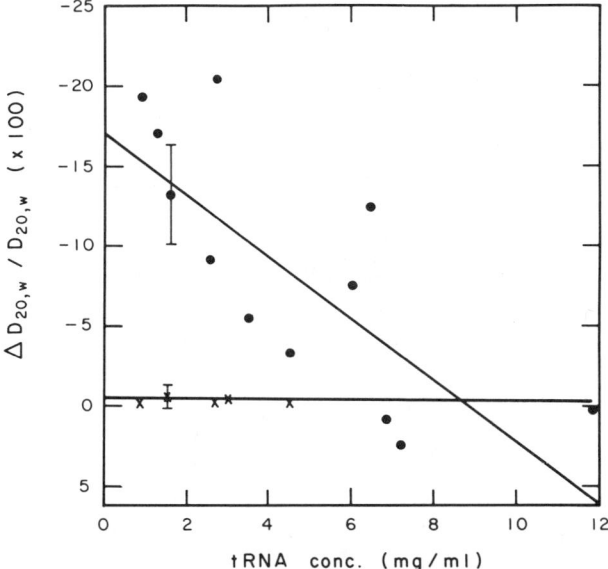

Figure 2 $\Delta D_{20,w}/D_{20,w}$ upon aminoacylation as a function of tRNA concentration. Experiments similar to those described in Fig. 1 were repeated at a variety of tRNA concentrations and the relative change in $D_{20,w}$ determined by a linear extrapolation to a starting point of pure, fully aminoacylated tRNA^{Phe}. The error bars signify ±1 S.D. determined from the linear extrapolation of data from $D_{20,w}$ vs the extent of aminoacylation at each tRNA concentration. The data at each Mg^{++} concentration were fit to a straight line by a least-squares technique. The experimental conditions are those of Fig. 1 with either 10 mM Mg^{++} (●) or 1 mM Mg^{++} (x).

increase in $D_{20,w}$ of 17 ± 2.7% and a slope of −1.9 ± 0.5%/(mg/ml). As indicated above, the $D_{20,w}$ values after complete deacylation were nearly identical to those obtained with nonacylated tRNA^{Phe} under the same conditions.

Although it seems most likely that the decrease in the relative change of $D_{20,w}$ with increasing tRNA concentration is due to electrostatic charge differences upon aminoacylation, it is also possible that the data reflect association of tRNA molecules (aminoacylated and/or nonaminoacylated tRNA). As before, the possibility of aggregation could be discounted because the diffusion constants and scattered intensity values obtained for both acylated and nonacylated tRNA in 1 and 10 mM Mg^{++} were consistent with the values expected for a monodisperse, monomeric sample. Thus, the changes in $\Delta D_{20,w}/D_{20,w}$ with tRNA concentration reflect electrostatic differences upon aminoacylation.

To quantitate conformational and electrostatic changes upon aminoacylation, it is useful to replot the data of Figure 2 as $D_{20,w}$ vs tRNA concentration. As indicated above, the slope in such a plot is proportional to the square of the average charge, whereas the intercept ($D^0_{20,w}$) is related to molecular conformation. When the diffusion constants for completely aminoacylated tRNA (extrapolated to 1800 pmoles/A_{260}) and deacylated tRNA (extrapolated to 0 pmole/A_{260}) are plotted vs tRNA concentration for 1 and 10 mM Mg^{++}, the following information was obtained. In 1 mM Mg^{++}, both aminoacylated tRNA and nonacylated tRNA exhibit the same $D^0_{20,w}$ ($[7.39 \pm 0.08] \times 10^{-7}$ cm²/sec) and slope ($[0.14 \pm 0.05] \times 10^{-7}$ [cm²/sec]/[mg/ml]). Using the value of the slope and Equation 1, the average molecular charge was calculated to be $11 \pm 4e^-$ for both, in good agreement with the value of $10e^-$ previously determined by light scattering for bulk *E. coli* tRNA and yeast tRNA^{Phe} under the same conditions (Olson et al. 1976).

By contrast, the aminoacylated and nonacylated species have different slopes and intercepts in 10 mM Mg^{++}. Aminoacylation results in a decrease in $D^0_{20,w}$ ($[6.84 \pm 0.10] - [5.8 \pm 0.2] \times 10^{-7}$ cm²/sec) and an increase in the slope ($[0.10 \pm 0.02]-[0.22 \pm 0.06] \times 10^{-7}$ [cm²/sec]/[mg/ml]). From the slope values, the average charge of the nonacylated species was estimated to be $10 \pm 2e^-$; for aminoacylated tRNA^{Phe} in 10 mM Mg^{++} the negative charge increases $5 \pm 2e^-$. The value obtained for nonacylated tRNA in 10 mM Mg^{++} is in close agreement with the value of $8e^-$ obtained by others in our laboratory for *E. coli* bulk tRNA under the same conditions (Olson et al. 1976). Thus, aminoacylation of tRNA^{Phe} results in no changes in charge or $D^0_{20,w}$ in 1 mM Mg^{++}, but an increase in negative charge and a decrease in $D^0_{20,w}$ are observed in 10 mM Mg^{++}.

Insight into the role of Mg^{++} ions in the conformational-electrostatic changes upon aminoacylation was obtained from the tRNA concentration dependence of $D_{20,w}$ changes at several Mg^{++} concentrations, all other conditions remaining the same. Data for each Mg^{++} concentration were fit to a straight line using n data points and the final results plotted as relative change in $D^0_{20,w}$ vs Mg^{++} concentration. This summary plot is shown in Figure 3. The data for 0.5 ($n = 3$) and 1.0 ($n = 5$) mM Mg^{++} give the same line, with no increase in $D^0_{20,w}$. The data for 2 ($n = 3$) and 3 ($n = 4$) mM Mg^{++} show increases in $D^0_{20,w}$ upon deacylation of $8.8 \pm 2.3\%$ and $11.7 \pm 2.7\%$, respectively. The data for 5 ($n = 6$) and 10 ($n = 12$) mM Mg^{++} produce the same line, with an increase in $D^0_{20,w}$ of $17 \pm 2.7\%$. The nature of the tRNA concentration dependence of scattered intensity changes observed with deacylation indicates that association is not responsible for the changes in $D^0_{20,w}$ at any of the Mg^{++} concentrations tested.

The data indicate cooperativity consistent with a two-state transition model, involving the binding of Mg^{++} ions. From a Hill analysis of the data of Figure 3 (in which the logarithm of Mg^{++} concentration is plotted against the logarithm of the ratio of the two putative conformers) it is

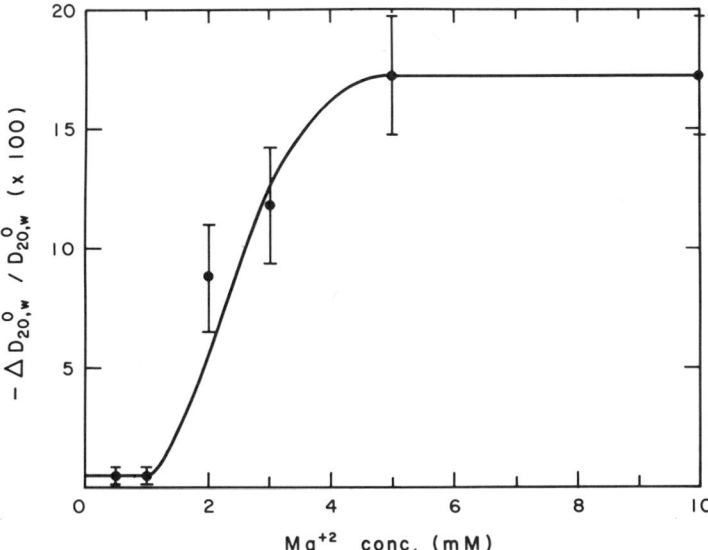

Figure 3 $\Delta D^0_{20,w}/D^0_{20,w}$ upon aminoacylation of tRNA^{Phe} as a function of Mg^{++} concentration. Each point represents the linear extrapolation to 0 tRNA concentration of data similar to those in Fig. 2, at various Mg^{++} concentrations. The error bars signify ±1 S.D. obtained by the least-squares technique. The experimental conditions are those of Fig. 2, with the exception that Mg^{++} concentrations were varied from 0.5 to 10 mM.

estimated that the transition involves the binding of 2–4 Mg^{++} ions. This estimate is somewhat imprecise because of errors in the determination of the percent changes in $D^0_{20,w}$, as well as inaccuracies inherent in calculations of the relative amounts of the individual conformers.

In summary, the changes observed in the diffusion constant ($D^0_{20,w}$) and in the average charge of yeast tRNA^{Phe} associated with aminoacylation depend on the Mg^{++} concentration and are not related to association/dissociation of the molecules. At low Mg^{++} concentrations (0.5 and 1.0 mM) no changes are observed; at higher Mg^{++} concentrations, however, an increase in the negative charge and a decrease in $D^0_{20,w}$ occur. Furthermore, the changes are dependent upon the Mg^{++} concentration in a cooperative manner, consistent with a two-state transition involving the binding of 2–4 additional Mg^{++} ions to the aminoacylated form.

EFFECTS OF SOLUTION CONDITIONS ON NONACYLATED tRNA

In work described previously (Olson et al. 1976) it was determined that nonaminoacylated *E. coli* tRNA and yeast tRNA^{Phe} can also undergo marked changes in diffusivity with seemingly small changes in solution conditions. These earlier results are shown in summary form in the upper panel in Figure 4. Here, the effects on $D^0_{20,w}$ of Mg^{++} and ionic strength are depicted.

For bulk *E. coli* tRNA dialyzed against 10 mM Mg^{++} there is a smooth transition to a slower-diffusing conformer as the ionic strength is reduced. However, an anomalous transition to a faster-diffusing conformer is

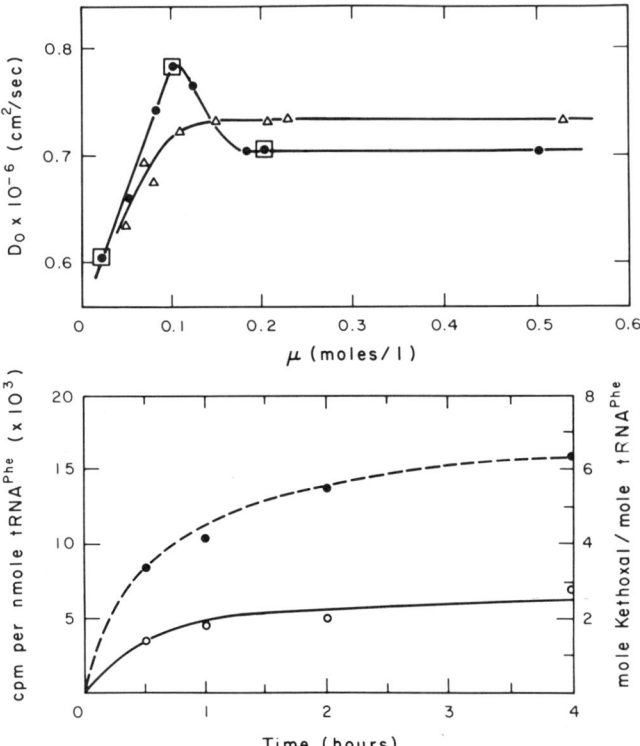

Figure 4 Analysis of solution effects on nonacylated tRNA by light scattering and chemical modification. (*Top*) Effect of ionic strength and magnesium on the diffusion constants of bulk *E. coli* tRNA^{Phe} and yeast tRNA^{Phe}. (Data from Olson et al. 1976.) $D^0_{20,w}$ values were obtained by extrapolation of $D_{20,w}$ vs tRNA-concentration data to 0 tRNA concentration. tRNA samples were dialyzed against 0.1 M Tris-Cl (pH 7.2) containing $MgCl_2$ and NaCl at the concentrations indicated; the ionic strength was adjusted with NaCl. (●) Unfractionated *E. coli* tRNA in 1 mM Mg^{++}; (△) unfractionated *E. coli* tRNA in 10 mM Mg^{++}; (▣) measurements obtained for both unfractionated *E. coli* tRNA and yeast tRNA^{Phe} in 1 mM Mg^{++}. (*Bottom*) Kinetics of kethoxal modification of yeast tRNA^{Phe}. tRNA samples were dialyzed against 0.01 M Tris-borate (pH 7.2) containing 1 or 10 mM Mg^{++} and NaCl ($\mu = 0.1$ M) and reacted with [^{3}H]kethoxal (5.55 mCi/mM; 18 mM) under the same conditions. The extent of modification was determined by sampling 7 μl of the reaction mixture at the times indicated and measuring the formation of labeled adduct by scintillation spectometry. (●) Data obtained in 1 mM Mg^{++}, 0.1 M ionic strength; (○) data obtained in 10 mM Mg^{++}, 0.1 M ionic strength.

observed for both bulk *E. coli* tRNA and yeast tRNA^{Phe} in 1 mM Mg^{++} as the ionic strength approaches 0.1 M. This "shrinkage" to a more compact form involves an increase in $D_{20,w}^0$ of approximately 12%. As the ionic strength is reduced further, $D_{20,w}^0$ values decrease in the same manner as for tRNA in 10 mM Mg^{++}. Because the increase in diffusivity in the lower magnesium condition occurs under solution conditions where secondary structure would be expected to be preserved, we conclude that this transition most likely occurs at the level of tertiary structure.

Independent evidence of a conformational change in the lower magnesium condition has been obtained from chemical-modification studies with yeast tRNA^{Phe}. The lower panel in Figure 4 shows the kinetics of modification of tRNA^{Phe} with kethoxal (β-ethoxy-α-ketobutyraldehyde), a guanine-specific reagent. The upper curve shows data obtained with tRNA dialyzed and modified in the solution condition that produces the fast-diffusing species, namely, 1 mM Mg^{++} and 0.1 M ionic strength. The results shown in the lower curve were obtained with sample dialyzed against solution containing 10 mM Mg^{++} but identical in ionic strength.

The data obtained for the slower-diffusing form show modification of two to three G residues per tRNA molecule. This result is in agreement with results reported by Litt (1969), who found that two G residues, 20 and 34, in the D and anticodon loops, were the first sites to be modified in tRNA^{Phe} in a high-salt, high-magnesium condition. By contrast, data obtained under solution conditions giving maximal diffusion-constant values show modification of five or six residues per molecule, indicating a structurally unique state. Thus, the solution conditions that promote the anomalous increase in $D_{20,w}^0$ alter the number of exposed reactive G residues. Data obtained at 0.25 M ionic strength in both 1 and 10 mM Mg^{++} were identical to those comprising the lower curve, suggesting a common structure for the slow-diffusing species. Work is currently in progress to identify specifically the sites of kethoxal modification under the various conditions.

Because both bulk *E. coli* tRNA and yeast tRNA^{Phe} exhibit the same transition to a more compact form in 1 mM Mg^{++}, it is logical to ask whether this behavior is typical of all tRNAs. To gain insight into the generality of the effect, the behaviors of four different single species of tRNAs were compared. tRNAs from different structural classes and one with special function were selected. Yeast tRNA^{Phe} was an obvious choice, since much is known about its crystal and solution structures. Two other class-1 tRNAs were also chosen. They were *E. coli* tRNA$_2^{Glu}$, which is very similar to yeast tRNA^{Phe} in size and function, and *E. coli* tRNA^{fMet}, which is functionally and structurally unique; tRNA^{fMet}, the initiator species, contains two unpaired bases at the top of the CCA stem. Finally, a class-2 tRNA, namely, *E. coli* tRNA$_1^{Leu}$, was selected. This species is about 15% larger in mass than the class-1 tRNAs, with the bulk of the extra nucleotides occurring in the large extra loop.

Figure 5 shows a plot of $D^0_{20,w}$ vs ionic strength at 1 and 10 mM Mg^{++} for all four tRNAs. At first glance, striking similarities are apparent for the noninitiator tRNAs (Fig. 5B, C, and D) and bulk tRNA (Fig. 4). Because of these similarities, it seems likely that the behavior of bulk tRNA represents an average of all the individual acceptor species. Features common to all include:

1. a decrease in the $D^0_{20,w}$ values at both Mg^{++} concentrations as the ionic strength is lowered below 0.1 M;
2. a transition to a faster-diffusing form in 1 mM Mg^{++} when the ionic strength is adjusted to 0.1 M;
3. $D^0_{20,w}$ values in 1 mM Mg^{++} lower at ionic strengths above 0.15 M than at $\mu = 0.1$ M.

Despite the qualitative similarities of the individual species, the data

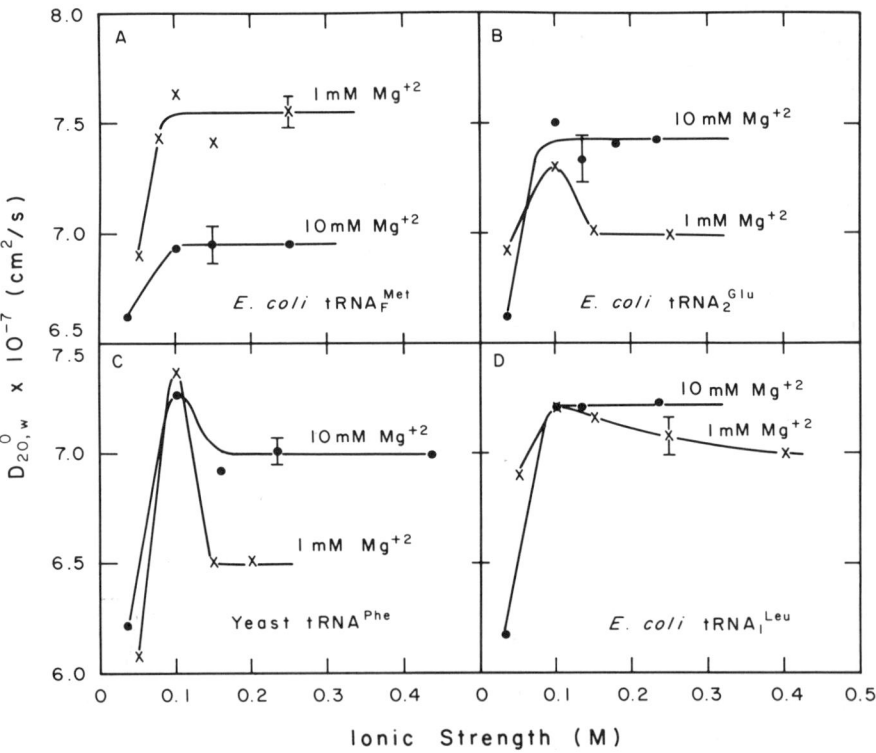

Figure 5 Effect of ionic strength and magnesium on the diffusion constants of four different tRNAs. $D^0_{20,w}$ values were derived for each solution condition from $D_{20,w}$ vs tRNA concentration. (x) Data obtained in 1 mM Mg^{++}; (●) data obtained in 10 mM Mg^{++}. (A) *E. coli* tRNA$^{fMet}_F$; (B) *E. coli* tRNA$^{Glu}_2$; (C) yeast tRNA^{Phe}; (D) *E. coli* tRNA$^{Leu}_1$.

also reveal the occurrence of unique features for each of the noninitiator tRNAs.

1. There are quantitatively different $D^0_{20,w}$ values for similar solution conditions.
2. The magnitude of the shrinkage in 1 mM Mg^{++} varies.
3. tRNA^{Phe} (Fig. 5C) exhibits qualitatively similar transitions in $D^0_{20,w}$ in 1 and 10 mM Mg^{++} at $\mu = 0.1$ M.

By far the most unique behavior is that of *E. coli* tRNA^{fMet} (Fig. 5A). This tRNA exhibits no shrinkage in 1 or 10 mM Mg^{++}. In addition, the diffusion constant in 1 mM Mg^{++} is larger than in 10 mM Mg^{++}, just the opposite of the situation with the other tRNAs. Thus, although the general behavior of most tRNAs is very similar, there are quantitative differences among them. Furthermore, in the case of at least one species, *E. coli* tRNA^{fMet}, the response to these solution conditions is quite unique.

DISCUSSION

The results presented here show a magnesium-dependent change to a slower-diffusing conformer of greater negative charge upon aminoacylation of yeast tRNA^{Phe}. In addition, the results with nonaminoacylated tRNA show an Mg^{++}-concentration-dependent and ionic-strength-dependent molecular shrinkage for several noninitiator tRNAs; the initiator species exhibits no such transition.

Laser light scattering yields information about the hydrodynamic shape and electrostatic charge of the macromolecule plus the associated ion atmosphere. To interpret the effective-charge and diffusion-constant measurements, it is necessary to consider the surrounding ion and hydration atmospheres, as these atmospheres can have important effects on the diffusivity. The tRNA molecule will have associated with it a hydration layer that includes the various pertinent counterions. In addition, some Mg^{++} may be bound to the tRNA in a rather tight site-specific manner. The effective charge then includes the intrinsic macromolecular charge plus the charge due to the associated Mg^{++} and other counterions. Some of these ions may exchange with the surrounding solution. Only those ions with exchange times that are long compared with the time required for the molecule to diffuse the distance $\lambda/(4\pi n \sin(\theta/2))$ will contribute to the experimentally determined charge (Ford 1972). Thus, for the experiments reported here, those ions with exchange times longer than about 10 μsec will be included in charge and diffusion-constant measurements. Nuclear magnetic resonance investigations of Eu^{3+} (Jones and Kearns 1974) and Mn^{++} (Chao and Kearns 1977) binding to tRNA have shown exchange rates ranging from 1 μsec to 15 msec. Since these ions have tRNA binding properties similar to Mg^{++}, it seems likely that at least some Mg^{++} ion binding can affect both the diffusion-constant and electrostatic-charge values.

We propose that aminoacylation results in a small conformational change resulting in changes in the ion atmosphere and Mg^{++} binding. Thus, changes in $D^0_{20,w}$ and electrostatic charge reflect an intrinsic molecular change plus the effect of altered hydration and ion atmospheres. This hypothesis is supported by results from two other investigations. Cohn and coworkers (1969), monitoring changes in the proton magnetic resonance (PMR) relaxation constant of water protons surrounding tRNA, found that aminoacylation of bulk *E. coli* tRNA and yeast tRNA[Phe] leads to an increase in Mn^{++} binding and a "looser" structure in the vicinity of binding. Ninio et al. (1972) found no intensity changes in small-angle X-ray scattering upon aminoacylation of *E. coli* tRNA[Val], indicating no large-scale structural changes. However, these same investigators found intensity differences at larger angles upon aminoacylation and they associated these changes with altered Mg^{++} binding and perturbation of the counterion distribution.

Several investigators have reported structural differences in tRNA upon aminoacylation, but much of the evidence is contradictory. The results of this investigation appear to provide at least a partial explanation for some of the discrepancies. It seems quite possible that a number of the negative findings are related to the solution conditions selected. In other cases the absence of an effect may be due to the use of a technique insensitive to the changes that do occur.

The results presented here also indicate a profound influence of solution conditions on the solution structure of nonaminoacylated tRNA. The effects observed are common to several, and perhaps most, tRNAs. For all tRNAs investigated to date by us, a marked decrease in $D^0_{20,w}$ is observed as the ionic strength is lowered below 100 mM. This effect is consistent with the polyanionic behavior of all tRNAs in which lowered ionic strength reduces the electrostatic shielding between phosphates, resulting in an overall expansion to reduce free energy.

The $D^0_{20,w}$ values obtained for the nonacylated tRNAs under identical solution conditions are rather similar. This is somewhat surprising in the case of the larger *E. coli* tRNA$_1^{Leu}$ species. The similarity in $D^0_{20,w}$ values may well reflect the dependence of light-scattering measurements on the hydro-dynamic rather than the molecular volume.

The anomalous shrinkage of noninitiator and bulk tRNAs at near-physiological solution conditions cannot be explained by simply invoking the polyelectrolytic character of tRNA. Furthermore, the kethoxal-modification results with yeast tRNA[Phe] indicate that a structurally distinct conformer occurs under these solution conditions. Because it is unlikely that changes in secondary structure occur at the high ionic strength and low temperature used in these experiments, it is reasonable to suggest that the changes observed reflect alterations that occur primarily at the level of the tertiary structure.

Although the number of similar properties of the individual and bulk tRNAs is striking, suggesting an average behavior for bulk tRNA, unique features also exist for each. Two alternative hypotheses can be advanced to explain the data. The first suggests that each tRNA undergoes unique structural transitions involving initial and final conformers that are different for each tRNA. In view of the known structural similarities among tRNAs this notion seems unlikely. The second hypothesis holds that different tRNAs undergo the same general transitions in response to changes in Mg^{++} concentration and ionic strength. The marked similarity among the noninitiator tRNAs supports the view that most tRNAs do undergo the same general transition with similar initial and final conformers. The small quantitative differences in the transition profiles for the various noninitiator tRNAs could reflect different equilibrium constants and ion requirements.

The results presented here suggest that most tRNAs behave similarly, but *E. coli* tRNA^{fMet} is the exception. The unique behavior of this tRNA precludes its involvement in structural transitions common to noninitiator tRNAs. The unique structural response of this tRNA to altered Mg^{++} concentration and ionic strength may be related to its unique biological function in protein synthesis and/or its structure, i.e., the absence of the terminal base pair in the CCA stem.

We have demonstrated conformational and electrostatic transitions of tRNA at near-physiological conditions as a result of aminoacylation and changes in solution conditions. It is tempting to suggest that these transitions could be important in regulating tRNA functions.

ACKNOWLEDGMENT

This work was supported by a grant from the National Science Foundation to M. J. F. and N. C. F.

REFERENCES

Chao, Y. H. and D. R. Kearns. 1977. Manganese (II) as a paramagnetic probe of the tertiary structure of transfer RNA. *Biochim. Biophys. Acta* **477**:20.

Chen, M. C., R. Geige, R. C. Lord, and A. Rich. 1975. Raman spectra and structure of yeast phenylalanine transfer RNA in the crystalline state and in solution. *Biochemistry* **14**:4385.

Cohn, M., A. Danchin, and M. Grunberg-Manago. 1969. Proton magnetic relaxation studies of manganous complexes of transfer RNA and related compounds. *J. Mol. Biol.* **39**:199.

Ford, N. C., Jr. 1972. Biochemical applications of laser Rayleigh scattering. *Chem. Scr.* **2:**193.

Jones, C. R. and D. R. Kearns. 1974. Investigation of the structure of yeast tRNA^{Phe} by nuclear magnetic resonance: Paramagnetic rare earth ion probes of structure. *Proc. Natl. Acad. Sci.* **71:**4237.

Ladner, J. E., A. Jack, J. D. Robertus, R. S. Brown, D. Rhodes, B. F. C. Clark, and A. Klug. 1975. Structure of yeast phenylalanine transfer RNA at 2.5 Å resolution. *Proc. Natl. Acad. Sci.* **72:**4414.

Litt, M. 1969. Structural studies on transfer ribonucleic acid. I. Labeling of exposed guanine sites in yeast phenylalanine transfer ribonucleic acid with kethoxal. *Biochemistry* **8:**3249.

Ninio, J., V. Luzzatti, and M. Yaniv. 1972. Comparative small-angle X-ray scattering studies on unacylated, acylated and cross-linked *Escherichia coli* transfer RNA$_1^{Val}$. *J. Mol. Biol.* **71:**217.

Olson, T., M. J. Fournier, K. H. Langley, and N. C. Ford, Jr. 1976. Detection of a major conformational change in transfer ribonucleic acid by laser light scattering. *J. Mol. Biol.* **102:**193.

Potts, R., M. J. Fourier, and N. C. Ford, Jr. 1977. Effect of aminoacylation on the conformation of yeast phenylalanine tRNA. *Nature* **268:**563.

Quigley, G. J., A. H.-J. Wang, N. C. Seeman, F. L. Suddath, A. Rich, J. L. Sussman, and S.-H. Kim. 1975. Hydrogen bonding in yeast phenylalanine transfer RNA. *Proc. Natl. Acad. Sci.* **72:**4866.

Rich, A. and U. L. RajBhandary. 1976. Transfer RNA: Molecular structure, sequence and properties. *Annu. Rev. Biochem.* **45:**805.

Robillard, G. T., C. E. Tan, F. Vosman, and H. J. C. Berendsen. 1976. Similarity of the crystal and solution structure of yeast tRNA^{Phe}. *Nature* **262:**363.

Schurr, J. M. 1977. Dynamic light scattering of biopolymers and biocolloids. *CRC Crit. Rev. Biochem.* **4:**371.

Aminoacyl-tRNA Synthetases

Structural Studies of Aminoacyl-tRNA Synthetases

Brian S. Hartley
Department of Biochemistry, Imperial College
London SW7 2AZ, England

Everyone admits that aminoacyl-tRNA synthetases are an important and interesting family of enzymes. By 1965, sequence and crystallographic studies had made it clear that the serine proteases were a family of enzymes with considerable sequence homologies and an amazing evolutionary conservation of tertiary structure. Dramatic differences in substrate specificity were grafted onto a common catalytic mechanism by only one or two amino acid changes in the substrate binding sites. If this were true of an apparently peripheral cluster of digestive enzymes, must it not be even more true of a family of enzymes that has a common catalytic mechanism and is under enormous selective pressure to conserve precise substrate specificities?

At that time, it had become clear that only a combination of the protein chemistry and X-ray crystallography of several aminoacyl-tRNA synthetases could provide definitive answers to such a question. However, unlike the pancreatic proteinases, these enzymes were not available in quantity in a pure form, since they represent a relatively minor fraction of the protein in most cells. The majority of early studies used animal enzymes, notably the tryptophanyl enzyme from beef pancreas, but as *Escherichia coli* had become the organism of choice for studying protein synthesis, workers naturally turned to purifying the aminoacyl-tRNA synthetases from this organism.

Purifying enzymes tends to be a frustrating occupation of low prestige, but it is remarkable how pioneering work in the field guides the future direction of biochemistry. The early purification studies of Berg and his collaborators (Baldwin and Berg 1966) led to the emergence of isoleucyl-tRNA synthetase as one of the *E. coli* enzymes most readily obtained pure in reasonable yield. Berg's selection of this enzyme to explore the molecular basis of the high specificity of these enzymes—a topic that had long intrigued theorists—led to the successful elucidation of editing mechanisms such as those discussed by Fersht (this volume). Unfortunately, however, isoleucyl-tRNA synthetase turned out to be a single chain of over 100,000 daltons, and this was enough in those early days to scare off both protein chemists and X-ray crystallographers.

My own group at Cambridge began to purify the methionyl-tRNA synthetase from *E. coli*, since an immediate question of local interest was whether a single enzyme catalyzed aminoacylation of both tRNA^{fMet} and tRNA^{mMet}, which had recently been discovered in our laboratory (Clark and Marcker 1966). It still seems remarkable to me that this enzyme fails to discriminate even kinetically between these two tRNAs (Bruton and Hartley 1968), which differ so considerably in sequence and which are so distinctly different in their reactions with the ribosome or with the transformylase. We always sought to work with the *E. coli* strain most popular with our tRNA colleagues; this was originally MRE600 (Heinrikson and Hartley 1967), but they switched to CA244 and so did we (Bruton and Hartley 1968). The two enzymes differ both in kinetic parameters and in amino acid composition, but since only the latter is a reputable *E. coli* K12, one could overlook this apparent species difference. A more serious problem arose when we discovered that our purification method, designed for convenient scale-up, yielded a stable form that was a proteolytic fragment of the native enzyme (Cassio and Waller 1968). As described by Blanquet et al. (this volume), Waller and his colleagues discovered strain EM20031 of *E. coli* K12, which superproduces methionyl-tRNA synthetase (Cassio et al. 1970), so we also switched to this latter strain for large-scale purification (Bruton et al. 1975) and sequencing.

I disinter this history because it hides a nasty problem. There appear to be appreciable differences in the sequences of methionyl-tRNA synthetases not only from MRE600, which is an ill-defined serotype, but also from CA244 and from EM20031, which are both reputedly K12 strains. Which, then, is the true *E. coli* enzyme? Are there such sequence differences between other *E. coli* enzymes within a single serotype such as *E. coli* K12 or are the methionyl-tRNA synthetases especially variable? If the latter is true, so much for the argument that these enzymes might be highly conserved by evolution.

Returning to the logistic question of obtaining a number of aminoacyl-tRNA synthetases in sufficient amounts for protein chemistry and X-ray crystallography, it became apparent that for economic reasons one ought to try to purify several of these enzymes simultaneously from a single, large-scale culture of *E. coli*. So was born the concept of a large-scale, multienzyme purification in which not only these enzymes but also others that were of interest would be isolated from the many distinct fractions that could be resolved by ion-exchange or hydroxyapatite chromatography. Pioneering work by Bruton and Atkinson led to routine procedures in which up to 70 kg of *E. coli* or *Bacillus stearothermophilus* (Atkinson et al. 1979) can be processed to yield as many as 20 different enzymes. In celebrating achievements in structural elucidation there is a tendency to play down the contribution of those who have struggled to purify the material. I will attempt to correct this tendency in this paper.

PRIMARY STRUCTURAL STUDIES

Gradually, a number of aminoacyl-tRNA synthetases were purified and their molecular weights and subunit structures determined. Table 1 shows the current picture, including results presented elsewhere in this volume. It is clear that these enzymes are not as similar as chymotrypsin, trypsin, and elastase; the sizes, number, and variety of peptide chains must surely squash such optimism. However, one can still hope that subfamilies with closely related structures may exist.

Results obtained from mapping and analysis with methionyl- and valyl-tRNA synthetases of *B. stearothermophilus* (Koch et al. 1974) and subsequently with isoleucyl-, leucyl-, and methionyl-tRNA synthetases of *E. coli* (Kula 1973; Bruton et al. 1974; Waterson and Konigsberg 1974) showed that all these enzymes with long polypeptide chains had extensive internal repeats of identical amino acid sequences. Therefore, one might legitimately expect two very similar domains of tertiary structure within each chain. Confirmation of this will require further adjudication by X-ray crystallographers.

A further attempt to systematize this family included study of whether two domains of 35,000–50,000 daltons are an essential requirement for a functional aminoacyl-tRNA synthetase. This speculation has been punctured by C. J. Bruton and L. A.-M. Cox (pers. comm.), who reported that cysteinyl-tRNA synthetase appears to be a unique sequence of about 500 residues that is functionally active as a monomer.

D. Kern et al. (pers. comm.) offer information that is more comforting to the seekers of simplicity. The glutamyl-tRNA synthetase of *E. coli* also turns out to be a monomer of unique sequence, with a molecular weight of 56,000; it was previously reported to be an $\alpha\beta$ dimer with 56,000 and 46,000 subunits (Lapointe and Söll 1972). Therefore, one needs only to consider three classes of subunit structure for these enzymes: α, α_2, and $\alpha_2\beta_2$.

Moreover, S. Robbe-Saul et al. (pers. comm.) have provided further evidence that the large chains in these enzymes contain extensive repeats. Both the α chain (73,000 daltons) and the β chain (63,000 daltons) of the yeast phenylalanyl-tRNA synthetase show extensive repeats and there is some evidence for repeats in the monomeric yeast arginyl-tRNA synthetase (75,000 daltons).

Hence it is tenable to hope that a similar structural domain of about 30,000 daltons may be common to all aminoacyl-tRNA synthetases. The smaller monomers (such as the cysteinyl- and glutamyl-tRNA synthetases) may contain only one such domain, the larger monomers and small dimers would contain two, the large dimers (such as methionyl-tRNA synthetase) would have four, and the huge tetramers (such as phenylalanyl-tRNA synthetase) could have up to eight.

Unfortunately, the limited sequence comparisons that are available fail

Table 1 Quaternary structures of purified aminoacyl-tRNA synthetases

Enzyme	Source	Quaternary structure	Reference
α_1 Class			
Arg	B. stearothermophilus	1 × 78,000	Parfait and Grosjean (1972)
Arg	E. coli	1 × 75,000	Marshall and Zamecnik (1969)
Arg	yeast	1 × 75,000	S. Robbe-Saul et al. (pers. comm.)
Arg	Neurospora crassa	1 × 85,000	Nazario and Evans (1974)
Cys	B. stearothermophilus	1 × 54,000	C. J. Bruton and L. A.-M. Cox (pers. comm.)
Gln	E. coli	1 × 70,000	Folk (1971)
Glu	E. coli	1 × 56,000	D. Kern et al. (pers. comm.)
Glu	Bacillus subtilis	1 × 65,000	D. Kern et al. (pers. comm.)
Ile	E. coli	1 × 110,000	Andt and Berg (1970)
Ile	B. stearothermophilus	1 × 110,000	Charlier and Grosjean (1972)
Leu	E. coli	1 × 100,000	Waterson and Konigsberg (1974)
Leu	B. stearothermophilus	1 × 110,000	Koch et al. (1974)
Val	E. coli	1 × 110,000	Berthelot and Yaniv (1970)
Val	B. stearothermophilus	1 × 110,000	Koch et al. (1974)
Val	yeast	1 × 113,000	Lagerkvist and Waldenstrom (1967)

α_2 Class

His	*E. coli*	$2 \times 43,000$	Kalousek and Konigsberg (1974)
Lys	*E. coli*	$2 \times 52,000$	Rymo et al. (1970)
Lys	yeast	$2 \times 72,000$	Rymo et al. (1970)
Met	*E. coli*	$2 \times 85,000$	Koch and Bruton (1974)
Met	*B. stearothermophilus*	$2 \times 85,000$	Mulvey and Fersht (1976)
Pro	*E. coli*	$2 \times 47,000$	Lee and Muench (1969)
Ser	*E. coli*	$2 \times 50,000$	Waterson and Konigsberg (1974)
Ser	yeast	$2 \times 60,000$	Hertz and Zachau (1973)
Trp	*E. coli*	$2 \times 37,000$	Joseph and Muench (1971)
Trp	*B. stearothermophilus*	$2 \times 37,000$	Koch et al. (1974)
Trp	yeast	$2 \times 50,000$	Hossein and Kallenbach (1974)
Trp	bovine pancreas	$2 \times 54,000$	Iborra et al. (1973)
Tyr	*E. coli*	$2 \times 45,000$	Chousterman and Chapeville (1973)
Tyr	*B. stearothermophilus*	$2 \times 45,000$	Koch et al. (1974)

$\alpha_2\beta_2$ Class

Gly	*E. coli*	$2 \times 33,000$	Ostrem and Berg (1974)
Gly	*E. coli*	$2 \times 80,000$	Ostrem and Berg (1974)
Phe	*E. coli*	$2 \times 37,000$	Fayat et al. (1974)
Phe	*E. coli*	$2 \times 98,000$	Fayat et al. (1974)
Phe	yeast	$2 \times 62,000$	Fasiolo et al. (1970)
Phe	yeast	$2 \times 72,000$	Fasiolo et al. (1970)

to reveal significant homology between different enzymes of this class. Winter et al. (this volume) report the sequence comparisons between the tryptophanyl-tRNA and tyrosyl-tRNA synthetases, which reduce at best to homology in position of two of the cysteine residues. C. J. Bruton and A. D. MacLachlan (unpubl.) have also carried out computer comparisons of these enzymes and of available sequences from the methionyl- and isoleucyl-tRNA synthetases of *E. coli*; again, there was no significant homology revealed.

Nevertheless, one cannot dismiss the possibility of common structural domains on these grounds alone. There are plenty of examples of precise conservation of structural homology that is not apparent in the primary structures. The tyrosyl- and tryptophanyl-tRNA synthetases do seem to have C-terminal domains of similar size that are resistant to proteolysis, and the repeats of identical sequence in the methionyl-, valyl-, leucyl-, isoleucyl-, and phenylalanyl-tRNA synthetase chains also imply repeats of small domains. In the latter case, however, we have another puzzle. If the two domains in methionyl-tRNA synthetase have conserved such high sequence homology, why is no sequence homology apparent between comparable domains in methionyl-, isoleucyl-, tyrosyl-, and tryptophanyl-tRNA synthetases?

The significance of the C-terminal repeated sequence in methionyl-tRNA synthetase is very doubtful. Bruton has shown (see Abstracts of European Molecular Biology Organization Workshop on tRNA, 1976) that although the tryptic fragment of methionyl-tRNA synthetase is fully active and stable, it contains only $1^1/_3$ repeat of sequence, about $^2/_3$ of the C-terminal repeat being removed in the redundant C-terminal peptides.

X-ray Crystallographic Studies

As was apparent from the outset, only X-ray crystallography will resolve the problems of the mechanism and evolution of the family of aminoacyl-tRNA synthetases. Progress in this direction has been rather disappointing. Relatively few laboratories have been able to prepare sufficient amounts of the pure enzymes to make possible extensive efforts to crystallize them. Even where this was possible, for example with pure aminoacyl-tRNA synthetases from *E. coli*, the enzymes resolutely refused to crystallize.

Some of the difficulty may have been due to the sensitivity of many of these enzymes to denaturation and proteolysis; a switch to the potentially more robust enzymes of *B. stearothermophilus* seems to have paid off. Even where crystals were obtained, for example of the lysyl-tRNA synthetase of yeast (Rymo et al. 1970), they were too small or too disordered to be useful for X-ray crystallography. In another case, crystals of the leucyl-tRNA synthetase of yeast (Chirikjian et al. 1972) proved to be crystals of a protein contaminant in the preparation. The two crystals currently under study are the tyrosyl-tRNA synthetase of *B. stearothermophilus* (Reid et al. 1973) and the tryptic fragment of methionyl-tRNA synthetase of *E. coli* (Zelwer et al. 1976), though we look forward to studying the findings of C. Carter (pers. comm.) on promising crystals of tryptophanyl-tRNA synthetase (*B. stearothermophilus*) and encouraging off-stage noises about some others.

The tyrosyl-tRNA synthetase of *B. stearothermophilus* remains the crystallographic front-runner (Irwin et al. 1976), but even with this there have been disappointments. As Rubin describes (J. R. Rubin and D. M. Blow, pers. comm.), only part of the chain is visible in the 2.7-Å electron-density maps; the rest of the molecule, amounting to 35% of the sequence, appears to be disordered in these crystals. Fortunately, the ordered part of the structure contains the binding sites for tyrosine, ATP, and aminoacyl adenylate. The intriguing result that both tyrosine and ATP bind at the tyrosine site when diffused separately into the crystals has obvious significance for mechanistic studies. There appear to be excellent prospects for building a reliable model of this part of the enzyme as soon as the latest X-ray results can be correlated with the latest sequence data.

Also avidly awaited are high-resolution electron-density maps of the monomeric tryptic fragment of methionyl-tRNA synthetase (*E. coli*) (Zelwer et al. 1976); here again there is good prospect of correlating these with the amino acid sequence. However, this structure will still only allow us to guess about the two-domain structure of repeated sequences in the chains of the native dimer. Structures and sequences of more aminoacyl-tRNA synthetases are clearly desirable.

Chemical Modification Studies

Active-site labeling proved a potent tool in recognizing serine protease families, but so far no such general method has emerged for aminoacyl-tRNA synthetases. Active-site-directed reagents based on amino acid analogs or derivatives attached to the appropriate tRNA can be devised (Bruton and Hartley 1970; Santi and Cunnion 1974), but these label only residues in the vicinity of the binding site. That *p*-nitrophenyl-carbamyl-methionyl-tRNA labels a specific lysine in methionyl-tRNA synthetase from *E. coli* K12 strain CA244 (Bruton and Hartley 1970) but not in methionyl-tRNA synthetase from *E. coli* K12 strain EM20031 (C. J. Bruton, pers. comm.) should be a warning. Photoaffinity labeling with analogs of ATP or aminoalkyl adenylates (Wetzel and Söll 1977) is also unlikely to identify catalytically active residues.

The exciting discovery by Kisselev et al. (this volume) that a unique carboxyl in the catalytic site of beef tryptophanyl-tRNA synthetase can be labeled with tryptophan or alkyl hydroxamates offers a possible tool for active-site labeling in all these enzymes. Surely these challenging observations will be followed up with other enzymes. Chemical modification is also likely to be useful in identifying residues in the enzyme that interact with tRNA. Photochemical cross-linking (Schoemaker and Schimmel 1974) has identified areas of the tRNA that appear to lie close to the enzyme, but the technique has obvious drawbacks for identifying residues in the enzyme that are so cross-linked. The reaction products are likely to be complex, since free radical intermediates are involved, and protein-sequencing techniques demand more material than nucleotide sequencing.

Another approach to identifying amino acid residues that interact with tRNA is competitive labeling of the complex. This technique is a modification of that of Kaplan et al. (1971), wherein amino groups in the enzyme or enzyme-tRNA complex are reacted with trace amounts of ^{3}H-labeled acetic anhydride of high specific radioactivity in the presence of a competing nucleophile such as phenylalanine. Under these conditions, the attacking nucleophile is always the native protein, since less than one residue per molecule is finally acetylated. The protein is then fully acetylated with unlabeled acetic anhydride and mixed with ^{14}C-labeled acetylated enzyme prior to digestion by proteases. The ^{3}H:^{14}C ratio in isolated peptides gives the relative reactivity of that amino group in the native enzyme or enzyme-tRNA complex. Amino groups that form salt bridges in the complex with the phosphate groups in the tRNA (or with carboxyl groups in the enzyme due to a change in tertiary or quaternary structure) will show reduced reactivity. An increase in reactivity can indicate breaking of a salt bridge in the free enzyme or a decrease in the pK_a of the amino group due to movement close to a positively charged residue in the complex.

In this way, Bosshard et al. (1978) compared the reactivities of amino groups in tyrosyl-tRNA synthetase (*B. stearothermophilus*) and in a complex of the dimeric enzyme with one molecule of the cognate tRNA^{Tyr}. Most of the amino groups showed comparable reactivities in the complex (relative reactivities of complex:enzyme of 0.9–1.2), but nine peptides showed reduced reactivities of 0.5–0.8. This is the range expected for a salt bridge formed in only one of the two subunits. One peptide showed a relative reactivity of less than 0.15 in the complex; this indicates that it forms salt bridges in both subunits, suggesting that a single tRNA interacts with the same residue on both subunits. Some of these residues with reduced reactivity have been located in the tyrosyl-tRNA synthetase sequence as described by Winter et al. (this volume).

Similar experiments have been carried out with the methionyl-tRNA synthetase of *E. coli* and its complexes with tRNA^{fMet} and tRNA^{mMet} under conditions where two molecules of tRNA would bind to the dimer

Table 2 Reactivities of amino groups in methionyl-tRNA synthetase of *E. coli* and its complexes with tRNA^{fMet} and tRNA^{mMet}

Peptide	Reactivity in complex:reactivity in enzyme	
	tRNA^{fMet} complex	tRNA^{mMet} complex
I B9	0.40	0.58
I A20 (N.t.[a])	0.32	0.94
III A5	0.84	3.44
V A9	0.38	0.50
VIII A25	1.81	1.98

The dimeric enzyme or its complexes with two molecules, tRNA^{fMet} and tRNA^{mMet}, were trace-labeled with ^{3}H-labeled acetic anhydride under competitive labeling conditions (Kaplan et al. 1971). After complete acetylation with unlabeled acetic anhydride, the proteins were mixed with ^{14}C-labeled acetylated enzyme and digested vigorously with trypsin. The ^{3}H:^{14}C ratio of a peptide from the complex compared with that obtained from the free enzymes indicates the relative reactivity of the N-terminal or ϵ-lysine amino group of over 60 peptides isolated; only 5 showed significant changes in reactivity in the complexes (<0.8 or >1.2). Partial sequence data have located I B9, I A20 (N-terminal), and V A9 in the known sequence of the tryptic fragment of methionyl-tRNA synthetase (M. Rangarajan et al., unpubl.).
[a]N-terminal.

of 85,000-dalton subunits (M. Rangarajan et al., in prep.). Most lysine-containing peptides showed relative reactivities of 0.8–1.2 in both complexes, but Table 2 shows five peptides with significantly changed reactivities in the complexes. Peptides I B9 and V A9 show reduced reactivities in both the tRNA^{fMet} and tRNA^{mMet} complexes. Peptide VIII A25 shows increased reactivity in both complexes, which suggests that it moves toward a positive group after the tRNA is bound. Most striking, however, is the behavior of the N-terminal residue (peptide I A20), which has considerably reduced reactivity in the tRNA^{fMet} complex but is unchanged in the tRNA^{mMet} complex. Conversely, peptide III A5 is normal in the tRNA^{fMet} complex. These clear distinctions are noteworthy when one recalls that the enzyme shows identical kinetics with both tRNAs (Bruton and Hartley 1968). So far only peptides I B9 and V A9 have been unambiguously located in the sequence of the 64,000-dalton tryptic fragment whose tertiary structure is eagerly awaited. These two are widely separated in the sequence and are not part of the repeating sequence in the fragment (Bruton et al. 1974).

CONCLUSIONS

Structural studies on aminoacyl-tRNA synthetases have so far provided more questions than answers. However, one can eliminate the naive view that selective pressure has conserved the sequence and structure of these enzymes as rigidly as with the serine protease family. One can indeed begin to question the dogma that these enzymes are likely to be conservative in evolutionary variance. It is generally supposed that protein synthesis is not controlled by the availability of a particular charged tRNA, but we have learned elsewhere in this text of how such a level of charging might control the translation of enzymes involved in amino acid biosynthesis (Umbarger 1980). In such circumstances the activity of a particular aminoacyl-tRNA synthetase would indirectly control the pool size for a particular amino acid. Thus, small changes in the K_m and V_{max} of the enzyme, appropriate for a particular ecological niche, would be selected. This would lead to greater-than-average species variability of these enzymes.

More sequences and more structures are the panacea for the waves of speculation that this field seems to engender. One must admit, however, that the intriguing speculations are a driving force for the hard work that such structural studies involve.

ACKNOWLEDGMENTS

I would like to thank all my present and former collaborators in this field, Yves Boulanger, Chris Bruton, Anne Dell, Alan Fersht, Ross Jakes, Gordon Koch, Minnie Rangarajan, Brian Reid, and Greg Winter, for allowing me to talk about the work that was truly theirs.

REFERENCES

Arndt, D. J. and P. Berg. 1970. Isoleucyl-tRNA synthetase is a single polypeptide chain. *J. Biol. Chem.* **245:**665.

Atkinson, A., G. T. Banks, C. J. Bruton, M. J. Comer, R. Rakes, A. Kamalagharan, A. R. Whitaker, and G. P. Winter. 1979. Large scale multienzyme isolation from *Bacillus stearothermophilus*. *Biochem. J.* (in press).

Baldwin, A. N. and P. Berg. 1966. Purification and properties of isoleucyl ribonucleic acid synthetase from *E. coli*. *J. Biol. Chem.* **241:**831.

Berthelot, F. and M. Yaniv. 1970. Presence of one polypeptide chain in valyl- and isoleucyl-tRNA synthetases from *E. coli*. *Eur. J. Biochem.* **16:**123.

Bosshard, H. R., G. L. E. Koch, and B. S. Hartley. 1978. The aminoacyl-tRNA synthetase-tRNA complex: Detection by differential labelling of lysine residues involved in complex formation. *J. Mol. Biol.* **119:**377.

Bruton, C. J. and B. S. Hartley. 1968. Sub-unit structure and specificity of methionyl-transfer-ribonucleic acid synthetase from *E. coli*. *Biochem. J.* **108:**281.

———. 1970. Chemical studies on methionyl-tRNA synthetase from *E. coli*. *J. Mol. Biol.* **52:**165.

Bruton, C. J., R. Jakes, and A. Atkinson. 1975. Gramme-scale purification of methionyl-tRNA and tyrosyl-tRNA synthetases from *E. coli*. *Eur. J. Biochem.* **59:**327.

Bruton, C. J., R. Jakes, and G. L. E. Koch. 1974. Repeated sequences in methionyl-tRNA synthetase from *E. coli. FEBS Lett.* **45:** 26.

Cassio, D. and J.-P. Waller. 1968. Étude de la méthionyl-tRNA synthétase d'*E. coli. Eur. J. Biochem.* **5:** 33.

Cassio, D., F. Lawrence, and D. A. Lawrence. 1970. Level of methionyl-tRNA synthetase in merodiploids of *E. coli* K12. *Eur. J. Biochem.* **15:** 331.

Charlier, J. and H. Grosjean. 1972. Isoleucyl-tRNA synthetase from *B. stearothermophilus*: Properties of the enzyme. *Eur. J. Biochem.* **25:** 163.

Chirikjian, J. G., H. T. Wright, and J. R. Fexo. 1972. Crystallization of leucyl-tRNA synthetase from bakers yeast. *Proc. Natl. Acad. Sci.* **69:** 1638.

Chousterman, S. and F. Chapeville. 1973. Tyrosyl-tRNA synthetase of *E. coli* B: Binding of various ligands. *Eur. J. Biochem.* **35:** 51.

Clark, B. F. C. and K. A. Marcker. 1966. The role of N-formyl-methionyl-sRNA in protein biosynthesis. *J. Mol. Biol.* **17:** 394.

Fasiolo, N., F. Befort, C. Bolloch, and J.-P. Ebel. 1970. Purification et quelques proprietes de la phenylalanyl-tRNA synthetase de la levure de boulangerie. *Biochim. Biophys. Acta* **217:** 305.

Fayat, G., S. Blanquet, P. Dessen, G. Batelier, and J.-P. Waller. 1974. The molecular weight and subunit composition of phenylalanyl-tRNA synthetase from *E. coli* K12. *Biochimie* **56:** 35.

Folk, W. R. 1971. Molecular weight of *E. coli* glutaminyl-tRNA synthetase and isolation of its complex with glutamine tRNA. *Biochemistry* **10:** 1728.

Heinrikson, R. L. and B. S. Hartley. 1967. Purification and properties of methionyl-transfer-ribonucleic acid synthetase from *E. coli. Biochem. J.* **105:** 17.

Hertz, H. S. and H. G. Zachau. 1973. Kinetic properties of phenylalanyl-tRNA and seryl-tRNA synthetases for normal substrates and fluorescent analogs. *Eur. J. Biochem.* **37:** 203.

Hossein, A. and N. R. Kallenbach. 1974. Purification and subunit structure of tryptophanyl-tRNA synthetase from baker's yeast. *FEBS Lett.* **45:** 202.

Iborra, F., M. Donizzi, and J. Labouesse. 1973. Tryptophanyl-tRNA synthetase from beef pancreas. *Eur. J. Biochem.* **39:** 275.

Irwin, M. J., J. Nyberg, B. R. Reid, and D. M. Blow. 1976. The crystal structure of tyrosyl-tRNA synthetase at 2.7Å resolution. *J. Mol. Biol.* **105:** 577.

Joseph, D. R. and K. H. Muench. 1971. Tryptophanyl-tRNA synthetase of *E. coli*: Purification of the enzyme and of tryptophan tRNA. *J. Biol. Chem.* **246:** 7602.

Kalousek, F. and W. H. Konigsberg. 1974. Purification and characterisation of histidinyl-tRNA synthetase of *E. coli. Biochemistry* **13:** 999.

Kaplan, H., K. J. Stevenson, and B. S. Hartley. 1971. Competitive labelling, a method for determining the reactivity of individual groups in proteins. *Biochem. J.* **124:** 289.

Koch, G. L. E. and C. J. Bruton. 1974. The subunit structure of methionyl-tRNA synthetase from *E. coli. FEBS Lett.* **40:** 180.

Koch, G. L. E., Y. Boulanger, and B. S. Hartley. 1974. Repeating sequences in aminoacyl-tRNA synthetases. *Nature* **249:** 316.

Kula, M.-R. 1973. Structural studies on isoleucyl-tRNA synthetase from *E. coli. FEBS Lett.* **35:** 299.

Lagerkvist, U. and J. Waldenstrom. 1967. Purification and some properties of valyl-tRNA synthetase from yeast. *J. Biol. Chem.* **242:** 3021.

Lapointe, J. and D. Söll. 1972. Glutamyl tRNA synthetase of *E. coli*: Purification and properties. *J. Biol. Chem.* **247:**4966.

Lee, M. and K. H. Muench. 1969. Prolyl-tRNA synthetase of *E. coli*: Purification and evidence for subunits. *J. Biol. Chem.* **244:**223.

Marshall, R. D. and P. C. Zamecnik. 1969. Some physical properties of lysyl- and arginyl-tRNA synthetases of *E. coli* B. *Biochim. Biophys. Acta* **181:**454.

Mulvey, R. S. and A. R. Fersht. 1976. Subunit interactions in the methionyl-tRNA synthetase of *B. stearothermophilus*. *Biochemistry* **15:**243.

Nazario, M. and J. A. Evans. 1974. Physical and kinetic studies of arginyl tRNA ligase of *Neurospora*. *J. Biol. Chem.* **249:**4934.

Ostrem, D. L. and P. Berg. 1974. Glycyl-tRNA synthetase from *E. coli*: Purification, properties and substrate binding. *Biochemistry* **13:**1338.

Parfait, R. and H. Grosjean. 1972. Arginyl-tRNA synthetase from *B. stearothermophilus*: Purification, properties and mechanism of action. *Eur. J. Biochem.* **30:**242.

Reid, B. R., G. L. E. Koch, Y. Boulanger, B. S. Hartley, and D. M. Blow. 1973. Crystallisation and preliminary X-ray diffraction studies on tyrosyl-tRNA synthetase from *B. stearothermophilus*. *J. Mol. Biol.* **80:**199.

Rymo, L., U. Lagerkvist, and A. Wonacott. 1970. Crystallisation of lysyl-tRNA synthetase from yeast. *J. Biol. Chem.* **245:**4308.

Santi, D. V. and S. O. Cunnion. 1974. Macromolecular affinity labelling agents: Reaction of *N*-bromoacetyl isoleucyl tRNA with isoleucyl-tRNA synthetase. *Biochemistry* **13:**481.

Schoemaker, H. J. P. and P. R. Schimmel. 1974. Photo-induced joining of a tRNA with its cognate aminoacyl-tRNA synthetase. *J. Mol. Biol.* **84:**503.

Umbarger, H. E. 1980. Comments on the role of aminoacyl-tRNA in the regulation of amino acid biosynthesis. In *Transfer RNA: Biological aspects* (ed. D. Söll et al.). Cold Spring Harbor Laboratory, Cold Spring Harbor, New York. (In press).

Waterson, R. M. and W. H. Konigsberg. 1974. Peptide mapping of aminoacyl-tRNA synthetases: Evidence for internal sequence homology in *E. coli* leucyl-tRNA synthetase. *Proc. Natl. Acad. Sci.* **71:**376.

Wetzel, R. and D. Söll. 1977. Analogs of methionyl-tRNA synthetase substrates containing photolabile groups. *Nucleic Acids Res.* **4:**1681.

Zelwer, C., J.-L. Risler, and C. Montheilhet. 1976. A low resolution model of crystalline methionyl-tRNA synthetase from *E. coli*. *J. Mol. Biol.* **102:**93.

Molecular Enzymology of Beef Pancreas Tryptophanyl-tRNA Synthetase

Lev L. Kisselev, Ol'ga O. Favorova, and Galina K. Kovaleva
Institute of Molecular Biology
Academy of Sciences of the USSR
Moscow V-334, USSR

Tryptophanyl-tRNA synthetase was the first aminoacyl-tRNA synthetase to be partially purified, the result of pioneering work done in Lipmann's laboratory (Davie et al. 1956). Later this enzyme became the subject of intensive study in the laboratory of Bernard and Julie Labouesse (Université de Bordeaux II) and in our laboratory. The parallel investigations of these two groups complement and enrich each other. Part of their work was summarized by Kisselev et al. (1978a).

The aim of this paper is to review briefly current results, including those of our colleagues (see Acknowledgments), obtained with tryptophanyl-tRNA synthetase.

ISOLATION AND CHARACTERIZATION OF THE TRYPTOPHANYL ENZYME

We have identified, isolated, and characterized a covalent derivative formed between the tryptophan residue and beef pancreas tryptophanyl-tRNA synthetase (Kovaleva et al. 1976, 1978a; Favorova et al. 1978). One mole or less of the tryptophan residues is covalently bound per mole of the dimeric enzyme. The tryptophan residue in this derivative is believed to be in the activated state because it (1) exchanges with exogenous tryptophan rather than with other amino acids, (2) reacts with NH_2OH yielding a tryptophanyl hydroxylamine, and (3) aminoacylates $tRNA^{Trp}$ in the absence of ATP (other tRNAs are not aminoacylated). The tryptophanyl enzyme is decomposed during incubation with AMP or pyrophosphate.

The energy-rich bond that makes it possible for the tryptophan in the tryptophanyl enzyme to be substituted by free tryptophan with a protonated amino group may be an anhydrous one (Bender 1960).

It is known that NH_2OH reacts with mixed anhydrides of carboxylic acids to yield the hydroxylamines of both. If the tryptophan residue forms an anhydrous bond with the enzyme, treatment of the tryptophanyl enzyme with NH_2OH can yield a protein hydroxylamine in addition to the accumulation of a tryptophanyl hydroxylamine. To reveal a possible coupled formation of the hydroxylamines of tryptophan and of the enzyme, we have treated the tryptophanyl enzyme containing 0.72 mole of the covalently bound tryptophan per mole of the enzyme with $^{14}CH_3ONH_2$ and isolated, by gel filtration on a Sephadex G-50 column, a labeled protein material containing 0.26 mole of the $^{14}CH_3ONH_2$ residue per mole of the enzyme. CM-cellulose chromatography of low-molecular-weight material obtained after gel filtration revealed the presence of 0.52 mole of tryptophanyl-O-[^{14}C]methyl hydroxylamine per mole of the enzyme.

However, in principle, labeling of the protein with $^{14}CH_3ONH_2$ could be the result of hydroxylaminolysis of asparagine-glycine or amide bonds of the protein (Bornstein and Balian 1970). To rule out this possibility, the following experiment was done. The $^{14}CH_3ONH_2$-treated protein was subjected to SDS-polyacrylamide gel electrophoresis. No scission of the polypeptide chains was observed. In addition, the ^{14}C-label incorporation in the control sample (the enzyme free from covalently bound tryptophan) was considerably lower than in the case of tryptophanyl enzyme (0.06 in comparison with 0.26 mole/mole of enzyme). Thus, we came to the conclusion that the reaction of the tryptophanyl enzyme with $^{14}CH_3ONH_2$ proceeds with simultaneous formation of the O-methyl hydroxylamines of the enzyme and tryptophan, the latter being preferential (0.56 and 0.2 mole/mole of the enzyme, respectively). These results indicate the presence of an anhydrous bond formed between the tryptophan moiety and the protein carboxylic group in the tryptophanyl enzyme.

The pH dependence of the stability of the tryptophanyl enzyme (Kovaleva et al. 1978a) is also in agreement with the chemical nature of the anhydrous bond (Bender 1960). In accord with the energy-rich character of the anhydrous bond, the tryptophanyl residue in the tryptophanyl enzyme is able to be transferred during denaturation with 8 M urea from the carboxylic to the sulfhydryl group of the enzyme (Moroz et al. 1979). The presence of covalently bound tryptophan on one of the enzyme's active sites hinders the formation or binding of aminoacyl adenylate on it (Favorova et al. 1978). As the content of covalently bound tryptophan never exceeded 1 mole per mole of the dimer enzyme, formation of the tryptophanyl enzyme involves half-of-the-sites reactivity, which was also observed in the binding of tRNA^{Trp} and tryptophanyl-tRNA (see below).

The activity of the enzyme treated with $^{14}CH_3ONH_2$ was compared with that of the tryptophanyl enzyme in the reactions of ATP-[^{32}P]pyrophosphate exchange and acylation of tRNA. Blocking of the carboxylic group of the protein had no effect on enzyme activity in the isotope exchange reaction but decreased enzyme activity in the aminoacylation

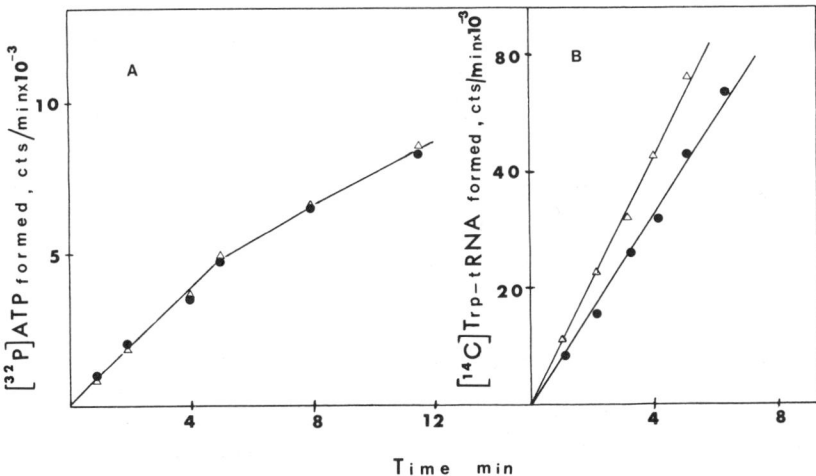

Figure 1 Comparison of the ATP-[^{32}P]pyrophosphate exchange (*A*) and aminoacylation (*B*) activities of the $^{14}CH_3ONH_2$-treated (●) and initial (△) tryptophanyl enzymes. ^{14}C-labeled enzyme contained 0.26 mole of the $^{14}CH_3ONH_2$ residue per mole of the protein.

reaction (Fig. 1). The degree of inactivation (about 28%) fitted in well with the molar ratio of the $^{14}CH_3ONH_2$ incorporation per mole of the enzyme. Although this correlation was observed at relatively low levels of protein modification, it was fully reproduced in several experiments and supported the idea that the protein carboxylic group involved in tryptophanyl-enzyme formation was essential for the tRNA acylation reaction. This conclusion is in agreement with the earlier observation that an activated tryptophan residue is directly transferred from the tryptophanyl enzyme to a specific tRNA^{Trp} molecule in the absence of ATP (Favorova et al. 1978).

The results obtained make it possible to consider this essential carboxylic group as a nucleophilic one that participates in the transfer of an activated amino acid moiety from an aminoacyl adenylate to a specific tRNA.

The existence of a covalent bond between aminoacyl-tRNA synthetase and a substrate amino acid, as well as its chemical nature, has been proved, as far as we are concerned, for the first time; however, further studies are necessary to elucidate the functional role of the tryptophanyl enzyme.

Formation of the acyl enzyme, an active intermediate, was demonstrated for a number of enzymes that, like aminoacyl-tRNA synthetase, activate a carboxyl group of a substrate using the energy of ATP hydrolysis (Anke and Spector 1975; Ray and Cronan 1976). ATP-citrate lyase, the citryl enzyme, is formed via an anhydride bond between citrate and the γ-carboxyl of glutamic acid (Suzuki 1971).

NUCLEOTIDE BINDING SITES

The aim of this section is to describe the data obtained with newly synthesized analogs of ATP used as reversible and irreversible inhibitors of the tryptophanyl-tRNA synthetase. Their structural formulas are presented in Figure 2.

Tryptophanyl-tRNA synthetase was irreversibly inhibited by most of the analogs tested so far (Kovaleva et al. 1978b). Dialysis followed by treating the enzyme with ATP did not restore the enzymatic activity in the ATP-[^{32}P]pyrophosphate exchange reaction. Graphical analysis according to the method of Kitz and Wilson (1968) allowed the calculation of the pseudomono-molecular constants for inhibition rates, and the data obtained are summarized in Table 1.

Obviously, the affinity of the I, II, III, IV, and VI analogs is high (K_i between 3.3×10^{-4} and 4×10^{-5} M) and exceeds the K_m value for ATP

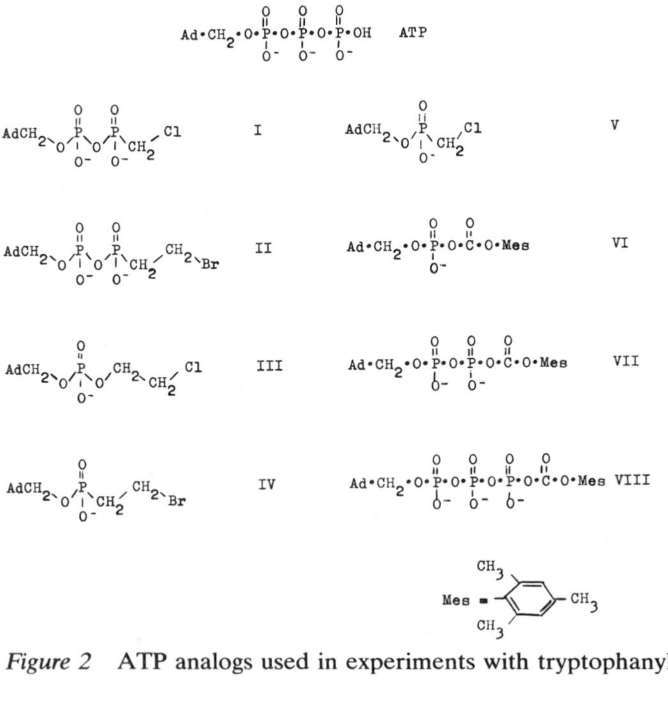

Figure 2 ATP analogs used in experiments with tryptophanyl-tRNA synthetase.

Table 1 ATP analogs, irreversible inhibitors of ATP-[^{32}P]pyrophosphate exchange, catalyzed by the tryptophanyl-tRNA synthetase

ATP analog (0.05–1 mM)	K_i, mM	k_2, min^{-1}	$t_{0.5}$, min
I	0.04	0.025	27.7
II	0.16	0.16	4.3
III	0.25	0.06	12.6
IV	0.09	0.042	16.5
V[a]			
VI	0.33	0.032	21.6
VII	0.15	0.03	21.0
VIII[b]			

[a]No irreversible inhibition.
[b]No inhibition.

(2×10^{-3} and 2×10^{-4} M for the exchange and aminoacylation reactions, respectively). This high affinity is easily explained in the case of the VI analog, which represents an acyl adenylate having a phosphoanhydrous bond oriented, with respect to the 5'-oxygen atom, similarly to the cognate aminoacyl adenylates known to manifest extremely high affinity to the synthetases.

The I, II, and III analogs containing P—C or P—O—C bonds, instead of P—O and P—O—P bonds, differ markedly from ATP in interatomic distances and angles in the polyphosphate chain but may resemble the ATP conformation in one of its transient states. Several researchers, including Santi et al. (1971) and Prasolov et al. (1975), observed that some of the ATP analogs they tested were noncompetitive with respect to ATP. The same type of reversible inhibition was observed with the IV analog (Fig. 3); the K_i constant of noncompetitive inhibition was found to be 10^{-3} M (Kovaleva et al. 1978b). Furthermore, ATP did not protect the enzyme against irreversible inhibition with the analog IV, although the reaction proceeds via formation of a reversible enzyme-inhibitor complex (Table 1).

All these observations lead to a suggestion that, in addition to the ATP binding sites located in the active centers of the dimeric enzyme and the tRNA-terminal A binding sites, the enzyme may contain nucleotide binding sites that are located outside the catalytic centers of the enzyme. This hypothesis has been formulated and tested experimentally by Nevinsky et al. (1979). It was observed that a wide variety of compounds, including AMP, GMP, CMP, ADP, GDP, and γ-p-azidoanilides of ADP, GDP, ATP, GTP, and CTP, activate the ATP-[^{32}P]pyrophosphate

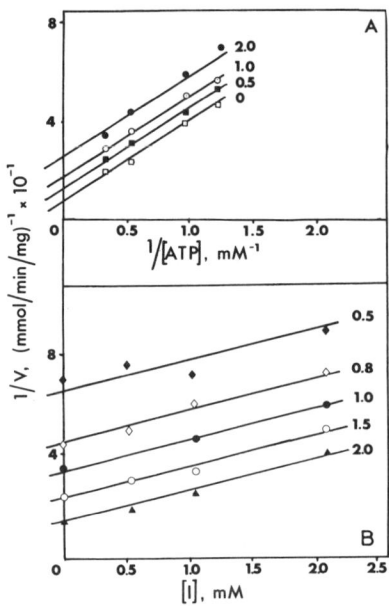

Figure 3 The reciprocal plots of the inhibition rates for ATP-[^{32}P]pyrophosphate exchange catalyzed by tryptophanyl-tRNA synthetase caused by the ATP analog IV against concentrations of ATP (*A*) and inhibitor (*B*). The ATP and inhibitor concentrations (in mM) are shown on the curves.

exchange catalyzed by tryptophanyl-tRNA synthetase. The optimum concentration for the activating effect lies within the range 1 to 5 μM for AMP, GMP, *p*-azidoanilides of ADP and GDP and within the range 10–50 μM for the other nucleotides. These values are 1–2 orders of magnitude lower than the K_m value for ATP. The results of one experiment of this type are shown in Figure 4.

The covalent attachment of the γ-*p*-azidoanilide of GMP to the enzyme abolishes the activating effect of nucleotides, leaving the enzymatic activity unchanged. Both ATP and GMP added to the enzyme separately block the covalent reaction with 2 moles of azido-ATP per mole of the enzyme. Protection by these nucleotides is additive: 4 moles of the analog are not covalently bound to the enzyme (Fig. 5).

All the aforementioned facts, as well as others (see Nevinsky et al. 1979), are consistent with the idea that there are two types of nucleotide binding sites on the enzyme; the first one includes the ATP binding sites of the two active centers and the second one is nonspecific with respect to ATP, both in terms of the number of phosphate groups and the nature of a heterocyclic base. These latter binding sites (probably two per dimeric enzyme) might be considered as effector sites that are located outside the active centers of the enzyme.

The discovery of effector sites in aminoacyl-tRNA synthetase opens a new area of research work on these intriguing enzymes and brings us closer to understanding the regulation of their activity in vivo.

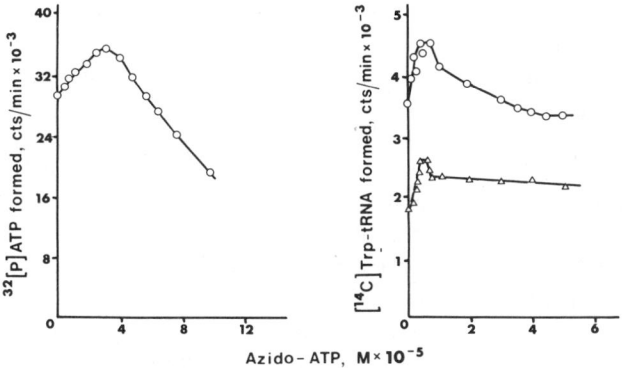

Figure 4 Dependence of the ATP-[^{32}P]pyrophosphate exchange (*left*) and tRNA aminoacylation (*right*) reactions on the azido-ATP concentration. Incubation time: 10 min (*left*) and 4 min (*right*); concentration of ATP: 0.25 mM (*left*) and 0.75 mM (*right*).

THE HALF-OF-THE-SITES REACTIVITY AND THE STEADY-STATE KINETIC MODEL FOR ENZYME FUNCTIONING

Investigation of the substrate binding and, independently, steady-state kinetic measurements, coupled with probabilistic treatment of the kinetic data, make it possible to construct a model for enzyme functioning (Kisselev et al. 1978b).

The binding of tRNA^{Trp} to the native tryptophanyl-tRNA synthetase is biphasic at pH 5.8 in the presence of Mg^{++}. Two tRNA molecules bind with $K_{dis} = 3.6 \times 10^{-8}$ M and $K_{dis} = 0.9 \times 10^{-6}$ M, respectively, pointing to a strong half-of-the-sites reactivity between the binding sites for tRNA (Akhverdyan et al. 1977). This observation opens a way for converting the enzyme with two active sites into a one-site enzyme. For this purpose,

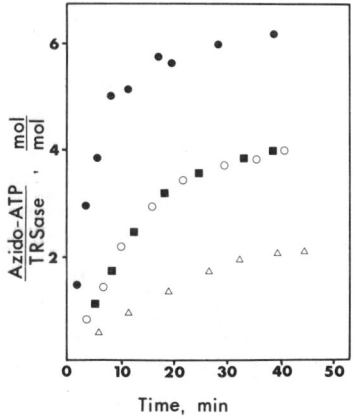

Figure 5 Kinetics of photoaddition of azido-[^{14}C]ATP in the presence of various ligands to the tryptophanyl-tRNA synthetase. Concentrations (in mM): azido-ATP, 0.1; GMP, 1.0; ATP; 1.0. (●) Azido-[^{14}C]ATP; (○) the same plus ATP; (■) the same plus GMP; (△) the same plus ATP and GMP.

N-chloroambucilyl-tryptophanyl-tRNA^{Trp} has been used, which is able to alkylate the enzyme.

Before the alkylation reaction, the tryptophanyl-tRNA analog binds noncovalently to the enzyme with $K_{dis} = 5.5 \times 10^{-8}$ M (pH 5.8, 10 mM Mg^{++}), similar to the binding of the first tryptophanyl-tRNA molecule ($K_{dis} = 4.8 \times 10^{-8}$ M). After sufficient incubation, a covalent bond is formed between N-chloroambucilyl-tryptophanyl-tRNA and the enzyme; 1 mole of the affinity reagent alkylates 1 mole of the protein. The alkylation reaction is completely inhibited in the presence of tRNA^{Trp}, whereas the tRNA free from the tRNA^{Trp} does not affect the rate of alkylation. The exhaustive alkylation of the enzyme partially inhibits the activity in the reaction of ATP-[^{32}P]pyrophosphate exchange and completely blocks the aminoacylation reaction of tRNA^{Trp}. Cleavage of the tRNA that is covalently bound to the protein restores both the exchange and aminoacylation reactions.

Thus, after modification of tryptophanyl-tRNA synthetase with the alkylating derivative of tryptophanyl-tRNA, the two-site enzyme converts into a one-site one that is inactive in the interaction with tRNA, due to half-of-the-sites reactivity, and has 50% activity in the tryptophan activation.

The approximately 50% reduction in the rate of the ATP-[^{32}P]pyrophosphate exchange reaction is observed when an excess of uncharged tRNA^{Trp} is added to the incubation mixture (Malygin et al. 1976). The inhibition curve allows one to calculate Hill's coefficient (n), which was found to be about 0.5. By contrast, if Hill's analysis is applied to the experimental dependencies of the rate of aminoacyl-tRNA formation on the tRNA concentration, the n values obtained from the graphs are about 2. The combination of interactions having opposite signs ($n < 1$ and $n > 1$) can be explained with the aid of the trigger model (Lazdunski et al. 1971) in the following way. Since we observe that n is about 2 in the tRNA aminoacylation reaction, it might be assumed that two tRNA molecules should be bound simultaneously to the enzyme in the course of the reaction. However, these molecules are in different states. tRNA^{Trp} bound to the enzyme first reacts with tryptophanyl adenylate, thus excluding the latter from the ATP-[^{32}P]pyrophosphate exchange. A second molecule of tRNA^{Trp} can bind to the enzyme, but its access to the adenylate is not permitted, and the latter continues to participate freely in the isotope exchange. This assumption explains why the ATP-[^{32}P]pryopophosphate exchange does not stop even under saturating tRNA concentrations. However, due to negative cooperativity among the tRNA molecules in the binding to the enzyme, tRNA^{Trp} contributes to the removal of tryptophanyl-tRNA^{Trp} from the enzyme, leading to the exchange of the states of the active centers. Consequently, we observe positive cooperativity in the tRNA aminoacylation reaction.

All the aforementioned considerations, as well as others, are reflected in the scheme (Fig. 6) for beef pancreas tryptophanyl-tRNA synthetase functioning (Kisselev et al. 1978b). There exist three routes for the aminoacylation reaction, having common and different stages. Route 1 presumably does not function in vivo, since it may proceed only at very low concentrations of tRNA. The ratio of routes 2 and 3 depends on the tRNA concentration and probably both of them may function in vivo.

A detailed description of the kinetic scheme is given by Kisselev et al. (1978b).

As shown above, the half-of-the-sites reactivity opens a way for converting the two-site enzyme into a one-site form: the latter, being a "simplified version" of the native enzyme, offers a unique possibility for investigating its kinetic characteristics and for comparing them with the nonmodified form. In principle, this is the way to elucidate the role of the active-site interactions in the catalytic mechanism.

The exchange reaction catalyzed by tryptophanyl-tRNA synthetase alkylated with N-chloroambucilyl-tryptophanyl-tRNA was studied by Zinoviev et al. (1977). It has been found that substrates bind to the one-site enzyme in the same order as to the two-site one: ATP is added first and tryptophan second. As modified tRNA is being fixed at one of the two sites, the equilibrium of the reaction is shifted toward the formation of an aminoacyl adenylate (5.8×10^5 M^{-1} as compared with 1.2×10^5 M^{-1} for the native enzyme). The parameters of the productive

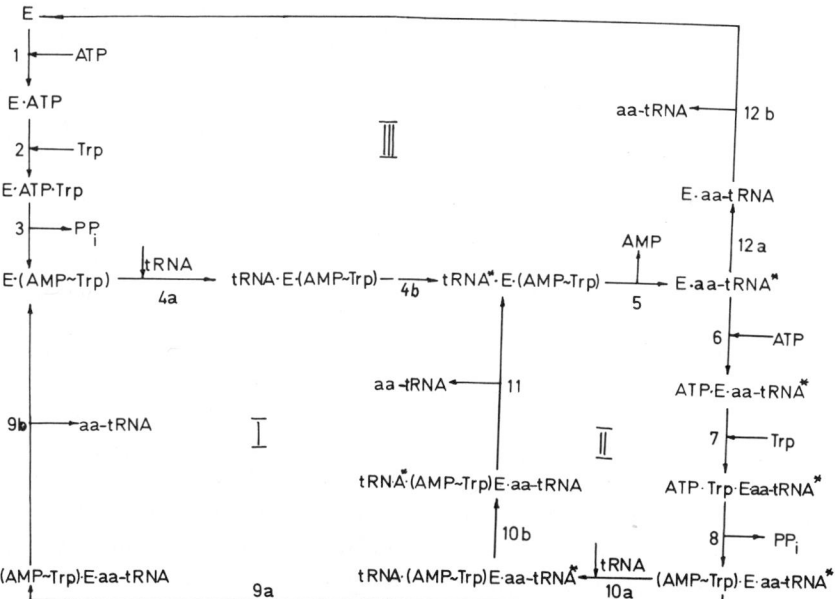

Figure 6 A scheme illustrating the functioning of beef pancreas tryptophanyl-tRNA synthetase (E). Trp, L-tryptophan; PPi, pyrophosphate; (AMP ~ Trp), tryptophanyl adenylate; aa-tRNA, tryptophanyl-tRNA; tRNA*, tRNA that has undergone conformational adjustment to the enzyme and thus is capable of accepting tryptophan.

reaction steps were found to be somewhat improved, whereas the opposite was true of the dead-end complexes.

The other way to produce a one-site enzyme is to use analogs of the other substrates, i.e., amino acid. It has been shown for valyl-tRNA synthetase (Frolova et al. 1973) and proved to be also true of some other synthetases (Silver and Laursen 1974; Rainey et al. 1976) that chloromethyl ketone derivatives of substrate amino acids are affinity-labeling reagents for the amino acid binding site. In the case of beef pancreas tryptophanyl-tRNA synthetase, it has been shown by Kovaleva et al. (1979) that the tryptophanyl chloromethyl ketone reacts at two enzyme active sites. However, under certain conditions in the absence of inorganic pyrophosphatase, only 1 mole of tryptophanyl adenylate is formed, leaving the other active site free. This half-of-the-sites reactivity opens a way to selective alkylation with the chloromethyl ketone of one of the two tryptophan binding sites. When the enzyme with one active site occupied by tryptophanyl adenylate is alkylated, the rate of modification of the nonoccupied active site turns out to be 1.5-fold higher than in the case of alkylation of the free enzyme and exceeds the rate of the tryptophanyl adenylate hydrolysis at the other active site by more than one order of magnitude.

On the other hand, the chloromethyl ketone of tryptophan located at the free active site promotes the hydrolysis of the tryptophanyl adenylate at the protected site, leading to its modification by the second molecule of chloromethyl ketone after prolonged incubation. Therefore, it was essential to stop the reaction properly to obtain a one-site enzyme.

The one-site enzyme prepared in this way is able to form 1 mole of adenylate if incubated in the presence of inorganic pyrophosphatase. The one-site enzyme has a K_m value for all the substrates similar to the native one, and V_{max} is twice as low in the ATP-[^{32}P]pyrophosphate exchange reaction.

The kinetic studies with this second type of one-site tryptophanyl-tRNA synthetase have been performed. Preliminary data have shown that some of the kinetic parameters were changed in the same way as in the enzyme alkylated with N-chloroambucilyl-tryptophanyl-tRNA, whereas the others resemble those of the native synthetase (Degtyaryev et al., in prep.).

The influence of the adenylate on the rate of alkylation reaction, on the one hand, and the influence of the covalently bound tryptophan analog on the adenylate hydrolysis, on the other, point once again to the mutual interactions between the active sites during catalysis, presumably via conformational changes of the interacting subunits.

ACKNOWLEDGMENTS

The authors are thankful to Professor V. A. Engelhardt for continuous encouragement of this work and to their colleagues at the Institute of

Molecular Biology (Moscow), the Institute of Organic Chemistry (Novosibirsk), and Moscow State University for their valuable contributions to the present work. Special thanks to Drs. D. G. Knorre, O. I. Lavrik, E. G. Malygin, G. A. Nevinsky, and V. V. Zinoviev from Novosibirsk; Drs. E. S. Severin, Z. Shabarova, V. Z. Akhvezdyan, L. L. Kochkina, and S. G. Moroz from Moscow; and Dr. R. Krauspe from Halle/Saale; their work contributed heavily to this discussion.

REFERENCES

Akhverdyan, V. Z., L. L. Kisselev, D. G. Knorre, O. I. Lavrik, and G. A. Nevinsky. 1977. Affinity labeling of tryptophanyl-transfer RNA synthetase. *J. Mol. Biol.* **113**:475.

Anke, H. and L. B. Spector. 1975. Evidence for an acetyl-enzyme intermediate in the action of acetyl-CoA synthetase. *Biochem. Biophys. Res. Commun.* **67**:767.

Bender, M. L. 1960. Mechanisms of catalysis of nucleophilic reactions of carboxylic acid derivatives. *Chem. Rev.* **60**:53.

Bornstein, P. and G. Balian. 1970. The specific non-enzymatic cleavage of bovine ribonuclease with hydroxylamine. *J. Biol. Chem.* **245**:4854.

Davie, E. W., V. V. Konigsberger, and F. Lipmann. 1956. The isolation of tryptophan-activating enzyme from beef pancreas. *Arch. Biochem. Biophys.* **65**:21.

Fasiolo, F. and A. R. Fersht. 1977. The aminoacyl adenylate mechanism in the aminoacylation reaction of yeast phenylalanyl-tRNA synthetase. *Eur. J. Biochem.* **85**:85.

Favorova, O. O., G. K. Kovaleva, S. G. Moroz, and L. L. Kisselev. 1978. Tryptophanyl-tRNA synthetase: Isolation and characterization of the tryptophanyl-enzyme. *Mol. Biol.* **12**:588.

Frolova, L. Yu., G. K. Kovaleva, M. B. Agalarova, and L. L. Kisselev. 1973. Irreversible inhibition of beef liver valyl-tRNA synthetase by an alkylating derivative of L-valine. *FEBS Lett.* **34**:213.

Kisselev, L. L., O. O. Favorova, and G. K. Kovaleva. 1978a. Tryptophanyl-tRNA synthetase from beef pancreas. *Methods Enzymol.* **59**:234.

Kisselev, L. L., E. G. Malygin, V. Z. Akhverdyan, and V. V. Zinoviev. 1978b. Mechanism of functioning of tryptophanyl-tRNA synthetase. *Dokl. Akad. Nauk SSSR* **238**:1475.

Kitz, R. and J. B. Wilson. 1968. Esters of methanesulfonic acid as irreversible inhibitors of acetylcholine-esterase. *J. Biol. Chem.* **237**:3245.

Knorre, D. G., E. G. Malygin, M. G. Slinko, V. I. Timoshenko, V. V. Zinoviev, L. L. Kisselev, L. L. Kochkina, and O. O. Favorova. 1974. Beef pancreas tryptophanyl-tRNA synthetase. Order of substrate binding in ATP-[^{32}P]pyrophosphate exchange reaction. *Biochimie* **56**:845.

Kochkina, L. L., V. Z. Akhverdyan, L. L. Kisselev, V. V. Zinoviev, and E. G. Malygin. 1976. Mechanism of reaction of tryptophanyl-tRNA formation catalysed by tryptophan: tRNA lygase. *Mol. Biol.* **10**:437.

Kovaleva, G. K., S. C. Degtyaryev, and O. O. Favorova. 1979. Affinity modification of tryptophanyl-tRNA synthetase by an alkylating analog of tryptophan. *Mol. Biol.* (in press).

Kovaleva, G. K., S. G. Moroz, O. O. Favorova, and L. L. Kisselev. 1978a.

Tryptophanyl-tRNA synthetase: Evidence for an anhydrous bond involved in the tryptophanyl enzyme formation. *FEBS Lett.* **95:**81.

Kovaleva, G. K., O. O. Favorova, S. G. Moroz, R. Krauspe, and L. L. Kisselev. 1976. An active derivative of tryptophan and tryptophanyl-tRNA synthetase. *Dokl. Akad. Nauk SSSR* **229:**492.

Kovaleva, G. K., L. L. Ivanov, I. A. Madoyan, O. O. Favorova, E. S. Severin, N. N. Gulyaev, L. A. Baranova, Z. A. Shabarova, N. I. Sokolova, and L. L. Kisselev. 1978b. Inhibition of tryptophanyl-tRNA synthetase by modifying ATP analogs. *Biokhimiya* **43:**525.

Lasdunski, M., C. Petitclerk, D. Chappelet, and C. Lasdunski. 1971. Flip-flop mechanisms in enzymology. A model: The alkaline phosphatase of *Escherichia coli. Eur. J. Biochem.* **20:**124.

Malygin, E. G., V. V. Zinoviev, F. Fasiolo, L. L. Kisselev, L. L. Kochkina, and V. Z. Akhverdyan. 1976. Interaction of aminoacyl-tRNA synthetases and tRNA: Positive and negative cooperativity of their active centers. *Mol. Biol. Rep.* **2:**445.

Moroz, S. G., R. Krauspe, G. K. Kovaleva, and O. O. Favorova. 1979. Characterization of the tryptophanyl enzyme obtained after denaturation on an active tryptophanylated derivative of the tryptophanyl-tRNA synthetase. *Bioorg. Khim.* (in press).

Nevinsky, G. A., O. I. Lavrik, O. O. Favorova, and L. L. Kisselev. 1979. Nucleotide binding sites of tryptophanyl-tRNA synthetase. *Bioorgan. Khim.* **5:**352.

Prasolov, V. S., A. M. Kritsyn, S. N. Michailov, and V. L. Florentiev. 1975. Influence of 9 (ω'-hydroxyalkyl)-adenines and their triphosphates on the ATP-[^{32}P]pyrophosphate exchange catalysed by tryptophanyl-tRNA synthetase. *Dokl. Akad. Nauk SSSR* **221:**1226.

Rainey, R., E. Holler, and M.-R. Kula. 1976. Labeling of L-isoleucine tRNA lygase from *Escherichia coli* with L-isoleucyl bromomethyl ketone. *Eur. J. Biochem.* **63:**419.

Ray, T. K. and J. E. Cronan. 1976. Activation of long chain fatty acids with acyl carrier protein: Demonstration of a new enzyme, acyl-acyl carrier protein synthetase in *Escherichia coli. Proc. Natl. Acad. Sci.* **73:**4374.

Santi, D. V., P. V. Danenberg, and K. A. Montgomery. 1971. Phenylalanyl-tRNA synthetase from *E. coli*. Analysis of the adenosine triphosphate binding site. *Biochemistry* **10:**4821.

Silver, J. and R. A. Laursen. 1974. Inactivation of aminoacyl-tRNA synthetases by amino acid chloromethylketones. *Biochim. Biophys. Acta* **340:**77.

Suzuki, F. 1971. Studies on adenosine triphosphate citrate lyase of rat liver. Binding site of citrate. *Biochemistry* **10:**2707.

Zinoviev, V. V., N. G. Rubtsova, O. I. Lavrik, E. G. Malygin, V. Z. Akhverdyan, O. O. Favorova, and L. L. Kisselev. 1977. Comparison of the ATP-[^{32}P]pyrophosphate exchange reactions catalysed by native (two-site) and chemically modified (one-site) tryptophanyl-tRNA synthetase. *FEBS Lett.* **82:**130.

Zinoviev, V. V., L. L. Kisselev, D. G. Knorre, L. L. Kochkina, E. G. Malygin, M. S. Slinko, V. I. Timoshenko, and O. O. Favorova. 1974. Kinetic scheme and kinetic parameters of ATP-[^{32}P]pyrophosphate exchange reaction catalysed by beef pancreas tryptophanyl-tRNA synthetase. *Mol. Biol.* **8:**380.

Editing Mechanisms in the Aminoacylation of tRNA

Alan R. Fersht*
Department of Chemistry
Imperial College of Science and Technology
London SW7 2AY England

There is a basic problem in protein biosynthesis in that there are insufficient differences in structure between certain amino acids to enable a sufficiently precise distinction between them by simple preferential binding of the correct substrate to the aminoacyl-tRNA synthetase. The classic example, anticipated by Pauling (1958), is the competition between valine and isoleucine for the isoleucyl-tRNA synthetase. Since the active site of the enzyme is large enough, or flexible enough, to accommodate isoleucine, it must also bind the smaller molecule of valine. The difference in binding energy between the two amino acids cannot be greater than that of the additional methylene group on the larger substrate, now known to be worth up to about 3 kcal/molecule, equivalent to a factor of 200–300. This sets the upper limit on the discrimination between valine and isoleucine by simple binding. Furthermore, this cannot be increased by invoking any of the theories of enzyme catalysis, such as "strain," "induced fit," and "nonproductive binding," or by the effects of a series of sequential steps or interacting active sites (Fersht 1974, 1977a). Yet, it has been found by Loftfield (1963; Loftfield and Vanderjagt 1972) that the overall error frequency for the misincorporation of valine for isoleucine in the synthesis of ovalbumin and globin is only 3×10^{-4}, and this includes any errors in transcription and translation.

The increase in specificity results from the evolution of an editing or proofreading function whereby the aminoacyl-tRNA synthetase has a hydrolytic activity that specifically destroys by deacylation any mis-aminoacylated product. Predicted by Crick, the editing mechanism was found experimentally by Baldwin, and Berg (Norris and Berg 1964; Baldwin and Berg 1966). The isoleucyl-tRNA synthetase activates valine to form an isolable, enzyme-bound valyl adenylate complex (although the value of k_{cat}/K_m for the reaction is about 150 times lower than that for the reaction with isoleucine). But, whereas the addition of tRNA[Ile] to the complex with isoleucyl adenylate leads to the formation of isoleucyl-

---

*Present address (1973–1979): Department of Biochemistry, Stanford University Medical Center, Stanford, California 94305.

tRNA^{Ile}, there is quantitative hydrolysis of the valyl adenylate on addition of tRNA^{Ile} to the enzyme-bound complex (Eq. 1).

$$E^{Ile} \xrightarrow[ATP]{Val} E^{Ile} \cdot Val \sim AMP + PPi \xrightarrow{tRNA^{Ile}} E^{Ile} + Val + AMP \qquad (1)$$

Thus, in the presence of tRNA^{Ile}, the isoleucyl-tRNA synthetase (E^{Ile}) is a valine-dependent ATP/pyrophosphatase as the amino acid is continuously activated and the error corrected.

THE REACTION PATHWAY: MISACYLATION/DEACYLATION

Subsequent to the observation that aminoacyl-tRNA synthetases possess weak hydrolytic activities toward their cognately charged tRNAs, it was found by Eldred and Schimmel (1972) that the isoleucyl-tRNA synthetase deacylates valyl-tRNA^{Ile} with a value of 0.02 sec^{-1} for k_{cat} at 3°C and by Yarus (1972a) that the phenylalanyl-tRNA synthetase deacylates isoleucyl-tRNA^{Phe} with a value of 2 sec^{-1} for k_{cat} at 37°C. It was thus speculated that the hydrolytic step in editing occurs after the transfer of amino acid to tRNA. This pathway has now been proved by the direct observation and trapping of the mischarged tRNAs in the rejection of threonine (Thr) and α-aminobutyrate by the valyl-tRNA synthetases from *Bacillus stearothermophilus, Escherichia coli,* and yeast (Eq. 2; E^{Val} is valyl-tRNA synthetase) (Fersht and Kaethner 1976a; Fersht and Dingwall 1979a,b).

$$E^{Val} \cdot Thr \sim AMP \cdot tRNA^{Val} \longrightarrow E^{Val} \cdot Thr\text{-}tRNA^{Val} + AMP$$
$$\downarrow 40 \text{ sec}^{-1} \qquad (2)$$
$$E^{Val} + Thr + tRNA^{Val}$$

The mischarged valyl-tRNA^{Ile} was not detected in the rejection of valine by the isoleucyl-tRNA synthetase, however, but this could be the result of a very rapid hydrolytic step (Fersht 1977b).

REACTION MECHANISM

Little is known about the actual reaction mechanism. The hydrolytic activity of the enzymes is either unaffected (Fersht and Kaethner 1976a) or only slightly modified (Schreier and Schimmel 1972; Yarus 1972a) when aminoacyl adenylate is bound to the active site, so there appears to be a separate and distinct hydrolytic site. The rate constant (turnover number) for the deacylation of α-aminobutyryl-tRNA^{Val} at 25°C by the valyl-tRNA synthetase is independent of pH between pH 5.8 and 7.8 at 50 sec^{-1} (Fersht and Dingwall 1979a,b).

From the observation that 2'-deoxytRNA^{Ile} (tRNA^{Ile} in which the terminal A is replaced by 2'-deoxyadenosine) is fully aminoacylated with valine by the isoleucyl-tRNA synthetase (von der Haar and Cramer 1975), it was suggested that the nonaccepting 3'-OH of the unmodified tRNA is chemically involved in the hydrolytic reaction (von der Haar and Cramer 1976; Igloi et al. 1977). This now seems unlikely, or at least not general, since we find that 3'-deoxytRNA^{Met} is as resistant to misacylation by the methionyl-tRNA synthetase as is tRNA^{Met} (Fersht and Dingwall 1979c).

GENERAL MECHANISMS FOR EDITING

Kinetic Proofreading

This elegant mechanism (Hopfield 1974) (Eq. 3) increases specificity without invoking an active hydrolytic site on the enzyme (E). (AA is amino acid.)

$$\begin{array}{c} E \cdot AA \xrightarrow{ATP\ \ PPi} E \cdot AA \sim AMP \xrightarrow{k_3} AA\text{-}tRNA \\ k_1 \updownarrow k_{-1} \qquad\qquad k_4 \updownarrow k_{-4} \\ E + AA \qquad\qquad E + AA \sim AMP \xrightarrow{k_h} AA + AMP \end{array} \qquad (3)$$

It was proposed that the complex with the incorrect aminoacyl adenylate would dissociate (via k_{-4}) and hydrolyze faster than the transfer (via k_3) to the tRNA. Since it has now been shown that the reaction pathway involves misacylation of the tRNA (Fersht and Kaethner 1976a) and that the dissociation of the isoleucyl-tRNA synthetase · valine $\sim$ AMP complex is very slow, the specific mechanism of Hopfield is not of importance (Fersht 1977b; Mulvey and Fersht 1977). If the discrimination in the first step of Equation 3 (k_1, k_{-1}) is f, the addition of the subsequent dissociation step can increase this only to f^2. Furthermore, the full increase can only occur at the expense of the correct substrate being substantially hydrolyzed via k_{-4} and k_h (Mulvey and Fersht 1977).

Hydrolytic Editing and Chemical Proofreading

Hydrolytic editing (Fersht and Kaethner 1976a) and chemical proofreading (von der Haar and Cramer 1976) are the same and are equivalent to the Yarus (1972a) and Schimmel (Eldred and Schimmel 1972; Schreier and Schimmel 1972) proposals. There is a specific enzyme-catalyzed deacylation of the mischarged tRNA. The increase in specificity is not limited to a factor f^2.

The Double-sieve Editing Mechanism

There is an almost bewildering array of misactivations of amino acids reported (summarized by Igloi et al. 1978). These may be rationalized, however, by one simple scheme that predicts both the nature of misactivations and when editing mechanisms are required. Furthermore, it enables a large range of amino acids to be edited by just the combination of one synthetic and one hydrolytic site. It is proposed (Fig. 1) that naturally occurring amino acids larger than the specific substrate are rejected at a tolerable level by simple steric exclusion from the aminoacylation site. Amino acids smaller than the specific substrate are accepted by the aminoacylation site and are activated, albeit at a lower rate because of the poorer fit. The steric exclusion principle is then used again, this time to exclude the specific substrate from the hydrolytic site but to accept the products of the smaller substrates. Substrates that are isosteric with the specific substrate, such as threonine competing with valine for the valyl-tRNA synthetase, cannot be rejected by size alone. In this case, the specific binding characteristics of the substrates must be invoked. For example, the hydrogen-bonding potential of the OH group of threonine could be used to draw it into a hydrophilic, hydrolytic site.

The hydrolytic and synthetic sites thus function as a pair of sieves, crudely sorting the substrates according to size. Superimposed upon this is a discrimination among the substrates on the grounds of steric fit: the closer the structure of the competing substrate to that of the correct substrate, the faster the activation.

The evidence available supports the double-sieve model (Fersht 1977a; Fersht and Dingwall 1979d): the only aminoacyl-tRNAs that are rapidly deacylated by aminoacyl-tRNA synthetases are those misacylated with amino acids that are smaller than, or isosteric with, the correct substrate; the most readily misactivated amino acids are those that are only slightly

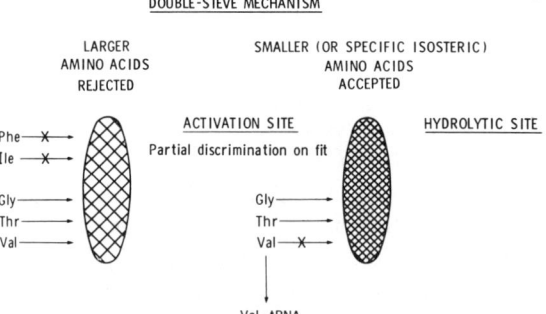

Figure 1 The double-sieve mechanism (modified from Fersht 1977a). Natural amino acids that are larger than the specific substrate are rejected at a tolerable level by steric exclusion (the first sieve). Isosteric and smaller amino acids are activated, but there is a discrimination by fit. Where necessary, there is a hydrolytic site that accepts the products of smaller or naturally occurring isosteric amino acids (the second sieve).

smaller than the correct substrate; larger naturally occurring amino acids are not activated by the enzymes at an appreciable rate.

In apparent contradiction to this, it has been widely thought that valyl-tRNA synthetases readily activate isoleucine. For example, the enzyme from *E. coli* has been reported to activate isoleucine with V_{max} 10% of that for valine (Loftfield and Eigner 1966) and the enzyme from yeast to activate isoleucine with V_{max} 57% of that for valine (Igloi et al. 1978). However, we have recently shown that these activities are caused mainly by residual traces of valine in preparations of isoleucine (Fersht and Dingwall 1979d). The valine was removed in the form of its hydroxamate by column chromatography after incubating the isoleucine with the valyl-tRNA synthetase, ATP, and hydroxylamine.

$$(CH_3)_2CHCH(NH_2)CO_2H \xrightarrow[ATP,H_2NOH]{E^{Val}} (CH_3)_2CHCH(NH_2)CONHOH \quad (4)$$

Over 97% of the isoleucine-stimulated pyrophosphate exchange activity is attributable to the impurity. The selectivity of the valyl-tRNA synthetase against isoleucine is at least 1×10^5. This is equivalent to at least 7 kcal/molecule of destabilization energy on cramming a methyl group into too small a cavity.

Although the enzymes have a high discrimination against naturally occurring amino acids, this may be reduced when faced with unnatural ones against which no defense has needed to be evolved. For example, the valyl-tRNA synthetase activates *O*-methylthreonine at an appreciable rate (Igloi et al. 1977), as does the methionyl synthetase activate ethionine (Trupin et al. 1966). But, true to the double-sieve model, the resultant misacylated tRNAs are not edited, the products being too large for the second sieve. The same applies to unnatural isosteres, but unnatural smaller amino acids are edited in the normal manner, consistent with the sorting-by-size principle.

THE GENERALITY OF EDITING MECHANISMS

It is predicted from the double-sieve model that editing is required whenever there are competing amino acids that are smaller than the correct substrate but sufficiently similar to be activated readily. Editing has been found so far with the isoleucyl- (Baldwin and Berg 1966), valyl- (Fersht and Kaethner 1976a; Igloi et al. 1977), and methionyl-tRNA synthetases (Fersht and Dingwall 1979c), in response to valine, threonine, and homocysteine, respectively. Editing will undoubtedly be found for the alanyl- (against glycine) and threonyl-tRNA synthetases (against serine and valine). The cysteinyl-tRNA synthetase, however, has such a high specificity for cysteine compared with serine and alanine that no editing is required and there is no evidence for its presence (Fersht and Dingwall 1979c). The same is probably true for the tyrosyl-tRNA synthetase (Igloi et al. 1978).

IS THE tRNA PROOFREAD?

Unlike the amino acid, the tRNA is such a complex structure that there should be adequate binding energy and structural variation for accurate discrimination without recourse to an editing mechanism. In a recent paper, however, Yamane and Hopfield (1977) calculate from the published association constants of tRNA^{fMet} and tRNA^{Phe} with the isoleucyl-tRNA synthetase from *E. coli* (Yarus 1972b; Mertes et al. 1972) that there are selectivities of only 3×10^3 and 2×10^4, respectively, for tRNA^{Ile}. But, the relevant kinetic quality for calculating selectivity is k_{cat}/K_m, not K_s or K_m alone (Fersht 1974, 1977a). Using k_{cat}/K_m, it is calculated that there is a selectivity for tRNA^{Ile} of 6.5×10^6 against tRNA^{fMet} and 2.5×10^7 against tRNA^{Phe}, values that are more than adequate for a tolerable error rate.

Yamane and Hopfield (1977) also claim that there is direct evidence for proofreading of the tRNA by the isoleucyl-tRNA synthetase in that 25 and 40 moles of ATP are hydrolyzed for each molecule of isoleucyl-tRNA^{fMet} and isoleucyl-tRNA^{Phe} formed, respectively. However, under the conditions of their experiments, the spontaneous rate of hydrolysis of isoleucyl-tRNA synthetase·isoleucine ~ AMP (Eq. 5) is some 25–100 times faster than the aminoacylation rates. The apparent proofreading is just an artifact of the slow reaction rates.

$$E^{Ile} \cdot Ile \sim AMP \xrightarrow[Ile + AMP]{} E^{Ile} \xrightarrow{ATP} E^{Ile} \cdot Ile \sim AMP \qquad (5)$$

(Turnover number for ATP consumption = 1.7×10^{-3} sec^{-1} at 25°C and pH 7.78 [Fersht and Kaethner 1976b; Mulvey and Fersht 1977]; isoleucylation rates of tRNA^{Met} [1 μM] and tRNA^{Phe} [2 μM] are estimated from the data of Yarus [1972b] and Mertes et al. [1972] to be 4×10^{-5} and 2×10^{-5} sec^{-1}, respectively.)

In vivo, the consumption of ATP would not occur because the isoleucyl-tRNA synthetase·isoleucine ~ AMP complex rapidly reacts with tRNA^{Ile}.

The suggestion that the deacylation of isoleucyl-tRNA^{Phe} catalyzed by phenylalanyl-tRNA synthetase (E^{Phe}) is a correction mechanism for the misrecognition of tRNA^{Phe} by the isoleucyl-tRNA synthetase (Eq. 6) (Yarus 1972a) now appears unlikely.

$$E^{Ile} + Ile + ATP + tRNA^{Phe} \rightarrow Ile\text{-}tRNA^{Phe} \atop \qquad\qquad\qquad\qquad\qquad\qquad\quad \downarrow E^{Phe} \atop \qquad\qquad\qquad\qquad\qquad\qquad Ile + tRNA^{Phe} \qquad (6)$$

First, Bonnet and Ebel (1974) have presented calculations that show that the reaction rate is too low to increase the accuracy appreciably. Second, this mechanism requires that the misacylated tRNA leave the first enzyme before editing. It is now known that there is sufficient elongation factor Tu in *E. coli* to sequester, and protect against hydrolysis, all the aminoacyl-tRNA that is released into solution (Mulvey and Fersht 1977). Thus, there appears to be no need, and no evidence, for the proofreading of the tRNA. The deacylation activity of phenylalanyl-tRNA synthetase probably corrects the misrecognition of amino acids by phenylalanyl-tRNA synthetase.

REFERENCES

Baldwin, A. N. and P. Berg. 1966. Transfer ribonucleic acid-induced hydrolysis of valyl adenylate bound to isoleucyl ribonucleic acid synthetase. *J. Biol. Chem.* **241:**839.

Bonnet, J. and J. P. Ebel. 1974. Correction of aminoacylation errors. Evidence for a nonsignificant role of the aminoacyl-tRNA synthetase catalysed deacylation of aminoacyl-tRNAs. *FEBS Letts.* **39:**259.

Eldred, E. W. and P. Schimmel. 1972. Rapid deacylation by isoleucyl transfer ribonucleic acid synthetase of isoleucine-specific transfer ribonucleic acid aminoacylated with valine. *J. Biol. Chem.* **247:**2961.

Fersht, A. R. 1974. Catalysis, binding and enzyme-substrate complementarity. *Proc. R. Soc. Lond. B* **187:**397.

———. 1977a. *Enzyme structure and mechanism* (see chapter 11). W. H. Freeman, San Francisco, California.

———. 1977b. Editing mechanisms in protein synthesis. Rejection of valine by the isoleucyl-tRNA synthetase. *Biochemistry* **16:**1025.

Fersht, A. R. and C. Dingwall. 1979a. Mechanism and specificity of aminoacyl-tRNA synthetases: "Double-sieve sorting." *FEBS Proc. Meet.* **52:**69.

———. 1979b. Establishing the misacylation/deacylation of tRNA pathway for the editing mechanism of prokaryotic and eukaryotic valyl-tRNA synthetases. *Biochemistry* **18:**1238.

———. 1979c. An editing mechanism for the methionyl-tRNA synthetase in the selection of amino acids in protein synthesis. *Biochemistry* **18:**1250.

———. 1979d. Evidence for the double-sieve editing mechanism for selection of amino acids in protein synthesis: Steric exclusion of isoleucine by valyl-tRNA synthetases. *Biochemistry* **18:**2627.

———. 1979e. Cysteinyl-tRNA synthetase from *Escherichia coli* does not need an editing mechanism for the rejection of serine and alanine. High binding energy of small groups in specific molecular interactions. *Biochemistry* **18:**1245.

Fersht, A. R. and M. Kaethner. 1976a. Enzyme hyperspecificity. Rejection of threonine by the valyl-tRNA synthetase by misacylation and hydrolytic editing. *Biochemistry* **15:**3342.

———. 1976b. Mechanism of aminoacylation of tRNA. Proof of the aminoacyl adenylate pathway for the isoleucyl- and tyrosyl-tRNA synthetases from *Escherichia coli* K12. *Biochemistry* **15:**818.

Hopfield, J. J. 1974. Kinetic proofreading: A new mechanism for reducing errors in biosynthetic processes requiring high specificity. *Proc. Natl. Acad. Sci.* **71:**4135.

Igloi, G. L., F. von der Haar, and F. Cramer. 1977. Hydrolytic action of

aminoacyl-tRNA synthetases from baker's yeast. "Chemical proofreading" of Thr-tRNA^{Val} and amino acid analogues. *Biochemistry* **16**:1969.
———. 1978. Aminoacyl-tRNA synthetases from yeast: Generality of chemical proofreading in the prevention of misaminoacylation of tRNA. *Biochemistry* **17**:3459.
Loftfield, R. B. 1963. The frequency of errors in protein biosynthesis. *Biochem. J.* **89**:82.
Loftfield, R. B. and E. A. Eigner. 1966. The specificity of enzymic reactions. Aminoacyl-soluble RNA ligases. *Biochim. Biophys. Acta* **130**:426.
Loftfield, R. B. and D. Vanderjagt. 1972. The frequency of errors in protein biosynthesis. *Biochem. J.* **128**:1353.
Mertes, M., M. A. Peters, W. Mahoney, and M. Yarus. 1972. Isoleucylation of transfer RNA$_f^{Met}$ (*E. coli*) by the isoleucyl-transfer RNA synthetase from *Escherichia coli*. *J. Mol. Biol.* **71**:671.
Mulvey, R. S. and A. R. Fersht. 1977. Editing mechanisms in aminoacylation of tRNA: ATP consumption and the binding of aminoacyl-tRNA by elongation factor Tu. *Biochemistry* **16**:4731.
Norris, A. T. and P. Berg. 1964. Mechanism of aminoacyl RNA synthesis: Studies with isolated aminoacyl adenylate complexes of isoleucyl tRNA synthetase. *Proc. Natl. Acad. Sci.* **52**:330.
Pauling, L. 1958. The probability of errors in the process of synthesis of protein molecules. In *Festschrift Arthur Stoll*, p. 597. Birkhäuser Verlag, Basel, Switzerland.
Schreier, A. A. and P. R. Schimmel. 1972. Transfer ribonucleic acid synthetase catalyzed deacylation of aminoacyl transfer ribonucleic acid in the absence of adenosine monophosphate and pyrophosphate. *Biochemistry* **11**:1582.
Trupin, J., H. Dickerman, M. Nirenberg, and M. Weissbach. 1966. Formylation of amino acid analogues of methionine sRNA. *Biochem. Biophys. Res. Commun.* **24**:50.
von der Haar, F. and F. Cramer. 1975. Isoleucyl-tRNA synthetase from baker's yeast: The 3'-hydroxyl of the 3'-terminal ribose is essential for preventing misacylation of tRNA^{Ile}-C-C-A with misactivated valine. *FEBS Lett.* **56**:215.
———. 1976. Hydrolytic action of aminoacyl-tRNA synthetases from baker's yeast: "Chemical proofreading" preventing acylation of tRNA with misactivated valine. *Biochemistry* **15**:4131.
Yamane, T. and J. J. Hopfield. 1977. Experimental evidence for kinetic proofreading in the aminoacylation of tRNA by synthetase. *Proc. Natl. Acad. Sci.* **74**:2246.
Yarus, M. 1972a. Phenylalanyl-tRNA synthetase and isoleucyl-tRNA^{Phe}: A possible verification mechanism for aminoacyl-tRNA. *Proc. Natl. Acad. Sci.* **69**:1915.
———. 1972b. Solvent and specificity. Binding and isoleucylation of phenylalanine transfer ribonucleic acid (*Escherichia coli*) by the isoleucyl transfer ribonucleic acid synthetase from *Escherichia coli*. *Biochemistry* **11**:2352.

The Tryptophanyl- and Tyrosyl-tRNA Synthetases from *Bacillus stearothermophilus*

Greg Winter and Gordon L. E. Koch
MRC Laboratory of Molecular Biology
Hills Road, Cambridge CB2 2QH, England

Anne Dell and Brian S. Hartley
Department of Biochemistry, Imperial College
London SW7 2AZ, England

Aminoacyl-tRNA synthetases are ancient enzymes, as evidenced by their central role in metabolism (Granick 1950). They may well have evolved their specificities for amino acid and tRNA as the early cell developed pathways for amino acid synthesis and as the first codons were assigned to each amino acid (Wong 1975). Similarities between synthetases of different amino acid specificity might therefore reflect the mechanisms of earliest evolution. At first sight, the aminoacyl-tRNA synthetases seem to be a diverse group of enzymes with a range of quaternary structures (α_1, α_2, and $\alpha_2\beta_2$) and subunit sizes (33,000–110,000 daltons) (see Table 1 of Hartley, this volume). Closer inspection reveals a loose class of large polypeptide chains centered on 100,000 daltons (and often monomeric) and a loose class of smaller chains centered on 50,000 daltons (and often dimeric). Peptide mapping, isolation of several tryptic peptides in greater than molar yield, and more detailed sequence studies have revealed that the large polypeptide chains contain areas of repeated sequence. (Kula 1973; Waterson and Konigsberg 1974; Koch et al. 1974). These larger chains could therefore have arisen by duplication and fusion of a common ancestral gene originally coding for the smaller chains. On the basis of this model, most synthetases would be composed of either a dimer of two identical chains or a monomer of fused domains. This arrangement would allow the binding of tRNA by contact of the pseudosymmetric regions on both tRNA and synthetase, each arm of the tRNA binding to corresponding portions of each subunit in the small dimeric enzymes or of each domain in the large, repeated-sequence enzymes (Kim 1975). This would presumably impose a conservation of sequence around the contacts. Repeated sequence in the monomeric large polypeptide chains might therefore comprise the tRNA contacts in each domain. In the dimeric (α_2) methionyl-tRNA synthetase from *Escherichia coli* (Bruton 1979) and the tetrameric ($\alpha_2\beta_2$) phenylalanyl-tRNA synthetase from yeast (Robbe-Saul et al. 1977), the conservation of repeated sequence might, alternatively, have been imposed by intersubunit contacts. Similarities among the leucyl-, isoleucyl-, valyl, methionyl-, and phenylalanyl-tRNA synthetases suggest that these five enzymes may have diverged from a repeated-sequence ancestor. Thus, each of these large, repeated-sequence enzymes loads a hydrophobic amino acid onto the 2' position of a tRNA with an A`at the discriminator position (fourth base from the 3' end [Crothers et al. 1972]) and a U in the middle of the codon (Wetzel 1978). The evolutionary relationships of the small subunit synthetases are more confusing, and only a comparison of the complete primary and tertiary structures of several of these enzymes can provide the definitive answer. We have therefore undertaken the purification and sequencing of crystallizable aminoacyl-tRNA synthetases.

Since aminoacyl-tRNA synthetases occur in very small amounts in bacterial cells, the purification procedures were scaled up to obtain the several hundred milligrams needed for sequencing. The procedure was optimized for the simultaneous purification of several synthetases and other enzymes (Bruton et al. 1975; Atkinson et al. 1978), and from 50 kg of wet *Bacillus stearothermophilus* cells it was possible to isolate about 500 mg of either tyrosyl- (Koch 1974) or tryptophanyl- (Winter 1976) tRNA synthetase. Both the tyrosyl (Reid et al. 1973) and the tryptophanyl enzymes (R. Jakes; C. Carter; both unpubl.) yielded crystals suitable for X-ray analysis and both were selected for protein sequencing. There seemed some prospect that a detailed structural comparison of these two enzymes would identify similar features, since both enzymes are dimers with subunits of comparable sizes (37,000 and 45,000 daltons) and a specificity for aromatic amino acids closely related in metabolism.

THE SEQUENCE STRATEGIES

The strategy used in sequencing the tryptophanyl-tRNA synthetase yielded numerous small peptides. The carboxymethylated enzyme was digested with trypsin and also with chymotrypsin, and the peptides fractionated by Sephadex, DEAE-cellulose, high-voltage paper electrophoresis, and paper chromatography. These peptides were sequenced in parallel by the dansyl-Edman method (Hartley 1970), which, although slow for sequencing a single peptide (~2 residues/day), becomes highly competitive with automated sequencing technologies if several peptides are sequenced together. A gas manifold was designed to purge 20 tubes simultaneously with nitrogen in the Edman coupling step (Edman 1949), and a new rapid method of hydrolyzing dansyl peptides was devised. The dansylated peptide is hydrolyzed with a mixture of equal volumes of hydrochloric acid and propionic acid at 165°C for 12 minutes. With these improvements it became possible to undertake two rounds of dansyl-Edman analysis on 20 peptides each day (Winter 1976). To ensure sequence accuracy and to assign amides, virtually all peptides were subdigested and the digest fractionated by high-voltage paper electrophoresis or paper chromatography. Each of these substituent peptides was characterized by dansylation, partial sequencing, amino acid analysis, and electrophoretic mobility (Offord 1966). Alternatively, subdigests were sequenced directly by mass-spectrometric mixture analysis, a method whereby the entire digest is loaded onto the probe of the mass spectrometer and the probe temperature is gradually increased. The mixture is thereby fractionated according to volatility, the sequence ions corresponding to each peptide component appearing at a characteristic temperature (Morris et al. 1971).

From the sequencer analysis and tryptic and chymotryptic digests, ten fragments internally overlapped by at least two residues could be constructed (Fig. 1). The one-residue overlaps between five of these fragments were allocated by noting which tryptic peptides were present in two cyanogen bromide (CNBr) peptides, and a further three fragments were overlapped after elastase and staphyloccocal protease digestion of two other CNBr peptides. The chain now lay in three pieces, which, assuming no outstanding sequence, must abut directly (Winter and Hartley 1977).

The sequence strategy for the tyrosyl enzyme was inspired by the particularly fast sequence determination of coelocanth triose phosphate isomerase (Kolb et al. 1974). In a two-step fragmentation, triose phos-

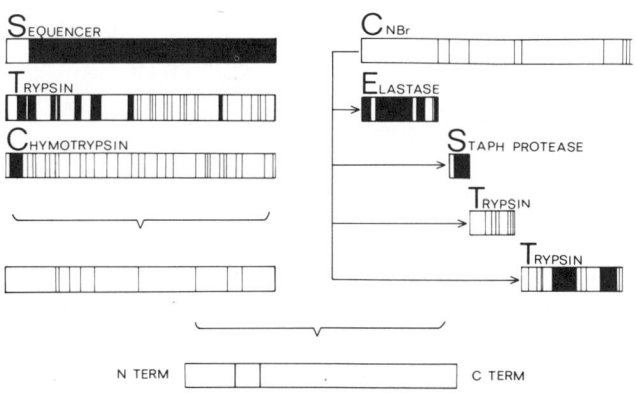

SEQUENCE STRATEGY FOR TRYPTOPHANYL tRNA SYNTHETASE

Figure 1 The white areas indicate those sequences deduced from enzymatic digests or sequencer analysis, the black areas indicate those sequences not so inferred, and the thin black lines indicate the sites of enzymatic or CNBr cleavage. Thus, to overlap the sequence, each black area or line from one digest must be straddled by a white area from another.

phate isomerase had been broken down into a small number of large fragments that were partly sequenced by a liquid-phase automated sequencer. Each of these fragments had then been subdigested and the remaining sequence attained by dansyl-Edman analysis. A CNBr digest of the carboxymethylated tyrosyl enzyme yielded four major fragments that could be completely separated by a denaturing sizing column (Fig. 2). The approximate sizes of these fragments were estimated as 27,000 daltons (fragment CNBr 4), 11,000 daltons (fragment CNBr 2), 5000 daltons (fragment CNBr 1), and 2000 daltons (fragment CNBr 3) from SDS-polyacrylamide gels. (The sizes of fragments CNBr 1 and CNBr 3 were deduced indirectly from the sizes of incompletely cleaved CNBr fragments.) There is some evidence for yet another CNBr fragment, probably small and located between fragments CNBr 3 and CNBr 4 (Bosshard et al. 1978). However, this fragment (CNBr 3a) has so far eluded capture. N-terminal sequencer analyses of tryptic and chymotryptic subdigests of CNBr fragments 1, 2, and 4 failed to complete the sequence. Subsequently, the enzyme has been cut at cysteine residues after cyanylation with 2-nitro-5-thiocyanatobenzoic acid and the cysteinyl fragments digested with elastase and staphyloccocal protease. The intact enzyme has also been digested with trypsin and a protease from *Armillaria mellea* and the acetylated enzyme with chymotrypsin. Two large runs of sequence can be constructed (Fig. 3), and most of the remaining sequence has been found in small peptides.

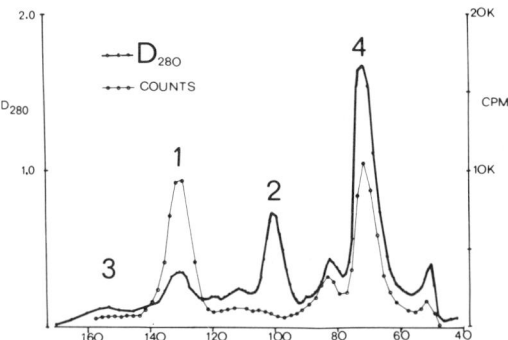

Figure 2 CNBr digest of the radioactively carboxymethylated tyrosyl-tRNA synthetase fractionated in 20% formic acid on a Bio-Gel P-100 column (100–200 mesh). Approximately 1 μmole digested enzyme was loaded in 3.3 ml 20% formic acid to a column, dimensions 2.5 × 88 cm with a flow rate of 10 ml hr^{-1}. Fractions of 2.5 ml were collected, the optical density measured at 280 nm (D$_{280}$), and radioactivity detected after adding 20-μl aliquots to a toluene, methoxyethanol, BBOT scintillant. The fragments are numbered according to their position in the sequence from the N terminal.

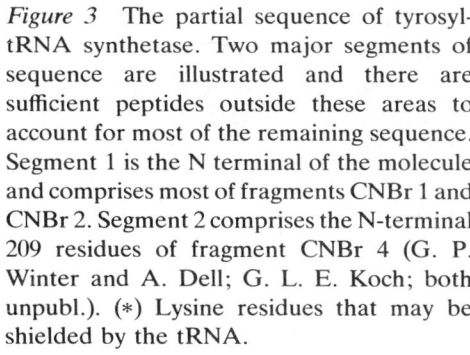

Figure 3 The partial sequence of tyrosyl-tRNA synthetase. Two major segments of sequence are illustrated and there are sufficient peptides outside these areas to account for most of the remaining sequence. Segment 1 is the N terminal of the molecule and comprises most of fragments CNBr 1 and CNBr 2. Segment 2 comprises the N-terminal 209 residues of fragment CNBr 4 (G. P. Winter and A. Dell; G. L. E. Koch; both unpubl.). (∗) Lysine residues that may be shielded by the tRNA.

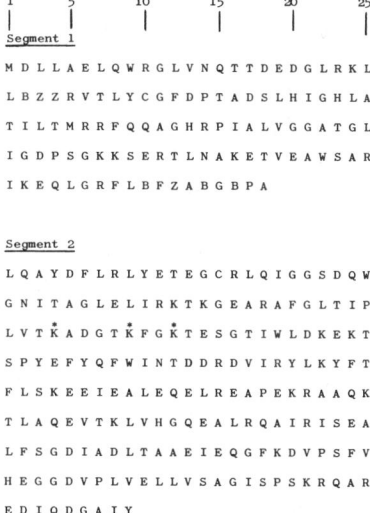

THE SEARCH FOR HOMOLOGY

Since we do not yet have the necessary crystallographic information, we have been searching for homology between the two enzymes in other ways.

Direct Comparison

The rules of evolution operating while synthetases evolved their amino acid specificities can scarcely be imagined. Presumably, with only a primitive translational apparatus, the relationship of genotype (in which species differences are stored) to phenotype (in which species differences are expressed and on which natural selection operates) must have been vague. Nevertheless, the requirement for correct three-dimensional structures of proteins in the cell may have been as necessary in that pre-Darwinian era as in the subsequent era of Darwinian evolution and, if so, would have imposed a fierce selection pressure on the synthetases. If a synthetase charges tRNA with the wrong amino acid, the resultant protein will have an altered, perhaps damaged, structure. Should a synthetase build such errors into another synthetase, the newly synthesized enzyme would then work less accurately and the proportion of errors would multiply rapidly. The extent of damage would depend on the amino acid inserted; whereas tyrosine might easily replace phenylalanine in a protein structure, it is unlikely that glycine would do so. Hence, although the divergence of these enzymes from any putative common ancestor must have occurred at the dawn of life, one might still hope to see such a relationship today at the level of primary structures. The frequencies with which amino acids replace each other throughout Darwinian evolution have been measured by comparing the sequences of the same protein from different organisms with the inferred ancestral sequence. A score matrix can thereby be constructed in which high scores are given to amino acid identities and low scores to infrequent substitutions. Each segment of one polypeptide chain A, with a total nA residues, can then be scored with respect to each segment of another polypeptide chain B, with a total nB residues, and the results expressed in a two-dimensional matrix ($nA \times nB$). Areas of similarity then appear in the matrix as a line of high scores (McLachlan 1971). When this technique was used to compare the two segments (Fig. 3) of the tyrosyl enzyme with the sequence of the tryptophanyl enzyme, absolutely no homology could be detected (A. D. McLachlan, unpubl.).

Key Residues

An alternative basis of comparison is to locate key residues or areas in the primary sequence and see whether these are conserved in both enzymes. Chemical modification at cysteine residues results in loss of activity for several aminoacyl-tRNA synthetases (Stern et al. 1966; Iaccarino and Berg 1969; Rouget and Chapeville 1971; Ostrem and Berg 1974; Kuehl et al. 1976; Rainey et al. 1976). The tryptophanyl-tRNA synthetase from *B. stearothermophilus* contains three cysteine residues per subunit and the tyrosyl-tRNA synthetase from *B. stearothermophilus* contains only two. It is interesting that the first cysteine in both sequences occurs 35 residues from the N terminal (Fig. 4). Other evidence suggests that the N-terminal portion of these molecules is critical; systematic alignments of tryptic peptides from the *E. coli* tryptophanyl-tRNA synthetase with the sequence of the enzyme from *B. stearothermophilus* show a highly conserved area at the N terminal (Fig. 5) (Winter et al. 1977). Partial proteolysis of the tryptophanyl-tRNA synthetase from beef pancreas suggests that this N-terminal area may be necessary for enzyme activity. Although the eukaryotic synthetase has larger subunits than the prokaryotic enzymes, these may be trimmed at their N terminal with

```
              5    10   15   20   25
TYR.  M D L L A E L Q W R G L V N Q T T D E D G L R K L
TRP.  M K T I F S G I Q P S G V I T I G N Y I G A L R Q


TYR.  L B Z Z R V T L Y C G F D P T A D S L H I G H L A
TRP.  F V E L Q H Z Y N C Y F C I V B Z H A I T V W Q D


TYR.  T I L T M R R F Q Q . . .
TRP.  P H E L R Q N I R R . . .
```

Figure 4 Alignment of the N-terminal portions of the tyrosyl- and tryptophanyl-tRNA synthetases (*B. stearothermophilus*). The first two cysteines of the tryptophanyl enzyme lie close together and the alignment can alternatively be based on the second cysteine.

Sequence in one letter code	Residue numbers
MKTI-FSGIQPSGVITIGNYIGALRQ	B.S. 1 - 25
IKPIVFXGAEPXGELXXGXYXGALR	E.C.
FVELQHZYNCYFCIVBZHAITVWQD	B.S. 26 - 50
CIVBZHAITVR	E.C.
PHELRQNIRRLAALYLAVGIDPTQA	B.S. 51 - 75
LACGIBZPK	E.C.
TLFIQSEVPAHAQAAWMLQCIVYIG	B.S. 76 - 100
WALBCYTYFG	E.C.
ELERMTQFKEKSAGKEAVSAGLLTY	B.S. 101 - 125
ELSR,MTZFK	E.C.
PPLMAADILLYNTDIVPVGEDQKQH	B.S. 126 - 150
LVPVGBZZK	E.C.
IELTRDLAERFDKRYGELFTIPEAR	B.S. 151 - 175
BIAZR,FBALYGZIFK	E.C.
IPKVGARIMSLVDPTKKMSKSDPNP	B.S. 176 - 200
IPK,SGAR MSK	E.C.
KAYITLLDDAKTIEKKIKSAVTDS-E	B.S. 201 - 225
IK AVTBSBZ	E.C.
GTIRYDKEAKPGISNLLNIYSTLSG	B.S. 226 - 250
PPVR	E.C.
QSIEELERQYEGKGYGVFKADLAQV	B.S. 251 - 275
SIPZLZZK	E.C.
VIETLRPIQERYHHWMESEELDRVL	B.S. 276 - 300
	E.C.
DEGAEKANRVASEMVRKMEQAMGLG	B.S. 301 - 325
BGAZK	E.C.
R.R.	B.S. 326 - 327
	E.C.

Figure 5 The sequence of tryptophanyl-tRNA synthetase (*B. stearothermophilus*) with conserved sequences from the tryptophanyl-tRNA synthetase (*E. coli*). Highly conserved areas are underlined.

elastase to give a fully active dimer of $2 \times 41,000$ daltons. The eukaryotic synthetase may therefore consist of an ancestral prokaryotic core with an N-terminal appendage. Proteolysis of the eukaryotic enzyme with trypsin removes about 20 residues from the N terminal of this core and results in total loss of pyrophosphate exchange activity (Lemaire et al. 1975).

The third cysteine (residue 95) of the tryptophanyl enzyme from *B. stearothermophilus*, like the second cysteine of the tyrosyl enzyme, lies well within the molecule. Both are situated at approximately 230 residues from the carboxyl terminal. Cysteine 95 may be located in, or near, the active site, as cyanylation of a single reactive cysteine in the homologous tryptophanyl-tRNA synthetase from *E. coli* destroys enzyme activity. This reactive cysteine apparently corresponds to cysteine 95 of the *B. stearothermophilus* enzyme (K. H. Muench, pers. comm.), although in the thermophile the cysteines remain buried to DTNB (R. Jakes, unpubl.). To summarize, there appears to be a matching of the cysteine locations in the tyrosyl and tryptophanyl enzymes that might be functionally important.

The binding of tRNA to aminoacyl-tRNA synthetase is salt-sensitive (see Loftfield 1972) and probably involves ionic interaction between basic residues on the protein and the phosphate groups of the tRNA. These points of contact may have been conserved throughout evolution and their identification in both the tyrosyl and tryptophanyl enzymes might highlight any evolutionary relationship between them. The binding of 1 molecule of tRNA^{Tyr} to tyrosyl-tRNA synthetase results in a decreased reactivity of particular lysine residues to acetic anhydride (Bosshard et al. 1978). These residues might therefore form direct salt links with the tRNA, although this alternatively could be due to new salt links within or between the protein subunits consequent on binding the tRNA. Three of these lysine residues have been identified in the primary sequence (Fig. 3) and at least two are close together. The changes in lysine reactivity in these residues were less than twofold; this is the maximum decrease expected if the tRNA were to interact with only one of the two identical subunits. The most dramatic change was a sevenfold decrease in reactivity for a lysine in the sequence Arg-Ile-Val-Thr-Gly-Met(Lys,Thr)Arg-Tyr (Bosshard et al. 1978); this peptide is thought to straddle CNBr 3 and the elusive CNBr 3a (G. L. E. Koch, unpubl. and not illustrated in Fig. 3). The sevenfold decrease in reactivity could therefore indicate that the tRNA binding site comprises the same portion of primary sequence in both subunits, entirely consistent with the proposal that tRNA and synthetase bind by recognition of pseudosymmetric regions on both molecules (Kim 1975). The corresponding areas of the tryptophanyl-tRNA synthetase have not yet been identified.

Partial Proteolysis

A completely different approach has been to probe the tertiary structure of both of these enzymes by partial proteolysis. Incubation of the tryptophanyl-tRNA synthetase with low levels of trypsin resulted in cleavage of the subunit to a protease-resistant fragment of 24,000 daltons and to smaller portions of about 6000 daltons, with concomitant loss of pyrophosphate exchange ability. The 24,000-dalton fragment was purified and the N terminal identified as residue 116 by automated sequencing. The carboxyl terminal of the fragment was shown to be the carboxyl terminal of the whole molecule by a CNBr digest; this portion of the chain produces several small, soluble CNBr peptides that are identified easily after high-voltage paper electrophoresis. The tryptic proteolysis apparently is not prevented by ATP or tryptophan singly but is prevented by ATP and tryptophan together (G. P. Winter, in prep.). Thus, in the inactive enzyme, surface loops at lysine 115 and one or more positions on the N-terminal side of this are attacked by trypsin; whereas in the catalytically active enzyme, they are not. Partial tryptic proteolysis of the tryptophanyl-tRNA synthetase from beef pancreas parallels that of the *B. stearothermophilus* enzyme. Trypsin attacks the core enzyme to give a protease-resistant fragment of 24,500 daltons (Prasolov et al. 1975), suggesting that both the eukaryotic and prokaryotic tryptophanyl-tRNA synthetases retain a protease-resistant carboxyl-terminal domain.

Incubation of the tyrosyl-tRNA synthetase with trypsin resulted in cleavage of the subunit to a protease-resistant fragment of 28,000 daltons and to smaller fragments, with concomitant loss of pyrophosphate exchange ability. The 28,000-dalton fragment could be trimmed with CNBr to give a 27,000-dalton fragment. Since this large CNBr fragment is known to be the carboxyl-terminal fragment, this demonstrates that like the tryptophanyl enzyme, the protease-resistant fragment corresponds to the carboxyl terminal of the molecule. The functional significance, if any, of these double-domain structures is open to speculation. Like with the tryptophanyl enzyme from *B. stearothermophilus*, the proteolysis is not prevented by ATP or tyrosine singly, but it is prevented by ATP and tyrosine together (G. P. Winter, in prep.). This resistance to proteases may reflect no more than a general tightening up of the enzyme structure upon the strong binding of the aminoacyl adenylate. Alternatively, it might be due to binding of ATP in the ATP site; when either ATP or tyrosine is soaked into crystals of the tyrosyl-tRNA synthetase, both ligands bind in the tyrosine cleft (Monteilhet and Blow 1978), but when ATP and tyrosine are soaked in together, the tyrosyl adenylate is formed in situ and the AMP moiety occupies a new site (J. R. Rubin and D. M. Blow, pers. comm.).

CONCLUSION

A comparison of the primary sequences of the tryptophanyl-tRNA synthetase and the tyrosyl-tRNA synthetase reveals little or no evidence of homology. Nevertheless, these enzymes do have features in common: one cysteine at 35 residues from the N terminal, another at about 230 residues from the carboxyl terminal, and a protease-resistant carboxyl-terminal domain comprising two-thirds of the molecule. This may reflect an underlying structural similarity.

ACKNOWLEDGMENTS

We would like to thank C. J. Bruton for providing crude fractions of the tyrosyl- and tryptophanyl-tRNA synthetases from the multienzyme preparation, for discussions, and for encouragement; A. D. McLachlan for running a computer program for assessing homologies; R. Jakes for purification details on the tryptophanyl-tRNA synthetase; and P. T. Jones, F. Northrop, M. J. Smith, and A. Kamalagharan for expert technical assistance on the tyrosyl-tRNA synthetase at various periods.

REFERENCES

Atkinson, A., G. T. Banks, C. J. Bruton, M. J. Comer, R. Jakes, A. Kamalagharan, A. R. Whitaker, and G. P. Winter. 1978. Large scale multienzyme isolation from *Bacillus stearothermophilus*. *Biochem. J.* (in press).

Bosshard, H. R., G. L. E. Koch, and B. S. Hartley. 1978. The aminoacyl-tRNA synthetase–tRNA complex: Detection by differential labelling of lysine residues involved in complex formation. *J. Mol. Biol.* **119**:377.

Bruton, C. J. 1979. Probing the substructure, evolution and interactions of aminoacyl tRNA synthetases. In *Nonsense mutations and tRNA suppressors.* (ed. J. Celis and J. D. Smith). Academic Press, New York. (In press.)

Bruton, C. J., R. Jakes, and A. Atkinson. 1975. Gram-scale purification of methionyl-tRNA and tyrosyl-tRNA synthetases from *Escherichia coli. Eur. J. Biochem.* **59**:327.

Crothers, D. M., T. Seno, and D. G. Söll. 1972. Is there a discriminator site in transfer RNA? *Proc. Natl. Acad. Sci.* **69**:3063.

Edman, P. 1949. A method for the determination of the amino acid sequence in peptides. *Arch. Biochem. Biophys.* **22**:475.

Granick, S. 1950. The structural and functional relationships between heme and chlorophyll. *Harvey Lect.* **44**:220.

Hartley, B. S. 1970. Strategy and tactics in protein chemistry. *Biochem. J.* **119**:805.

Iaccarino, M. and P. Berg. 1969. Requirement of sulfhydryl groups for the catalytic and tRNA recognition functions of isoleucyl-tRNA synthetase. *J. Mol. Biol.* **42**:151.

Kim, S. H. 1975. Symmetry recognition hypothesis model for tRNA binding to aminoacyl-tRNA synthetase. *Nature* **256**:679.

Koch, G. L. E. 1974. Tyrosyl transfer ribonucleic acid synthetase from *Bacillus stearothermophilus*. Preparation and properties of the crystallizable enzyme. *Biochemistry* **13**:2307.

Koch, G. L. E., Y. Boulanger, and B. S. Hartley. 1974. Repeating sequences in aminoacyl-tRNA synthetases. *Nature* **249**:316.

Kolb, E., J. I. Harris, and J. Bridgen. 1974. Triose phosphate isomerase from the coelacanth. *Biochem. J.* **137**:185.

Kuehl, G. V., M. Lee, and K. H. Muench. 1976. Tryptophanyl transfer ribonucleic acid synthetase of *Escherichia coli. J. Biol. Chem.* **251**:3254.

Kula, M. R. 1973. Structural studies on isoleucyl-tRNA synthetase from *E. coli. FEBS Lett.* **35**:299.

Lemaire, G., C. Gros, S. Epely, M. Kalinski, and B. Labouesse. 1975. Multiple

forms of tryptophanyl-tRNA synthetase from beef pancreas. *Eur. J. Biochem.* **51:**237.
Loftfield, R. B. 1972. The mechanism of aminoacylation of transfer RNA. *Prog. Nucleic Acid Res. Mol. Biol.* **12:**87.
McLachlan, A. D. 1971. Tests for comparing related amino-acid sequences. Cytochrome C and cytochrome C551. *J. Mol. Biol.* **61:**409.
Monteilhet, C. and D. M. Blow. 1978. Binding of tyrosine, adenosine triphosphate and analogues to crystalline tyrosyl transfer RNA synthetase. *J. Mol. Biol.* **122:**407.
Morris, H. R., D. H. Williams, and R. P. Ambler. 1971. Determination of the sequences of protein-derived peptides and peptide mixtures by mass spectrometry. *Biochem. J.* **125:**189.
Offord, R. E. 1966. Electrophoretic mobilities of peptides on paper and their use in the determination of amide groups. *Nature* **211:**591.
Ostrem, D. L. and P. Berg. 1974. Glycyl transfer ribonucleic acid synthetase from *Escherichia coli*; purification, properties and substrate binding. *Biochemistry* **13:**1338.
Prasolov, V. S., O. O. Favorova, G. V. Margulis, and L. L. Kisselev. 1975. Limited proteolysis of the tryptophanyl-tRNA synthetase. *Biochim. Biophys. Acta* **378:**92.
Rainey, P., E. Holler, and M. R. Kula. 1976. Labelling of L-isoleucine tRNA ligase from *Escherichia coli* with L-isoleucyl-bromomethyl ketone. *Eur. J. Biochem.* **63:**419.
Reid, B. R., G. L. E. Koch, Y. Boulanger, B. S. Hartley, and D. M. Blow. 1973. Crystallization and preliminary X-ray diffraction studies on tyrosyl-transfer RNA synthetase from *Bacillus stearothermophilus*. *J. Mol. Biol.* **80:**199.
Robbe-Saul, S., F. Fasiolo, and Y. Boulanger. 1977. Phenylalanyl-tRNA synthetase from bakers' yeast. Repeated sequence in the two sub-units. *FEBS Lett.* **84:**57.
Rouget, P. and F. Chapeville. 1971. Leucyl-tRNA synthetase. Two forms of the enzyme: Role of sulfhydryl groups. *Eur. J. Biochem.* **23:**452.
Stern, R., M. Deluca, A. M. Mehler, and W. D. McElroy. 1966. Role of sulfhydryl groups in activating enzymes, properties of *Escherichia coli* lysine-transfer ribonucleic acid synthetase. *Biochemistry* **5:**126.
Waterson, R. M. and W. H. Konigsberg. 1974. Peptide mapping of aminoacyl tRNA synthetases: Evidence for internal sequence homology in *Escherichia coli* leucyl-tRNA synthetase. *Proc. Natl. Acad. Sci.* **71:**376.
Wetzel, R. 1978. Aminoacyl-tRNA synthetase families and their significance to the origin of the genetic code. *Origins Life* (in press).
Winter, G. P. 1976. "Sequence of tryptophanyl tRNA synthetase." Ph.D. thesis, Cambridge University, Cambridge, England.
Winter, G. P. and B. S. Hartley. 1977. The amino acid sequence of tryptophanyl tRNA synthetase from *Bacillus stearothermophilus*. *FEBS Lett.* **80:**340.
Winter, G. P., B. S. Hartley, A. D. McLachlan, M. Lee, and K. H. Muench. 1977. Sequence homologies between the tryptophanyl tRNA synthetases of *Bacillus stearothermophilus* and *Escherichia coli*. *FEBS Lett.* **82:**348.
Wong, J. T. F. 1975. A co-evolution theory of the genetic code. *Proc. Natl. Acad. Sci.* **72:**1909.

Mechanism of Aminoacyl-tRNA Synthetases: Recognition and Proofreading Processes

Friedrich Cramer, Friedrich von der Haar, and Gabor L. Igloi
Abteilung Chemie
Max-Planck-Institut für experimentelle Medizin
D-3400 Göttingen, Federal Republic of Germany

Over a decade has elapsed since it became clear that tRNA is specifically aminoacylated by the cognate aminoacyl-tRNA synthetase to provide the correct aminoacyl-tRNA as a building block for protein biosynthesis (Loftfield 1972; Goddard 1978). Despite many attempts, both theoretical and practical, to understand the pathways leading to the final, error-free product formation, there have been until recently only a few experimental facts with which to approach this all-important aspect of aminoacyl-tRNA synthetase function.

The problem is conveniently divisible into two parts. Concerning the small substrate, the amino acid, structural differences may be envisaged as playing a part in the recognition. On the other hand, with regard to tRNA recognition, initial studies were aimed at determining structural differences within the group of ligands and were, perhaps with hindsight, predictably unsuccessful in the search for a recognition region unique to each tRNA. In recent years, partly by abandoning preconceived ideas and recognition theories, great progress has been made in achieving a clearer, semimolecular picture of the processes involved. These new approaches, with their associated jargon (i.e., proofreading, editing, mopping up, triggering, etc.), have revived the interest in the general field of accuracy with a rapid increase in new data that need to be integrated into the framework of previously known facts.

We have recently reviewed the field of aminoacyl-tRNA synthetase specificity up to the beginning of 1977 (Igloi and Cramer 1978). We now discuss new results and interpretations with the hope of obtaining a more up-to-date survey of this important area.

RECOGNITION OF tRNA AS A SEQUENCE OF EVENTS

Attempts to assign a recognition site on the tRNA corresponding to a target in the binding pocket of the aminoacyl-tRNA synthetase have so far been unsuccessful. We therefore considered the possibilities arising out of a sequential recognition of a specific tRNA. This alternative approach was initiated by two experimental observations:

1. $tRNA^{Phe}$-CC, lacking the invariant 3′-terminal AMP, inhibits the aminoacylation by phenylalanyl-tRNA synthetase from baker's yeast only very weakly, even if present in a large excess over the substrate $tRNA^{Phe}$-CCA. This weak inhibition is not of a competitive type (von der Haar and Gaertner 1975).
2. $tRNA^{Phe}$-CCF, having the A analog formycin at its 3′ terminal instead of A, is aminoacylated with the same K_m as $tRNA^{Phe}$-CCA but with an approximately 20-fold reduction in rate (von der Haar and Gaertner 1975). The rate of aminoacylation of mixtures of $tRNA^{Phe}$-CCF and $tRNA^{Phe}$-CCA is, however, completely different from what one would expect if the aminoacylation of both of these substrates passed through the same rate-limiting step (Fig. 1).

Both of these unexpected results could be explained by the following scheme for the aminoacylation reaction. (E^{Phe} is phenylalanyl-tRNA synthetase.)

$$E^{Phe} + tRNA^{Phe} \rightleftarrows [E^{Phe} \cdot tRNA^{Phe}]^1 \rightleftarrows [E^{Phe} \cdot tRNA^{Phe}]^2 \rightarrow E + \text{Product} \quad (1)$$
$$\text{step 1} \quad\quad \text{I} \quad\quad \text{step 2} \quad\quad \text{II}$$

$tRNA^{Phe}$, irrespective of whether an A or a formycin is positioned at the 3′ terminal, is bound in a rapid equilibrium to form a catalytically incompetent complex (I). To pass into the catalytically competent complex (II), a conformational transition has to be induced by the 3′-terminal A. This conformational transition is more efficiently induced by the A than by the formycin at the 3′ terminal and, consequently, the $tRNA^{Phe}$-CCA is kinetically selected against $tRNA^{Phe}$-CCF during aminoacylation, as demonstrated in the experiment shown in Figure 1.

This explanation was supported by further studies with a set of $tRNA^{Phe}$-CCN, where N was 2′-deoxyadenosine (2′dA), 3′-deoxyadenosine (3′dA), oxidized-reduced A ($A_{oxi\text{-}red}$), and oxidized-reduced formycin ($F_{oxi\text{-}red}$) (von der Haar and Gaertner 1975). Furthermore, using rapid kinetic techniques, Maass and his colleagues found that $tRNA^{Phe}$-CCA does, in fact, induce a conformational transition in the enzyme (Eq. 1, step 2), whereas $tRNA^{Phe}$-CC does not (Krauss et al. 1977). This equilibrium (Eq. 1, step 2) is such that it has little influence on the thermodynamics of binding of the respective $tRNA^{Phe}$, the association constant of $tRNA^{Phe}$-CCA being 8×10^5 liters/mole and that of $tRNA^{Phe}$-CC being 2×10^5 liters/mole (Krauss et al. 1977).

The study with the 3′-terminal modified tRNA-CCN was then extended to other systems (von der Haar and Cramer 1978a). Whereas for the phenylalanyl-tRNA synthetase the $tRNA^{Phe}$-CC-2′dA was a competitive inhibitor, for valyl-, seryl-, and threonyl-tRNA synthetases—all from baker's

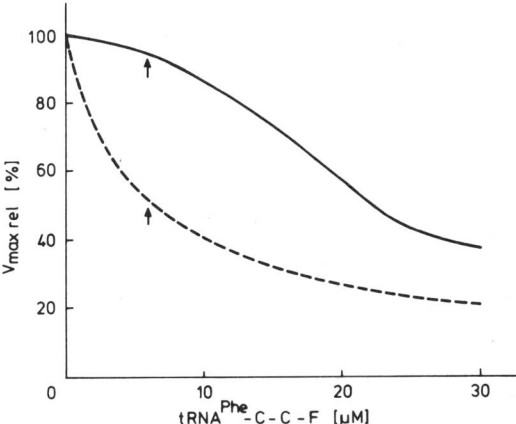

Figure 1 Velocity of aminoacylation of a mixture of tRNA^{Phe}-CCA and tRNA^{Phe}-CCF by phenylalanyl-tRNA synthetase. To a constant concentration of tRNA^{Phe}-CCA (6.4 μM) increasing amounts of tRNA^{Phe}-CCF were added, and the velocity of aminoacylation of these mixtures with phenylalanyl-tRNA synthetase (1.2 nM) was determined (———). Experimental data are compared to a steady-state kinetic calculation (----) assuming that two substrates with identical K_m but different V_{max} compete for one enzyme. (↑) The point where the tRNA^{Phe} species are present in equimolar amounts.

yeast—the tRNAs carrying their respective nonaccepting 2'dA or 3'dA at the 3' terminal showed the same type of inhibition as did tRNA^{Phe}-CC in the phenylalanine system. Since in all cases the tRNA-CCN bound as well as tRNA-CCA (von der Haar and Cramer 1978b), the conclusion is that a situation analogous to that outlined for the phenylalanine system in Equation 1 exists. However, in these examples the induction of the conformational transition must be intimately related to the accepting hydroxyl of the 3' terminal. This situation prevents further analysis of the type applied to the phenylalanyl-tRNA synthetase, because the accepting and triggering functions of the tRNA are identical (von der Haar and Cramer 1978a).

A much more productive analysis could be carried out in the case of the isoleucyl-tRNA synthetase from baker's yeast (von der Haar and Cramer 1978a). By analogous arguments we came to the conclusion that this enzyme transfers the isoleucine to the 2'-OH of the 3'-terminal A, whereas the nonaccepting 3'-OH is of major importance in inducing the conformational transition. Furthermore, we observed that (1) tRNA^{Ile}-CC-3'dA shows severe substrate inhibition, whereas tRNA^{Ile}-CCA does not; (2) the substrate inhibition exhibited by tRNA^{Ile}-CC-3'dA can be suppressed by tRNA^{Ile}-CC; and (3) valylation of tRNA^{Ile}-CC-3'dA by isoleucyl-tRNA synthetase (von der Haar and Cramer 1975) is inhibited by tRNA^{Ile}-CCA in a strictly noncompetitive pattern.

From this it was concluded that isoleucyl-tRNA synthetase, a single-chain enzyme of 115,000 m.w., possesses two binding sites for tRNA^{Ile} that work in an alternating manner. This cooperative behavior is triggered by the nonaccepting 3'-OH of the 3'-terminal A (von der Haar and Cramer 1978a).

For the isoleucyl-tRNA synthetase, we were also able to demonstrate that the conformational transition induced by the 3'-terminal A, or parts thereof, is of utmost importance in the maintenance of the specificity of the reaction. Isoleucyl-tRNA synthetase is only able to misaminoacylate tRNA^{Phe}-CCA to a negligible extent, if at all. By contrast, tRNA^{Phe}-CC-3'dA, lacking the nonaccepting 3'-OH responsible for induction of the conformational transition, is extensively misaminoacylated (von der Haar and Cramer 1978a).

This vital 3'-OH therefore seems to play a dual role. On the cognate tRNA^{Ile}-CCA, it enhances the fidelity of the aminoacylation (see Recognition of Amino Acid as a Two-step Process); on the noncognate tRNA^{Phe}-CCA, it prevents misaminoacylation.

A question arising from the above is whether the selection of the cognate tRNA out of the population of all tRNAs leads to further conformational changes. To investigate this we studied the misaminoacylation of tRNA^{Ile}-CCA by valyl- and phenylalanyl-tRNA synthetases and its inhibition by the cognate tRNA^{Phe}-CCN and tRNA^{Val}-CCN, respectively. The N in these cases was chosen on the basis that it was unable to induce the conformational transition described in Equation 1 (von der Haar and Cramer 1978b). Hence, any such hypothetical selection step should occur prior to the conformational transition given in Equation 1. The striking observation for phenylalanyl-tRNA synthetase is that minute amounts of those tRNA^{Phe}-CCN for which inhibition of aminoacylation of tRNA^{Phe}-CCA is not detectable almost completely inhibit misaminoacylation of noncognate tRNA^{Ile}-CCA, despite the fact that tRNA^{Ile}-CCA and tRNA^{Phe}-CCA exhibit the same K_m during aminoacylation with phenylalanyl-tRNA synthetase (von der Haar and Cramer 1978b). Identical results were found with valyl-tRNA synthetase. From this we concluded that the conformational transition indicated in Equation 1 must be preceded by a conformational transition leading to selection of cognate over noncognate tRNAs, at least in the case of the phenylalanyl- and valyl-tRNA synthetases studied (von der Haar and Cramer 1978b).

The above interpretations may shed some light on some previous anomalous observations concerning the specific aminoacylation of tRNA. These observations may be summarized as follows. Phenylalanyl-tRNA synthetase from baker's yeast possesses two binding sites for tRNA^{Phe} with identical K_{ass} (Krauss et al. 1976) that, under the conditions of the experiments in question, were independent. Using the same conditions, however, only one binding site could be entered by the noncognate tRNA^{Tyr} from *Escherichia coli* (Krauss et al. 1976) or tRNA^{Ile} from baker's yeast (von der Haar and Cramer 1978b). Furthermore, in the affinity elution system the cognate tRNA^{Phe}-CCA is also only bound to one site (von der Haar 1976a), whereas under yet another set of conditions the second site of the phenylalanyl-tRNA synthetase·tRNA^{Phe} is occupied in preference to the unoccupied sites of excess free enzyme (von der Haar 1978). Taking into account the postulate that there is a conformational transition induced by tRNA^{Phe}-CC, it would seem reasonable to suggest that a corresponding transition is responsible for the complex behavior observed. This phenomenon might also explain the type of cooperativity between both sites during phenylalanylation of tRNA^{Phe}-CCA (Fasiolo et al. 1976).

In conclusion, according to our present view, the specificity of aminoacylation of tRNA is not merely controlled by the thermodynamics of binding of the individual tRNA. Instead, after binding of the tRNA, the enzyme·tRNA complex must pass through a cataract of distinct conformational transitions as outlined in scheme 1 (Fig. 2), which includes one further step deduced from the different mode of binding of ATP (von der Haar and Gaertner 1975). The conversion of one state to the other is achieved by the induction of conformational transitions by unique elements on the cognate tRNA. One of these elements is the chemically invariant 3′ A, which, as we pointed out earlier (Cramer et al. 1969), must exhibit steric variability to accomplish its different actions in different tRNAs.

RECOGNITION OF AMINO ACID AS A TWO-STEP PROCESS

Qualitative Considerations

The selection of an amino acid by a particular aminoacyl-tRNA synthetase purely on the basis of thermodynamic considerations is rather unspecific. Indeed, it is to be anticipated from our current knowledge that any amino acid analog bearing only superficial resemblance to the natural substrate has a good chance of associating with the activating enzyme (Flossdorf and Kula 1973; Leporo et al. 1975). This lack of binding selectivity extends to the D-isomer series, where it is sometimes found that the unnatural enantiomer is bound to a certain extent (Calendar and Berg 1966; Owens and Bell 1970).

The process of activation, on the other hand, imposes more stringent requirements on the amino acid or analog. For example, amines or amino alcohols may be tightly bound to the enzyme (Holler et al. 1973) but lack the carboxyl group needed for the formation of a mixed anhydride with AMP. Evolutionary processes may also be involved in determining which

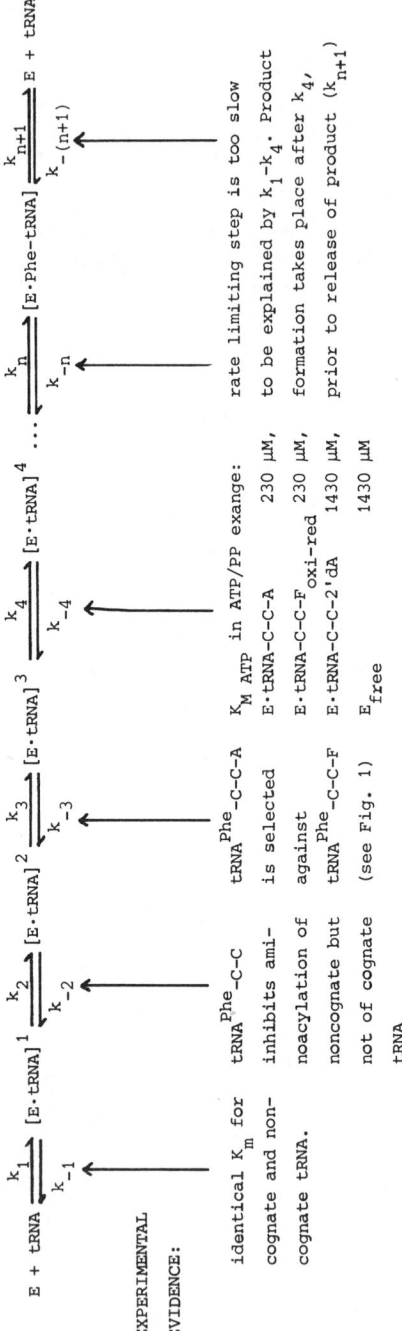

Figure 2 Scheme 1: Present view of aminoacylation cataract for phenylalanyl-tRNA synthetase from baker's yeast.

analog becomes activated. For instance, the proline analog azetidine-2-carboxylic acid is activated by prolyl-tRNA synthetase from *E. coli* but is inert toward the same enzyme from *Polygonatum multiflorum* in which it is a metabolite (Peterson and Fowden 1965). Within the group of naturally occurring activable L-amino acids, however, there does seem to be a more sensitive binding discrimination on the basis of structural criteria (Igloi et al. 1978). How effective is this discrimination in terms of the overall fidelity of protein biosynthesis? Early theoretical considerations by Pauling (1958) established an upper limit of 5% for the accuracy with which two similar amino acids, such as isoleucine and valine, could be distinguished by interaction with a polypeptide. More recent calculations have reduced this value to 0.7% (De Maeyer 1976) and, as described below, the highly specialized synthetase binding site is able to realize a binding discrimination to at least this extent. Nevertheless, the standard that has been reached in the evolution of the architecture of the binding site is not sufficient to ensure faultless translation of the genetic material.

The catalytic cycle of the aminoacyl-tRNA synthetase does not end, however, at the activation stage and, as was demonstrated by Baldwin and Berg many years ago, the addition of the cognate tRNA^{Ile} to a misactivated isoleucyl-tRNA synthetase · valine-AMP complex brings about the breakdown and release of the wrong amino acid (Baldwin and Berg 1966). Therefore, the noncognate amino acid does not enter the protein-synthesizing machinery as valyl-tRNA^{Ile} and the fidelity of the system is maintained. Clearly, there has been an amplification of specificity by use of a second discriminating step subsequent to the initial binding/activation.

There are several possibilities in the reaction sequence for where this second step (proofreading) may be located. Some of those for which theoretical or experimental evidence has been obtained are depicted in scheme 2 (Fig. 3). The theoretical treatment of Hopfield (1974) demands a rejection of the wrong amino acid purely on thermodynamic grounds and prior to transfer to the tRNA. The unlikelihood of this taking place in practice has been amply discussed (von der Haar 1977; Fersht 1977a; Igloi and Cramer 1978). We favor the chemical proofreading route in which we propose (von der Haar 1976b; von der Haar and Cramer 1976; Igloi et al. 1977) the involvement of a transient transfer of the incorrect amino acid from the adenylate to the 2'-OH or 3'-OH of the 3'-terminal A of the tRNA cognate to the enzyme. There follows an enzyme-catalyzed hydrolysis of the ester bond, the mechanism of which depends on the structure of the amino acid concerned. In the case of the misactivation of valine by isoleucyl-tRNA synthetase, a water molecule takes the place of the missing methyl group and enzymatic activation of this water for ester bond cleavage occurs through the absolutely essential 3'-OH (Fig. 4). The valyl-tRNA synthetase, on the other hand, makes use of an existing

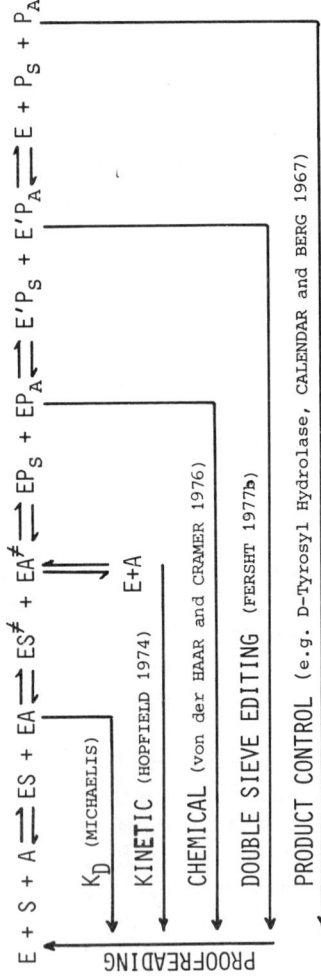

E = Aminoacyl-tRNA synthetase

S = Cognate amino acid

A = Non-cognate amino acid

$ES^{\neq}$, $EA^{\neq}$ = Activated enzyme·amino acid complex (e.g. E · aminoacyladenylate)

EP_S, EP_A, $E'P_S$, $E'P_A$ = Enzyme·product complex (e.g. E · aminoacyl-tRNA) in various conformational states E and E'

Figure 3 Scheme 2: A highly schematic representation of the theoretical location and the experimentally determined location of the discrimination steps.

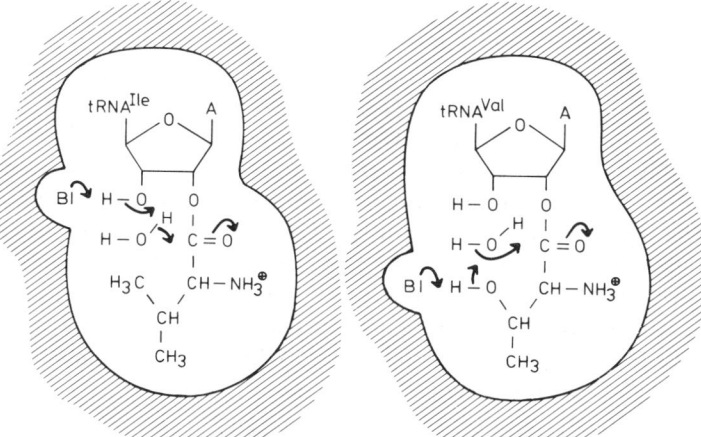

Figure 4 Mechanism of proofreading by isoleucyl- and valyl-tRNA synthetases.

amino acid functionality rather than a size criterion to achieve the same end. In this case, a water molecule in the binding site is activated through the β-OH group of threonine, which is transiently linked to the 2'-OH of tRNA^{Val}. The 3'-OH is not essential for correction in this case. We have elucidated these mechanisms using amino acid analogs and modified tRNAs bearing 3'dA and 3'-amino-3'dA at the 3' terminals. In the case of the deoxy derivative, no 2' to 3' isomerization can take place. The particular use of the amino nucleoside is described below. Using unmodified substrates, an indication for the transient transfer of an amino acid to tRNA has been obtained by the AMP/PPi-independent hydrolysis of aminoacyl-tRNAs during aminoacylation, which causes an overall nonstoichiometric ATP hydrolysis due to an aminoacylation-deacylation cycle.

Fersht, using a rapid sampling method, has been able to detect directly the transient formation of threonyl-tRNA^{Val} and its concomitant enzymatic destruction (Fersht and Kaethner 1976). In another example, he simulated the ester hydrolysis by measuring the rate of breakdown of an isolated misaminoacylated tRNA in the presence of a large excess of cognate synthetase (Fersht 1977a). The very fast rate of this process is not observed in our systems when a similar reaction is carried out under catalytic conditions of excess tRNA over enzyme. The relatively slow turnover under catalytic conditions probably reflects a series of slow rate-determining steps subsequent to the rapid hydrolysis. By decreasing the tRNA:enzyme ratio, we can also achieve a single turnover state in which there is an initial burst of hydrolysis followed by the slow release of the amino acid from aminoacyl-tRNA (F. von der Haar, unpubl.).

Fersht considers the problem of amino acid selection as one based on size alone and proposes a double-sieve editing model to take this into account (Fersht 1977b). The suggestion is that through movement of the amino acid at the 3' terminal of tRNA it can either occupy a site selecting against amino acids larger than the true substrate or a site selecting against smaller amino acids. The movement of the amino acid may be associated with 2' to 3' migration or, since we have shown that in the valyl-tRNA synthetase/threonine case the 3'-OH is not essential for proofreading, with movement of the tRNA terminal as a whole. This model, as it stands, must be considered an oversimplification, since, for example, isosteric valine and threonine are distinguished by valyl-tRNA synthetase and other cases are documented in which amino acids larger than the natural substrate have become attached to tRNA (Igloi et al. 1977, 1978). We feel that a less generalized mechanism of discrimination, making use of the properties, be they structural or functional, of the individual amino acids, is more appropriate for such an individualistic set of enzymes as the aminoacyl-tRNA synthetases.

Quantitative Aspects

Valine, misactivated by isoleucyl-tRNA synthetase, can be transferred to tRNA^{Ile}-CC-3'NH$_2$A, bearing a terminal 3'-amino-3'dA, without being subjected to the chemical proofreading step described above. There are two reasons this modification blocks the correcting activity. First, the 3'-amino group is not able to replace the 3'-OH in its essential role during the proofreading process (von der Haar and Cramer 1976). Second, release of the isoleucyl-tRNA^{Ile}-CC-3'NH$_2$A or valyl-tRNA^{Ile}-CC-3'NH$_2$A from the enzyme results in the isomerizing transacylation from the accepting 2'-OH to the 3'-amino group of the 3'-terminal A and a stable amide is formed. This is not subject to further hydrolytic breakdown. The application of such a modification allows one to measure the valylation of tRNA^{Ile}-CC-3'NH$_2$A in competition with its isoleucylation; therefore, an estimate of the contribution of differences in binding to the specificity of aminoacylation is obtained. We find that the error rate for activation and transfer of valine in competition with isoleucine by isoleucyl-tRNA synthetase is 1 in 225 ± 25 (Table 1).

The additional contribution to accuracy of the chemical proofreading step was determined for the isoleucyl-tRNA synthetase from *E. coli* by Hopfield and colleagues (Hopfield et al. 1976). Although these authors interpreted their results differently, the real nature of the experiment as a test for the efficiency of proofreading was evaluated by reinterpretation by one of us (von der Haar 1977). They determined the amount of ATP consumed during the interaction of isoleucyl-tRNA synthetase with valine and tRNA^{Ile}-CCA in comparison to the small amount of valyl-tRNA^{Ile}-CCA escaping the proofreading event.

Table 1 Formation of ^{14}C-labeled isoleucyl-tRNA^{Ile}-CC-3'NH$_2$A

^{14}C-labeled isoleucine (mM)	Valine (mM)	Valine/isoleucine	Isoleucylation (%)
0.02	0	0	100
0.02	0.2	10	96
0.02	1	50	80
0.02	2	100	67
0.02	3	150	56
0.02	4	200	50

Formation of ^{14}C-labeled isoleucyl-tRNA-CC-3'NH$_2$A was determined in the presence of increasing amounts of unlabeled valine. The purity of the unlabeled valine with respect to isoleucine contamination was checked by running the same set of experiments with tRNA^{Ile}-CCA. At the highest concentration of valine, a radioactive dilution of less than 3% was determined. This is equivalent to an impurity of <1 unlabeled isoleucine in 4000 valine molecules.

For the enzyme from *E. coli*, Hopfield et al. found a specificity of 1 valylation per 270 proofreading steps. We have repeated this experiment with the isoleucyl-tRNA synthetase from baker's yeast. Despite severe experimental difficulties, mainly due to having to determine a small amount of valyl-tRNA^{Ile}-CCA against a large background of labeled valine and ATP/AMP, our data point to a specificity of 800 ± 200 chemical proofreading events per release of 1 valyl-tRNA^{Ile}-CCA (F. von der Haar, unpubl.). In other words, chemical proofreading by the yeast enzyme is three to four times more efficient than by the enzyme from *E. coli*.

Thus, the activation and transfer of valine has an accuracy of 225 ± 25 and chemical proofreading has an error rate of 800 ± 200. The overall error rate for the consecutive steps is equal to the product of the individual error rates. Hence, according to the best of our presently available data, the overall error rate for selection of valine against isoleucine by isoleucyl-tRNA synthetase from baker's yeast is about 1 in 180,000.

REFERENCES

Baldwin, E. A. and P. Berg. 1966. Transfer ribonucleic acid-induced hydrolysis of valyladenylate bound to isoleucyl ribonucleic acid synthetase. *J. Biol. Chem.* **241:** 839.

Calendar, R. and P. Berg. 1966. The catalytic properties of tyrosyl ribonucleic acid synthetases from *E. coli* and *Bacillus subtilis*. *Biochemistry* **5:** 1690.

———. 1967. D-Tyrosyl RNA: Formation, hydrolysis and utilization for protein synthesis. *J. Mol. Biol.* **26:** 39.

Cramer, F., V. A. Erdmann, F. von der Haar, and E. Schlimme. 1969. Structure and reactivity of tRNA. *J. Cell. Physiol.* (suppl. 1) **74:** 163.

De Maeyer, L. C. M. 1976. Energiebedarf der molekularen Informationsübertragung. *Ber. Bunsen Ges.* **80:** 1189.

Fasiolo, F., J. P. Ebel, and M. Lazdunski. 1976. Non equivalence of the sites of yeast phenylalanyl-tRNA synthetase during catalysis. *Eur. J. Biochem.* **73:** 7.

Fersht, A. 1977a. Editing mechanisms in protein synthesis. Rejection of valine by the isoleucyl-tRNA synthetase. *Biochemistry* **16:** 1025.

———. 1977b. *Enzyme structure and mechanism.* W. H. Freeman, London.

Fersht, A. and M. M. Kaethner. 1976. Enzyme hyperspecificity. Rejection of threonine by the valyl-tRNA synthetase by misacylation and hydrolytic editing. *Biochemistry* **15:** 3342.

Flossdorf, J. and M. R. Kula. 1973. Ultracentrifuge studies on binding of aliphatic amino acids to isoleucyl-tRNA synthetase from *E. coli* MRE 600. *Eur. J. Biochem.* **36:** 534.

Goddard, J. 1978. The structures and functions of transfer RNA. *Prog. Biophys. Mol. Biol.* **32:** 233.

Holler, E., P. Rainey, A. Orme, E. L. Bennet, and M. Calvin. 1973. On the active site topography of isoleucyl-transfer ribonucleic acid synthetase of *E. coli* B. *Biochemistry* **12:** 1150.

Hopfield, J. J. 1974. Kinetic proofreading: A new mechanism for reducing errors in biosynthetic processes requiring high specificity. *Proc. Natl. Acad. Sci.* **71:** 4135.

Hopfield, J. J., T. Yamane, V. Yue, and S. M. Coutts. 1976. Direct experimental evidence for kinetic proofreading in aminoacylation of tRNA^{Ile}. *Proc. Natl. Acad. Sci.* **73:** 1164.

Igloi, G. L. and F. Cramer. 1978. Interaction of aminoacyl-tRNA synthetases and their substrates with a view to specificity. In *Transfer RNA* (ed. S. Altman), p. 294. MIT Press, Cambridge, Massachusetts.

Igloi, G. L., F. von der Haar, and F. Cramer. 1977. Hydrolytic action of aminoacyl-tRNA synthetase from baker's yeast: Chemical proofreading of Thr-tRNA^{Val} by valyl-tRNA synthetase studied with modified tRNA^{Val} and amino acid analogues. *Biochemistry* **16:** 1696.

———. 1978. Aminoacyl-tRNA synthetases from yeast: The generality of chemical proofreading in the prevention of misaminoacylation of tRNA. *Biochemistry* **17:** 3459.

Krauss, G., D. Riesner, and G. Maass. 1976. Mechanism of discrimination between cognate and noncognate tRNAs by phenylalanyl-tRNA synthetase from yeast. *Eur. J. Biochem.* **68:** 81.

———. 1977. Mechanism of tRNA recognition: Role of terminal A. *Nucleic Acids Res.* **4:** 2253.

Leporo, G. C., P. DiNatale, L. Guarini, and F. DeLorenzo. 1975. Histidyl-tRNA synthetase from *Salmonella typhimurium*: Specificity in the binding of histidine analogues. *Eur. J. Biochem.* **56:** 369.

Loftfield, R. B. 1972. The mechanism of aminoacylation of transfer RNA. *Prog. Nucleic Acid Res. Mol. Biol.* **12:** 87.

Owens, S. and F. Bell. 1970. Specificity of the valyl ribonucleic acid synthetase from *E. coli* in the binding of valine analogues. *J. Biol. Chem.* **245:** 5515.

Pauling, L. 1958. The probability of errors in the process of synthesis of protein molecules. In *Festschrift Arthur Stoll*, p. 597. Birkhäuser Verlag, Basel.
Peterson, P. J. and L. Fowden. 1965. Purification, properties and comparative specificities of the enzyme prolyl-transfer ribonucleic acid synthetase from *Phaseolus aureus* and *Polygonatum multiflorum*. *Biochem. J.* **97:**112.
von der Haar, F. 1976a. Phenylalanyl-tRNA synthetase from baker's yeast: Specificity and quantitation of affinity elution with tRNA. *Hoppe-Seyler's Z. Physiol. Chem.* **357:**819.

———. 1976b. Korrekturschritte bei der Aktivierung der Aminosäuren zur Proteinbiosynthese. *Naturwissenschaften* **63:**519.

———. 1977. Enzyme specificity resulting from proofreading events. *FEBS Lett.* **79:**225.

———. 1978. The ligand-induced solubility shift in salting out chromatography. A new affinity technique, demonstrated with phenylalanyl- and isoleucyl-tRNA synthetase from baker's yeast. *FEBS Lett.* **94:**371.

von der Haar, F. and F. Cramer. 1975. Isoleucyl-tRNA synthetase from baker's yeast: The 3'-hydroxyl group of the 3'-terminal ribose is essential for preventing misacylation of $tRNA^{Ile}$-C-C-A with misactivated valine. *FEBS Lett.* **56:**215.

———. 1976. Hydrolytic action of aminoacyl-tRNA synthetases from baker's yeast: "Chemical proofreading" preventing acylation of $tRNA^{Ile}$ with misactivated valine. *Biochemistry* **15:**4131.

———. 1978a. Seryl-, threonyl-, valyl- and isoleucyl-tRNA synthetase from baker's yeast: Role of the 3' terminal adenosine in the dynamic recognition of tRNA. *Biochemistry* **17:**3139.

———. 1978b. Valyl- and phenylalanyl-tRNA synthetase from baker's yeast: Recognition of tRNA results from a multistep process, as indicated by inhibition of aminoacylation with modified tRNA. *Biochemistry* **17:**4509.

von der Haar, F. and E. Gaertner. 1975. Phenylalanyl-tRNA synthetase from baker's yeast: Role of the 3'-terminal adenosine of $tRNA^{Phe}$ in enzyme substrate interaction studied with 3' modified $tRNA^{Phe}$ species. *Proc. Natl. Acad. Sci.* **72:**1378.

Methionyl-tRNA Synthetase from *Escherichia coli*: Structure-Function Relationships of a Dimeric Enzyme with Repeated Sequences

Sylvain Blanquet, Philippe Dessen, and Guy Fayat
Laboratoire de Biochimie
Laboratoire Associé n° 240 du Centre National de la Recherche Scientifique
Ecole Polytechnique
91128 Palaiseau Cedex, France

Aminoacyl-tRNA synthetases display a variety of molecular structures. For instance, in *Escherichia coli* the molecular weights of native aminoacyl-tRNA synthetases range from 44,000 in the case of cysteinyl-tRNA synthetase to 265,000 in the case of phenylalanyl-tRNA synthetase (Fayat et al. 1974). Moreover, depending on the enzyme considered, the synthetases consist of a single polypeptide chain or result from the assembly of identical or nonidentical subunits. Structures of the α, α_2, and $\alpha_2\beta_2$ types can be distinguished (for a review, see Ofengand 1977). Until recently, this complex pattern of structures has defied any attempt to find a unifying trait within this class of isofunctional enzymes. In the past few years it has been shown that several aminoacyl-tRNA synthetases exhibit extensive sequence duplication inside polypeptide chains. Those enzymes that have so far been shown to possess this property are composed of one or more 100,000-m.w. polypeptide chain. On the basis of this observation, the bacterial aminoacyl-tRNA synthetases can be schematically grouped into a minimum of four nonexclusive classes based on protomer number and degree of sequence duplication.

The first class groups the enzymes composed of two identical subunits that do not have internal repeats. The molecular weights of the protomers are in the range of 35,000–55,000. The second class are monomeric enzymes of about 100,000 m.w. with extensive duplication (Kula 1973; Koch et al. 1974; Waterson and Konigsberg 1974; Bruton 1975). The third class is represented by the bacterial methionyl-tRNA synthetase. This enzyme is a dimer ($2 \times 85,000$) composed of identical subunits resembling the monomer in class 2, i.e., sequence duplicated (Bruton et al. 1974). Finally, a fourth class includes the $\alpha_2\beta_2$ enzymes. Yeast phenylalanyl-tRNA synthetase, an $\alpha_2\beta_2$ structure of 270,000 m.w. ($[2 \times 73,000] + [2 \times 62,000]$), has recently been reported to contain repeated sequences in both the α and β subunits (Robbe-Saul et al. 1977).

A unifying pattern of structure thus emerges according to which the majority of known synthetases appear to be composed of at least two "domains" of about 40,000–50,000 m.w. These domains either correspond to each protomer of the dimers in class 1 or are covalently bound within the large protomers of classes 2 and 3. These large protomers may have arisen through duplication and fusion of adjacent genes coding for normal-sized subunits, such as those in class 1. Consequently, this raises the important issue of the functional significance of such two-domain structures. Two types of answers have been proposed.

In the case of tyrosyl-tRNA synthetase, which belongs to class 1, it has been shown that the subunits exhibit anticooperativity for substrate binding to the tyrosine-adenylating sites, whereas both subunits are able to synthesize tyrosyl adenylate at different rates (Jakes and Fersht 1975). By extrapolating this property to the synthetases that possess the sequence-duplication feature, Fersht (1975) has proposed a general mechanism for enhancement of the catalytic rate in the esterification of tRNA. This mechanism, which involves two interacting domains, is based on the utilization, during a single aminoacylation reaction cycle, of the binding energies of the substrates from both of the adenylate sites.

On the other hand, based on an internal pseudosymmetry appearing in the three-dimensional structure of yeast tRNA^{Phe} and on the two-domain structure of the synthetases, a general symmetry-recognition hypothesis of tRNAs by the enzymes has been proposed (Kim 1975). According to this hypothesis, the pseudo-twofold symmetry of an aminoacyl-tRNA synthetase is recognized by the pseudo-twofold symmetry of tRNA. It must be emphasized that such a two-point binding process should also lead to better discrimination between cognate and noncognate tRNAs by spacing apart their free energy of binding.

In discussing these hypotheses, this short review will focus on the structure-function relationship of methionyl-tRNA synthetase from *E. coli*, an enzyme composed of two large protomers, each having the double-domain structure.

STRUCTURE

Subunits

Native methionyl-tRNA synthetase from *E. coli* has molecular weight of about 175,000 (Heinrikson and Hartley 1967; Lemoine et al. 1968). Polyacrylamide gel electrophoresis in SDS and quantitative carboxyl-terminal determinations have shown that the native enzyme is a homologous dimer composed of two 85,000-m.w. subunits (Koch and Bruton 1974; Fayat et al. 1974).

Primary Structure

It was shown in the past that the native enzyme can be converted into four 45,000-m.w. peptides by treatment with 8 M urea (Lemoine et al. 1968; Cassio and Waller 1971) or maleic anhydride (Bruton and Hartley 1970), both at pH 9. Moreover, amino acid sequence studies and fingerprinting clearly indicate that methionyl-tRNA synthetase has repeated sequences in different portions of the polypeptide chain (Bruton et al. 1974). These findings suggest that the 85,000-m.w. subunit may contain a high degree of duplication and that the four similar peptides observed earlier may have resulted from chain cleavage of each subunit into two similar fragments.

Limited Proteolysis

Several studies have indicated that native methionyl-tRNA synthetase is highly susceptible to limited proteolysis, leading to enzymatically active, irreversibly modified fragments. The modification produced by trypsin has been investigated in detail (Cassio and Waller 1971). Proteolysis of the native enzyme by trypsin leads to the release of enzymatically inactive fragments corresponding to 25% of the original protein, with concomitant dissociation of the original dimer into 2 moles of enzymatically active modified monomer. The latter, henceforth referred to as trypsin-modified methionyl-tRNA synthetase, has a molecular weight of 64,000. This modified enzyme has been shown to be a single polypeptide chain with the same amino-terminal sequence as the native enzyme (Koch and Bruton 1974). Thus, it must arise by the removal of about 200 amino acids from the carboxyl terminal of the native molecule, which clearly results in the dissociation of the subunits. Most remarkable is the fact that despite the loss of nearly 25% of the polypeptide chain, the derivative retains full specificity toward methionine and tRNA^{Met} and has unimpaired activity in both the activation and aminoacylation reactions (Cassio and Waller 1971; Blanquet et al. 1973, 1974; Lawrence et al. 1973; Hyafil et al. 1976).

Tertiary Structure

By contrast to the native enzyme, trypsin-modified methionyl-tRNA synthetase has been successfully crystallized (Waller et al. 1971). A 4-Å electron-density map has been calculated (Zelwer et al. 1976). The molecule appears as an elongated ellipsoid with overall dimensions of $90 \times 43 \times 43$ Å. It is built of two parts separated by a large cleft. The volume of one of these domains is approximately twice that of the other. This observation is consistent with the hypothesis that each protomer of the native enzyme might be composed of two nuclei, each of 45,000 m.w. Only one of these nuclei would be attacked by trypsin, since the amino-terminal residues are the same in the native protomer and in the tryptic fragment. One can therefore expect this fragment to be composed of one intact nucleus and one proteolyzed nucleus, thus leading to two domains with a volume ratio of about 2 to 1. At this stage, the 4-Å Fourier map did not reveal clear structural homologies between these two domains. This could also reflect that a major part of the duplicated residues has been split off by the proteolysis. Indeed, close to 150 amino acid residues out of the 200 contained in peptides that are present twice in the native protomer are released upon tryptic digestion. Only one duplicated peptide, containing 47 amino acid residues, remains (Bruton 1979).

Small-angle X-ray-scattering experiments performed on solutions of the trypsin-modified enzyme indicate structural parameters that can account for an equivalent prolate ellipsoid of revolution having an axial ratio of 2.3 and a maximum length of 90 Å, with a creviced surface (Gulik et al. 1976). These dimensions are in excellent agreement with the crystallographic data. In the case of the native enzyme, small-angle X-ray- and neutron-scattering experiments (Gulik et al. 1976; Dessen et al. 1978) have been performed. The radius of gyration (43–44 Å) is large when compared with that of the modified enzyme. This indicates that the native enzyme is more anisometric than the modified enzyme. The distribution of scattered intensity is bimodal, suggesting the presence of two or more parts loosely connected together. The volumes of the modified (90,000 Å^3) and native (244,000 Å^3) enzymes are in the ratio of their molecular weights, thus indicating that in both cases all or most of the polypeptide chains are folded into globular particles. Finally, the small-angle X-ray experiments reveal a distribution of chords that could also indicate the association within the native enzyme of at least two globular parts, probably the two protomers, linked together by a fairly flexible hinge involving a limited region of the structure. In view of the mode of attack by trypsin on the native dimeric enzyme, these findings strongly suggest that the two subunits in the native enzyme are linked together chiefly through interactions involving the carboxyl-terminal end of each subunit. This idea is also supported by the observation that in the course of the trypsin attack, an intermediary monomeric enzymatic species with a molecular weight of 80,000–90,000 occurs (Cassio and Waller 1971). The size of this monomer is very similar to that of each subunit of the native enzyme, therefore suggesting that the release upon proteolysis of a very small number of amino acid residues from the carboxyl-terminal end is enough to abolish the association of subunits.

SUBSTRATE BINDING

Adenylation Site

The interactions of methionyl-tRNA synthetase with various ligands involved in the methionine activation reaction have been investigated by a variety of methods, including equilibrium dialysis, absorbance difference spectroscopy, and, most often, protein fluorescence.

The dimeric native enzyme possesses two equivalent sites for methionine (Blanquet et al. 1972; Fayat and Waller 1974), methionyl adenylate (Hyafil et al. 1976), or methioninyl adenylate (Blanquet et al. 1972), the latter being a stable and unreactive structural analog of methionyl adenylate (Cassio et al. 1967). On the other hand, the native enzyme has two distinct classes of ATP sites (Fayat and Waller 1974). Out of a total of four ATP sites per mole of native enzyme, two appear not to be involved in adenylate formation, as attested to by their continued ability to bind ATP in the presence of an excess of methioninyl adenylate. Their function is presently unknown. The remaining two ATP sites are blocked by methioninyl adenylate and therefore correspond to the catalytic sites. These sites are equivalent with respect to ATP-Mg^{++} binding (Blanquet et al. 1974; Hyafil et al. 1976). However, in the absence of the divalent ion, a complex anticooperative pattern of ATP binding to the catalytic sites appears (Fayat and Waller 1974). The monomeric trypsin-modified enzyme has one catalytic site that binds a unique molecule of methionine, methionyl adenylate, ATP-Mg^{++}, or ATP (Fayat and Waller 1974; Hyafil et al. 1976). The noncatalytic ATP site that was observed on each protomer of the native enzyme has disappeared upon proteolysis. In view of the closely similar functional parameters of the native and trypsin-modified forms of the enzyme, it can be concluded that these additional ATP sites have no significant influence on the activation and aminoacylation reactions.

tRNA^{Met} Binding

The interaction of methionyl-tRNA synthetase with its cognate tRNAs^{Met} is accompanied by variations in tryptophan fluorescence of large amplitude (Bruton and Hartley 1970; Blanquet et al. 1973). The method has thus provided a sensitive tool for defining optimal interaction conditions and binding parameters of tRNA (Blanquet et al. 1973). It is shown that native methionyl-tRNA synthetase can only bind a single molecule of tRNA^{fMet} or tRNA^{mMet} in normal conditions (10 mM $MgCl_2$). Despite extensive differences in their primary structures (Dube et al. 1968; Cory et al. 1968), tRNA^{fMet} and tRNA^{mMet} induce similar quenching of the enzyme fluorescence. However, their affinity constants differ by a factor of three in favor of tRNA^{mMet}. Evidence for the existence of a second tRNA binding site on the dimeric native synthetase is provided by the fact that raising the $MgCl_2$ concentration to 40 mM abolishes the anticooperative character of the binding of tRNA^{Met} by promoting the opening of a second binding site (Blanquet et al. 1973; Dessen et al. 1978). This property indicates the existence of two potentially equivalent tRNA sites on the dimeric methionyl-tRNA synthetase.

The monomeric trypsin-modified methionyl-tRNA synthetase has only one binding site for tRNA^{fMet} or tRNA^{mMet} whatever the magnesium concentration (Blanquet et al. 1973). When compared in conditions where the intact dimeric synthetase exhibits two equivalent tRNA binding sites, the binding parameters of tRNAs^{Met} to the modified enzyme are found to be identical to those for the native enzyme. This result supports the view that the trypsin-modified enzyme, in spite of extensive structural modifications, has kept an intact tRNA binding site and remains a functionally representative monomeric model of the native dimeric enzyme.

STRUCTURE-FUNCTION RELATIONSHIPS

Activation of Methionine

The complete rate equation for the methionine-dependent isotopic ATP-PPi exchange reaction at equilibrium catalyzed by native or trypsin-modified methionyl-tRNA synthetases has been determined (Blanquet et al. 1974). The method does not differentiate between the two forms of the enzyme. Under the same standard conditions, spectrofluorometric analyses at equilibrium of the reactions of methionine with the enzyme·ATP-Mg^{++} complex, of ATP-Mg^{++} with the enzyme·methionine complex, and of PPi-Mg^{++} with the enzyme·methionyl-adenylate complex have been performed (Blanquet et al. 1974). The results obtained provide additional evidence for the similarity between the two forms of the enzyme. Both approaches indicate a stoichiometry of two and one catalytic sites per mole of enzyme for native and trypsin-modified methionyl-tRNA synthetases, respectively. Fluorescence stopped-flow analysis of the kinetics of the reversible adenylation reaction has also been performed (Hyafil et al. 1976). This study demonstrates that each form of the enzyme catalyzes adenylate synthesis according to a single scheme involving a unique set of active sites. The equilibrium constants and rate values fitting this scheme have been determined. They demonstrate that each of the subunits of the dimer displays the same thermodynamic parameters as those of the modified monomer. It is concluded that the native enzyme has two independent active sites for adenylate synthesis, the catalytic properties of which are unaffected by subunit assembly or by losing a major part of the peptides present twice in each native protomer.

On the other hand, the methionine-dependent isotopic ATP-PPi exchange reaction has been shown to be sensitive to the occupation of the tRNA binding sites (Jacques and Blanquet 1977). One bound tRNA per dimer decreases the exchange velocity by 50%, whereas one bound tRNA per trypsin-modified monomer decreases it by about 100%. Therefore, the anticooperative binding to the dimer of a single tRNA molecule, which potentially induces asymmetry between the subunits, is able to reveal an asymmetry between the two adenylating sites of the native enzyme. Such coupling of the tRNA binding site to its corresponding amino acid activating site might be an important feature of the mechanisms insuring specific aminoacylation of tRNA.

Also contributing to the specificity of the amino acid activation reaction is the coupling between ligands, i.e., the change in the dissociation of one ligand upon binding of another ligand, which has been observed within the adenylation site of methionyl-tRNA synthetase (Fayat and Waller 1974; Blanquet et al. 1975a,b; Fayat et al. 1977a). From these detailed studies it is concluded that the binding of methionine to its specific enzyme triggers the coupling. Thus, ensuing from the specificity of amino acid binding is a free energy of coupling with the ATP-Mg^{++} binding, which is utilized to overcome the geometric and entropy requirements for aminoacyl adenylate synthesis. In other words, this coupling energy, instead of providing free energy of binding, is consumed to drive the catalysis (Jencks 1975).

Interestingly, the application to affinity chromatography of the couplings between ligands within the activation site of methionyl-tRNA synthetase has provided the basis for a new purification procedure for methionyl-tRNA synthetase, which leads to a 200-fold purification in one chromatographic step (Fayat et al. 1977b).

Coupling properties similar to those of methionyl-tRNA synthetase have emerged from the study of other aminoacyl-tRNA synthetases (Kosakowski and Holler 1973; Holler et al. 1975). This could indicate the involvement of common mechanisms in the specific amino acid activation reaction catalyzed by this class of enzymes. In this context it is noteworthy that several bacterial aminoacyl-tRNA synthetases have recently been shown to contain within their catalytic site an essential lysine residue (Fayat et al. 1978). This lysine can be affinity labeled with the 2',3'-dialdehyde derivative of ATP obtained by periodate oxidation. The conservation of a lysine residue within the amino acid activating site of synthetases may indeed reflect a general catalytic mechanism characteristic of these enzymes.

Aminoacylation of tRNA

In the reactions leading to the aminoacylation of $tRNA^{fMet}$ and $tRNA^{mMet}$, small differences in the behavior of the two enzymatic forms of methionyl-tRNA synthetase have been observed (Lawrence et al. 1973). For instance, excess ATP exhibits an inhibitory effect on the aminoacylation of $tRNA^{fMet}$ catalyzed by the native enzyme, whereas no significant inhibition occurs with the modified enzyme. This observation could be accounted for by the existence of four ATP sites on the native enzyme and only one on the modified enzyme (Fayat and Waller 1974). On the other hand, the Michaelis parameters for the aminoacylation reactions of $tRNA^{fMet}$ and $tRNA^{mMet}$ catalyzed by native or modified enzyme have been compared at their respective optimal magnesium concentrations. The only observed difference is a ratio of the rate values for the reactions between a given tRNA and the two enzymes, which tends to be closer to 4 than to 2. This result contrasts with that obtained for the methionine-dependent ATP-PPi isotopic exchange reaction where the enzymes exhibit turnover numbers within the ratio of their subunit number.

A lower rate of aminoacylation upon trypsin treatment could arise either through subtle structural modifications of the catalytic sites or merely through functional uncoupling of the two enzyme protomers. Clearly, as emphasized above, the proteolytic treatment, which eliminates about three-quarters of the detected sequence redundancy within each protomer, has no noticeable effect on the activation and tRNA binding sites. It has also been verified that the trypsin modification does not affect the specificity of the enzyme-catalyzed aminoacylation reaction toward $tRNA^{fMet}$ with respect to bulk tRNA. These facts lead to the puzzling conclusion that either redundancy has no obvious functional significance or that the only duplicate sequence remaining after the digestion is still enough to keep the enzyme properties intact.

Although it does not affect the fundamental properties of the enzyme, the limited trypsin digestion monomerizes the dimeric native methionyl-tRNA synthetase and abolishes the anticooperative character of tRNA binding. To acquire deeper insight into the functional significance of such an anti-cooperativity, the kinetics of the binding of $tRNA^{fMet}$ to the native and modified enzymes were examined by stopped-flow analysis (Blanquet et al. 1976). Advantage was taken of the pronounced effect of methionine-specific tRNAs on the fluorescence of the enzyme. Interestingly, analysis of the results reveals that in the case of the dimer, the rate of exchange between strongly enzyme-bound tRNA and free competing tRNA increases as the concentration of the competing tRNA rises, instead of occurring at a limiting rate imposed by the kinetic dissociation constant of the enzyme-tRNA complex. By contrast, in the case of the modified enzyme the exchange proceeds normally to a limiting rate independent of the competing tRNA concentration in the solution. Interpretation of this observation requires that one assume the existence of a transient enzymatic complex in which both the "strong" and "weak" anticooperative tRNA sites are occupied. Exchange between tRNA molecules within the preferred binding site on the dimer should occur via this transient complex. Due to the faster rate of dissociation of tRNA from this transient complex, the rate of exchange increases with the concentration of the transient complex, and therefore with the concentration of free competing tRNA. The assumption of such a complex tends to indicate that despite the anticooperativity, both subunits of the dimer can symmetrically bind a tRNA molecule. Therefore, the anticooperativity does not arise from an asymmetric assembly of two enzyme subunits. Rather, it is the strong binding of one tRNA that promotes a decrease in the affinity for the second tRNA.

Small-angle neutron-scattering studies of the binding of initiator tRNA^{Met} to native methionyl-tRNA synthetase provide further information on the structural changes of the enzyme involved in the anticooperative formation of this transient complex (Dessen et al. 1978). In 10 mM MgCl$_2$, a condition where the anticooperativity is insured, the binding of the first tRNA molecule induces a conformational change of the enzyme with concomitant inhibition of the accessibility to the second site. This change is a contraction of the enzyme. Upon fixing the second tRNA, the protomers move apart and the conformation of the enzyme dimer is more open. Combined with the observation in 50 mM MgCl$_2$ (where the two tRNA sites have very similar affinities) that the conformation of the enzyme is open even in the absence of tRNA and remains open upon binding one or two tRNA, the latter observations indicate that the hindrance to binding a second tRNA in 10 mM MgCl$_2$ arises from the constrained conformation of the one tRNA·enzyme complex. It is easy to conceive that on the dimer, the positioning of two bulky tRNA molecules requires a good distance between sites. In 10 mM MgCl$_2$, the anticooperativity is expressed because the opening of the conformation of the enzyme, required for the binding of a second tRNA molecule, is achieved at the expense of the free energy of binding of this second tRNA.

The rationale for the observed variations of the radius of gyration of the enzyme may involve conformational changes within each protomer upon binding the tRNA, especially since in preliminary experiments the radius of gyration of the monomeric trypsin-modified methionyl-tRNA synthetase has been observed to decrease slightly upon saturation by tRNA^{fMet} (S. Blanquet et al., in prep.). However, in the case of the dimer the scattering data suggest that the observed changes are dominated by subunit rearrangements affecting the angle that the enzyme protomers form at their hinge. Reinforcing this idea, the neutron data also show that every time the enzyme structure is opened either by anticooperatively binding a second tRNA or by raising the MgCl$_2$ concentration, a concomitant tendency for the dimer to dissociate occurs. This fits with the idea that the movement of the protomers relative to each other couples with the stability of the dimer.

Reciprocation between strong and weak tRNA sites on methionyl-tRNA synthetase can take place when both tRNA sites are anticooperatively occupied. According to the above structural observations, the opening of the structure of the enzyme expresses a symmetry between the anticooperative sites, which could be required for the triggering of the reciprocation (Levitzki et al. 1971). However, the partial dissociation of the dimer upon anticooperatively binding the second tRNA could also account for the reciprocation. Upon dissociation of the dimer followed by reassociation of the protomers, each bound tRNA molecule would have an equal probability of occupying the strong tRNA binding site on the dimer. Whatever the mechanism may be, both possibilities are compatible with the observation that the rate of exchange of enzyme-bound tRNA depends on the concentration of the free competing tRNA (Blanquet et al. 1976).

CONCLUDING REMARKS

The functional comparison of the intact dimeric methionyl-tRNA synthetase with its derived monomeric fragment raises as many questions as it has resolved. For instance, the dimeric structure of native methionyl-tRNA synthetase may be understood in terms of regulation. It can relate the rate of exchange of bound tRNA molecules to the number of free tRNA molecules. On the basis of this observation and on the condition that the release of acylated tRNA is rate-limiting in the aminoacylation reaction, a plausible mechanism can be envisaged where release of aminoacyl-tRNA from its complex with the enzyme is controlled by the level of free unacylated tRNA.

On the other hand, except that it dissociates the dimer and therefore uncouples the subunits, the proteolysis of one-fourth of each subunit does not alter the specificity and catalytic properties of the enzyme. This leads to the following puzzling remarks. Each identical subunit of the intact native enzyme contains an important number of repeated sequences. Such a redundancy could have arisen through fusion between adjacent genes coding for the same polypeptide chain. The conservation during evolution of such a large extent of redundancy within the subunits of methionyl-tRNA synthetase supposes a quite powerful selection pressure. This pressure should reflect the importance of the redundancy in functional terms. About three-quarters of the peptides present twice in each subunit are lost through proteolytic digestion, yet, despite this, the resulting modified enzyme is still active. An identical behavior is found in the case of methionyl-tRNA synthetase from *Bacillus stearothermophilus*, a 180,000-m.w. dimeric enzyme that also contains the sequence-duplication feature. It is converted by subtilisin into an active 64,000-m.w. monomer (Koch et al. 1974; Mulvey and Fersht 1976). There again, no differences emerge from a comparison of the adenylation reaction catalyzed by either the intact dimer or its derived monomer (S. Blanquet et al., in prep.). Each protomer of the dimer behaves exactly as the modified monomer. Moreover, due to the rather high stability of the enzymes from *B. stearothermophilus*, it has been possible to compare the rates of renaturation of the intact dimer and of the modified monomer after denaturation by 5 M guanidine. The monomer recovers its activity by dilution as well as, or better than, the dimer. This observation strongly supports the idea that the part of duplicated sequence that has been lost in each protomer upon subtilisin digestion did not contain any program for the building up of an active three-dimensional structure of the enzyme.

In the case of *E. coli*, as well as *B. stearothermophilus*, no more than one catalytic site per subunit of methionyl-tRNA synthetase could be detected. In addition, no more than one methionine can be bound per polypeptide chain. Therefore, there is no evidence that the existence of repeated sequences within each subunit of methionyl-tRNA synthetase is reflected by binding or reaction at a second site per polypeptide chain.

The assumption that repeated sequences within subunits originate from gene duplication followed by fusion of adjacent genes indicates that the process leading to covalent fusion of domains has either abolished or altered one of the original active sites per subunit. The extra ATP site that is reported in the case of the *E. coli* enzyme and that disappears upon trypsin treatment could reflect remains of this lost active site. It remains possible that the duplication is essential for the formation of two structural domains that would, in turn, generate as large a cleft as the one shown by the low-resolution three-dimensional map of the modified enzyme from *E. coli*. Inside the cleft, the active adenylating site, or the tRNA binding site, should be created with participation of both domains. The neutron-scattering data have indicated that the tRNA center of mass should lie very close to the center of mass of the protomer of the native enzyme with which it associates (Dessen et al. 1978). This supports the idea of a $tRNA^{fMet}$ molecule encompassing the subunit to which it binds, rather than cross-bridging between enzyme subunits. The tRNA molecule could, for instance, fill up the cleft between the two domains, thus satisfying the symmetry recognition hypothesis of Kim (1975).

It is expected that the imminent availability of the 2.4-Å map of trypsin-modified methionyl-tRNA synthetase (J. L. Risler and C. Zelwer, pers. comm.), combined with affinity labeling and chemical studies aimed at probing the adenylate and tRNA binding sites (Fayat et al. 1978; Bruton 1979; S. Blanquet et al., in prep.), will provide answers to these questions.

ACKNOWLEDGMENTS

We are pleased to thank Jean-Pierre Waller for his constant interest, support, and encouragement during the course of our work. We are also indebted to him for critical reading of the manuscript. The authors are grateful to Chris J. Bruton for kindly furnishing a manuscript of his work before publication. The authors' work has been supported, in part, by grants from the Commissariat à l'Energie Atomique.

REFERENCES

Blanquet, S., P. Dessen, and M. Iwatsubo. 1976. Anticooperative binding of bacterial and mammalian initiator $tRNA^{Met}$ to methionyl-tRNA synthetase from *Escherichia coli*. *J. Mol. Biol.* **103:** 765.

Blanquet, S., G. Fayat, and J. P. Waller. 1974. The mechanism of action of methionyl-tRNA synthetase from *Escherichia coli*. Mechanism of the amino acid activation reaction catalyzed by the native and the trypsin-modified enzymes. *Eur. J. Biochem.* **44:**343.

———. 1975a. The amino acid activation reaction catalyzed by methionyl-transfer RNA synthetase. Evidence for synergistic coupling between the sites for methionine, adenosine and pyrophosphate. *J. Mol. Biol.* **94:**1.

Blanquet, S., M. Iwatsubo, and J. P. Waller. 1973. The mechanism of action of methionyl-tRNA synthetase from *Escherichia coli*. I. Fluorescence studies on tRNA^{Met} binding as a function of ligands, ions and pH. *Eur. J. Biochem.* **36:**213.

Blanquet, S., G. Fayat, M. Poiret, and J. P. Waller. 1975b. The mechanism of action of methionyl-tRNA synthetase from *Escherichia coli*. Inhibition by adenosine and 8-aminoadenosine of the amino acid activation reaction. *Eur. J. Biochem.* **51:**567.

Blanquet, S., G. Fayat, J. P. Waller, and M. Iwatsubo. 1972. The mechanism of reaction of methionyl-tRNA synthetase from *Escherichia coli*. Interaction of the enzyme with ligands of the amino acid-activation reaction. *Eur. J. Biochem.* **24:**461.

Bruton, C. J. 1975. The infrastructure of valyl transfer ribonucleic acid synthetase from yeast. *Biochem. J.* **147:**191.

———. 1979. Probing the sub-structure, evolution and interactions of aminoacyl-tRNA synthetases. In *Non-sense mutations and tRNA suppressors* (ed. J. D. Smith and J. Celis). Academic Press, London. (In press.)

Bruton, C. J. and B. S. Hartley. 1970. Chemical studies on methionyl-tRNA synthetase from *Escherichia coli*. *J. Mol. Biol.* **52:**165.

Bruton, C. J., R. Jakes, and G. L. E. Koch. 1974. Repeated sequences in methionyl-tRNA synthetase from *E. coli*. *FEBS Lett.* **45:**26.

Cassio, D. and J. P. Waller. 1971. Modification of methionyl-tRNA synthetase by proteolytic cleavage and properties of the trypsin-modified enzyme. *Eur. J. Biochem.* **20:**283.

Cassio, D., F. Lemoine, J. P. Waller, E. Sandrin, and R. A. Boissonas. 1967. Selective inhibition of aminoacyl ribonucleic acid synthetases by aminoalkyl adenylates. *Biochemistry* **6:**827.

Cory, S., K. A. Marcker, S. K. Dube, and B. F. C. Clark. 1968. Primary structure of a methionine transfer RNA from *Escherichia coli*. *Nature* **220:**1039.

Dessen, P., S. Blanquet, G. Zaccai, and B. Jacrot. 1978. Anticooperative binding of initiator tRNA^{Met} to methionyl-transfer RNA synthetase from *Escherichia coli*: Neutron scattering studies. *J. Mol. Biol.* **126:**293.

Dube, S. K., K. A. Marcker, B. F. C. Clark, and S. Cory. 1968. Nucleotide sequence of N-formyl-methionyl transfer RNA. *Nature* **218:**223.

Fayat, G. and J. P. Waller. 1974. The mechanism of action of methionyl-tRNA synthetase from *Escherichia coli*. Equilibrium-dialysis studies on the binding of methionine, ATP and ATP-Mg^{2+} by the native and trypsin-modified enzymes. *Eur. J. Biochem.* **44:**335.

Fayat, G., M. Fromant, and S. Blanquet. 1977a. Couplings between the sites for methionine and adenosine-5' triphosphate in the amino acid activation reaction catalyzed by trypsin modified methionyl-transfer RNA synthetase from *Escherichia coli*. *Biochemistry* **16:**2570.

———. 1978. Aminoacyl-tRNA synthetases: Affinity labeling of the ATP binding site by 2',3'-ribose oxidized ATP. *Proc. Natl. Acad. Sci.* **75:**2088.

Fayat, G., M. Fromant, D. Kahn, and S. Blanquet. 1977b. Affinity chromatography on agarose-hexyl-adenosine-5'-phosphate of methionyl-tRNA synthetase from *Escherichia coli*. Application of the couplings between the methionine and ATP sites. *Eur. J. Biochem.* **78:**333.

Fayat, G., S. Blanquet, P. Dessen, G. Batelier, and J. P. Waller. 1974. The molecular weight and subunit composition of phenylalanyl-tRNA synthetase from *Escherichia coli* K-12. *Biochimie* **56:**35.

Fersht, A. R. 1975. Demonstration of two active sites on a monomeric aminoacyl-tRNA synthetase. Possible roles of negative cooperativity and half-of-the-sites reactivity in oligomeric enzymes. *Biochemistry* **14:**5.

Gulik, A., C. Monteilhet, P. Dessen, and G. Fayat. 1976. Small-angle X-ray and light scattering study of native and trypsin-modified methionyl-tRNA synthetase from *Escherichia coli*. *Eur. J. Biochem.* **64:**295.

Heinrikson, R. L. and B. S. Hartley. 1967. Purification and properties of methionyl-transfer ribonucleic acid synthetase from *Escherichia coli*. *Biochem. J.* **105:**17.

Holler, E., B. Hammer-Raber, T. Hanke, and P. Bartman. 1975. The catalytic mechanism of amino acid:tRNA ligases. Synergism and formation of the ternary enzyme-amino-acid-ATP complex. *Biochemistry* **14:**2496.

Hyafil, F., Y. Jacques, G. Fayat, M. Fromant, P. Dessen, and S. Blanquet. 1976. Methionyl-tRNA synthetase from *Escherichia coli*: Active stoichiometry and stopped-flow analysis of methionyl-adenylate formation. *Biochemistry* **15:**3678.

Jacques, Y. and S. Blanquet. 1977. Interrelation between transfer RNA and amino acid activating sites of methionyl-transfer RNA synthetase from *Escherichia coli*. *Eur. J. Biochem.* **79:**433.

Jakes, R. and A. R. Fersht. 1975. Tyrosyl-tRNA synthetase from *Escherichia coli*. Stoichiometry of ligand binding and half-of-the-sites reactivity in aminoacylation. *Biochemistry* **14:**3344.

Jencks, W. P. 1975. Binding, energy, specificity and enzymic catalysis. The Circe effect. *Adv. Enzymol.* **43:**219.

Kim, S. H. 1975. Symmetry recognition hypothesis model for tRNA binding to aminoacyl-tRNA synthetase. *Nature* **256:**679.

Koch, G. L. E. and C. J. Bruton. 1974. The subunit structure of methionyl-tRNA synthetase from *Escherichia coli*. *FEBS Lett.* **40:**180.

Koch, G. L. E., Y. Boulanger, and B. S. Hartley. 1974. Repeating sequences in aminoacyl-tRNA synthetases. *Nature* **249:**316.

Kosakowski, H. M. and E. Holler. 1973. Phenylalanyl-tRNA synthetase from *Escherichia coli* K10. Synergistic coupling between the sites for binding of L-phenylalanine and ATP. *Eur. J. Biochem.* **38:**274.

Kula, M. R. 1973. Structural studies on isoleucyl-tRNA synthetase from *E. coli*. *FEBS Lett.* **35:**299.

Lawrence, F., S. Blanquet, M. Poiret, M. Robert-Gero, and J. P. Waller. 1973. The mechanism of action of methionyl-tRNA synthetase. 3. Ion requirements and ionic parameters of the ATP-PP exchange and methionine-transfer reactions catalyzed by the native and trypsin-modified enzymes. *Eur. J. Biochem.* **36:**234.

Lemoine, F., J. P. Waller, and R. van Rapenbusch. 1968. Studies on methionyl-transfer RNA synthetase. 1. Purification and some properties of methionyl transfer RNA synthetase from *Escherichia coli* K12. *Eur. J. Biochem.* **4:**213.

Levitzki, A., W. B. Stallcup, and D. E. Koshland, Jr. 1971. Half-of-the-sites reactivity and the conformational states of cytidine triphosphate synthetase. *Biochemistry* **10**:3371.

Mulvey, R. S. and A. R. Fersht. 1976. Subunit interactions in the methionyl-tRNA synthetase of *Bacillus stearothermophilus*. *Biochemistry* **15**:243.

Ofengand, J. 1977. tRNA and aminoacyl-tRNA synthetases. In *Molecular mechanisms of protein biosynthesis* (ed. H. Weissbach and S. Pestka), p. 7. Academic Press, New York.

Robbe-Saul, S., F. Fasiolo, and Y. Boulanger. 1977. Phenylalanyl-tRNA synthetase from baker's yeast: Repeated sequences in the two subunits. *FEBS Lett.* **84**:57.

Waller, J. P., J. L. Risler, C. Monteilhet, and C. Zelwer. 1971. Crystallisation of trypsin-modified methionyl-tRNA synthetase from *Escherichia coli*. *FEBS Lett.* **16**:186.

Waterson, R. M. and W. H. Konigsberg. 1974. Peptide mapping of aminoacyl-tRNA synthetases: Evidence for internal sequence homology in *Escherichia coli* leucyl-tRNA synthetase. *Proc. Natl. Acad. Sci.* **71**:376.

Zelwer, C., J. L. Risler, and C. Monteilhet. 1976. A low resolution model of crystalline methionyl-transfer RNA synthetase from *Escherichia coli*. *J. Mol. Biol.* **102**:93.

Recognition of tRNAs by Proteins

Similarities in the Structural Organization of Complexes of tRNAs with Aminoacyl-tRNA Synthetases and the Mechanism of Recognition

Paul R. Schimmel
Department of Biology
Massachusetts Institute of Technology
Cambridge, Massachusetts 02139

tRNAs are specifically recognized and discriminated by several protein systems during the events of protein synthesis. For example, the aminoacyl-tRNA synthetases discriminate between the various specific tRNAs and correctly match each amino acid with its cognate tRNA. The elongation factor Tu (EF-Tu) has a strong preference for aminoacylated vs nonacylated tRNA but does not discriminate between the various acylated tRNA species. The *Escherichia coli* transformylase formylates charged methionine-specific tRNA^{fMet} but not charged tRNA^{mMet} or many other charged tRNA species. Even within a given system of protein-tRNA interactions there are some extraordinary features in the recognition process. For example, the aminoacyl-tRNA synthetases not only distinguish between tRNA species, but they also show a preference, which varies from enzyme to enzyme, for either the 2'- or 3'-OH as the initial site of aminoacylation.

In this paper and those by Schulman, Ebel et al., and Hecht (all this volume), the question of recognition and how it is achieved are discussed. This paper focuses mainly on the recognition of tRNAs by aminoacyl-tRNA synthetases. Ebel et al. and Schulman also consider this issue. Some of the other interesting questions are taken up by Schulman and by Hecht.

The problem of how synthetases recognize tRNAs has attracted a great many investigators. Apart from the obvious importance of the problem, its popularity is not hard to understand. Because tRNAs are low-molecular-weight, well-characterized nucleic acids, they are attractive, defined objects for experimental study. In addition, with such a large number of different synthetases and tRNAs available, it is possible to cross-compare results in any one system with those in several others. In principle, the possibilities for cross-comparisons should greatly aid in solving the recognition problem.

In spite of the many advantages of studying the synthetase-tRNA systems, and in spite of the efforts of a large number of laboratories, we still do not understand how synthetases differentiate between tRNA molecules. One difficulty is that in some cases the binding of a noncognate tRNA to a synthetase can be quite strong. Although the interaction of the cognate tRNA with an enzyme is generally strongest (under optimal conditions the association constant is as high as 10^8–10^9 M^{-1} [Yarus and Berg 1967; Lam and Schimmel 1975]), some noncognate interactions have stability constants within one to two orders of magnitude of that of the cognate ones (Ebel et al. 1973; Roe et al. 1973). Thus, at the level of binding, recognition is not simply an all-or-none phenomenon. Instead, discrimination is also achieved at the level of the maximal velocity for aminoacylation; that is, even if an enzyme binds to a noncognate tRNA, the velocity of misacylation is very slow (Ebel et al. 1973; Roe et al. 1973). This means that a twofold filter, or dual discrimination, operates on selecting the right tRNA—one discriminates at the level of binding, the other operates at the level of catalysis. Finally, it should be noted that it is possible to induce a synthetase to misacylate a large variety of tRNAs. This is accomplished by the use of special reaction conditions, such as buffers containing minor proportions of organic solvents (Giegé et al. 1971, 1972, 1974; Yarus 1972; Kern et al. 1972; Yarus and Mertes 1973; Ebel et al. 1973). What this means is that the recognition process is subtle and delicate.

Synthetases appear to have widely different subunit structures. There are four basic subunit types, which we designate as α, α_2, α_4, and $\alpha_2\beta_2$ (Kisselev and Favorova 1974; Söll and Schimmel 1974; Schimmel and Söll 1979; see also Appendix IV). This lack of uniformity in structure makes it difficult to compare results on one enzyme with those on another because such comparisons can be misleading.

In spite of these difficulties, significant progress has been made. There are three major areas on the tRNA structure that a number of investigations have indicated are important for synthetase-tRNA interactions. These areas are the acceptor end, the D stem, and the anticodon (Schimmel 1977). These regions are designated in Figure 1, which shows a cloverleaf structure and a three-dimensional structure of tRNA, as deduced by X-ray crystallography.

In the discussion below, the lines of evidence supporting this conclusion are reviewed briefly. Following this, some additional data are considered and some ideas on how recognition is achieved are discussed. A main conclusion is that there are some common features and similarities in the organization of synthetase-tRNA complexes and that proper recognition involves the precise alignment of the 3' terminal of the tRNA within the catalytic groups on the enzyme; slight distortions from a precise alignment mean that the bound tRNA cannot be aminoacylated, a mechanism by which synthetases avoid aminoacylation of noncognate tRNAs that

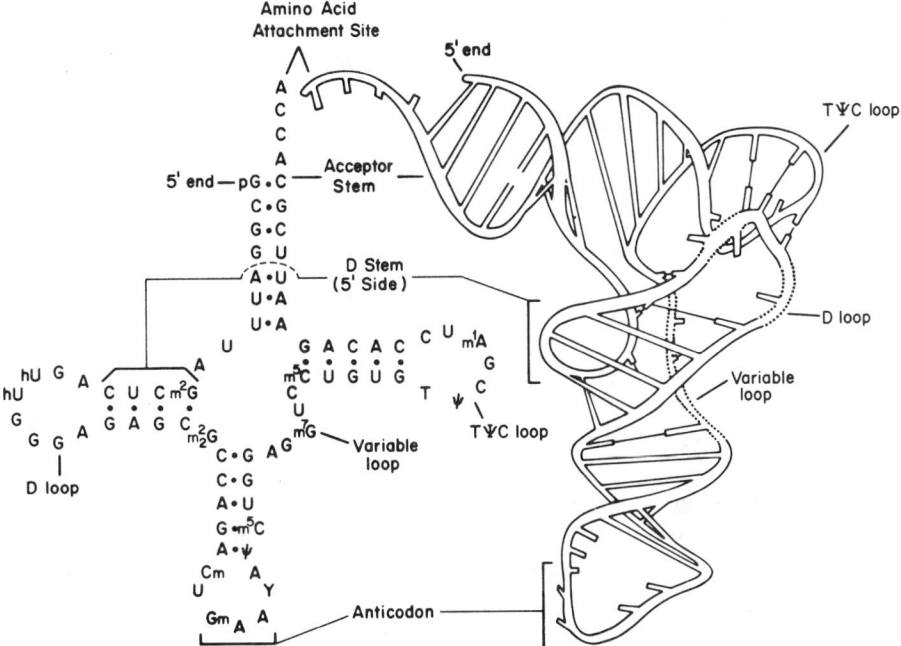

Figure 1 Sequence and cloverleaf structure of yeast tRNA^Phe (*left*) with a schematic illustration of three-dimensional structure (*right*; Kim et al. 1974b). In the schematic (*right*), the ribose-phosphate backbone is a continuous tube, whereas cloverleaf base pairs are indicated by crossbars and thin lines join bases that have tertiary interactions. Major landmarks are indicated on each structure. (Reprinted, with permission, from Schimmel 1979.)

nevertheless bind. These ideas lead to the notion that amino acid substitutions in the tRNA binding cleft may enable a synthetase to align its catalytic groups precisely with the 3′ terminal of a bound noncognate tRNA and thereby carry out a misacylation. This line of thought has led to the successful isolation of a mutant aminoacyl-tRNA synthetase with altered tRNA recognition. The recognition properties of the mutant enzyme and the implications for our understanding of the recognition problem are also discussed.

ACCEPTOR TERMINAL

It is not surprising that the acceptor terminal is somehow involved in the synthetase-tRNA interactions and recognition. After all, the amino acid is attached to the 3′ acceptor terminal; this means that the enzyme must come into close contact with the acceptor end of the molecule. This is easy to demonstrate. For example, Dickson and Schimmel (1975) incubated an aminoacyl-tRNA with ribonuclease A in the presence and absence of bound synthetase. In the absence of synthetase, the nuclease rapidly removes the aminoacyl moiety from the rest of the tRNA by cleaving after one of the cytidylic acid residues at the 3' end of the chain. However, in the presence of bound synthetase the action of the nuclease is greatly supressed in this area of the tRNA, indicating that this section is blocked by the bound synthetase.

Genetic studies have clearly shown the significance of the acceptor stem in synthetase-tRNA interactions. Table 1 summarizes a number of tRNA species that have altered recognition properties. Five different base changes in the acceptor stem have been produced in su^+3 tRNA^{Tyr} (Shimura et al. 1972; Hooper et al. 1972; Smith and Celis 1973). Each of these mutations enables the tRNA to be misacylated with glutamine. Thus, these mutations clearly affect the recognition process itself. Chemical modification studies using the *E. coli* initiator tRNA^{Met} have been carried out by Schulman and coworkers. They have found that modification of either of two bases in the acceptor stem results in inactivation of the tRNA (for a summary, see Schulman and Pelka 1977; Schulman, this volume).

Finally, arguments have been presented to suggest that, at least in some situations, the fourth base from the 3' end is an important discriminator in synthetase-tRNA interactions (Crothers et al. 1972).

D STEM

One of the earliest systematic studies of recognition took advantage of heterologous aminoacylations. Dudock et al. (1971) and Roe et al. (1973) observed that pure yeast phenylalanyl-tRNA synthetase can aminoacylate a number of noncognate tRNA species that derive from a different cell origin, such as *E. coli* or wheat germ. It was shown that 11 pure tRNA species can be aminoacylated to high extents in the heterologous systems. (In this study, there are no special reaction conditions used to induce misacylations.) The sequences of these heterologous tRNAs were compared with that of yeast tRNA^{Phe}. A major feature these tRNAs have in common with each other is the occurrence of a particular sequence of bases in the D stem.

A second piece of evidence involving the D stem comes from photochemical cross-linking studies. Under the action of UV light it is possible to cross-link together synthetase and tRNA that are bound in a complex. In six different complexes involving noninitiator tRNAs, the region in or around the 5' side of the D stem has been found to cross-link to bound synthetase (Schoemaker and Schimmel 1974; Budzik et al. 1975; Schoemaker et al. 1975; Schimmel 1977). Thus, there seems to be little doubt that this region is a common contact site for various synthetase-tRNA complexes.

ANTICODON

A priori, one might reason that the anticodon is the best candidate for a recognition site on tRNA. Although the anticodon is clearly important in some systems (see below), it is apparent from the foregoing that other areas also play a role in synthetase-tRNA interactions. Moreover, the anticodon appears to be of little or no significance in some systems. For example, Thiebe et al. (1972) showed that the entire anticodon may be removed from yeast tRNA^{Phe}, and yet the phenylalanyl-tRNA synthetase is still capable of aminoacylating the modified tRNA. The su^+3 tRNA^{Tyr} is aminoacylated normally by tyrosyl-tRNA synthetase. This tRNA is an amber suppressor by virtue of having the first base of the anticodon change from Q (which is a derivative of G) to C (Abelson et al. 1970).

But there are situations in which the anticodon is clearly important for recognition (see Table 1). For example, the mutation of the third base in the anticodon of *E. coli* tRNA$^{Gly}_2$ or tRNA$^{Gly}_3$ results in a greatly decreased rate of aminoacylation with glycyl-tRNA synthetase (see Carbon and Squires 1971). Also, Yaniv et al. (1974) showed that mutation of the second base in the anticodon of *E. coli* tRNA^{Trp} enables this tRNA to be misacylated with glutamine. In addition, H. Ozeki et al. (pers. comm.) reported that $\psi 37 \to A37$ and $\psi 37 \to C37$ mutations at the 3' end of the anticodon loop of a tRNA^{Gln} result in the misacylation of this tRNA with an unknown amino acid.

In addition to these genetic studies, oligonucleotide hybridization (Schimmel et al. 1972), photochemical cross-linking (Schimmel 1977), and isotope-labeling studies (Schoemaker and Schimmel 1976) have shown that in some systems the anticodon loop comes into close contact with the surface of the synthetase. In chemical modification studies of tRNA^{fMet} it has been shown that modification of either of the first two bases of the anticodon results in an inactive tRNA species (see Schulman and Pelka 1977).

What these studies clearly show is that in some systems the anticodon, and possibly some contiguous bases in the anticodon loop, is somehow involved in the mechanism by which synthetases differentiate between tRNAs. Of course, these studies do not pin down the mechanism by which the anticodon contributes to the recognition process, but they do point to the importance of the anticodon in some systems.

LOCATION OF ACCEPTOR END, D STEM, AND ANTICODON IN THREE-DIMENSIONAL STRUCTURE

The three areas shown to be important for synthetase-tRNA interactions, at least in a number of systems, are indicated in the illustration of the three-dimensional structure of tRNA in Figure 1. It is clear that the amino acid acceptor terminal, 5' side of the D stem, and the anticodon are all located on the same side of the tRNA structure. These regions

Table 1 Mutations in *E. coli* tRNAs that affect synthetase interactions

tRNA	Mutation	Location in structure	Effect of mutation	References
tRNA$_1^{Gly}$		altered anticodon	greatly decreased rate of aminoacylation with glycyl-tRNA synthetase	Carbon and Squires (1971)
tRNA$_2^{Gly}$	C36 → U36	third base in anticodon	greatly decreased rate of aminoacylation with glycyl-tRNA synthetase	Carbon and Squires (1971)
tRNA$_3^{Gly}$	C35 → U35	second base in anticodon	greatly decreased rate of aminoacylation with glycyl-tRNA synthetase	Carbon and Squires (1971)
su^+3 tRNA^{Tyr}	A82 → G82	acceptor stem	misacylation with glutamine	Shimura et al. (1972), Hooper et al. (1972), Smith and Celis (1973)
su^+3 tRNA^{Tyr}	C81 → A81	acceptor stem	misacylation with glutamine	Hooper et al. (1972), Smith and Celis (1973)

su^+3 tRNA^{Tyr}	C81→U81	acceptor stem	misacylation with glutamine	Hooper et al. (1972), Smith and Celis (1973)
su^+3 tRNA^{Tyr}	G2→A2	acceptor stem	misacylation with glutamine	Hooper et al. (1972), Smith and Celis (1973)
su^+3 tRNA^{Tyr}	G1→A1	acceptor stem	misacylation with glutamine	Smith and Celis (1973)
su^+7 tRNA^{Tyr}	C35→U35	second base in anticodon	misacylation with glutamine	Yaniv et al. (1974)
su^+2 tRNA$_2^{Gln}$	ψ37→A37	anticodon loop	misacylation with unknown amino acid	H. Ozeki et al. (pers. comm.)
su^+2 tRNA$_2^{Gln}$	ψ37→C37	anticodon loop	misacylation with unknown amino acid	H. Ozeki et al. (pers. comm.)
su^+2 tRNA$_2^{Gln}$	ψ37→A37/ G29→A29		misacylation with tryptophan	H. Ozeki et al. (pers. comm.)
su^+2 tRNA$_2^{Gln}$	ψ38→C38	anticodon loop-stem	misacylation with tryptophan	H. Ozeki et al. (pers. comm.)

Adapted from Schimmel and Söll (1979).

occur along the inner part of the roughly L-shaped molecule. Therefore, it seems likely that the enzymes bind for the most part along and around the inner part of the L-shaped structure. Some enzymes may extend all the way from the acceptor terminal to the anticodon; others may only reach part of the way and not make contact with the anticodon. In those cases, we would expect the anticodon not to be significant for recognition (see above); in other cases, however, the enzyme extends far enough to reach this part of the molecule so that it can play a role.

Table 1 gives a tabulation of mutations in various tRNAs that affect the synthetase interactions. These mutations are clustered in the anticodon region and near the acceptor end. They occur in the general vicinity of the area believed crucial for synthetase interactions.

As discussed above, a variety of studies have suggested that the three areas mentioned above play a part in the synthetase-tRNA interactions, at least in a number of systems. This indicates that there are common features to the synthetase-tRNA complexes, as already mentioned. Two other lines of evidence also suggest that there are similar structural features. First, photochemical cross-linking data have been obtained on cognate and noncognate synthetase-tRNA complexes (Budzik et al. 1975; Schoemaker et al. 1975). The conclusion from these studies is that noncognate complexes share certain features in common with cognate complexes. Thus, from a structural viewpoint, the noncognate complexes bear an overall similarity to the cognate ones. Second, as mentioned above, misacylations are easy to induce by using special reaction conditions. This implies that only subtle changes in the structure of a tRNA and, additionally or alternatively, an enzyme are sufficient to induce misrecognition. This is easy to visualize if, under normal circumstances, the noncognate tRNA already binds close to the proper mode for aminoacylation; the special reaction conditions merely provide a subtle conformational distortion that enables aminoacylation to occur. Thus, from these observations, and from the data discussed above, it appears that in spite of the wide variety of subunit structures of synthetases, the tRNA complexes are fairly similar in organization. The question then arises as to how, within the context of a roughly similar structural organization, do the synthetases differentiate among the various tRNA species.

SCHEME FOR RECOGNITION

The above facts can be incorporated into a simple model for the recognition process. The tRNA is envisioned as an L-shaped structure that is made up of a number of elements that are important for recognition. These structural elements may be comprised of one or more bases or ribose-phosphate groups. The idea is that these elements come close to the surface of the synthetase when the complex is formed. But the crucial question is whether the amino acid attachment site then fits into the catalytic site of the enzyme in a stereo-chemically precise way. If this occurs, aminoacylation can take place.

It is generally believed that the three-dimensional structures of various tRNAs share much in common with that of yeast tRNA^{Phe} (Kim et al. 1974a,b; Robertus et al. 1974). This means that the folding of the various tRNA species will approximate that of yeast tRNA^{Phe}, particularly in giving them a roughly L-shaped conformation. We envision that synthetases are designed to accommodate this structure and, accordingly, that their binding sites must bear much in common to accommodate the basic overall tRNA shape. What this means is that synthetases probably have a basic framework of amino acids that is designed to accommodate the general tRNA structure while superimposed upon it are residues at a few critical positions that determine the specificity of the aminoacylation process.

Thus, we envision that a synthetase has receptor sites for residues at specific positions in the cognate tRNA structure and that these receptor sites precisely fit with specific structural elements of the cognate tRNA. These structural elements are probably spatially dispersed in the tRNA molecule, so that elements far apart in the structure may play a critical role in binding to the synthetase. This situation contrasts with the recognition of specific sequences of adjacent nucleotides such as is found for the recognition of sites in DNA by restriction enzymes (Nathans and Smith 1975) or by repressors such as the *lac* repressor (Gilbert et al. 1973).

In the case of a noncognate interaction, the enzyme may fit well with the corresponding structural elements on the noncognate tRNA, except at one or a few positions. The mismatch at one of these positions could distort the bound noncognate tRNA just enough so that the 3' terminal does not fit into the catalytic groups on the enzyme. Consequently, even though a certain degree of binding does occur (although it is apt to be weaker, as is often observed, because of steric mismatches between the noncognate tRNA and protein), no aminoacylation can occur.

In the cognate case, the enzyme accommodates all of the relevent elements on the tRNA, including the one or more tRNA sites that cannot fit in the noncognate case. As a result, the tRNA is oriented perfectly on the surface of the synthetase so that the amino acid acceptor terminal rests properly in the catalytic site. This would enable aminoacylation to occur.

By use of these simple ideas it is possible to explain a variety of the complex features of the recognition problem. For example, the observation that the introduction of DMSO or dioxane in minor proportions can lead to extensive misacylations can be explained by simply postulating that some small structural change in the enzyme or tRNA occurs so that the 3' terminal of the noncognate tRNA is now brought into the proper spatial orientation with respect to the catalytic groups on the enzyme. Indeed, Yarus (1972) showed that the tRNA conformation is perturbed by small proportions of organic solvent mixed in aqueous solutions. The possibility for extensive misacylations under special solution conditions is viewed as a direct consequence of the similar structural features of the various synthetase-tRNA complexes, including the noncognate ones, and of the crucial dependence of the rate of aminoacylation on the exact positioning of the acceptor terminal in the catalytic site in the enzyme.

It should be emphasized that this model does not imply that there should be a correlation between the strength of a noncognate synthetase-tRNA interaction and the V_{max} for misacylation. The crucial factor for aminoacylation is the orientation of the 3' terminal with respect to the catalytic site. Binding strength is determined by a variety of interactions between groups on the enzyme and tRNA, such as electrostatic, hydrogen, and hydrophobic bonds. Even if a tRNA is not properly oriented for efficient catalysis on the surface of the enzyme, one can imagine that in the misoriented configuration the tRNA and enzyme can have favorable nonspecific interactions that would enhance the stability of the complex, or at least compensate for some of the unfavorable interactions.

It has not been assumed that the tRNA or synthetase undergoes a substantial conformational change upon complex formation. There is no strong evidence for such a change; nor is there compelling evidence ruling it out. However, if a conformational change should occur, the basic idea is the same. In the end, the crucial question is whether the 3' terminal of the tRNA aligns properly with the catalytic site on the enzyme.

MUTANT SYNTHETASES WITH ALTERED RECOGNITION CHARACTERISTICS

The scheme discussed above suggests that it should be possible to isolate mutant synthetases that misacylate tRNAs. The point is that if an interaction between a tRNA contact site and the corresponding site on the synthetase is unfavorable or sterically disallowed so that aminoacylation cannot occur, then the right amino acid substitution in the synthetase at that site should be able to relieve the constraint.

We set out to isolate mutant aminoacyl-tRNA synthetases with altered tRNA recognition characteristics. Recombinant DNA methods were used to insert the alanyl-tRNA synthetase gene on a plasmid, and mutagenesis of the plasmid was carried out followed by a transformation and selection procedure. The details are to be described elsewhere (D. LaDage and P. R. Schimmel, in prep.).

The goal was to obtain a mutant alanyl-tRNA synthetase that would attach alanine onto tRNA^{Ile}. A clone was obtained with a phenotype that suggested that such a mutant was present. Alanyl-tRNA synthetase was partially purified from this clone and tested out.

The aminoacylation of tRNA^{Ile} with mutant and wild-type alanyl-tRNA synthetases was tested. With the wild-type enzyme, no alanyl-tRNA^{Ile} is made. With the mutant enzyme, a slow synthesis of mischarged tRNA^{Ile} occurs. This demonstrates that a change in recognition specificity has been introduced.

The mutant enzyme also aminoacylates tRNA^{Ala} and does so much more efficiently than it aminoacylates tRNA^{Ile}. The question arises as to whether the mischarging of tRNA^{Ile} is specific, or whether a number of tRNAs can be aminoacylated by the mutant enzyme. It turns out that both tRNA^{Ala} and tRNA^{Ile} have a number of similarities. Each belongs to the largest class of tRNA structures that have 4 base pairs in the D stem and 5 bases in the variable loop (Sprinzl et al. 1978). And each has the same sequence of bases in the D stem and the same base (an A) at the fourth position from the 3' terminal (the site proposed as a discriminator). Two questions arise. Will another tRNA species with similar structural features also be misaminoacylated? How will the enzyme react toward a tRNA species with quite different structural features?

The species tRNA^{Val} bears similarities to both tRNA^{Ala} and tRNA^{Ile} in the aspects described above. Thus, this tRNA is a good one to check with respect to the first question. A quite different species is tRNA^{Tyr}. This tRNA has only 3 base pairs in the D stem, a large variable loop of 13 bases, and no sequence homologies with tRNA^{Ala} in the region of the D stem (Sprinzl et al. 1978). Clearly, to check on the second question, it is of interest to investigate this tRNA species.

From experiments with tRNA^{Ile}, tRNA^{Val}, and tRNA^{Tyr}, it is clear that the misaminoacylation is specific for tRNA^{Ile}. The amount of aminoacylation of tRNA^{Val} and of tRNA^{Tyr} is negligibly small and of questionable significance. It appears that the mutant enzyme has highly specific recognition characteristics.

These results indicate that the tRNA recognition sites of aminoacyl-tRNA synthetases are plastic and can be altered through amino substitutions to give enzymes with altered specificities. This fits well with the idea that there is much in common in the structural organization of the various synthetase-tRNA complexes and that specificity is achieved by the subtle effects of the placement of a few amino acids in the general binding cleft. According to the ideas discussed above, we imagine that the amino acid substitution introduced into alanyl-tRNA synthetase enables tRNA^{Ile} to fit more precisely so that the 3' terminal now lands in the right orientation at the catalytic site. The exquisite sensitivity and specificity of this system are shown by the fact that the alterations in alanyl-tRNA synthetases that enable misacylation of tRNA^{Ile} do not facilitiate aminoacylation of either tRNA^{Val} or tRNA^{Tyr}. If wholesale misaminoacylations were to occur, the effects would undoubtedly be lethal.

It should also be mentioned that H. Ozeki et al. (pers. comm.) appear to have identified genetically a glutaminyl-tRNA synthetase with altered tRNA recognition, although biochemical characterizations have not yet been carried out.

CONCLUSIONS

At this point we have a fair idea of how aminoacyl-tRNA synthetases bind to a tRNA. And we also have a reasonable working hypothesis to explain the subtle features of the recognition process (see above). The recently discovered ability to isolate a mutant aminoacyl-tRNA synthetase with altered recognition characteristics argues strongly that the recognition sites on the enzymes are closely similar and that specificity is achieved by the proper placement of perhaps only a few critical residues in a general tRNA binding cleft. At present, there is considerable need for a crystal of a synthetase-tRNA complex that is suitable for X-ray diffraction analysis. When this is available, some of the ideas discussed above will be brought into sharper focus and possibly altered.

REFERENCES

Abelson, J. N., M. L. Gefter, L. Barnett, A. Landy, R. L. Russell, and J. D. Smith. 1970. Mutant tyrosine transfer ribonucleic acids. *J. Mol. Biol.* **47**:15.

Budzik, G. P., S. S. M. Lam, H. J. P. Schoemaker, and P. R. Schimmel. 1975. Two photo-cross-linked complexes of isoleucine specific transfer ribonucleic acid with aminoacyl transfer ribonucleic acid synthetases. *J. Biol. Chem.* **250**:4433.

Carbon, J. and C. Squires. 1971. Studies on genetically altered transfer RNA species in *Escherichia coli. Cancer Res.* **31**:663.

Crothers, D. M., T. Seno, and D. G. Söll. 1972. Is there a discriminator site in transfer RNA? *Proc. Natl. Acad. Sci.* **69**:3063.

Dickson, L. A. and P. R. Schimmel. 1975. Structure of transfer RNA-aminoacyl transfer RNA synthetase complexes investigated by nuclease digestion. *Arch. Biochem. Biophys.* **67**:638.

Dudock, B., C. DiPeri, K. Scileppi, and R. Reszelbach. 1971. The yeast phenylalanyl-transfer RNA synthetase recognition site: The region adjacent to the dihydrouridine loop. *Proc. Natl. Acad. Sci.* **68**:681.

Ebel, J. P., R. Giegé, J. Bonnet, D. Kern, N. Befort, C. Bollack, F. Fasiolo, J. Gangloff, and G. Dirheimer. 1973. Factors determining the specificity of the tRNA aminoacylation reaction. Non-absolute specificity of tRNA-aminoacyl-tRNA synthetase recognition and particular importance of the maximal velocity. *Biochimie* **55**:547.

Giegé, R., D. Kern, and J. P. Ebel. 1972. Incorrect aminoacylations catalysed by *E. coli* valyl-tRNA synthetase. *Biochimie* **54:**1245.

Giegé, R., D. Kern, J. P. Ebel, and R. Taglang. 1971. Incorrect heterologous aminoacylation of various yeast tRNAs catalysed by *E. coli* valyl-tRNA synthetase. *FEBS Lett.* **15:**281.

Giegé, R., D. Kern, J. P. Ebel, H. Grosjean, S. De Henau, and H. Chantrenne. 1974. Incorrect aminoacylations involving tRNAs or valyl-tRNA synthetase from *Bacillus stearothermophilus*. *Eur. J. Biochem.* **45:**351.

Gilbert, W., N. Maizels, and A. Maxam. 1973. Sequences of controlling regions of the lactose operon. *Cold Spring Harbor Symp. Quant. Biol.* **38:**845.

Hooper, M. L., R. L. Russell, and J. D. Smith. 1972. Mischarging in mutant tyrosine transfer RNAs. *FEBS Lett.* **22:**149.

Kern, D., R. Giegé, and J. P. Ebel. 1972. Incorrect aminoacylations catalysed by the phenylalanyl- and valyl-tRNA synthetases from yeast. *Eur. J. Biochem.* **31:**148.

Kim, S.-H., F. L. Suddath, G. J. Quigley, A. McPherson, J. L. Sussman, A. H.-J. Wang, N. C. Seeman, and A. Rich. 1974a. Three-dimensional tertiary structure of yeast phenylalanine transfer RNA. *Science* **185:**435.

Kim, S.-H., J. L. Sussman, F. L. Suddath, G. J. Quigley, A. McPherson, A. H.-J. Wang, N. C. Seeman, and A. Rich. 1974b. The general structure of transfer RNA molecules (base stacking/hydrogen bonding/tRNA conformation). *Proc. Natl. Acad. Sci.* **71:**4970.

Kisselev, L. L. and O. O. Favorova. 1974. Aminoacyl-tRNA synthetases: Some recent results and achievements. *Adv. Enzymol.* **40:**141.

Lam, S. S. M. and P. R. Schimmel. 1975. Equilibrium measurements of cognate and noncognate interactions between aminoacyl transfer RNA synthetases and transfer RNA. *Biochemistry* **14:**2775.

Nathans, D. and H. O. Smith. 1975. Restriction endonucleases in the analysis and restructuring of DNA molecules. *Annu. Rev. Biochem.* **44:**273.

Robertus, J. D., J. E. Ladner, J. T. Finch, D. Rhodes, R. S. Brown, B. F. C. Clark, and A. Klug. 1974. Structure of yeast phenylalanine tRNA at 3 Å resolution. *Nature* **250:**546.

Roe, B., M. Sirover, and B. Dudock. 1973. Kinetics of homologous and heterologous aminoacylation with yeast phenylalanyl transfer ribonucleic acid synthetase. *Biochemistry* **12:**4146.

Schimmel, P. R. 1977. Approaches to understanding the mechanism of specific protein-transfer RNA interactions. *Accts. Chem. Res.* **10:**411.

———. 1979. Recent results on how aminoacyl transfer RNA synthetases recognize specific transfer RNAs. *Mol. Cell. Biochem.* (in press).

Schimmel, P. R. and D. Söll. 1979. Aminoacyl-tRNA synthetases. General features and recognition of transfer RNAs. *Annu. Rev. Biochem.* **48:**601.

Schimmel, P. R., O. C. Uhlenbeck, J. B. Lewis, L. A. Dickson, E. W. Eldred, and A. A. Schreier. 1972. Binding of complementary oligonucleotides to free and aminoacyl transfer ribonucleic acid synthetase bound transfer ribonucleic acid. *Biochemistry* **11:**642.

Schoemaker, H. J. P. and P. R. Schimmel. 1974. Photo-induced joining of a transfer RNA with its cognate aminoacyl-transfer RNA synthetase. *J. Mol. Biol.* **84:**503.

———. 1976. Isotope labeling of free and aminoacyl transfer RNA synthetase-bound transfer RNA. *J. Biol. Chem.* **251:**6823.

Schoemaker, H. J. P., G. P. Budzik, R. Giegé, and P. R. Schimmel. 1975. Three photo-cross-linked complexes of yeast phenylalanine specific transfer ribonucleic acid with aminoacyl transfer ribonucleic acid synthetases. *J. Biol. Chem.* **250:**4440.

Schulman, L. H. and H. Pelka. 1977. Structural requirements for aminoacylation of *Escherichia coli* formylmethionine transfer RNA. *Biochemistry* **16:**4256.

Shimura, Y., H. Aono, H. Ozeki, A. Sarabhai, H. Lamfrom, and J. Abelson. 1972. Mutant tyrosine tRNA of altered amino acid specificity. *FEBS Lett.* **22:**144.

Smith, J. D. and J. E. Celis. 1973. Mutant tyrosine transfer RNA that can be charged with glutamine. *Nature* **243:**66.

Söll, D. and P. R. Schimmel. 1974. Aminoacyl tRNA synthetases. *Enzyme* **10:**489.

Sprinzl, M., F. Grüter, and G. H. Gauss. 1978. Collection of published tRNA sequences. *Nucleic Acids Res.* **5:**r15.

Thiebe, R., K. Harbers, and H. G. Zachau. 1972. Aminoacylation of fragment combinations from yeast tRNA^{Phe}. *Eur. J. Biochem.* **26:**144.

Yaniv, M., W. R. Folk, P. Berg, and L. Soll. 1974. A single mutational modification of a tryptophan-specific transfer RNA permits aminoacylation by glutamine and translation of the codon UAG. *J. Mol. Biol.* **86:**245.

Yarus, M. 1972. Binding of isoleucyl transfer ribonucleic acid by isoleucyl transfer ribonucleic acid synthetase: Solvents, the strength of interaction and a proposed source of specificity. *Biochemistry* **11:**2050.

———. 1972. Solvent and specificity. Binding and isoleucylation of phenylalanine transfer ribonucleic acid (*Escherichia coli*) by isoleucyl transfer ribonucleic acid synthetase from *Escherichia coli*. *Biochemistry* **11:**2352.

Yarus, M. and M. Berg. 1967. Recognition of tRNA by aminoacyl tRNA synthetases. *J. Mol. Biol.* **28:**479.

Yarus, M. and M. Mertes. 1973. The variety of intraspecific misacylations carried out by isoleucyl transfer ribonucleic acid synthetase of *Escherichia coli*. *J. Biol. Chem.* **248:**6744.

Chemical Approaches to the Study of Protein-tRNA Recognition

LaDonne H. Schulman
Department of Developmental Biology and Cancer
Division of Biology, Albert Einstein College of Medicine
Bronx, New York 10461

Chemical modification methods can be used to study the structural requirements for recognition of tRNAs by proteins, provided that alteration of a specific site in the tRNA can be correlated with the presence or absence of an effect on the protein-tRNA interaction. This can often be accomplished by selection of a reagent that attacks a unique target in the tRNA, such as the 3' or 5' terminal, or a specific minor base. The amount of structure-function information that can be obtained using such highly selective reagents is obviously limited to a very small number of sites. To probe other internal regions of tRNA structure, reagents can be used that modify major bases. Such reagents normally alter a number of sites simultaneously, making it necessary to carry out additional steps to determine which of the modifications produced are significant in terms of their effect on a particular biological activity. In this paper, I will describe the general methods used in my laboratory to obtain specific information on the structural requirements for recognition of *Escherichia coli* tRNAs[Met] by a number of *E. coli* proteins following modification of the tRNAs with reagents that alter the structure of major bases.

CONFORMATIONAL EFFECTS ON BIOLOGICAL ACTIVITY

Recognition of tRNAs by proteins is expected to involve interactions between essential ligands located at specific sites in the three-dimensional structure of each macromolecule. Such interactions can be destroyed either by altering the conformation of the tRNA in a manner that changes the spatial orientation of a required functional group, or by direct modification of an essential ligand. Conformational effects can often be eliminated or confined to the immediate environment of the modified base when chemical modifications are carried out under conditions that maintain the compact, native tRNA structure. Under such conditions, the sites of reaction are ordinarily limited to accessible bases not involved in secondary or tertiary structure interactions. For example, modification of *E. coli* tRNA[fMet] with 3 M sodium bisulfite (pH 6.0) in 10 mM $MgCl_2$ at 25°C leads to deamination of only 6 of the 26 C residues in the tRNA

(Goddard and Schulman 1972). High-resolution nuclear magnetic resonance (NMR) spectroscopy shows that the modified tRNA retains all of the secondary and tertiary NH hydrogen bonds of unmodified tRNA^{fMet} (Fig. 1). In addition, a new UA base pair is formed at the end of the acceptor stem of the modified tRNA as a result of the bisulfite-catalyzed C→U conversion at the 5' terminal. By contrast, the NMR spectra of two tRNA^{fMet} isomers that differ by a single base at position 47 in the variable loop show a complex difference spectrum (Daniel and Cohn 1976). The isomer containing m^{7}G47 has been shown to have a more stable, ordered structure and the NMR spectrum indicates the presence of additional hydrogen bonds, which give rise to several new peaks and produce shifts in the peak positions of a number of other resonances. The complexity of the difference spectrum resulting from a single base change in a region of the tRNA involved in tertiary structure serves to illustrate the sensitivity of NMR spectroscopy to small conformational changes and to emphasize the simplicity of the difference spectrum of unmodified and bisulfite-modified tRNA^{fMet}.

CHEMICAL REAGENTS

Use of reaction conditions that maintain the native structure of the tRNA during chemical modification not only limits conformational effects to

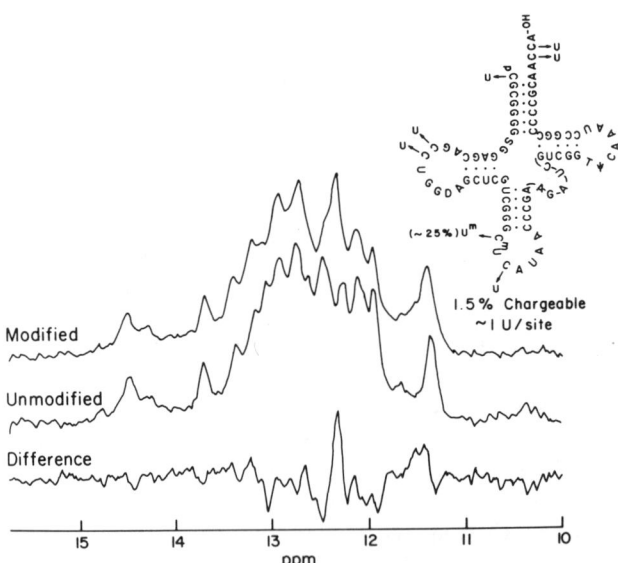

Figure 1 360-MHz proton NMR spectra of modified and unmodified *E. coli* tRNA$_3^{fMet}$. The modified tRNA was treated with 3 M sodium bisulfite (pH 6), 10 mM MgCl$_2$ at 25°C for 24 hr as described by Goddard and Schulman (1972). Spectra were recorded in 0.01 M sodium cacodylate (pH 7), 0.1 M NaCl, 0.015 M MgCl$_2$, 1 mM EDTA at 35°C (R. Romer et al., unpubl.).

local structural regions, but also greatly reduces the number of affected sites, simplifying subsequent analysis of structure-function data. The complexity of chemical-modification reactions is further reduced by using reagents that have chemical selectivity for one, or possibly two, of the major bases. It is also desirable to use reagents that do not greatly increase the size of the nucleotide base in order to avoid adverse steric effects on protein-tRNA interactions.

Sodium bisulfite has proven to be the most useful of the chemical reagents we have used for structure-function studies. Reaction conditions can be selected that result only in the conversion of exposed C residues in the tRNA to U residues (Hayatsu 1976). Thus, the product is a naturally occurring base with well-defined properties. Sodium bisulfite can also be used to modify exposed U residues selectively under other reaction conditions (Hayatsu 1976). The product is a U-bisulfite adduct that is nonplanar and has altered base-stacking and hydrogen-bonding properties. Treatment of tRNAs with chloroacetaldehyde leads to simultaneous modification of exposed A and C residues (Schulman and Pelka 1976). G residues can be selectively modified by photo-oxidation in the presence of methylene blue (Simon and van Vunakis 1964). Although ordered structure has been reported to retard the rate of modification of G in tRNA (Kuwano et al. 1968), G residues in the double-stranded D-stem region and a G residue near the 5' terminal of *E. coli* tRNA^{fMet} were found to be extensively modified under conditions in which G residues in the D loop, the anticodon stem, the TψC stem, and the remainder of the acceptor stem were unreactive (Schulman 1971). As no other purified tRNA has been studied, it is not clear whether the reactivity of G residues in the D stem is a unique property of tRNA^{fMet}. The modifications lead to a local disruption of secondary structure in the D stem but do not lead to simultaneous loss of ordered structure in other parts of the molecule (L. H. Schulman, unpubl.).

RATE OF MODIFICATION AND CHANGE IN BIOLOGICAL ACTIVITY

When a chemical reagent introduces multiple modifications, the structural changes must occur independently of one another to allow identification of primary events that alter protein-tRNA interactions. An indication of the suitability of a given chemical reagent for use in structure-function studies is obtained by investigating the rate of change of biological activity. Observation of pseudo first-order reaction kinetics indicates that the change results from a single-hit process. A lag in the rate of change of activity shows that the effective modification is dependent on a prior reaction at another site. This could result, for example, if modification of one residue increased the chemical reactivity of another or if a change in biological activity required alteration of two sites within the same molecule.

SEPARATION OF MODIFIED tRNA MOLECULES ON THE BASIS OF BIOLOGICAL ACTIVITY

To distinguish modifications that affect protein-tRNA interactions from those that do not, the modified molecules must be fractionated on the basis of the biological activity of interest. In the case of the aminoacylation reaction, we have used the phenoxyacetylation procedure (Gillam et al. 1968) to separate modified molecules that can still be aminoacylated by *E. coli* methionyl-tRNA synthetase from those that cannot. Partially modified tRNA is allowed to react with the enzyme, which attaches methionine to the active molecules. After derivatization, these molecules are separated from inactive, deacylated molecules on benzoylated DEAE-cellulose and structural analyses are performed (Schulman 1970, 1971, 1972; Schulman and Goddard 1973; Schulman and Pelka 1977a). Samples that contain a mixture of modified tRNA molecules that are aminoacylated at significantly different rates can also be separated by this procedure into fractions having normal and altered aminoacylation kinetics following reaction with a limiting amount of enzyme (Schulman and Pelka 1977b).

Modified tRNA molecules capable of forming a stable ternary complex with bacterial elongation factor Tu (EF-Tu) and GTP have been separated from inactive modified molecules by gel filtration on Sephadex G-100 at 4°C (Schulman et al. 1974). Modified tRNA molecules that are attacked by peptidyl-tRNA hydrolase have been separated from those that are resistant to hydrolysis by this enzyme by chromatography on benzoylated DEAE-cellulose (Schulman and Pelka 1975). The deacylated (active) tRNA fraction readily separates from unreacted N-substituted aminoacyl-tRNAs by elution with a salt gradient.

RECOGNITION OF *E. COLI* METHIONINE tRNAs BY *E. COLI* METHIONYL-tRNA SYNTHETASE

We have studied the effect of modifications at 25 sites in tRNA^{fMet} on the ability of the tRNA to be aminoacylated by methionyl-tRNA synthetase (Fig. 2). These data, as well as relevant results from other laboratories, have recently been summarized elsewhere (Schulman and Pelka 1977a).

Following separation of active and inactive molecules and structural analysis, the ability of each tRNA fraction to interact with methionyl-tRNA synthetase is investigated. The rate of aminoacylation of active modified tRNA is measured, as well as the ability of inactive, modified tRNA to inhibit aminoacylation. Data obtained from such studies have revealed a wide variety of effects of different modifications on biological activity; many modifications have no detectable effect on the ability of the tRNA to interact with the enzyme, whereas others alter the

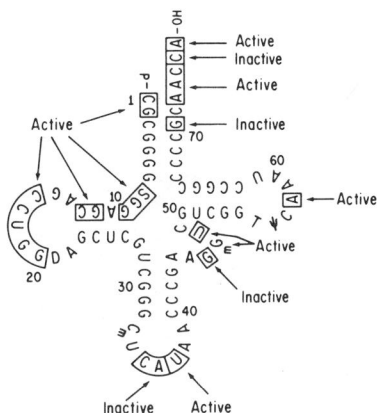

Figure 2 Sites of modified nucleotides in *E. coli* tRNA^{fMet} that have been studied in this laboratory. (Active) Residues that have been modified without loss of methionine-acceptor activity; (Inactive) sites where modification has resulted in complete loss of methionine acceptance. (Reprinted, with permission, from Schulman and Pelka 1977a.)

K_m and/or V_{max} for aminoacylation. A small number of modifications prevent recognition of the tRNA by the enzyme, and at least one modification causes loss of activity by interfering with the catalytic step in the reaction.

Modifications affecting protein-tRNA recognition have been found in three structural regions: the anticodon, the variable loop, and the acceptor-stem region. A summary of the structural changes in the anticodon bases of the *E. coli* tRNAs^{Met} and their effects on methionine acceptor activity is given in Figure 3. The data indicate that C35 is an essential ligand of methionyl-tRNA synthetase and point to the N3 position of the C base as a probable binding site for the enzyme.

Modification of A36 in the middle of the anticodon with chloroacetaldehyde leads to loss of methionine-acceptor activity. This may be due to a requirement for interaction of the enzyme with N1 or the exocyclic-amino group of this purine base; however, a negative effect of this modification on interaction of methionyl-tRNA synthetase with the adjacent base (C35) cannot be ruled out. Modifications at the 3′ end of the anticodon have not been found to inactivate tRNA^{fMet}; thus, U37 is not an essential base for recognition of the tRNA by methionyl-tRNA synthetase.

Photo-oxidation of G46 in the variable loop inactivates tRNA^{fMet}; however, the structural complexity of this region and the possible participation of G46 in tertiary structure interactions complicate interpretation of this result. Considerable information is available on the effect of structural modifications in the acceptor-stem region on aminoacylation (Fig. 4). The overall data indicate that this part of the molecule contains some essential structural element for recognition of tRNA^{fMet} by methionyl-tRNA synthetase and suggest a specific role for G71 in the interaction of the tRNA with this protein.

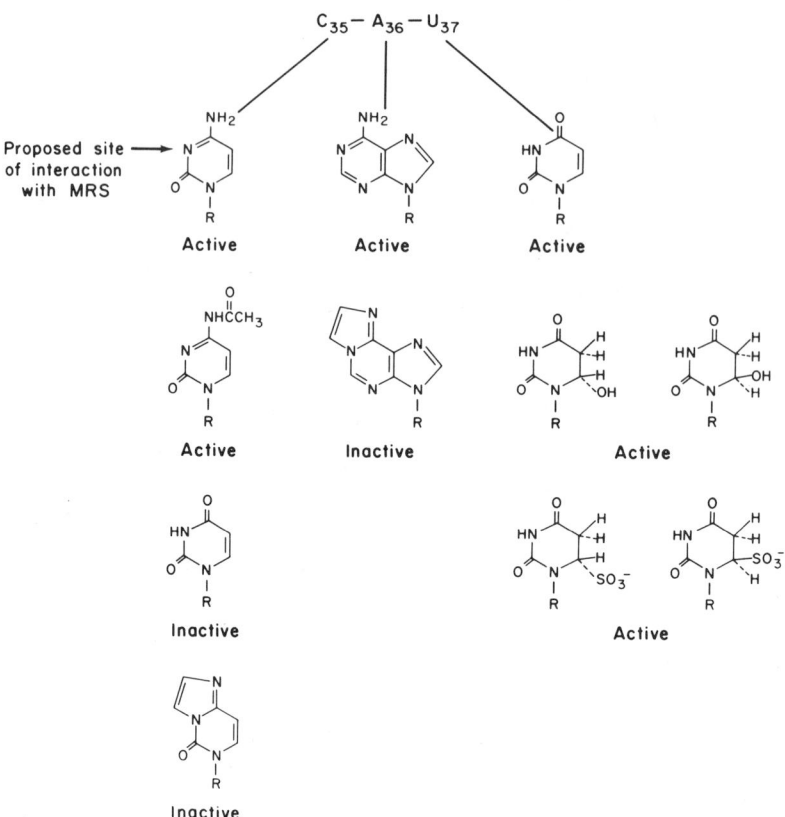

Figure 3 Effects of structural changes in the anticodon bases of *E. coli* tRNAs^{Met} on methionine-acceptor activity. (Reprinted, with permission, from Schulman and Pelka 1977a.)

COMPARISON OF PRIMARY-STRUCTURE AND CHEMICAL-MODIFICATION DATA FOR AMINOACYLATION OF tRNAs BY *E. COLI* METHIONYL-tRNA SYNTHETASE

It has been common practice to compare primary-sequence data for tRNAs aminoacylated by the same enzyme in an attempt to deduce the structural requirements for tRNA-synthetase recognition. It is not yet clear, however, whether a composite structure of the nucleotide bases common to all tRNA substrates of a given enzyme contains all of the critical protein-recognition sites. It is possible that certain essential interactions require a specific functional group that is common to two types of nucleotide bases, such as the carboxyl group located at the C2 position in both U and C. Purines and pyrimidines that share common functional groups might also be able to substitute for each other, provided that other structural changes were present elsewhere in the molecule that allowed the essential ligand to be properly positioned with respect to the

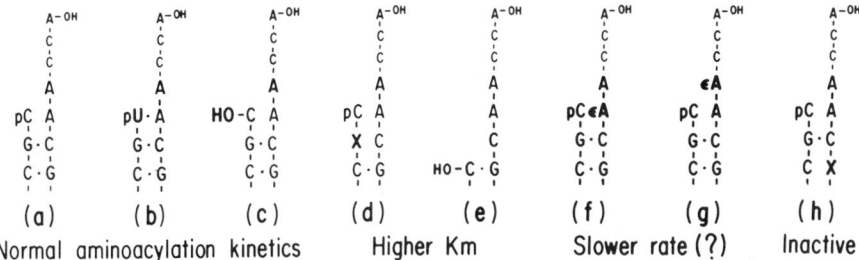

Figure 4 Effects of structural changes in the acceptor-stem region of *E. coli* tRNA^{fMet} on methionine-acceptor activity. Sites of modifications are indicated by heavy letters. (X) Photo-oxidized G and ϵA is 1,N^6-ethenoadenosine. The data are taken from Schulman and Goddard (1973) (*b*); L. H. Schulman and H. Pelka (unpubl.) (*c*); Schulman (1971) (*d*); Seno et al. (1971) and Seno and Sano (1971) (*e*); Schulman and Pelka (1977a) (*f, g*); and Schulman (1972) (*h*). Modification of A73 (*f*) or A74 (*g*) appears to alter the rate of aminoacylation, as judged by the fact that the sites are less extensively modified in the isolated active molecules following fractionation of the modified tRNA by the phenoxyacetylation procedure. The possibility of an increase in the rate of the enzymatic deacylation reaction for such modified molecules has not been excluded, however.

enzyme. Comparisons of sequence data and structural modification data should provide a clearer picture of the structural basis for recognition of tRNAs by aminoacyl-tRNA synthetases than either type of data alone.

The *E. coli* initiator and noninitiator tRNAs^{Met} and the initiator tRNAs from yeast and mammalian species are all aminoacylated by *E. coli* methionyl-tRNA synthetase with similar kinetics (Simsek et al. 1974). More recently, the structures of the initiator tRNAs^{Met} from *B. subtilis* (Yamada and Ishikura 1975), the blue green alga *Anacystis nidulans* (Ecarot-Charrier and Cedergren 1976), *Mycoplasma* (Walker and Raj-Bhandary 1978), wheat germ (Ghosh et al., pers. comm.), and the cytoplasmic and mitochondrial initiator tRNAs of *Neurospora crassa* (Gillum et al. 1977; Heckman et al. 1978) have been reported. All of these tRNAs are good substrates for *E. coli* methionyl-tRNA synthetase. The composite primary structure of the *E. coli* tRNAs^{Met} and the sequenced initiator tRNAs from both prokaryotic and eukaryotic organisms is shown in Figure 5. The composite contains only 19 nucleotides, of which 11 are common to all tRNAs. The size of the D loop varies between 7 and 9 nucleotides, and the variable loop contains only 4 rather than 5 nucleotides in the case of the *N. crassa* mitochondrial initiator tRNA. For comparison, the sites of modifications in *E. coli* tRNA^{fMet} leading to loss of methionine-acceptor activity are shown in Figure 5. Four of the five sites of inactivating modifications correspond to sites of identical nucleotides in the composite. The site of inactivation in the variable loop is a G residue in all of the tRNAs in the composite except

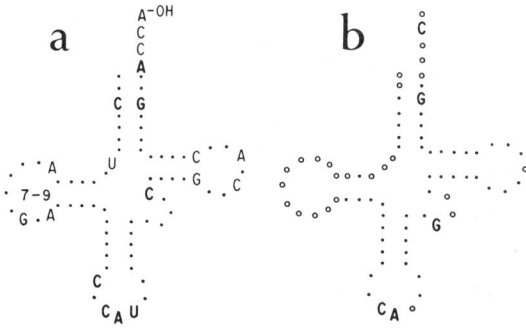

Figure 5 Comparison of primary structure and chemical modification data for aminoacylation of tRNAs by *E. coli* methionyl-tRNA synthetase. (*a*) Composite structure of tRNAs recognized by *E. coli* methionyl-tRNA synthetase. Heavy letters indicate nucleotides uniquely common to the methionine tRNAs. Light letters indicate nucleotides that are also common to other tRNAs. (·) Nucleotides not common to the methionine tRNAs. (*b*) Sites of modifications resulting in loss of methionine acceptor activity. Heavy letters indicate sites where chemical modifications result in inactivation of methionine acceptance. (○) Sites that can be nucleolytically excised or chemically modified without loss of methionine acceptance; (·) sites for which experimental data is not available (Schulman and Pelka 1977a).

yeast tRNA^{fMet}, which has a U residue at the same position. Eight of the nucleotide bases in the composite have been chemically modified or nucleolytically excised without inactivation of *E. coli* tRNA^{fMet}; however, certain structural alterations at three of these sites (U37, A74, and A77) have been found to affect the kinetics of aminoacylation of this tRNA (Schulman and Pelka 1977a).

It should be pointed out that noninitiator tRNAs^{Met} from several mammalian species contain all of the nucleotides in the composite structure of tRNAs recognized by *E. coli* methionyl-tRNA synthetase, yet they are not aminoacylated by this enzyme. Thus, the composite may not contain all of the essential recognition sites or the mammalian noninitiator tRNAs^{Met} may contain other structural features that inhibit binding of the *E. coli* methionyl-tRNA synthetase to recognition sites that are accessible in prokaryotic and eukaryotic initiator tRNA species.

RECOGNITION OF INITIATOR tRNAs BY *E. COLI* METHIONYL-tRNA TRANSFORMYLASE AND BACTERIAL INITIATION FACTOR 2

Following chemical modification, the fraction of structurally altered tRNA molecules that retains the ability to accept an amino acid can be used to investigate the structural requirements for recognition by other proteins that utilize aminoacylated tRNA as a substrate. In the case of the initiator tRNA, modified methionyl-tRNA^{fMet} can be tested for its ability to be formylated by *E. coli* methionyl-tRNA transformylase and modified formylmethionyl-tRNA^{fMet} can be tested for its ability to be recognized by bacterial initiation factor 2 (IF-2) and form a ribosomal initiation complex in a factor-dependent reaction.

A composite of primary-sequence data and chemical-modification data for tRNAs that can be formylated by *E. coli* methionyl-tRNA transformylase is shown in Figure 6. Thirty-one nucleotides are common to all prokaryotic and eukaryotic initiator tRNAs that are good substrates for this enzyme (Gillum et al. 1977; Heckman et al. 1978), including 11 nucleotides that are also common to other tRNA species. We have identified 15 sites in *E. coli* tRNA^{fMet} where chemical modifications can occur without loss of recognition of the tRNA by the transformylase (Schulman 1970, 1971, 1972; Schulman and Goddard 1973). These include six sites where identical nucleotides occur in all tRNAs that are good substrates for the enzyme (G2, U8, G12, G20, U37, and C75). We have not yet identified any site where chemical modification specifically inactivates tRNA^{fMet} with respect to its ability to be formylated.

Purified bacterial IF-2 forms a stable GTP-independent binary complex with *E. coli* formylmethionyl-tRNA^{fMet} in the absence of Mg^{++} ions (Majumdar et al. 1976). The structural requirements of the initiator tRNA for formation of this complex have been studied using modified formylmethionyl-tRNA^{fMet} molecules having structural alterations at 20 different sites (Sundari et al. 1976). The modified tRNAs were found to have the same affinity for IF-2 as unmodified formylmethionyl-tRNA^{fMet}. Additional studies showed that the binary complex with IF-2 can be formed with any tRNA structure covalently attached to an N-substituted methionine group. These results indicate that IF-2 alone is not capable of distinguishing the nucleotide sequence of tRNAs and that it selects the initiator tRNA by recognizing the formylmethionine moiety. The presently available data suggest that the role of IF-2 in formation of the ribosomal

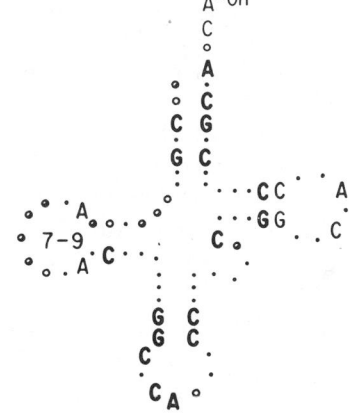

Figure 6 Composite of primary-sequence data and chemical-modification data for tRNAs recognized by *E. coli* methionyl-tRNA transformylase. Heavy letters indicate nucleotides uniquely common to tRNAs that can be formylated by *E. coli* methionyl-tRNA transformylase. Light letters indicate nucleotides that are also common to other tRNAs. (·) Nucleotides that are not common to the formylatable tRNAs^{Met}; (●) sites in *E. coli* tRNA^{fMet} that have been chemically modified without loss of formate-acceptor activity; (○) modifications of nucleotides that are part of the composite structure of tRNAs recognized by the *E. coli* formylating enzyme.

initiation complex is to stabilize the interaction of formylmethionyl-tRNA^{fMet} with the ribosome by binding to both the ribosomal particle and the formylmethionyl group of the tRNA (Sundari et al. 1976). The ribosome itself has an inherent specificity for the initiator tRNA sequence (Rudland et al. 1969), and initiation-factor-dependent ribosome binding of formylmethionyl-tRNA^{fMet} is sensitive to a variety of structural alterations in the tRNA (Sundari et al. 1976, 1977).

RECOGNITION OF AMINOACYL-tRNAs BY BACTERIAL EF-Tu

Bacterial EF-Tu binds aminoacyl-tRNAs to the acceptor site on the ribosome during polypeptide-chain elongation. Factor-dependent ribosome binding occurs through formation of an intermediate ternary complex consisting of EF-Tu, GTP, and aminoacyl-tRNA (Lucas-Lenard and Lipmann 1971). No complex is formed with deacylated or N-substituted aminoacyl-tRNAs. In addition, E. coli methionyl-tRNA^{fMet} is unable to form a stable ternary complex (Ono et al. 1968). This discrimination against the bacterial initiator tRNA is due to the presence of a unique structural feature in the tRNA itself, as all other aminoacyl-tRNAs that participate in protein synthesis form a stable EF-Tu·GTP·AA-tRNA complex. (AA is amino acid.)

One unique structural feature of prokaryotic tRNA^{fMet} species is the presence of a non-hydrogen-bonded base at the 5' terminal (Rich and RajBhandary 1976). Treatment of E. coli tRNA^{fMet} with sodium bisulfite results in conversion of the unpaired 5'-terminal C to U and formation of a normal UA base pair at the end of the acceptor stem (Goddard and Schulman 1972). This modification was found to have no effect on the ability of the tRNA to be aminoacylated or formylated (Schulman and Goddard 1973) or to initiate protein synthesis (Sundari et al. 1977). It was therefore of interest to determine the effect of this structural change on recognition of the tRNA by those proteins capable of discriminating against the initiator tRNA species.

The biologically active fraction of bisulfite-modified tRNA^{fMet} was found to form a ternary complex with EF-Tu and GTP (Schulman and Her 1973). The C→U base change in the 3'-terminal CCA sequence reduced the affinity of the modified tRNA for the protein, as indicated by the fact that the maximum level of binding was increased following removal of the damaged UCA sequence by limited digestion with snake venom phosphodiesterase and resynthesis of a normal CCA sequence using tRNA nucleotidyl transferase (Schulman et al. 1974). The ternary complex formed between EF-Tu, GTP, and bisulfite-modified methionyl-tRNA^{fMet} having a fully base-paired acceptor stem and a repaired CCA terminal had properties analogous to the ternary complexes formed with other aminoacyl-tRNAs. In addition to the C→U base change at the 5' terminal, this tRNA was also partially modified at two C residues in the D loop. Several lines of evidence indicated that the modification at the 5' terminal was responsible for the newly acquired ability of EF-Tu to recognize the bisulfite-treated tRNA. First, the maximum fraction of modified methionyl-tRNA^{fMet} that could be bound in the ternary complex in the presence of excess EF-Tu quantitatively corresponded to the fraction of the tRNA that contained 5'-terminal U and was about twice as great as the fraction having a C→U base change in the D loop. Second, when a mixture of methionyl-tRNA^{fMet} having 5'-terminal U and 5'-terminal C was bound to excess EF-Tu and GTP and the ternary complex was separated from unbound methionyl-tRNA^{fMet} by Sephadex G-100 gel filtration, the fraction bound to the protein was heavily enriched in a 5'-terminal U (Schulman et al. 1974).

The requirement for a fully base-paired acceptor stem for recognition of methionyl-tRNA^{fMet} suggested a possible role for the 5'-phosphate group in the formation of stable EF-Tu·GTP·AA-tRNA complexes. Removal of the 5'-phosphate from several normal AA-tRNAs was found to drastically reduce their affinity for EF-Tu, supporting the idea that a specific spatial orientation of the terminal phosphate greatly enhances binding of tRNA substrates to this protein.

RECOGNITION OF N-ACYLAMINOACYL-tRNAs BY PEPTIDYL-tRNA HYDROLASE

Peptidyl-tRNA hydrolase catalyzes the hydrolysis of N-acylaminoacyl-tRNAs and peptidyl-tRNAs to free tRNA and N-acyl amino acid or peptide. Unsubstituted aminoacyl-tRNAs are resistant to attack. In addition, *E. coli* formylmethionyl-tRNA^{fMet} is resistant to hydrolysis by this enzyme (Kössel and RajBhandary 1968; Vogel et al. 1968). Treatment of tRNA^{fMet} with sodium bisulfite, followed by aminoacylation and formylation of the biologically active fraction, yields modified formylmethionyl-tRNA^{fMet}, which is a good substrate for the hydrolase (Schulman and Pelka 1975). The products of the enzymatic cleavage are free tRNA^{fMet} and formylmethionine. To determine whether the base change at the 5' terminal is responsible for the change in activity, a mixture of formylmethionyl-tRNA^{fMet} having 5'-terminal U and 5'-terminal C was incubated with the enzyme for various times and the unreacted starting material was separated from hydrolyzed, deacylated tRNA^{fMet}. Structural analysis of the remaining unhydrolyzed formylmethionyl-tRNA^{fMet} showed that the enzyme had preferentially attacked those molecules containing the 5'-terminal U. The unusual structure of the acceptor stem of *E. coli* tRNA^{fMet} thus plays a critical role in maintaining the viability of the organism by preventing enzymatic cleavage of the formylmethionyl group from the bacterial initiator tRNA. This is in keeping with the results of primary-sequence studies that have shown that all prokaryotic initiator tRNA species that initiate protein synthesis with formylated methionyl-tRNA^{fMet} have an unpaired base at the 5' terminal, whereas eukaryotic initiator tRNAs and prokaryotic initiator tRNAs that utilize unformylated methionyl-tRNA^{fMet} for initiation have fully base-paired acceptor stems (Rich and RajBhandary 1976).

ACKNOWLEDGMENTS

The work from this laboratory has been supported by grants from the American Chemical Society (NP-19) and the National Institutes of Health (GM-16995).

REFERENCES

Daniel, W. E., Jr. and M. Cohn. 1976. Changes in tertiary structure accompanying a single base change in transfer RNA. Proton magnetic resonance and aminoacylation studies of *Escherichia coli* tRNA$_{\text{f1}}^{\text{Met}}$ and tRNA$_{\text{f3}}^{\text{Met}}$ and their spin-labeled (S^{4}U$_8$) derivatives. *Biochemistry* **15**:3917.

Ecarot-Charrier, B. and R. J. Cedergren. 1976. The preliminary sequence of tRNA$_\text{f}^{\text{Met}}$ from *Anacystis nidulans* compared with other initiator tRNAs. *FEBS Lett.* **63**:287.

Gillam, I., D. Blew, R. C. Warrington, M. von Tigerstrom, and G. M. Tener. 1968. A general procedure for the isolation of specific transfer ribonucleic acids. *Biochemistry* **7**:3459.

Gillum, A. M., L. I. Hecker, M. Silberklang, S. D. Schwartzbach, U. L. RajBhandary, and W. E. Barnett. 1977. Nucleotide sequence of *Neurospora crassa* cytoplasmic initiator tRNA. *Nucleic Acids Res.* **4**:4109.

Goddard, J. P. and L. H. Schulman. 1972. Conversion of exposed cytidine residues to uridine residues in *Escherichia coli* formylmethionine transfer ribonucleic acid. *J. Biol. Chem.* **247**:3864.

Hayatsu, H. 1976. Bisulfite modification of nucleic acids and their constituents. *Prog. Nucleic Acid Res. Mol. Biol.* **16**:75.

Heckman, J. E., L. I. Hecker, S. D. Schwartzbach, W. E. Barnett, B. Baumstark, and U. L. RajBhandary. 1978. Structure and function of initiator methionine tRNA from the mitochondria of *Neurospora crassa*. *Cell* **13**:83.

Kössel, H. and U. L. RajBhandary. 1968. Studies on polynucleotides. LXXXVI. Enzymatic hydrolysis of N-acylaminoacyl-transfer RNA. *J. Mol. Biol.* **35**:539.

Kuwano, M., Y. Hayashi, H. Hayashi, and K. Miura. 1968. Photochemical modification of transfer RNA and its effect on aminoacyl RNA synthesis. *J. Mol. Biol.* **32**:659.

Lucas-Lenard, J. and F. Lipmann. 1971. Protein biosynthesis. *Annu. Rev. Biochem.* **40**:409.

Majumdar, A., K. K. Bose, N. K. Gupta, and A. J. Wahba. 1976. Specific binding of *Escherichia coli* chain initiation factor 2 to fMet-tRNA$_\text{f}^{\text{Met}}$. *J. Biol. Chem.* **251**:137.

Ono, Y., A. Skoultchi, A. Klein, and P. Lengyel. 1968. Peptide chain elongation:

Discrimination against the initiator transfer RNA by microbial amino acid polymerization factors. *Nature* **220:**1304.
Rich, A. and U. L. RajBhandary. 1976. Transfer RNA: Molecular structure, sequence and properties. *Annu. Rev. Biochem.* **45:**805.
Rudland, P. S., W. A. Whybrow, K. A. Marcker, and B. F. C. Clark. 1969. Recognition of bacterial initiator tRNA by initiation factors. *Nature* **222:**750.
Schulman, L. H. 1970. Structure and function of *E. coli* formylmethionyl tRNA. I. Effect of modification of pyrimidine residues on aminoacyl synthetase recognition. *Proc. Natl. Acad. Sci.* **66:**507.
―――. 1971. Structure and function of *Escherichia coli* formylmethionine transfer RNA. II. Effect of modification of guanosine residues on aminoacyl synthetase recognition. *J. Mol. Biol.* **58:**117.
―――. 1972. Structure and function of *Escherichia coli* formylmethionine transfer RNA: Loss of methionine acceptor activity by modification of a specific guanosine residue in the acceptor stem of formylmethionine transfer RNA from *Escherichia coli*. *Proc. Natl. Acad. Sci.* **69:**3594.
Schulman, L. H. and J. P. Goddard. 1973. Loss of methionine acceptor activity resulting from a base change in the anticodon of *Escherichia coli* formylmethionine transfer ribonucleic acid. *J. Biol. Chem.* **248:**1341.
Schulman, L. H. and M. O. Her. 1973. Recognition of altered *E. coli* formylmethionine transfer RNA by bacterial T factor. *Biochem. Biophys. Res. Commun.* **51:**275.
Schulman, L. H. and H. Pelka. 1975. The structural basis for the resistance of *Escherichia coli* formylmethionyl transfer ribonucleic acid to cleavage by *Escherichia coli* peptidyl transfer ribonucleic acid hydrolase. *J. Biol. Chem.* **250:**542.
―――. 1976. Location of accessible bases in *Escherichia coli* formylmethionine transfer RNA as determined by chemical modification. *Biochemistry* **15:**5769.
―――. 1977a. Structural requirements for aminoacylation of *Escherichia coli* formylmethionine transfer RNA. *Biochemistry* **16:**4256.
―――. 1977b. Alteration of the kinetic parameters for aminoacylation of *Escherichia coli* formylmethionine transfer RNA by modification of an anticodon base. *J. Biol. Chem.* **252:**814.
Schulman, L. H., H. Pelka, and R. M. Sundari. 1974. Structural requirements for recognition of *Escherichia coli* initiator and non-initiator transfer ribonucleic acids by bacterial T factor. *J. Biol. Chem.* **249:**7102.
Seno, T. and K. Sano. 1971. Kinetical study on the reconstitution of methionine-acceptor activity from fragments of *Escherichia coli* tRNA^{fMet} with a deletion in the dihydrouridine region or the amino acid acceptor stem. *FEBS Lett.* **16:**180.
Seno, T., K. Sano, and T. Katsura. 1971. Reconstitution of methionine-acceptor activity from fragments of *E. coli* tRNA^{fMet} with pCpGp deleted from the 5'-terminus. *FEBS Lett.* **12:**137.
Simon, M. I. and H. van Vunakis. 1964. The dye-sensitized photooxidation of purine and pyrimidine derivatives. *Arch. Biochem. Biophys.* **105:**197.
Simsek, M., U. L. RajBhandary, M. Boisnard, and G. Petrissant. 1974. Nucleotide sequence of rabbit liver and sheep mammary gland cytoplasmic initiator transfer RNAs. *Nature* **247:**518.
Sundari, R. M., H. Pelka, and L. H. Schulman. 1977. Structural requirements of

Escherichia coli formylmethionyl transfer ribonucleic acid for ribosome binding and initiation of protein synthesis. *J. Biol. Chem.* **252**:3941.

Sundari, R. M., E. A. Stringer, L. H. Schulman, and U. Maitra. 1976. Interaction of bacterial initiation factor 2 with initiator tRNA. *J. Biol. Chem.* **251**:3338.

Vogel, Z., A. Zamir, and D. Elson. 1968. On the specificity and stability of an enzyme that hydrolyzes N-substituted aminoacyl-transfer RNAs. *Proc. Natl. Acad. Sci.* **61**:701.

Walker, R. T. and U. L. RajBhandary. 1978. The nucleotide sequence of formylmethionine tRNA from *Mycoplasma mycoides* sp. capri. *Nucleic Acids Res.* **5**:57.

Yamada, Y. and H. Ishikura. 1975. Nucleotide sequence of initiator tRNA from *Bacillus subtilis*. *FEBS Lett.* **54**:155.

Interaction between tRNA and Aminoacyl-tRNA Synthetase in the Valine and Phenylalanine Systems from Yeast

Jean-Pierre Ebel, Michel Renaud, André Dietrich,
Franco Fasiolo, Gérard Keith, Olga O. Favorova,*
Slava Vassilenko,† Mireille Baltzinger, Ricardo Ehrlich,
Pierre Remy, Jacques Bonnet, and Richard Giegé
Laboratoire de Biochimie
Institut de Biologie Moléculaire et Cellulaire du CNRS
67084 Strasbourg Cedex, France

Much work has been devoted to studies of the interaction between tRNAs and aminoacyl-tRNA synthetases. The aim was to understand the mechanisms that govern the specificity of the tRNA aminoacylation reaction (for recent general reviews, see Kisselev and Favorova 1974; Söll and Schimmel 1974; Goddard 1977; Ofengand 1977). In the early days, indirect approaches to this problem were used; recently, direct approaches have been introduced.

The indirect approaches included perturbing the primary structure of tRNA by chemical or genetic means, excising small sequences in tRNA and correlating these perturbations with the activity of the modified molecules, or comparing the sequences of different tRNAs recognized by the same aminoacyl-tRNA synthetase. The direct approaches include measuring the inhibition of the tRNA · aminoacyl-tRNA synthetase interaction caused by various factors (e.g., oligonucleotides or ionic strength) or studying the shielding of some areas of the tRNA by the aminoacyl-tRNA synthetase against RNases, oligonucleotides, or chemicals; they also include measurements by physical means of the thermodynamic parameters involved in the interaction between the aminoacyl-tRNA synthetase and intact or fragmented tRNA and the determination of the contact zones between the two macromolecules by cross-linking methods (for additional details, see reviews by Loftfield 1972; Schimmel 1977; Smith 1977). The interpretation of the results of most of these approaches has been greatly facilitated by knowledge of the structures of tRNA (Rich and RajBhandary 1976; Dirheimer et al., this volume) and aminoacyl-tRNA synthetase (Irwin

Present addresses: *Institute of Molecular Biology, Academy of Sciences of the USSR, Moscow B-312, USSR; and †Institute of Organic Chemistry, Academy of Sciences of the USSR, V. Siberian Department, Novosibirsk, USSR.

et al. 1976; Zelwer et al. 1976; Winter and Hartley 1977). At this stage, however, this problem is far from being solved, especially since many of these studies focused only on tRNA.

In this paper we report new results from our laboratory obtained with two tRNA·aminoacyl-tRNA synthetase systems from *Saccharomyces cerevisiae*, the valine and phenylalanine systems, using several direct approaches. On the tRNA side, we studied the contact zones with the aminoacyl-tRNA synthetase by UV photochemical cross-linking experiments and by studying the protection of the tRNA by the aminoacyl-tRNA synthetase against digestion by a structure-specific RNase from *Naja oxiana* cobra venom. On the aminoacyl-tRNA synthetase side, we investigated the localization of tRNA^{Phe} on the subunits of phenylalanyl-tRNA synthetase by cross-linking methods and the gross organization of the tRNA^{Val}·valyl-tRNA synthetase complex by neutron small-angle scattering. Finally, the comparison of interaction and aminoacylation parameters of intact or modified tRNA^{Phe} will allow us to propose a structural model that accounts for the specific aminoacylation of a tRNA by its cognate aminoacyl-tRNA synthetase.

DETERMINATION OF REGIONS WITHIN tRNA INTERACTING WITH AMINOACYL-tRNA SYNTHETASES

Photochemical Cross-linking between tRNAs and Aminoacyl-tRNA Synthetases

This approach was first used in the field of tRNA·aminoacyl-tRNA synthetase by Schimmel and his colleagues (Schimmel 1977) for six different complexes. Our experiments extend the investigations of that group in the sense that (1) results concerning a new system (tRNA^{Val}·valyl-tRNA synthetase) are presented and (2) a new methodology is used, comparable in its principles to that of the Schimmel group but differing by the irradiation conditions, the isolation procedure of the complexes, and the identification of the cross-linked oligonucleotides.

The irradiation of enzyme·tRNA complexes was performed with monochromatic light at 248 nm, which gave a favorable ratio of enzyme inactivation to cross-linking in both systems studied (Fig. 1). Indeed, we found that the inactivation of the aminoacyl-tRNA synthetases measured after irradiation is mainly due to their specific joining with tRNA and not to UV-light-induced side effects, as cross-links do not appear in the presence of salt at high concentration and no significant joining of tRNA with bovine serum albumin is found.

The determination of the regions within tRNA^{Val} or tRNA^{Phe} that are cross-linked to the cognate aminoacyl-tRNA synthetase was done by performing an RNase T1 digestion of the cross-linked tRNA·protein

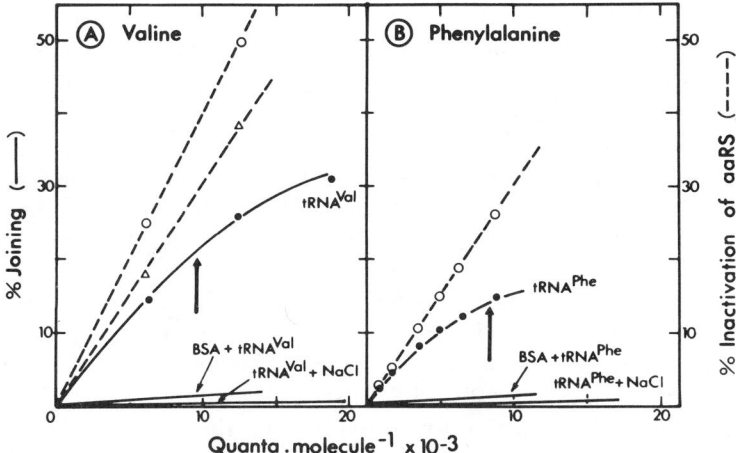

Figure 1 Behavior of the valine (*A*) and phenylalanine (*B*) tRNA·aminoacyl-tRNA synthetase systems upon irradiation at 248 nm. Yeast valyl- and phenylalanyl-tRNA synthetases, and their cognate tRNAs, were prepared by conventional methods. tRNA^{Val} and tRNA^{Phe} were ^{3}H-labeled in vitro. 1 to 1 complexes between the aminoacyl-tRNA synthetase (8×10^{-6} molecule) and tRNA were irradiated at 4°C at 248 nm for various times in 50 mM phosphate buffer (pH 7.2), 10 mM MgCl$_2$. The illumination system used, equipped with a 2-kilowatt Hg-Xe lamp, was set in such a way that the bandwidth of the emitted UV radiation was 3 nm. The percent joining of protein to tRNA was measured by filtration of the incubation mixtures on nitrocellulose membranes in the presence of 6 M urea. (●) Joining of tRNA to aminoacyl-tRNA synthetase; (———) joining of tRNA to bovine serum albumin or of tRNA to aminoacyl-tRNA synthetase in the presence of 1 M NaCl; (△) inactivation of complex formation; (○) inactivation of aminoacylation capacity; (→) conditions that have subsequently been used in this work.

complexes and then determining which T1 oligonucleotides were linked to the aminoacyl-tRNA synthetase by identification of those missing in the digest. Typically, for one analysis, 1 mg of [^{3}H]tRNA, mixed with the adequate amount of aminoacyl-tRNA synthetase, was irradiated with about 10^4 quanta × molecule^{-1} UV light at 248 nm. The irradiated material was then dialyzed against 6 M urea and the denatured covalent complexes were separated from free tRNA by high-voltage electrophoresis on sucrose gradients. After RNase T1 digestion of the complexes and of the free irradiated tRNAs (as a control), the oligonucleotides were analyzed on DEAE-cellulose columns. This methodology allowed us to get information both from the optical density and from the radioactivity of the oligonucleotides. Worth noticing is the necessity to analyze, as an internal control, the fraction of tRNA remaining non-cross-linked after irradiation in the presence of the enzyme, as sometimes the yield of some oligonucleotides was lowered. In some cases this was due to nonspecific photo-induced splits. The results showed, in both systems, six oligonucleotides significantly cross-linked to the enzyme, but to variable extents. Their localization in the cloverleaf sequence of yeast tRNA^{Val} or tRNA^{Phe} is given in Figure 2.

Some additional results were obtained in the case of the phenylalanine system at the level of the anticodon region. Upon irradiation of the complex at 313 nm, where only the Y base of the anticodon loop absorbs light, a specific cross-link of tRNA^{Phe} to phenylalanyl-tRNA synthetase occurs. This bond, however, appears to be rather unstable and its stability depends upon the handling conditions. The existence of this cross-link, however, is in agreement with the joining at 248 nm of the Y-base oligonucleotide of tRNA^{Phe} to phenylalanyl-tRNA synthetase (see Fig. 2B) and is a good indication for the proximity of the Y base to the enzyme during the interaction.

Other experiments also revealed the instability of UV-induced bonds. Decreasing amounts of cross-linked material, depending on the handling conditions, were observed as a function of time, and the percent of cross-links increased or decreased from one experiment to another to the same extent for all oligonucleotides. These observations suggest an explanation for the apparent discrepancy concerning the number of fragments of tRNA linked to the aminoacyl-tRNA synthetase found by us (six RNase T1 fragments) and by Schimmel and coworkers, who reported only three cross-linked oligonucleotides in all systems (Budzik et al. 1975; Schoemaker et al. 1975). Indeed, the percent joining measured is relevant to the ability of some regions within tRNA to react with the aminoacyl-tRNA synthetase, but it also reflects the more or less strong stability of the covalent bonds formed. For particularly unstable covalent bonds, the residual cross-links may fall under the limit of uncertainty of the method. We believe that the methodology we have developed is more favorable to the formation and/or to the preservation of a cross-link than that worked out by Schimmel and coworkers. It is worth mentioning that the six cross-linked oligonucleotides we have characterized in the yeast phenylalanine system correspond exactly to the set of fragments described by Schimmel's group (Schimmel 1977) in the cross-linking of yeast tRNA^{Phe} with three different enzymes (yeast phenylalanyl-, *E. coli* isoleucyl-, and *E. coli* valyl-tRNA synthetases). Furthermore, the oligonucleotides of various tRNAs interacting with yeast valyl-tRNA synthetase are located at similar positions either in the cognate yeast tRNA^{Val} (our results) or in the noncognate yeast tRNA^{Phe} and *E. coli* tRNA^{Ile} (Schimmel 1977).

For the interpretation of these results we have to keep in mind that the absence of cross-links does not imply the absence of interactions and that a cross-link near a G residue could prevent the action of RNase T1 and thus lead to the apparent cross-linking of the adjacent oligonucleotide.

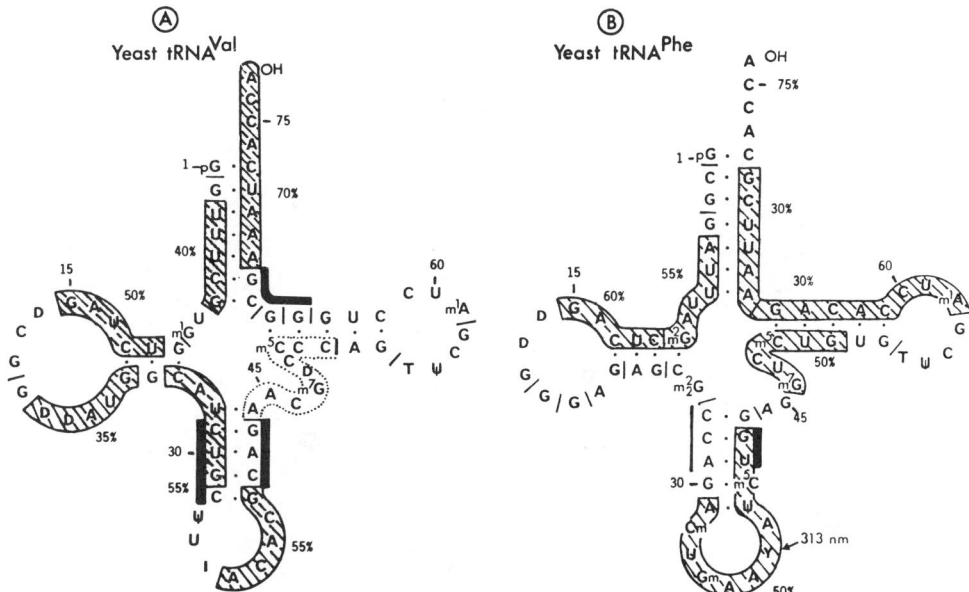

Figure 2 Cloverleaf structures of yeast tRNA^{Val} (A) and tRNA^{Phe} (B) with contact areas with valyl- and phenylalanyl-tRNA synthetases, as revealed by cross-linking and nuclease-digestion experiments. Cross-linking experiments: The RNase T1 oligonucleotides in tRNA^{Val} (Bonnet et al. 1974) and tRNA^{Phe} (Raj-Bhandary and Chang 1968) cross-linked to aminoacyl-tRNA synthetase are enclosed in full lines and hatched. Oligonucleotides were considered cross-linked only if their percent joining to the aminoacyl-tRNA synthetase (indicated on figures) was significantly higher than the errors that were estimated as ±15-20% joining. No information about oligonucleotide A44-C52 in tRNA^{Val} (enclosed in dotted lines) could be obtained as a consequence of an unspecific split occurring at C52 after hydrolysis of the irradiated complex. The percent joining was calculated by comparing the areas of the oligonucleotide peaks obtained after DEAE-cellulose column chromatography of RNase T1 hydrolyzates of free irradiated tRNA and of the corresponding tRNA engaged in the covalent complex. It should be recalled that joined oligonucleotides are those that are missing in the hydrolyzates of the covalent complexes. More details will be published elsewhere (M. Renaud et al., in prep.). (←313) tRNA^{Phe} region cross-linked to phenylalanyl-tRNA synthetase upon irradiation at 313 nm. Nuclease digestion experiments: Regions protected (or nonprotected) by aminoacyl-tRNA synthetase against digestion by cobra venom RNase are bordered in the cloverleaf structure by heavy (or light) lines. Experiments were conducted with tRNAs first dephosphorylated and then 5'-end-labeled with [γ-^{32}P]ATP in the presence of T4 polynucleotide kinase according to a combination of methods described by Donis-Keller et al. (1977) and Silberklang et al. (1977a). Incubation of tRNA or of tRNA · aminoacyl-tRNA synthetase complexes with cobra venom RNase was performed in 10 mM phosphate buffer (pH 7.0), 10 mM MgCl$_2$ at 0°C and the radioactive fragments arising from the digestions were analyzed on 15% polyacrylamide slab gels. Precise experimental conditions will be described elsewhere (O. O. Favorova et al., in prep.).

Nuclease Digestion of tRNA in the Absence or Presence of Aminoacyl-tRNA Synthetase

A few experiments on nuclease digestion of tRNA·aminoacyl-tRNA synthetase complexes have already been performed with RNases that preferentially attack single-stranded regions (Hörz and Zachau 1973; Schimmel 1977); tRNAs were found to be strongly protected by their cognate aminoacyl-tRNA synthetase.

In this work, we studied the digestion of yeast tRNA^{Val} and tRNA^{Phe}, free or in the presence of the corresponding aminoacyl-tRNA synthetase, using an RNase from the *Naja oxiana* cobra venom, which preferentially cuts RNAs in structured regions, without any base specificity (Vassilenko and Ryte 1975). To establish the sites of attack of tRNAs by this RNase, the digestions were performed on [5'-^{32}P]tRNAs. The radioactive 5' fragments arising from these hydrolyses were analyzed by the rapid gel-sequencing technique as described by Donis-Keller et al. (1977).

In the absence of aminoacyl-tRNA synthetase, the two tRNAs are cut predominantly within the anticodon stem on both the 5' and 3' sides. Furthermore, as to tRNA^{Val}, major cuts were also found in a zone belonging to the amino acid acceptor and the TψC stems.

Kinetics of digestion of tRNA^{Val} and tRNA^{Phe} with the cobra enzyme showed rather constant patterns of hydrolysis. This is a very favorable situation as compared with the action of other nucleases such as T1 or pancreatic RNases where the prolonged digestion finally leads to the total digestion of the tRNA molecule.

The pattern of 5'-labeled fragments arising from digestions on the tRNAs in the presence of their cognate aminoacyl-tRNA synthetase clearly showed in both tRNAs tested the almost complete disappearance of some cuts that took place within the free tRNAs. The total disappearance either of one or of two cuts that occurred in the anticodon stem in free tRNAs was observed (Fig. 2). A likely explanation is a strong protection of this region by the aminoacyl-tRNA synthetase. In yeast tRNA^{Phe} only one of the cuts disappears, the one located on the 3' side of this stem. In yeast tRNA^{Val} all of the cuts on the 5' and 3' sides of the anticodon stem disappear. These results show for yeast tRNA^{Phe} an asymmetric protection by the cognate aminoacyl-tRNA synthetase, suggesting an asymmetric recognition of the anticodon stem by phenylalanyl-tRNA synthetase. On the contrary, for yeast tRNA^{Val} the protection of the anticodon stem seems to be more complete, suggesting that in this tRNA the anticodon stem is more buried in the valyl-tRNA synthetase. In tRNA^{Val} another protected region was found located at the corner between the amino acid acceptor and TψC stems (Fig. 2). It is also important to mention the disappearance in the presence of valyl- and phenylalanyl-tRNA synthetases of some minor cuts that were often observed in the absence of aminoacyl-tRNA synthetase, but further experiments will be necessary to decide whether these results can be interpreted in terms of protection.

Finally, we can mention that uncharged and charged tRNAs^{Phe} gave the same hydrolysis pattern when either free or engaged in the complex with phenylalanyl-tRNA synthetase.

To conclude, it is worth noticing that some regions of tRNAs found protected against digestion by the cobra RNase in the presence of their cognate aminoacyl-tRNA synthetase were also found cross-linked to the enzymes. This is particularly the case with regions within the anticodon stem in both tRNA^{Val} and tRNA^{Phe}. However, some discrepancies exist between the two approaches: for instance, the 3' side of the anticodon stem in tRNA^{Val} is protected against nuclease digestion but is not found cross-linked to valyl-tRNA synthetase. These differences could be due to the proper limitations of the two approaches. In the case of the nuclease-digestion approach, they may be related to steric hindrance and to difficulty in explaining the disappearance of a cut either by protection or conformational change; in the case of UV cross-linking, they may be due to the absence of chromophores or to joining occurring in adjacent regions.

Relationship between the Structure of tRNA^{Phe} and Its Ability to Be Recognized and Aminoacylated by Phenylalanyl-tRNA Synthetase

In the two approaches above, the tRNA was investigated as a whole. In this part we describe the recognition and aminoacylation of modified yeast tRNA^{Phe}, either by dissection or by chemical excision of minor bases. These experiments have been conducted using the cognate yeast phenylalanyl-tRNA synthetase. The modified tRNAs, listed in Figure 3, were compared with tRNA^{Phe} as to their affinity for phenylalanyl-tRNA synthetase, their aminoacylation parameters, and their melting behavior.

The affinities for yeast phenylalanyl-tRNA synthetase of all these molecules are shown in Figure 3. The values indicate that the single removal of m^{7}G from tRNA^{Phe} has no effect on the strength of interaction. As to the three-quarter fragment, its affinity for phenylalanyl-tRNA synthetase, as compared with that of intact tRNA^{Phe}, drops by a factor of 7, as long as it contains m^{7}G. This is in agreement with another experiment that shows that this fragment is a competitive inhibitor in tRNA^{Phe} aminoacylation with an inhibition constant increased about sixfold compared with the K_m of tRNA^{Phe} in the acylation reaction. This affinity drops ten times more when the fragment is deprived of m^{7}G, which seems rather large when one considers that only one base has been excised, as compared with the effect of the removal of a 17-nucleotide fragment.

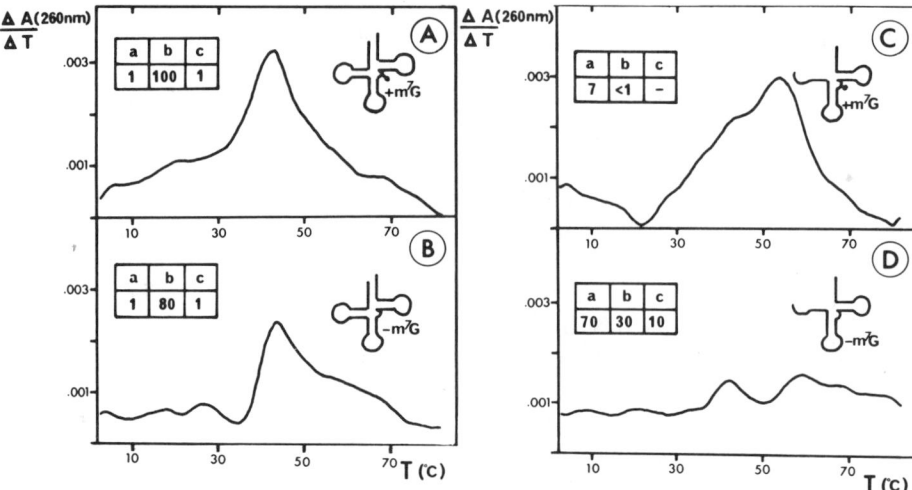

Figure 3 Affinities and aminoacylation parameters of yeast tRNA^{Phe}, intact or modified for phenylalanyl-tRNA synthetase, and derivated melting curves of the various tRNA^{Phe} forms. Three-quarter tRNA^{Phe} was prepared by cleavage with lead acetate at the D level (Werner et al. 1976). Affinities were measured by filtration on nitrocellulose membranes and dissociation constant values calculated by competition experiments with oxidized-reduced [^{3}H]tRNA^{Phe}. Aminoacylation assays were performed in standard conditions using saturating amounts of tRNAs so that results can be expressed in relative V_{max} values. Aminoacylation extents at the plateau of less than 100% (intact and three-quarter tRNA^{Phe} [$-$m^{7}G]) most likely result from the occurrence of inactive molecules in the incubation mixture. The melting curves were obtained by the method developed by Reiss et al. (1974). The experiments were performed at low ionic strength in the complete absence of Mg^{++}. Before melting, samples were submitted to a denaturation-renaturation process. To increase the signal-to-noise ratio, experimental data were submitted to a smoothing procedure (in this case, Fourier transform smoothing) so that the mean squared error of the reconstituted signal is approximately 5×10^{-8}. The melting profiles are of native yeast tRNA^{Phe} (*A*), tRNA^{Phe} ($-$m^{7}G) (*B*), three-quarter tRNA$^{Phe}_{18-76}$ (*C*), and three-quarter tRNA$^{Phe}_{18-76}$ ($-$m^{7}G) (*D*). (a) Ratio of dissociation constant of intact tRNA to dissociation constant of modified tRNA, the dissociation constant of intact tRNA^{Phe} for phenylalanyl-tRNA synthetase being 0.2 μM; (b) aminoacylation extent (%) at the plateau; (c) ratio of V_{max} of intact tRNA to V_{max} of modified tRNA. Three-quarter tRNA$^{Phe}_{18-76}$ (+m^{7}G) is not aminoacylated.

In Figure 3 some aminoacylation parameters of these forms derived from tRNA^{Phe} are also shown. Of particular importance is the recovery of the chargeability of the three-quarter fragment after removal of the m^{7}G (this will be discussed below), although its affinity for phenylalanyl-tRNA synthetase is significantly reduced. The observation of the three-dimensional structure of tRNA^{Phe} (Jack et al. 1976; Quigley and Rich 1976) suggests a possible explanation for this latter point. Indeed, in the intact tRNA the m^{7}G residue is one of the seven bases implicated in the stabilization of the tertiary structure, whereas in the three-quarter fragment it looks like a bolt in the zone appearing as the hinge between the two remaining helical regions (TψC and anticodon stem), and thus could play a key role in stabilizing and maintaining the structure of the fragment. Thus, the decrease of the affinity can be the consequence of an important alteration of the structure of the tRNA provoked by the removal of this strategic residue. Melting experiments support this interpretation. Figure 3 A, B, and C show thermal denaturation profiles of intact tRNA^{Phe}, tRNA^{Phe} ($-$m^{7}G), and three-quarter tRNA^{Phe} ($+$m^{7}G), respectively. Although differences can be seen, these profiles are characteristic of the maintenance of an important tertiary and secondary structure (R. Ehrlich et al., in prep.). By contrast, Figure 3D shows a completely different melting profile, which can be interpreted by a collapse of the structure of three-quarter tRNA^{Phe} ($-$m^{7}G).

GROSS LOCALIZATION OF tRNAs ON AMINOACYL-tRNA SYNTHETASES

Localization of tRNA^{Val} on Yeast Valyl-tRNA Synthetase

Yeast valyl-tRNA synthetase is a 130,000-dalton monomeric enzyme that contains duplicated sequences (Kern et al. 1975) and appears to have an extended structure, as evidenced by neutron scattering. The localization of tRNA^{Val} in the complex with the enzyme was deduced from neutron-scattering contrast-variation experiments (Zaccai et al. 1979). They showed that the tRNA center of mass lies close to that of the complex and that the tRNA is closer to it than the protein, so that the enzyme appears to surround the tRNA. This association is accompanied by a conformational change of the enzyme, which becomes more contracted in the presence of tRNA^{Val}, especially at low ionic strength.

Localization of tRNA^{Phe} on Yeast Phenylalanyl-tRNA Synthetase

Yeast phenylalanyl-tRNA synthetase is a tetrameric enzyme ($\alpha_2\beta_2$) of 260,000 daltons (Fasiolo et al. 1974). The localization of tRNA^{Phe} binding sites among the constitutive subunits was undertaken using two different methods: formation of covalent bonds between enzyme and tRNA^{Phe} upon UV irradiation and formation of a Schiff's base between enzyme and tRNA^{Phe} after periodate oxidation of the latter.

The UV cross-linking of phenylalanyl-tRNA synthetase to tRNA^{Phe} was conducted by irradiation of the complex at 248 nm, 313 nm (Y-base absorption band), and 335 nm in the case of a chemically modified tRNA^{Phe} in which U residues were statistically transformed into s^{4}U residues (this modification was carried out by Dr. Lapidot, The Hebrew University of

Jerusalem). In the two latter cases, the decrease of the acylation activity of the enzyme is strictly equal to the amount of tRNA bound, the ATP-PPi exchange reaction being unaffected. The experiments showed that in all three cases, the tRNA was almost exclusively linked to the β subunit (m.w. 63,000) of phenylalanyl-tRNA synthetase.

As to the reaction of periodate-oxidized tRNA with phenylalanyl-tRNA synthetase, here too the 3'-terminal A of tRNA^{Phe} is found to covalently bind almost exclusively to the β subunit. The covalent complex formation proceeds with a good yield, and it is specific for the tRNA binding site, since addition of an excess of native tRNA completely protects the enzyme. Moreover, a stoichiometric inactivation of the acylation activity is observed, whereas the ATP-PPi exchange activity remains unaffected.

DISCUSSION
Structural Basis of tRNA · Aminoacyl-tRNA Synthetase Interaction

Two direct approaches, photochemical cross-linking between tRNA and aminoacyl-tRNA synthetase and nuclease digestion of the noncovalent complex, have been used to understand the interaction between the two types of macromolecules. They allowed us to define tRNA regions close to the enzyme as pictured in Figure 2.

At this point, it is worth comparing these results with previous observations in the literature. For tRNA^{Val}, on the basis of valylation experiments of modified or dissected tRNAs, some of the regions we defined in this work had already been regarded as important, in particular the 3'-end region and the anticodon loop (Mirzabekov et al. 1969; Chambers et al. 1973). Moreover, mischarging experiments allowed us to emphasize the importance of the D arm for tRNA^{Val} aminoacylation (Kern et al. 1972). There are also remarkable sequence similarities between yeast tRNA^{Val} and the tRNA-like 3' end of turnip yellow mosaic virus, which are valylated by yeast valyl-tRNA synthetase with similar K_m and V_{max} values (Giegé et al. 1978). These homologies occur in the 3'-end region of the amino acid acceptor stem and in the anticodon region, where the anticodon itself, as well as the whole stem, is conserved in the cloverleaf models proposed by Briand et al. (1977) and Silberklang et al. (1977b).

Concerning tRNA^{Phe}, as seen above our results confirm and extend those of Schimmel's group (Schimmel 1977). Moreover, they are consistent with many data in the literature (reviewed by Goddard 1977). For instance, Hörz and Zachau (1973) found that phenylalanyl-tRNA synthetase protects the D and anticodon loops against RNase T1 digestion. Furthermore, on the basis of sequence comparisons, the D stem and the extra loop were described as important for the aminoacylation (Dudock et al. 1971; Kern et al. 1972).

However, the difficulty of interpreting, in terms of recognition sites, results arising from indirect approaches, such as the study of aminoacylation of modified tRNAs or the sequence comparison of tRNAs aminoacylated by the same aminoacyl-tRNA synthetase, must be emphasized. The direct approaches, such as cross-linking or nuclease-digestion experiments, give results that are, in this respect, much less ambiguous, although they have their proper limitations.

As for the number of interaction sites, our cross-linking data led us to define, in both systems investigated, six contact areas. We have already discussed the apparent discrepancy of these results with those of Schimmel's group. It is of interest to compare the number of six oligonucleotides we found bound to the aminoacyl-tRNA synthetase with the number of ionic interaction sites (Bonnet et al. 1975a). We found five to six ionic sites for the intact tRNA and four for a three-quarter tRNA^{Phe} deleted of the 5' quarter containing two of the cross-linked oligonucleotides. If this correspondence is not coincidental, this would mean that tRNAs and aminoacyl-tRNA synthetases interact only at a few points.

One can now ask the question as to the relative contribution of each of these interaction regions. This has been approached through dissection experiments. The removal of seven or nine nucleotides from the 3' end of yeast tRNA^{Phe} and tRNA^{Val}, respectively, did not significantly affect the affinities of these tRNAs for their cognate aminoacyl-tRNA synthetase (Bonnet et al. 1975b). This could be interpreted as meaning that this region does not have an essential role in the interaction process. But this could also be said for all the other parts of the tRNA^{Phe} molecule. For instance, according to excision experiments, removal of fragments 1-17 (Bonnet et al. 1975a), 18-37 (Beltchev and Grunberg-Manago 1970), or even 38-70 (Hörz and Zachau 1973) leaves molecules still able to interact strongly with phenylalanyl-tRNA synthetase.

These data, together with the results described above, lead to the conclusion, first suggested by Zachau's group, that there must be several sites of interaction between tRNAs and aminoacyl-tRNA synthetase and that these sites are independent in the sense that the interaction of each of them does not need the presence of the others. However, these sites are obviously spatially related by a three-dimensional definite tRNA structure, as shown in Figure 4. In this three-dimensional picture, the interacting regions are concentrated in both the valine and phenylaline systems at the inside of the two branches that form the L-shaped structure of tRNA, in accordance with the model proposed by Rich and Schimmel (1977). The valyl- and phenylalanyl-tRNA synthetases would span the distance of approximately 80 Å from the anticodon to the 3' terminal, with several contact points with the inside of the L-shaped tRNA. The fact that this geometry was found in all systems studied up to now strongly suggests that it represents a general model for tRNA · aminoacyl-tRNA synthetase interaction.

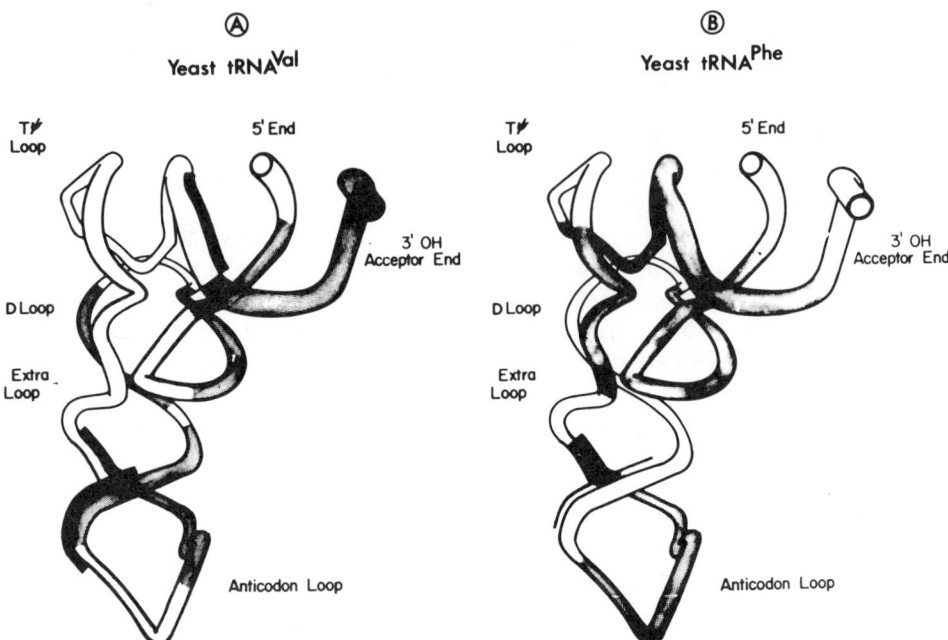

Figure 4 Three-dimensional structure of a tRNA molecule with regions of yeast tRNA^{Val} (*A*) and tRNA^{Phe} (*B*) cross-linked to yeast valyl- and phenylalanyl-tRNA synthetases and protected within the cognate tRNA·aminoacyl-tRNA synthetase complex against cleavage by cobra venom RNase. The cross-linked regions are indicated by shaded areas and the protected regions are indicated by heavy lines in sketches of the three-dimensional structure of a tRNA molecule. (Reprinted, with permission, from Kim et al. 1973. Copyright 1973 by the American Association for the Advancement of Science.) The region within tRNA^{Val} that is nonprotected against RNase digestion in the presence of valyl-tRNA synthetase is indicated by a light line.

Concerning the contact points of the enzyme with the tRNA, partial answers can be given by physical methods. As to the valine system, we have used neutron-scattering methods to get structural parameters on the valyl-tRNA synthetase molecule when it is engaged in complexes with tRNAs. (For a discussion of the method, see Jacrot 1976.) It has been found (Zaccaï et al. 1979) that tRNA^{Val} is bound close to the center of mass of the complex and that tRNA induces a conformational change in valyl-tRNA synthetase, which appears more contracted in the complex. This conformational change is ionic-strength-dependent, the enzyme being the most contracted at low ionic strength (radius of gyration decreasing from 39 to 35 Å); it takes place only when the tRNA (cognate and noncognate) interacts strongly with the enzyme. This happens, for example, with yeast tRNA^{Asp} (Bonnet and Ebel 1975). However, for other RNAs that interact less strongly and in a less specific manner (for example, tRNA$_3^{Leu}$ or 5S

RNA), no conformational change was found. It is of interest to relate these observations to conformational changes that have been observed by fluorescence studies for cognate tRNA · aminoacyl-tRNA synthetase complexes and did not occur in noncognate ones (Rigler et al. 1976; Krauss et al. 1976) or in systems where the tRNA is deprived of the 3'-terminal A (Krauss et al. 1977). Our own sedimentation experiments (Fasiolo et al. 1974) can also be explained by conformational changes of phenylalanyl-tRNA synthetase.

In addition, if we consider that valyl-tRNA synthetase possesses an extended structure, likely made up of two domains, as suggested by sequence duplication (Kern et al. 1975; Bruton 1975), one can draw a tentative model (Fig. 5) for the complex valyl-tRNA synthetase · tRNA^{Val} in which tRNA^{Val} would be held between the two domains of the aminoacyl-tRNA synthetase so that these domains come closer together. The tRNA would cover a large part of the aminoacyl-tRNA synthetase molecule because of the relative sizes of the two macromolecules, which are similar, although their molecular weights are quite different. In this model, the complex presents some character of symmetry and resembles the one proposed by Kim (1975) for tRNA · aminoacyl-tRNA synthetase recognition based on the pseudosymmetry existing between the two branches of the L-shaped tRNA. The 3'-terminal A would interact with the catalytic site located in one domain of valyl-tRNA synthetase in a manner symmetrical to that of the anticodon that would interact with another site located in the second domain of the enzyme.

One may question the generality of the model proposed for the

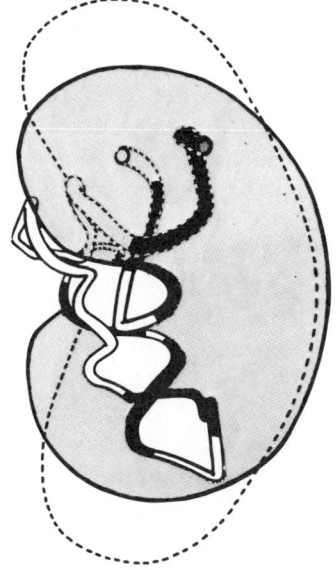

Figure 5 A possible model for yeast tRNA^{Val} · valyl-tRNA synthetase interaction. This model takes into account neutron-scattering, cross-linking, and RNase-digestion-protection experiments; it accounts for the sequence duplication in valyl-tRNA synthetase (Bruton 1975; Kern et al. 1975) and for the symmetry recognition hypothesis for tRNA binding to aminoacyl-tRNA synthetase (Kim 1975). (The model is reprinted, with permission, from Kim 1973. Copyright 1973 by the American Association for the Advancement of Science.) The regions of tRNA joined to valyl-tRNA synthetase are dark; the extended structure of valyl-tRNA synthetase is indicated with a broken line; the contracted enzyme form, interacting with tRNA, is outlined in full lines and is shaded. This model is not unique, particularly the shape of the enzyme, but it is the simplest available to explain our results.

tRNA^{Val}·valyl-tRNA synthetase interaction. On the tRNA side, the situation seems similar for all systems (see Schimmel, this volume; Rich and Schimmel 1977), but the model proposed here could be different for other aminoacyl-tRNA synthetases, especially for the dimeric and tetrameric ones. For the tetrameric $\alpha_2\beta_2$ phenylalanyl-tRNA synthetase, we have also to keep in mind that this aminoacyl-tRNA synthetase has two tRNA binding sites, so that one β subunit can interact either with a complete tRNA or symmetrically with two halves of tRNA. Whatever might be the real situation, in each case one can give a functional significance for the sequence duplication found in the β subunit (Robbe-Saul et al. 1977).

Relationship between Structural Features and tRNA Aminoacylation

Concerning the relationship between the structural requirements discussed above and the aminoacylation ability, it can be thought that the spatial organization of the interaction sites will lead to the optimal aminoacylation of tRNA by allowing a correct adjustment of the 3'-terminal A in the catalytic site. Experiments reported here bring some information about this problem. For instance, in the three-quarter tRNA^{Phe} the preserved domains still allow a strong interaction with the enzyme, but some constraints prevent the correct adjustment of this 3'-terminal A in the catalytic site of phenylalanyl-tRNA synthetase (Fig. 4). The excision of m^{7}G results in an important recovery of the acylation capacity while the affinity drops. This can be explained by the loosening of the remaining three-dimensional structure (melting experiments of Fig. 4), which removes some of the constraints, thus allowing the 3'-terminal A to again reach the catalytic site.

Other important results were provided by aminoacylation experiments of 3' halves of tRNA (Wübbeler et al. 1975) and of the tRNA-like 3' end of turnip yellow mosaic virus RNA, which is valylated as well as tRNA^{Val} by yeast valyl-tRNA synthetase (Giegé et al. 1978). In both cases, conformational similarities with intact tRNA should exist, although the overall three-dimensional structure of these RNAs is obviously different from that of the normal substrate of the aminoacyl-tRNA synthetase. It follows that molecules exhibiting very large differences compared to the tRNA structure are not necessarily inactive. On the contrary, minute changes in the primary structure of tRNA can lead to a complete loss of activity. This occurs, for instance, after point mutations, chemical modifications, or small excisions in tRNA (for a review, see Ofengand 1977; Goddard 1977).

These last results confirm that the aminoacylation rate cannot be directly correlated to the strength of interaction. This had first been suggested by studies in our laboratory (Kern et al. 1972; Giegé et al. 1974) on mischarging reactions, which were later confirmed by others (Roe et al. 1973; von der Haar and Cramer 1978). Furthermore, the correct adjustment of the 3'-terminal A on the enzyme, which we considered as determining the V_{max} in our concept of kinetic specificity of tRNA aminoacylation (Ebel et al. 1973), can be achieved, on the basis of the present studies, only by part of the independent (in the sense discussed above) interaction domains of tRNA. The strong effect of some minute modifications of tRNA can thus be understood by a slight spatial disorganization of the molecule, sufficient to prevent the correct positioning of the terminal A.

CONCLUSIONS

Our data and concepts, from this report and from previously published work (Ebel et al. 1973; Bonnet and Ebel 1975; Bonnet et al. 1975a; Giegé et al. 1974, 1978), together with those already presented by others (Thiebe et al. 1972; Wübbeler et al. 1975; Freist and Cramer 1976; Krauss et al. 1977; Ofengand 1977; Rich and Schimmel 1977; Schimmel 1977) can be summarized as follows:

1. A large part of the energy of interaction between tRNA and aminoacyl-tRNA synthetase is due to electrostatic forces; however, the number of ionic sites seems to be limited.
2. Most of the odd nucleosides in tRNA do not seem to be directly involved in the aminoacylation function of tRNA.
3. Large areas of the tRNA, located at the inside of the two branches of the L-shaped tRNA molecule, are in close proximity to the aminoacyl-tRNA synthetase.
4. Yeast tRNA^{Val} is located in a central and symmetrical manner with respect to the monomeric valyl-tRNA synthetase that undergoes a conformational change. Yeast tRNA^{Phe} is located on the β subunits of the tetrameric phenylalanyl-tRNA synthetase; this interaction may also lead to conformational changes that could be induced by the 3'-terminal A.
5. The tRNAs carry several independent domains that can interact with the aminoacyl-tRNA synthetase. The role of the three-dimensional structure is to maintain these domains in a geometry leading to the most effective positioning of the 3'-terminal A, thus allowing an optimal aminoacylation reaction. These domains, however, are not all required for the aminoacylation reaction, since fragments can be charged. This enables us to understand why the strength of interaction between tRNA and aminoacyl-tRNA synthetase is not correlated, in most cases, with the aminoacylation activity. This extends the general concept of kinetic specificity of tRNA aminoacylation that we previously proposed (Ebel et al. 1973).

Most of these conclusions were derived from studies of tRNA · aminoacyl-tRNA synthetase complexes performed on the tRNA side, and thus the present picture of these complexes is obviously incomplete. Although similar structural investigations on the enzyme side should allow a better understanding of the tRNA·aminoacyl-tRNA synthetase interaction, it is likely that in the future a deeper insight will come from a crystallographic study of a tRNA·aminoacyl-tRNA synthetase complex.

ACKNOWLEDGMENTS

We wish to thank the Institut Laue Langevin, Grenoble, and the Institut du Radium, Orsay, for experimental facilities and B. Jacrot, P. Morin, G. Zaccaï (Institut Laue Langevin), D. Moras, J. C. Thierry (Université Louis Pasteur, Strasbourg), D. Fréchet (Institut de Biologie Moléculaire et Cellulaire, Strasbourg), J. Gabarro-Arpa, and C. Reiss (Institut du Radium, Orsay) for fruitful discussions and collaboration in neutron-scattering and $tRNA^{Phe}$ melting experiments. We very much enjoyed many stimulating discussions with Drs. Y. Boulanger and D. Kern (Institut de Biologie Moléculaire et Cellulaire).

REFERENCES

Beltchev, B. and M. Grunberg-Managno. 1970. Preparation of a pG-fragment from $tRNA^{Phe}$ yeast by chemical scission at the dihydrouracil and inhibition of $tRNA^{Phe}_{yeast}$ charging by this fragment when combined with the -CCA half of this tRNA. *FEBS Lett.* **12**:24.

Bonnet, J. and J.-P. Ebel. 1975. Influence of various factors on the recognition specificity of tRNA's by yeast valyl-tRNA synthetase. *Eur. J. Biochem.* **58**:193.

Bonnet, J., M. Renaud, J. P. Raffin, and P. Remy. 1975a. Quantitative study of the ionic interactions between yeast $tRNA^{Val}$ and $tRNA^{Phe}$ and their cognate aminoacyl-tRNA ligases. *FEBS Lett.* **53**:154.

Bonnet, J., N. Befort, C. Bollack, F. Fasiolo, and J.-P. Ebel. 1975b. Study of the role of the acceptor stem in the interactions between tRNAs and aminoacyl-tRNA synthetases. *Nucleic Acids Res.* **2**:211.

Bonnet, J., J.-P. Ebel, G. Dirheimer, L. P. Shershneva, A. I. Krutilina, T. V. Venkstern, and A. A. Bayev. 1974. The corrected nucleotide sequence of valine tRNA from baker's yeast. *Biochimie* **56**:1211.

Briand, J. P., G. Jonard, H. Guilley, K. Richards, and L. Hirth. 1977. Nucleotide sequence (n = 159) of the amino acid accepting 3'-OH extremity of turnip-yellow-mosaic virus RNA and the last portion of its coat-protein cistron. *Eur. J. Biochem.* **72**:453.

Bruton, C. J. 1975. The infrastructure of valyl-transfer ribonucleic acid synthetase from yeast. *Biochem. J.* **147**:191.

Budzik, G. P., S. S. M. Lam, H. J. P. Schoemaker, and P. R. Schimmel. 1975. Two photo-cross-linked complexes of isoleucine specific transfer ribonucleic acid with aminoacyl-transfer ribonucleic acid synthetases. *J. Biol. Chem.* **250**:4433.

Chambers, R. W., S. Aoyagi, Y. Furukawa, H. Zawadzka, and O. S. Bhanot. 1973. Inactivation of valine acceptor activity by a C–U missence change in the anticodon of yeast valine transfer ribonucleic acid. *J. Biol. Chem.* **248**:5549.

Donis-Keller, H., A. M. Maxam, and W. Gilbert. 1977. Mapping adenines, guanines and pyrimidines in RNA. *Nucleic Acids Res.* **4:**2527.

Dudock, B., C. Diperi, K. Scileppi, and R. Reszelbach. 1971. The yeast phenylalanyl-transfer RNA synthetase recognition site: The region adjacent to the dihydrouridine loop. *Proc. Natl. Acad. Sci.* **68:**681.

Ebel, J.-P., R. Giegé, J. Bonnet, D. Kern, N. Befort, C. Bollack, F. Fasiolo J. Gangloff, and G. Dirheimer. 1973. Factors determining the specificity of the tRNA aminoacylation reaction. Non-absolute specificity of tRNA-aminoacyl-tRNA synthetase recognition and particular importance of the maximal velocity. *Biochimie* **55:**547.

Fasiolo, F., P. Remy, J. Pouyet, and J.-P. Ebel. 1974. Yeast phenylalanyl-tRNA synthetase. Molecular weight and interaction with tRNA^{Phe} phenylalanine. *Eur. J. Biochem.* **50:**227.

Freist, W. and F. Cramer. 1976. A binary code model for substrate recognition of aminoacyl-tRNA ligases. *J. Theor. Biol.* **58:**401.

Giegé, R., J. P. Briand, R. Mengual, J.-P. Ebel, and L. Hirth. 1978. Valylation of the two RNA components of turnip-yellow mosaic virus and specificity of the tRNA aminoacylation reaction. *Eur. J. Biochem.* **84:**251.

Giegé, R., D. Kern, J.-P. Ebel, H. Grosjean, S. De Henau, and H. Chantrenne. 1974. Incorrect aminoacylations involving tRNAs or valyl-tRNA synthetase from *Bacillus stearothermophilus*. *Eur. J. Biochem.* **45:**351.

Goddard, J. P. 1977. The structures and functions of transfer RNA. *Prog. Biophys. Mol. Biol.* **32:**233.

Hörz, W. and H. G. Zachau. 1973. Complexes of aminoacyl-tRNA synthetases with tRNAs as studied by partial nuclease digestion. *Eur. J. Biochem.* **32:**1.

Irwin, M. J., J. Nyborg, B. R. Reid, and D. M. Blow. 1976. The crystal structure of tyrosyl-transfer RNA synthetase at 2.7 Å resolution. *J. Mol. Biol.* **105:**577.

Jack, A., J. E. Ladner, and A. Klug. 1976. Crystallographic refinement of yeast phenylalanine transfer RNA at 2.5 Å resolution. *J. Mol. Biol.* **108:**619.

Jacrot, B. 1976. The study of biological structures by neutron scattering from solution. *Rep. Prog. Phys.* **39:**911.

Kern, D., R. Giegé, and J.-P. Ebel. 1972. Incorrect aminoacylations catalysed by the phenylalanyl- and valyl-tRNA synthetases from yeast. *Eur. J. Biochem.* **31:**148.

Kern, D., R. Giegé, S. Robbe-Saul, Y. Boulanger, and J.-P. Ebel. 1975. Complete purification and studies on the structural and kinetic properties of two forms of yeast valyl-tRNA synthetases. *Biochimie* **57:**1167.

Kim, S.-H. 1975. Symmetry recognition hypothesis model for tRNA binding to aminoacyl-tRNA synthetase. *Nature* **256:**5519.

Kim, S.-H., G. J. Quigley, F. L. Suddath, A. McPherson, D. Sneden, J. J. Kim, J. Weinzierl, and A. Rich. 1973. Three dimensional structure of yeast phenylalanine transfer RNA: Folding of the polynucleotide chain. *Science* **179:**285.

Kisselev, L. L. and O. O. Favorova. 1974. Aminoacyl-tRNA synthetases: Some recent results and achievements. *Adv. Enzymol.* **40:**141.

Krauss, G., D. Riesner, and G. Maass. 1976. Mechanism of discrimination between cognate and non-cognate tRNAs by phenylalanyl-tRNA synthetase from yeast. *Eur. J. Biochem.* **68:**81.

———. 1977. Mechanism of tRNA-synthetase recognition: Role of terminal A. *Nucleic Acids Res.* **4:**2253.

Loftfield, R. B. 1972. The mechanism of aminoacylation of transfer RNA. *Prog. Nucleic Acid Res. Mol. Biol.* **12:**87.

Mirzabekov, A. D., L. Y. Kazarinova, D. Lastity, and A. A. Bayev. 1969. Enzymatic aminoacylation of dissected molecules of baker's yeast valine tRNA 1. *FEBS Lett.* **3:**268.

Ofengand, J. 1977. tRNA and Aminoacyl-tRNA synthetases. In *Molecular mechanisms of protein biosynthesis* (ed. H. Weissbach and S. Petsko), p. 7. Academic Press, New York.

Quigley, G. J. and A. Rich. 1976. Structural domains of transfer RNA molecules. *Science* **194:**796.

RajBhandary, U. L. and S. H. Chang. 1968. Yeast phenylalanine transfer ribonucleic acid: Partial digestion with ribonuclease T_1 and derivatives of the total primary structure. *J. Biol. Chem.* **243:**598.

Reiss, C., F. Michel, and J. Gabarro. 1974. An apparatus for studying the thermal transitions of nucleic acids at high resolution. *Anal. Biochem.* **62:**499.

Rich, A. and U. L. RajBhandary. 1976. Transfer RNA: Molecular structure, sequence and properties. *Annu. Rev. Biochem.* **45:**805.

Rich, A. and P. R. Schimmel. 1977. Structural organization of complexes of transfer RNAs with aminoacyl transfer RNA synthetases. *Nucleic Acids Res.* **4:**1649.

Rigler, R., U. Pachmann, R. Hirsch, and H. G. Zachau. 1976. On the interaction of seryl-tRNA synthetase with $tRNA^{Ser}$. A contribution to the problem of synthetase tRNA recognition. *Eur. J. Biochem.* **65:**307.

Robbe-Saul, S., F. Fasiolo, and Y. Boulanger. 1977. Phenylalanyl-tRNA from baker's yeast. Repeated sequences in the two subunits. *FEBS Lett.* **84:**57.

Roe, B., M. Sirover, and B. Dudock. 1973. Kinetics of homologous and heterologous aminoacylation with yeast phenylalanyl transfer ribonucleic acid synthetase. *Biochemistry* **12:**4146.

Schimmel, P. R. 1977. Approaches to understanding the mechanism of specific protein-transfer RNA interactions. *Accts. Chem. Res.* **10:**411.

Schoemaker, H. J. P., G. P. Budzik, R. Giegé, and P. R. Schimmel. 1975. Three photo-cross-linked complexes of yeast phenylalanine specific transfer ribonucleic acid synthetases. *J. Biol. Chem.* **250:**4440.

Silberklang, M., A. M. Gillum, and U. L. RajBhandary. 1977a. The use of nuclease P_1 in sequence analysis of end group labeled RNA. *Nucleic Acids Res.* **4:**4091.

Silberklang, M., A. Prochiantz, A. L. Haenni, and U. L. RajBhandary. 1977b. Studies on the sequence of the 3'-terminal region of turnip-yellow mosaic virus RNA. *Eur. J. Biochem.* **72:**464.

Smith, I. 1977. Genetics of the translational apparatus. In *Molecular mechanisms of protein biosynthesis* (ed. H. Weissbach and S. Petsko), p. 627. Academic Press, New York.

Söll, D. and P. R. Schimmel. 1974. Aminoacyl-tRNA synthetases. In *The enzymes* (ed. P. D. Boyer), vol. 10, p. 489. Academic Press, New York.

Thiebe, R., K. Harbers, and H. G. Zachau. 1972. Aminoacylation of fragment combinations from yeast $tRNA^{Phe}$. *Eur. J. Biochem.* **26:**144.

Vassilenko, S. K. and V. C. Ryte. 1975. Isolation of highly purified ribonuclease from cobra (*Naja oxiana*) venom. *Biokhimya* **40:**578.

von der Haar, F. and F. Cramer. 1978. Valyl- and phenylalanyl-tRNA synthetase from baker's yeast: Recognition of transfer RNA results from a multistep process, as indicated by inhibition of aminoacylation with modified transfer RNA. *Biochemistry* **17**:4509.

Werner, C., B. Krebs, G. Keith, and G. Dirheimer. 1976. Specific cleavages of pure tRNAs by plumbous ions. *Biochim. Biophys. Acta* **432**:161.

Winter, G. P. and B. S. Hartley. 1977. The amino acid sequence of tryptophanyl-tRNA synthetase from *Bacillus stearothermophilus*. *FEBS Lett.* **80**:340.

Wübbeler, W., C. Lossow, F. Fittler, and H. G. Zachau. 1975. Amino acid incorporation into tRNA fragments and into heterologous combination of fragments. *Eur. J. Biochem.* **59**:405.

Zaccaï, G., P. Morin, B. Jacrot, D. Moras, J. C. Thierry, and R. Giegé. 1979. Interaction of valyl-tRNA synthetase with RNAs and conformational changes of the enzyme. *J. Mol. Biol.* **129**:483.

Zelwer, C., J. L. Risler, and C. Monteilhet. 1976. A low-resolution model of crystalline methionyl-transfer RNA synthetase from *Escherichia coli*. *J. Mol. Biol.* **102**:93.

2'-OH vs 3'-OH Specificity in tRNA Aminoacylation

Sidney M. Hecht
Department of Chemistry
Massachusetts Institute of Technology
Cambridge, Massachusetts 02139

Although the aminoacylation of tRNA and participation of the aminoacylated species in the partial reactions of protein biosynthesis have been studied in detail for a number of years, until recently little data were available concerning the (2' or 3') positional isomer of tRNA that participated in each of these reactions. The source of ambiguity has been the aminoacyl-tRNA itself, which is believed to exist in solution as a rapidly equilibrating mixture of 2'- and 3'-*O*-aminoacylated molecules. Thus, whereas the activation of a given tRNA has been thought to involve attachment of the amino acid to a single hydroxyl group at the 3' terminal of the tRNA, early efforts at trapping the initially formed isomer prior to equilibration were unsuccessful (Feldmann and Zachau 1964; Wolfenden et al. 1964; McLaughlin and Ingram 1965). Equilibration of the formed aminoacyl-tRNAs (Fig. 1) has also precluded the use of single isomers of

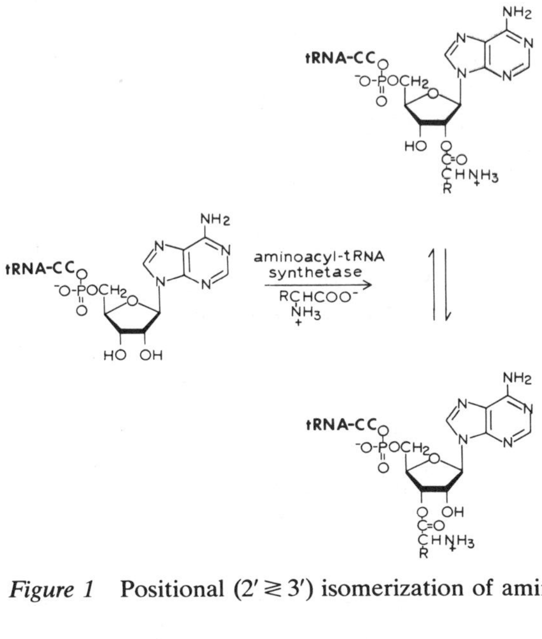

Figure 1 Positional ($2' \geq 3'$) isomerization of aminoacyl-tRNAs.

the aminoacylated species to probe the partial reactions subsequent to aminoacylation.

In 1973, two laboratories described the preparation of tRNAs terminating in isomeric, modified adenine nucleotides (Sprinzl et al. 1973; Hecht et al. 1973), and such positionally defined, isomeric tRNAs were shown to be of utility for studying certain of the individual transformations in the overall process of peptide bond formation (Sprinzl and Cramer 1973; Fraser and Rich 1973). Attention has now focused on tRNAs terminating in 2'- and 3'-deoxyadenosine, which are good analogs of unmodified tRNA to a reasonable level of refinement and have been useful in identifying the positional isomers of aminoacyl-tRNA that participate in tRNA activation, the peptidyl transferase reaction, etc. This paper summarizes the data pertinent to positional specificities in the aminoacylation of individual tRNA isoacceptors and the binding of the aminoacylated species to elongation factor Tu (EF-Tu).

MODIFIED tRNAs ARE PREPARED USING CTP (ATP):tRNA NUCLEOTIDYL TRANSFERASE

The preparation of tRNA species 2 and 3 (Fig. 2) is effected by reconstruction of "abbreviated" tRNA (tRNA-CC_{OH}) with the deoxynucleotide of interest. The abbreviated tRNA itself can be prepared conveniently by treating unmodified tRNA (fractionated or unfractionated) with purified venom exonuclease (Zubay and Takanami 1964) and monitoring the extent of hydrolysis carefully in terms of loss of amino acid acceptance. In this fashion, hydrolysis can be confined to the last three nucleotides in the acceptor stem, permitting subsequent reconstitution to tRNA-CC_{OH} by incubation of the mixture of venom-treated tRNAs with CTP (ATP):tRNA nucleotidyl transferase and CTP.

Abbreviated tRNA can also be prepared chemically by treatment of the tRNA with periodate to form the 3'-terminal dialdehyde, followed by treatment with a primary amine that affords tRNA-CCp (Khym and Uziel 1968; Tal et al. 1972; Uziel 1973). Additional treatment with alkaline phosphatase yields tRNA-CC_{OH}. This procedure has the obvious advantage of removing a single nucleotide from each tRNA, thus

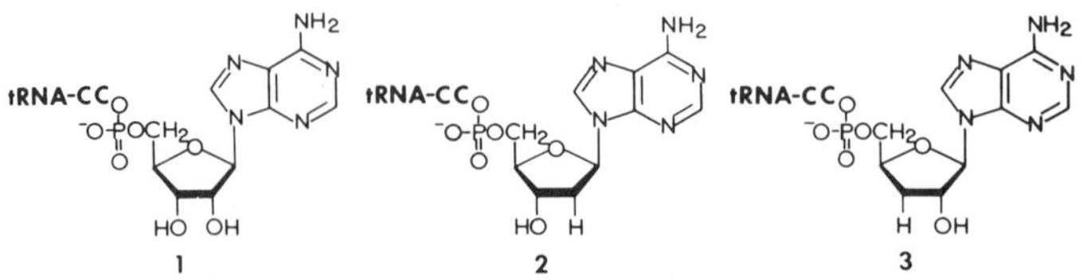

Figure 2 tRNA species 1–3.

obviating the need for reconstitution with Cs. The stepwise chemical degradation also tends to give more reproducible results than those obtained with venom exonuclease. A limitation of the chemical procedure, however, is that it causes an irreversible diminution of several amino acid acceptor activities for reasons other than oxidation of the 3'-terminal nucleoside (Tal et al. 1972; Sprinzl and Cramer 1975; Hecht 1977).

Reconstruction of tRNAs 2 and 3 from tRNA-CC$_{OH}$ is effected most conveniently by incubation of the latter in the presence of yeast CTP(ATP):tRNA nucleotidyl transferase and the deoxynucleoside 5'-triphosphate of interest. Although incorporation proceeds much less quickly than that observed for ATP, by the use of substantial amounts of deoxynucleotides and enzyme, as well as an extended period of incubation, tRNAs 2 and 3 can be obtained routinely in 45–65% yields (Sprinzl and Cramer 1975; Hecht and Chinault 1976; Alford et al. 1979). Separation of tRNA species 2 and 3 from tRNA-CC$_{OH}$ can be achieved easily by chromatography on DBAE-cellulose (Rosenberg et al. 1972; McCutchan et al. 1975; Alford et al. 1979).

Another procedure for the preparation of tRNA species 2 and 3 has involved incubation of tRNA-CC$_{OH}$ with the appropriate deoxynucleoside 5'-diphosphates in the presence of polynucleotide phosphorylase (Hecht et al. 1974); because the enzyme might be expected to effect multiple additions of 2'-deoxyadenosine diphosphate to tRNA-CC$_{OH}$ (Hawley et al. 1978), this nucleotide was utilized as its 3'-O-α-methoxyethyl derivative. Although the yields of tRNAs 2 and 3 obtained to date by this method have been modest (Hecht et al. 1974), more recent developments in the addition of nucleotides to oligonucleotide primers (Gillam and Smith 1974; Gillam et al. 1974; Sninsky et al. 1974) may facilitate progress in this area.

A third approach that can be utilized for the conversion of tRNA-CC$_{OH}$ to tRNAs 2 and 3 derives from the observation that RNA ligase can add single nucleotides to the 3' terminals of acceptor oligo- and polynucleotides and that a wide range of substrates can be transferred (Kikuchi et al. 1978; England and Uhlenbeck 1978; Barrio et al. 1978), apparently via the intermediacy of dinucleoside diphosphates formed by initial adenylylation of the substrates (Kaufmann and Littauer 1974; England et al. 1977). B. L. Alford and S. M. Hecht (unpubl.) have recently shown that tRNA-CC$_{OH}$ can function as an efficient acceptor for RNA ligase and that by the use of dinucleoside bisphosphates 4 and 5 (Fig. 3) as donor substrates, tRNAs 2 and 3 can be prepared in yields up to 26%. Further development of this technique might provide an attractive alternative to the use of CTP(ATP):tRNA nucleotidyl transferase for the preparation of modified tRNAs.

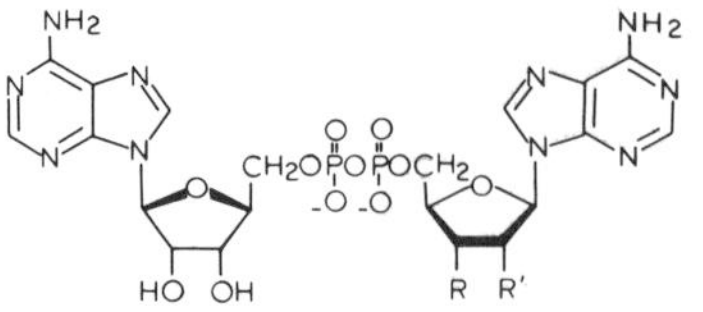

Figure 3   Nucleotide species 4 and 5.

4 R=OCH(CH₃)OCH₃, R'=H
5 R=H, R'=OH

tRNAs TERMINATING IN DEOXYADENOSINE ARE SUBSTRATES FOR THEIR COGNATE AMINOACYL-tRNA SYNTHETASES

Only Single Isomers of Most Isoacceptors Can Be Activated

Sprinzl and Cramer (1973) compared the aminoacylation of yeast tRNA^{Phe} species 1–3 by yeast phenylalanyl-tRNA synthetase. Aminoacylation of the modified tRNA (3) terminating in 3'-deoxyadenosine occurred with the same apparent K_m (2.8 μM) as that of unmodified tRNA and with a similar V_{max} (0.18 μmole/min vs 0.28 μmole/min for species 1), whereas tRNA^{Phe} 2 was not a substrate for the same enzyme. These observations are consistent with the interpretation that yeast phenylalanyl-tRNA synthetase normally aminoacylates its cognate tRNA on the 2'-OH group. Additional supportive evidence, which also extended this finding to other systems, was obtained for modified *Escherichia coli* tRNA^{Phe} terminating in 3'-amino-3'-deoxyadenosine (Fraser and Rich 1973) and for *E. coli*, yeast, and rat liver tRNAs^{Phe} treated successively with sodium periodate and sodium borohydride (Ofengand et al. 1974).

Determination of the initial positions of aminoacylation of each of the tRNA isoacceptors was first carried out by modification of unfractionated *E. coli* tRNAs (Sprinzl and Cramer 1975; Hecht and Chinault 1976). Attempted activation of each of the isoacceptors present in the mixtures of modified tRNAs was effected using an unfractionated *E. coli* aminoacyl-tRNA synthetase preparation and the appropriate radiolabeled amino acid; Hecht and Chinault (1976) also added each of the noncognate (unlabeled) amino acids to the incubation mixture to suppress possible misacylations (Ebel et al. 1973; Giegé et al. 1974). The aminoacylation of *E. coli* tRNAs 1–3 with [³H]tryptophan revealed that the tRNA species (3) terminating in 3'-deoxyadenosine was aminoacylated at essentially the same rate and to the same extent as unmodified *E. coli* tRNA, whereas the mixture of tRNAs terminating in 2'-deoxyadenosine accepted little, if any, tryptophan relative to a control assay lacking tRNA. Remarkably, activation of the same tRNA samples in the presence of [³H]serine gave the opposite result, namely, that species 1 and 2, but not 3, accepted serine. Assay of additional isoacceptors revealed that the arginyl-, glutamyl-, isoleucyl-, leucyl-, methionyl-, and valyl-tRNA synthetases also utilized exclusively tRNA species 3 as substrate, i.e., that they aminoacylated their cognate tRNAs solely on the 2'-OH group. Other aminoacyl-tRNA synthetases that utilized solely the 3'-OH groups for tRNA aminoacylation included the activating enzymes specific for alanine, glycine, histidine, lysine, proline, and threonine (Table 1).

Results complementary to those described above were obtained by Fraser and Rich (1975), who studied the aminoacylation of *E. coli* tRNAs terminating in 2'-amino-2'-deoxyadenosine and 3'-amino-3'-deoxyadenosine. Activation of these tRNAs could occur in two ways, namely, by direct acylation of the amino moieties or aminoacylation of the hydroxyl groups followed by O→N acyl migration; in fact, both isomers of many of the isoaccepting activities were activated, complicating interpretation of the data. Nevertheless, by considering the relative rates of aminoacylation of the two isomers and comparing the results obtained for tRNAs 2 and 3, it was possible to attribute the faster activation process to O acylation and the slower to N acylation. When the aminoacylation results were compared in this context with those obtained for tRNAs 2 and 3, good agreement was obtained (Fraser and Rich 1975; Hecht 1977).

The aminoacylation experiments described above were carried out using ATP as an energy source and with a mixture of aminoacyl-tRNA synthetases contaminated to some extent with other enzyme activities. Some preparations included detectable CTP(ATP):tRNA nucleotidyl

Table 1 Initial positions of aminoacylation of tRNAs

2'-OH[a]	3'-OH[a]	2'- and 3'-OH[a]	Variable
Arg	Ala	Asn	Trp
Glu[b]	Gln[c]	Asp[d]	
Ile	Gly	Cys	
Leu	His	Tyr	
Met	Lys		
Phe	Pro		
Val	Ser		
	Thr		

These assignments were made using unfractionated tRNAs. Repetition of the experiments with many of the purified isoacceptors has led to the same conclusions in each case, although some of the ostensibly nonsubstrate tRNAs (e.g., *E. coli* tRNA^{Trp} 2) do act as less efficient substrates for their cognate aminoacyl-tRNA synthetases.
[a]Hydroxyl group at 3' terminal of tRNA that is aminoacylated.
[b]Based on the activation of *E. coli* tRNAs^{Glu} 1–3.
[c]Based on the activation of yeast tRNAs^{Gln} 1–3.
[d]*E. coli* tRNA^{Asp} 2 and calf liver tRNA^{Asp} 3 were substrates for activation; aminoacylation of the isomeric tRNAs^{Asp} in each case was uncertain.

transferase activity, which could have effected the conversion of tRNA species 2 or 3 to 1 prior to aminoacylation. In fact, whereas only single isomeric forms of most isoacceptors were used as substrates, suggesting that the exchange process was not important under the reaction conditions, *E. coli* asparaginyl-, cysteinyl-, and tyrosyl-tRNA synthetases activated tRNA species 2 and 3. Therefore, to establish that the observed aminoacylations did involve modified tRNAs, Hecht and Chinault (1976) repeated each of the successful activation experiments utilizing as an energy source the deoxynucleoside 5'-triphosphate corresponding to the deoxynucleoside at the 3' terminal of the substrate tRNA.

Although ATP was a more efficient energy source than the deoxynucleotides in a number of cases, no qualitative differences in tRNA activation were found. The possibility that the apparent aminoacylation of both isomers of some *E. coli* tRNAs was due to the presence of two or more isoaccepting tRNAs having different positional specificities cannot be excluded. However, for yeast tRNA^{Asn} and tRNA^{Cys} the derived modified tRNAs (2 and 3) were each aminoacylated to the same extent as the unmodified species, indicating activation of the same molecules at both positions. It should be noted that although both hydroxyl groups in such tRNAs can be utilized for aminoacylation, one is generally used much more efficiently than the other. Presumably, this means that one position is normally preferred for activation but in its absence the other suffices.

Positional Specificities Have Been Conserved during Evolution

The finding that most *E. coli* aminoacyl-tRNA synthetases utilized a single, preferred hydroxyl moiety for activation of their cognate tRNAs was not unanticipated, as the use of one hydroxyl group might be expected to increase the overall specificity of the aminoacylation process. However, since the positional ($2' \gtrless 3'$) equilibration of the aminoacyl moiety on the formed aminoacyl-tRNA is believed to be much more rapid than the overall process of protein biosynthesis (Griffin et al. 1966), it was thought that no significance was attached to the use of a particular hydroxyl group and that, for example, the initial position of aminoacylation of any given tRNA isoacceptor might well change during evolution.

The study of the initial position of tRNA aminoacylation was therefore extended to yeast (Sprinzl and Cramer 1975; Hecht and Chinault 1976) and calf liver tRNAs (Chinault et al. 1977). As a result of these studies, it was found that the initial position of aminoacylation of any given isoacceptor was conserved during evolution, with the exception that *E. coli* tRNA^{Trp} was aminoacylated on the 2'-OH group, whereas yeast and calf liver tRNAs were aminoacylated on the 3'-OH group. Recent experiments utilizing purified *E. coli* and yeast tRNAs^{Trp} 1–3 and purified tryptophanyl-tRNA synthetases from the same organisms have established that tRNA^{Trp} 2 and 3 from *E. coli* and yeast can both be aminoacylated, but that the preferred substrates are indeed *E. coli* tRNA^{Trp} 3 and yeast tRNA^{Trp} 2 (B. L. Alford et al., in prep.). It should also be noted that the initial positions of aminoacylation of tRNA^{Glu} and tRNA^{Gln} have been measured successfully only in the *E. coli* and yeast systems, respectively, and that the results obtained with yeast tRNA^{Gln} 2 and 3 did not suggest the same positional specificity as those obtained with *E. coli* and wheat germ tRNAs terminating in 2'(3')-amino-2'(3')-deoxyadenosine (Fraser and Rich 1975; Julius et al. 1979).

The change in the preferred position of aminoacylation of tRNA^{Trp} during evolution, and possibly also of tRNA^{Gln}, indicated that other tRNAs should also have been able to undergo analogous transformations. Therefore, the observation that the positional specificities for activation of most isoacceptors remained the same suggested that some selective pressure had functioned to maintain this specificity. Since no advantage would seem to derive from the use of a particular hydroxyl group, at least in terms of the ability of the derived aminoacyl-tRNA to function in protein biosynthesis, the nature of this selective pressure has been a matter of some conjecture.

Possible Bases for the Conservation of Positional Specificity

Cramer and his coworkers (von der Haar and Cramer 1975, 1976; Igloi et al. 1977) have suggested that the individual hydroxyl groups at the 3' terminal of tRNA have evolved specific functions in the deacylation of incorrectly activated aminoacyl-tRNAs. The experimental evidence initially offered in support of this thesis derived from the earlier observation by Baldwin and Berg (1966) that *E. coli* isoleucyl-tRNA synthetase effected the dissociation of valyl adenylate, either by direct hydrolysis or by rapid deacylation of valyl-tRNA^{Ile}. The (transient) accumulation of valyl-tRNA^{Ile} was not observed. Repetition of the experiment with yeast isoleucyl-tRNA synthetase (von der Haar and Cramer 1975) gave analogous results with regard to both dissociation of the valyl adenylate and accumulation of yeast valyl-tRNA^{Ile}. Remarkably, in the presence of yeast tRNA^{Ile} 3, valine was stably transferred to the tRNA by the homologous isoleucyl-tRNA synthetase. This suggested that both *E. coli* and yeast isoleucyl-tRNA synthetases effected dissociation of the valyl adenylate by hydrolysis of valyl-tRNA^{Ile} species 1 and also that this process was operative only when the 2'-OH group was present at the 3' terminal of tRNA^{Ile}. Specific chemical proofreading mechanisms were proposed for yeast valyl-tRNA^{Ile} and threonyl-tRNA^{Val}, and the hydrolytic capacities of a number of aminoacyl-tRNA synthetases were investigated (von der Haar and Cramer 1976; Igloi et al. 1977). Clearly, the need to maintain a chemical proofreading function in parallel with the ability to aminoacylate their cognate tRNAs could limit the variety of acceptable modifications of certain aminoacyl-tRNA synthetases during evolution and thereby constrain these species to the use of a specific hydroxyl group for aminoacylation.

Although the existence of a chemical proofreading mechanism could provide a satisfactory explanation for the conservation of (ostensibly randomly chosen) initial positions of tRNA activation, it would bear no obvious relevance to perceived patterns of positional specificity among different tRNA isoacceptors. Therefore, the observation by Hecht (1977) that such a pattern did exist suggested the need for an additional or alternate explanation for conservation of initial positions of tRNA aminoacylation during evolution. The correlation noted by Hecht was related to an earlier observation by Woese et al. (1967) that the arrangement of amino acids within the genetic code is nonrandom. Specifically, it was noted that all of the amino acids having U as their second codon base were lipophilic in nature, whereas those having C as the second base were relatively nonpolar (Fig. 4). Since lipophilic amino acids are generally thought to serve a structural rather than a functional role in proteins, it was reasoned that the occasional interchange of such species would likely have little effect on protein function. Therefore, by having such amino acids grouped together in the genetic code, the potentially deleterious effects of point mutations in the DNA, or mistakes in transcription or translation at the first or third codon positions, would be minimized. Even an alteration at the second codon position, which is generally regarded as the most secure component of codon-anticodon interaction, would result only in the substitution of a nonpolar amino acid in place of the intended lipophilic species.

A potential weakness in this scheme is that it would not compensate for errors in tRNA aminoacylation. One might therefore anticipate the existence of some additional mechanism at this level for suppressing the

U - 2nd base - C

1st base	3rd base U	C	A	G
U	phe	phe	leu	leu
C	leu	leu	leu	leu
A	ile	ile	ile	met
G	val	val	val	val

1st base	3rd base U	C	A	G
U	ser	ser	ser	ser
C	pro	pro	pro	pro
A	thr	thr	thr	thr
G	ala	ala	ala	ala

Figure 4 Arrangement of the genetic code according to the second codon bases.

formation of nonfunctional proteins. One possible way to accomplish this would be to effect the activation of structurally related amino acids by aminoacyl-tRNA synthetases that functioned by similar mechanisms, such that the occasional misacylation of a given tRNA would likely occur with an amino acid structurally similar to the cognate amino acid. In fact, in the absence of their cognate tRNAs, certain aminoacyl-tRNA synthetases bind efficiently to several noncognate tRNAs and utilize these species as substrates. The existence of such families of tRNAs has been recognized (Yarus and Mertes 1973; Ebel et al. 1973; Giegé et al. 1974), as has the fact that they include the lipophilic and nonpolar amino acids in Figure 4 (Hecht 1977); the term secondary cognition has been suggested for this process, as it would tend to complement the primary recognition mechanism between a given tRNA and its cognate activating enzyme. Additional evidence for the functional nature of the secondary cognition process may be inferred from the observations that the aminoacyl-tRNA synthetases whose amino acids have codons containing U as the second base acylate their cognate tRNAs specifically on the 2′-OH group and misacylate noncognate tRNAs at the same position, whereas those enzymes whose amino acids are specified by codons having C in the corresponding position all utilize the 3′-OH groups for aminoacylation (Table 2) (Hecht 1977; Alford and Hecht 1978).

Table 2 Correlation between initial position of tRNA aminoacylation and second base of codon specifying each isoacceptor

Hydroxyl group[a]	Codon assignments of individual tRNAs			
	XUY	XCY	XAY	XGY
2′-OH	Ile Leu Met Phe Val		Glu	Arg
3′-OH		Ala Pro Ser Thr	Gln His Lys	Gly Ser
2′- and 3′-OH			Asn Asp Tyr	Cys Trp

The assignments of initial position of aminoacylation are based on the data from Table 1 (Sprinzl and Cramer 1975; Hecht and Chinault 1976; Chinault et al. 1977).
[a]Hydroxyl group at 3′ terminal of tRNA that is aminoacylated.

ISOMERIC AMINOACYL-tRNAs ARE ACCESSIBLE VIA CHEMICAL AMINOACYLATION

Studies of the positional specificities of the partial reactions subsequent to aminoacylation would all involve the use of isomeric aminoacyl-tRNAs (7 and 8; Fig. 5) for direct comparison with the respective unmodified species. Although both isomers of some aminoacyl-tRNAs are accessible by enzymatic activation of tRNAs 2 and 3 (Table 1), only single isomers of most modified isoacceptors can be aminoacylated, with the result that few isomeric aminoacyl-tRNAs have been available for study. An apparently general solution to this problem has recently been described by Hecht and his coworkers (1978).

The new technique is based on the observation by England et al. (1977) that certain dinucleoside bisphosphates are substrates for T4 RNA ligase, which transfers a single nucleotide from each of these species to the 3' terminal of an appropriate acceptor oligomer. By the use of chemically preaminoacylated derivatives of these dinucleoside bisphosphates (e.g., 9; Fig. 6), it has been possible to prepare, e.g., yeast phenylalanyl-tRNA^{Phe} species 7 and methionyl-tRNA^{iMet} 7 in yields of 45% and 91%, respectively, as well as a substantial variety of misacylated tRNAs of type 6. The key to the success of this procedure was the use of the O-nitrophenylsulfenyl protecting group, which served to stabilize the aminoacyl moiety, thus allowing an extended period of incubation in the presence of the ligase, and also provided a lipophilic group that permitted purification of each of the formed nitrophenylsulfenyl-aminoacyl-tRNAs (Fig. 7) by chromatography on BD-cellulose.

BOTH ISOMERS OF AMINOACYL-tRNA ARE BOUND BY EF-Tu

That both isomers of aminoacyl-tRNA are bound by EF-Tu at reasonable concentrations of the factor was established by Hecht et al. (1977) and later confirmed by Sprinzl et al. (1977). Of somewhat more concern, though, has been the question of the possible preference for a single positional isomer of aminoacyl-tRNA at limiting concentrations of EF-Tu. Although ostensibly this would seem not to be physiologically rele-

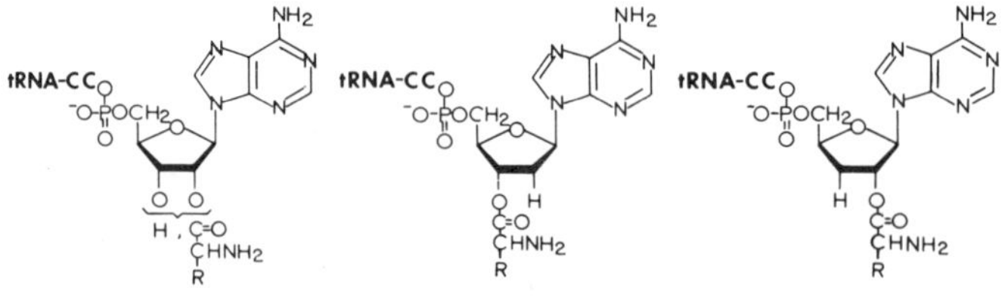

Figure 5 tRNA species 6, 7, and 8.

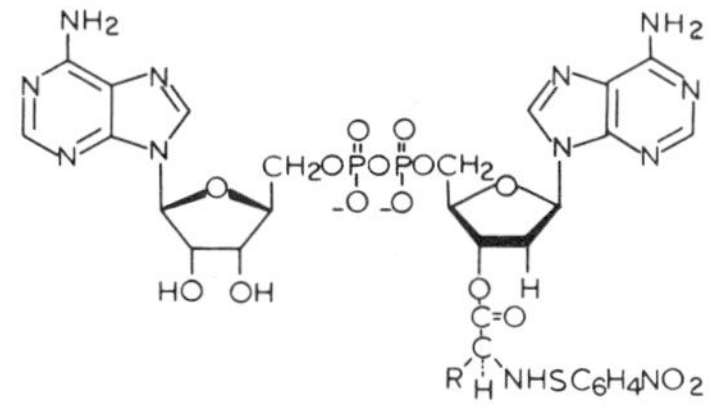

Figure 6 Nucleotide species 9.

vant, as high concentrations of EF-Tu are present in *E. coli* (Furano 1975), on the basis of the greater affinity for the factor of yeast 2'-*O*-tyrosyl-tRNA^{Tyr}, as compared with the 3' isomer, Sprinzl et al. (1977) suggested that the "significantly higher efficiency" of binding of such 2'-*O*-aminoacylated species would result in conversion of all of the aminoacyl-tRNAs to this positional isomer after initial ternary complex formation. This postulated positional preference was included as a key feature of a proposed scheme descriptive of the positional specifities of the individual partial reactions leading to peptide bond formation (Sprinzl 1977; Sprinzl and Wagner, this volume). The lack of an obvious preference for a single positional isomer of aminoacyl-tRNAs during our initial study (Hecht et al. 1977) prompted a more detailed investigation of the problem.

Gel filtration, on Ultrogel AcA 44 resin, of the ternary complex resulting from incubation of guanylylimidodiphosphate, equimolar amounts of *E. coli* ^{3}H-labeled 2'-deoxy-3'-*O*-tyrosyl-tRNA^{Tyr} and ^{14}C-labeled 2'-*O*-tyrosyl-3'-deoxy-tRNA^{Tyr}, and a limiting amount of EF-Tu resulted in inclusion within the complex of 42% of the input 2'-*O*-tyrosyl-tRNA but only 27% of the 3'-*O*-aminoacylated species. This was entirely consistent with the findings of Sprinzl et al. (1977) for the positional isomers of yeast tyrosyl-tRNA^{Tyr}, which had been obtained under the same conditions. However, repetition of the gel filtration experiments with the *E. coli* tyrosyl-tRNAs in the presence of GTP resulted in inclusion in the ternary complex of a greater amount of the 3'-*O*-tyrosylated species (61%) than of its 2'-*O*-aminoacylated isomer (42%). It should be noted that all of these results were quite reproducible, as was the additional reversal of the apparent positional preference noted

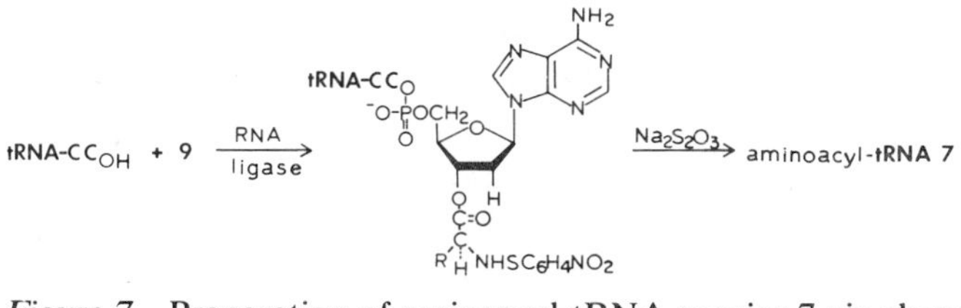

Figure 7 Preparation of aminoacyl-tRNA species 7 via chemical aminoacylation.

when the latter experiment was analyzed by gel filtration on Sephadex G-100. The extreme sensitivity of the measured positional preference to the experimental conditions employed was also noted for isomeric yeast tRNAs^{Phe} and tRNAs^{iMet} and for isomeric *E. coli* tRNAs^{Lys} and *E. coli* and yeast tRNAs^{Trp} (data not shown). In sharp contrast to this sensitivity was the greater affinity for EF-Tu · GTP of the respective unmodified tRNAs under all of the conditions employed. For example, direct competition of equimolar amounts of yeast ^{14}C-labeled 2'-O-phenylalanyl-3'-deoxy tRNA^{Phe} and unmodified ^{3}H-labeled 2'(3')-O-phenylalanyl-tRNA^{Phe} for the same (limited) amount of EF-Tu · GTP resulted in inclusion of 65% of the unmodified species, but only 15% of the 2'-O-phenylalanyl-tRNA, in the ternary complex. Consistent with earlier results (Hecht et al. 1977) it appeared that the possible difference in affinities for EF-Tu between positional isomers was generally much smaller than the difference between either of the isomers and the respective unmodified aminoacyl-tRNA.

The affinities of the isomeric tRNAs for EF-Tu · GTP were also measured in terms of the ability of a limited amount of the factor to diminish the rate of chemical hydrolysis of the isomeric aminoacyl-tRNAs. These experiments were carried out using the smallest amount of EF-Tu · GTP that would provide a reliably detectable increment in the lifetime of the aminoacyl-tRNAs; deacylation curves were determined in each case and the 45-minute time points were compared relative to the extent of deacylation observed in each case in the absence of EF-Tu·GTP. When equimolar amounts of *E. coli* tyrosyl-tRNAs 2 and 3 were incubated in the presence of a limiting amount of EF-Tu · GTP, the extent of deacylation of species 2 after 45 minutes was diminished 1.29-fold; that of tRNA 3 was reduced 1.18-fold. The respective numbers for yeast tRNAs^{Phe} 2 and 3 were 1.23 and 1.55, however, and no clear pattern could be discerned among six isomeric aminoacyl-tRNAs. When the same experiments were carried out in the presence of guanylylimidodiphosphate, a similar lack of a common positional preference among the various isoacceptors was observed.

As in the case of the gel filtration experiments, the competition between modified and unmodified aminoacyl-tRNAs was more instructive. Thus, deacylation of yeast 2'(3')-O-phenylalanyl-tRNA^{Phe} and 2'-O-phenylalanyl-3'-deoxy tRNA^{Phe} in an incubation mixture in the presence of a limited amount of EF-Tu·GTP resulted in a 4.32-fold diminution in the extent of deacylation of the unmodified species after 45 minutes but only a 1.42-fold reduction in the extent of deacylation of the modified phenylalanyl-tRNA. The analogous experiment was carried out with unmodified and modified (2'-O-tyrosyl) *E. coli* tRNAs^{Tyr} and resulted in 8.91- and 1.06-fold reductions in the extent of chemical hydrolysis of the two species, respectively. These results are in good agreement with those previously obtained that utilized unfractionated mixtures of tRNAs (Hecht 1977).

The small and variable differences observed between positional isomers of individual aminoacyl-tRNAs, as compared with the larger differences typically observed between either of the isomeric species and the respective unmodified aminoacyl-tRNAs, indicate that EF-Tu has no obvious preference for either of these positionally defined analogs of aminoacyl-tRNA. Given the large difference in the affinity of EF-Tu·GTP for individual tRNA isoacceptors (Hecht et al. 1977; Hecht 1977; Pingoud et al. 1977), it is not surprising that the factor fails to exhibit a clearly defined preference for a common positional isomer of each aminoacyl-tRNA. It seems reasonable to conclude that although the studies with the modified tRNAs (2 and 3) have established convincingly that both positional isomers of aminoacyl-tRNA can bind to EF-Tu, it is less clear that the modified species represent adequate analogs of aminoacyl-tRNA at the level of refinement needed to establish the possible positional preference of the factor. If support for the postulate (Sprinzl 1977; Sprinzl and Wagner, this volume) of positional preference is accessible at all with such modified species, it could probably derive only from direct measurement of the association constants (Pingoud et al. 1977) of several different isomeric aminoacyl-tRNAs in comparison with the respective unmodified species. It should be noted, however, that even if one positional isomer of aminoacyl-tRNA can be shown to have a somewhat greater affinity for EF-Tu · GTP, it does not follow directly that isomerization would result following ternary complex formation or that the difference in affinities would be sufficient to result in the production of a single isomeric aminoacyl-tRNA within the complex.

REFERENCES

Alford, B. and S. M. Hecht. 1978. 2'-Versus 3'-OH specificity in tRNA aminoacylation. Further support for the "secondary cognition" proposal. *J. Biol. Chem.* **253:**4844.

Alford, B. L., A. C. Chinault, S. O. Jolly, and S. M. Hecht. 1979. Preparation of tRNA's terminating in 2'- and 3'-deoxyadenosine. *Methods Enzymol.* **59:**121.

Baldwin, A. N. and P. Berg. 1966. Transfer ribonucleic acid-induced hydrolysis of valyladenylate bound to isoleucyl ribonucleic acid synthetase. *J. Biol. Chem.* **241:**839.

Barrio, J. R., M. del Carmen, G. Barrio, N. J. Leonard, T. E. England, and O. C. Uhlenbeck. 1978. Synthesis of modified nucleoside 3',5'-bis-phosphates and their incorporation into oligoribonucleotides with T4 RNA ligase. *Biochemistry* **17:**2077.

Chinault, A. C., K. H. Tan, S. M. Hassur, and S. M. Hecht. 1977. Initial position of aminoacylation of individual *Escherichia coli*, yeast and calf liver transfer RNA's. *Biochemistry* **16:**766.

Ebel, J.-P., R. Giegé, J. Bonnet, D. Kern, N. Befort, C. Bollack, F. Fasiolo, J. Gangloff, and G. Dirheimer. 1973. Factors determining the specificity of the tRNA aminoacylation reaction. Non-absolute specificity of tRNA-aminoacyl-tRNA synthetase recognition and particular importance of the maximal velocity. *Biochimie* **55**:547.

England, T. E. and O. C. Uhlenbeck. 1978. Enzymatic oligoribonucleotide synthesis with T4 RNA ligase. *Biochemistry* **17**:2069.

England, T. E., R. I. Gumport, and O. C. Uhlenbeck. 1977. Dinucleoside pyrophosphates are substrates for T4-induced RNA ligase. *Proc. Natl. Acad. Sci.* **74**:4839.

Feldmann, H. and H. G. Zachau. 1964. Chemical evidence for the 3'-linkage of amino acids to s-RNA. *Biochem. Biophys. Res. Commun.* **15**:13.

Fraser, T. H. and A. Rich. 1973. Synthesis and aminoacylation of 3'-amino-3'-deoxy transfer RNA and its activity in ribosomal protein synthesis. *Proc. Natl. Acad. Sci.* **70**:2671.

―――. 1975. Amino acids are not all initially attached to the same position on transfer RNA molecules. *Proc. Natl. Acad. Sci.* **72**:3044.

Furano, A. V. 1975. Content of elongation factor Tu in *Escherichia coli*. *Proc. Natl. Acad. Sci.* **72**:4780.

Giegé, R., D. Kern, J.-P. Ebel, H. Grosjean, S. DeHenau, and H. Chantrenne. 1974. Incorrect aminoacylations involving tRNA's or valyl-tRNA synthetase from *Bacillus stearothermophilus*. *Eur. J. Biochem.* **45**:351.

Gillam, S. and M. Smith. 1974. Enzymatic synthesis of deoxyribo-oligonucleotides of defined sequence. Properties of the enzyme. *Nucleic Acids Res.* **1**:1631.

Gillam, S., K. Waterman, M. Doel, and M. Smith. 1974. Enzymatic synthesis of deoxy ribo-oligonucleotides of defined sequence. Deoxy ribo-oligonucleotide synthesis. *Nucleic Acids Res.* **1**:1649.

Griffin, B. E., M. Jarman, C. B. Reese, J. E. Sulston, and D. R. Trentham. 1966. Some observations relating to acyl mobility in aminoacyl soluble ribonucleic acids. *Biochemistry* **5**:3638.

Hawley, D. M., J. J. Sninsky, G. N. Bennett, and P. T. Gilham. 1978. Activity of polynucleotide phosphorylase with nucleoside diphosphates containing sugar ring modifications. *Biochemistry* **17**:2082.

Hecht, S. M. 1977. Participation of isomeric tRNA's in the partial reactions of protein biosynthesis. *Tetrahedron* **33**:1671.

Hecht, S. M. and A. C. Chinault. 1976. Position of aminoacylation of individual *Escherichia coli* and yeast tRNA's. *Proc. Natl. Acad. Sci.* **73**:405.

Hecht, S. M., J. W. Kozarich, and F. J. Schmidt. 1974. Isomeric phenylanalyl-tRNA's. Position of the aminoacyl moitey during protein biosynthesis. *Proc. Natl. Acad. Sci.* **71**:4317.

Hecht, S. M., B. L. Alford, Y. Kuroda, and S. Kitano. 1978. "Chemical aminoacylation" of tRNA's. *J. Biol. Chem.* **253**:4517.

Hecht, S. M., K. H. Tan, A. C. Chinault, and P. Arcari. 1977. Isomeric aminoacyl-tRNA's are both bound by elongation factor Tu. *Proc. Natl. Acad. Sci.* **74**:437.

Hecht, S. M., S. D. Hawrelak, J. W. Kozarich, F. J. Schmidt, and R. M. Bock. 1973. Chemical modifications of transfer RNA species. Transfer RNA's terminating in 2'- and 3'-O-methyladenosine. *Biochem. Biophys. Res. Commun.* **52**:1341.

Igloi, G. L., F. von der Haar, and F. Cramer. 1977. Hydrolytic action of aminoacyl-tRNA synthetases from baker's yeast. "Chemical proofreading" of thr-tRNA^{Val} by valyl-tRNA synthetase studied with modified tRNA^{Val} and amino acid analogues. *Biochemistry* **16**:1696.

Julius, D. J., T. H. Fraser, and A. Rich. 1979. The isomeric specificity of aminoacylation of wheat germ tRNA and the specificity of interaction of elongation factor Tu with aminoacyl-tRNA. *Biochemistry* **18**:604.

Kikuchi, Y., F. Hishinuma, and K. Sakaguchi. 1978. Addition of mononucleotides to oligoribonucleotide acceptors with T4 RNA ligase. *Proc. Natl. Acad. Sci.* **75**:1270.

Kaufmann, G. and U. Z. Littauer. 1974. Covalent joining of phenylalanine transfer ribonucleic acid half-molecules by T4 RNA ligase. *Proc. Natl. Acad. Sci.* **74**:3741.

Khym, J. X. and M. Uziel. 1968. The use of cetyltrimethylammonium cation in terminal sequence analyses of ribonucleic acids. *Biochemistry* **7**:422.

McCutchan, T. F., P. T. Gilham, and D. Söll. 1975. An improved method for the purification of tRNA by chromatography on dihydroxyboryl substituted cellulose. *Nucleic Acids Res.* **2**:853.

McLaughlin, C. S. and V. M. Ingram. 1965. Chemical studies on amino acid acceptor ribonucleic acids. IV. Position of the amino acid residue in aminoacyl s-RNA: Chemical approach. *Biochemistry* **4**:1442.

Ofengand, J., S. Chládek, G. Robilard, and J. Bierbaum. 1974. Enzymatic acylation of oxidized-reduced transfer ribonucleic acid by *Escherichia coli*, yeast and rat liver synthetases occurs almost exclusively at the 2'-hydroxyl. *Biochemistry* **13**:5425.

Pingoud, A., C. Urbanke, G. Krauss, F. Peters, and G. Maass. 1977. Ternary complex formation between elongation factor Tu, GTP and aminoacyl-tRNA: An equilibrium study. *Eur. J. Biochem.* **78**:403.

Rosenberg, M., J. L. Wiebers, and P. T. Gilham. 1972. Studies on the interactions of nucleotides, polynucleotides and nucleic acids with dihydroxyboryl-substituted celluloses. *Biochemistry* **11**:3623.

Sninsky, J. J., G. N. Bennett, and P. T. Gilham. 1974. "Single addition" and "transnucleotidation" reactions catalyzed by polynucleotide phosphorylase. Effect of enzymatic removal of inorganic phosphate during reaction. *Nucleic Acids Res.* **1**:1665.

Sprinzl, M. 1977. 2',3'-Isomerization of aminoacyl-tRNA during EF-Tu dependent binding to the ribosomal A-site. 11th *FEBS Proc. Meet.* Copenhagen, August 1977. Abstracts A2-6 164.

Sprinzl, M. and F. Cramer. 1973. Accepting site for aminoacylation of tRNA^{Phe} from yeast. *Nat. New Biol.* **245**:3.

———. 1975. Site of aminoacylation of tRNA's from *Escherichia coli* with respect to the 2' or 3'-hydroxyl group of the terminal adenosine. *Proc. Natl. Acad. Sci.* **72**:3049.

Sprinzl, M., M. Kucharzewski, J. B. Hobbs, and F. Cramer. 1977. Specificity of elongation factor Tu from *Escherichia coli* with respect to attachment of the amino acid to the 2' or 3'-hydroxyl group of the terminal adenosine of tRNA. *Eur. J. Biochem.* **78**:55.

Sprinzl, M., K. H. Scheit, H. Sternbach, F. von der Haar, and F. Cramer. 1973. *In vitro* incorporation of 2'-deoxyadenosine and 3'-deoxyadenosine into yeast tRNA^{Phe} using tRNA nucleotidyltransferase and properties of tRNA^{Phe}-C-C-2'dA and tRNA^{Phe}-C-C-3'dA. *Biochem. Biophys. Res. Commun.* **51**:881.

Tal, J., M. P. Deutscher, and U. Z. Littauer. 1972. Biological activity of *Escherichia coli* tRNA^{Phe} modified in its C-C-A terminus. *Eur. J. Biochem.* **28**:478.

Uziel, M. 1973. Periodate oxidation and amine-catalyzed elimination of the terminal nucleoside from adenylate or ribonucleic acid. Products of overoxidation. *Biochemistry* **12**:938.

von der Haar, F. and F. Cramer. 1975. Isoleucyl-tRNA synthetase from baker's yeast: The 3'-hydroxyl group of the 3'-terminal ribose is essential for preventing misacylation of tRNA^{Ile}-C-C-A with misactivated valine. *FEBS Lett.* **56**:215.

———. 1976. Hydrolytic action of aminoacyl-tRNA synthetases from baker's yeast: "Chemical proofreading" preventing acylation of tRNA^{Ile} with misactivated valine. *Biochemistry* **15**:4131.

Woese, C. R., D. H. Dugre, S. A. Dugre, M. Kondo, and W. C. Saxinger. 1967. On the fundamental nature and evolution of the genetic code. *Cold Spring Harbor Symp. Quant. Biol.* **31**:723.

Wolfenden, R., D. H. Rammler, and F. Lipmann. 1964. On the site of esterification of amino acids to soluble RNA. *Biochemistry* **3**:329.

Yarus, M. and M. Mertes. 1973. The variety of intraspecific misacylations carried out by isoleucyl transfer ribonucleic acid synthetase of *Escherichia coli*. *J. Biol. Chem.* **248**:6744.

Zubay, G. and M. Takanami. 1964. Observations on the configuration of nucleotides near the 3'-hydroxy end of adapter RNA. *Biochem. Biophys. Res. Commun.* **15**:207.

tRNA Interactions with Ribosomes

tRNA-Ribosome Interactions

Charles R. Cantor
Barth Laboratory
Departments of Chemistry and Biological Sciences
Columbia University,
New York, New York 10027

The bacterial ribosome is 100 times the molecular weight of tRNA. Ribosomes from eukaryotes are even larger. Thus, a detailed description of ribosome structure and the mechanism of protein synthesis could dwarf a book about tRNA (for recent reviews, see Nomura et al. 1974; Weissbach and Pestka 1977; Kurland 1977; Brimacombe et al. 1978; Grunberg-Manago et al. 1978). Here we shall focus on ribosome-tRNA interactions from the myopic point of view of the tRNA. Details of ribosome structure not related directly to tRNA function will be ignored as much as possible. For example, we shall stress what happens to a tRNA when it binds to a ribosome, but not what happens to the ribosome. Almost all examples will be drawn from *Escherichia coli*. However, most generalizations probably apply accurately to other ribosomes.

The essential problem in studying tRNA-ribosome interactions is that far more is known about tRNA structure and dynamics than about ribosomes. The known structure and properties of tRNAs raise numerous interesting questions about how tRNA may function on the ribosome. However, it seems far easier to frame such questions than to design definitive experimental approaches for answering them. Some of the most powerful techniques available for the study of an isolated tRNA cannot be used on tRNA-ribosome complexes because of their large size, heterogeneity, and instability. The situation is analogous to the study of enzyme-substrate interactions in the absence of crystals of the enzyme-substrate complex. One has to hope that available knowledge about the substrate will be applicable on the enzyme and one has to exploit the greater ease of manipulating the structure of the substrate in planning experimental approaches.

Few hard facts are available about the structure of tRNAs on the ribosome. No high-resolution structural information exists on ribosomal tRNA binding sites. Virtually no equilibrium and kinetic data are available about tRNA-ribosome interactions under physiological conditions, and such physicochemical studies are in their infancy even under less realistic conditions. In spite of this, an impressive number of generalizations can be made about the likely modes of tRNA function on the ribosome. Few inconsistencies exist in the large set of biochemical and low-resolution structural data available. In this paper, we will summarize what is known and then indicate important aspects of tRNA behavior on the ribosome that demand further inquiry. Experimental tools now exist that should make such explorations highly productive. These will be discussed and a few examples of recent progress will be given.

THE RIBOSOME AS SEEN BY tRNA

An Outline of Protein Synthesis

tRNA Binding Sites

The current picture of the mechanism of protein synthesis places a number of constraints on the behavior of tRNA on the ribosome. It requires the existence of at least two tRNA binding sites that can be occupied simultaneously. These are usually called the peptidyl site and the aminoacyl site. During the elongation cycle of protein synthesis, the peptidyl site can be occupied by deacylated or peptidyl-tRNA, whereas the aminoacyl site is occupied by aminoacyl- or peptidyl-tRNA (see Fig. 1). However, other states of ribosome-bound tRNA exist, as shown in Table 1. Some of these are probably just in vitro constructions that have no significant biological role. However, others are important in function. For example, occupation of the aminoacyl site by deacylated tRNA is the signal that triggers the stringent response (Cashel and Gallant 1974; Block and Haseltine 1974).

The existence of additional distinct ribosome binding sites with functional roles in protein synthesis has occasionally been suggested (Swan et al. 1970; Lake 1977). Indeed, it would simplify our understanding of a number of the steps in protein synthesis if more discrete tRNA locations

Table 1 tRNA binding configurations on 70S ribosomes

tRNA		
peptidyl site	aminoacyl site	Biological role
Peptidyl	—	posttranslocation
Peptidyl	aminoacyl	peptidyl transferase substrate
Deacylated	peptidyl	pretranslocation
Deacylated	—	termination
Peptidyl	deacylated	stringent response
Deacylated	deacylated	stringent response
Deacylated	aminoacyl	no known role
Aminoacyl	aminoacyl	in vitro peptidyl transferase substrate
Aminoacyl	—	no known role
—	—	no known role

on the ribosome could be found. However, evidence to date supports only the occurrence of minor variants of the aminoacyl and peptidyl sites. For example, the initiator tRNA, formylmethionyl-tRNA^{fMet}, may occupy a site that overlaps parts of the aminoacyl and peptidyl sites before ultimately reaching the peptidyl site (Revel 1977). Similarly, aminoacyl-tRNA may pass through a site (sometimes called the recognition site) that only partially overlaps the aminoacyl site before ultimately reaching the aminoacyl site (Haenni and Lucas-Lenard 1968; Skoultchi et al. 1970). In both cases, affinity labeling and other biochemical studies suggest that the differences between these sites and the normal aminoacyl and peptidyl sites most probably involve a displacement in the position of the 3' ends of the tRNAs (Hauptmann et al. 1974; Johnson et al. 1977).

Initiation

In *E. coli*, as far as we know, tRNA entry into the ribosome always requires the presence of a protein factor (see Fig. 1). For example, initiation factor 2 (IF-2), a protein that exists in several forms with molecular weights between 85,000 and 117,000, will promote formylmethionyl-tRNA^{fMet} binding to 30S ribosomes in the presence of AUG (Grunberg-Manago et al. 1978). IF-2 will bind to 30S subunits in the absence of tRNA or message (Revel 1977). It will also form a binary complex with formylmethionyl-tRNA^{fMet} under some conditions (Sundari et al. 1976) and some evidence exists for ternary complexes of IF-2, GTP, and formylmethionyl-tRNA^{fMet}. However, it is not clear whether a tRNA-containing binary or ternary complex is an actual intermediate in initiation. The situation is complicated by the presence of two other initiation factors, IF-1 (9500 m.w.) and IF-3 (22,000 m.w.). Kinetic studies should eventually help sort out their roles (see, for example, Gualerzi et al. 1977).

Elongation

During the elongation cycle of protein synthesis, aminoacyl-tRNA binding to the aminoacyl site is promoted by elongation factor Tu (EF-Tu), a 44,000-m.w. protein. An EF-Tu·GTP complex binds tightly to 70S ribosomes. This complex also binds tightly to aminoacyl-tRNA (Miller and Weissbach 1977). Recently it has been shown that a covalent aminoacyl-tRNA·EF-Tu·GTP ternary complex will bind to 70S ribosomes (Johnson et al. 1978). This suggests that the ternary complex is almost certainly capable of serving as an intermediate in protein synthesis. However, it has not yet been proved that this intermediate is obligatory.

Translocation of peptidyl-tRNA from the aminoacyl site to the peptidyl site and release of deacylated tRNA from the peptidyl site are promoted by elongation factor G (EF-G). This is an 80,000-m.w. protein that, like EF-Tu and IF-2, appears to function as a GTP complex (Brot 1977). Most available evidence indicates that the direct site of EF-G function is the

50S moiety of the 70S ribosome. However, it has not been established whether or not EF-G will promote translocation in the absence of the 30S particle. No evidence exists that EF-G interacts directly with tRNA in the absence of ribosomes. EF-G will bind to 70S particles in the absence of tRNA. It is not known with certainty whether EF-G directly contacts tRNA while bound to the ribosome.

The mechanism of EF-G-promoted translocation remains almost a complete mystery. Most suggestions involve a concerted release of deacylated tRNA and a shift of the peptidyl-tRNA from the aminoacyl site to the peptidyl site. No compelling evidence exists for an intermediate state with an empty peptidyl site and a filled aminoacyl site. In fact, as shown in Table 1, no method is known for preparing a ribosome with tRNA in the aminoacyl site without simultaneous occupancy of the peptidyl site. A number of biochemical studies have suggested that the aminoacyl site may not even exist unless the peptidyl site contains a bound tRNA (de Groot et al. 1971). Thus, a concerted movement of the tRNAs during translocation is an attractive model. In support of this idea, it has been shown that EF-G will not cause release of a deacylated tRNA from the peptidyl site if the aminoacyl site is empty (Lucas-Lenard and Haenni 1969; Modolell et al. 1973).

It is pretty clear that the relative movement of the mRNA in translocation is the result of tRNA movement and not vice versa (Rich 1974). A frameshift suppressor tRNA^{Gly} isolated by Riddle and Carbon (1973) has an extra nucleotide in its anticodon loop. Apparently this tRNA can read four bases on the mRNA instead of three. Translocation of the tRNA moves the message four bases and alleviates the effect of the frameshift mutation. Thus, the ribosome must be programmed to shift a tRNA during translocation rather than make a rigid three-base shift of a message.

Termination

Termination of protein synthesis occurs when protein release factors bind to the ribosome in response to the appearance of a termination codon in the aminoacyl site (Caskey 1977). Release factor 1 (RF-1) and release factor 2 (RF-2) are codon-specific factors, and one may speculate that they are the protein analogs of tRNAs. With molecular weights of about 45,000 each, they could occupy a volume not much bigger than that of tRNA. A third factor, RF-3, acting as a GTP complex, assists the binding of RF-1 or RF-2. This results in hydrolysis of the peptidyl-tRNA, leaving a deacylated tRNA bound at the peptidyl site. How the ultimate removal of this tRNA is accomplished is unknown.

The overall scheme of bacterial protein synthesis, highlighting the reactions that involve tRNA, is shown in Figure 1. Ignoring a few aspects not resolved completely, the following generalizations emerge.

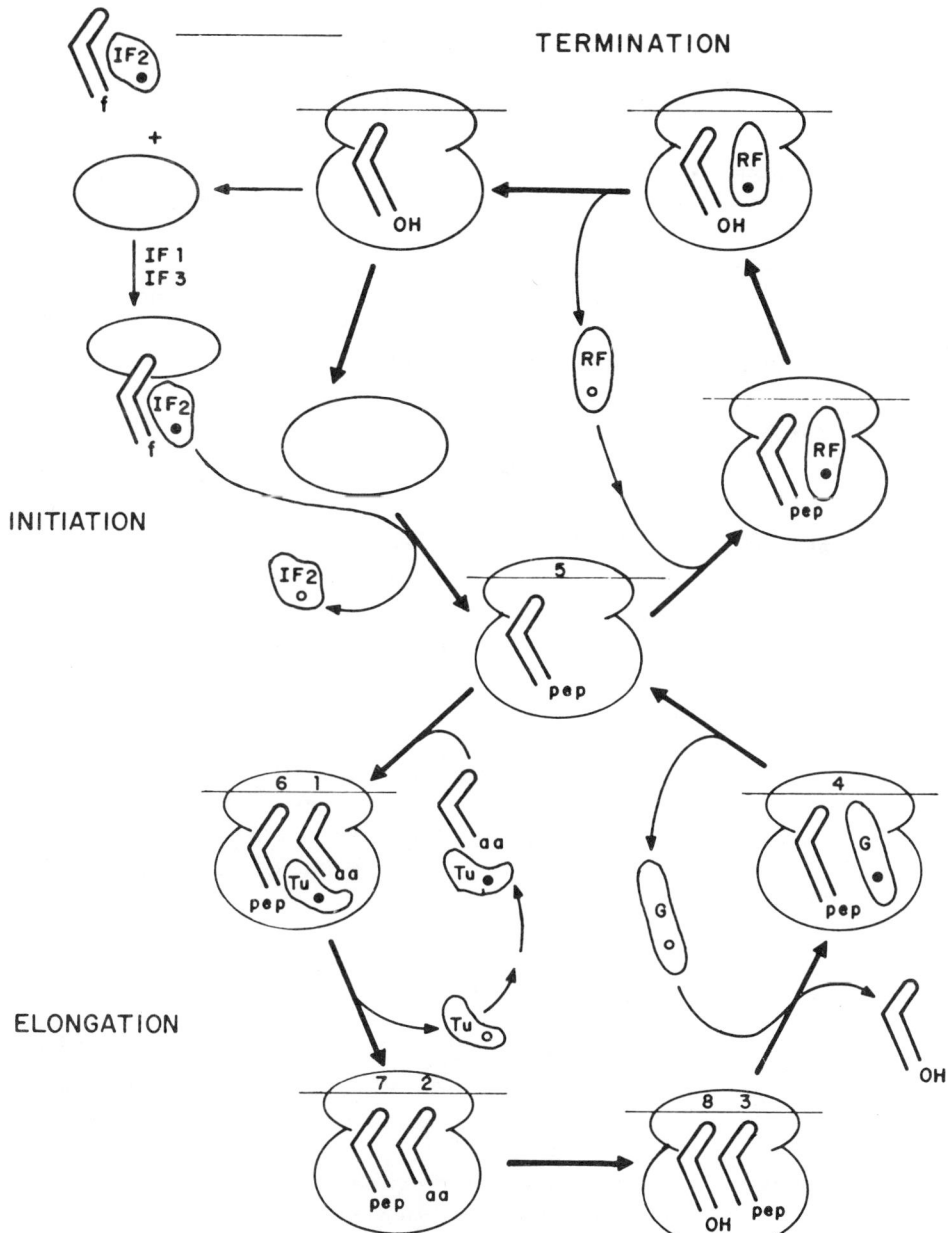

Figure 1 An outline of the current accepted scheme for protein biosynthesis in *E. coli*. At least 12 distinct tRNA-ribosome interactions are involved, including 8 in the elongation cycle (numbered in the order in which they occur for one tRNA in this cycle), 3 in termination, and 1 in initiation. (●) Factor-bound GTP; (○) factor-bound GDP. Some of the details of initiation and of EF-Tu recycling are omitted.

1. The entry of tRNA into any site always requires a GTP-containing protein factor. Probably the tRNA is presented to the site as an actual complex with the factor.
2. Release of the protein factor must occur before the newly bound tRNA can function. This release always requires the hydrolysis of the GTP bound to the factor, which thus leaves the ribosome as a GDP complex.
3. tRNA always enters sites on the ribosome with an aminoacylated 3' terminal. Removal of tRNA from the ribosome always occurs as the departure of deacylated tRNA from the peptidyl site.
4. Each tRNA that participates in protein synthesis elongation is involved in at least eight discrete ribosome interactions, as shown schematically in Figure 1. Thus, questions such as, "What is the structure of ribosome-bound tRNA?" are really rather complex ones.

Nature and Location of the tRNA Binding Sites

The *E. coli* 70S ribosome contains three rRNAs and about 54 different proteins. The 30S subunit consists of the 16S rRNA, with 1541 nucleotides, and 21 proteins, each present in one copy. These proteins are named S1, S2, . . . , S21. The 50S subunit consists of the 23S rRNA, with about 3000 nucleotides, the 5S rRNA, with 120 nucleotides, and 34 proteins called L1, L2, . . . , L34. Of these proteins, all except L7 and L12 are found in a single copy. L7 is simply N-acetyl-L12, and these two proteins together occur in a total of four copies. The idea that the symmetry implied by multiple copies is somehow used in recognizing the pseudosymmetry seen in the tRNA structure (Kim 1978) is intriguing but unproven.

Much is known about the primary structure of these ribosomal components, but little is known about their tertiary structure except that many of the proteins appear to be quite elongated (Brimacombe et al. 1978).

Affinity Labeling

The most direct approach for the identification of those ribosomal components that come into contact with bound tRNA is affinity labeling. Chemically or photochemically reactive analogs of tRNA are constructed that still retain partial or full biological activity. These are bound to one of the tRNA sites on the ribosome and, after incubation or activation, any covalent tRNA-ribosome products are isolated and identified. The advantages and some of the pitfalls of this technique are described in a number of recent review articles (Johnson and Cantor 1977; Cooperman 1978; Pellegrini and Cantor 1977). Several detailed examples of tRNA affinity-labeling studies can be found in Kuechler and Ofengand and in Johnson (both this volume).

The most convincing test of whether a covalent tRNA-ribosome reaction results from tRNA bound in a correct functional mode is a direct link between function and the reaction. Can the tRNA function, even in a limited sense, after it has become covalently attached? Or, can some function be shown as absolutely prerequisite to the covalent attachment? Quite a few tRNA affinity analogs have passed these stringent tests.

A representative survey of some of the ribosomal components that may be directly involved in tRNA binding sites is given in Table 2. All the results listed as direct were obtained using affinity analogs of tRNA. The large number of components that have been located near the 3' end of ribosome-bound tRNA is in part a reflection of the large number of different analogs that have been studied by many groups. Chemical manipulation of the 3' end of tRNA is far easier than that of other sites. Note, however, that a single 3'-end affinity analog in a pure binding site can react covalently with a number of different components. Since each of the reaction products meets the stringent tests described above, one must conclude that many different components of the ribosome approach to within a few angstroms of the 3' end of bound tRNA. It is also probable that some flexibility exists either in the 3' end of tRNA or the ribosome itself, which lenghtens the apparent reach of the chemically reactive moieties of the affinity analogs.

Less-direct Methods

Table 2 also contains a list of ribosomal components of tRNA binding sites identified by less direct measurements. Most of these are the result of affinity-labeling studies that employed analogs of either mRNA codons or antibiotics, such as puromycin, that are clearly tRNA mimics. In addition, a large number of ribosomal components that can affect tRNA binding or function have been implicated by indirect methods. These include effects of chemical modification of ribosome components on subsequent function, interference in function by immunoglobulins specific to ribosome components, genetic studies, and the in vitro reassembly of ribosomes missing various components. Most of the components seen by

Table 2 Ribosome components that contact bound tRNAs

tRNA region	Direct evidence	Less direct evidence
Anticodon	16S rRNA	S1, S4, S12, S18, S21
Other	16S rRNA	5S rRNA
Aminoacyl end	S3, S13, S14, 23S rRNA, L2, L11, L16, L18, L24, L27, L32-33	S6, L23

See Grunberg-Manago et al. (1978) for a more detailed listing.

affinity labeling are also detected by these other methods. Probably many of the others indirectly implicated are actually components of the tRNA binding sites. We cannot be sure, however, because in a structure as complex as the ribosome, a chemical perturbation in one region may be felt at others. A number of examples of such potential long-range interactions on the ribosome exist (see, for example, Langlois et al. 1977; Saltzman and Apirion 1976). Because of this, it is better to be conservative about the assignment of binding-site components.

The results in Table 2 do not discriminate between components of the aminoacyl site and the peptidyl site. Affinity-labeling studies do show clear differences between these two sites at several regions of the tRNA. For example, L2 is most probably a peptidyl-site component, whereas L16 is predominantly an aminoacyl-site component. However, in many cases, although each affinity reagent shows some site-specificity, different reagents do not lead to consistent results. It is probably fairly accurate to conclude that the two tRNAs are so near each other at both the anticodon and aminoacyl ends that many ribosomal components span both binding sites. Insufficient affinity-labeling data exist on the remainder of the tRNA structure to be sure about other parts of the binding sites.

Interactions with Single Components

It would simplify studies of tRNA function considerably if significant interactions between tRNAs and individual or small groups of ribosomal components could be found. Such studies are in their infancy, but a number of interesting findings have emerged. Dahlberg et al. (1978) have observed a strong, specific complex between 23S rRNA and tRNA^{fMet} when the rRNA is extracted from cells by a hot SDS procedure. The physiological significance of this complex is unclear because there is no evidence for it when 50S ribosomes are similarly extracted. There are also indications of a complex between formylmethionyl-tRNA^{fMet} and the 16S rRNA when 30S initiation complexes are extracted with cold SDS at high magnesium concentrations (P. Thammana, unpubl.). The instability of this complex makes detailed characterization difficult.

Complexes between tRNA and individual ribosomal proteins have been observed by chromatography of proteins on a column containing covalently attached tRNA. Unfortunately, the proteins that complex with tRNA appear to depend on the length of the spacers between tRNA and the resin. With a long spacer, eight 50S proteins stick tightly to tRNA (Ustav et al. 1977, 1978). Most are proteins implicated by affinity labeling as located near the aminoacyl terminal. An alternative way to approach the identification of components required for tRNA binding is to start with intact 70S particles and see how much can be removed while retaining tRNA binding ability. For example, isolated 50S subunits still have two tRNA binding sites; 30S subunits have at least one message-specific site. Ribosomal particles can be prepared by various treatments with salts and denaturants. Usually they are assayed by their ability to rebind tRNA or perform other limited protein-synthesis functions. However, when 70S particles containing bound oligo-phenylalanyl-tRNA^{Phe} are extracted with SDS at high magnesium, the 30S particle and a substantial amount of the 50S proteins are removed, whereas half of the tRNA is retained in a core particle (P. Thammana, unpubl.). Such systems could prove useful in identifying additional tRNA binding components.

General Conclusions

A number of generalizations about ribosome binding sites for tRNAs emerge from the results shown in Table 2 and the large amount of structural information currently available for the *E. coli* ribosome.

The tRNA binding sites contain both rRNA and proteins. It is not possible to say what fraction of the binding surface is made up of each, but rRNAs appear to contact tRNA in numerous places.

The aminoacyl ends of tRNA are bound predominantly on the 50S subunit. The components of the ribosome responsible for catalyzing peptidyl transfer are located exclusively on this subunit. The isolated 50S particle can carry out peptidyl transfer.

The anticodon ends of tRNA appear to be exclusively in contact with the 30S subunit. No strong evidence for direct contact between the mRNA and the 50S subunit exists, although, curiously, mRNAs can stimulate the binding of tRNA to the isolated 50S subunit (Jonak and Rychlik 1970). The anticodon must protrude a fair distance from the surface of the 50S subunit, as complexes between two tRNAs with complementary anticodons bind to the 50S subunit stronger than individual tRNAs (P. Thammana, unpubl.).

Virtually all of the 50S components implicated in tRNA contact have a portion that lies in a small region of the 50S subunit as seen by immuno-electron microscopy (Brimacombe et al. 1978). Virtually all of the 30S components implicated in tRNA contact have a portion that lies in one section of the 30S particle. Unfortunately, the resolution of current electron microscopic techniques is not really sufficient to define the precise arrangement of components and the tRNA binding sites. However, the agreement between affinity labeling and electron microscopy is really encouraging.

The tRNA binding sites must lie between the subunits of the 70S particle (see Fig. 2). The two regions described above face toward each other into a central cavity bounded by the subunit interface. This location is consistent with the resistance of ribosome-bound tRNA to nuclease digestion (Kuechler et al. 1972). How tRNAs enter or leave this cavity is a

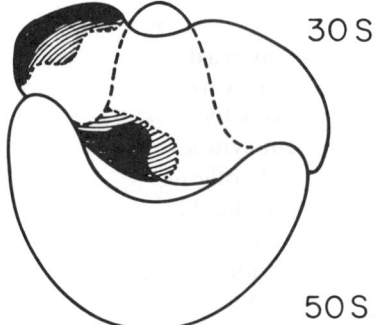

Figure 2 A model for the structure of *E. coli* 70S ribosomes based on immuno-electron microscopy data (Stöffler and Wittmann 1977). Shaded areas indicate locations of proteins implicated by affinity labeling as situated near the 3' end of tRNA on the 50S subunit and near the anticodon on the 30S subunit. The two bound tRNAs apparently are sandwiched between the two subunits.

major unsolved puzzle in our picture of protein synthesis. In some models of the 70S particle the cavity is joined to the surface of the 70S particle by holes large enough for a tRNA to pass. In others, it would appear that the subunits would have to swing open and allow tRNA binding and release, an idea originally suggested by Spirin (1970).

Nearly all of the ribosomal components implicated in translocation reside on the 50S subunit. This suggests that translocation is predominantly mediated by this subunit. However, a critical test of this idea is lacking. Few, if any, 50S components involved in translocation appear to contact tRNA directly. Perhaps EF-G mediates translocation by directly pushing tRNA itself, while anchored to the 50S subunit.

Interaction of tRNA with Other Ribosome-bound Species

While on the ribosome, tRNAs can interact with many other species. It is of interest to examine whether these interactions are different from those that can occur in the absence of the ribosome. For example, an interaction too weak to cause appreciable association in solution might still be strong enough to affect the properties of ribosome-bound tRNAs. Alternatively, the constraints imposed on a tRNA while bound to the ribosome might eliminate the possibility of interaction with some other molecules.

tRNA-mRNA Interactions

Although there is little direct proof, the indirect evidence that tRNA in the aminoacyl site is in contact with the message is overwhelming. The message-directed specific binding of aminoacyl-tRNA to the aminoacyl site is, of course, one of the principal assays that was used to determine the genetic code. Similar evidence indicates that the binding of tRNA in configurations that at least partially resemble aminoacyl-site binding is codon-specific. This includes deacylated tRNA binding and the resulting stringent response, EF-Tu·tRNA·GMPP(CH_2)P binding that presumably corresponds to occupancy of the recognition site, and initiator tRNA binding to the 30S subunit.

Two recent experiments indicate that codon-anticodon interaction is a feature of bound tRNA and not just a transient stage in the binding. Direct irradiation of poly(U)-programmed, aminoacyl-site-bound yeast tRNA^{Phe} yields a covalent tRNA-mRNA complex (Kuechler and Ofengand, this volume). Fluorescence studies on a derivative of this same tRNA in which a proflavine replaces the Y base show that when deacylated tRNA is bound at the aminoacyl site, the presence of poly(U) causes a perturbation of the environment of the proflavine (R. H. Fairclough et al., in prep.).

Although virtually all models of protein synthesis show an interaction between the message and the anticodon of tRNA bound in the peptidyl site, evidence for this interaction has only recently been obtained. It was reasonable to suspect contact with message because binding of deacylated tRNA to ribosomes is codon-dependent (Levin 1970) and deacylated tRNA has a strong preference for peptidyl-site binding (de Groot et al. 1971). The same kinds of cross-linking and fluorescence results described above for the aminoacyl site are observed when tRNA^{Phe} is bound at the peptidyl site. This strongly suggests that there is an mRNA-tRNA interaction in both sites.

Probing further one can ask whether both tRNA anticodons can simultaneously contact adjacent mRNA codons on the ribosome. Isolation of a double cross-link would answer this question definitely. Until that has been accomplished, one must be content with less-direct evidence. Occupation of the aminoacyl site occurs with no apparent change in the codon-anticodon interaction in the peptidyl site, as judged by the fluorescence of the adjacent hypermodified base in tRNA^{Phe} (R. H. Fairclough and C. R. Cantor, in prep.). Furthermore, corresponding positions of the anticodons of two simultaneously bound tRNA^{Phe} molecules are 18 Å apart. This is exactly the distance expected for a continuous helix formed by two adjacent codon-anticodon triplets (R. H. Fairclough and C. R. Cantor, in prep.). The energy transfer measurements that allowed this distance to be measured also provide the first really direct proof that two tRNAs can simultaneously occupy one ribosome. Under most conditions, ribosome binding stoichiometries are so poor that attempts to fill both sites usually result in a total of less than one tRNA per ribosome in the sample.

tRNA-tRNA Interactions

A variety of tRNA dimers in solution are known, but most of these do not seem to be appropriate models for tRNA interactions on the ribosome (Crothers and Cole 1978). No cross-linking between adjacent ribosome-bound tRNAs has been reported. The evidence of direct tRNA-tRNA contact is indirect. Binding to the aminoacyl site appears to require prior occupancy of the peptidyl site (Table 1). Exchange of peptidyl-site-bound tRNA is blocked by occupancy of the aminoacyl site (N. Farber et al., unpubl.). The simplest interpretation of these results is that the tRNA in the peptidyl site actually provides some of the binding contacts that constitute the aminoacyl site. However, less-direct interactions cannot be excluded.

What regions of the two adjacent tRNAs are most likely in contact? It is difficult to construct models of tRNAs bound to adjacent mRNA codons without making direct contacts between the tRNA anticodon loops (Lake 1977). Fluorescence studies support the idea that the hypermodified base of tRNA in the aminoacyl site comes into contact with the anticodon loop of the tRNA in the peptidyl site (R. H. Fairclough and C. R. Cantor, in prep.). Binding of 3' fragments of aminoacyl-tRNA to the aminoacyl site of the ribosome requires the presence of tRNA in the peptidyl site (Haselkorn and Rothman-Denes 1973). This suggests the possibility of a direct interaction between the 3' ends of the two bound tRNAs.

tRNA-Factor Interactions

Aminoacyl-tRNA enters the aminoacyl site as an EF-Tu·tRNA·GTP ternary complex. If the nonhydrolyzable analog $GMPP(CH_2)P$ is substituted for GTP, binding still occurs, but the EF-Tu is now retained on the ribosome. Presumably it is still in direct contact with the tRNA. In this complex, transfer of the peptide chain from the tRNA in the peptidyl site is blocked (Shorey et al. 1971). However, puromycin is able to bind and participate in peptidyl transfer (Modolell and Vazquez 1973). Thus, the effect of EF-Tu·tRNA interaction on the ribosome is apparently to hold the 3' end of the tRNA away from the peptidyl transferase center. Further indirect evidence of EF-Tu·tRNA interaction on the ribosome is the substantial overlap between ribosomal proteins known to contact tRNA and those known to contact bound EF-Tu (San José et al. 1976; Fabian 1976). The most direct evidence for this interaction is the ability of GTP bound to a covalent complex of ϵ bromoacetyl lysyl-tRNA^{Lys} and EF-Tu to undergo ribosome-promoted hydrolysis (Johnson et al. 1978).

Although no direct evidence exists, it is fairly likely that initiation factors (at least IF-2) contact tRNA^{fMet} on the 30S subunit. The principal evidence is the similarity of proteins believed to contact the tRNA and proteins known to contact initiation factors (for a review, see Grunberg-Manago et al. 1978). Similar evidence suggests the possibility of contact between tRNA and EF-G (Acharya et al. 1973). It is less clear at present whether there is any direct contact between ribosome-bound tRNA and release or stringent factors.

tRNA–Small-molecule Interactions

GTP hydrolysis plays an almost ubiquitous role in tRNA functions on the ribosome. Most experimental results available favor the notion that every protein factor has its own GTP binding site. It is not known whether tRNA ever contacts GTP directly on the ribosome.

As isolated from cells, tRNA contains bound polyamines. In solution, polyamines and divalent cations bind strongly to tRNA and affect its structure. Spermine and Mg^{++} cations are present in crystals of yeast $tRNA^{Phe}$ and the locations of two spermines and four Mg^{++} ions have been found in the crystal (Quigley et al. 1978). Are such cations still present on tRNA when it binds to ribosomes or are they replaced by positively charged side chains of ribosomal proteins? Elevated magnesium concentrations promote extremely efficient tRNA binding to the ribosome (Grajevskaja et al. 1972). It is not quite clear how this binding should be interpreted, as it requires no mRNA or factors, although it is site-specific. However, it suggests that the direct reaction between tRNA and ribosomes should be written as:

$$a Mg^{++} \cdot tRNA + b Mg^{++} + c Mg^{++} \cdot 70S \leftrightarrow tRNA \cdot (a+b+c) Mg^{++} \cdot 70S \quad (1)$$

Whether the extra magnesiums present in the tRNA-ribosome complex are bound to tRNA, ribosome, or both is unknown. However, if there is extensive contact between tRNA and rRNA, it seems reasonable that some new Mg^{++} binding sites will be created.

tRNA AS SEEN BY THE RIBOSOME

Aspects of tRNA Essential for Function

The variety of tRNA species that can all function in the same ribosomal sites is truly impressive. With rare exceptions, eukaryotic and prokaryotic tRNAs can function on both eukaryotic and prokaryotic ribosomes. Thus, some features of tRNA-ribosomal interaction must be as universal as the genetic code itself.

By examining variations in the structures of tRNAs competent in protein synthesis, one can identify parts or features of tRNA that may be essential for function on the ribosome. At the level of the nucleotide sequence it is clear that all but a few residues of tRNA vary (Sprinzl et al. 1978; RajBhandary et al., this volume). Thus, for function in both the peptidyl and aminoacyl sites, a precise base in a precise place is not required, except possibly at a few positions. These include the 3'-CCA end, 1 base pair of the TψC stem and part of the TψC loop, a few residues in the D loop, U8, G10, and the U on the 5' side of the anticodon. All of these except the first and the last are involved in extensive tertiary interactions within the known structure of yeast tRNA^{Phe}. Changes in some of these features by chemical modification or base-analog incorporation still often leave partial functional ability.

It is reasonable to suggest that some of these conserved regions are required not simply because of their role in tRNA tertiary structure interactions, but also because, on the ribosome, partial disruption of the tertiary structure may free these residues to interact with ribosomal components. The possible TψC-loop interaction with 5S rRNA has received the most attention (for a review, see Möller et al., this volume). A possible role for the U in the anticodon loop has recently been suggested (Lake 1977). It remains to be proved whether any of these residues interact with 16S rRNA, 23S rRNA, or the corresponding RNA species in mitochondrial, chloroplast, and eukaryotic cytoplasmic ribosomes. Another possibility is that conserved residues are used for interaction between the two ribosome-bound tRNAs. Such interactions would place especially strong evolutionary constraints on tRNA sequence. All tRNAs would have to be able to form these interactions with all other tRNAs.

A striking feature of tRNA structure is the rigid maintenance of most features of the cloverleaf pattern of base-paired stems and intervening loops in the face of enormous sequence variation. Thus, it is reasonable to infer that the size of the anticodon loop and stem, the size of the TψC loop and stem, the size of the aminoacyl stem, and certain tertiary-structure features that connect these regions are essential for some of the normal functions of tRNAs on the ribosome. Significantly, tRNAs that do not have to participate in the complete ribosome cycle sometimes show alterations in these features. Variants include prokaryotic and eukaryotic initiator tRNAs that never have to occupy the normal aminoacyl site and tRNAs involved in cell-wall biosynthesis that apparently do not interact with ribosomes at all.

Once more is known about tRNA binding sites and tRNA structures on the ribosome, it will be easier to interpret the constant features implied by available tRNA sequences. However, it is probably safe to infer that some of these features are required simply to insure the right distance and orientation between the anticodon and the aminoacyl end. The extent to which this is flexible and is adjusted by ribosome binding is unknown. However, the ability of ribosomes to tolerate, albeit with reduced function, tRNAs with an extra C next to the 3' A (Thang et al. 1974) and to tolerate extensive substitution of 5' fluorouridine for uracil and its derivatives (Ofengand et al. 1974) implies that there must be some flexibility in binding. The still unexplained mechanism by which a G24→ A24 mutation in tRNA^{Trp} leads to a change in codon recognition may be another indication of tRNA flexibility (Hirsh and Gold 1971; Buckingham and Kurland 1977). Other constant features must be involved in insuring acceptable binding strengths to the ribosome and recognition by the various factors and by any ribosomal components directly involved in translocation. The time seems ripe for a systematic approach to the synthesis of tRNA variants, whose altered function could shed considerable light on the mechanism of protein synthesis.

Unfortunately, we do not yet know the tertiary structure of any tRNA with a large extra loop. Since such tRNAs function fully on the ribosome, the extra loop presumably is oriented away from any contact sites between the tRNA and either the ribosome or the adjacent tRNA. The extra loop is not near the three regions of tRNA currently implicated in ribosome contacts: the aminoacyl end, the anticodon, and the TψC loop. In yeast tRNA^{Phe}, residues in or near the extra loop may be as exposed when bound to the peptidyl site as when free in solution (N. Farber, unpubl.). Thus, the ribosome may not directly contact tRNA near this region.

Functional Discrimination at the Aminoacyl End

It has long been debated whether esterification or other modifications of the 3' end of tRNA leads to significant conformational changes (Crothers and Cole 1978). In protein synthesis, the state of the aminoacyl end has profound effects on function. Most factors and the ribosomal sites themselves are able to discriminate between different esterified derivatives.

Nonenzymatic tRNA Binding

At low magnesium in the absence of deacylated tRNA, peptidyl-tRNA binds specifically to the peptidyl site. However, deacylated tRNA is a reasonably strong competitor for this site. At high magnesium, excess deacylated tRNA will completely displace peptidyl-tRNA from the peptidyl site. But under these conditions, peptidyl-tRNA binds to the aminoacyl site much stronger than deacylated tRNA. Thus, one can produce, in a very high yield, a ribosomal configuration, with deacylated tRNA in the peptidyl site and peptidyl-tRNA in the aminoacyl site, which is presumably the same as the pretranslocation state (de Groot et al. 1971). How magnesium alters the site specificity of the ribosome is unknown. Its effect is easier to rationalize if, on the ribosome, deacylated tRNA and peptidyl-tRNA adopt different structures. If this were the case, it would be possible that peptidyl transfer is directly coupled to tRNA conformational changes.

The apparent binding strength of different species of deacylated or acylated tRNAs varies considerably (Peters 1977; B. Wells, unpubl.). However, few quantitative measurements of tRNA-ribosome binding are available and the significance of these differences is not yet clear.

Enzymatic tRNA Binding

EF-Tu specifically recognizes features at the aminoacyl end of tRNA (Miller and Weissbach 1977). It apparently requires a base pair between the 5' terminal and the acceptor stem (Schulman et al. 1974). EF-Tu requires a 5'-terminal phosphate for efficient ternary complex formation with tRNA and GTP. Aminoacyl-tRNA is complexed much stronger than deacylated tRNA (Schulman et al. 1974). The amino group itself is not required, as its replacement by a hydroxyl still allows some ternary complex formation (Fahnestock et al. 1972). By contrast, N-acetylphenylalanyl-tRNA or N-acetyl-lysyl-tRNA are very poor substrates for EF-Tu (Ravel et al. 1967; Johnson et al. 1977).

IF-2 recognizes the initiator tRNA in *E. coli* primarily by the fact that it is formylated. Experiments on complex formation between tRNAs and IF-2 show that deacylated tRNA and aminoacyl-tRNA bind weakly, whereas any N-blocked methionyl-tRNA tested binds strongly, as do several other N-blocked aminoacyl-tRNAs (Sundari et al. 1976). However, IF-dependent tRNA binding to ribosomes is much stronger with formylmethionyl-tRNA^{fMet} than with any other N-blocked aminoacyl-tRNAs (Rudland et al. 1969). Thus, the ribosome itself must have a specific preference for some of the unique features of initiator tRNA.

Peptidyl Transferase

The peptidyl-transferase activity of ribosomes appears to be quite unspecific. In addition to the obvious ability to use essentially an infinite variety of peptidyl-tRNAs as donors, and all 20 aminoacyl-tRNAs as acceptors, this ribosome activity will use compounds like lactyl-tRNA and phenyl-lactyl-tRNA in which a hydroxyl group is substituted for the amino group so that the ultimate product is a polyester rather than a polypeptide (Fahnestock and Rich 1971). Numerous tRNA fragments will serve as substrates, as will antibiotics like puromycin (Harris and Pestka 1977). However, tRNA with an extra C next to the 3' end is a poor substrate (Thang et al. 1974). Interestingly, ethanol and water can substitute for aminoacyl-tRNA, resulting in hydrolysis of peptidyl-tRNA or the synthesis of an ethylester (Caskey 1977).

The versatility of the peptidyl transferase is reasonable, considering that it must be able to handle three different types of aminoacyl-site substrates during protein synthesis: aminoacyl-tRNA and iminoacyl-tRNA (Prolyl) during elongation, and water during termination.

Translocation

EF-G appears to have a definite requirement for certain 3'-end tRNA structures (see Grunberg-Manago et al. 1978). These are summarized in Table 3. Peptidyl-tRNA in the peptidyl site or aminoacyl-tRNA in the aminoacyl site inhibits the binding of EF-G to the ribosome and also GTP hydrolysis. Translocation appears to require a peptidyl-tRNA in the aminoacyl site and deacylated tRNA in the peptidyl site. In other words, EF-G specifically recognizes the pretranslocation state.

Stringent Response

The synthesis of pppGpp by the stringent factor-ribosome complex requires a deacylated tRNA bound at the aminoacyl site in response to an mRNA codon. Apparently the peptidyl site can contain peptidyl-tRNA or deacylated tRNA. But the aminoacyl site requirements for stringent response are quite strict. tRNAs with aminoacylated, shortened, oxidized, oxidized and reduced, or phosphorylated 3' terminals are poor substrates. Only a lengthened 3' terminal can be tolerated (Chinali et al. 1978).

Fine Distinctions at the 3' End

In addition to all of the specificity described above, ribosomes and other components of the protein-synthesizing systems are sometimes quite particular about whether the amino acid is esterified to the 2'-OH or 3'-OH of the terminal A. The natural aminoacyl species can migrate between the two positions fairly easily. However, studies of 3' analogs of tRNA in which this migration was blocked reveal that peptidyl transfer may require a 3'-esterified amino acid even though ribosome binding does not require one. (For recent reviews of these complex and interesting studies, see Hecht [1977] and Sprinzl and Wagner [this volume].)

Table 3 Effects of tRNA on EF-G function

Ribosome configuration		
peptidyl site	aminoacyl site	EF-G action
Deacylated tRNA	peptidyl-tRNA	translocation, GTPase
Peptidyl-tRNA	—	binding and GTPase activity of EF-G inhibited
Deacylated tRNA	—	binding and GTPase but no translocation
Deacylated tRNA	aminoacyl-tRNA	binding and GTPase activity of EF-G inhibited
—	—	binding, GTPase

The ability of so many protein factors to make fine distinctions about the nature of substitutions at the 3' end of tRNA on the ribosome is striking. It is known that most or all of these factors have overlapping binding sites so that only one can be bound to the ribosome at a time. The simplest explanation is that each of the binding sites includes regions immediately adjacent to the 3' end of bound tRNA. This appears to adequately account for what is known about the structure of ribosome-tRNA complexes at the current level of resolution. However, no data compel this explanation and it remains possible that less-direct mechanisms apply.

STRUCTURE OF tRNA ON THE RIBOSOME

Some Evidence Suggests an Altered Structure

A great deal of effort has been devoted to the study of tRNA structure in the crystal and in solution. This is summarized elsewhere (Kim; Crothers; both this volume). Several basic conclusions have emerged. The crystal structure of yeast tRNA^{Phe} appears to be an excellent representation of the structure of this tRNA and probably all other tRNAs in aqueous solution at room temperature in the presence of Mg^{++} or other multivalent cations. Some structural changes may occur as salt and magnesium are varied or an amino acid or peptide is added, but there is little evidence thus far for major structural alterations at near-physiological conditions. However, the existence of many minor structural variations in the anticodon loops and other regions is indicated. The functional significance of any of these variations is unknown.

When subjected to more violent environmental perturbations, such as elevated temperature, magnesium removal, or low salt, tRNAs can undergo major structural changes (Crothers and Cole 1978). The energy required to drive these changes under physiological conditions would be large but, in principle, binding of a tRNA to a ribosome might involve large enough energies to promote such changes and still retain an acceptable binding affinity. On the other hand, the kinetic barriers to such major structural changes would be considerable, and the safest guess at present would have to be that, at most, minor changes in tRNA structure accompany ribosome binding. By this is meant minor thermodynamic changes. If tRNA contains any easily deformed hinge regions, the gross appearance of the bound molecule could be very different than in the crystal.

Reid (1977) has suggested that during aminoacylation tRNA may fold into a U-shaped structure bringing the anticodon loop and aminoacyl ends into close proximity. Such a structure has been suggested under some conditions in solution by inelastic light scattering (Olson et al. 1976) and indeed was an early speculative model for tRNA structure (Cantor et al. 1966). This kind of structural change is very easily examined by energy transfer between dyes at the anticodon loop and the aminoacyl end. Fluorescence studies of deacylated tRNA in solution (Beardsley and Cantor 1970; Yang and Soll 1974) and on the ribosome (B. Wells, unpubl.) make it very unlikely that a U-shaped structure contributes significantly. However, these studies do not exclude the possibility of U-shaped structures for aminoacyl- or peptidyl-tRNAs.

Whether or not major changes occur in tRNA conformation during normal protein synthesis, some species of tRNA can exist in more than one conformation capable of ribosome binding. For example, Kirillov and Odinzov (1978) have recently described two interconvertible conformations of *E. coli* phenylalanyl-tRNA^{Phe} with very different affinities for codon-dependent binding to 70S ribosomes. The high-affinity form has an association constant of 10^{11} per molecule, whereas the low-affinity form binds about 10^5-fold weaker. The activation energy for conversion from the low- to high-affinity form is 35 kcal per molecule. It will be of interest to learn more about the structural differences between these two conformations and their modes of binding to the ribosome. The binding differences are visible only with 70S particles and not with 30S particles, suggesting that the anticodon loop and stem regions are identical in the two structures.

A number of regions of tRNA are candidates for possible structure changes upon ribosome binding.

Anticodon

Numerous studies in solution suggest the possibility of other stable anticodon structures in addition to the 3′ stack seen in the crystal and probably in solution under conditions similar to those used for crystallization (Langlois et al. 1975). Some results support the existence of a 5′ stack (Eisinger and Spahr 1973; Yoon et al. 1975; Kan et al. 1975). Others indicated more complex stacking arrangements (Lee and Tinoco 1970). Multiple anticodon environments are clearly indicated by fluorescence studies (Beardsley et al. 1970; Rigler et al. 1977). What is the relevance of all of these observations for the structure of the anticodon on the ribosome? They indicate that latered structures are energetically feasible, but not that they actually occur.

On the ribosome, tRNA in both aminoacyl and peptidyl sites is in interaction with mRNA. Some information on the nature of mRNA-anticodon interactions is available by examining the fluorescence of the Y base, or dyes substituted in this position, when tRNA^{Phe} is bound to ribosomes in the presence and absence of poly(U). The results clearly indicate that the hypermodified base is affected by the presence of message in both the aminoacyl and peptidyl sites, but in very different ways (R. H. Fairclough et al., in prep.). In neither site does the ribosome itself cause much spectral perturbation compared with the tRNA free in solution. In neither site is the effect of message at all like what is seen when poly(U) interacts with tRNA in the absence of ribosomes. When these results are considered along with distances and angles measured between the anticodons, the following picture emerges. In the aminoacyl site in the presence of mRNA the anticodon loop structure is (as viewed by the hypermodified base) different from that of tRNA free in solution (R. H. Fairclough and C. R. Cantor, in prep.). In the peptidyl site in the absence of mRNA the anticodon structure of tRNA is apparently the same as seen for free tRNA in solution. However, its complex with message is different.

A number of interesting speculations have been put forth to explain how tRNAs in the aminoacyl and peptidyl sites simultaneously base pair with the message (Fuller and Hodgson 1967; Woese 1970; Sundaralingam et al. 1975; Lake 1977; Kim 1978). Almost none of these fit with the small amount of hard structural information currently available (R. H. Fairclough and C. R. Cantor, in prep.). However, minor variations on some of the models would clearly bring them close enough to observed properties of the hypermodified base to stay in contention. What is badly needed is more experimental data. This seems most likely to come from specific chemical modification and cross-linking.

TψC-D Loop Tertiary Contacts

In crystals of yeast tRNA^{Phe} extensive tertiary interactions occur between bases in the TψC and D loops. These interactions appear to persist in solution, as the residues involved are inaccessible to a variety of chemical and physical probes. Some evidence exists that binding of a message to the anticodon in the absence of ribosomes triggers a conformational charge in tRNA leading to exposure of residues in the TψC loop. This includes changes in chemical reactivity (Wagner and Garrett 1978), oligonucleotide binding (Schwarz and Gassen 1977), and fluorescence (Robertson et al. 1977). However, nuclear magnetic resonance studies (Geerdes et al. 1978) and other oligonucleotide-binding studies (J. Liesch and O. C. Uhlenbeck, in prep.) do not show such a change. (For further comments, see Möller et al. [this volume].)

Additional studies support the idea that such a conformational change may occur on the ribosome. Oligonucleotide binding to a tRNA·30S·mRNA complex suggests that the TψC loop is accessible in this complex (Schwar et al. 1976). Several results discussed earlier suggest that codon interactions stimulate tRNA binding to isolated 50S subunits. Much indirect evidence exists that the GTψC sequence interacts with 5S rRNA on the 50S ribosome (Erdmann 1976). This interaction is suggested to occur in the aminoacyl site, and the D-loop residues that pair with the

TψC loop in solution may also be involved in direct ribosome interactions (Sprinzl et al. 1976).

Oligonucleotides that compete for tRNA-binding-dependent, aminoacyl-site functions do not compete for corresponding peptidyl-site functions. This suggests that the D and TψC loops of tRNA in the peptidyl site are not involved in direct ribosome interactions. However, tritium-exchange studies of peptidyl-site-bound, deacylated tRNA^{Phe} indicate clearly that the D and TψC loops are more accessible when ribosome bound (without an occupied aminoacyl site) than when free in solution (N. Farber, unpubl.). In this regard, it is interesting that addition of TψCG sequence actually stimulated the binding of tRNA to the peptidyl site (Sprinzl et al. 1976).

The simplest picture consistent with all of these findings is that in both peptidyl and aminoacyl sites, some of the tertiary contacts seen between D and TψC loops in tRNA crystals are disrupted. The disruption probably does not occur in the EF-Tu·GTP ternary complex (Crothers and Cole 1978) but may occur either when the complex is bound to ribosomes or when EF-Tu·GDP leaves. On the ribosome, the exposed residues are involved in interactions important for function, but the exact nature of these interactions (whether to 5S rRNA, 23S rRNA, ribosomal proteins, or even the other bound tRNA, which, after all, has the complementary exposed sequences) remains to be proved. Additional chemical-modification, cross-linking, and tritium-exchange studies of ribosome-bound tRNA should be productive measures to resolve these questions. It is interesting to speculate that the free energy stored in ribosome-bound tRNA by the disrupted tertiary contacts may eventually assist the removal of tRNA from the ribosome.

Miniloop (U8)

tRNAs cross-linked between s^{4}U8 and C13 appear to be fully functional in protein synthesis (Yaniv et al. 1971). This means that any structural changes that occur on the ribosome must not require a separation of these two residues. This constraint still leaves open many possibilities for structural changes near s^{4}U8. Indeed, several fluorescent derivatives of s^{4}U8 in several tRNA species consistently show a pattern of large spectral changes upon ribosome binding (A. Johnson, unpubl.). Even though fluorescence changes per se are not proof of a structural change, their occurrence in such a large number of different samples is most easily explained in this way.

Functional Aspects of tRNA Structural Changes

The results described above indicate that tRNA structural changes may play a role in tRNA function on the ribosome (Crothers and Cole 1978).

However, it remains difficult to pin down the steps where tRNA structural changes dominate. Furthermore, there is a real possibility that tRNA binding induces structural changes in the ribosome. For example, several fluorescent, ribosomal protein derivatives are altered by tRNA binding (Perrin and Pochon 1975; C. C. Lee and B. Wells, unpubl.). Really very little is known about the details of tRNA function on the ribosome. A sketch of some of the items that remain to be explained is given below.

Codon Recognition

It was recognized first by Crick (1966) that the rules for codon-anticodon interaction on the ribosome may be different from base pairing in a long, perfect, double helix. Whether this is directly due to tRNA structure or whether it reflects the interaction of ribosomal components is still not clear. Recently there have been suggestions that suppressor tRNAs may respond to only two bases (Buckingham and Kurland 1977), and even that two-base coding may be more general (Lagerkvist 1978).

Little quantitative data exist on the equilibrium thermodynamics or kinetics of tRNA-mRNA interactions on the ribosome. Although much is known about these interactions in solution, it may be very hazardous to extrapolate this knowledge to the ribosome in view of the possible structural changes discussed earlier. What information is available suggests that under some conditions (e.g., uncharged tRNA) not much stabilization of the cognate tRNA binding actually results from interactions with the correct message, although noncognate is weaker than cognate binding in the presence of message (Table 4). However, under other conditions, especially aminoacyl-tRNA in the aminoacyl site, there is a substantial effect of mRNA on the binding constant. The true function of the message is probably to exclude the binding of the wrong tRNA as much as to promote the binding of the right one.

The thermodynamic data available for codon-anticodon interactions in solution do not readily explain the low error frequencies observed (Loftfield and Vanderjagt 1972; Edelmann and Gallant 1977) for in vivo protein synthesis. Nor do they explain how various antibiotics and mutations increase the observed error frequencies, although such effects are potentially caused by rather simple mechanisms (Ninio 1974). Several complex and interesting kinetic (Hopfield 1974; Ninio 1975) and equilibrium (Kurland et al. 1975) binding schemes have been proposed to improve the degree of specific codon recognition. All of these, in different ways, use the difference between correct and incorrect tRNA binding in more than one step to multiply the extent of discrimination. The small amount of available data is consistent with some kind of proofreading but is insufficient to establish a mechanism (Thompson and Stone 1977).

Here it is worth pointing out that what is already known about the

Table 4 Equilibrium binding of tRNA to 70S *E. coli* ribosomes

		Binding constant (per molecule)		
Sample	Poly(U)	peptidyl site	aminoacyl site	Reference
tRNA^{bulk}	+	1.4×10^7		S. V. Kirillov et al. (in prep.)[a]
tRNA^{bulk}	−	1.4×10^7		S. V. Kirillov et al. (in prep.)[a]
tRNA^{Phe}	+	3.0×10^8	1×10^7	M. A. Peters and M. Yarus (unpubl.)[b]
tRNA^{Phe}	−	1.4×10^7	$< 10^6$	M. A. Peters and M. Yarus (unpubl.)[b]
tRNA^{Phec}	+	1.0×10^9	3×10^7	R. H. Fairclough et al. (in prep.)[d]
tRNA^{Phec}	−	1.0×10^9	3×10^7	R. H. Fairclough et al. (in prep.)[d]
ϕ-lac-tRNA^{Phee}	+	9.0×10^7	3×10^7	M. A. Peters and M. Yarus (unpubl.)[b]
ϕ-lac-tRNA^{Phec}	−	$\sim 2.0 \times 10^7$	$\leqslant 10^6$	M. A. Peters and M. Yarus (unpubl.)[b]
Phenylalanyl-tRNA^{Phe} HA[f]	+		$> 1 \times 10^{11}$	S. V. Kirillov et al. (in prep.)[a]
Phenylalanyl-tRNA^{Phe} LA[g]	+		5×10^6	S. V. Kirillov et al. (in prep.)[a]
Phenylalanyl-tRNA^{Phe} HA	−		3×10^8	S. V. Kirillov et al. (in prep.)[a]
Phenylalanyl-tRNA^{Phe} LA	−		2×10^6	S. V. Kirillov et al. (in prep.)[a]
tRNA^{Ile}	+	8.0×10^7	8×10^5	M. A. Peters and M. Yarus (unpubl.)[b]
tRNA^{Val}	+	4.0×10^5	1×10^6	M. A. Peters and M. Yarus (unpubl.)[b]
tRNA^{fMet}	+	1.0×10^6	5×10^6	M. A. Peters and M. Yarus (unpubl.)[b]
tRNA^{Arg}	+	2.0×10^5	6×10^5	M. A. Peters and M. Yarus (unpubl.)[b]
tRNA^{Glu}	+	8.0×10^5	1×10^6	M. A. Peters and M. Yarus (unpubl.)[b]

tRNAs are from *E. coli*.
[a] At 0°C in 20 mM Mg^{++}.
[b] At 7°C in 25 mM Mg^{++}.
[c] tRNA^{Phe} from yeast.
[d] At 30°C in 25 mM Mg^{++}.
[e] ϕ-lac is phenyl-lactyl.
[f] HA is high-affinity conformation.
[g] LA is low-affinity conformation.

process of protein synthesis offers several untested possible stages for proofreading. We do not really know at which, if any, stage proofreading occurs. For example, does EF-Tu discriminate against a ribosome with an incorrectly base-paired peptidyl-tRNA in the peptidyl site? Or does EF-G discriminate against a ribosome with an incorrectly base-paired peptidyl-tRNA in the aminoacyl site? Either of these selections could lead to aborted synthesis if a protein chain picked up an error. Can the peptidyl transferase reaction be reversed? Can translocation be reversed? No evidence exists for these possibilities, but no decisive evidence seems to exist against them, either. If any of these processes are coupled to conformational changes in tRNA, it is easy to see how mispairing might be felt by regions of tRNA far from the anticodon.

FUTURE DIRECTIONS

If further progress is to be made on understanding tRNA function on the ribosome, improved kinetic and thermodynamic data on tRNA-ribosome interactions are vitally needed. Such information has been difficult to obtain because until recently most ribosome preparations contained too large a percentage of inactive particles. However, there are no serious obstacles in the way of obtaining such data now (see, for example, Wintermeyer et al., this volume). A host of fluorescent tRNA, message, and ribosome derivatives are available for direct physical studies. Even filter-binding assays are quite suitable for binding studies, when used with proper precautions. For example, we have recently shown that the conveniently slow rate of exchange of peptidyl-site-bound, deacylated tRNA^{Phe} is proportional to the fraction of the time that the aminoacyl site is unoccupied (N. Farber et al., unpubl.). This will allow an accurate determination of the binding constant of various tRNAs to the aminoacyl site by simple filter-binding measurements. Because the peptidyl-site component will be examined, the results will not be perturbed by variation in the kinetics of interaction at the aminoacyl site.

The role of tRNA or the role of tRNA structure in regulating protein synthesis or processes coupled to protein synthesis remains to be elucidated. However, the indications that the kinetics of ribosome movement on nascent mRNA may effect transcription termination are indeed fascinating (Lee and Yanofsky 1977). They suggest numerous possible regulatory functions for tRNA species through the kinetics of their interactions with ribosomes.

ACKNOWLEDGMENTS

I wish to thank R. Amils, B. Clark, P. Cole, R. Fairclough, N. Farber, J. Hildebrandt, A. Johnson, C. C. Lee, E. Matthews, P. Schimmel,

P. Thammana, and B. Wells for helpful discussions and permission to cite unpublished results. I am also grateful to M. A. Petters, M. Yarus, and S. Bressler for making available much useful data on tRNA binding. Studies in this laboratory were supported by grants from the U.S. Public Health Service and the National Science Foundation.

REFERENCES

Acharya, A. S., P. B. Moore, and F. M. Richards. 1973. Crosslinking of elongation factor EF-G to the 50S ribosomal subunit of *Escherichia coli*. *Biochemistry* **12**:3108.

Beardsley, K. and C. R. Cantor. 1970. Studies of transfer RNA tertiary structure by singlet-singlet energy transfer. *Proc. Natl. Acad. Sci.* **65**:39.

Beardsley, K., T. Tao, and C. R. Cantor. 1970. Studies on the conformation of the anticodon loop of phenylalanine transfer ribonucleic acid. Effect of environment on the fluorescence of the Y-base. *Biochemistry* **9**:3524.

Block, R. and W. Haseltine. 1974. In vitro synthesis of ppGpp and pppGpp. Cellular regulation of guanosine tetraphosphate and guanosine pentaphosphate. In *Ribosomes* (ed. M. Nomura et al.), p. 747. Cold Spring Harbor Laboraty, Cold Spring Harbor, New York.

Brimacombe, R., G. Stöffler, and H. G. Wittmann. 1978. Ribosome structure. *Annu. Rev. Biochem.* **47**:217.

Brot, N. 1977. Translocation. In *Molecular mechanisms of protein biosynthesis* (ed. H. Weissbach and S. Pestka), p. 375. Academic Press, New York.

Buckingham, R. H. and C. G. Kurland. 1977. Codon specificity of UGA suppressor tRNA^{Trp} from *Escherichia coli*. *Proc. Natl. Acad. Sci.* **74**:5496.

Cantor, C. R., S. R. Jaskunas, and I. Tinoco, Jr. 1966. Optical properties of ribonucleic acids predicted from oligomers. *J. Mol. Biol.* **20**:39.

Cashel, M. and J. Gallant. 1974. Cellular regulation of guanosine tetraphosphate and guanosine pentaphosphate. In *Ribosomes* (ed. M. Nomura et al.), p. 733. Cold Spring Harbor Laboratory, Cold Spring Harbor, New York.

Caskey, T. 1977. Peptide chain termination. In *Molecular mechanisms of protein biosynthesis* (ed. H. Weissbach and S. Pestka), p. 443. Academic Press, New York.

Chinali, G., R. Liov, and J. Ofengand. 1978. Role of the aminoacyl end of transfer RNA in the allosteric control of guanosine pentaphosphate synthesis by the stringent factor ribosome complex of *Escherichia coli*. *Biochemistry* **17**:2761.

Cooperman, B. S. 1977. Affinity labeling studies on *Escherichia coli* ribosomes. In *Bioorganic chemistry IV* (ed. E. Van Tamelan), p. 81. Academic Press, New York.

Crick, F. H. C. 1966. Codon-anticodon pairing: The wobble hypothesis. *J. Mol. Biol.* **19**:548.

Crothers, D. M. and P. E. Cole. 1978. Conformational changes of tRNA. In *Transfer RNA* (ed. S. Altman), p. 196. MIT Press, Cambridge, Massachusetts.

Dahlberg, J. E., C. Kintner, and E. Lund. 1978. Specific binding of tRNA$_f^{Met}$ to 23S rRNA of *Escherichia coli*. *Proc. Natl. Acad. Sci.* **75**:1071.

de Groot, N., A. Panet, and Y. Lapidot. 1971. The binding of purified Phe-tRNA and peptidyl tRNA^{Phe} to *Escherichia coli* ribosomes. *Eur. J. Biochem.* **23:**523.

Edelmann, P. and J. Gallant. 1977. Mistranslation in *Escherichia coli*. *Cell* **10:**131.

Eisinger, J. and P. F. Spahr. 1973. Binding of complementary pentanucleotides to the anticodon loop of transfer RNA. *J. Mol. Biol.* **73:**131.

Erdmann, V. A. 1976. Structure and function of 5S and 5.8S RNA. *Prog. Nucleic Acid Res.* **18:**45.

Fabian, U. 1976. Identification of proteins located in the neighborhood of the binding site for the elongation factor EF-Tu on *Escherichia coli* ribosomes. *FEBS Lett.* **71:**256.

Fahnestock, S. and A. Rich. 1971. Ribosome-catalyzed polyester formation. *Science* **173:**340.

Fahnestock, S., H. Weissbach, and A. Rich. 1972. Formation of a ternary complex of phenyllactyl-tRNA with transfer factor Tu and GTP. *Biochim. Biophys. Acta* **269:**62.

Fuller, W. and A. Hodgson. 1967. Conformation of the anticodon loop in tRNA. *Nature* **215:**817.

Geerdes, H. A. M., J. H. Van Boom, and C. W. Hilbers. 1978. Codon-anticodon interaction in yeast tRNA^{Phe}. *FEBS Lett.* **88:**27.

Grajevskaja, R. A., E. M. Saminski, and S. E. Bresler. 1972. Interaction of transfer RNA with ribosomes in the absence of messenger. *Biochem. Biophys. Res. Commun.* **46:**1106.

Grunberg-Manago, M., R. J. Buckingham, B. S. Cooperman, and J. W. B. Hershey. 1978. Structure and function of the translation machinery. *Symp. Soc. Gen. Microbiol.* **28:**27.

Gualerzi, C., G. Risuleo, and C. L. Pon. 1977. Initial rate kinetic analysis of the mechanism of initiation complex formation and the role of initiation factor IF-3. *Biochemistry* **16:**1684.

Haenni, A. and J. Lucas-Lenard. 1968. Stepwise synthesis of a tripeptide. *Proc. Natl. Acad. Sci.* **61:**1363.

Harris, R. J. and S. Pestka. 1977. Peptide bond formation. In *Molecular mechanisms of protein biosynthesis* (ed. H. Weissbach and S. Pestka), p. 413. Academic Press, New York.

Hauptmann, R., A. P. Czernilofsky, H. O. Voorma, G. Stöffler, and E. Kuechler. 1974. Identification of a protein at the ribosomal donor site. *Biochem. Biophys. Res. Commun.* **56:**331.

Haselkorn, R. and L. B. Rothman-Denes. 1973. Protein synthesis. *Annu. Rev. Biochem.* **42:**397.

Hecht, S. M. 1977. Participation of isomeric tRNAs in the partial reactions of protein biosynthesis. *Tetrahedron* **33:**1671.

Hirsch, D. and L. Gold. 1971. Translation of the UGA triplet in vitro by tryptophan transfer RNAs. *J. Mol. Biol.* **58:**459.

Hopfield, J. J. 1974. Kinetic proofreading. A new mechanism for reducing errors in biosynthetic processes requiring high specificity. *Proc. Natl. Acad. Sci.* **71:**4135.

Johnson, A. E. and C. R. Cantor. 1977. Affinity labeling of multicomponent systems. *Methods Enzymol.* **46:**180.

Johnson, A. E., R. H. Fairclough, and C. R. Cantor. 1977. Some approaches for the study of ribosome-tRNA interactions. In *Nucleic acid-protein recognition* (ed. H. J. Vogel), p. 469. Academic Press, New York.

Johnson, A. E., D. L. Miller, and C. R. Cantor. 1978. A functional covalent complex of elongation factor Tu and an analog of lysyl-tRNA. *Proc. Natl. Acad. Sci.* **79**:3078.

Jonak, J. and I. Rychlik. 1970. Role of messenger RNA in binding of peptidyl transfer RNA to 30S and 50S ribosomal subunits. *Biochim. Biophys. Acta* **199**:421.

Kan, L.-S., P. O. P. Ts'o, F. von der Haar, M. Sprinzl, and F. Cramer. 1975. Proton magnetic resonance studies on the conformation of the hexanucleotide, G$_m$pApApYpApψp, and related fragments from the anticodon loop of baker's yeast phenylalanine transfer ribonucleic acid. *Biochemistry* **14**:3278.

Kim, S.-H. 1978. Crystal structure of yeast tRNA^{Phe}. Its correlation to the solution, structure and functional implications. In *Transfer RNA* (ed. S. Altman), p. 248. MIT Press, Cambridge, Massachusetts.

Kirillov, S. V. and V. B. Odinzov. 1978. The interconversion of conformers of phenylalanyl tRNA with different affinity to 70S ribosomes of *Escherichia coli*. *Nucleic Acids Res.* **5**:1501.

Kuechler, E., K. Bauer, and A. Rich. 1972. Protein synthesis with ribonuclease digested ribosomes. *Biochim. Biophys. Acta* **277**:615.

Kuriki, Y. and A. Kaji. 1967. Specific binding of sRNA to ribosomes during polypeptide synthesis and the nature of peptidyl sRNA. *J. Mol. Biol.* **25**:407.

Kurland, C. G. 1977. Structure and function of the bacterial ribosome. *Annu. Rev. Biochem.* **46**:173.

Kurland, C. G., R. Rigler, M. Ehrenberg, and C. Blomberg. 1975. Allosteric mechanism for codon-dependent tRNA selection on ribosomes. *Proc. Natl. Acad. Sci.* **72**:4245.

Lagerkvist, U. 1978. Two out of three—An alternative method of codon reading. *Proc. Natl. Acad. Sci.* **75**:1759.

Lake, J. A. 1977. Aminoacyl-tRNA binding at the recognition site is the first step of the elongation cycle of protein synthesis. *Proc. Natl. Acad. Sci.* **74**:1903.

Langlois, R., S.-H. Kim, and C. R. Cantor. 1975. A comparison of the fluorescence of the Y base of yeast tRNA^{Phe} in solution and in crystals. *Biochemistry* **14**:2554.

Langlois, R., C. R. Cantor, R. Vince, and S. Pestka. 1977. Interaction between the erythromycin and chloramphenicol binding sites on the *Escherichia coli* ribosome. *Biochemistry* **16**:2349.

Leder, P., A. Bernardi, D. Livingston, B. Lund, D. Roufa, and L. Skogerson. 1970. Protein biosynthesis: Studies using synthetic and viral mRNAs. *Cold Spring Harbor Symp. Quant. Biol.* **34**:411.

Lee, C.-H. and I. Tinoco, Jr. 1977. Studies of the conformation of modified dinucleoside phosphates containing 1,N^6-ethenoadenosine and 2'-O-methylcytidine by 360-MHz ^{1}H nuclear magnetic resonance spectroscopy. Investigation of the solution conformations of dinucleoside phosphates. *Biochemistry* **16**:5403.

Lee, F. and C. Yanofsky. 1977. Transcription termination at Trp operon attenuator of *Escherichia coli* and *Salmonella typhimurium*. RNA secondary structure and regulation of termination. *Proc. Natl. Acad. Sci.* **74**:4365.

Levin, J. G. 1970. Codon-specific binding of deacylated transfer ribonucleic acid to ribosomes. *J. Biol. Chem.* **245:** 3195.
Loftfield, R. B. and D. Vanderjagt. 1972. Frequency of errors in protein biosynthesis. *Biochem. J.* **128:** 1353.
Lucas-Lenard, J. and A. L. Haenni. 1969. Release of tRNA during peptide chain elongation. *Proc. Natl. Acad. Sci.* **63:** 93.
Miller, D. L. and H. Weissbach. 1977. Factors involved in the transfer of aminoacyl tRNA to the ribosome. In *Molecular mechanisms of protein biosynthesis* (ed. H. Weissbach and S. Pestka), p. 324. Academic Press, New York.
Modolell, J. and D. Vazquez. 1973. Inhibition by aminoacyl tRNA of elongation factor G-dependent binding of guanosine nucleotide to ribosomes. *J. Biol. Chem.* **245:** 458.
Modolell, J., B. Cabrer, and D. Vazquez. 1973. Interaction of elongation factor G with *N*-acetylphenylalanyl transfer RNA-ribosomes complexes. *Proc. Natl. Acad. Sci.* **70:** 3561.
Ninio, J. 1974. A semiquantitative treatment of missense and nonsense suppression in the strA and ram ribosomal mutations of *Escherichia coli*. Evaluation of molecular parameters of translation in vivo. *J. Mol. Biol.* **84:** 297.
———. 1975. Kinetic amplification of enzyme discrimination. *Biochimie* **57:** 587.
Nomura, M., A. Tissieres, and P. Lengyel, eds. 1974. *Ribosomes.* Cold Spring Harbor Laboratory, Cold Spring Harbor, New York.
Ofengand, J., J. Bierbaum, J. Horowitz, C. N. Ou, and M. Ishaq. 1974. Protein synthetic ability of *Escherichia coli* valine transfer RNA with pseudouridine, ribothymidine, and other uridine-derived residues replaced by 5-florouridine. *J. Mol. Biol.* **88:** 313.
Olson, T., M. J. Fournier, K. H. Langley, and N. C. Ford. 1976. Detection of a major conformational change in transfer ribonucleic acid by laser light scattering. *J. Mol. Biol.* **102:** 103.
Pellegrini, M. and C. R. Cantor. 1977. Affinity labeling of ribosomes. In *Molecular mechanisms of protein biosynthesis* (ed. H. Weissbach and S. Pestka), p. 203. Academic Press, New York.
Perrin, M. and F. Pochon. 1975. Fluorescent labeling of *Escherichia coli* ribosomal sulfhydryl groups. *Eur. J. Biochem.* **57:** 319.
Peters, M. A. 1977. "Ribosome binding of tRNA." Ph.D. thesis, University of Colorado, Boulder.
Quigley, G. J., M. M. Teeter, and A. Rich. 1978. Structural analysis of spermine and magnesium binding to yeast phenylalanine transfer RNA. *Proc. Natl. Acad. Sci.* **75:** 64.
Ravel, J. M., R. C. Shorey, and W. Shive. 1967. Evidence for a guanine nucleotide aminoacyl tRNA complex as an intermediate in the enzymatic transfer of tRNA to ribosomes. *Biochem. Biophys. Res. Commun.* **29:** 68.
Reid, B. R. 1977. Synthetase-tRNA recognition. In *Nucleic acid-protein recognition* (ed. H. J. Vogel), p. 375. Academic Press, New York.
Revel, M. 1977. Initiation of messenger RNA translation into protein and some aspects of its regulation. In *Molecular mechanisms of protein biosynthesis* (ed. H. Weissbach and S. Pestka), p. 248. Academic Press, New York.
Rich, A. 1974. How transfer RNA may move inside the ribosome. In *Ribosomes* (ed. M. Nomura et al.), p. 871. Cold Spring Harbor Laboratory, Cold Spring Harbor, New York.

Riddle, D. L. and J. Carbon. 1973. Frameshift suppression: A nucleotide addition to the anticodon of a glycine transfer RNA. *Nat. New Biol.* **242:** 230.

Rigler, R., M. Ehrenberg, and W. Wintermeyer. 1977. Structural dynamics of tRNA—A fluorescence relaxation study of tRNA$^{Phe}_{yeast}$. *Mol. Biol. Biochem. Biophys.* **24:** 219.

Robertson, J. M., M. Kahn, W. Wintermeyer, and H. G. Zachau. 1977. Interactions of yeast tRNA^{Phe} with ribosomes from yeast and *Escherichia coli. Eur. J. Biochem.* **72:** 117.

Rudland, P. S., W. A. Whybrow, K. A. Marcker, and B. F. C. Clark. 1969. Recognition of bacterial tRNA by initiation factors. *Nature* **222:** 750.

Saltzman, L. and D. Apirion. 1976. Binding of erythromycin to the 50S ribosomal subunit is affected by alterations in the 30S ribosomal subunit. *Mol. Gen. Genet.* **143:** 301.

San José, C., C. G. Kurland, and G. Stöffler. 1976. The protein neighborhood of ribosome-bound elongation factor Tu. *FEBS Lett.* **71:** 133.

Schulman, L. H., H. Pelka, and R. M. Sundari. 1974. Structural requirements for recognition of *Escherichia coli* initiation and non-initiation transfer ribonucleic acids by bacterial T factor. *J. Biol. Chem.* **249:** 7102.

Schwarz, U. and H. G. Gassen. 1977. Codon-dependent rearrangement of the tertiary structure of tRNA^{Phe} from yeast. *FEBS Lett.* **78:** 267.

Schwarz, U., H. M. Manzel, and H. G. Gassen. 1976. Codon-dependent rearrangement of the three-dimensional structure of phenylalanine tRNA, exposing the T-ψ-C-G sequence for binding to the 50S ribosomal subunit. *Biochemistry* **15:** 2482.

Shorey, R. L., J. M. Ravle, and W. Shive. 1971. The effect of guanylyl-5′-methylene diphosphonate on binding of aminoacyl transfer ribonucleic acid to ribosomes. *Arch. Biochem. Biophys.* **146:** 110.

Shulman, R. G., C. W. Hilbers, and D. L. Miller. 1974. Nuclear magnetic resonance studies of protein-tRNA interactions. I. The elongation factor Tu·GTP-aminoacyl tRNA complex. *J. Mol. Biol.* **90:** 601.

Skoultchi, A., Y. Ono, J. Waterson, and P. Lengyel. 1970. Peptide chain elongation: Indication for the binding of an amino acid polymerization factor, guanosine 5′-triphosphate·aminoacyl transfer ribonucleic acid complex to the messenger-ribosome complex. *Biochemistry* **9:** 508.

Spirin, A. S. 1970. A model of the functioning ribosome: Locking and unlocking of the ribosome subparticles. *Cold Spring Harbor Symp. Quant. Biol.* **34:** 197.

Sprinzl, M., F. Gruter, and D. H. Gauss. 1978. Collection of published tRNA sequences. *Nucleic Acids Res.* **5:** 215.

Sprinzl, M., T. Wagner, S. Lorenz, and V. A. Erdmann. 1976. Regions of tRNA important for binding to the ribosomal A site. *Biochemistry* **15:** 3031.

Stöffler, G. and H. G. Wittmann. 1977. Primary structure and three-dimensional arrangement of proteins within the *Escherichia coli* ribosome. In *Molecular mechanisms of protein biosynthesis* (ed. H. Weissbach and S. Pestka), p. 117. Academic Press, New York.

Sundaralingam, M., T. Brennan, N. Yathindra, and T. Ichikawa. 1975. Stereochemistry of messenger RNA (codon)-transfer RNA (anticodon) interaction in the ribosome during peptide bond formation. In *Structure and conformation of nucleic acids and protein-nucleic acid interactions* (ed. M. Sundaralingam and S. T. Rao), p. 101. University Park Press, Baltimore, Maryland.

Sundari, R. M., E. A. Stringer, L. H. Schulman, and U. Maitra. 1976. Interaction of bacterial initiation factor 2 with initiator tRNA. *J. Biol. Chem.* **251**:3338.

Swan, D., G. Sander, E. Bermek, W. Kramer, T. Kreuzer, C. Arglebe, R. Zollner, K. Eckert, and H. Matthei. 1970. On the mechanism of coded binding of aminoacyl-tRNA and ribosomes: Number and properties of sites. *Cold Spring Harbor Symp. Quant. Biol.* **34**:179.

Thang, M. N., L. Dondon, and B. Rether. 1974. Structural change of the phe-tRNA$^{Phe}_{(CCCA)}$ and the effect on the rate of peptide formation. *FEBS Lett.* **40**:67.

Thompson, R. C. and P. J. Stone. 1977. Proofreading of the codon-anticodon interaction on ribosomes. *Proc. Natl. Acad. Sci.* **74**:198.

Ustav, M., M. Saarma, A. Lind, and R. Villems. 1978. The domain for transfer ribonucleic acid binding to the *Escherichia coli* ribosomes. *FEBS Lett.* **87**:315.

Ustav, M., R. Villems, M. Saarma, and A. Lind. 1977. The interaction of transfer ribonucleic acid with 50S ribosomal subunit proteins. *FEBS Lett.* **83**:353.

Wagner, R. and R. A. Garrett. 1978. Chemical evidence for a codon-induced change of tRNA conformation. *FEBS Lett.* **85**:291.

Weissbach, H. and S. Pestka, eds. 1977. *Molecular mechanisms of protein biosynthesis.* Academic Press, New York.

Woese, C. 1970. Molecular mechanics of translation: A reciprocating ratchet mechanism. *Nature* **226**:817.

Yaniv, M., A. Chestier, F. Gros, and A. Favre. 1971. Biological activity of irradiated tRNA^{Val} containing a 4-thiouridine-cytosine dimer. *J. Mol. Biol.* **58**:381.

Yang, C.-H. and D. Söll. 1974. Studies of transfer RNA tertiary structure by singlet-singlet energy transfer. *Proc. Natl. Acad. Sci.* **71**:2838.

Yoon, K., D. H. Turner, and I. Tinoco, Jr. 1975. The kinetics of codon-anticodon interaction in yeast phenylalanine transfer RNA. *J. Mol. Biol.* **99**:507.

Ribosome Structure and tRNA Binding Sites

James A. Lake
Molecular Biology Institute and Department of Biology
University of California, Los Angeles
Los Angeles, California 90024

During the last decade, knowledge of ribosome structure has evolved from a 100-Å resolution structure obtained by three-dimensional reconstruction (Lake and Slayter 1970, 1972; Unwin 1977) to current models for the *Escherichia coli* 70S ribosome at 30- to 40-Å resolution (Lake 1976; Tischendorf et al. 1976) derived by immunoelectron microscopy (for a review, see Lake 1978). This progress has resulted, in large part, from the convergence of different approaches toward studying ribosome structure and function, including biochemical, genetic, immunological, neutron-diffraction, and electron microscopic techniques (for a review, see Nomura et al. 1974). This paper reviews the relationship between ribosome structure and function with special reference to tRNA binding and the molecular events of protein synthesis.

ANTIBODY MAPPING STUDIES

Small Subunits, Large Subunits, and Monomeric Ribosomes

The structures of the small and large subunits of the *E. coli* ribosome are illustrated in Figure 1. The small subunit (Fig. 1a,b) is divided into two unequal parts by an indentation and a region of accumulated negative stain (Lake et al. 1974b). The parts are referred to as the upper one-third and the lower two-thirds. A region of the subunit, called the platform, extends from the lower two-thirds of the small subunit and forms a cleft between it and the upper one-third (Lake and Kahan 1976). Recent tilting experiments have determined the absolute hand of the 30S subunit (the correct enantiomorph is shown in Fig. 1) and have confirmed its asymmetric properties (Leonard and Lake 1979).

The large subunit, like the small subunit, is asymmetric (Fig. 1c,d) (Lake et al. 1974a; Lake 1976). It consists of a central protuberance (Fig. 1c, frame 1) and protrusions inclined approximately 50° from the central protuberance. One of these, the L7/L12 stalk, is at the right. In an orientation approximately orthogonal to this (shown in right frames of Fig. 1c,d), the large subunit is characterized by a notch on the upper surface.

In the monomeric ribosome the small subunit is positioned asymmetrically on the large subunit, as shown in Figure 2. The platform of the

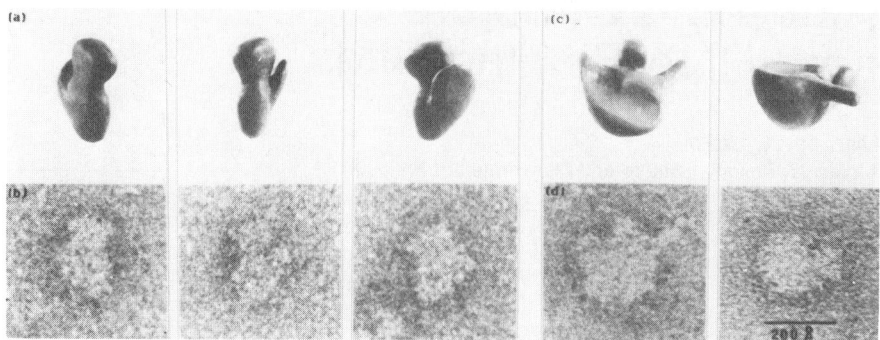

Figure 1 A comparison of three-dimensional models in different orientations (*top*) with electron micrographs (*bottom*). (*a*, *b*) 30S subunits; (*c*, *d*) 50S subunits. (Adapted from Lake 1976.)

small subunit is in contact with the large subunit (Lake 1976) so that the one-third–two-thirds partition of the small subunit is approximately aligned with the notch of the large subunit.

Using these three-dimensional structures, preliminary locations of most ribosomal proteins have been mapped in collaboration with L. Kahan (Department of Physiological Chemistry, University of Wisconsin) and W. Strycharz and M. Nomura (Institute for Enzyme Research, University of Wisconsin) and their locations have raised puzzling questions about protein synthesis. In this paper, most proteins that are described have been both implicated in important functional events and well documented by our immune mapping experiments.

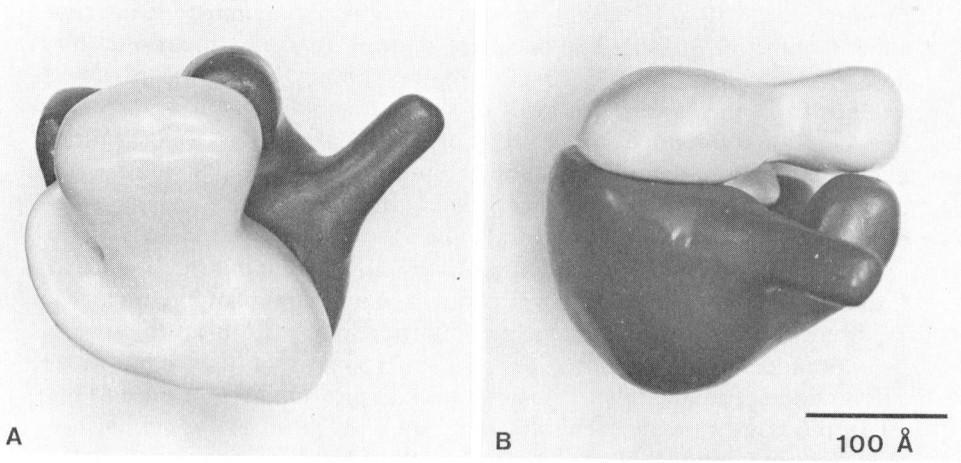

Figure 2 Model of the 70S *E. coli* ribosome showing relative orientations of the subunits. The 30S subunit is light; the 50S subunit is dark. (Adapted from Lake 1976.)

Exterior, Small-subunit Proteins

The orientation of the small subunit on the large subunit makes an operational distinction between 30S protein locations on the exterior (i.e., the side away from the interface) of the subunit and those at the interface. In the view of the subunit shown in Figure 2B, the 30S subunits are characterized by a platform and a cleft. In this projection, exterior proteins map on the side opposite the platform. They are shown schematically on the surface of the 70S ribosome in Figure 3.

It was unexpected to find so many proteins clustered on the 30S exterior surface, as most events occurring during elongation were thought to happen at the subunit interface, and it was even more unexpected to find that all five of these proteins have been implicated in tRNA binding and recognition interactions.

Omission of small-subunit proteins S8, S10, or S14 eliminates tRNA binding, and mutation of S5 or S12 alters tRNA binding.

The importance of proteins S8, S10, and S14 was suggested in single-component reconstitution experiments (Nomura et al. 1970) when an almost absolute requirement for them was found in poly(U)-dependent polyphenylalanine synthesis. When further investigated by poly(U)-directed phenylalanyl-tRNA binding, these three proteins were found to be almost absolutely required for tRNA binding. (Since subunits with S8 omitted sedimented at 20–25S, the reduced tRNA binding observed in the absence of S8 probably represents an indirect effect of structural disorder caused by lack of S8 binding.) Omission of proteins S10 and especially S14 caused only minor changes in subunit sedimentation and, hence, may directly participate in tRNA binding.

The two remaining exterior proteins, S5 and S12, have been implicated in tRNA recognition by misreading studies. Streptomycin and its effects on misreading have been characterized extensively (for a review, see Gorini 1974). Among cultures surviving treatment with streptomycin, two classes of mutants are found. These are streptomycin-resistant and strep-

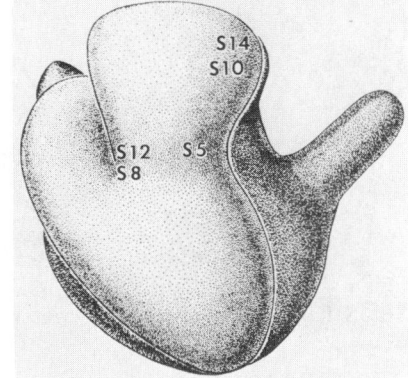

Figure 3 Schematic diagram of the locations of proteins mapping on the exterior surface of the 30S subunit. The entire 70S ribosome is illustrated to emphasize the locations of the proteins.

tomycin-dependent strains. Protein S12 is altered in both resistant and dependent strains, and revertants from streptomycin dependence, able to grow in the presence or absence of streptomycin, have altered S5 or S4. (Mutations in ribosomal proteins have been reviewed by Nomura [1970].) By observing the degree of restriction of suppression of an amber codon with related amber and ocher suppressors, Gorini (1970) concluded that the streptomycin-resistance mutation (S12) involved an interaction with some part of the tRNA other than the anticodon. Similarly, protein S5 is altered by the mutation to spectinomycin resistance and this mutation can restore the suppressor activity of restricted streptomycin-resistant strains. Thus, mutations in proteins S5 and S12 alter tRNA recognition. Modifications in both affect streptomycin-induced misreading and alterations in both can modify (i.e., restrict [S12] or restore [S5]) the suppressor activity of mutant strains.

Proteins on the 30S Platform

The platform is a region of the small subunit with a unique shape and location that seems to be associated with a special structural role. This evolutionarily conserved feature is present in such diverse prokaryotes and eukaryotes as mammals, e.g., rat liver and rabbit reticulocytes (Emanuilov et al. 1978), and *E. coli* (Lake and Kahan 1975). In the 70S ribosome, one surface of the platform faces the cytoplasm, whereas the other faces the large subunit. Because of its thinness (~30 Å), it is possible for a protein to be exposed on both the exterior and the interface surfaces of the platform. Proteins S1, S11, and S18 map on the platform at sites summarized in two views of the subunit in Figure 4. The location of S1 should be considered preliminary.

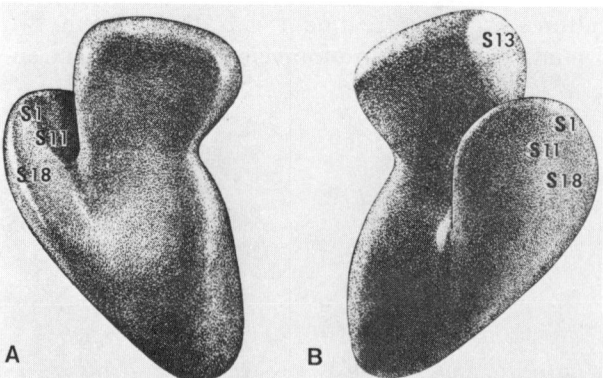

Figure 4 Diagrammatic representation of the locations of ribosomal proteins S1, S11, S18, and S13 shown on the external (*A*) and interface (*B*) surfaces of the 30S subunit.

Several lines of experimentation have linked these proteins to mRNA binding. Single-component omission experiments (Nomura et al. 1970) demonstrated that streptomycin-induced misreading is insensitive to the omission of most proteins with two exceptions, S11 and S12. In the absence of S11, misreading (i.e., incorporation of isoleucine, tyrosine, and serine in a poly[U]-directed system) was increased by a factor of 6.5. Decreased misreading in the absence of S12 was consistent with an S12-tRNA interaction at a region other than the codon-anticodon site. Since the omission of no other protein significantly increased the error frequency, it was concluded that S11 may be participating directly in the selection of the correct tRNA, i.e., in the codon-anticodon interaction (Nomura 1970).

Proteins S1 and S18 have been cross-linked to mRNA using a variety of affinity-labeling analogs (these have been comprehensively reviewed by Cooperman [1978]). The reagents employed have included poly(U), UUUU, UGA, GUUU, electrophilic derivatives of AUG, AAAAAAAA, poly(s^4U), and poly(Br^5U). The most frequently labeled protein is S18 and the second most frequently labeled protein is S1 (other labeled proteins include S4, S12, and S21). Proteins S1 and S18 are consistently labeled by messenger affinity analogs and are likely to be near the codon-anticodon interaction site.

Additional evidence placing S1 near the codon comes from the suggestion (Shine and Delgarno 1975) that the 3' end of RNA can make between 4 and 7 base pairs with a leader sequence located about 10 nucleotides before (i.e., on the 5' side of) the initiation codon, from the experiments supporting this suggestion (Steitz and Jakes 1975; Dunn et al. 1978), and from the isolation of protein S1 complexed to this base-pairing region (Dahlberg and Dahlberg 1975). Consistent with this, dimethyl As (located on the 16S RNA near the leader-sequence pairing region) have been mapped (Politz and Glitz 1977) on the platform by immuno-electron microscopy.

Thus, a considerable body of evidence exists to suggest that the platform proteins S1, S11, and S18 are close to the site of codon-anticodon interaction and we conclude that the platform is probably the anticodon binding site.

Initiation-factor Binding Region

The final 30S protein to be discussed, S13, is also shown in Figure 4B. Protein S13 maps on the interface side of the subunit on the upper one-third opposite the platform (Lake and Kahan 1975).

The location of protein S13, taken together with the mapping of those proteins mapping on the platform and the location of S12, suggests a binding site for initiation factors 1, 2, and 3 (IF-1, IF-2, and IF-3). IF-3, required for the binding of natural mRNA to the 30S subunit, has been efficiently cross-linked to S7 (van Duin et al. 1975), and to S11, S12, S13, and S19 (Heimark et al. 1976). IF-2 stimulates the binding of formyl-methionyl-tRNA^{fMet} to the 30S subunit and catalyzes the hydrolysis of GTP on the 70S ribosome (reviewed by Lucas-Lenard and Lipmann 1971; Haselkorn and Rothman-Denes 1973). In the presence of IF-1 and IF-3 it has been efficiently cross-linked to S1, S2, S11, S13, S14, and S19, with the strongest immunological reactions being against S13 and S19 and the next strongest against S11 and S12 (Bollen et al. 1975). (IF-1 has been cross-linked to S1, S12, S13, and S19 [Langberg et al. 1977] in the presence of IF-2 and IF-3.) Consideration of the results of both cross-linking and protein localization suggests that IF-3, IF-2, and IF-1 are positioned across the groove between S13 and S12 on the one-third side and S11 on the platform side in the general region indicated in Figure 5. (Our preliminary mapping of S7 is also consistent with this interpretation.) The binding of IF-3 is consistent with the location of the codon-anticodon interaction; the binding of IF-2, in turn, provides an indication of the binding site for formyl-methionyl-tRNA^{fMet}.

Localization of L7/L12

Ribosomal proteins L7 and L12 have several unique properties (for reviews, see Nomura et al. 1974; Weissbach and Pestka 1977). Proteins L7 and L12 have identical amino acid sequences except that the amino terminal of L7 is acetylated, whereas that of L12 is not. L7 and L12 are the only ribosomal proteins present in multiple copies, and the total number of L7 and L12 copies per ribosome is most likely four. Both proteins are intimately involved in elongation factor Tu (EF-Tu), in elongation factor G (EF-G)-dependent GTP hydrolysis, and in IF-2-dependent GTP hydrolysis. Hence, localization of the L7/L12 stalk pro-

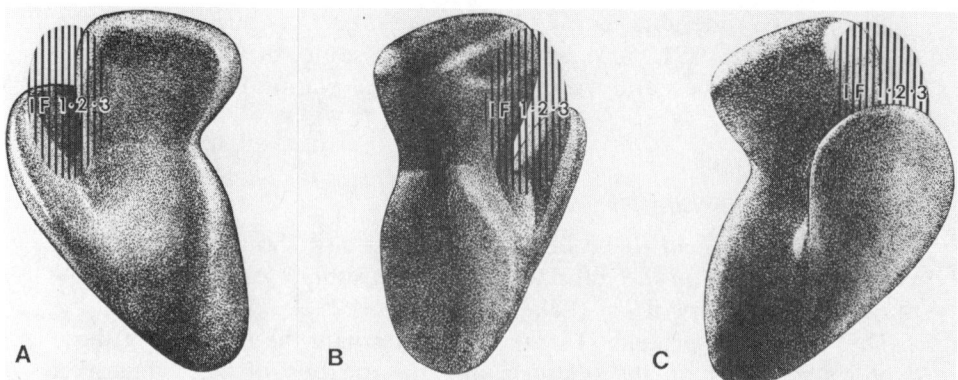

Figure 5 Locations deduced for IF-1, IF-2, and IF-3 on the exterior surface (*A*), in the cleft (*B*), and at the interface (*C*) of the 30S subunit.

vides information about the ribosome binding sites of factors EF-Tu, EF-G, and IF-2. Recently, using several approaches, including doubly labeled subunits, Strycharz et al. (1978) demonstrated that L7 and L12 map exclusively on the elongated appendage called the L7/L12 stalk.

The relationship of proteins L7 and L12 to the monomeric 70S ribosome is shown in Figure 6. For reference, some small-subunit protein locations that have been previously discussed are shown. The large subunit is illustrated in the conformation principally observed in fields of dissociated 50S subunits (Fig. 6) and in the conformation principally observed in fields of 70S ribosomes (Fig. 6). In either conformation, however, L7 and L12 are quite distant from the convex edge (at the left) of the small subunit. The L7/L12 stalk location at the 30S/50S subunit interface is consistent with the IF-2 binding site discussed in the section on the Initiation-factor Binding Region. However, since IF-2 as well as EF-Tu and EF-G is relatively large, factor binding sites could be positioned over a sizable region of the 70S ribosome and will be consistent with the reported factor-L7/L12 stalk cross-links (EF-G, Acharya et al. 1973; EF-Tu, San José et al. 1977).

50S Proteins Located at the Peptidyl Transferase

The locations of four additional large subunit proteins, L17, L21, L23, and L27, are shown schematically in Figure 7. At present, there is little information relating either L17 or L21 to functional events occurring on the 50S subunits.

Protein L27 is a likely candidate to be near the peptidyl transferase. It is among the proteins labeled by modified aminoacyl-tRNA affinity labels

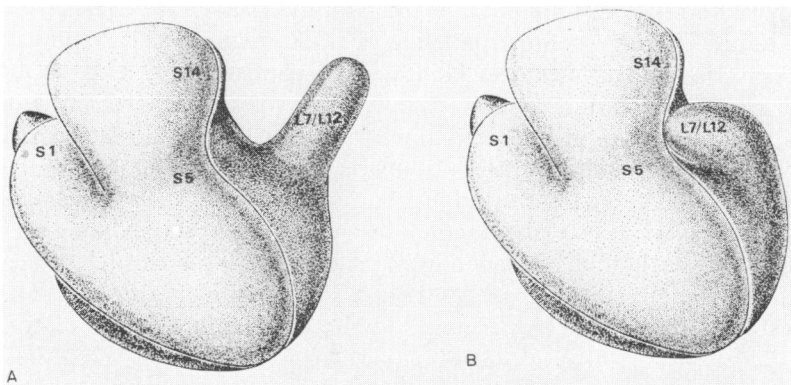

Figure 6 The location of the L7/L12 stalk relative to the small subunit. (*A*) The morphology of the large subunit corresponding to the major population observed in fields of large subunits. (*B*) The morphology principally observed in fields of 70S ribosomes. For reference, the locations of a few small-subunit proteins are labeled.

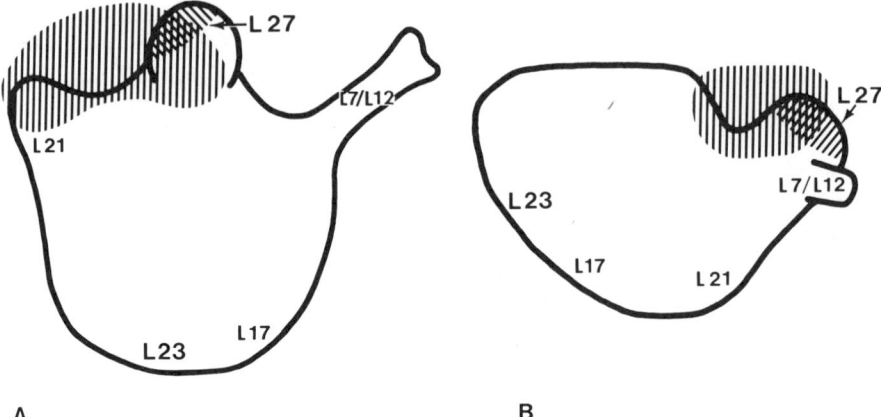

Figure 7 Suggested location of the peptidyl transferase shown in the quasi-symmetric (A) and asymmetric (B) projections of the 50S subunit. The region of vertical stripes corresponds to the possible region of contact of the amino acid acceptor end of the tRNA assuming the codon-anticodon interaction occurs on the platform. The diagonal stripes indicate the localization of protein L27. The location of L23 is also indicated, as are the locations of the L7/L12 stalk, L17, and L21.

bound to the peptidyl site and to the aminoacyl site (for reviews, see Traut et al. 1974; Cooperman 1978). The most common affinity labels for peptidyl-site studies have been derivatives of N-acyl-phenylalanyl-tRNA^{Phe}. Using these affinity labels, the principal labeled proteins are L2, L11, L18, and L27. Under conditions where aminoacyl-site labeling is favored, the predominantly labeled proteins are L2, L16, and L27. The third suggestion that L27 might be at the peptidyl transferase comes from acyl-methionyl-tRNA^{fMet} affinity labeling of ribosomes, where L2 and L27 have been the major product. Hence, L27 consistently seems to be labeled near the peptidyl transferase by affinity analogs of initiator tRNAs and of tRNAs in both the aminoacyl and the peptidyl sites. By contrast, L23 is not significantly labeled by peptidyl-tRNA analogs, but it is labeled with radioactive puromycin. Rather unexpectedly, however, small-subunit proteins were also labeled, the most significant reaction being with S14. This suggests that the L23 labeling might be occurring somewhere other than at the peptidyl transferase, although the possibility that L27 was spuriously labeled with an electrophilic affinity label cannot be eliminated.

RIBOSOME FUNCTION

Functional Regions of the Ribosome

Given the locations of the ribosomal proteins discussed in this paper, one can reasonably start to ask where the functional regions of the ribosome are and, in particular, where the tRNA binding sites are located. The process of combining data on the functional roles of ribosomal proteins is subject to interpretation regarding which experiments are significant and to what extent structural inferences can be obtained from them. Nevertheless, the ability to map protein locations in three dimensions at 30- to 40-Å resolution makes it possible to start to answer specific questions about the structural events of protein synthesis.

In the section entitled Proteins on the 30S Platform, evidence was discussed that suggested general regions for codon-anticodon interaction and for the initiation-factor binding sites. Based on (1) localizations of proteins S1, S11, and S18 presented, (2) the binding site deduced for IF-3, (3) the localization of m_6^2A (Politz and Glitz 1977), and (4) the localization of eukaryotic initiation factor 3 (eIF-3) in native eukaryotic small subunits (Emanuilov et al. 1978), the codon-anticodon interaction probably occurs on the platform of the 30S subunit.

The region of the 30S subunit to which IF-1, IF-2, and IF-3 have been cross-linked is indicated in Figure 5 and the protein locations and the cross-linking studies involved in deriving this site are also described. The location of IF-2 is relevant because formylmethionyl-tRNA^{fMet} binding to the 30S subunit is dependent upon binding of both the initiator codon and IF-2. For the anticodon of a tRNA in the aminoacyl or the peptidyl site to contact the codon on the platform, the tRNA can only contact a limited region of the 50S subunit. Using the constraints of tRNA size and shape, it was suggested (Lake 1976) that the peptidyl transferase might be located within the area indicated by vertical striations in Figure 7, and a flexing of the tRNA at the TψC corner, as suggested (Richter et al. 1973), would not greatly modify this region. The location of L27 is also indicated by diagonal striping. This protein is one of the major molecules labeled by both peptidyl- and aminoacyl-tRNA affinity labels and is presumed to be near the peptidyl transferase. Hence, the crosshatched region in Figure 7 where both regions overlap is suggested as a possible peptidyl transferase site.

This location of the peptidyl transferase, which is separated from proteins L7 and L12, is consistent with biochemical evidence suggesting the separation. Removal of L7 and L12, for example, does not alter the peptidyl-transferase activity of the cores (Hamel et al. 1972). Large-subunit cores with L7, L12, L10, and L11 removed are active in the fragment reaction (Howard and Gordon 1974; Ballesta and Vazquez 1974). The erythromycin binding site, thought to be near the peptidyl transferase, is approximately 70 Å from a labeled region of the L7/L12 stalk (Langlois et al. 1976). Inhibition of both EF-G binding and EF-Tu-associated GTPase activity, but not of peptidyl-transferase activity, by binding of the antibiotic siomycin (Modolell et al. 1971) also suggests a separation of the L7/L12 stalk from the peptidyl-transferase center.

For a tRNA to simultaneously contact the codon, the IF-2 binding region, and the peptidyl transferase, it must be positioned in the aminoacyl or the peptidyl site approximately as shown in Figure 8. For comparison, the approximate locations of the external proteins are also marked. The separation of the aminoacyl- and peptidyl-tRNA binding sites from these proteins is quite remarkable in view of the considerable experimental evidence suggesting their role in tRNA binding. It could, perhaps, be argued that these proteins might alter tRNA binding indirectly or that, although they might extend into the tRNA binding region, they might not be sufficiently exposed to permit antibody labeling. The simplest alternative, however, is that they directly interact with tRNAs. Certainly in the case of protein S5 at least, the information obtained from neutron diffraction makes the extension of a significant portion of S5 well beyond a diameter of 40 Å unlikely (Engelman et al. 1975; Langer et al. 1978). Whereas a single tRNA binding protein might be easily explained, five proteins strongly implicated in tRNA binding that map outside the expected aminoacyl-site region are much more difficult to understand.

The cluster of proteins that is involved in tRNA binding but is located on the exterior of the 30S subunit suggested that another tRNA binding site might exist. As a result, a reconsideration of the elongation cycle led to the conclusion that during protein synthesis initial codon recognition occurs with the aminoacyl-tRNA bound in a structurally distinct recognition tRNA binding site (Lake 1977). The concept of the recognition site is supported by the fundamental experiments defining the elongation cycle, quite independently of the results of our antibody-labeling studies.

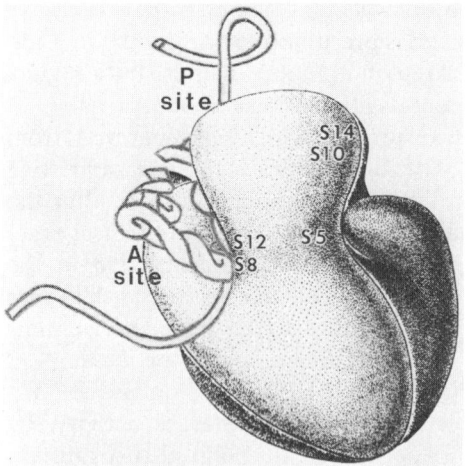

Figure 8 Model of the monomeric ribosome showing suggested aminoacyl- and peptidyl-tRNA binding sites.

Initial Codon Recognition Occurs at the Recognition tRNA Binding Site

The broad outlines of the chain elongation cycle of protein synthesis are quite well understood (for reviews, see Leder 1973; Kaziro 1978). Typically, the major events that occur are summarized as (1) codon recognition, (2) peptide-bond formation, and (3) translocation. In Figure 9A, a cycle of elongation has been completed and the $n + 1$ codon is available to be read by the correct tRNA. In the first step of the next elongation cycle (initial codon recognition; Fig. 9B), the codon is recognized and the correct aminoacyl-tRNA · EF-Tu · GTP complex is bound by the ribosome. In the second step (peptide-bond formation; Fig. 9C), the GTP is cleaved to GDP and Pi, the EF-Tu · GDP complex and Pi are released from the ribosome, the peptide bond is completed, and the $n + 1$ tRNA occupies the aminoacyl site. The third event, translocation, refers to a switching of the peptidyl-tRNA from the aminoacyl site (Fig. 9C) to the peptidyl site (Fig. 9D). This step utilizes EF-G and the cleavage of GTP and results in the release of EF-G, GDP, Pi, and the uncharged tRNA.

Each of the three states shown in Figure 9, B, C, and D, is a functionally distinct state (and, therefore, a physically distinct state) of the elongation cycle. The three states are functionally distinguishable according to the following criteria. If peptidyl-tRNA is reactive with puromycin it is said to be in the peptidyl site (Fig. 9D) and if peptidyl-tRNA is not reactive with puromycin then it is in the aminoacyl site (Fig. 9C). In a similar way, with an aminoacyl-tRNA in the recognition site (i.e., in the presence of both EF-Tu and GMPP[CH$_2$]P, the nonhydrolyzable analog of GTP), peptidyl transfer does not occur, yet the aminoacyl site of the peptidyl transferase is vacant because the peptidyl-tRNA in the peptidyl site is puromycin-reactive (Lucas-Lenard et al. 1970). Just as the aminoacyl- and peptidyl-tRNA binding sites refer to tRNAs bound to the ribosome in defined physical states, the term recognition site describes the distinct physical state of the aminoacyl-tRNA when bound to the ribosome during initial codon recognition.

The Recognition Site Is Probably on the External Surface of the Small Subunit

The existence of the recognition site rests on the current biochemical understanding of the elongation cycle, quite different data from that provided by our mapping studies. Information on the elongation cycle itself provides few clues to the location of the recognition site other than to suggest that the recognition and aminoacyl sites are distinct. Distinct may refer to only a separation of a few angstroms between the sites or could refer to a separation of many angstroms. The fact that puromycin an release the peptidyl-tRNA while an aminoacyl-tRNA is bound to the

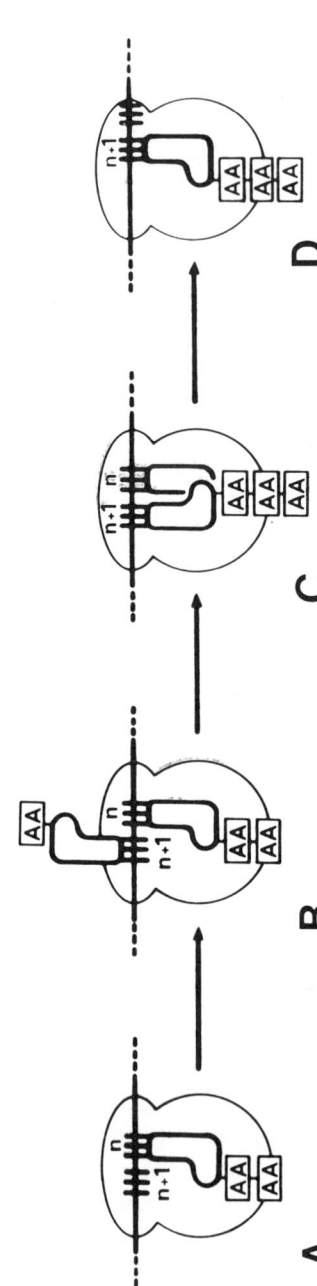

Figure 9 The elongation cycle of protein synthesis. (*A*) Peptidyl-tRNA in the peptidyl site; (*B*) aminoacyl-tRNA in the recognition site; (*C*) peptidyl-tRNA in the aminoacyl site; (*D*) peptidyl-tRNA in the peptidyl site. (Reprinted, with permission, from Lake 1977.)

recognition site implies that the CCA end of the tRNA bound to the recognition site is separated from the aminoacyl site by at least 5–10 Å.

A probable location for the recognition site is the cluster of five tRNA binding proteins found on the exterior of the 30S subunit. If the recognition site is located as shown in Figure 10, then an aminoacyl-tRNA bound to the recognition site can contact (or be near) external proteins S5, S8, S10, S12, and S14 while its anticodon is simultaneously base pairing with the codon. This would be consistent with the previously discussed effects of these proteins on tRNA binding. In addition, the D stem of a tRNA in the recognition site would be near proteins S5 and S12, consistent with their roles in restricting the suppression of nonsense mutations and with the role of a single base substitution in the D stem in converting tRNA^{Trp} into a nonsense suppressor tRNA (Hirsh 1970).

EF-Tu has been cross-linked to the L7/L12 stalk (San José et al. 1977) in the presence of GMPP(CH$_2$)P (i.e., under conditions probably corresponding to recognition-site binding). In addition, other evidence, such as the lack of EF-Tu binding to L7/L12 stripped cores (for a review, see Kaziro 1979), suggests an interaction of the L7/L12 stalk with EF-Tu. Thus, the recognition site as illustrated in Figure 10 on the exterior side of the small subunit is consistent with its proximity to proteins L7 and L12. However, Ef-Tu is relatively large so that other

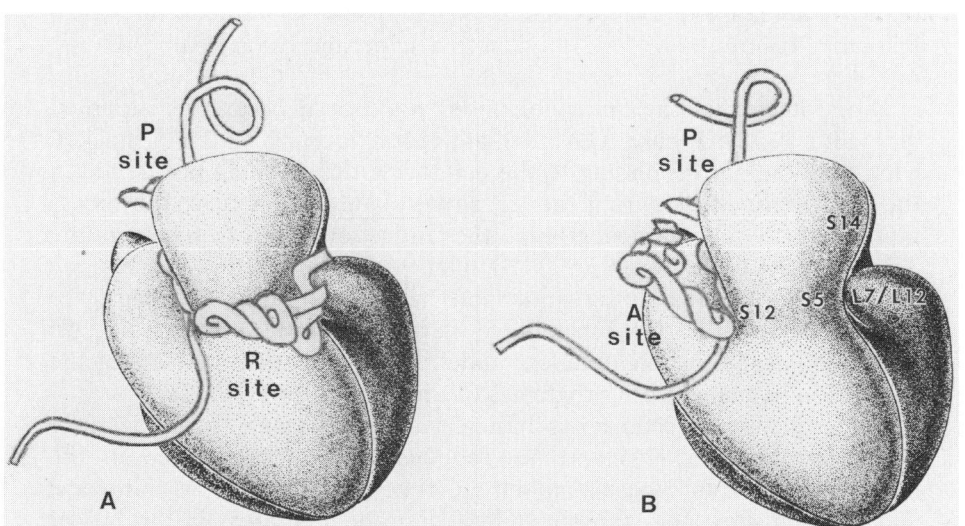

Figure 10 Suggested locations of the tRNA binding sites on the ribosome. (*A*) Recognition-site binding (corresponding to Fig. 9B); (*B*) aminoacyl-site binding (corresponding to Fig. 9C). The locations of exterior proteins are marked in *B*. The tRNAs are approximately drawn to scale.

locations on the small subunit could also be consistent with the reported factor-L7/L12 stalk cross-link.

A Model for Switching from the Recognition Site to the Aminoacyl Site

If the recognition tRNA site is located on the exterior of the 30S subunit, then a mechanism must exist for switching the tRNA from the recognition site to the aminoacyl site. This tRNA movement to the aminoacyl site must be consistent with the observation (Gupta et al. 1971; Thach and Thach 1971) that no measurable messenger movement, relative to the ribosome, occurs during recognition- to aminoacyl-site switching and that the message is shifted three nucleotides only during the aminoacyl- to peptidyl-site switching.

Evidence suggests that the essential element of the switching mechanism could involve the tRNA itself. The anticodon loop is thought to assume two conformations, referred to as the 5′- and the 3′-stacked conformations. In the 5′ stack, the five bases at the 5′ end of the loop are stacked in a quasi-helical manner. In the 3′ stack, the five bases at the 3′ end of the anticodon loop are stacked. The 3′-stacked conformation has been demonstrated by X-ray diffraction of yeast tRNA^{Phe} in both the orthorhombic (Kim et al. 1974) and monoclinic crystal forms (Robertus et al. 1974). The 5′-stacked configuration has been suggested by exploration of anticodon-loop base pairing with complementary polynucleotides (Eisinger and Spahr 1973; Uhlenbeck 1972) and by temperature-jump kinetics (Yoon et al. 1975). Definitive experiments are still lacking, however.

The switching in the anticodon loop would then occur as diagramed in Figure 11. In Figure 11A, the anticodon loop is in the 5′-stacked conformation, corresponding to the aminoacyl-tRNA being positioned in the recognition site. During the transition of the tRNA from the recognition site to the aminoacyl site, the anticodon loop switches to the 3′-stacked conformation (Fig. 11B). Note that during this process the mRNA (in black) and the anticodon have not moved, although the acceptor end of the tRNA has moved a large distance (see Fig. 10). An interesting aspect of this detailed model is that essential elements of the translation machinery are properties of the tRNA (Crick 1968).

The switching mechanism is basically different from the model proposed by Woese (1970) and also applies to a different step in the elongation cycle. In Woese's model, the switching between the aminoacyl and peptidyl sites, i.e., translocation, is described, whereas in this model the switching between the recognition and aminoacyl sites is described. Also, the tRNA remains fixed on the ribosome in the model of Woese and the mRNA moves, whereas in this model the message remains fixed and the tRNA moves. Recent evidence (cited by Nierhaus 1978) is consistent with the Woese model.

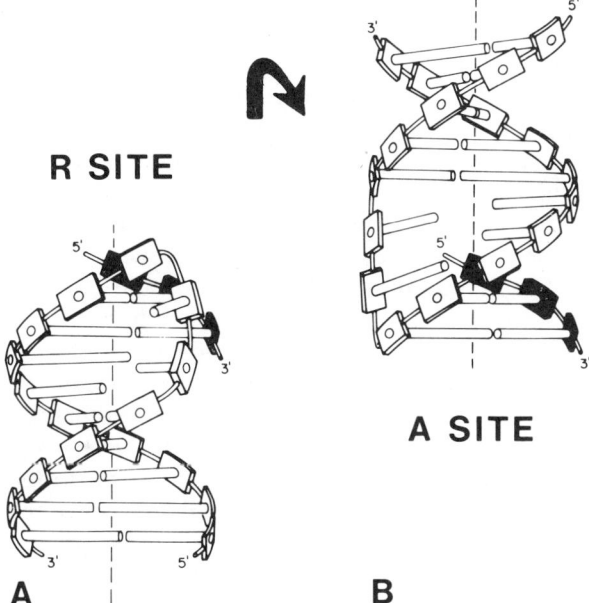

Figure 11 Diagrammatic representation of the switching of the anticodon loop from the 5'-stacked conformation (A) to the 3'-stacked conformation (B). Note that during the switching the codon (black) and the anticodon remain fixed, although the acceptor stem of the tRNA (not shown) moves a considerable distance. (Reprinted, with permission, from Lake 1977.)

The Recognition Process

The ideas presented here have been developed from the point of view of protein locations and ribosomal function, but they fit in well with current ideas on fidelity in protein synthesis. In particular, the recognition site provides the structural site needed for initial tRNA recognition in translation models in which the codon is read twice.

It is thought that the difference in binding energy between a correct and an incorrect codon-anticodon triplet is not sufficient to explain the accuracy obtained in protein synthesis. In general, off rates measured from complexes of tRNAs having complementary anticodons are two orders of magnitude greater than rates measured from tRNAs with anticodons having two out of three complementary bases (Grosjean et al. 1978), and it was suggested that the ribosome must amplify this small difference in off rates to reduce errors. Most mechanisms for doing this involve a two-step process, an initial reversible reading step (presumably corresponding to recognition-site binding) followed by a second selection involving an irreversible branch-point reaction (Hopfield 1974; see also Fersht 1977). Ninio (1974) found experimental support for a two-step recognition process in his analysis of the data of Gorini (1970) on the efficiency of missense and nonsense suppressors in streptomycin-resistant *E. coli* strains. Experimental support for a two-step model involving an irreversible branch point has recently been demonstrated by Thompson and Stone (1977). Hence, our current ideas on ribosome structure are converging, in a potentially productive manner, with the current understanding of the fidelity of protein synthesis and with the understanding of the role of factors in protein synthesis.

In conclusion, the localization of ribosome proteins is providing data that were unobtainable a few years ago. We now know where most ribosomal proteins, including the multiple-copy proteins L7 and L12, are located, and we have a fairly accurate idea of the location of the codon-anticodon interaction site and of the aminoacyl- and peptidyl-tRNA binding sites. It is known that the existence of a recognition site is necessary, given the current understanding of the elongation cycle, and a location for it on the exterior of the small subunit has been suggested. Just as the aminoacyl-tRNA and peptidyl-tRNA sites describe tRNAs bound to the ribosome in well-defined physical states, the recognition site describes the functionally and physically distinct state of the aminoacyl-tRNA bound to the ribosome during initial codon recognition. The concept of the recognition site is supported by the basic ideas of the elongation cycle, and obviously its existence is not dependent on the correctness of the details of the molecular model for switching from the recognition site to the aminoacyl site.

ACKNOWLEDGMENTS

I thank M. Hwang, A. Kost, and S. Pressman-Nelson for excellent electron microscopy and photography, and L. Kahan, W. Strycharz, and M. Nomura for helpful comments on this paper. Supported by grants from the National Science Foundation and from the National Institute of General Medical Sciences.

REFERENCES

Acharya, A. S., P. B. Moore, and F. M. Richards. 1973. Cross-linking of elongation factor EF-G to the 50S ribosomal subunit of *Escherichia coli*. *Biochemistry* **12**:3108.

Ballesta, J. P. G. and D. Vazquez. 1974. Activities of ribosomal cores deprived of proteins L7, L10, L11 and L12. *FEBS Lett.* **48**:266.

Bollen, A., R. L. Heimark, A. Cozzone, R. R. Traut, J. W. B. Hershey, and L. Kahan. 1975. Crosslinking of initiation factor IF-2 to *Escherichia coli* 30S ribosomal proteins with dimethyl suberimidate. *J. Biol. Chem.* **250**:4310.

Cooperman, B. S. 1978. Affinity labeling studies on *Escherichia coli* ribosomes. *Bioorg. Chem.* (suppl.) **4**:81.

Crick, F. H. C. 1968. The origin of the genetic code. *J. Mol. Biol.* **38**:367.

Dahlberg, A. and J. Dahlberg. 1975. Binding of ribosomal protein S1 of *Escherichia coli* to the 3' end of 16S rRNA. *Proc. Natl. Acad. Sci.* **72:** 2940.

Dunn, J. J., E. Buzash-Pollert, and W. F. Studier. 1978. Mutations of bacteriophage T7 that affect initiation of synthesis of the gene 0.3 protein. *Proc. Natl. Acad. Sci.* **75:** 2741.

Eisinger, J. and P.-F. Spahr. 1973. Binding of complementary pentanucleotides to the anticodon loop of transfer RNA. *J. Mol. Biol.* **73:** 131.

Emanuilov, I., D. D. Sabatini, J. A. Lake, and C. Freienstein. 1978. Localization of eukaryotic initiation factor 3 on native small ribosomal subunits. *Proc. Natl. Acad. Sci.* **75:** 1389.

Engelman, D. M., P. B. Moore, and B. P. Schoenborn. 1975. Neutron scattering measurements of separation and shape of proteins in 30S ribosomal subunit of *Escherichia coli*: S2-S3, S5-S8, S3-S7. *Proc. Natl. Acad. Sci.* **72:** 3888.

Fersht, A. 1977. *Enzyme structure and mechanism.* W. H. Freeman, San Francisco.

Gorini, L. C. 1970. The contrasting role of *str A* and *ram* gene products in ribosomal functioning. *Cold Spring Harbor Symp. Quant. Biol.* **34:** 101.

———. 1974. Streptomycin and misreading of the genetic code. In *Ribosomes* (ed. M. Nomura et al.), p. 791. Cold Spring Harbor Laboratory, Cold Spring Harbor, New York.

Grosjean, H. T., S. de Henau, and D. M. Crothers. 1978. On the physical basis for ambiguity in genetic coding interactions. *Proc. Natl. Acad. Sci.* **75:** 610.

Gupta, S. L., J. Waterson, M. L. Sopori, S. M. Weissman, and P. Lengyel. 1971. Movement of the ribosome along the messenger ribonucleic acid during protein synthesis. *Biochemistry* **10:** 4410.

Hamel, E., M. Koka, and T. Nakamoto. 1972. Requirement of an *Escherichia coli* 50S ribosomal protein component for effective interaction of the ribosome with T and G factors and with guanosine triphosphate. *J. Biol. Chem.* **247:** 805.

Haselkorn, R. and L. B. Rothman-Denes. 1973. Protein synthesis. *Annu. Rev. Biochem.* **42:** 397.

Heimark, R. L., L. Kahan, K. Johnston, J. W. B. Hershey, and R. R. Traut. 1976. Cross-linking of initiation factor IF3 to proteins of the *Escherichia coli* 30S ribosomal subunit. *J. Mol. Biol.* **105:** 219.

Hirsh, D. 1970. Tryptophan tRNA of *Escherichia coli*. *Nature* **228:** 57.

Hopfield, J. J. 1974. Kinetic proofreading: A new mechanism for reducing errors in biosynthetic processes requiring high specificity. *Proc. Natl. Acad. Sci.* **71:** 4135.

Howard, G. A. and J. Gordon. 1974. Peptidyl transferase activity of ribosomal particles lacking protein L11. *FEBS Lett.* **48:** 271.

Kaziro, Y. 1978. The role of guanosine 5'-triphosphate in polypeptide chain elongation. *Biochim. Biophys. Acta* **505:** 95.

Kim, S.-H., F. L. Suddath, G. J. Quigley, A. McPherson, J. L. Sussman, A. H. J. Wang, N. C. Seeman, and A. Rich. 1974. Three-dimensional tertiary structure of yeast phenylalanine transfer RNA. *Science* **185:** 435.

Lake, J. A. 1976. Ribosome structure determined by electron microscopy of *Escherichia coli* small subunits, large subunits and monomeric ribosomes. *J. Mol. Biol.* **105:** 131.

———. 1977. Aminoacyl-tRNA binding at the recognition site is the first step of the elongation cycle of protein synthesis. *Proc. Natl. Acad. Sci.* **74:** 1903.

———. 1978. Electron microscopy of specific proteins: Three dimensional mapping of ribosomal proteins using antibody labels. In *Advanced techniques in biological electron microscopy II* (ed. J. Koehler), p. 173. Springer-Verlag, Berlin.

Lake, J. A. and L. Kahan. 1975. Ribosomal proteins S5, S11, S13 and S19 localized by electron microscopy of antibody-labeled subunits. *J. Mol. Biol.* **99:** 631.

Lake, J. A. and H. S. Slayter. 1970. Three dimensional structure of the chromatoid body (ribosome helix) of *Entamoeba invadens*. *Nature* **227:** 1032.

———. 1972. Three dimensional structure of the chromatoid body (ribosome) helix of *Entamoeba invadens*. *J. Mol. Biol.* **66:** 271.

Lake, J. A., Y. Nonomura, and D. D. Sabatini. 1974a. Ribosome structure as studied by electron microscopy. In *Ribosomes* (ed. M. Nomura et al.), p. 543. Cold Spring Harbor Laboratory, Cold Spring Harbor, New York.

Lake, J. A., M. Pendergast, L. Kahan, and M. Nomura. 1974b. Localization of *E. coli* ribosomal proteins S4 and S14 by electron microscopy of antibody-labeled subunits. *Proc. Natl. Acad. Sci.* **71:** 4688.

Langberg, S., L. Kahan, R. Traut, and J. W. B. Hershey. 1977. Binding of protein synthesis initiation factor IF1 to 30S ribosomal subunits: Effects of other initiation factors and identification of proteins near the binding site. *J. Mol. Biol.* **117:** 307.

Langer, J. A., D. M. Engelman, and P. B. Moore. 1978. Neutron-scattering studies of the ribosome of *E. coli*, a provisional map of the location of proteins S3, S4, S5, S7, S8 and S9 in the 30S subunit. *J. Mol. Biol.* **119:** 463.

Langlois, R., C. C. Lee, C. R. Cantor, R. Vince, and S. Pestka. 1976. The distance between two functionally significant regions of the 50S *Escherichia coli* ribosome: The erythromycin binding site and proteins L7/L12. *J. Mol. Biol.* **106:** 297.

Leder, P. 1973. The elongation reactions in protein synthesis. *Adv. Protein Chem.* **27:** 213.

Leonard, K. R. and J. A. Lake. 1979. Ribosome structure—Hand determination by electron microscopy of 30S subunits. *J. Mol. Biol.* **129:** 155.

Lucas-Lenard, J. and F. Lipmann. 1971. Protein biosynthesis. *Annu. Rev. Biochem.* **40:** 409.

Lucas-Lenard, J., P. Tao, and A. Haenni. 1970. Further studies on bacterial polypeptide elongation. *Cold Spring Harbor Symp. Quant. Biol.* **34:** 455.

Modolell, J., B. Cabrer, A. Parmeggiani, and D. Vazquez. 1971. Inhibition by siomycin and thiostrepton of both aminoacyl-tRNA and factor G binding to ribosomes. *Proc. Natl. Acad. Sci.* **68:** 1796.

Nierhaus, K. 1978. The ribosome: Still a knotty problem. *Nature* **24:** 743.

Ninio, J. 1974. A semi-quantitative treatment of missense and nonsense suppression in the *str A* and *ram* ribosomal mutants of *Escherichia coli*. Evaluation of some molecular parameters of translation *in vivo*. *J. Mol. Biol.* **84:** 297.

Nomura, M. 1970. Bacterial ribosome. *Bacteriol. Rev.* **34:** 228.

Nomura, M., A. Tissières, and P. Lengyel, eds. 1974. *Ribosomes* (see pp. 711–731). Cold Spring Harbor Laboratory, Cold Spring Harbor, New York.

Nomura, M., S. Mizushima, M. Ozaki, P. Traub, and C. V. Lowry. 1970. Structure and function of ribosomes and their molecular components. *Cold Spring Harbor Symp. Quant. Biol.* **34:** 49.

Politz, S. M. and D. G. Glitz. 1977. Ribosome structure: Localization of N^6,N^6-dimethyladenosine by electron microscopy of a ribosome-antibody complex. *Proc. Natl. Acad. Sci.* **74:**1468.

Richter, D., V. A. Erdmann, and M. Sprinzl. 1973. Specific recognition of GTψC loop (loop IV) of tRNA by 50S ribosomal units. *Nat. New Biol.* **246:**132.

Robertus, J. D., J. E. Ladner, J. T. Finch, D. Rhodes, R. S. Brown, B. F. C. Clark, and A. King. 1974. Structure of yeast phenylalanine tRNA at 3 Å resolution. *Nature* **250:**546.

San José, C., C. G. Kurland, and G. Stöffler. 1977. The protein neighborhood of ribosome-bound elongation factor Tu. *FEBS Lett.* **71:**133.

Shine, J. and L. Delgarno. 1975. Determinant of cystron specificity in bacterial ribosomes. *Nature* **254:**34.

Steitz, J. A. and K. Jakes. 1975. How ribosomes select initiator regions in mRNA: Base pair formation between the 3' terminus of 16S rRNA and the mRNA during initiation of protein synthesis in *Escherichia coli. Proc. Natl. Acad. Sci.* **72:**4734.

Strycharz, W. A., M. Nomura, and J. A. Lake. 1978. Ribosomal proteins L7/L12 localized at a single region of the large subunit by immune electron microscopy. *J. Mol. Biol.* **126:**123.

Thach, S. S. and R. E. Thach. 1971. Translocation of messenger RNA and "accommodation" of fMet-tRNA. *Proc. Natl. Acad. Sci.* **68:**1791.

Thompson, R. C. and P. J. Stone. 1977. Proofreading of the codon-anticodon interaction on ribosomes. *Proc. Natl. Acad. Sci.* **74:**198.

Tischendorf, G. W., B. Tesche, and G. Stöffler. 1976. The structural organization of the proteins on the *Escherichia coli* ribosome, as determined by immunoelectron microscopy. In *Proceedings of the 6th European Congress on Electron Microscopy*, Jerusalem (ed. B. Shaul), vol. 2, p. 524. Tal. Int. Publ., Jerusalem.

Traut, R. R., R. L. Heimark, T.-T. Sun, J. W. B. Hershey, and A. Bollen. 1974. Protein topography of ribosomal subunit from *Escherichia coli*. In *Ribosomes* (ed. M. Nomura et al.), p. 543. Cold Spring Harbor Laboratory, Cold Spring Harbor, New York.

Uhlenbeck, O. C. 1972. Complementary oligonucleotide binding to transfer RNA. *J. Mol. Biol.* **65:**25.

Unwin, P. T. N. 1977. Three dimensional model of membrane-bound ribosomes obtained by electron microscopy. *Nature* **269:**118.

van Duin, J., C. G. Kurland, J. Dandon, and M. Grunberg-Managi. 1975. Near neighbors of IF3 bound to 30S ribosomal subunits. *FEBS Lett.* **59:**287.

Weissbach, H. and S. Pestka, eds. 1977. *Molecular mechanisms of protein biosynthesis*. Academic Press, New York.

Woese, C. 1970. Molecular mechanics of translation: A reciprocating ratchet mechanism. *Nature* **226:**817.

Yoon, K., D. H. Turner, and I. Tinoco. 1975. The kinetics of codon-anticodon interaction in yeast phenylalanine transfer RNA. *J. Mol. Biol.* **99:**507.

Affinity Labeling of tRNA Binding Sites on Ribosomes

Ernst Kuechler
Institute of Biochemistry
University of Vienna, Austria

James Ofengand
Roche Institute of Molecular Biology
Nutley, New Jersey 07110

During the past several years a variety of chemically reactive derivatives of tRNA have been synthesized for use as affinity labels for amino acid-tRNA ligases, ribosomes, and factors involved in protein biosynthesis. Particularly in the study of *Escherichia coli* ribosomes, progress has been rather rapid. Besides affinity-label derivatives of tRNA, reactive derivatives of oligo- and polynucleotides have been constructed to label the ribosomal site for mRNA binding. Furthermore, a number of small-molecular-weight substances, such as antibiotics and GTP, have been chemically modified to obtain specific reagents for their respective ribosomal binding sites. The field of affinity labeling of ribosomes has been reviewed before with emphasis on various aspects of this problem (Cantor et al. 1974; Pellegrini and Cantor 1977; Zamir 1977; Cooperman 1978; Kuechler 1978). Here we want to concentrate primarily on affinity-labeling studies dealing with the ribosomal binding site for tRNA and on the interaction of tRNA with mRNA. The results will be correlated with the existing data on the structural organization of the ribosome as obtained by electron and immuno-electron microscopy and used to outline the functional domains on the ribosome for different regions of the tRNA.

AFFINITY- AND PHOTOAFFINITY-LABEL DERIVATIVES OF tRNA

Position of the Affinity-labeling Group

Attachment of an affinity-labeling group to aminoacyl-tRNA requires special precautions to insure single-site specificity. In most studies, advantage has been taken of the existence of the single, highly reactive α-amino group on the aminoacyl moiety of tRNA. Thus, affinity reagents have usually been bound via an amide bond with the α-amino group. Ribosomes tolerate rather extensive modifications at this site of aminoacyl-tRNA without reduction in binding, probably because the group attached to the aminoacyl residue occupies the same space as the growing peptide chain. Thus, the relatively large and bulky groups commonly used for photoaffinity labeling can be attached to the aminoacyl moiety of tRNA without apparent loss in binding specificity. Additionally, identification of the labeled ribosomal component is simplified by the use of radioactive amino acids. Following the affinity-labeling reaction, the bond between the aminoacyl residue and the tRNA is cleaved and the radioactive amino acid remains as a tag on the labeled ribosomal component. This procedure also eliminates a potential problem with some tRNAs, which is the concomitant attachment of the affinity label to the aliphatic amino group of acp^3U, which is found at position 47 in several *E. coli* tRNAs. Although cross-linking to the ribosome can occur in this situation (Table 1, reagent XIV), it is not detected because the radioactive amino acid tag is removed from the tRNA.

Acylated aminoacyl-tRNAs have binding properties similar to peptidyl-tRNA and bind preferentially to the peptidyl site of the ribosome. Due to the blocked α-amino group, the derivatives are unable to form ternary complexes with elongation factor Tu (EF-Tu) and GTP and cannot be bound enzymatically to the aminoacyl site. However, derivatives of the initiator formylmethionyl-tRNA^{fMet} are recognized by initiation factor 2 (IF-2) and can be bound to the peptidyl site in response to natural mRNA (Hauptmann et al. 1974; Sonenberg et al. 1976).

Nonenzymatic binding to the aminoacyl site has been performed by preincubation of the ribosome with uncharged tRNA at high magnesium ion concentration (de Groot et al. 1971; Watanabe 1972). Under these conditions, the uncharged tRNA binds tightly to the peptidyl site. Upon subsequent incubation, the affinity-label derivative can then bind only to the aminoacyl site (Eilat et al. 1974b). Enzymatic labeling of the aminoacyl site has been achieved recently by use of lysyl-tRNA substituted at the ϵ-amino group. This derivative is recognized by EF-Tu because of its free α-amino group and is bound specifically to the aminoacyl site (Johnson et al. 1977; Johnson, this volume).

Other sites that have been specifically substituted with affinity probes are s^4U8 and acp^3U47 in certain *E. coli* tRNAs. These locations and the stereochemical relationship of probes placed at these sites to the rest of the tRNA molecule are illustrated in Figure 1, using as examples reagent XII on s^4U and reagent VII on acp^3U. Probes at these sites do not prevent nonenzymatic peptidyl-site binding, EF-Tu-dependent aminoacyl-site binding, or initiation-factor-dependent binding of formylmethionyl-tRNA (Ofengand et al. 1977), nor do they affect the ability of such modified tRNAs to be recognized by stringent factor (Ofengand and Liou 1978). Consequently, tRNAs modified at these positions can be used to compare the interaction of tRNA at both the peptidyl and aminoacyl sites. These positions are particularly suited to such studies because they are located in the central region of the tRNA, which is that part expected to show the maximum movement during translocation.

XII fits readily into the tRNA structure with its reactive group pointing outward. It cannot be extended beyond the overall perimeter of the tRNA. VII (or its analog XIV) is located in a more exposed region and has considerably greater flexibility. The two probes are so placed that they scan opposite surfaces of the plane defined by the L-shaped tRNA. They are expected to give complementary results.

Finally, naturally occurring modified bases in the anticodon loop of some tRNAs can be photo cross-linked upon irradiation with UV light. Examples are the cmo^{5}U and mo^{5}U residues in the 5' position of the anticodon of certain tRNAs from *E. coli* and *Bacillus subtilis*, respectively, which can be photo cross-linked to 16S RNA (Ofengand et al. 1979) and the wybutine base in tRNA^{Phe} from yeast. This base is situated next to the 3' end of the anticodon (Fig. 1). It has been shown to photoreact on the ribosome with poly(U) as mRNA (A. J. M. Matzke et al., unpubl.). The occurrence of natural photoreactive nucleotides in the anticodon loop of tRNA provides a particularly favorable situation. Specific chemical modifications, such as the attachment of photoreactive groups to this part of the molecule, would be chemically difficult and would likely affect codon recognition and, in some cases, aminoacylation.

Ribosomal Components Labeled

The affinity labels employed are usually alkylating or acylating reagents (Table 1). The most extensively studied reagents are the bromoacetyl group (II) (Pellegrini et al. 1972) and the *p*-nitrophenoxycarbonyl group (IV) (Czernilofsky and Kuechler 1972). Although both react with nucleophilic residues, their specificities are somewhat different. Reagent II reacts preferentially with sulfhydryl groups, but can also alkylate histidine, hydroxyl, and amino groups in proteins. In addition, it can react with RNA. Bromoacetyl-phenylalanyl-tRNA (IIa) labeled preferentially large-subunit proteins L2 and L27 when the tRNA derivative was bound at the ribosomal peptidyl site (Oen et al. 1973; Pellegrini et al. 1974). Upon nonenzymatic binding to the aminoacyl site, large-subunit protein L16 was labeled (Eilat et al. 1974b). However, in other experiments (Breitmeyer and Noller 1976), RNA was labeled. Eilat et al. (1974a) prepared a series of bromoacetyl derivatives of peptidyl-tRNA (IIc) containing between 1 and 16 glycine residues attached to phenylalanyl-tRNA^{Phe}. These were used to map the site of the growing peptide chain on the ribosome. At short chain length, L2 was labeled predominantly. With increasing chain length, L16, L26/27, L32/33, and L24 became labeled. When the bromoacetyl group was placed on the ϵ-NH$_2$ of lysyl-tRNA (IId), mostly 23S RNA was labeled from the peptidyl site and L27 from the aminoacyl site (Johnson, this volume). Peptidyl-site labeling was much greater than aminoacyl-site labeling.

Reagent IV preferentially acylates amino groups, but it can also react

Table 1 Affinity and photoaffinity reagents attached to tRNA

Reagent		Site of attachment
—C(O)—(CH$_2$)$_3$—C$_6$H$_4$—N(CH$_2$CH$_2$Cl)$_2$	(I)	α-NH$_2$, phenylalanyl-tRNA
—C(O)—CH$_2$—Br	(II)	(a) α-NH$_2$, phenylalanyl-tRNA
		(b) α-NH$_2$, methionyl-tRNA^{fMet}
		(c) α-NH$_2$, (Gly)$_n$-phenylalanyl-tRNA (n = 1–16)
		(d) ε-NH$_2$, lysyl-tRNA
—C(O)—CH$_2$—I	(III)	α-NH$_2$, phenylalanyl-tRNA
—C(O)—O—C$_6$H$_4$—NO$_2$	(IV)	(a) α-NH$_2$, phenylalanyl-tRNA
		(b) α-NH$_2$, methionyl-tRNA^{fMet}
—C(O)—C(N$_2$)—COO—C$_2$H$_5$	(V)	α-NH$_2$, phenylalanyl-tRNA
—C(O)—CH$_2$—C$_6$H$_4$—O—C$_6$H$_3$(NO$_2$)—N$_3$	(VI)	α-NH$_2$, phenylalanyl-tRNA
(a) —C(O)—CH$_2$—NH—C$_6$H$_3$(NO$_2$)—N$_3$	(VII)	(a) α-NH$_2$, phenylalanyl-tRNA
		(b) α-NH$_2$, acp^{3}U of *E. coli* tRNA^{Phe}

Photoaffinity	Identified components		Reference
	major	minor	
−	23S RNA		Bochkareva et al. (1971, 1973)
−	L2, L27	L14-L17	Pellegrini et al. (1972, 1974) Oen et al. (1973)
	L16 (aminoacyl site) 23S RNA		Eilat et al. (1974b) Breitmeyer and Noller (1976)
−	L2, L27	L6, L14, L20	Sopori et al. (1974)
−	L2, L26/27, L32/33, L24 (with increasing chain length)	L16	Eilat et al. (1974a)
−	23S RNA (peptidyl site)	L2, L13/14/15	Johnson (this volume)
−	L27 (aminoacyl site)		
−	23S RNA (A*UUUAG)		Yukioka et al. (1975, 1977)
−	L15, L27	L2, L16, L34	Czernilofsky and Kuechler (1972), Czernilofsky et al. (1974), Bauer et al. (1975), E. Kuechler and G. Stöffler (unpubl.)
−	L15, L27		Hauptmann et al. (1974), Collatz et al. (1976)
+	23S RNA		Bispink and Matthaei (1973)
+	L11, L18	L2, L5, L27, L33	Hsiung et al. (1974)
+	L11, L18	L2, L5, L27, L33	Hsiung and Cantor (1974)
+	50S, 30S (peptidyl site)		F.-L. Lin and J. Ofengand (unpubl.)

Table 1 (*Continued*)

Reagent		Site of attachment
[structure: –C(=O)–C₆H₃(NO₂)–N₃]	(VIII)	α-NH$_2$, phenylalanyl-tRNA
[structure: –C(=O)–CH(NHtBoc)–CH$_2$–C$_6$H$_4$–N$_3$]	(IX)	(a) α-NH$_2$, phenylalanyl-tRNA (b) α-NH$_2$, methionyl-tRNA^{fMet} (c) α-NH$_2$, (Gly)$_n$-phenylalanyl-tRNA (n = 2, 4)
[structure: –C(=O)–CH$_2$CH$_2$–C$_6$H$_4$–C(=O)–C$_6$H$_5$]	(X)	α-NH$_2$, phenylalanyl-tRNA
[structure: –C(=O)–CH$_2$–CH$_2$–C(=O)–C$_6$H$_5$]	(XI)	α-NH$_2$, phenylalanyl-tRNA
[structure: –CH$_2$–C(=O)–C$_6$H$_4$–N$_3$]	(XII)	(a) SH, s^{4}U8 of *E. coli* tRNA^{Phe} (b) SH, s^{4}U8 of *E. coli* tRNA$_1^{Val}$ (c) SH, s^{4}U8 of *E. coli* tRNA^{fMet}
[structure: –CH$_2$–C(=O)–O–CH$_2$–C(=O)–C$_6$H$_4$–N$_3$]	(XIII)	(a) SH, s^{4}U8 of *E. coli* tRNA^{Phe}, tRNA$_1^{Val}$ (b) tRNA^{fMet}
[structure: –C(=O)–(CH$_2$)$_5$–NH–C$_6$H$_3$(NO$_2$)–N$_3$]	(XIV)	α-NH$_2$, acp^{3}U47 of *E. coli* tRNA^{Phe}

The arrow indicates the site of reaction. (*) Modified nucleotide residue; (SH) sulfhydryl; (TR) tight couple ribosomes (Noll et al. 1973); (WR) washed ribosomes (Schwartz et al. 1975); (I site) initiation-factor-dependent binding site.

	Identified components		Reference
Photoaffinity	major	minor	
+	23S RNA S3, S7, S14 (on 30S)	50S proteins	Girshovich et al. (1974)
+	23S RNA (18S fragment)		Sonenberg et al. (1975)
+	23S RNA (18S fragment)		Sonenberg et al. (1976)
+	23S RNA (18S fragment)		Sonenberg et al. (1977)
+	23S RNA (18S fragment)		Barta et al. (1975)
+	23S RNA (18S fragment)		Kuechler and Barta (1977)
+	30S proteins (aminoacyl site, TR)		Hsu et al. (1978), L. Hsu et al. (unpubl.)
+	30S subunit (aminoacyl site, TR)		Hsu et al. (1978), L. Hsu et al. (unpubl.)
+	30S, 50S proteins 16S RNA (aminoacyl site, WR)		Schwartz et al. (1975)
+	30S subunit (I site, WR)	50S subunit	Schwartz et al. (1976)
+	30S subunit (aminoacyl site, TR)		Hsu et al. (1978), L. Hsu et al. (unpubl.)
+	30S subunit (I site, WR)		Schwartz et al. (1976)
+	50S subunit, 30S protein(s) (peptidyl site)		F.-L. Lin and J. Ofengand (unpubl.)

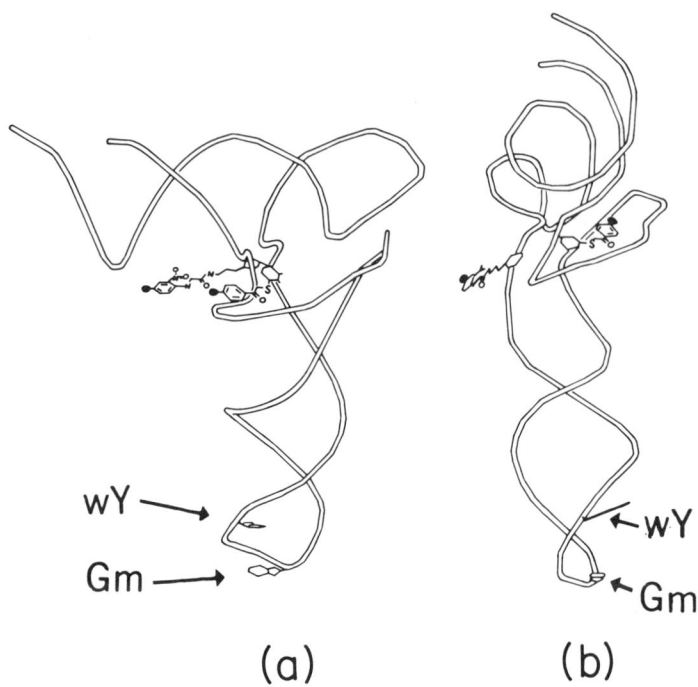

Figure 1 Stereochemistry of reagents VII and XII when attached to the body of tRNA. Three views of the structure of yeast tRNA[Phe] (Kim et al. 1974) with reagent VII attached to U47 and reagent XII attached to U8 as models for the addition of XIV to acp^{3}U47 and of XII to s^{4}U8. The continuous cylinder shows the ribose phosphate backbone. Reagents VII and XII, drawn to scale, are shown in their maximally extended form. Other less-extended conformations are also possible as a result of bond rotations. The azide group (●) is not to scale. The 5′-anticodon base, Gm34, and wybutine at position 37 are shown in *a* and *b*.

with the hydroxyl group. During the reaction, *p*-nitrophenol is cleaved off and the cross-link is formed via the remaining carbonyl group. Therefore, with this reagent more than with II the two reacting sites must be in extremely close contact to allow the formation of a covalent bond. Labeling occurred predominantly on L15 and L27 (Czernilofsky et al. 1974). There was no reaction with rRNA. L2, which was a major labeled component when IIa was used, was found as a minor component with IVa.

The same affinity reagents were attached to methionyl-tRNA[fMet] with results similar to those with the phenylalanyl-tRNA derivatives except that labeling of minor components was reduced (Hauptmann et al. 1974; Sopori et al. 1974). This may be due to a higher degree of specificity in

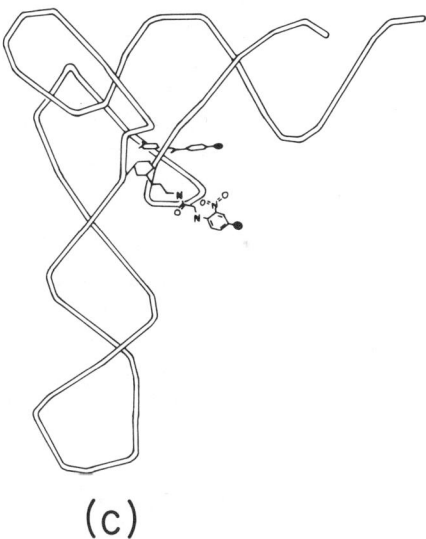

(c)

Figure 1 (*Continued*)

the presence of initiation factors. The difference in the labeling patterns observed between reagents IIa,b and IVa,b may be due to the difference in length or to the different nucleophile specificity.

Other alkylating tRNAs investigated were the N-chlorambucyl- (I) and iodoacetyl- (III) phenylalanyl-tRNAs. Both of these tRNAs reacted mainly with 23S RNA.

Most of the photoaffinity-label derivatives substituted at the α-amino group of aminoacyl-tRNA also reacted primarily with 23S RNA. They include the ethyldiazomalonyl group (V), various aromatic azides (VIII, IX), and the aromatic ketones (X, XI). In all cases in which it was studied, labeling occurred in the 18S fragment derived from the 3' part of 23S RNA. So far, sequence analysis has been performed only for the site labeled by III. The sequence $A^xUUUUAG$, in which A^x indicates the alkylated residue, is also situated in the 3' part of 23S RNA (Yukioka et al. 1977).

The only exceptions to photoaffinity labeling of 23S RNA are p-(2-nitro-4-azidophenoxy)phenylacetyl-phenylalanyl-tRNA (VI), which predominantly reacted with L1, L11, L18, and L22, and N-(2-nitro-4-azidophenyl)glycyl-phenylalanyl-tRNA (VIIa), which labeled L11 and L18 (Hsiung et al. 1974; Hsiung and Cantor 1974). This result is unexpected, as the length of both of these reagents falls within the range covered by VIII and IX, which have similar or identical reactive ends. Reagents VI and VIIa labeled quite different proteins from those that reacted with either reagent II or IV. This may be due to different reactivity patterns or to their differences in length.

Some of these reagents have also been applied to affinity labeling of eukaryotic ribosomes. Czernilofsky et al. (1977) used reagent IV attached to yeast phenylalanyl-tRNA to label rat liver ribosomes, and Perez-Gosalbez et al. (1978) used reagents III and IV attached to yeast phenylalanyl-tRNA to label yeast ribosomes.

To identify ribosomal components near regions of the tRNA other than the aminoacyl end, the aromatic azides XII and XIII were attached to the S atom of s^4U8 of several *E. coli* tRNAs. As illustrated in Figure 1, this location provides different information from that described above because it is near the junction of the two helical stems. Most work has been done with XIIa, the phenylalanyl-tRNA derivative (Hsu et al. 1978; L. Hsu et al., unpubl.). When this tRNA was enzymatically bound to the aminoacyl site, only labeling of 30S proteins was found. The 50S subunit and 16S RNA did not react. Cross-linking required not only mRNA (poly[U]) and EF-Tu, but also the hydrolysis of GTP. Replacement of GTP by the nonhydrolyzable analog GDPCP completely blocked cross-linking, although it only decreased noncovalent binding by half. This result could be interpreted in one of several ways.

1. There is a distinct recognition site, different from the aminoacyl site, that requires GTP hydrolysis for entry (Lake 1977).
2. The ribosome or tRNA has a different conformation when EF-Tu is bound in the complex.
3. Ribosome-bound EF-Tu interferes directly with the cross-linking reaction.

Increasing the Mg^{++} concentration increased the cross-linking efficiency, suggesting that the second possibility may be the correct interpretation.

Cross-linking was aminoacyl-site-specific, as translocation of the tRNA to the peptidyl site with EF-G also abolished cross-linking. It seems that close contact between 30S proteins and the central fold of the tRNA can only occur once the tRNA is correctly bound (i.e., with hydrolysis of GTP) into the aminoacyl site.

Similar specific binding to the 30S subunit was also observed with tRNA^{Val} (XIIb) and with the longer probe (XIII) on both tRNA^{Phe} and tRNA^{Val}. The labeling of the 50S subunit with XIIb observed previously (Schwartz et al. 1975) was not detected when tight couple ribosomes were used. This observation emphasizes that ribosomes in subtly different conformational states may give different cross-linking results.

When reagents XII and XIII were attached to tRNA^{fMet} (XIIc and XIIIb) to study interaction at the initiator tRNA binding sites on 70S ribosomes (Schwartz et al. 1976), the 30S subunit was the primary target in a reaction strongly dependent on initiation factors. This result should be compared to the failure of XIIa to cross-link when translocated to the peptidyl site. This result might be explained by appropriate differences in the conformations of phenylalanyl-tRNA and formylmethionyl-tRNA when on the ribosome, but the possibility should also be considered that the initiator tRNA binding site on 70S ribosomes is not identical to the peptidyl site. Further studies will be needed to clarify this situation. None of the components cross-linked to reagent XII or XIII have been identified as yet. The 30S proteins linked to XIIa are currently under analysis.

Attachment of 6-(2-nitro-4-azidophenylamino)caproate (XIV) to the acp^{3}U47 of *E. coli* tRNA^{Phe} was used to probe the region of the ribosome opposite to that which reacted with XII and XIII (see Fig. 1). At the peptidyl site, XIV reacted with both 50S and 30S subunits. Recall that the complementary probe XIIa was unreactive at the peptidyl site. The yield and distribution of the cross-link between subunits depended on the presence or absence of IF-2, and on whether GTP or GDPCP was used. In all cases, the cross-linked tRNA was still capable of reaction with puromycin. All of the 30S cross-linking was to protein. The 50S cross-linking site is not yet clear. When the shorter probe VIIb was substituted for XIV at the peptidyl site, cross-linking similar to XIV could be detected, but with lower yield. Characterization of the cross-linked ribosomal components has not been done.

Specificity of the Labeling Reaction

The significance of an affinity-labeling experiment on a complicated organelle like the ribosome depends critically on the control experiments used to check the specificity of the reaction. Fortunately, affinity labeling with tRNA derivatives has allowed this specificity to be examined both by stimulatory and inhibitory effects of various ribosomal ligands. The heterogeneity of the usual ribosome preparations has posed a serious problem. Normal preparations of ribosomes are only 20–50% active in protein synthesis. Thus, a considerable fraction is either totally inactive or is blocked in some step of the translational cycle. To insure that only the active fraction is labeled, the affinity reaction has frequently been performed with a molar excess of ribosomes over tRNA. Thus, only those ribosomes having a high binding constant should bind the tRNA.

The basic criteria used for affinity labeling at correct tRNA binding sites were the dependence of the reaction on the presence of the cognate mRNA and the inhibition by underivatized aminoacyl-tRNA (Bauer et al. 1975). However, such controls do not specify the site of labeling, nor do they specify that the labeled ribosomes were capable of peptide-bond formation.

Labeling at the peptidyl site has frequently been checked by the addition of puromycin. Puromycin reacts with peptidyl-tRNA bound at the peptidyl site. The reaction is catalyzed by peptidyl transferase and results in the release of peptidyl-puromycin from the ribosome. Affinity labeling at the peptidyl site of active ribosomes should therefore be strongly inhibited by adding the antibiotic either before or together with the aminoacyl-tRNA derivative (Czernilofsky et al. 1974; Kuechler 1976; Schwartz and Ofengand 1978). A further test for peptidyl-site labeling that was applicable to derivatives of initiator tRNA was the dependence on the presence of IF-2 (Schwartz et al. 1976; R. Hauptmann, unpubl.). The most unambiguous proof for affinity labeling at the peptidyl site of functional ribosomes has been the demonstration that the covalently attached affinity label can still participate in peptide-bond formation (Pellegrini et al. 1974; Hsiung and Cantor 1974; Hsiung et al. 1974; Sopori et al. 1974; Sonenberg et al. 1977; Ofengand et al. 1979). Evidence for affinity labeling at the aminoacyl site has been provided by the dependence of the reaction on EF-Tu and GTP in the presence of an excess of deacylated tRNA (Schwartz et al. 1975; Johnson et al. 1977; L. Hsu et al., unpubl.); by tetracycline inhibition (Eilat et al. 1974b); by insensitivity to puromycin; in suitable cases, by inhibition as a result of translocation to the peptidyl site (L. Hsu et al., unpubl.); and by incorporation of a peptidyl donor from the peptidyl site into covalent linkage with the ribosome (L. Hsu et al., unpubl.).

INTERACTION BETWEEN tRNA AND THE mRNA-RIBOSOME COMPLEX

Affinity and Photoaffinity Derivatives of mRNA Analogs

Most of the information about the site of codon-anticodon interaction comes from affinity-labeling studies with analogs of mRNA. For this purpose, derivatives of oligonucleotides and polynucleotides have been prepared (Table 2). The first polynucleotide tested was poly-4-thiouridylic acid (XX) (Fiser et al. 1974). It resembles poly(U) in its physicochemical properties and codes for poly(phenylalanine) in an in vitro protein-synthesizing system. Upon irradiation at 330 nm, the C—S bond is activated. It reacts with oxygen to yield an unstable product that subsequently reacts with a neighboring nucleophilic group resulting in cleavage of the C—S bond. XX was found to react with the small-subunit proteins S1, S18, and S21, and also with 16S RNA (Fiser et al. 1975b). Irradiation resulted in a rapid inhibition of poly(phenylalanine) synthesis. By contrast, ribosomal complexes containing poly(U) were refractory to irradiation at 330 nm. The specificity of the reaction was further shown by the demonstration that preincubation of ribosomes with phage R17 RNA in the presence of initiation factors and formylmethionyl-tRNA inhibited both binding and photoaffinity labeling with XX (Fiser et al. 1977).

Table 2 Polynucleotide and oligonucleotide derivatives used as affinity and photoaffinity labels

Derivative	Photoaffinity	Identified components		Reference
		major	minor	
ApUpG / UpGpA (XV)	—	S4, Cys-10 of S18 (on 70S) S4, S12, S18 (on 30S) S18 (on 70S) S4, S18 (on 30S)	S21 (on 70S) S11, S13, S21 (on 30S)	Pongs and Lanka (1975), Yaguchi et al. (1978) Pongs et al. (1975a) Pongs and Rossner (1976)
(XVI)	—	16S RNA	30S proteins	Budker et al. (1973)
(XVII)	—	S1	S18	Pongs et al. (1976)

Table 2 (*Continued*)

Derivative	Photoaffinity	Identified components major	Identified components minor	Reference
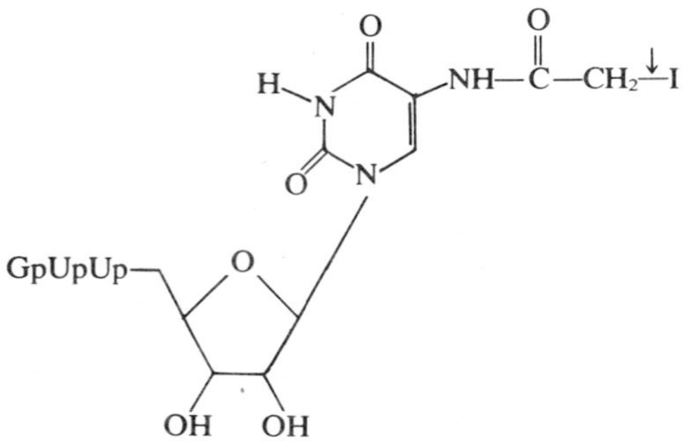 (XVIII)	—	S1, S18		Lührmann et al. (1976)
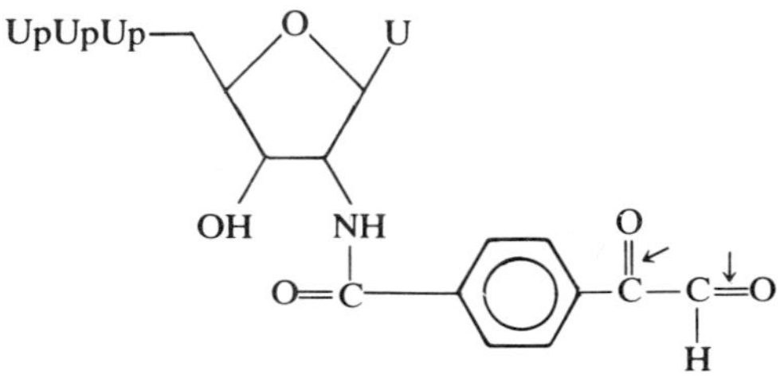 (XIX)	—	16S RNA (segment I)		Wagner and Gassen (1975), Wagner et al. (1976)

426

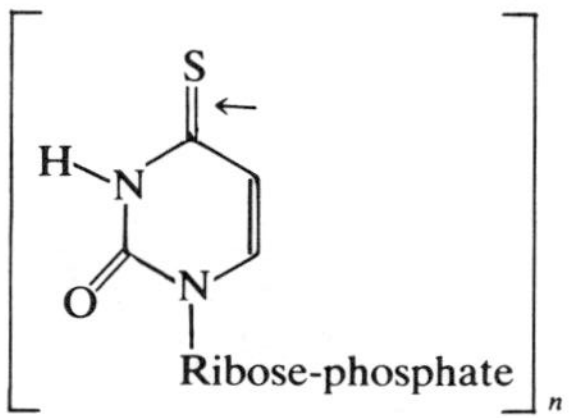

(XX) + S1, S18, S21
16S RNA
Fiser et al. (1974, 1975b, 1977)

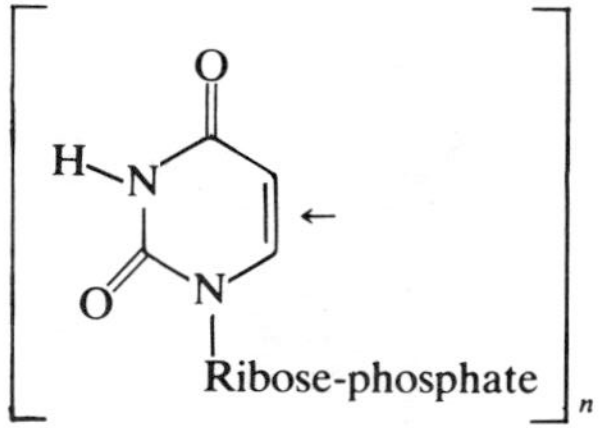

(XXI) + 16S RNA
S1
Schenkman et al. (1974)
Fiser et al. (1975a),
Margaritella and Kuechler (1978)

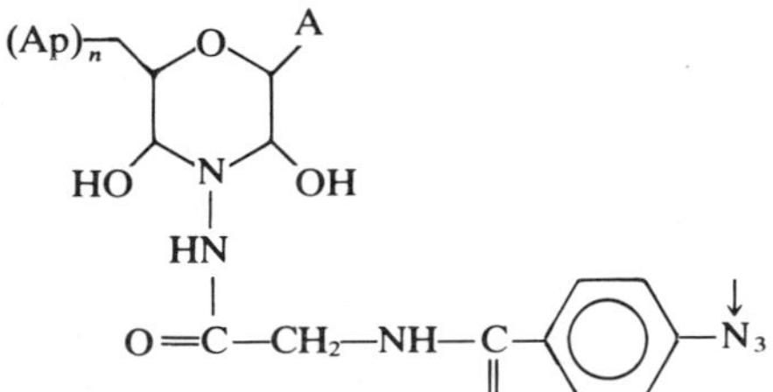

(XXII) + S3, S5
16S RNA
Towbin and Elson (1978)

($n = 5$–7)

The arrow indicates the site of reaction.

Similar experiments have been carried out with poly(U) (XXI), which photoreacts upon irradiation at 254 nm, with 16S RNA, and with S1 (Schenkman et al. 1974; Fiser et al. 1975a). Labeling on S1 was strongly stimulated by the addition of tRNA^{Phe} (Margaritella and Kuechler 1978).

Several derivatives of oligonucleotides have been tested as affinity labels. pApUpG and pUpGpA were substituted at the 5'-phosphate with a bromoacetamidophenyl residue (XV). When these derivatives were incubated with 70S ribosomes, S18 and a small amount of S21 were labeled. Under certain conditions S4 was also labeled. With 30S ribosomal subunits, S4, S12, and S18 were found to be labeled (Pongs and Lanka 1975; Pongs et al. 1975a; Pongs and Rossner 1976); the site of reaction in S18 was Cys-10 (Yaguchi et al. 1978). Attachment of the affinity label at the 3' end of (Ap)$_7$A was achieved by forming the p-[N-(2-chloroethyl)-N-(methyl)amino]benzylidene derivative (XVI) with the 2'-OH, 3'-OH groups of the terminal ribose. In this case, primarily 16S RNA was labeled (Budker et al. 1973). Alternatively, 5-aminouridylic acid or 2'-aminouridylic acid was added to the 3' terminal of a trinucleotide. The reactive moiety was then attached to the amino group of the aminouridine. ApUpGp-5(NH$_2$)U and GpUpUp-5(NH$_2$)U substituted with bromoacetyl (XVII) or iodoacetyl (XVIII) residues, respectively, were shown to react with S1 and S18 (Pongs et al. 1976; Lührmann et al. 1976). The benzoylglyoxal derivative of UpUpUp-2'-(NH$_2$)U (XIX) was found to react with segment I of the 16S RNA (Wagner and Gassen 1975; Wagner et al. 1976). Recently, Elson and coworkers have reported another affinity reagent derived from oligo(A). N-[4-azidobenzoyl]glycylhydrazide attached to the dialdehyde of oligo(A) (XXII) (Towbin and Elson 1978) is similar to XVI in that it also is a derivative of the 3'-terminal ribose with similar stereochemistry, except that it is a photoreactive derivative; it labeled S3, S5, and 16S RNA.

The specificity of these labeling reactions was shown by the demonstration that ribosomes labeled with these oligonucleotides were primed to bind their cognate aminoacyl-tRNA. The labeled ribosomes showed decreased binding and translation of poly(U) and the other mRNAs. In some instances it was also demonstrated that the presence of the cognate tRNA stimulated the affinity labeling. This is the most direct proof of the specificity of the reaction.

Photo Cross-linking from Wybutine to mRNA and Translocation

It has recently been discovered that phenylalanyl-tRNA^{Phe} from yeast can be photo cross-linked to poly(U) when ribosomal complexes are irradiated at 320 nm (A. J. M. Matzke et al., unpubl.). The cross-linked tRNA-poly(U) product can be isolated by chromatography on oligo(dA)-cellulose. Several lines of evidence indicate that the photoreaction occurs at the wybutine base at position 37 of yeast tRNA^{Phe}, i.e., next to the 3' side of the anticodon (Fig. 1).

1. Upon digestion of the cross-linked phenylalanyl-tRNA^{Phe}·poly(U) complex with RNase T1 only the tRNA fragment from the anticodon loop was retained on an oligo(dA)-cellulose column.
2. Total hydrolysis of this tRNA fragment yielded all the nucleotides surrounding the wY in the correct stoichiometric ratio except for the wY that was missing. Instead, a new spot appeared on the chromatogram.
3. tRNA^{Phe} from *E. coli*, which does not contain the wybutine, was not photo cross-linked to poly(U) under these conditions.

The efficiency of cross-linking was very similar when acetylphenylalanyl-tRNA^{Phe} was bound nonenzymatically at the aminoacyl site (Modolell et al. 1973) or at the peptidyl site. When peptidyl-site-bound acetyl-[^{3}H]phenylalanyl-tRNA^{Phe} was irradiated and subsequently treated with puromycin, there was a drastic decrease in the amount of [^{3}H]phenylalanine bound to poly(U). Thus, acetylphenylalanyl-tRNA^{Phe} cross-linked to poly(U) at the peptidyl site is still capable of peptide-bond formation. On the other hand, acetyl-[^{3}H]phenylalanyl-tRNA cross-linked at the aminoacyl site was refractory to puromycin treatment unless EF-G and GTP were added to promote translocation, in which case the [^{3}H]phenylalanine was released from linkage to the poly(U). This result clearly demonstrates that acetylphenylalanyl-tRNA cross-linked from wY to poly(U) at the aminoacyl site can still be translocated efficiently to the peptidyl site.

Several conclusions concerning the mechanism of translocation can be drawn from these experiments.

1. tRNA and mRNA are able to move together during translocation.
2. tRNA and mRNA are closely associated both at the aminoacyl site and at the peptidyl site.
3. The relative arrangement of the wybutine base with respect to the mRNA is similar both at the aminoacyl site and at the peptidyl site.

Two basically different hypotheses concerning the mechanism of translocation have been suggested (Woese 1970; Rich 1974). Woese proposed that translocation is brought about by a change in the configuration of the anticodon loop from a 5'-stacked form to a 3'-stacked form. During this structural transition, mRNA and tRNA would perform a 180° rotation with respect to each other. In the 5'-stacked form the anticodon would be stacked onto the U on the 5' side of the anticodon. The loop would make the turn between the anticodon and the wybutine. In the 3'-stacked form the wybutine is stacked onto the anticodon. This is the structure found in the tRNA^{Phe} crystal (Kim et al. 1973, 1974; Robertus et al. 1974). Thus, the orientation of the wybutine with respect to the mRNA is very different when the 5'-stacked and the 3'-stacked forms are compared (Woese 1970; Lake 1977). In the 5'-stacked form the wybutine points away from the mRNA, whereas in the 3'-stacked form it would point toward the mRNA.

Due to steric inhibition resulting from different orientations of the wybutine, the efficiencies of photo cross-linking with the mRNA should be expected to be different when comparing tRNA in the aminoacyl site and in the peptidyl site. Such a difference was, however, not observed. Furthermore, it is difficult to imagine how the tRNA-poly(U) complex cross-linked at the wybutine could still perform the 180° rotation from a 5'-stacked to a 3'-stacked form during translocation. Thus, the results of photo cross-linking between tRNA^{Phe} and poly(U) do not support the model of translocation suggested by Woese.

Instead, the data fit a model of translocation in which the relative orientation of the wybutine with respect to the mRNA is more or less conserved during translocation. For example, in the model suggested by Rich (1974) the tRNAs both at the peptidyl site and the aminoacyl site have the same structure as in the crystal. The experimental results are in agreement with this model but they cannot be taken as proof, as the molecular mechanism of the photo cross-linking is not yet known. More structural information will be needed before conclusions about the absolute configuration of the anticodon loop on the ribosome can be drawn.

Photo Cross-linking from (c)mo⁵U34 to 16S RNA

Certain *E. coli* tRNAs can be photo cross-linked to 16S RNA when ribosomal peptidyl-site complexes are irradiated at 310–325 nm (Schwartz and Ofengand 1978; Ofengand et al. 1978, 1979). Cross-linking required the presence of the cognate mRNA. Up to 70% of the bound tRNA could be cross-linked. The reaction was peptidyl-site-specific, as more than 80% of the ribosome-attached tRNA could still react with puromycin. Moreover, the same aminoacyl-tRNA bound to the aminoacyl site (by EF-Tu) did not cross-link.

The reactive base in the tRNA was deduced to be (c)mo^{5}U34, the 5'-anticodon base (see Fig. 1). The argument, given in detail elsewhere (Ofengand et al. 1979), is based on (1) a comparison of nucleotides at a given position in the reactive tRNAs, *E. coli* tRNA$_1^{Val}$ and tRNA$_1^{Ser}$ and *B. subtilis* tRNA^{Val} and tRNA^{Thr}, and in the inactive tRNAs, *E. coli* 5-fluorouracil-substituted tRNA$_1^{Val}$ (Horowitz et al. 1974), tRNA$_2^{Val}$, tRNA^{Phe}, and tRNA^{fMet}; (2) the lack of involvement of the s^{4}U residue; (3) the single-hit kinetics of cross-linking; and (4) the linking of each of the reactive tRNAs to the same-size fragment of the 16S RNA. Since both mo^{5}U and cmo^{5}U were equally well cross-linked, the COOH group cannot play any essential role in the mechanism of cross-linking.

The site of attachment to 16S RNA was localized to an approximately 100-nucleotide fragment of the 8S piece (Zimmermann et al. 1979). The 8S piece is the 3'-third of the molecule. The two RNAs are unlikely to be linked by a ribosomal protein spacer because the complex is resistant to proteases, and both the characteristics of formation and photolysis of the cross-link are inconsistent with known nucleotide-amino acid photo cross-linking reactions. The presence of an mRNA spacer has not been ruled out entirely, but since GpUpU can serve as well as poly(U_2, G), the putative pair of linkage points can be, at most, 2 nucleotides apart.[1]

Cross-linking was not quenched by free radical scavenging agents and was not affected by the presence or absence of oxygen or by prephotolysis. Most importantly, the cross-link could be cleaved by 254-nm irradiation with complete regeneration of the reactants. Thus, the reactants could subsequently be cross-linked again. The sum of these properties are characteristic of the formation of pyrimidine-pyrimidine cyclobutane dimers.

Should this be the correct structure, then the 5'-anticodon base of tRNA, which is located at the tip of the molecule, must be very close (<4 Å) to a pyrimidine base somewhere in the 3'-third of the 16S RNA and may even be stacked upon it. Such an interaction need not interfere with codon-anticodon recognition, if indeed this occurs directly in the peptidyl site, but rather could extend the anticodon stack into the 16S RNA and stabilize the tRNA-ribosome interaction. Such a stabilization might be important for maintenance of the proper alignment of peptidyl-tRNA on the ribosome, particularly if codon-anticodon base pairing is incomplete or absent. The presence of another tRNA in the aminoacyl site would not complicate this model. It is possible, by suitable manipulation of the anticodon loops and/or diester link between mRNA codons, for a base from 16S RNA to stack on the 5'-anticodon base at the peptidyl site even though it would now be located in the center of a 6-bp codon-anticodon complex.

A list of the cross-linking results from minor bases around the anticodon is given in Table 3. Recently, Abdurashidova et al. (1978) reported preliminary experiments using 254-nm irradiation to randomly cross-link tRNA to the aminoacyl and peptidyl sites. With suitable controls for ribosome survival, site specificity, and identification of the tRNA bases involved, this approach should also be informative.

LOCATION OF FUNCTIONAL SITES ON THE RIBOSOME

Enough information now exists about the structural organization of the *E. coli* ribosome to ask how the accumulated data of Tables 1–3 can be used to locate the peptidyl transferase center and the decoding region on a

---

[1]*Note added in proof*: The absence of mRNA has now been shown.

Table 3 Photo-induced cross-linking of tRNA to the ribosome-mRNA complex

tRNA	Site in tRNA	Ribosome-mRNA complex		Reference
		binding site	labeling site	
E. coli tRNA$_1^{Val}$ and tRNA$_1^{Ser}$	cmo^{4}U34	peptidyl	16S RNA (8S fragment)	Schwartz and Ofengand (1978), Ofengand et al. (1978, 1979), Zimmermann et al. (1979)
Bacillus subtilis tRNA^{Val} and tRNA^{Thr}	mo^{5}U34	peptidyl	16S RNA (8S fragment)	Ofengand et al. (1979), Zimmermann et al. (1979)
Yeast tRNA^{Phe}	wybutine at position 37	aminoacyl, peptidyl	mRNA (poly[U])	A. J. M. Matzke et al. (unpubl.)

ribosomal particle. However, before proceeding, one must recognize that there is still no universal agreement about the structure of even the exhaustively studied *E. coli* ribosome. Figure 2 shows three models of this ribosome with a tRNA added to give some dimension to our quest for the tRNA binding sites.

Only affinity-labeling data, as given in Tables 1 and 2, have been used in the following summary. There is a large body of additional data (reviewed by Stöffler and Wittmann 1977), based on indirect evidence of various kinds, that supports the identifications found by affinity labeling. However, since the methods used were not always able to distinguish between direct and indirect effects, they have not been considered here. One point that must be mentioned, however, is that affinity labeling and immuno-electron microscopy of ribosomal proteins, the technique used for localization of the proteins on the ribosome, do not necessarily detect the same region of a protein. If the protein is elongated, as many are (multiple antigenic sites will be noted in Figs. 3 and 4), the affinity-labeled site and the antigenic site may be far apart. Given this situation, a localization becomes more convincing when at least one antigenic site for each of a set of affinity-labeled proteins is observed to cluster in a single region.

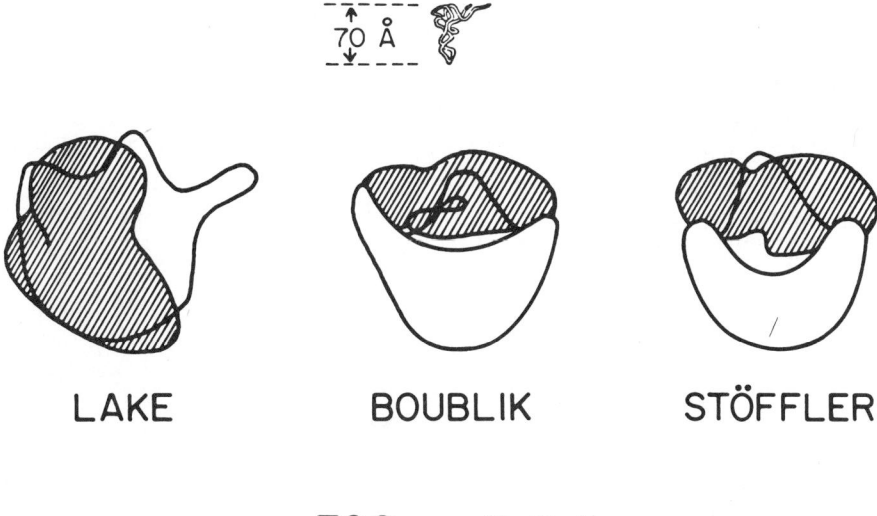

Figure 2 Proposed structures for the *E. coli* 70S ribosome. The three structural diagrams are taken from Lake (1978), M. Boublik et al. (1977 and pers. comm.), and Brimacombe et al. (1978). The tRNA structure is the crystal structure of yeast tRNA^{Phe} (Kim et al. 1974). All are drawn to the same scale.

Peptidyl Transferase Center

The affinity probes of the peptidyl transferase center listed in Table 1 reacted mostly with the 50S subunit. All of the major proteins listed in the table are shown in Figure 3. To be counted as a true minor protein, two listings were arbitrarily required in order to reduce the frequency of false positive results. Only large-subunit proteins L5 and L14 qualified (shown in Fig. 3 as broken circles). Additional affinity-labeling information concerning the region of the peptidyl transferase center has been obtained by the use of suitably modified antibiotics (reviewed by Pellegrini and Cantor 1977; Cooperman 1978). Only two studies, both of which used puromycin or analogs (Pongs et al. 1975b; Jaynes et al. 1978), have identified additional proteins, which are indicated by underlining. It is clear that the peptidyl transferase center is localized to the central cavity and left arm (Fig. 3a), as indicated by the shading.

The location of the 18S fragment of the 23S RNA, and specifically the sequence labeled by Yukioka et al. (1977), is not known, although the 18S fragment does complex with L1, L5, L18, and L23 (Branlant et al. 1977), which is in agreement with its being a frequent target of affinity probes. It is also interesting to note that the two short probes (II and IV) labeled primarily L2, L16, and L27, which are clustered on the left arm of the 50S chair, whereas the two longer probes (VI and VII) labeled L1, L11, and L18, which are clustered in the seat region.

Decoding Site

The mRNA analogs listed in Table 2 labeled nine proteins of the 30S subunit, as well as the 16S RNA. Two of them, S11 and S13, have not been included in Figure 4, as they were minor proteins found in only one case and were not reactive in 70S ribosomes. The proteins (shown inside

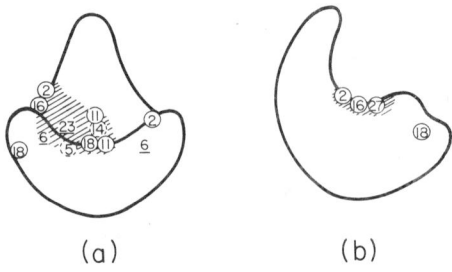

Figure 3 Proteins on the surface of the *E. coli* 50S subunit that are in the vicinity of the peptidyl transferase center. (*a*) Front view; (*b*) left-side view. Location of the proteins and overall shape of the particle is according to Brimacombe et al. (1978). Circled numbers represent proteins near the peptidyl transferase center according to the data of Table 1. Solid circles are major proteins and dashed circles are minor proteins. The criteria for selection are given in the text. Underlined numbers denote proteins affinity labeled by puromycin or an analog (Pongs et al. 1975b; Jaynes et al. 1978). The shaded area denotes the putative peptidyl transferase center.

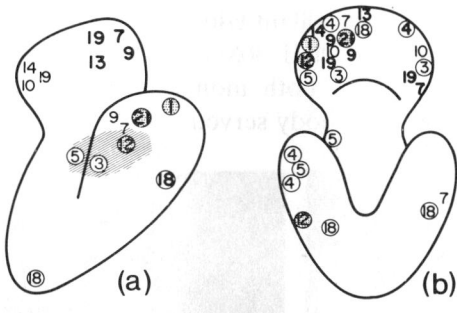

Figure 4 Proteins on the surface of the *E. coli* 30S subunit that are in the vicinity of mRNA or are near the 3' segment of the 16S RNA. Location of the proteins and overall shape of the particles are from (*a*) Lake (1978) and Winkelmann and Kahan (1978) and (*b*) Brimacombe et al. (1978) as derived by immunoelectron microscopy. Circled numbers represent proteins near mRNA according to the data of Table 2. Only major proteins have been considered. Plain numbers represent proteins associated with the 3'-third of the 16S RNA on the basis of protein-RNA binding studies (Zimmermann 1974), analysis of ribonucleoprotein fragments (Morgan and Brimacombe 1972; Rinke et al. 1977), and both chemical and photochemical cross-linking (Kenner 1973; Czernilofsky et al. 1975; Rinke et al. 1976; see also Stöffler and Wittmann 1977). Stippled circles represent proteins near to both mRNA and the 3'-third of 16S RNA. Boldface type signifies proteins on the surface toward the viewer and light type indicates those on the surface away from the viewer. The approximate location of the m_2^6A residues near the 3' end of the 16S RNA is shown by shading (Politz and Glitz 1977).

circles in Fig. 4) cluster on the 30S subunit in both the Lake and Stöffler models. The Stöffler model (Fig. 4b), which has more identified proteins, also illustrates well the problem of multiple antigenic sites. Initially, it appears as though the circled proteins (Fig. 4) are scattered throughout the particle. However, on closer inspection, one can find at least one antigenic site for each of the circled proteins clustered in the upper left crest. In the Lake model, the cluster appears to be localized on and around the platform. This region should be the prime candidate for the decoding site. As indicated in Table 3, the tRNA anticodon (and thus the decoding site) is in close contact with some portion of the 3'-third of 16S RNA. In agreement with that fact, proteins known to be associated with the 3'-third of 16S RNA are also observed to cluster in the region of the mRNA marker proteins. In particular, S1, S12, and S21, which possess both properties, are grouped together.

Localization of the Decoding Site by Immunoelectron Microscopy

Keren-Zur et al. (1979) have recently succeeded in visualizing the vicinity of the decoding site by electron microscopy by taking advantage of the ability to attach tRNA to the 30S subunit via its anticodon (see Table 3). The procedure used was acylation of the α-amino position of valine on valyl-tRNA with the dinitrophenyl group, linkage to the ribosome by irradiation, and, after isolation of the 30S subunits, formation of 30S dimers by reaction with antidinitrophenyl antibody. The dimers, which were 1:1 covalent complexes of tRNA and 30S, could then be separated from unreacted 30S by centrifugation. Upon examination by electron microscopy, both monomers and dimers were found (Fig. 5). The Y-shaped antibody served as a marker for the aminoacyl end of the tRNA,

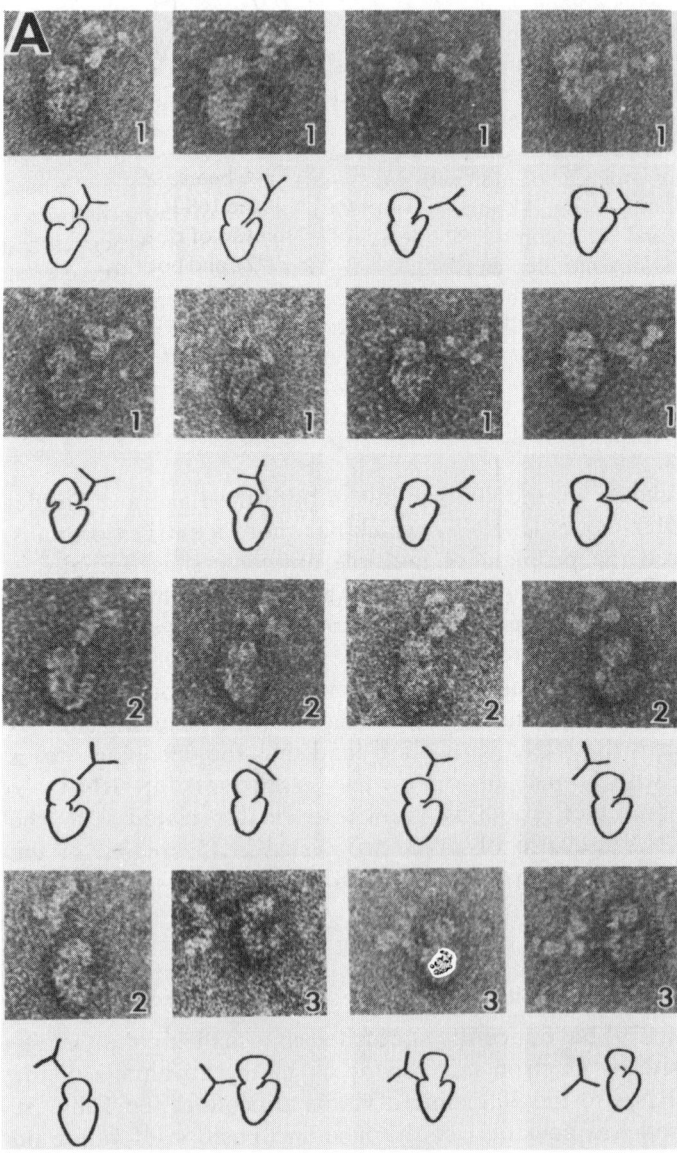

Figure 5 Electron micrographs of 30S-tRNA covalent complexes with antidinitrophenyl antibody (Keren-Zur et al. 1979). The antidinitrophenyl antibody labels the aminoacyl end of the tRNA. (*A*) Monomers with interpretive drawings.

which is about 80 Å from the actual site of cross-linking, assuming that the crystal structure of tRNA is preserved under these conditions (1 mM Mg^{++}). Multiple attachment sites of the antibody were found, distributed as shown in Figure 6. Most antibody was attached to area 1, near to the cleft and large projection, although some was also found attached to areas

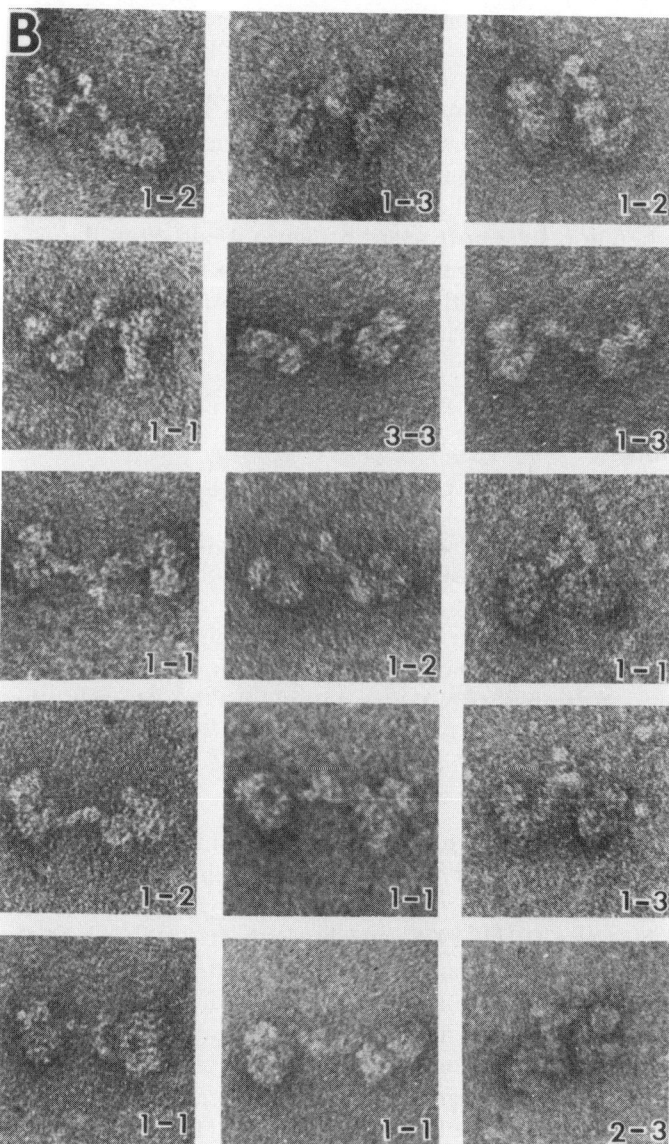

Figure 5 (continued) (*B*) Dimers of 30S subunits with attached antidinitrophenyl antibody. The numbers on each electron micrograph identify the sites of antibody attachment as defined in Fig. 6. Magnification, 520,000× in each case.

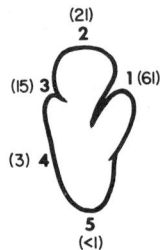

Figure 6 A schematic drawing of the 30S *E. coli* ribosomal subunit according to Keren-Zur et al. (1979). The attachment sites of the antibody to the subunit are denoted as 1–5. The percentage distribution of bound antibody is given in brackets. A total of 500 antibody-30S complexes were examined, 240 as antibody-30S monomers and 130 as antibody-30S dimers.

2 and 3. By extrapolating the 80-Å distance between the antigenic site and the point of cross-linking from each of the three major points of attachment, an approximate region of intersection was obtained. This area, in the cleft region, should be the decoding site. This result agrees well with the affinity-labeling results portrayed in Figure 4, although the fit is somewhat better to the assignments of Lake than to those of Stöffler. However, it should be stressed that the resolution achieved so far is too low to make a definite statement. Antigens closer to the cross-linking site should improve the resolution.

CONCLUDING REMARKS

From the studies done so far, four hard facts have emerged.

1. The decoding site (and thus the anticodons of both tRNAs) is located in the upper one-third and possibly near the cleft of the 30S subunit (Figs. 4 and 5).
2. The peptidyl transferase center (and thus the aminoacyl end) runs from the center of the seat of the 50S to the left arm (Fig. 3).
3. The s^4U residues of the two tRNAs are within 20–30 Å of each other (A. E. Johnson et al., unpubl.).
4. The s^4U region of the aminoacyl-site-bound tRNA is within 9 Å of the 30S subunit (Table 1; L. Hsu et al., unpubl.).

The first two facts localize the two ends of both the aminoacyl- and peptidyl-site tRNAs, whereas the latter two place certain limits on their orientation within the constraints of the first two facts. For example, the models proposed by Lake (1977, 1978) and by Rich (1974) in which the central regions of the two tRNAs at the aminoacyl and peptidyl sites are widely separated appear inconsistent with fact 3. Indeed, virtually the only way in which two tRNAs can be arranged on a ribosome so that their aminoacyl ends, their anticodons, and their s^4U sites are close to each other is in some type of side-by-side arrangement. If this is the case, fact 4 strongly implies that the s^4U side (see Fig. 1) of the tRNA bound at the aminoacyl site should be on the side away from the other tRNA so it can react with the 30S subunit.

In conclusion, the agreement among affinity-labeling data from a variety of sources and the correspondence with completely independent findings encourages one to believe that this approach will continue to be useful for mapping the topographical features of ribosomal binding sites for tRNA. In particular, affinity probes such as XII, XIII, and XIV placed at the central part of the tRNA should be particularly useful for determining the orientation of tRNA in the aminoacyl and peptidyl sites. Thus, there is every reason to believe that the affinity-labeling approach, in conjunction with immunoelectron microscopy, will eventually provide the experimental basis for determining how two tRNAs fit on a 70S ribosome.

ACKNOWLEDGMENTS

The expenses of the research carried out at the Institute of Biochemistry, University of Vienna, were met by a grant from the Austrian Fonds zur Forderung der wissenschaftlichen Forschung.

REFERENCES

Abdurashidova, G. G., M. F. Turchinsky, T. A. Salikhov, C. A. Aslanov, and E. I. Budowsky. 1978. Identification of proteins interacting with tRNA in the A and P sites of *E. coli* ribosome. *Bioorg. Khim.* **4**:982.

Barta, A., E. Kuechler, C. Branlant, J. Sri Widada, A. Krol, and J.-P. Ebel. 1975. Photoaffinity labeling of 23S RNA at the donor-site of the *Escherichia coli* ribosome. *FEBS Lett.* **56**:170.

Bauer, K., A. P. Czernilofsky, and E. Kuechler. 1975. Affinity-labeling of *Escherichia coli* ribosomes by a derivative of phenylalanyl-tRNA. A critical test for the specificity. *Biochim. Biophys. Acta* **395**:146.

Bispink, L. and H. Matthaei. 1973. Photoaffinity labeling of 23S rRNA in *Escherichia coli* ribosomes with poly(U)-coded ethyl-2-diazo-malonyl-Phe-tRNA. *FEBS Lett.* **37**:291.

Bochkareva, E. S., V. G. Budker, A. S. Girshovich, D. G. Knorre, and N. M. Teplova. 1971. An approach to specific labeling of ribosome in the region of peptidyltransferase center using *N*-acylaminoacyl tRNA with an active alkylating grouping. *FEBS Lett.* **19**:121.

———. 1973. Specific chemical modification of the ribosome close to the peptidyl transferase center. *Mol. Biol.* **7**:278.

Boublik, M., W. Hellmann, and A. K. Kleinschmidt. 1977. Size and structure of *E. coli* ribosomes by electron microscopy. *Cytobiologie* **14**:293.

Branlant, C., J. Sri Widada, A. Krol, and J.-P. Ebel. 1977. Studies on the primary structure of the ribosomal 23S RNA of *Escherichia coli*. II. A characterisation and an alignment of 24 sections spanning the entire molecule and its application to the localisation of specific fragments. *Nucleic Acids Res.* **4**:4323.

Breitmeyer, J. B. and H. F. Noller. 1976. Affinity labeling of specific regions of 23 S RNA by reaction of *N*-bromoacetyl-phenylalanyl-transfer RNA with *Escherichia coli* ribosomes. *J. Mol. Biol.* **101**:297.

Brimacombe, R., G. Stöffler, and H. G. Wittmann. 1978. Ribosome structure. *Annu. Rev. Biochem.* **47:**217.

Budker, V. G., A. S. Girshovich, N. I. Grineva, G. G. Karpova, D. G. Knorre, and N. D. Kobets. 1973. Specific chemical modification of the ribosome close to the mRNA-binding center. *Dokl. Akad. Nauk SSSR* **211:**725.

Cantor, C. R., M. Pellegrini, and H. Oen. 1974. Affinity labeling techniques for examining functional sites of ribosomes. In *Ribosomes* (ed. M. Nomura et al.), p. 573. Cold Spring Harbor Laboratory, Cold Spring Harbor, New York.

Collatz, E., E. Kuechler, G. Stöffler, and A. P. Czernilofsky. 1976. The site of reaction on ribosomal protein L27 with an affinity label derivative of tRNA$_f^{Met}$. *FEBS Lett.* **63:**283.

Cooperman, B. S. 1978. Affinity labeling studies on *Escherichia coli* ribosomes. *Bioorg. Chem.* (suppl.) **4:**81.

Czernilofsky, A. P. and E. Kuechler. 1972. Affinity label for the tRNA binding site on the *Escherichia coli* ribosome. *Biochim. Biophys. Acta* **272:**667.

Czernilofsky, A. P., C. G. Kurland, and G. Stöffler. 1975. 30S ribosomal proteins associated with the 3′-terminus of 16S RNA. *FEBS Lett.* **58:**281.

Czernilofsky, A. P., E. Collatz, G. Stöffler, and E. Kuechler. 1974. Proteins at the tRNA binding sites of *Escherichia coli* ribosomes. *Proc. Natl. Acad. Sci.* **71:**230.

Czernilofsky, A. P., E. Collatz, A. M. Gressner, I. G. Wool, and E. Kuechler. 1977. Identification of the tRNA-binding sites on rat liver ribosomes by affinity labeling. *Mol. Gen. Genet.* **153:**231.

de Groot, N., A. Panet, and Y. Lapidot. 1971. The binding of purified Phe-tRNA and peptidyl-tRNA^{Phe} to *Escherichia coli* ribosomes. *Eur. J. Biochem.* **23:**523.

Eilat, D., M. Pellegrini, H. Oen, Y. Lapidot, and C. R. Cantor. 1974a. A chemical mapping technique for exploring the location of proteins along the ribosome bound peptide chain. *J. Mol. Biol.* **88:**831.

Eilat, D., M. Pellegrini, H. Oen, N. de Groot, Y. Lapidot, and C. R. Cantor. 1974b. Affinity labeling the acceptor site of the peptidyl transferase centre of the *Escherichia coli* ribosome. *Nature* **250:**514.

Fiser, I., P. Margaritella, and E. Kuechler. 1975a. Photoaffinity reaction between polyuridylic acid and protein S1 on the *Escherichia coli* ribosome. *FEBS Lett.* **52:**281.

Fiser, I., K. H. Scheit, and E. Kuechler. 1977. Poly (4-thiouridylic acid) as messenger RNA and its application for photoaffinity labeling of the ribosomal mRNA binding site. *Eur. J. Biochem.* **74:**447.

Fiser, I., K. H. Scheit, G. Stöffler, and E. Kuechler. 1974. Identification of protein S1 at the messenger RNA binding site of the *Escherichia coli* ribosome. *Biochem. Biophys. Res. Commun.* **60:**1112.

―――. 1975b. Proteins at the mRNA binding site of the *Escherichia coli* ribosome. *FEBS Lett.* **56:**226.

Girshovich, A. S., E. S. Bochkareva, V. A. Kramarov, and Yu. A. Ovchinnikov. 1974. *E. coli* 30S and 50S ribosomal subparticle components in the localization region of the tRNA acceptor terminus. *FEBS Lett.* **45:**213.

Hauptmann, R., A. P. Czernilofsky, H. O. Voorma, G. Stöffler, and E. Kuechler. 1974. Identification of a protein at the ribosomal donor-site by affinity labeling. *Biochem. Biophys. Res. Commun.* **56:**331.

Horowitz, J., C.-N. Ou, M. Ishaq, J. Ofengand, and J. Bierbaum. 1974. Isolation and partial characterization of *Escherichia coli* valine transfer RNA with uridine and uridine-derived residues replaced by 5-fluorouridine. *J. Mol. Biol.* **87:**301.

Hsiung, N. and C. R. Cantor. 1974. A new simpler photoaffinity analogue of peptidyl tRNA. *Nucleic Acids Res.* **1:**1753.

Hsiung, N., S. A. Reines, and C. R. Cantor. 1974. Investigation of the ribosomal peptidyl transferase center using a photoaffinity label. *J. Mol. Biol.* **88:**841.

Hsu, L., M. Keren-Zur, and R. Liou. 1978. Photoaffinity labeling of the ribosomal A site with base modified tRNA. *Fed. Proc.* **37:**1405.

Jaynes, E. N., Jr., P. G. Grant, G. Giangrande, R. Wieder, and B. S. Cooperman. 1978. Photoinduced affinity labeling of the *Escherichia coli* ribosome puromycin site. *Biochemistry* **17:**561.

Johnson, A. E., R. H. Fairclough, and C. R. Cantor. 1977. Some approaches for the study of ribosomal-tRNA interactions. In *Nucleic acid-protein recognition* (ed. H. J. Vogel), p. 469. Academic Press, New York.

Kenner, R. A. 1973. A protein-nucleic acid crosslink in 30S ribosomes. *Biochem. Biophys. Res. Commun.* **51:**932.

Keren-Zur, M., M. Boublik, and J. Ofengand. 1979. Localization of the decoding region on the 30S *E. coli* ribosomal subunit by affinity immunoelectron microscopy. *Proc. Natl. Acad. Sci.* **76:**1054.

Kim, S.-H., G. J. Quigley, F. L. Suddath, A. McPherson, D. Sneden, J. J. Kim, J. Weinzierl, and A. Rich. 1973. Three-dimensional structure of yeast phenylalanine transfer RNA. Folding of the polynucleotide chain. *Science* **179:**285.

Kim, S.-H., F. L. Suddath, G. J. Quigley, A. McPherson, J. L. Sussman, A. H. J. Wang, N. C. Seeman, and A. Rich. 1974. Three-dimensional tertiary structure of yeast phenylalanine transfer RNA. *Science* **185:**435.

Kuechler, E. 1976. Chemical methods of studying ribosome structure. *Angew. Chem. Int. Ed. Engl.* **15:**533.

―――. 1978. Affinity labels for tRNA and mRNA binding sites on ribosomes. In *Theory and practice in affinity techniques* (ed. P. V. Sundaram and F. Eckstein). Academic Press, New York. (In press.)

Kuechler, E. and A. Barta. 1977. Aromatic ketone derivatives of aminoacyl-tRNA as photoaffinity labels for ribosomes. *Methods Enzymol.* **46:**676.

Lake, J. A. 1977. Aminoacyl-tRNA binding at the recognition site is the first step of the elongation cycle of protein synthesis. *Proc. Natl. Acad. Sci.* **74:**1903.

―――. 1978. Electron microscopy of specific proteins: 3-Dimensional mapping of ribosomal proteins using antibody labels. In *Advanced techniques in biological electron microscopy. II.* (ed. J. K. Koehler), p. 173. Springer-Verlag, Berlin.

Lührmann, R., H. G. Gassen, and G. Stöffler. 1976. Identification of the 30S ribosomal proteins at the decoding site by affinity labeling with a reactive oligonucleotide. *Eur. J. Biochem.* **66:**1.

Margaritella, P. and E. Kuechler. 1978. Specificity of the photocrosslinking between poly(U) and protein S1 on the *Escherichia coli* ribosome. *FEBS Lett.* **88:**131.

Modolell, J., B. Cabrer, and D. Vazquez. 1973. The interaction of elongation factor G with N-acetylphenylalanyl transfer RNA·ribosome complexes. *Proc. Natl. Acad. Sci.* **70:**3561.

Morgan, J. and R. Brimacombe. 1972. A series of specific ribonucleoprotein fragments from the 30S subparticle of *Escherichia coli* ribosomes. *Eur. J. Biochem.* **29**:542.

Noll, M., B. Hapke, M. H. Schreier, and H. Noll. 1973. Structural dynamics of bacterial ribosomes. I. Characterization of vacant couples and their relation to complexed ribosomes. *J. Mol. Biol.* **75**:281.

Oen, H., M. Pellegrini, D. Eilat, and C. R. Cantor. 1973. Identification of 50S proteins at the peptidyl-tRNA binding site of *Escherichia coli* ribosomes. *Proc. Natl. Acad. Sci.* **70**:2799.

Ofengand, J. and R. Liou. 1978. Ability of modified forms of phenylalanine tRNA to stimulate guanosine pentaphosphate synthesis by the stringent factor-ribosome complex of *E. coli*. *Nucleic Acids Res.* **5**:1325.

Ofengand, J., R. Liou, J. Kohut III, I. Schwartz, and R. A. Zimmermann. 1979. Covalent crosslinking of tRNA to the ribosomal P site. Mechanism and site of reaction in tRNA. *Biochemistry* **18**. (In press.)

Ofengand, J., I. Schwartz, G. Chinali, S. S. Hixson, and S. H. Hixson. 1977. Photoaffinity-probe-modified tRNA for the analysis of ribosomal binding sites. *Methods Enzymol.* **46**:683.

Ofengand, J., I. Schwartz, R. A. Zimmermann, S. M. Gates, and R. Liou. 1978. Photochemical crosslinking of unmodified acetylvalyl-tRNA to 16S RNA at the ribosomal P site. *Fed. Proc.* **37**:1658.

Pellegrini, M. and C. R. Cantor. 1977. Affinity labeling of ribosomes. In *Molecular mechanisms of protein biosynthesis* (ed. H. Weissbach and S. Pestka), p. 203. Academic Press, New York.

Pellegrini, M., H. Oen, and C. R. Cantor. 1972. Covalent attachment of a peptidyl-transfer RNA analog to the 50S subunit of *Escherichia coli* ribosomes. *Proc. Natl. Acad. Sci.* **69**:837.

Pellegrini, M., H. Oen, D. Eilat, and C. R. Cantor. 1974. The mechanism of covalent reaction of bromoacetyl-phenylalanyl-transfer RNA with the peptidyl-transfer RNA binding site of the *Escherichia coli* ribosome. *J. Mol. Biol.* **88**:809.

Perez-Gosalbez, M., D. Vazquez, and J. P. G. Ballesta. 1978. Affinity labeling of yeast ribosomal peptidyl transferase. *Mol. Gen. Genet.* **163**:29.

Politz, S. M. and D. G. Glitz. 1977. Ribosome structure: Localization of N^6,N^6-dimethyladenosine by electron microscopy of a ribosome-antibody complex. *Proc. Natl. Acad. Sci.* **74**:1468.

Pongs, O. and E. Lanka. 1975. Affinity labeling of the ribosomal decoding site with an AUG-substrate analog. *Proc. Natl. Acad. Sci.* **72**:1505.

Pongs, O. and E. Rossner. 1976. Comparison of the reactions of chemically reactive analogs of U-G-A and of A-U-G with ribosomes of *Escherichia coli*. *Nucleic Acids Res.* **3**:1625.

Pongs, O., G. Stöffler, and R. W. Bald. 1976. Location of protein S1 of *Escherichia coli* ribosomes at the A-site of the codon binding site. Affinity labeling studies with a 3'-modified A-U-G analog. *Nucleic Acids Res.* **3**:1635.

Pongs, O., G. Stöffler, and E. Lanka. 1975a. The codon binding site of the *Escherichia coli* ribosome as studied with a chemically reactive AUG analog. *J. Mol. Biol.* **99**:301.

Pongs, O., R. Bald, V. A. Erdmann, and E. Reinwald. 1975b. Studies on active sites of ribosomes with haloacetylated antibiotic analogs. In *Topics in infectious diseases* (ed. J. Drews and F. E. Hahn), vol. 1, p. 179. Springer-Verlag, Wien.

Rich, A. 1974. How transfer RNA may move inside the ribosome. In *Ribosomes* (ed. M. Nomura et al.), p. 871. Cold Spring Harbor Laboratory, Cold Spring Harbor, New York.

Rinke, J., A. Ross, and R. Brimacombe. 1977. Characterization of RNA fragments obtained by mild nuclease digestion of 30S ribosomal subunits from *Escherichia coli*. *Eur. J. Biochem.* **76:** 189.

Rinke, J., A. Yuki, and R. Brimacombe. 1976. Studies on the environment of protein S7 within the 30S subunit of *Escherichia coli* ribosomes. *Eur. J. Biochem.* **64:** 77.

Robertus, J. D., J. E. Ladner, J. T. Finch, D. Rhodes, R. S. Brown, B. F. C. Clark, and A. Klug. 1974. Structure of yeast phenylalanine tRNA at 3Å resolution. *Nature* **250:** 546.

Schwartz, I. and J. Ofengand. 1978. Photochemical crosslinking of unmodified acetylvalyl-tRNA to 16S RNA at the ribosomal P site. *Biochemistry* **17:** 2524.

Schwartz, I., E. Gordon, and J. Ofengand. 1975. Photoaffinity labeling of the ribosomal A site with S-(p-azidophenacyl)valyl-tRNA. *Biochemistry* **14:** 2907.

Schwartz, I., R. Tejwani, and J. Ofengand. 1975. Photoaffinity labeling of the initiator tRNA binding site on *E. coli* ribosomes. *Int. Congr. Biochem. Abstr.* **10:** 121.

Schenkman, M. L., D. C. Ward, and P. B. Moore. 1974. Covalent attachment of a messenger RNA to the *Escherichia coli* ribosome. *Biochim. Biophys. Acta* **353:** 503.

Sonenberg, N., M. Wilchek, and A. Zamir. 1975. Identification of a region in 23S rRNA located at the peptidyl transferase center. *Proc. Natl. Acad. Sci.* **72:** 4332.

―――. 1976. Photo-affinity labeling of 23S RNA by an analog of fMet-tRNA$_f^{Met}$. *Biochem. Biophys. Res. Commun.* **72:** 1534.

―――. 1977. Mapping of 23S rRNA at the ribosomal peptidyl-transferase center by photoaffinity labeling. *Eur. J. Biochem.* **77:** 217.

Sopori, M., M. Pellegrini, P. Lengyel, and C. R. Cantor. 1974. Affinity labeling of *Escherichia coli* ribosomal proteins with an analog of the natural initiator tRNA. *Biochemistry* **13:** 5432.

Stöffler, G. and H. G. Wittmann. 1977. Primary structure and three-dimensional arrangement of proteins within the *Escherichia coli* ribosome. In *Molecular mechanisms of protein biosynthesis* (ed. H. Weissbach and S. Pestka), p. 117. Academic Press, New York.

Towbin, H. and D. Elson. 1978. A photoaffinity labeling study of the messenger RNA-binding region of *Escherichia coli* ribosomes. *Nucleic Acids Res.* **5:** 3389.

Wagner, R. and H. G. Gassen. 1975. On the covalent binding of mRNA models to the part of the 16S RNA which is located in the mRNA binding site of the 30S ribosome. *Biochem. Biophys. Res. Commun.* **65:** 519.

Wagner, R., H. G. Gassen, Ch. Ehresmann, P. Stiegler, and J.-P. Ebel. 1976. Identification of a 16S RNA sequence located in the decoding site of 30S ribosomes. *FEBS Lett.* **67:** 312.

Watanabe, S. 1972. Interaction of siomycin with the acceptor site of *Escherichia coli* ribosomes. *J. Mol. Biol.* **67:** 443.

Winkelmann, D. and L. Kahan. 1978. Accessibility of the antigenic determinants of ribosomal protein S4 on the 30S ribosomal subunit and assembly intermediates. *Fed. Proc.* **37:** 1739.

Woese, C. 1970. Molecular mechanisms of translation: A reciprocating ratchet mechanism. *Nature* **226:**817.

Yaguchi, M., E. Lanka, B. Dworniczak, H. H. Kiltz, and O. Pongs. 1978. Identification of cysteine-10 of protein S18 as part of the mRNA-binding site of *Escherichia coli* ribosomes by affinity-labeling studies with a chemically reactive A-U-G analog. *Eur. J. Biochem.* **92:**243.

Yukioka, M., T. Hatayama, and S. Morisawa. 1975. Affinity labeling of the ribonucleic acid component adjacent to the peptidyl recognition center of peptidyl transferase in *Escherichia coli* ribosomes. *Biochim. Biophys. Acta* **390:**192.

Yukioka, M., T. Hatayama, and K. Omori. 1977. Nucleotide sequence of a region in 23S RNA adjacent to peptidyl transferase catalytic center of *Escherichia coli* ribosomes. *Eur. J. Biochem.* **73:**449.

Zamir, A. 1977. Affinity labeling of ribosomal functional sites. *Methods Enzymol.* **46:**621.

Zimmermann, R. A. 1974. RNA-protein interactions in the ribosome. In *Ribosomes* (ed. M. Nomura et al.), p. 225. Cold Spring Harbor Laboratory, Cold Spring Harbor, New York.

Zimmermann, R. A., S. M. Gates, I. Schwartz, and J. Ofengand. 1979. Covalent cross-linking of tRNA to the ribosomal P site. Site of reaction in 16S RNA. *Biochemistry* **18**. (In press.)

Studies on tRNA Conformation and Ribosome Interaction with Fluorescent tRNA Derivatives

Wolfgang Wintermeyer, James M. Robertson, Hermann Weidner,* and Hans G. Zachau
Institut für Physiologische Chemie
Physikalische Biochemie und Zellbiologie der Universität München
8000 München 2, Federal Republic of Germany

Specific interactions of tRNA with aminoacyl-tRNA synthetases and with mRNA-programmed ribosomes are key steps in protein biosynthesis. It is the precision of these two steps that determines the accuracy of codon-dependent amino acid incorporation into protein. Structure-function relationships of the common reactant of the two steps, the tRNA, have been studied extensively by using a variety of approaches (see summaries by Rich and RajBhandary 1976; Schimmel 1977; Holbrook et al. 1978). Although a detailed picture of the structure of tRNA and an understanding of the various steps of protein synthesis have emerged, little is known concerning the interactions of tRNA on the molecular level. Furthermore, the conformational changes of the tRNA, which are indicated by a number of studies, have not yet been fully established.

Fluorescence spectroscopy has proved to be useful in both thermodynamic and kinetic studies of conformational changes and interactions of macromolecules. In addition, information may be obtained about the environment and the mobility of the fluorophor (Cantor and Tao 1971). In this paper, experiments on the use of fluorescent tRNA derivatives in tRNA-conformation and ribosome-interaction studies are reviewed. Studies on the interaction of such derivatives with synthetases have been published (Pachmann et al. 1973) and will not be included here.

PREPARATION AND CHARACTERIZATION OF FLUORESCENT tRNA DERIVATIVES

A number of fluorescent compounds have been introduced into tRNA by covalent attachment to odd bases carrying unique functional groups (see, e.g., Yang and Söll 1974) or to the aminoacyl moiety of aminoacylated tRNA (see, e.g., Lynch and Schimmel 1974). Since the attachment of a bulky group may impair the functions of the tRNA, we have developed an alternative

*Present address: Siemens AG, 8000 München 70, Federal Republic of Germany.

procedure by which odd bases are replaced with fluorescent dyes that possess either a primary amino or a hydrazino group. The procedure has been used to insert proflavine (Prf) or ethidium (Etd) in the place of wybutine, dihydrouracil, or 7-methylguanine in tRNA^{Phe} and in the dihydrouracil position of tRNA^{Ser} from yeast (Wintermeyer and Zachau 1971, 1974; Wintermeyer et al. 1979). The hydrazine incorporation seems to be generally applicable (Schleich et al. 1978), whereas the condensation products with some basic aromatic and particularly aliphatic amines were found to hydrolyze easily or to undergo chain scission by β-elimination (Philippsen et al. 1968; Wintermeyer et al. 1972).

Introducing an extrinsic fluorescent group implies some disturbance of the macromolecule under study. High biological activity is probably the most sensitive criterion for the preservation of a near-native structure of a tRNA. The proflavine and ethidium derivatives of tRNA^{Phe} exhibited good or excellent activities in the assay systems of aminoacylation with synthetases from yeast and *Escherichia coli* (Wintermeyer and Zachau 1971 and in prep.) and of poly(U)-dependent binding and poly(Phe) synthesis in the ribosomal systems from *E. coli* (Wintermeyer and Zachau 1971, 1975a), yeast (Robertson et al. 1977), and rabbit reticulocytes (Odom et al. 1975).

In addition to the activity measurements, direct structural studies have been performed to strengthen the statement that the tRNA-dye compounds are present in a near-native conformation. High-resolution nuclear magnetic resonance of low-field protons engaged in base-pairing has shown that the insertion of proflavine into tRNA^{Phe} at the position vacated by excision of wybutine restored the original spectrum of unmodified tRNA^{Phe} (Wong et al. 1975). Wybutine fluorescence was used as a parameter for comparing tRNA$^{Phe}_{Etd16/17}$ and unmodified tRNA^{Phe}; a very similar response to changes of the Mg^{++} concentration in the range 0–15 mM was observed for the two tRNAs.

In another group of experiments, the partial ribonuclease (RNase) T1 digestion patterns of tRNA^{Phe} and tRNA$^{Phe}_{Etd16/17}$ were compared at 20 mM Mg^{++}. The patterns were the same, indicating that in both tRNAs cleavage at G57 occurred only after G20 and G19 had been excised (Harbers et al. 1972). This result is expected from the crystal structure.

In conclusion, the comparison of the structures and activities of the tRNA^{Phe}-dye derivatives with those of unmodified tRNA^{Phe} indicates that the insertion of the dyes does not disturb essential elements of the native structure.

CONFORMATIONS OF tRNA^{Phe} IN SOLUTION

The crystallographic determination of the structure of yeast tRNA^{Phe} (Quigley et al. 1975; Ladner et al. 1975) had a great impact on the investigation of the solution structure of the molecule. Particular attention has been given to the question of whether the conformation in the crystal also prevails in solution. Evidence obtained from enzymatic degradation experiments (Zachau et al. 1972) and, more recently, from dynamic light scattering (Olson et al. 1976) indicates that tRNA^{Phe} exists in more than one conformation in solution. We have demonstrated that m^{7}G in the extra loop of yeast tRNA^{Phe} becomes more exposed to attack by NaBH$_4$ when the ionic strength is increased, suggesting some structural flexibility around this residue (Wintermeyer and Zachau 1975b).

More detailed studies on the solution structure of tRNA^{Phe} have been possible by the use of tRNA^{Phe}-ethidium derivatives. It turned out that the fluorescence properties of these compounds are rather sensitive to changes in the solution conditions. Addition of MgCl$_2$ to Mg^{++}-free solutions of both tRNA$^{Phe}_{Etd37}$ and tRNA$^{Phe}_{Etd16/17}$ led to a strong quenching of the ethidium fluorescence; the effect was saturated only around 20 mM Mg^{++} (Rigler et al. 1977; M. Ehrenberg et al., in prep.). No such quenching was observed with ethidium-containing tRNA fragments. These observations immediately suggest Mg^{++}-induced conformational changes of the tRNA. An analysis of the fluorescence lifetimes and the chemical relaxation kinetics has shown that there is an equilibrium of at least two conformational states of the tRNA molecule that are present in roughly equal proportions at Mg^{++} concentrations around 5 mM. Although independent Mg^{++}-induced conformational changes in the anticodon and D regions cannot be excluded, the striking similarity of the effects suggests that the conformational changes in the two loop regions are coupled and represent a transition of the tertiary structure of the whole tRNA molecule.

An involvement of tertiary-structure changes in the observed transitions is further indicated by the finding that the quenching effect of Mg^{++} is quantitatively reversed upon excision of G19 and G20 from the D loop of tRNA$^{Phe}_{Etd16/17}$ by partial digestion with RNase T1 (Fig. 1). As mentioned above, the excision destabilizes the tertiary interactions be-

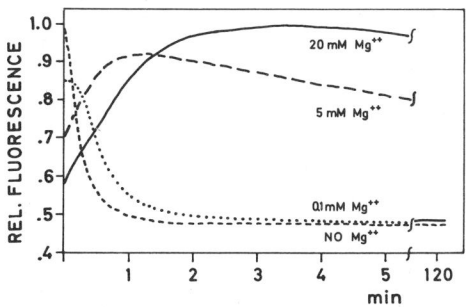

Figure 1 Digestion of tRNA$^{Phe}_{Etd16/17}$ with RNase T1 at different Mg^{++} concentrations. Ethidium fluorescence was recorded in a Perkin-Elmer MPF-2A spectrofluorometer. Disc electrophoretic analysis showed that in the presence of 20 mM Mg^{++} after 4 min nearly all of the tRNA was cleaved specifically after G18. Same ordinate for the four samples.

tween the D loop and the TψC loop. Because of further digestion, the rate of which depends on the Mg^{++} concentration, the relative fluorescence intensity of the clipped molecule has to be obtained by extrapolation of the fluorescence decrease to zero time. A value close to 1 is obtained irrespective of the Mg^{++} concentration, indicating that the fluorescence of the clipped molecule is not dependent on the Mg^{++} concentration and equals that of the intact tRNA$^{Phe}_{Etd16/17}$ in the absence of Mg^{++}. This result clearly indicates that the conformation of the tRNA^{Phe} that dominates at low Mg^{++} concentration is more open with respect to tertiary-structure interactions than the conformation that is favored at high Mg^{++} concentrations. The latter probably represents the conformation of the tRNA in the crystal.

There are numerous reports that describe the strong influence of Mg^{++} on the transitions of tRNAs from denatured or partially melted forms into the folded state. By contrast to these studies, the Mg^{++} effects described here have been observed under conditions in which the tRNA can be regarded as being in the native state, i.e., where good activity in biochemical assay systems can be demonstrated.

THE POSSIBILITY OF INDUCED CONFORMATIONAL CHANGES WITHIN tRNA^{Phe}

The strong complexes formed by tRNAs with complementary anticodons, observed first by Eisinger (1971), have been investigated rather intensively (see, e.g., Grosjean et al. 1978). The complexes of the fluorescent tRNA^{Phe} derivatives with tRNA$^{Glu}_2$ from *E. coli* seemed to be well suited for investigating the question of whether the binding of the complementary sequence to the anticodon of tRNA^{Phe} changes the conformation of the anticodon loop and possibly the D loop.

In the complex of tRNA$^{Phe}_{Etd37}$ with tRNA$^{Glu}_2$, the emission of ethidium was quenched and substantially red-shifted (Fig. 2). No change was observed upon addition of two other tRNAs. Apparently, the interaction of the complementary anticodons induces a structure of the anticodon loop in

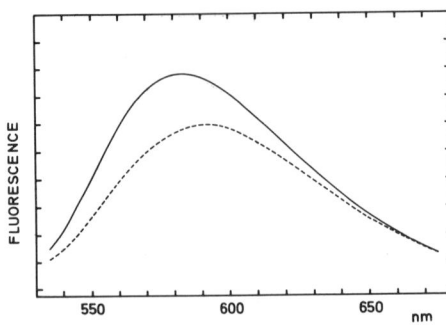

Figure 2 Uncorrected emission spectra of free tRNA$^{Phe}_{Etd37}$(———) and its complex with tRNA$^{Glu}_2$ (----).

which the stacking interactions of the dye with the neighboring adenines are diminished, thus increasing its exposure to the solvent. The dye does not interfere with the complex formation, as indicated by a binding constant of 3×10^5 liters/mole (24°C), which is close to that reported for the complex of $tRNA_2^{Glu}$ with unmodified $tRNA^{Phe}$.

Binding of $tRNA_2^{Glu}$ to $tRNA_{Etd16/17}^{Phe}$ enhanced the fluorescence without a spectral shift (Fig. 3). Addition of $tRNA_2^{Ser}$ (yeast) and $tRNA_1^{Val}$ (*E. coli*) caused only small effects; $tRNA^{fMet}$ (*E. coli*) had no effect. These observations indicate a structural change of the D loop of $tRNA_{Etd16/17}^{Phe}$ that is induced by the binding of $tRNA_2^{Glu}$ at the anticodon. The small fluorescence change observed upon addition of the other two tRNAs might be explained by a weak interaction of the CGA sequences in the TψC loops of the two tRNAs with the complementary ψCG sequence in the TψC loop of $tRNA_{Etd16/17}^{Phe}$. These are the only complementary trinucleotide sequences located in single-strand regions of cloverleaf models of those tRNAs. Consistent with this explanation is the failure of $tRNA^{fMet}$ (which has a CAA sequence instead) to induce a fluorescence change when added to $tRNA_{Etd16/17}^{Phe}$. If the proposed interaction of the TψC loops of two tRNAs exists, it should also be expected in the $tRNA^{Phe}$-$tRNA_2^{Glu}$ pair in addition to the interaction of the anticodons. However, preliminary titration data indicate that the latter interaction is stronger than the former one and causes most of the fluorescence change seen in Figure 3.

The effect on the fluorescence of $tRNA_{Etd16/17}^{Phe}$ of adding $tRNA^{Phe}$ was investigated in some detail. Complexes also seem to be formed here (Fig. 3). As indicated by titration experiments, the complex is rather stable, with a dissociation constant in the micromolar range. A self-association of $tRNA^{Phe}$ cannot be explained by an interaction of the TψC loops, as the presence of m^1A in the CGm^1A sequence of $tRNA^{Phe}$ prevents the formation of the third base pair that is probably required for a strong complex to be formed. One can imagine, however, that the anticodon of $tRNA^{Phe}$, GmAA, forms three base pairs with the complementary TψC sequence in the TψC loop of $tRNA_{Etd16/17}^{Phe}$. Obviously, such an interaction may also lead to $tRNA^{Phe}$ aggregates higher than dimers, and it should be mentioned that in

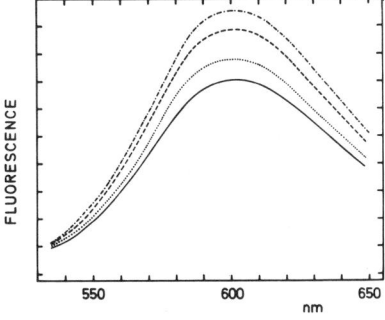

Figure 3 Uncorrected emission spectra of $tRNA_{Etd16/17}^{Phe}$. The figure gives $tRNA_{Etd16/17}^{Phe}$ alone (———) and after addition of $tRNA_2^{Ser}$ (baker's yeast) or $tRNA_1^{Val}$ (*E. coli*) (· · · ·), $tRNA_2^{Glu}$ (*E. coli*) (----), and $tRNA^{Phe}$ (yeast) (· - · -).

experiments not related to those described here, we have observed aggregates containing three molecules of tRNA^{Phe}. They were formed during NaBH$_4$ reduction of tRNA$^{Phe}_{Etd37}$ in up to 30% yield and could be isolated by gel filtration. The material moved as a single band of the size of a trimeric tRNA aggregate on electrophoresis in urea-containing polyacrylamide gels and was not dissociated by heating in the presence of urea. We concluded that at some stage during reduction or reoxidation, trimeric tRNA aggregates had been covalently linked in some unknown way. The gel electrophoretic patterns of the oligonucleotides obtained by total digestion of the material with RNase T1 or pancreatic RNase were consistent with the idea that the ethidium-containing dodecanucleotide from the anticodon region had been cross-linked to another oligonucleotide. The tRNA$^{Phe}_{Etd}$ aggregates could not be aminoacylated. Returning to the present results, a trimeric aggregate in which one molecule of tRNA$^{Phe}_{Etd16/17}$ and two molecules of tRNA^{Phe} form a circular structure by anticodon–TψC-loop associations is an attractive speculation to explain the observed stability of the complex.

To explain the experimental results presented above, we had to postulate two novel interactions between tRNA molecules, both of which involve base pairing to the TψC sequence of tRNA^{Phe}. This implies that these bases, which in the crystals are engaged in intramolecular base pairs, have to become available for intermolecular base pairing in solution. The equilibrium of two tRNA conformations that differ in the interactions between the D loop and the TψC loop (see above) is consistent with this proposal. Trapping an open form by base pairing to an added tRNA would shift the equilibrium, thus explaining the observed fluorescence changes. The effect of tRNA$^{Glu}_2$ on the fluorescence of tRNA$^{Phe}_{Etd16/17}$ may be due to the same conformational transition, which in this case, however, is coupled to a conformational transition of the anticodon loop. A codon-dependent rearrangement of tRNA^{Phe} that results in an exposed TψCG sequence has been suggested (Schwarz et al. 1974; Schwarz and Gassen 1977) on the basis of oligonucleotide binding data. Our results support and extend this hypothesis.

tRNA-RIBOSOME INTERACTIONS

The programmed ribosome in vivo selects the correct aminoacyl-tRNA with about 10^4 times higher probability than the wrong one. Experiments in ribosome-free model systems (such as the tRNA-tRNA pairs described above) have shown that complexes formed by the interaction of complementary trinucleotide sequences at the best are only about 100 times more stable than noncognate ones (see, e.g., Grosjean et al. 1978). The role of the ribosome in codon-anticodon recognition might well account for the missing factor of 100 in specificity. Several models have been proposed that describe possible mechanisms by which proofreading on the ribosome might occur. Some of these models involve conformational transitions of an initially unspecific tRNA-ribosome complex and lead to the desired preference for the correct tRNA (Ninio 1973; Kurland et al. 1975). There are only very few thermodynamic and essentially no kinetic data available that would allow one to decide whether the codon-dependent binding of tRNA to the decoding site on the ribosome proceeds by a single- or a multistep mechanism.

To contribute data relevant to these questions, we have used the fluorescent tRNA derivatives and have started to study the thermodynamics and kinetics of tRNA-ribosome complex formation. At first, we investigated the complexes of the nonaminoacylated $tRNA^{Phe}$-dye derivatives with poly(U)-programmed ribosomes; some results with aminoacylated tRNAs are also reported.

Equilibrium Studies

The emission of $tRNA^{Phe}_{Etd37}$ is substantially increased and slightly blue-shifted upon addition of *E. coli* ribosomes in the presence of poly(U) under conditions where the tRNA is bound to the peptidyl site (Fig. 4). Apparently, the probe in the complex is located in a rather hydrophobic environment where it is rigidly fixed, probably by stacking on top of the codon-anticodon double helix. The latter conclusion was reached from polarization measurements that showed an extensive immobilization of the dye in the complex (Robertson et al. 1977). The spectral change seen in Figure 4 was not observed when poly(U) was not present or when it was replaced by poly(A), although the polarization measurement showed that a tRNA-ribosome complex was formed even under those conditions. Analogous polarization measurements with the ribosome complexes of $tRNA^{Phe}_{Etd16/17}$ have revealed that, in addition, the D loop of $tRNA^{Phe}$, which in the absence of poly(U) had some mobility, was rigidly fixed in the presence of poly(U). To explain this effect of the codon-anticodon inter-

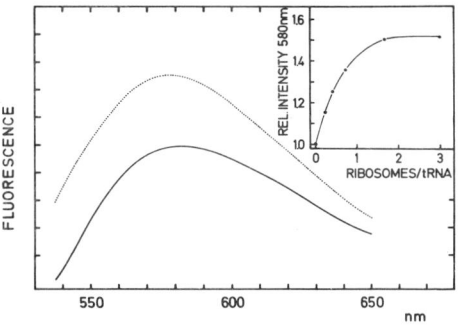

Figure 4 Complex formation of $tRNA^{Phe}_{Etd37}$ with poly(U)-programmed *E. coli* ribosomes in the presence of 20 mM Mg^{++}. Emission spectra were measured at 24°C upon addition of increasing amounts of ribosomes (tight couples according to Noll et al. [1973]). Spectra for free tRNA (———) and after saturation with ribosomes (· · · ·) are given and the titration curve (*inset*) was determined from the spectra.

action on the D loop we considered the possibility of a codon-dependent rearrangement of the tRNA molecule (of the type discussed above) by which the D loop is made available for a direct contact with the ribosome. Results very similar to those described here have been obtained with yeast ribosomes.

The quantitative evaluation of the titration curves revealed that 0.5–1 mole of tRNA was bound per mole of ribosomes, with a binding constant in the order of 10^8 mole^{-1} (24°C, 20 mM Mg^{++}). The apparent binding constants were not appreciably dependent on the presence of poly(U), although the stability against dissociation of the complexes was very different. In the absence of poly(U), the tRNA$_{Etd37}^{Phe}$-ribosome complex dissociated rapidly upon addition of nonlabeled tRNA^{Phe} (faster than 15 sec), whereas in the presence of poly(U), dissociation was extremely slow.

The use of proflavine as the fluorescent probe next to the anticodon of tRNA^{Phe} has revealed additional features of the tRNA complexes with *E. coli* ribosomes (R. H. Fairclough et al., in prep.). The spectroscopic analysis has shown that the emission spectra of the peptidyl-site- and aminoacyl-site-bound tRNA$_{Prf37}^{Phe}$ are different, making possible the separate evaluation of the binding parameters. The affinity of the tRNA for the two sites differs; the association constants (7°C, 25 mM Mg^{++}) were determined to be 10^9 mole^{-1} (peptidyl site) and 3×10^7 mole^{-1} (aminoacyl site), respectively. Addition of poly(U) did not change the binding constants. It did change the spectra of tRNA$_{Prf37}^{Phe}$ in the two sites, however, indicating that the anticodon is in contact with the poly(U) in both the aminoacyl site and the peptidyl site. Continuing these experiments, we recently found similar results with yeast ribosomes. The behavior of the proflavine label in the D loop was quite different; tRNA$_{Prf16/17}^{Phe}$ gave no indications for spectral differences when bound to the aminoacyl or peptidyl site of either *E. coli* or yeast ribosomes.

The results reported so far have been obtained with nonaminoacylated tRNAs. To approximate more closely the physiological situation, it is necessary to work with aminoacyl-tRNA, although in binding experiments at 20 mM Mg^{++} both aminoacylated and nonaminoacylated

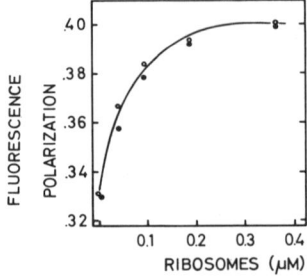

Figure 5 Complex formation of tRNA$_{Prf37}^{Phe}$ (○) and phenylalanyl-tRNA$_{Prf37}^{Phe}$ (●) with poly(U)-programmed yeast ribosomes as measured by fluorescence polarization at 500 nm.

tRNA^{Phe}-dye derivatives behave rather similarly. As an example, Figure 5 shows the comparison of phenylalanyl-tRNA$^{Phe}_{Prf37}$ and tRNA$^{Phe}_{Prf37}$ in titrations with yeast ribosomes. There is, however, evidence indicating that at low Mg^{++} concentrations binding of the aminoacylated species is significantly weaker, whereas the nonaminoacylated species still binds strongly.

Kinetic Studies

To follow the kinetics of tRNA binding to ribosomes we have performed stopped-flow experiments monitoring the fluorescence changes of tRNA^{Phe}-dye compounds upon formation of the complexes. Rapid mixing of ethidium- and proflavine-labeled tRNA^{Phe} with an excess of ribosomes from *E. coli* or yeast in the absence of poly(U) resulted in a small fluorescence change that was faster than the dead-time of the stopped-flow apparatus (~2 msec). The presence of poly(U) completely changed the kinetics. Now four slow steps could be resolved (Table 1) in addition to the unresolved rapid step; the two slowest steps exhibited only very small amplitudes (1–2% of the total signal). Practically identical relaxation times were observed when the binding of tRNA$^{Phe}_{Prf37}$ and tRNA$^{Phe}_{Prf16/17}$ to *E. coli* ribosomes was measured. The use of yeast ribosomes introduced some quantitative changes, but the qualitative picture was not altered. The ethidium label did not show the step in the 100-msec range. When a fluorescent tRNA^{Ser}-proflavine derivative was used, only the unresolved fast step was observed with poly(U)-programmed ribosomes. Variation of the concentration of tRNA, ribosomes, or both (see Table 1) did not change the resolved relaxation times; measurements at lower concentrations, where such changes would be expected, cannot be carried out yet. Analogous experiments with phenylalanyl-tRNA$^{Phe}_{Prf37}$ and poly(U)-programmed ribosomes from *E. coli* have revealed the same relaxation spectrum as observed with the deacylated tRNAs.

Table 1 Results of kinetic experiments at 20°C

tRNA	Ribosomes (μM)	Relaxation times (sec ±20%)			
tRNA$^{Phe}_{Prf37}$	*E. coli* (0.6)	0.14	2.3	20	46
tRNA$^{Phe}_{Prf16/17}$		0.12	2.4	15	37
tRNA$^{Phe}_{Etd37}$ [a]	*E. coli* (0.35)	—	1.0	8	50
tRNA$^{Phe}_{Prf37}$	yeast (1.0)	0.25	1.0	7	100

The tRNA concentrations varied between 0.03 and 1.5 μM, and ribosome concentrations varied between 0.03 and 0.6 μM without significant effects on the observed relaxation times.
[a]Measured at 10°C.

The kinetic data reported here suggest that the binding of the tRNA to a functional ribosomal site occurs in several steps. In the unresolved, probably diffusion-controlled association step, a tRNA-ribosome complex is formed in a codon-independent manner that is easily reversible. In the following steps, which depend on the presence of the proper codon, the initial complex undergoes transitions until the final state is reached. This is qualitatively what is expected from the proofreading models mentioned above, in which the codon-anticodon interaction acts as an allosteric signal triggering conformational transitions of the tRNA-ribosome complex. At present there are not enough data to develop a consistent model that correlates the kinetic steps with biochemically defined functional states of the tRNA-ribosome complex. An explanation can be offered only for the first resolved step, which in the experiments with $tRNA_{Prf37}^{Phe}$ was seen as a quenching of fluorescence. In the equilibrium experiments, a lower fluorescence was observed only when $tRNA_{Prf37}^{Phe}$ was bound to the ribosomal aminoacyl site. This raises the possibility that this step represents transient binding of the tRNA to the aminoacyl site, implying that also in the highly simplified system studied here the tRNA reaches the peptidyl site via the aminoacyl site.

The rates of the slow steps need some further comment. Obviously, these rates are much too slow in view of the in vivo rate of protein synthesis. It has to be kept in mind, however, that the experiments reported here have been performed in the absence of elongation factors. To approach the in vivo rates one certainly has to work with conditions that are closer to the physiological ones, i.e., to study the factor-dependent binding of aminoacyl-tRNA. Studies on the influence of the elongation factors Tu (EF-Tu) and 1 (EF-1) from *E. coli* and yeast, respectively, and on the equilibrium and kinetics of formation of tRNA-ribosome complexes are presently being carried out. Nevertheless, we feel that the present results, particularly the finding of the same relaxation spectrum for proflavine labels located at two different positions of the tRNA, indicate that conformational transitions of the tRNA molecule are involved.

CONCLUDING REMARKS

After characterization of the fluorescent $tRNA^{Phe}$ derivatives, groups of experiments on conformational states and ligand-induced conformational changes have been described and related to results of studies on tRNA-ribosome interaction. Under physiological conditions with respect to ionic strength, Mg^{++} concentration, and temperature, $tRNA^{Phe}$ seems to exist in two conformations. The conformation favored by Mg^{++} concentrations of 15 mM and higher is probably close to the one present in the crystals; the other one, favored at lower Mg^{++} concentrations, appears to be different with respect to the conformation of the anticodon loop and the tertiary-structure interactions between the D and TψC loops. The latter type of conformation also seems to be favored when the anticodon or the TψCG region of the fluorescent tRNA derivatives interacts with complementary sequences of other tRNAs.

The investigation of complexes of anticodon-loop- and D-loop-labeled tRNAs[Phe] with ribosomes in both equilibrium and kinetic studies has provided evidence suggesting that the tRNA changes its conformation when it is bound to the ribosome in a codon-dependent manner. From stopped-flow experiments it is concluded that binding of the tRNA to a functional ribosomal site occurs in several steps. The existence of different conformational states of tRNA on the ribosome and their close relation to those of tRNA free in solution are likely but they still have to be fully established.

From the present evidence and from related results of other laboratories a consistent picture emerges: the tRNA may act as an allosteric molecule, the conformational transitions of which are triggered by binding or removal of ligands such as Mg^{++}, the complementary codon, or nucleotide sequences competing with tertiary-structure base pairs. Operating these conformational switches may constitute an important means to initiate the transitions between different functional states of the tRNA in the elongation cycle.

ACKNOWLEDGMENT

These studies were supported by Deutsche Forschungsgemeinschaft, Sonderforschungsbereich 51.

REFERENCES

Cantor, C. R. and T. Tao. 1971. Application of fluorescence techniques to the study of nucleic acids. In *Procedures in nucleic acid research* (ed. G. L. Cantoni and D. R. Davies), vol. 2, p. 31. Harper and Row, New York.

Eisinger, J. 1971. Complex formation between transfer RNAs with complementary anticodon loops. *Biochem. Biophys. Res. Commun.* **43:**854.

Grosjean, H. J., S. de Henau, and D. M. Crothers. 1978. On the physical basis for ambiguity in genetic coding. *Proc. Natl. Acad. Sci.* **75:**610.

Harbers, K., R. Thiebe, and H. G. Zachau. 1972. Preparation and characterization of fragments from yeast tRNA[Phe]. *Eur. J. Biochem.* **26:**132.

Holbrook, S. R., J. L. Sussman, R. W. Warrant, and S.-H. Kim. 1978. Crystal structure of yeast phenylalanine transfer RNA-structural features and functional implications. *J. Mol. Biol.* **123:**631.

Kurland, C. G., R. Rigler, M. Ehrenberg, and C. Blomberg. 1975. Allosteric mechanism for codon-dependent tRNA selection on ribosomes. *Proc. Natl. Acad. Sci.* **72:**4248.

Ladner, J. E., A. Jack, J. D. Robertus, R. S. Brown, D. Rhodes, B. F. C. Clark, and A. Klug. 1975. Structure of yeast phenylalanine transfer RNA at 2.5 Å resolution. *Proc. Natl. Acad. Sci.* **72:** 4414.

Lynch, D. C. and P. R. Schimmel. 1974. Cooperative binding of magnesium to transfer ribonucleic acid studied by a fluorescent probe. *Biochemistry* **13:** 1841.

Ninio, J. 1973. Recognition in nucleic acids and the anticodon families. *Prog. Nucleic Acid Res. Mol. Biol.* **13:** 301.

Noll, M., B. Hapke, and H. Noll. 1973. Structural dynamics of bacterial ribosomes. II. Preparation and characterization of ribosomes and subunits active in the translation of natural messenger RNA. *J. Mol. Biol.* **80:** 519.

Odom, O. W., B. Hardesty, W. Wintermeyer, and H. G. Zachau. 1975. Efficient polyphenylalanine synthesis with proflavine and ethidium labeled tRNA^{Phe} from yeast in the reticulocyte ribosomal system. *Biochim. Biophys. Acta* **378:** 159.

Olson, T., M. J. Fournier, K. H. Langley, and N. C. Ford, Jr. 1976. Detection of a major conformational change in transfer ribonucleic acid by laser light scattering. *J. Mol. Biol.* **102:** 193.

Pachmann, U., E. Cronvall, R. Rigler, R. Hirsch, W. Wintermeyer, and H. G. Zachau. 1973. On the specificity of interactions between transfer ribonucleic acids and aminoacyl-tRNA synthetases. *Eur. J. Biochem.* **39:** 265.

Philippsen, P., R. Thiebe, W. Wintermeyer, and H. G. Zachau. 1968. Splitting of phenylalanine specific tRNA into half molecules by chemical means. *Biochem. Biophys. Res. Commun.* **33:** 922.

Quigley, G. J., A. H. J. Wang, N. C. Seeman, F. L. Suddath, A. Rich, J. L. Sussman, and S.-H. Kim. 1975. Hydrogen bonding in yeast phenylalanine transfer RNA. *Proc. Natl. Acad. Sci.* **72:** 4866.

Rich, A. and U. L. RajBhandary. 1976. Transfer RNA: Molecular structure, sequence, and properties. *Annu. Rev. Biochem.* **45:** 805.

Rigler, R., M. Ehrenberg, and W. Wintermeyer. 1977. Structural dynamics of tRNA—A fluorescence relaxation study of tRNA$^{Phe}_{yeast}$. *Mol. Biol. Biochem. Biophys.* **24:** 219.

Robertson, J. M., M. Kahan, W. Wintermeyer, and H. G. Zachau. 1977. Interactions of yeast tRNA^{Phe} with ribosomes from yeast and *E. coli*. *Eur. J. Biochem.* **72:** 117.

Schimmel, P. R. 1977. Approaches to understanding the mechanism of specific protein-tRNA interactions. *Accts. Chem. Res.* **10:** 411.

Schleich, H. G., W. Wintermeyer, and H. G. Zachau. 1978. Replacement of wybutine by hydrazines and its effect on the active conformation of yeast tRNA^{Phe}. *Nucleic Acids Res.* **5:** 1701.

Schwarz, U. and H. G. Gassen. 1977. Codon-dependent rearrangement of the tertiary structure of tRNA^{Phe} from yeast. *FEBS Lett.* **78:** 267.

Schwarz, U., R. Lührmann, and H. G. Gassen. 1974. On the mRNA-induced conformational change of aa-tRNA exposing the TψCG sequence for binding to the 50S ribosomal subunit. *Biochem. Biophys. Res. Commun.* **56:** 807.

Wintermeyer, W. and H. G. Zachau. 1971. Replacement of Y base, dihydouracil and 7-methylguanine in tRNA by artificial odd bases. *FEBS Lett.* **18:** 214.

———. 1974. Replacement of odd bases in tRNA by fluorescent dyes. *Methods Enzymol.* **29:** 667.

———. 1975a. Characterization of fluorescent derivatives of tRNA^{Phe} by experiments in the ribosomal system. *Mol. Biol.* **9:** 49.

———. 1975b. Tertiary structure interactions of 7-methylguanosine in yeast tRNA^{Phe} as studied by borohydride reduction. *FEBS Lett.* **58:** 306.

Wintermeyer, W., H. G. Schleich, and H. G. Zachau. 1979. Incorporation of amines or hydrazines into tRNA replacing wybutine or dihydrouracil. *Methods Enzymol.* **59:** 110.

Wintermeyer, W., R. Thiebe, and H. G. Zachau. 1972. Amine catalyzed cleavage of phenylalanine tRNA after base elimination. *Hoppe-Seyler Z. Physiol. Chem.* **353:** 1625.

Wong, K. L., D. R. Kearns, W. Wintermeyer, and H. G. Zachau. 1975. NMR investigation of the effect of selective modifications in the anticodon loop on the conformation of yeast transfer RNA^{Phe}. *Biochim. Biophys. Acta* **395:** 1.

Yang, C. H. and D. Söll. 1974. Studies of transfer RNA tertiary structures by singlet-singlet energy transfer. *Proc. Natl. Acad. Sci.* **71:** 2838.

Zachau, H. G., R. E. Streeck, and U. J. Hanggi. 1972. Conformational states of transfer ribonucleic acids. In *Gene expression and its regulation* (ed. F. T. Kenney et al.), p. 217. Plenum Press, New York.

Codon-induced Structural Transitions in tRNA

Achim Möller, Ulrike Manderschied, Rolf Lipecky, Sabine Bertram, Marion Schmitt, and Hans Günter Gassen
Institut für Organische Chemie und Biochemie
Technische Hochschule Darmstadt
D-6100 Darmstadt, Federal Republic of Germany

With the structure of an RNA known in one conformation, one may ask whether this RNA molecule may change its conformation during its biological function (Kurland et al. 1975). Such could occur during aminoacylation (Dvorak et al. 1976, 1978), in the formation of the ternary complex aminoacyl-tRNA·EF-Tu·GTP (Pingoud et al. 1978), or during the codon-directed binding of aminoacyl-tRNA to the ribosome (Schwarz et al. 1976; Schwarz and Gassen 1977). (EF-Tu is elongation factor Tu.)

In this paper we restrict ourselves to structural transitions that occur during the codon-directed binding of either initiator or elongator tRNA to the ribosome.

Conformational changes in the anticodon-loop region of the tRNA were postulated by Woese (1970). In an extension of the Fuller and Hodgson (1967) model, he proposed a transition between the 3'-stacked and the 5'-stacked conformation as a basic mechanism for mRNA translocation. Recently, Urbanke and Maass (1978) measured the temperature dependence of the fluorescence of the "Y-base" in yeast tRNA^{Phe}. A slow structural transition characterized by a monomolecular all-or-none effect was found. These authors, too, discuss as a possible explanation for their data a transition from the 3'-stacked (the more stable one) to the 5'-stacked anticodon structure (Urbanke and Maass 1978). Similar experiments were performed earlier by Yoon et al. (1975), who arrived at similar conclusions.

Although C32 and U33 are not available for binding in the crystal lattice of tRNA^{Phe}, UUCA and the pentamer UUCAG are bound more effectively to the anticodon region of the tRNA than UUC. Thus, the anticodon conformation could change on binding the complementary oligonucleotides (Eisinger and Spahr 1973; Pongs et al. 1973).

If one accepts a conformational flexibility in the anticodon region of the tRNA, the question arises whether a transition in this region could also lead to changes in other parts of the tRNA molecule.

When Uhlenbeck et al. (1974) measured the binding of complementary oligonucleotides to the native or denatured forms of tRNA$_3^{Leu}$ from yeast, they found that in the native form only the anticodon and the CCA end were accessible to oligonucleotide binding, whereas in the denatured form the D loop and stem were unfolded, making the anticodon region no longer accessible. A correlation between the D-loop structure and the conformation of the anticodon loop was further substantiated by Cameron and Uhlenbeck (1973), who showed that excision of the Y base from tRNA^{Phe} changed the oligonucleotide binding pattern in the D loop. These experiments can be interpreted in terms of a dynamic tRNA structure in which changes in one part of the molecule lead to altered properties at another site.

How can conformational transitions in the tRNA molecule be related to the codon-directed binding of tRNA to the ribosome?

The base complementarity between the constant tetranucleotide TψC(G,A) in tRNA and the CGAA in 5S RNA led to the suggestion that tRNA is bound to the 50S subunit via the four base pairs (Brownlee et al. 1967; Erdmann et al. 1973). This theory was supported by inactivation of the 50S subunit by N-oxidation of the As in the CGAA sequence within the 5S RNA and by the inhibition of aminoacyl-tRNA binding to 70S ribosomes by TψCG (Ofengand and Henes 1969; Erdmann et al. 1976). Furthermore, experimental evidence for this hypothesis has been presented showing that CGAA—an oligonucleotide complementary to the TψCG region—binds to tRNA only in the presence of the codon (Schwarz et al. 1976).

In the following we review the experimental evidence leading to the conclusion that the tRNA undergoes a structural transition when it binds to the codon. Furthermore, we discuss the stabilization of the codon-anticodon complex by CGAA and the effect of EF-Tu on this transition. In comparing the codon-directed binding of elongator and initiator tRNAs to 30S and 70S ribosomes, we present experimental evidence that the constant U next to the anticodon may play a crucial role in the codon-directed binding of the initiator tRNA to the 30S ribosome.

THE CODON-INDUCED STRUCTURAL TRANSITION OF tRNA AS FOLLOWED BY CGAA BINDING TO THE TψCG LOOP

We have presented experimental evidence that the capacity of TψCG in tRNA^{Phe} to bind to the 50S subunit is controlled by codon-anticodon interaction. Codon-anticodon complex formation and binding of the complementary oligonucleotide CGAA parallel each other in varying the concentrations of 30S and EF-Tu·GTP (Schwarz et al. 1976).

As can be seen in Table 1, optimal CGAA binding can be achieved only in the complete system 30S·UUUUUUUU·phenylalanyl-tRNA·EF-Tu·GTP. If the amount of CGAA bound is related to the amount of

Table 1 Specificity of CGAA binding at 0°C (pH ~ 7.5)

Components	CGAA bound (pmole)
30S	1.2
30S + EF-Tu·GTP	0.1
30S + phenylalanyl-tRNA^{Phe}·EF-Tu·GTP	1.1
U_8^a + phenylalanyl-tRNA^{Phe}·EF-Tu·GTP	4.4
30S + U_8 + phenylalanyl-tRNA^{Phe}	4.6
30S + U_8 + phenylalanyl-tRNA^{Phe}·EF-Tu·GTP	10.7

$^a U_8$ is UUUUUUUU.

ribosomes in the system, then only 10–20% of the bound phenylalanyl-tRNA^{Phe} is active in CGAA binding.

BINDING OF CGAA TO tRNA IN THE NONRIBOSOMAL SYSTEM

To discriminate between the possibilities that the rearrangement of tRNA is due to the binding of the tRNA to the ribosome or is a consequence of codon-anticodon complex formation, the binding of CGAA to the tRNA in the nonribosomal system was followed.

Figure 1 shows the Mg^{++} dependence of CGA binding to tRNA^{Lys} in the presence of poly(A). In the poly(A)·lysyl-tRNA^{Lys} system, we replaced CGAA by CGA, as the tetranucleotide binds to the anticodon of tRNA^{Lys}. With this tRNA, an Mg^{++} dependence of CGA binding is obtained with a point of inflection at 7 mM. About 5% of the input tRNA binds the CGA. The percentage of unfolded tRNA cannot be increased with higher concentrations of either CGA or the codon.

CGAA INFLUENCES THE STABILITY OF THE CODON-ANTICODON COMPLEX

An obvious question is whether the codon-anticodon complex is stabilized by the presence of CGAA.

The binding of the codon to its cognate tRNA with or without CGAA was examined with yeast tRNA^{Phe} and *Escherichia coli* tRNA^{Lys}. As can be seen in Table 2, CGAA has a small, but positive effect on codon-anticodon complex formation in both cases. In evaluating these data, it should be noted that only a 20-fold excess of CGAA over tRNA was used in these experiments. With tRNA^{Lys}, the data must be analyzed further, because, as mentioned before, CGAA competes with oligoadenylates for the anticodon region of the tRNA^{Lys}. At higher concentrations (0.2 mM), CGAA shows about 60% efficiency in lysyl-tRNA binding to 70S ribosomes

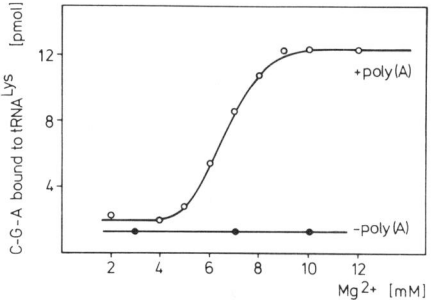

Figure 1 Mg^{++} dependence of the binding of ^{3}H-labeled CGA to *E. coli* tRNA^{Lys} in the nonribosomal system at 0°C (pH ~ 7.5).

as compared with AAA as a codon. In the nonribosomal system, the number of binding sites for AAAA obtained from a Scatchard plot drops from 1.0 to 0.8 in the presence of CGAA.

EFFECTIVE BINDING OF OLIGONUCLEOTIDES TO THE ANTICODON OF A tRNA WITHOUT STIMULATION OF AMINOACYL-tRNA BINDING TO 30S RIBOSOMES

It seems to be a reasonable assumption that the high efficiency of UUCA and AAAA in complex formation with their respective tRNAs should be reflected in a better stimulation of aminoacyl-tRNA binding to the ribosomes, i.e., either in the amount of tRNA bound or in the amount of oligonucleotide needed to obtain saturation. When the coding properties of UUCA were examined, a surprising effect emerged (Möller et al. 1978). In spite of its high association constant with tRNA^{Phe}, UUCA is practically inactive in the stimulation of phenylalanyl-tRNA binding to

Table 2 Influence of CGAA on the formation of the codon-anticodon complex at 0°C (pH ~ 7.2)

Complex	Oligonucleotide	Association constant ($\times 10^3$ mole^{-1})	
		−	+
UUUU·tRNA^{Phe}	CGAA	1.7	3.7
UUCC·tRNA^{Phe}	CGAA	6.3	9.6
AAA·tRNA^{Lys}	CGAA	37.0	63.0
AAAA·tRNA^{Lys}	CGAA	173.0	273.0
AAA·tRNA^{Lys}	GUAA	37.0	8.3
AAA·tRNA^{Lys}	CCCC	37.0	37.0

30S ribosomes. A similar effect was found for AAAA/lysyl-tRNA and GUAA/valyl-tRNA (Fig. 2). With 70S ribosomes, the effect is less pronounced, but UUC is still a more effective codon than UUCA.

EF-Tu·GTP SHOWS NO DIRECT EFFECT ON CODON-ANTICODON INTERACTION

Since we have shown (Table 1) that the ternary complex aminoacyl-tRNA·EF-Tu·GTP binds CGAA at lower codon concentrations than free tRNA, we examined the influence of EF-Tu on the stability of codon-anticodon complex formation (Table 3). Under our conditions, no major change in the association constant between free tRNA and the ternary complex was detected. Interestingly, the binding constants with tetranucleotides terminating in A were reduced in the presence of EF-Tu·GTP. This would again support the assumption that EF-Tu·GTP, in addition to the codon, stabilizes one defined tRNA conformation.

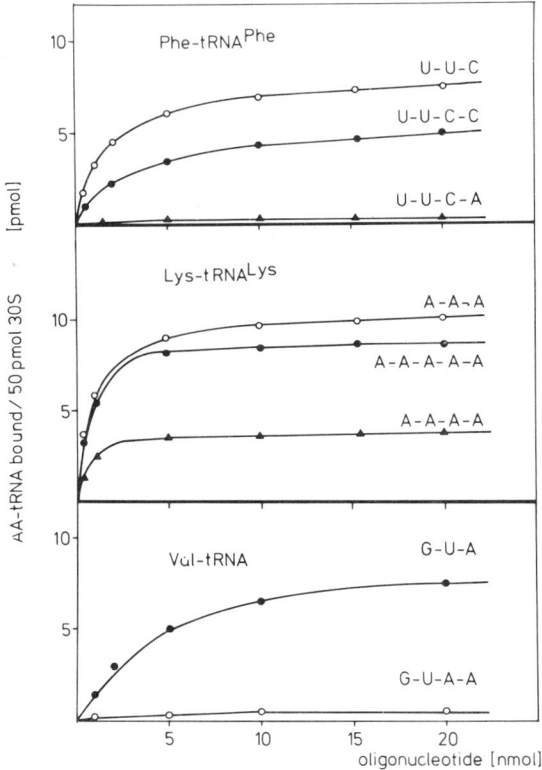

Figure 2 Oligonucleotide-dependent binding of ^{3}H-labeled aminoacyl-tRNA to 30S ribosomes at 37°C (pH ~ 7.2).

Table 3 Binding of oligonucleotides to the aminoacyl-tRNA and to the ternary complex aminoacyl-tRNA · EF-Tu · GTP at 0°C (pH ~ 7.2)

Complex	Association constant ($\times 10^3$ mole^{-1})
UUCA · phenylalanyl-tRNA^{Phe}	41.0
UUCA · phenylalanyl-tRNA^{Phe} · EF-Tu · GTP	25.0
UUUU · phenylalanyl-tRNA^{Phe}	2.7
UUUU · phenylalanyl-tRNA^{Phe} · EF-Tu · GTP	3.4
AAA · lysyl-tRNA^{Lys}	13.0
AAA · lysyl-tRNA^{Lys} · EF-Tu · GTP	17.0
AAAA · lysyl-tRNA^{Lys}	41.0
AAAA · lysyl-tRNA^{Lys} · EF-Tu · GTP	12.0

INITIATOR tRNA MAY RECOGNIZE A TETRANUCLEOTIDE CODON DURING THE 30S INITIATION COMPLEX FORMATION

Since we presented evidence before that tetranucleotides with a terminal A do not stimulate the binding of aminoacyl-tRNA to 30S, we repeated these experiments comparing AUG and AUGA in their capacity to guide the binding of formylmethionyl-tRNA^{fMet} to 30S and 70S ribosomes (Manderschied et al. 1978). By contrast to the elongation system, AUGA stimulates the binding of formylmethionyl-tRNA^{fMet} to 30S ribosomes as well as or even better than AUG or AUGU. With 70S ribosomes, the situation is reversed: AUG is more effective than AUGA (Fig. 3). The AUGA-mediated binding to 30S requires the presence of saturating amounts of IF-2 · GTP. The dominant influence of IF-2 · GTP is evident in the case of acetylvalyl-tRNA binding to 30S ribosomes (Fig. 4). In the absence of IF-2 · GTP, this pseudoinitiator tRNA does not bind in

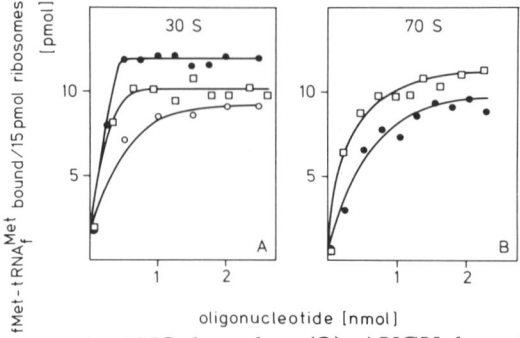

Figure 3 AUG-dependent (○), AUGU-dependent (□), and AUGA-dependent (●) ^{3}H-labeled formylmethionyl-tRNA^{fMet} binding to 30S and 70S ribosomes at 20°C (pH ~ 7.4).

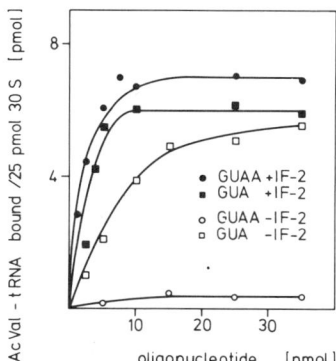

Figure 4 GUA- and GUAA-dependent ³H-labeled acetylvalyl-tRNA binding to 30S ribosomes in the presence or absence of initiator factors at 20°C (pH ~ 7.4). (Reprinted, with permission, from Manderschied et al. 1978.)

response to GUAA, whereas in its presence the GUAA concentration required to obtain maximal acetylvalyl-tRNA binding is very low.

Since the initiator codon AUG in natural mRNAs is often followed by a codon starting with G (Koper-Zwarthoff et al. 1977), we compared the stimulatory effects of AUGG, AUGA, and AUGU (Fig. 5). It may be seen from Figure 5 that the AUGG-mediated binding of formylmethionyl-tRNA^{fMet} is completely dependent on the presence of IF-2. If the factor-independent binding of formylmethionyl-tRNA^{fMet} is substracted, AUGG is the best codon, followed by AUGA and AUGU (Fig. 5).

Using AUGUUU as a template, the prebinding of formylmethionyl-tRNA inhibits the phenylalanyl-tRNA binding on the 30S level, whereas phenylalanyl-tRNA binding with 70S ribosomes was stimulated (Fig. 6). Together with the findings of others, this led us to assume that the formylmethionyl-tRNA · IF-2 · GTP complex recognizes an initiator

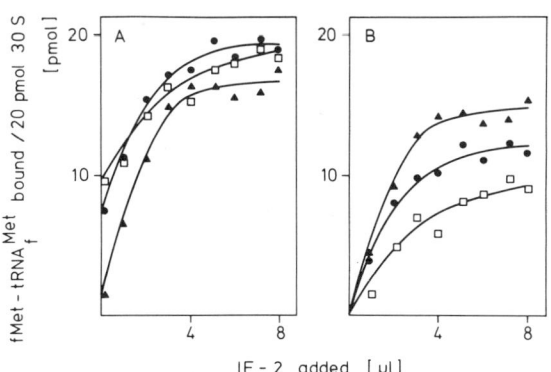

Figure 5 IF-2 dependence of ³H-labeled formylmethionyl-tRNA^{fMet} binding to 30S ribosomes at saturating amounts of oligonucleotides at 20°C (pH ~ 7.4). (▲) AUGG; (●) AUGA; (□) AUGU. (Reprinted, with permission, from Manderschied et al. 1978.)

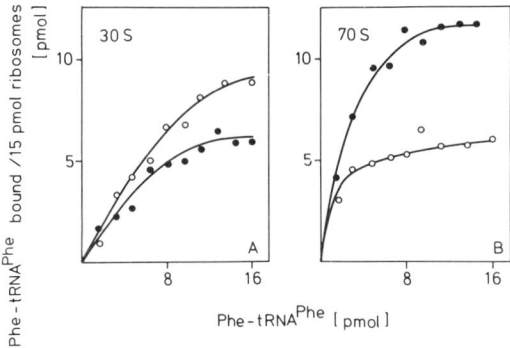

Figure 6 AUGUUU-coded [^{3}H]phenylalanyl-tRNA^{Phe} binding to 30S and 70S in the presence (●) or absence (○) of formylmethionyl-tRNA^{fMet}·IF-2·GTP at 0°C (pH ~ 7.4).

tetranucleotide, preferably AUGA or AUGG (Castel et al. 1977; Grunberg-Manago and Gros 1977; Taniguchi and Weissmann 1978). This interaction serves to prevent the binding of an elongator tRNA to the 30S ribosome by blocking the first base of the elongation codon. With the removal of IF-2·GDP from the 70S initiation complex, the trinucleotide interaction to form AUG·formylmethionyl-tRNA takes place and the formation of the first aminoacyl-tRNA·EF-Tu·GTP complex can occur. The postulated switch from tetranucleotide to trinucleotide interaction with tRNA^{fMet} would require a positional flexibility of the constant U next to the wobble nucleoside in the anticodon.

DISCUSSION

R. D. Wells first presented evidence that long-range transitions do occur in nucleic acids by examining the binding of drugs to DNA models (Burd et al. 1975). He named the stabilization of one region of a DNA helix by a region 50 Å away "telestability."

Obviously, the three-dimensional structure of any macromolecule is changed to some degree when a ligand is bound. As long as we do not know the final structure of the liganded macromolecule from X-ray analysis, we can only identify induced structural transitions by altered biochemical properties of the complex.

Codon-anticodon complex formation triggers a conformational change in the tRNA tertiary structure (Schwarz et al. 1976). This structural transition was monitored by the binding of CGAA to the now accessible TψCG region of the liganded tRNA. In the ribosomal system, this sequence most likely binds to the C43-GA-A46 sequence in the 5S RNA of the 50S subunit. All our binding studies were done by equilibrium dialysis. Geerdes et al. (1978) detected no significant structural change, i.e., disruption of hydrogen bonds, in the tRNA^{Phe}·UUCA complex on the basis of nuclear magnetic resonance measurements. In evaluating their results, however, one has to keep in mind that, at least in our hands, under optimal conditions only 5–10% of the tRNA is in the open conformation. M. Sprinzl (pers. comm.) investigated the tRNA^{Phe}·tRNA^{Glu} couple using the same technique. He, too, found no change in the interactions between the D and the TψCG loops following anticodon-anticodon complex formation. Removal of the Y base from tRNA^{Phe}, however, caused a destabilization in the T54 and m$_2^2$G26 region, which points to long-range effects transmitted from the anticodon loop to the TψCG region 60 Å away. Using fluorescence spectroscopy, Wintermeyer and colleagues (Robertson et al. 1976) were able to show a moderate change in the quantum yield with a dye located in the D loop on binding tRNA^{Phe} to poly(U)-programmed ribosomes. Wagner and Garrett (1978) examined the reactivity of the Gs in tRNA^{Phe} and tRNA^{Lys} with a kethoxal type of reagent in the absence and presence of the cognate codon. The kinetics of G modification showed that in the tRNA-codon complexes a number of oligonucleotides, including TψCG, were modified at a higher rate. Their results strongly point toward an equilibrium between the two tRNA conformers, the ratio of which depends on the presence of the codon.

We are concerned with the bad stoichiometry obtained in the binding of CGAA to the tRNA-codon complex. Even with 30S ribosomes, we never found more than 20% of the tRNA active in CGAA binding. Increasing the codon concentrations above the millimolar range caused artificial binding, especially of tetranucleotides, to parts of the tRNA other than the anticodon. Furthermore, one has to keep in mind that we lack the 50S subunit in our experiments, which may act as a scavenger for the unfolded tRNA structure in the native system.

Since the presence of the codon increases the binding of CGAA to the tRNA, the reverse experiment was undertaken, the stabilization of codon-anticodon interaction by a constant amount of CGAA. A small but positive effect was found for both tRNA species. Since CGAA competes with AAA and AAAA for the anticodon of tRNA^{Lys}, the actual binding constant A$_n$·tRNA^{Lys} (where n = no. of As) may even be further increased by CGAA. To exclude the possibility that the influence of CGAA on codon-anticodon interaction depends on an unspecific binding of an oligonucleotide to some parts of the tRNA, CCCC was used as a control. CCCC shows no effect on codon-anticodon complex formation, whereas GUAA is bound to the anticodon region of tRNA^{Lys}.

Our experiments are hampered by the low codon-tRNA association constants. Therefore, we used tetranucleotides terminating in A, which show a 10–20-fold higher association constant as compared with the corresponding trinucleotides. Their effectiveness in binary complex formation, however, is not related to their coding activity. Since AAAAA shows binding and coding properties similar to AAA and since UUCA has no coding activity over a broad Mg^{++} concentration range, we do not assume that the effect of the terminal A is due to a higher population of the stacked conformer in the tetranucleotides. More likely, tetranucleotides terminating in A stabilize an inactive tRNA conformation by an interaction of that terminal A with U33 of the tRNA. This would interrupt the hydrogen bond N3H from U33 to OP of A36, which keeps the U33 inside the anticodon loop.

No difference was found when the binding of a codon to tRNA alone or to the ternary complex EF-Tu·GTP·aminoacyl-tRNA was investigated. However, it is of interest that the association constants between tRNA and tetranucleotides terminating in A are reduced in the presence of EF-Tu·GTP. Together with the finding that EF-Tu·GTP increases CGAA binding to $tRNA^{Phe}$ in the ribosomal system (Table 1), these data support the idea that EF-Tu·GTP stabilizes a tRNA conformation in which the U33 is not accessible to artificial oligonucleotide binding.

Interestingly, the oligonucleotide-stimulated binding of initiator tRNA shows the opposite. In this case, AUGA and AUGG are highly effective in stimulating the binding of the ternary complex formylmethionyl-$tRNA^{fMet}$·IF-2·GTP to the 30S ribosome. IF-2 is needed in stoichiometric amounts in this binding reaction. The crucial effect of the initiation factor becomes even more evident in the experiments shown in Figures 3 and 4. When, in valyl-tRNA, the α-amino group of the valyl residue is acetylated, a complex of acetylvalyl-tRNA with IF-2·GTP can be formed. As a consequence, GUAA, which is inactive in valyl-$tRNA^{Val}$ binding, becomes an effective codon for the binding of the ternary complex acetylvalyl-tRNA·IF-2·GTP to the 30S ribosome.

In the 30S initiation complex, IF-2 should stabilize a conformation of $tRNA^{fMet}$ in which U33 is available for complex formation with the tetranucleotide AUGN. This tetranucleotide interaction would simultaneously block the first nucleotide in the following elongation codon, preventing the binding of an elongator tRNA to the 30S initiation complex (Manderschied et al. 1978). Removal of IF-2·GDP from the 70S initiation complex reduces formylmethionyl-tRNA binding to the usual triplet interaction and makes the first elongator codon accessible for aminoacyl-tRNA binding. Since not all elongator triplets following the AUG start with A or G (only ~80%), the type of the first elongator codon may have, in addition, some regulatory function in translation.

The experiments discussed support a structural flexibility of the elongator as well as of the initiator tRNA during their codon-directed binding to the ribosome. Additional changes of the tRNA conformation may occur in defined steps during protein synthesis as postulated by Kurland et al. (1975).

Finally, we wish to discuss the effect of a flexible tRNA structure on the codon-directed binding of aminoacyl-tRNA to ribosomes and the recognition process in general. Grosjean et al. (1976) showed that the major difference in complex formation between cognate and noncognate tRNAs is found in the dissociation rate constants, which are 10^3 times those for the noncognate tRNA. We postulate from our data that a codon-tRNA complex on the ribosome must exist long enough to allow a transition in the tRNA structure resulting in the exposition of the TψCG region as a binding site for the 50S subunit. Thus, a selective type of binding—different anticodons in all tRNAs—would be transferred via a structural change of a macromolecule to a stable binding site common to all tRNAs. The lifetime of the correct codon-anticodon complex must be greater than the time necessary for the rearrangement of the tRNA structure. The required time for the structural transition may be an intrinsic property of each tRNA, possibly regulated by the three-dimensional structure and the rare nucleosides.

SUMMARY

Codon-anticodon complex formation changes the tRNA structure from the nonbinding-type to the ribosome-binding-type structure. This conformational change is monitored by the binding of ^{3}H-labeled CGAA to either the 30S·codon·tRNA complex or the codon-tRNA complex. The codon-anticodon complex is stabilized by CGAA, whereas EF-Tu·GTP shows only a minor effect. Tetranucleotides terminating in A show a high association constant toward their cognate tRNA, but, surprisingly, a low codon activity in binding an elongator tRNA to the 30S ribosome. However, AUGA represents an excellent codon for the binding of the formylmethionyl-tRNA^{fMet}·IF-2·GTP to the 30S ribosome. Thus, the constant U next to the anticodon plays a crucial role in the codon-directed binding of an aminoacyl-tRNA to the ribosome.

ACKNOWLEDGMENTS

The work presented was supported by grants from the Deutsche Forschungsgemeinschaft and the Fonds der Chemischen Industrie.

REFERENCES

Brownlee, G. G., F. Sanger, and B. G. Barrell. 1967. Nucleotide sequence of 5S-ribosomal RNA from *Escherichia coli. Nature* **215**:735.

Burd, J. F., J. E. Larson, and R. D. Wells. 1975. Further studies on telestability in DNA. *J. Biol. Chem.* **250**:6002.

Cameron, V. and O. C. Uhlenbeck. 1973. Removal of Y-37 from tRNA$^{Phe}_{yeast}$ alters oligomer binding to two loops. *Biochem. Biophys. Res. Commun.* **50**:635.

Castel, A., B. Kraal, P. R. M. Kerklaan, J. Klok, and L. Bosch. 1977. Initiation of polypeptide synthesis with various NH_2-blocked aminoacyl-tRNAs under the direction of alfalfa mosaic virus RNA 4. *Proc. Natl. Acad. Sci.* **74:** 5509.

Dvorak, D. J., C. Kidson, and R. C. Chin. 1976. Aminoacyl-tRNA conformation. *J. Biol. Chem.* **251:** 6730.

Dvorak, D. J., C. Kidson, and D. J. Winzor. 1978. Conformational changes in tRNA. Consequences of aminoacylation and codon-anticodon recognition. *FEBS Lett.* **90:** 187.

Eisinger, J. and P. F. Spahr. 1973. Binding of complementary pentanucleotides to the anticodon loop of transfer RNA. *J. Mol. Biol.* **73:** 131.

Erdmann, V. A., M. Sprinzl, and O. Pongs. 1973. The involvement of 5S RNA in the binding of tRNA to ribosomes. *Biochem. Biophys. Res. Commun.* **54:** 942.

Erdmann, V. A., J. R. Horne, O. Pongs, J. Zimmermann, and M. Sprinzl. 1976. 5S RNA-protein complex: Enzymatic activities and interaction with TpψpCpGp. *First Symposium on Ribosomes and Ribonucleic Acid Metabolism* (ed. J. Zelinka and J. Balan), p. 363. Slovak Academy of Sciences, Bratislava.

Fuller, W. and A. Hodgson. 1967. Conformation of the anticodon loop in tRNA. *Nature* **215:** 817.

Geerdes, H. A. M., J. H. Van Boom, and C. W. Hilbers. 1978. Codon-anticodon interaction in yeast tRNA^{Phe}. *FEBS Lett.* **88:** 27.

Grosjean, H. J., D. G. Söll, and D. M. Crothers. 1976. Studies of the complex between transfer RNAs with complementary anticodons. *J. Mol. Biol.* **103:** 499.

Grunberg-Manago, M. and F. Gros. 1977. Initiation mechanism of protein synthesis. *Prog. Nucleic Acid Res. Mol. Biol.* **20:** 209.

Koper-Zwarthoff, E. C., R. E. Lockhard, U. L. RajBhandary, B. Alzner-De Weerd, and J. E. Bol. 1977. Nucleotide sequence of 5' terminus of alfalfa mosaic virus RNA$_4$ leading into coat protein cistron. *Proc. Natl. Acad. Sci.* **74:** 5504.

Kurland, C. G., R. H. Rigler, M. Ehrenberg, and C. Blomberg. 1975. Allosteric mechanism for codon-dependent tRNA selection on ribosomes. *Proc. Natl. Acad. Sci.* **72:** 4248.

Manderschied, U., S. Bertram, and H. G. Gassen. 1978. Initiator-tRNA recognizes a tetranucleotide codon during the 30S initiation complex formation. *FEBS Lett.* **90:** 162.

Möller, A., U. Schwarz, R. Lipecky, and H. G. Gassen. 1978. Effective binding of oligonucleotides to the anticodon of a tRNA without stimulation of tRNA binding to 30S ribosomes. *FEBS Lett.* **89:** 263.

Ofengand, J. and C. Henes. 1969. The function of pseudouridylic acid in transfer ribonucleic acid. *J. Biol. Chem.* **244:** 6241.

Pingoud, A., C. Urbanke, G. Krauss, F. Peters, and G. Maass. 1978. The binding of kirromycin to elongation factor Tu. *Eur. J. Biochem.* **86:** 153.

Pongs, O., R. Bald, and E. Reinwald. 1973. On the structure of yeast tRNA^{Phe}. *Eur. J. Biochem.* **32:** 117.

Robertson, J. M., M. Kahan, W. Wintermeyer, and H. G. Zachau. 1976. Interaction of yeast tRNA^{Phe} with ribosomes from yeast and *Escherichia coli*. *Eur. J. Biochem.* **72:** 117.

Schwarz, U. and H. G. Gassen. 1977. Codon-dependent rearrangement of the tertiary structure of tRNA^{Phe} from yeast. *FEBS Lett.* **78:** 267.

Schwarz, U., H. M. Menzel, and H. G. Gassen. 1976. Codon-dependent rearrangement of the three-dimensional structure of phenylalanine tRNA, exposing the TψCG sequence for binding to the 50S ribosome. *Biochemistry* **15**:2484.

Taniguchi, T. and C. Weissmann. 1978. Site-directed mutations in the initiator region of the bacteriophage Qβ coat cistron and their effect on ribosome binding. *J. Mol. Biol.* **118**:533.

Uhlenbeck, O. C., J. G. Chirikjian, and J. R. Fresco. 1974. Oligonucleotide binding to the native and denatured conformers of yeast transfer RNA$_3^{Leu}$. *J. Mol. Biol.* **89**:495.

Urbanke, C. and G. Maass. 1978. A novel conformational change of the anticodon region of tRNA^{Phe} (yeast). *Nucleic Acids Res.* **5**:1551.

Yoon, K., D. Turner, and J. Tinocco. 1975. The kinetics of codon-anticodon interaction in yeast phenylalanine transfer RNA. *J. Mol. Biol.* **99**:507.

Wagner, R. and R. A. Garrett. 1978. Chemical evidence for a codon-induced change of tRNA conformation. *FEBS Lett.* **85**:291.

Woese, C. 1970. Molecular mechanics of translation: A reciprocating ratchet mechanism. *Nature* **226**:817.

Role of the 2',3'-Isomerization of Aminoacyl-tRNA during Ribosomal Protein Synthesis

Mathias Sprinzl and Thomas Wagner
Abteilung Chemie
Max-Planck-Institut für Experimentelle Medizin
D-3400 Göttingen, Federal Republic of Germany

The structure of aminoacyl-tRNA is not yet established. Moreover, changes of tRNA conformation were suggested to take place after the attachment of the amino acid to the tRNA during the aminoacylation reaction (Potts et al. 1977) and because of codon-dependent binding of aminoacyl-tRNA to ribosomes (Schwarz et al. 1976). Apart from these suggested changes in the tertiary structure during the functional cycle of aminoacyl-tRNA, an obvious structural transformation can occur on the 3' terminal of the molecule to which the amino acid is attached. The presence of a vicinal *cis*-diol function on the 3'-terminal A allows a migration of the aminoacyl residue between the 2'- and the 3'-OH groups (Fig. 1). The rate of this isomerization has not yet been determined. An estimate, based on experiments with model substances, suggests that the halftime for isomerization of an average aminoacyl-tRNA in neutral buffered medium at 37°C is about 1.8×10^{-4} seconds (Griffin et al. 1966). Little attention has been paid to the possible significance of the existence of the isomers, i.e., 2'-aminoacyl-tRNA and 3'-aminoacyl-tRNA, in the elucidation of the mechanism of protein biosynthesis.

In recent years aminoacyl-tRNA species have become available in which the aminoacyl migration can be prevented by a modification of the 3'-terminal A residue. Thus, if the amino acid is linked to a hydroxyl group in which a vicinal *cis* hydroxyl is missing or to an amino group via an amide bond (Fig. 2), a migration of the amino acid cannot take place.

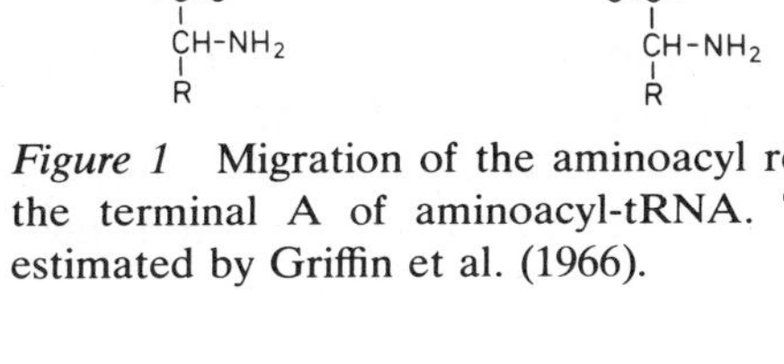

Figure 1 Migration of the aminoacyl residue between the 2' and 3' positions of the terminal A of aminoacyl-tRNA. The halftime of the isomerization was estimated by Griffin et al. (1966).

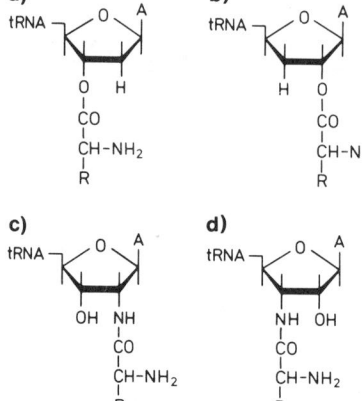

Figure 2 Structure of the 3′-terminal aminoacyl-A residue of the nonisomerizable aminoacyl-tRNA species. (a) Aminoacyl-tRNA-2′-deoxyadenosine; (b) aminoacyl-tRNA-3′-deoxyadenosine; (c) N-aminoacyl-tRNA-2′NH_2A; (d) N-aminoacyl-tRNA-3′NH_2A.

Such tRNA species with a terminating 2′-deoxyadenosine, 3′-deoxyadenosine, 2′-deoxy-2′-aminoadenosine, and 3′-deoxy-3′-aminoadenosine, respectively, were prepared by incorporation of the appropriate nucleoside-5′-phosphates using AMP(CMP)tRNA nucleotidyl transferase (Fraser and Rich 1974; Chinault et al. 1977b; Sprinzl et al. 1977b).

In this paper we discuss recent results from our laboratory based on the utilization of "nonisomerizable" aminoacyl-tRNA species in the particular steps of the elongation process. It is evident that for two possible structures of the aminoacyl-tRNA, either of the two species can be utilized in the interaction with other macromolecules provided that a spontaneous transacylation can occur. The questions that then arise are: which isomer is used in each particular step, does an aminoacyl migration take place during the different steps of the elongation process, and what is the significance of such structural variations for the mechanism of the recognition process?

AMINOACYLATION OF tRNA

In principle, during enzymatic aminoacylation either of the two hydroxyl groups of the 3′-terminal A of the tRNA can be esterified. Although the aminoacyl-tRNA synthetase might be expected to be specific for one of the hydroxyl groups, the site of primary attachment cannot be determined by a chemical analysis of the products of this enzymatic reaction due to aminoacyl migration.

Using tRNA species in which the absence of the *cis*-diol function on the 3′-terminal A precludes isomerization, the site of enzymatic aminoacylation can be elucidated. Thus, if the 2′-OH group is missing and the tRNA is still aminoacylated, it is obvious that the aminoacylation must have occurred on the 3'-OH. If the K_m and V_{max} values for the aminoacylation of such deoxy species are similar to those of native tRNA, it can be assumed that the initial position of aminoacylation in the unmodified tRNA is that which is aminoacylated in the modified tRNA species. This approach, developed in our laboratory (Sprinzl and Cramer 1973), was later utilized for the determination of the 2',3' specificity of synthetases from *Escherichia coli* (Sprinzl and Cramer 1974), yeast, and calf liver (Chinault et al. 1977a). According to the given criteria, i.e., aminoacylation of a particular 2'-deoxy- or 3'-deoxy-tRNA, the synthetases from all three sources show the same specificity. About half of the synthetases aminoacylate the tRNA when the terminal 2'-OH group is missing and these are classified as 3'-specific; others, the 2'-specific synthetases, aminoacylate only the 3'-deoxy-tRNAs. Some synthetases, such as tyrosyl-tRNA synthetase, can utilize both deoxy-tRNAs; however, the rate of aminoacylation is higher for tRNA^{Tyr}-3'-deoxyadenosine.

A different approach for the determination of the site of aminoacylation was used by Fraser and Rich (1974). These authors incorporated 2'-deoxy-2'-aminoadenosine or 3'-deoxy-3'-aminoadenosine into the 3' terminal of tRNA and determined the extent of aminoacylation of these modified species. With the assumption that the aminoacyl residue can be transferred enzymatically only to a hydroxyl group of the 3'-terminal A, these modified tRNAs could be used for the determination of the 2',3' specificity of synthetases. For instance, a 2'-specific synthetase would aminoacylate only the tRNA-3'NH$_2$A via the 2'-OH group. Aminoacylation of such a tRNA would then result in a 3'-N-aminoacyl-tRNA, since the amino acid migrates to the 3' position, forming a stable amide bond. The opposite should be true for 3'-specific synthetases, where the aminoacylation of the 3'-OH group of tRNA-2'NH$_2$A would result in a stable 2'-N-aminoacyl-tRNA. In Table 1 some representative results obtained for the aminoacylation of modified amino- and deoxy-tRNA species are summarized. The following conclusions can be drawn from these investigations:

1. The determination of the site of aminoacylation of the native tRNAs with respect to the 2'- or 3'-OH group of the terminal A, using different modified tRNA species, is ambiguous. For example, yeast phenylalanyl-tRNA synthetase is clearly a 2'-specific enzyme if the deoxy-tRNA^{Phe} species are tested but it appears to be unspecific if the amino derivatives of tRNA^{Phe} are employed. On the other hand, yeast tyrosyl-tRNA synthetase is unspecific with respect to the deoxy-tRNA species but 2'-specific with respect to the amino-tRNA species. The determination of the 2',3' specificity must therefore be seen in relation to the pair of terminal-ribose-modified tRNAs used.

Table 1 Aminoacylation of 3′-modified tRNA species using purified aminoacyl-tRNA synthetases

tRNA			Aminoacyl-tRNA synthetase[a]		Maximal aminoacylation (pmoles/A$_{260}$ unit)	K_m (μM)	V_{max} (rel)
source	specificity	3′-terminal base 76	specificity	source			
Yeast	Phe	A	Phe	yeast	1460	280	100
Yeast	Phe	2′dA[b]	Phe	yeast	20	—	—
Yeast	Phe	3′dA	Phe	yeast	1420	2.86	64
Yeast	Phe	2′NH$_2$A	Phe	yeast	1520	3.80	40
Yeast	Phe	3′NH$_2$A	Phe	yeast	1580	2.20	85
E. coli	Phe	A	Phe	E. coli	1500	0.5	100
E. coli	Phe	2′dA	Phe	E. coli	80	—	—
E. coli	Phe	2′dA	Tyr	yeast	1175	n.d.[c]	—
E. coli	Phe	3′dA	Phe	E. coli	1450	0.5	0.125
E. coli	Phe	3′dA	Tyr	yeast	1625	n.d.	—
E. coli	Phe	2′NH$_2$A	Phe	E. coli	60	—	—
E. coli	Phe	2′NH$_2$A	Phe	yeast	640	n.d.	—
E. coli	Phe	3′NH$_2$A	Phe	E. coli	1450	0.5	125
E. coli	Phe	3′NH$_2$A	Phe	yeast	1490	n.d.	—
Yeast	Tyr	A	Tyr	yeast	1630	1.7	100
Yeast	Tyr	2′dA	Tyr	yeast	1460	4.5	18
Yeast	Tyr	3′dA	Tyr	yeast	1420	1.7	75
Yeast	Tyr	2′NH$_2$A	Tyr	yeast	40	—	—
Yeast	Tyr	3′NH$_2$A	Tyr	yeast	1480	2.22	63
E. coli	Lys	A	Lys	E. coli	1450	6.4	100
E. coli	Lys	A	Phe	yeast	1550	5.5	100
E. coli	Lys	2′dA	Lys	E. coli	1420	8.3	68
E. coli	Lys	2′dA	Phe	yeast	20	—	—
E. coli	Lys	3′dA	Lys	E. coli	30	—	—
E. coli	Lys	3′dA	Phe	yeast	1510	2.1	16.8

[a]Aminoacylation was performed with the amino acid cognate to the given synthetase.
[b]dA is deoxyadenosine.
[c]Not determined.

2. The rate of aminoacylation of some deoxy-tRNAs is strongly reduced. For example, yeast tRNA^{Phe}-3′-deoxyadenosine is aminoacylated by a homologous yeast phenylalanyl-tRNA synthetase with a rate that is 64% of the rate for the native tRNA. Using the same modification in the *E. coli* system, the rate is decreased 800-fold as compared with the unmodified tRNA. This finding suggests that not only the removal of the hydroxyl group, but also changes in the conformation of the terminal ribose ring have to be considered when this type of modified tRNA species is used.

3. The specificity for one of the hydroxyl groups during the aminoacylation reaction is determined by the structure of the synthetase active site and not by a structural feature of the tRNA. This can be concluded from different misaminoacylation experiments given in Table 1. Phenylalanyl-tRNA synthetase from yeast, which aminoacylates only tRNA^{Phe}-3'-deoxyadenosine, phenylalanylates tRNA^{Lys}-3'-deoxyadenosine from *E. coli*, although the cognate lysyl-tRNA synthetase is able to aminoacylate only tRNA^{Lys}-2'-deoxyadenosine. Similar examples are given for the misaminoacylation of modified *E. coli* tRNA^{Phe} species by phenylalanyl- or tyrosyl-tRNA synthetase from yeast.

The aminoacyl-tRNA species listed in Table 1 were prepared on a preparative scale and used to elucidate the 2',3' specificity of elongation factor Tu (EF-Tu) and ribosomes from *E. coli*.

FORMATION OF EF-Tu·GTP·AMINOACYL-tRNA TERNARY COMPLEXES

We showed previously that the EF-Tu from *E. coli* recognizes both tyrosyl-tRNA^{Tyr} species from yeast; the one with the amino acid attached to the 2' position, as well as the one in which the amino acid is linked to the 3' position of the terminal A. From gel filtration experiments with 3'-tyrosyl-tRNA^{Tyr}-2'-deoxyadenosine·EF-Tu·GMPP(NH)P and 2'-tyrosyl-tRNA^{Tyr}-3'-deoxyadenosine·EF-Tu·GMPP(NH)P ternary complexes, we could, however, conclude that the 2'-tyrosyl-tRNA^{Tyr} species forms more stable ternary complexes (Sprinzl et al. 1977a).

Recently Pingoud et al. (1977) elaborated on a method for the determination of binding constants of aminoacyl-tRNA·EF-Tu·GTP ternary complexes. These authors measured the effect of EF-Tu·GTP on the rate of spontaneous hydrolysis of the amino acid from the aminoacyl-tRNA. Using this method, a quantitative difference between the two isomeric tyrosyl-tRNAs^{Tyr} in their interaction with EF-Tu·GTP could be determined (Table 2). The results indicate that the 2' aminoacyl-tRNA species binds about six times more strongly to EF-Tu·GTP than the 3' aminoacylated isomer. Taking into account that the concentration of EF-Tu in the *E. coli* cell (Furano 1975) is about 1000-fold higher than the measured apparent dissociation constant of the tyrosyl-tRNA^{Tyr}-A·EF-Tu·GTP ternary complex, this difference should not have an apparent biological significance. Clearly, EF-Tu·GTP is able to bind both isomers of tyrosyl-tRNA^{Tyr}. However, the fact that the association constant for the native tyrosyl-tRNA^{Tyr} is almost identical with that for 2'-tyrosyl-tRNA^{Tyr}-3'-deoxyadenosine suggests that the native tyrosyl-tRNA^{Tyr} is complexed with EF-Tu·GTP predominantly as a 2' aminoacyl isomer. This would be especially the case if a migration of the aminoacyl residue could take place while the aminoacyl-tRNA is bound to EF-Tu·GTP, allowing its accommodation to the more favorable 2' binding site.

Table 2 Association constants for the interaction of native and modified tyrosyl-tRNAs^{Tyr} from yeast with *E. coli* EF-Tu·GTP

tRNA	Association constant (mole^{-1})
Tyrosyl-tRNA^{Tyr}-A	7×10^6
Tyrosyl-tRNA^{Tyr}-2'-deoxyadenosine	0.5×10^6
Tyrosyl-tRNA^{Tyr}-3'-deoxyadenosine	3×10^6

The data were obtained by measurements of the protection of the hydrolysis of tyrosyl-tRNA^{Tyr} species by EF-Tu·GTP according to the method of Pingoud et al. (1977).

From experiments made with one particular aminoacyl-tRNA, it is not possible to generalize for all aminoacyl-tRNA species. The higher affinity of 2' aminoacyl-tRNA to EF-Tu·GTP as compared to the 3' isomer may be an intrinsic property of tyrosyl-tRNA^{Tyr}. It is possible that all tRNAs carrying an aromatic amino acid have a similar specificity, whereas others are more unspecific with regard to the binding of 2' or 3' isomers to EF-Tu·GTP (Hecht et al. 1977).

The use of nonisomerizable aminoacyl-tRNAs for the determination of the 2',3' specificity of aminoacyl-tRNA·EF-Tu·GTP ternary complexes may even be misleading. It is possible that not a defined isomer but some intermediate structure of aminoacyl-tRNA is the proper substrate for EF-Tu·GTP. This can be suggested from the observation that both phenylalanyl-tRNA^{Phe}-2'NH$_2$A and phenylalanyl-tRNA^{Phe}-3'NH$_2$A in which the phenylalanine is attached via an amide bond to the 2' or 3' position, respectively, are inactive in ternary-complex formation (Sprinzl et al. 1977a). The high rigidity of the linking amide bond in these modified tRNAs probably prevents the accommodation of the amino acid in its binding site. In the case of the deoxy aminoacyl-tRNAs, the flexibility of the ester bond by which the amino acid is linked to the tRNA allows such an accommodation, provided that the amino acid side chain is contributing to this interaction.

EF-Tu-DEPENDENT BINDING OF NONISOMERIZABLE AMINOACYL-tRNAs TO PROGRAMMED RIBOSOMES

The aminoacyl-tRNA-binding or codon-recognition step results in the addition of the appropriate aminoacyl-tRNA to the peptidyl-tRNA·mRNA·ribosome complex. It is somewhat difficult to investigate this process in an in vitro system using a single purified aminoacyl-tRNA, because significant simplifications with regard to the mRNA used and the occupation of the ribosomal peptidyl site must be made to perform such experiments. The 2',3' specificity of aminoacyl-tRNAs in the EF-Tu-dependent binding reaction to poly(U)-programmed ribosomes can be investigated by using phenylalanyl-tRNA^{Phe}-3'-deoxyadenosine, tyrosyl-tRNA^{Phe}-3'-deoxyadenosine (2' aminoacylated species), phenylalanyl-tRNA^{Phe}-3'NH$_2$A, and tyrosyl-tRNA^{Phe}-2'-deoxyadenosine (3' aminoacylated species).

Chinali et al. (1974) demonstrated that the EF-Tu-dependent binding of phenylalanyl-tRNA^{Phe}-3'-deoxyadenosine to *E. coli* ribosomes proceeds at a rate and to an extent similar to that of the native phenylalanyl-tRNA^{Phe}. These authors also showed that the enzymatic binding of phenylalanyl-tRNA^{Phe}-3'-deoxyadenosine induces the EF-Tu-dependent hydrolysis of GTP. However, only a very slow formation of the new peptide bond was observed after the binding of this tRNA.

Experiments demonstrating the differences between the rate of binding and the rate of peptide-bond formation using phenylalanyl-tRNA^{Phe}-3'-deoxyadenosine are shown in Figure 3. The binding of both the native and the modified tRNAs is completed within several seconds and GTP hydrolysis is not required for this process. However, only the phenylalanyl-tRNA^{Phe}-A that was bound in the presence of GTP is able to act as an acceptor of the peptidyl residue. Neither the phenylalanyl-tRNA^{Phe}-3'-deoxyadenosine nor the native phenylalanyl-tRNA^{Phe} bound in the presence of a nonhydrolyzable GTP analog, GMPP(CH$_2$)P, is able to participate in the peptidyl-transfer reaction. It can be concluded that after the EF-Tu-mediated binding of aminoacyl-tRNA to programmed ribosomes, the following reaction steps have to take place: (1) the hydrolysis of GTP to GDP, which is followed by the dissociation of EF-Tu·GDP from the ribosome (Yokosawa et al. 1973), and (2) the transacylation of the amino acid from the 2' to the 3' position.

The question of whether the attachment of the amino acid to the 2'

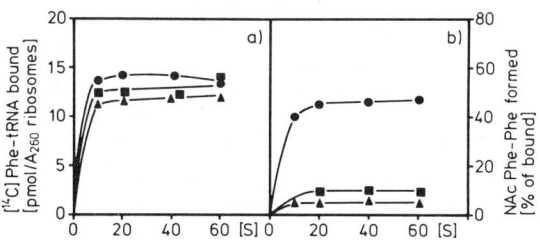

Figure 3 (*a*) EF-Tu-dependent binding of phenylalanyl-tRNA^{Phe}-A to 70S ribosomes from *E. coli* in the presence of GTP (●) or GMPP(CH$_2$)P (■), and phenylalanyl-tRNA^{Phe}-3'-deoxyadenosine in the presence of GTP (▲). (*b*) Subsequent dipeptide formation with peptidyl-site-bound *N*-acetylphenylalanyl-tRNA^{Phe} (symbols same as in *a*). The EF-Tu-dependent binding was monitored by Millipore filtration. The dipeptide formation was determined by chromatographic analysis according to Chinali et al. (1974).

position is obligatory in the EF-Tu-dependent binding of aminoacyl-tRNA to ribosomes is still open. Due to the limitation that poly(U)- or poly(A)-programmed ribosomes have to be used for in vitro assays, only nonisomerizable aminoacylated tRNA^{Phe} or tRNA^{Lys} species can be investigated. Since phenylalanyl-tRNA^{Phe}-2'-deoxyadenosine cannot be prepared by enzymatic aminoacylation (Table 1), the 3' aminoacylated phenylalanyl-tRNA^{Phe}-3'NH$_2$A was utilized in the elongation-factor-dependent binding experiment (Baksht et al. 1976). This tRNA was considerably less active in the binding reaction as compared with native phenylalanyl-tRNA^{Phe} or 2' aminoacylated phenylalanyl-tRNA^{Phe}-3'-deoxyadenosine. However, this difference probably reflects the fact that the charged amino tRNAs interact only weakly or not at all with EF-Tu·GTP (Sprinzl et al. 1977a).

More recently we compared the activity of lysyl-tRNA^{Lys}-2'-deoxyadenosine and phenylalanyl-tRNA^{Lys}-3'-deoxyadenosine in the EF-Tu-dependent binding to poly(A)-programmed ribosomes (Table 3). Under conditions where the respective preformed aminoacyl-tRNA·EF-Tu·GTP ternary complexes were in excess over the ribosomes, more phenylalanyl-tRNA^{Lys}-3'-deoxyadenosine could be bound as compared with lysyl-tRNA^{Lys}-2'-deoxyadenosine. Since the nature of the amino acid side chain of the aminoacyl-tRNA might have an influence on the binding of aminoacyl-tRNA to EF-Tu·GTP (Pingoud et al. 1977), we investigated this binding reaction over a wide range of ternary-complex concentrations. The results are consistent with the interpretation that the preference for phenylalanyl-tRNA^{Lys}-3'-deoxyadenosine in the EF-Tu-dependent binding is due to the attachment of the amino acid to the 2'-OH and not due to its structure.

The interaction of phenylalanyl-tRNA^{Lys}-3'-deoxyadenosine with the ribosomal aminoacyl site is also qualitatively different from that of lysyl-tRNA^{Lys}-2'-deoxyadenosine. Although the 2' species is able to bind to a

Table 3 Binding of aminoacyl-tRNA^{Lys} species to 70S ribosomes

Aminoacyl-tRNA^{Lys} species	pmoles/A$_{260}$ unit of ribosomes		
	+ EF-Tu	− EF-Tu	stimulation
Lysyl-tRNA^{Lys}-A	17.3	5.5	11.8
Phenylalanyl-tRNA^{Lys}-A	16.1	2.9	13.2
Lysyl-tRNA^{Lys}-2'-deoxyadenosine	8.5	4.6	3.9
Phenylalanyl-tRNA^{Lys}-3'-deoxyadenosine	18.3	4.0	14.3

The binding reaction was performed at an Mg^{++} concentration of 10 mM, according to the method of Wagner and Sprinzl (1978). The concentration of aminoacyl-tRNA^{Lys} is 1.1 µM and that of EF-Tu·GTP is 10.7 µM. Preformed ternary complexes are present in a 1.5-fold M excess over the 70S ribosomes.

higher extent, it can be replaced by native lysyl-tRNA^{Lys} in the presence of EF-Tu·GTP. The lysyl-tRNA^{Lys}-2'-deoxyadenosine does not show this effect. The reversibility of the binding reaction in the ribosomal decoding process is a necessary requirement for an efficient codon-specific selection of aminoacyl-tRNA (Thompson and Stone 1977). In the systems tested, only the 2' isomer of aminoacyl-tRNA fulfills this requirement (Chinali et al. 1974; T. Wagner and M. Sprinzl, unpubl.).

From the investigation of the nonisomerizable aminoacyl-tRNA species (Fraser and Rich 1973; Hecht et al. 1974; Chinali et al. 1974), as well as the nonisomerizable aminoacylated-tRNA fragments (Ringer et al. 1976), it is evident that the ribosomal aminoacyl site is able to accommodate both isomers of the aminoacyl-tRNA. The function of the 3' binding site can be rationalized as the site where the amino acid has to be localized to accept the peptidyl residue (Nathans and Neidle 1963; Fraser and Rich 1973). The function of the ribosomal binding site for an aminoacyl residue attached to the 2' position of aminoacyl-tRNA must be in an intermediate step of the decoding process prior to the formation of a new peptide bond (Ringer et al. 1976).

CONCLUSIONS

Our present knowledge on the participation of the isomeric 2' or 3' aminoacyl-tRNA in the elongation process is summarized in Figure 4. Both isomers can be formed by aminoacylation, depending upon the specificity of the particular synthetase. Furthermore, spontaneous migration of the aminoacyl residue leads to a mixture of isomers, provided that the aminoacyl-tRNA is free in solution. EF-Tu·GTP is able to interact with both 2' and 3' aminoacyl-tRNA isomers. Although such a mechanism is not yet supported by any experimental evidence, the observed unspecificity of EF-Tu·GTP could be required to allow a direct catalysis of the aminoacyl-tRNA release from both 2'- and 3'-aminoacylating aminoacyl-tRNA synthetases.

In the next step, in which a codon-specific aminoacyl-tRNA·EF-Tu·GTP ternary complex is selected by the programmed ribosome from the pool of the available ternary complexes, probably a 2' aminoacyl-tRNA isomer is utilized. This suggestion is based on the following observations. First, the binding of the ternary complex to the programed ribosomes is more efficient if the 2' isomer of the aminoacyl-tRNA is used rather than the 3' isomer. Second, in the binding process the selection of the proper aminoacyl-tRNA must take place in a reversible step. The 2' aminoacyl isomer of the aminoacyl-tRNA fulfills this requirement.

It is established that only the 3' isomer of the aminoacyl-tRNA bound to the ribosomal aminoacyl site can act efficiently as an acceptor of the

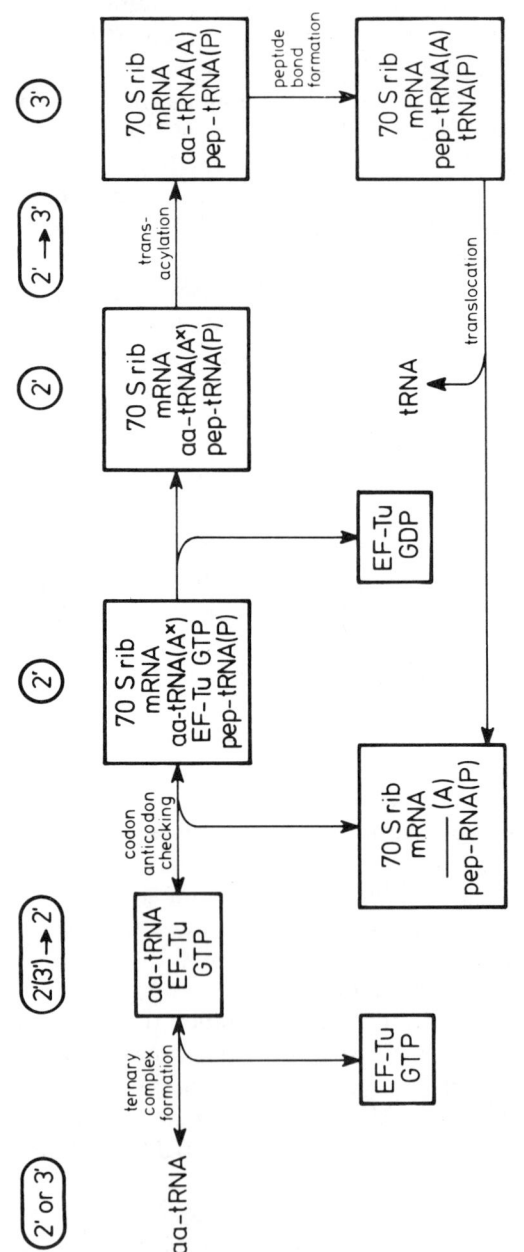

Figure 4 Isomers of the aminoacyl-tRNA (aa-tRNA) in the steps of the elongation process. The position of the aminoacyl residue with respect to the attachment on the 2'-OH or 3'-OH of the terminal A is given above the boxes that indicate the particular intermediate.

peptidyl residue (Fraser and Rich 1973; Chinali et al. 1974). Our finding that the 2′ aminoacylated tRNA species participates in the binding to ribosomes that is EF-Tu·GTP-dependent therefore implies that a migration of the amino acid from the 2′ to the 3′ position of the terminal A of tRNA takes place prior to the peptidyl-transfer reaction. Since it has been demonstrated previously that the 2′ aminoacylated phenylalanyl-tRNA^{Phe}-3′-deoxyadenosine can trigger the EF-Tu-dependent hydrolysis of GTP to GDP, we suggest that the 2′ to 3′ transaminoacylation takes place after GTP hydrolysis and EF-Tu·GDP release.

The significance of the 2′ to 3′ transaminoacylation during the binding of aminoacyl-tRNA to programmed ribosomes becomes apparent from the following consideration. If the α-amino group of the amino acid attached to the tRNA is brought into the vicinity of the carbonyl group of the ester linkage by which the peptidyl residue is bound to the peptidyl-tRNA, a very fast reaction leading to the formation of a new peptide bond should be expected. This can be deduced from the halftime of migration of the amino acid on the aminoacyl-tRNA, which is probably shorter than 10^{-4} seconds (Griffin et al. 1966). In this case, the optimal arrangement of the vicinal hydroxyl group facilitates the transfer of the amino acid. It can be expected that its transfer to a more nucleophilic amino group will take place at an even higher rate.

In the process of binding of aminoacyl-tRNA to ribosomes, two principal functions must be accomplished. First, the α-amino group of the amino acid must be brought to the position that permits the reaction with the ester carbonyl of the peptidyl-tRNA. Second, codon-specific selection of the appropriate aminoacyl-tRNA from the pool of all species has to take place. Considering the very fast peptide-bond formation, if the first point is fulfilled, it is clear that the two processes must be separated during the binding interaction. In other words, an efficient selection of the proper aminoacyl-tRNA can take place only under conditions where there is enough time for the noncognate aminoacyl-tRNA to dissociate from the ribosome before the peptide-bond formation takes place.

The scheme presented in Figure 4 provides a possible pathway for such a mechanism. At the stage where the 2′ aminoacyl-tRNA·EF-Tu·GTP ternary complex is bound to the ribosomes, the amino acid is in its unreactive 2′ position and the entire ternary complex can still easily dissociate. The proper codon-anticodon matching triggers, in an as yet unknown mechanism, the hydrolysis of GTP and the release of EF-Tu·GDP. In the absence of EF-Tu, the aminoacyl residue moves spontaneously to the reactive 3′ position and participates in the de facto irreversible peptidyl-transfer reaction. According to this model there is a sufficient time lag available during the residence time of the ternary complex on the ribosome for the selection of the proper aminoacyl-tRNA before the irreversible peptide-bond formation takes place. EF-Tu-controlled migration of the aminoacyl residue of aminoacyl-tRNA from the nonreactive 2' to the reactive 3' position during the binding to ribosomes is thus necessary for an error-free translation. However, it has to be pointed out that if a dynamic structure with respect to 2',3'-isomerization on the aminoacyl-tRNA is important in the selection process, the use of nonisomerizable tRNA species can only give indirect hints on the mechanism. Clearly, a method that allows the monitoring of the transacylation on the native aminoacyl-tRNA is required for a proof.

ACKNOWLEDGMENTS

We are indebted to F. Cramer for his encouragement, interest, and continuous support during the progress of this work. We thank F. von der Haar, H. Faulhammer, H. Sternbach, E. Graeser, and M. Kucharzewski for their helpful cooperation.

REFERENCES

Baksht, E., N. de Groot, M. Sprinzl, and F. Cramer. 1976. Properties of tRNA species modified in the 3'-terminal ribose moiety in an eucaryotic ribosomal system. *Biochemistry* **15**:3639.

Chinali, G., M. Sprinzl, A. Parmeggiani, and F. Cramer. 1974. Participation in protein biosynthesis of transfer ribonucleic acids bearing altered 3'-terminal ribosyl residues. *Biochemistry* **13**:3001.

Chinault, A. C., K. H. Tan, S. M. Hassur and S. M. Hecht. 1977a. Initial position of aminoacylation of individual *Escherichia coli*, yeast, and calf liver transfer RNAs. *Biochemistry* **16**:766.

Chinault, A. C., J. W. Kozarich, S. M. Hecht, F. J. Schmidt, and R. M. Bock. 1977b. Preparation of *Escherichia coli* tRNAs terminating in modified nucleosides by the use of CTP(ATP):tRNA nucleotidyltransferase and polynucleotide phosphorylase. *Biochemistry* **16**:756.

Fraser, T. H. and A. Rich. 1973. Synthesis and aminoacylation of 3'-amino-3'-deoxy transfer RNA and its activity in ribosomal protein synthesis. *Proc. Natl. Acad. Sci.* **70**:2671.

———. 1974. Amino acids are not all initially attached to the same position on transfer RNA molecules. *Proc. Natl. Acad. Sci.* **72**:3044.

Furano, A. V. 1975. Content of elongation factor Tu in *Escherichia coli*. *Proc. Natl. Acad. Sci.* **72**:4780.

Griffin, B. E., H. Jarman, C. B. Reese, J. E. Sulston, and D. R. Trentham. 1966. Some observations relating to acyl mobility in aminoacyl soluble ribonucleic acids. *Biochemistry* **5**:3638.

Hecht, S. M., J. W. Kozarich, and F. J. Schmidt. 1974. Isomeric phenylalanyl-tRNAs. Position of the aminoacyl moiety during protein biosynthesis. *Proc. Natl. Acad. Sci.* **71**:4317.

Hecht, S. M., K. H. Tan, A. C. Chinault, and P. Arcari. 1977. Isomeric aminoacyl-tRNAs are both bound by elongation factor Tu. *Proc. Natl. Acad. Sci.* **74**:437.

Nathans, D. and A. Neidle. 1963. Structural requirements for puromycin inhibition of protein synthesis. *Nature* **197**:1076.

Potts, R., M. J. Fournier, and N. C. Ford. 1977. Effect of aminoacylation on the conformation of yeast phenylalanine tRNA. *Nature* **268**:563.

Pingoud, A., C. Urbanke, G. Krauss, F. Peters, and G. Maass. 1977. Ternary complex formation between elongation factor Tu, GTP and aminoacyl-tRNA: An equilibrium study. *Eur. J. Biochem.* **78**:403.

Ringer, D., S. Chladek, and J. Ofengand. 1976. Enzymatic binding of aminoacyl transfer ribonucleic acid to ribosomes: The study of binding sites of 2' and 3' isomers of aminoacyl transfer ribonucleic acid. *Biochemistry* **15**:2759.

Schwarz, U., H. M. Menzel, and H. G. Gassen. 1976. Codon-dependent rearrangement of the three-dimensional structure of phenylalanine tRNA, exposing the T-Ψ-C-G- sequence for binding to the 50S ribosomal subunit. *Biochemistry* **15**:2484.

Sprinzl, M. and F. Cramer. 1973. Accepting site for aminoacylation of tRNA^{Phe} from yeast. *Nat. New Biol.* **245**:3.

———. 1974. Site of aminoacylation of tRNAs from E. coli with respect to the 2'- or 3'-hydroxyl group of the terminal adenosine. *Proc. Natl. Acad. Sci.* **72**:3049.

Sprinzl, M., M. Kucharzewski, J. B. Hobbs, and F. Cramer. 1977a. Specificity of elongation factor Tu from *E. coli* with respect to attachment of the amino acid to the 2'- or 3'-hydroxyl group of the terminal adenosine of tRNA. *Eur. J. Biochem.* **78**:55.

Sprinzl, M., H. Sternbach, F. von der Haar, and F. Cramer. 1977b. Enzymatic incorporation of ATP and CTP analogues into the 3' end of tRNA. *Eur. J. Biochem.* **81**:579.

Thompson, R. C. and P. J. Stone. 1977. Proofreading of the codon-anticodon interaction on ribosomes. *Proc. Natl. Acad. Sci.* **74**:198.

Wagner, T. and M. Sprinzl. 1978. Enzymatic binding of aminoacyl-tRNA to *Escherichia coli* ribosomes using modified tRNA species and tRNA fragments. *Methods Enzymol.* **60**:615.

Yokosawa, H., N. Inoue-Yokosawa, K. I. Arai, M. Kawakita, and Y. Kaziro. 1973. The role of guanosine triphosphate hydrolysis in elongation factor Tu-promoted binding of aminoacyl transfer ribonucleic acid to ribosomes. *J. Biol. Chem.* **248**:375.

Analogs of Lysyl-tRNA as Probes of Ribosome and Elongation Factor Tu Structure and Function

Arthur E. Johnson
Department of Chemistry
University of Oklahoma
Norman, Oklahoma 73019

tRNA is the substrate in nearly all of the partial reactions of protein biosynthesis. Therefore, the most direct means of investigating the structural and functional states of the ribosome and associated macromolecules is to monitor the tRNA. With this approach, it is also possible to correlate directly a structural state with a particular functional state, as the functional state of a tRNA in a ribosomal complex can be determined by biochemical assay.

Selective observation of the tRNA is complicated by the presence of rRNA and mRNA in the ribosomal complex. Techniques that measure an intrinsic property of RNA are generally unsuitable because a tRNA constitutes only about 1% of the total RNA in the complex. This problem is circumvented by covalently attaching to a tRNA a tag or probe moiety that is easily observable above the background signal of the system. This paper describes our work using a unique class of aminoacyl-tRNA analogs that have probe moieties attached to the side chain of the lysine in lysyl-tRNA.

THE PROBE ATTACHMENT SITE

Probes attached to (or in place of) various unusual bases in the tRNA have provided information about each of the tRNA loop regions and their environments. At the aminoacyl end of the molecule, probes have usually been attached to the α-amino nitrogen of the amino acid. These modified tRNAs are analogs of peptidyl-tRNA, however, and cannot interact with elongation factor Tu (EF-Tu; I will confine my discussion to the *Escherichia coli* system) because of the blocked α-amino group (Miller and Weissbach 1977). Thus, certain states of the ribosomal complex, such as the EF-Tu·GMPP(NH)P-dependent recognition conformation (Johnson et al. 1977; Lake 1977), are not accessible for study using these analogs.

My approach is to attach a probe to the side chain of the amino acid. In this location, the probe does not interfere with the interaction between the aminoacyl-tRNA and the EF-Tu. As a result, the aminoacyl regions of the EF-Tu·aminoacyl-tRNA·GTP complex and all of the EF-Tu-dependent states of the ribosomal complex can be investigated.

Side-chain-modified lysyl-tRNA was prepared primarily because the reaction of the probe reagent at other sites on the tRNA was minimal and did not interfere with tRNA function and because the probe is linked to the lysine through a stable amide bond (Johnson et al. 1976). In addition, mRNA requirements for in vitro assays are satisfied by commercially available poly(rA) or easily synthesized oligonucleotides, such as $AUGA_n$. An advantage in using this type of aminoacyl-tRNA analog is that radioactive probe moieties need not be synthesized. Instead, probes are covalently attached to radioactive amino acids that are commercially available. On the other hand, these probes are susceptible to hydrolysis from the tRNA, as they are ultimately attached to the tRNA through the aminoacyl-ester bond.

CHEMICAL MODIFICATION OF THE ϵ-AMINO GROUP OF LYSYL-tRNA

Aminoacyl-tRNA synthetases will not attach substantially altered amino acids to their cognate tRNAs. Lysyl-tRNA synthetase, for example, does not recognize N^ϵ-acetyllysine as a legitimate substrate (Johnson et al. 1976). Side-chain modifications must therefore be done chemically using enzymatically prepared aminoacyl-tRNA as one of the reagents. This in turn necessitates that the reaction conditions be mild enough to avoid significant hydrolysis of the aminoacyl ester bond.

When lysyl-tRNA in a tetrahydrofuran:aqueous mixture (1:4) is exposed for a short time (15 sec) to a high apparent pH ($\sim$11) in the presence of an N-hydroxysuccinimide ester of, e.g., bromoacetic acid, the lysyl-tRNA is preferentially acylated at the ϵ-amino group (Johnson et al. 1976). Despite the exposure to high pH, the recovery of lysyl-tRNA specific activity (pmoles lysine/A_{260} unit tRNA) is greater than 80%. Lysines that have been acylated at the α-amino group can be enzymatically hydrolyzed from the tRNA using peptidyl-tRNA hydrolase. Following incubation with this enzyme, the percentage of N^ϵ-labeled lysyl-tRNA in the solution is 85–93% (Johnson et al. 1976 and unpubl.).

For most purposes, it is essential that the probe be located at only one site on the aminoacyl-tRNA. Several tRNAs, including E. coli tRNA^{Lys} (Chakraburtty et al. 1975) contain the X base, 3-(3-amino-3-carboxypropyl)uridine (Friedman et al. 1974; Ohashi et al. 1974), and this base can react with N-hydroxysuccinimide esters (Friedman 1972, 1973; Nauheimer and Hedgcoth 1974; Schiller and Schechter 1977; A. E. Johnson, unpubl.). The reaction conditions used to modify the ϵ-amino group of lysine only slightly diminish the ability of the tRNA^{Lys} to be aminoacylated (Johnson et al. 1976) and to participate in protein synthesis (discussed below). Therefore, reaction of the probe with a functionally important tRNA base is minimal. However, the extent to which these reaction conditions acylate tRNA has not been determined directly. In the affinity-labeling experiments using N^ϵ-bromoacetyl-lysyl-tRNA (ϵBrAcLys-tRNA), only the lysine was radioactive, and therefore only covalent reactions involving the bromoacetyl moiety on the lysine were detected (A. E. Johnson and C. R. Cantor, in prep.). On the other hand, to prepare the fluorescent analog N^ϵ-(N-methylanthraniloyl)-lysyl-tRNA, it is necessary to acetylate the X base of tRNA^{Lys} using N-acetoxysuccinimide before exposing the tRNA to the N-hydroxysuccinimide ester of N-methylanthranilic acid (A. E. Johnson, unpubl.).

FUNCTIONAL ACTIVITY OF N^ϵ-MODIFIED LYSYL-tRNA

The primary prerequisite for a useful analog of aminoacyl-tRNA is that it retains its functional activity. N^ϵ-modified lysyl-tRNA interacts normally with EF-Tu in terms of both ternary-complex formation and ternary-complex interaction with ribosomes. The aminoacyl ester bond of ϵBrAcLys-tRNA is protected from hydrolysis in the presence of EF-Tu and GTP, demonstrating the formation of the ϵBrAcLys-tRNA·EF-Tu·GTP complex (Johnson et al. 1978). The cross-linking of ϵBrAcLys-tRNA to EF-Tu, which is discussed below, also requires ternary-complex formation. The binding of N^ϵ-modified lysyl-tRNAs to ribosomes is EF-Tu-dependent, as well as message-dependent and message-specific (Johnson et al. 1977 and unpubl.). Peptide-bond formation between N,N-diacetyl-lysyl-tRNA (DiAcLys-tRNA) in the peptidyl site and ϵBrAcLys-tRNA in the aminoacyl site has been observed in in vitro incubations using poly(rA) as the message (Johnson et al. 1977). The analog can also serve as a donor in transpeptidation, as peptidyl-site-bound DiAcLys-ϵBrAcLys-tRNA reacts with puromycin. In addition, translocation of ϵBrAcLys-tRNA from the aminoacyl to the peptidyl site is elongation factor G (EF-G)-dependent (see below).

The most stringent test of the functional activity of N^ϵ-modified lysyl-tRNA is whether it can compete with unmodified aminoacyl-tRNA in a protein-synthesizing system containing natural mRNA. As an assay medium we used a rabbit reticulocyte, cell-free, protein-synthesizing system that was nearly as active as the intact cell for up to 20 minutes, synthesized 10–15 complete chains of globin per ribosome during a 60-minute incubation, and contained the full in vivo complement of reticulocyte tRNA^{Lys} (Woodward et al. 1974). When $E.$ $coli$ lysyl-tRNA and $E.$ $coli$ ϵAcLys-tRNA were added to parallel lysate incubations, the N^ϵ-acetyllysine was incorporated 82% as well as the unmodified lysine (Johnson et al. 1976). Further analysis demonstrated that the

N^ϵ-acetyllysine was not only incorporated into complete globin chains, but was present in tetrameric hemoglobin (Johnson et al. 1976).

The N^ϵ-modified lysyl-tRNAs are therefore fully competent from a functional standpoint. Particularly convincing is their ability to interact successfully with heterologous ribosomes and natural message and to successfully compete with unmodified reticulocyte lysyl-tRNA. Equally important, ribosomes and elongation factors appear to accommodate the elongated lysine side chain with little difficulty.

STRUCTURAL STUDIES UTILIZING εBrAcLys-tRNA

εBrAcLys-tRNA is an excellent choice for an affinity label because its functional activity is virtually unperturbed by the bromoacetyl moiety. The analog therefore associates normally with the EF-Tu·GTP and ribosomal complexes. This automatically places the probe in a functional aminoacyl-tRNA binding site. A nearby nucleophilic residue of EF-Tu, rRNA, or ribosomal protein can then attack the electrophilic bromoacetyl group and be identified through the radioactive lysine to which it becomes covalently linked.

AFFINITY LABELING OF EF-Tu

The association of εBrAcLys-tRNA, EF-Tu, and GTP to form a ternary complex results in the covalent cross-linking of EF-Tu and the aminoacyl-tRNA analog (Fig. 1). The yield of covalent product is exceptionally high, typically 40–50% of the εBrAcLys-tRNA added to the incubation (Johnson et al. 1978). No covalent bond is formed between EF-Tu and εBrAcLys-tRNA in the absence of GTP (the phosphoenolpyruvate-free incubation in Fig. 1). In addition, N,N-dibromoacetyl-lysyl-tRNA (DiBrAcLys-tRNA), a reactive analog of peptidyl-tRNA that does not form a ternary complex (Johnson et al. 1977), does not react covalently with EF-Tu (Fig. 1). These two controls show that there is no nonspecific covalent reaction between EF-Tu and tRNA-bound bromoacetyl moieties. The addition of unmodified lysyl-tRNA dramatically reduces the amount of cross-linking (Johnson et al. 1978) and further indicates that the covalent reaction is binding-site-specific.

The reactive EF-Tu residue is accessible and less than 9 Å (the fully extended length of the N^ϵ-bromoacetyllysine side chain) along the surface of the protein from the normal binding site of the α-carbon of the aminoacyl-tRNA's amino acid in the ternary complex. The EF-Tu nucleophile is not covered by either the aminoacyl-tRNA or the GTP in the ternary complex. The reactive EF-Tu amino acid has not yet been identified, but only one (perhaps two) tryptic peptide is alkylated by the εBrAcLys-tRNA (D. L. Miller and A. E. Johnson, unpubl.). Knowing the

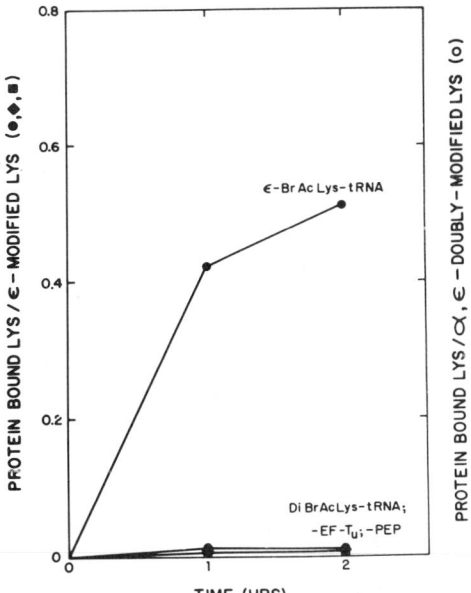

Figure 1 Affinity labeling of EF-Tu by εBrAcLys-tRNA. Aliquots were assayed for lysine moieties covalently attached to protein by measuring the hot (10 min, 80°C) trichloroacetic-acid-insoluble cpm. Ordinate values for these incubations, complete (●), without EF-Tu (◆), and without phosphoenolpyruvate (■), were calculated as a function of the amount of εBrAcLys-tRNA originally present. Data obtained from a complete incubation that contained DiBrAcLys-tRNA instead of εBrAcLys-tRNA are given (○). Ordinate values were calculated in this case as a function of the amount of DiBrAcLys-tRNA originally present and corrected for the cross-linking expected from the εBrAcLys-tRNA impurity in the DiBrAcLys-tRNA solution. (Reprinted, with permission, from Johnson et al. 1978.)

cross-linked EF-Tu residue will assist greatly in locating the binding site of the aminoacyl-tRNA's amino acid in the EF-Tu crystal structure, once it is sufficiently resolved to allow identification of individual EF-Tu residues.

FUNCTIONAL ACTIVITY OF CROSS-LINKED TERNARY COMPLEX

Since the εBrAcLys-tRNA·EF-Tu·GTP complex may have been trapped by the cross-linking reaction in a nonfunctional conformation, it is desirable to demonstrate that the cross-linked complex is functional. Following purification by gel filtration, the cross-linked ternary complex was assayed for binding to ribosomes by both filter binding and Sepharose gel chromatography (Johnson et al. 1978). The covalent ternary complex exhibited binding that was both message-dependent and message-specific.

Its affinity for the ribosome·poly(rA) complex was similar to that of the noncovalent ternary complex (Johnson et al. 1978).

Normally in bacterial protein synthesis, the GTP is hydrolyzed following ternary complex association with the ribosome (Haselkorn and Rothman-Denes 1973). The resulting EF-Tu·GDP and Pi are released from the ribosome, and the aminoacyl-tRNA is positioned in the aminoacyl site to react with the peptidyl-tRNA in the peptidyl site. To determine whether the cross-linking interfered with the GTP hydrolysis, covalent ternary complexes were prepared with either [γ-^{32}P]GTP or [^{14}C]GTP (Johnson et al. 1978). Gel filtration of an incubation containing ribosomes, poly(rA), and cross-linked ϵBrAc-[^{3}H]Lys-tRNA·EF-Tu·[γ-^{32}P]GTP showed that ^{3}H label, but no ^{32}P label, eluted with the ribosomes in the void volume (fractions 8–11, Fig. 2). In an incubation containing [^{14}C]GTP instead of [γ-^{32}P]GTP, both ^{3}H label and ^{14}C label were found associated with ribosomes (Fig. 3). Thus, the GTP is hydrolyzed following association of the cross-linked ternary complex with the ribosome, but the newly formed GDP remains bound to the ribosome because the cross-linking prevents the release of EF-Tu·GDP.

The cross-linked complex therefore appears to function normally, except for its inability to dissociate. This means that the EF-Tu nucleophile is not required for binding to the ribosome or for GTP hydrolysis. In addition, the immobilization of the lysine side chain does not interfere with normal function. This suggests that EF-Tu and aminoacyl-tRNA interactions with the ribosome involve regions of the ternary complex removed from the site of cross-linking.

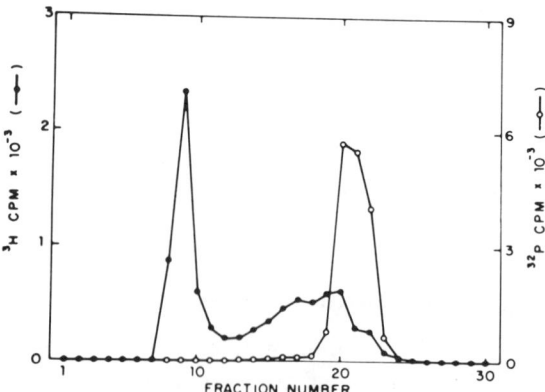

Figure 2 Gel-filtration analysis of the binding to poly(rA)-programmed ribosomes of cross-linked EF-Tu·ϵBrAc-[^{3}H]Lys-tRNA·[γ-^{32}P]GTP. A 100-μl binding incubation was layered on a 0.6-cm diameter × 8.5 cm Sepharose 6B column and three-drop (0.13 ml) fractions were assayed for radioactivity. (Reprinted, with permission, from Johnson et al. 1978.)

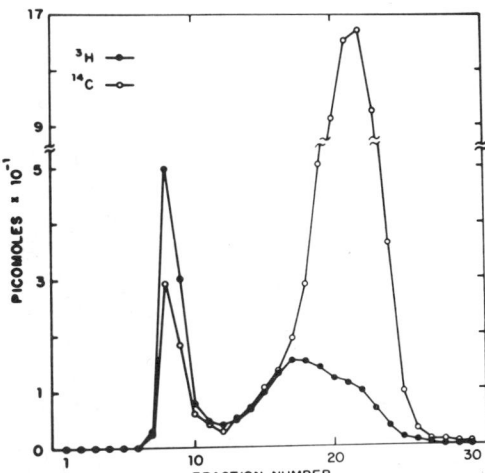

Figure 3 Gel-filtration analysis of the binding to poly(rA)-programmed ribosomes of cross-linked EF-Tu·εBrAc-[^{3}H]Lys-tRNA·[^{14}C]GTP. This incubation was treated as in Fig. 2. (Reprinted from Johnson et al. 1978.)

This functional covalent complex provides a unique opportunity to investigate both protein-nucleic acid interactions and ternary-complex interactions with the ribosome. For example, since complex dissociation has so far stymied attempts to crystallize protein-nucleic acid complexes (except for nucleosome core particles; Finch et al. 1977), the nondissociable nature of this cross-linked complex greatly improves the chances of obtaining a crystal structure of an aminoacyl-tRNA in a functional protein-nucleic acid complex. The structural and functional consequences of restricting amino acid attachment to either the 2'-OH or the 3'-OH of the terminal A can also be investigated.

AFFINITY LABELING OF RIBOSOMES: SITE SPECIFICITY

Affinity-labeling experiments involving multicomponent complexes are subject to several complications, as we have discussed at length elsewhere (Johnson and Cantor 1977). The primary concern is whether the covalent reaction occurred while the reactive tRNA was in a functional binding site. It is also desirable in the ribosomal system to determine in which functional site the tRNA was bound when the covalent reaction occurred.

The key observation in affinity-labeling experiments with ribosomes is that photoreactive and electrophilic tRNA analogs react covalently with ribosomal components with a very low efficiency (Johnson and Cantor 1977). This fact makes unambiguous interpretation of the data difficult because of heterogeneity in the ribosomal population (composition, conformational state, and activity) and the possibility of multiple tRNA binding sites (including nonspecific sites). For example, except for εBrAcLys-tRNA, only analogs of peptidyl-tRNA have been used to investigate the topography of the aminoacyl ends of the aminoacyl and peptidyl sites (for a review, see Pellegrini and Cantor 1977). The inability of these analogs to interact with EF-Tu means that their binding to the ribosome is nonenzymatic and is split between the aminoacyl and peptidyl sites (Eilat et al. 1974b; Pellegrini et al. 1974). The latter is also true of analogs of formylmethionyl-tRNA that exhibit initiation-factor-dependent binding to ribosomes (Sopori et al. 1974). In both cases, 60–80% of the analogs react with puromycin and are, by definition, in the peptidyl site (Pellegrini et al. 1974; Sopori et al. 1974). But the fraction of ribosome-bound analog that attaches covalently to ribosomal components is usually less than 10% (Johnson and Cantor 1977) so it cannot be concluded unambiguously that the covalent reaction occurred in the peptidyl site.

It is also difficult to determine whether a probe reacts covalently with the ribosome while it is in the binding conformation indicated by functional analysis. Because the function assay (e.g., dipeptide formation) and the covalent reaction do not occur simultaneously, it is possible during the time interval between the two for the tRNA to be translocated from one site to another. (Note that this is also the case in photoaffinity-labeling experiments.) We have found, in fact, that even the dipeptide approach to site identification can be misleading and must be interpreted cautiously (A. E. Johnson and C. R. Cantor, in prep.).

ELONGATION-FACTOR-DEPENDENT AFFINITY LABELING OF RIBOSOMES WITH εBrAcLys-tRNA

To avoid the objections and uncertainties inherent in a ribosomal affinity-labeling experiment, it is necessary to demonstrate that the same tRNA that reacted covalently with the ribosome was bound in a functional manner to a particular functional site on a functional ribosome at the time of covalent reaction. All of these conclusions are important in view of the heterogeneity in ribosomes and tRNA binding sites. These strict requirements were successfully met using elongation factors to position εBrAcLys-tRNA in a particular binding site. This was possible, of course, only because εBrAcLys-tRNA interacts normally with elongation factors and ribosomes.

When bound to the ribosome-mRNA complex with EF-Tu and a nonhydrolyzable analog of GTP (e.g., GMPP[NH]P), an aminoacyl-tRNA cannot participate in transpeptidation and the EF-Tu and GTP analog remain bound to the ribosome (Skoultchi et al. 1970). We have termed this the recognition binding site or conformation (Johnson et al. 1977). Lake (1977) has since independently proposed a detailed model of such a site, which he also termed recognition. When bound to the ribosomal complex with EF-Tu and GTP, the aminoacyl-tRNA is positioned in the aminoacyl site. The addition of EF-G to such complexes results, in the presence of GTP, in the translocation of the tRNA from the aminoacyl to the peptidyl site. The elongation cycle will be halted at this point if the incubation contains fusidic acid. This antibiotic permits only one round of translocation by stabilizing the binding of EF-G to the ribosome, thereby preventing the EF-G from acting catalytically (Bodley et al. 1970) and also preventing further EF-Tu·aminoacyl-tRNA·GTP association with the ribosome (Cabrer et al. 1972; Miller 1972; Richman and Bodley 1972; Richter 1972, 1973). In the absence of fusidic acid, protein chain elongation will proceed.

Ribosomal complexes that contained ϵBrAcLys-tRNA in each of its functional states were prepared by incubating salt-washed ribosomes, poly(rA), and ϵBrAcLys-tRNA with the appropriate pure elongation factors and either GMPP(NH)P or GTP. The binding incubations contained 7 mM Mg^{++} because at that concentration the ϵBrAcLys-tRNA binding to ribosomes is EF-Tu-dependent. By doing the experiment in this way, EF-Tu screens both the ϵBrAcLys-tRNA molecules and the ribosomal complexes for active particles, and thus the system itself selects only functional macromolecules and macromolecular complexes.

Ribosomal RNA was assayed for covalent reaction with ϵBrAcLys-tRNA by sedimentation through a sucrose gradient containing SDS after dissociation from ribosomal proteins in 1% SDS (A. E. Johnson and C. R. Cantor, in prep.). Radioactive N^{ϵ}-bromoacetyllysine cosedimented with 23S rRNA (and therefore was covalently attached to it) only in those incubations that contained EF-G (Fig. 4). The total cross-linking of ϵBrAcLys-tRNA to 23S rRNA was reduced by about 50% in the presence of fusidic acid (3 mM). Essentially no covalent reaction occurred with rRNA when ϵBrAcLys-tRNA was in the recognition or aminoacyl site. Various control incubations showed that the 23S rRNA reaction with ϵBrAcLys-tRNA was dependent upon EF-Tu, poly(rA), and an intact bromoacetyl moiety, and that it did not result from covalent reaction through a bromoacetyl-labeled X base.

The dependence of the covalent reaction upon EF-G demonstrates that the cross-linking to 23S rRNA occurs when ϵBrAcLys-tRNA is in the peptidyl site. The dependence of the cross-linking on both EF-Tu and EF-G also shows that the covalent reaction occurs on functional ribosomes with functional ϵBrAcLys-tRNA. The addition of [^{14}C]lysyl-tRNA to ribosomal complexes that had already covalently reacted with ϵBrAc-[^{3}H]Lys-tRNA resulted in ^{14}C radioactivity cosedimenting with 23S rRNA (A. E. Johnson and C. R. Cantor, in prep.). This means that the same ϵBrAcLys-tRNA that reacts covalently with 23S rRNA can still participate in peptide-bond formation and therefore must be bound in a functional manner to a functional peptidyl site. Since the reactive 23S

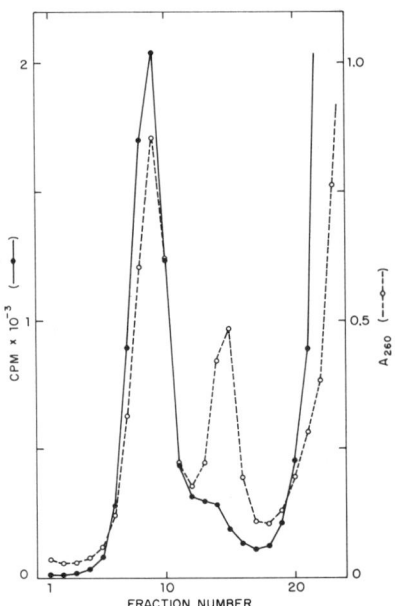

Figure 4 Covalent reaction of εBrAcLys-tRNA with 23S rRNA. The incubation (37°C, 40 min, 0.28 ml) contained: 7 mM magnesium acetate; 50 mM Tris-HCl (pH 7.4); 75 mM NH$_4$Cl; 75 mM KCl; 33 μM GTP; 1.05 A$_{260}$ units of poly(rA); 6.5 A$_{260}$ units of salt-washed *E. coli* 70S ribosomes; 9.7 A$_{260}$ units of unacylated, unfractionated tRNA; 15 μg of EF-Tu·GDP; and 2.4 A$_{260}$ units of εBrAc-[^{3}H]Lys-tRNA (100 pmoles of [^{3}H]lysine; 82% εBrAcLys-tRNA; 4790 cpm/pmole lysine). EF-G (13.6 μg) and GTP (to 1 mM) were added at 15 min. Following incubation in 1% SDS, 0.1 M LiCl for 20 min at 22°C, the solution was layered on a 5–20% sucrose gradient in 0.1% SDS, 0.1 M LiCl and centrifuged for 16 hr at 20°C and 24,500 rpm in an SW 41 rotor. Eight-drop fractions were assayed for radioactivity and absorbance at 260 nm.

rRNA nucleophile is not accessible from the aminoacyl site, the aminoacyl end of the peptidyl site must differ structurally from that of the aminoacyl site, and thus the aminoacyl and peptidyl sites are distinct both structurally and functionally.

These conclusions are supported by an analysis of εBrAcLys-tRNA cross-linking to ribosomal proteins using two-dimensional gel electrophoresis (A. E. Johnson and C. R. Cantor, in prep.). The amount of radioactivity in the gel near large-subunit proteins L13, L14, and L15 is dramatically increased in those incubations containing EF-G. The covalent reaction with L2 is also increased by EF-G action, whereas L27 reacts covalently with εBrAcLys-tRNA in the absence of EF-G. Thus, reactive residues of L13, L14, and/or L15 and of L2 are located next to the aminoacyl end of the peptidyl site, and a reactive residue of L27 appears to be located adjacent to the aminoacyl end of the aminoacyl site. These results confirm the tentative assignments made in previous investigations (for a review, see Pellegrini and Cantor 1977), except in the case of L27. This protein is the primary target of electrophilic tRNA analogs as diverse as bromoacetyl-(glycyl)$_{12}$-phenylalanyl-tRNA (Eilat et al. 1974a), bromoacetylphenylalanyl-tRNA (Pellegrini et al. 1974), and εBrAcLys-tRNA (A. E. Johnson and C. R. Cantor, in prep.). Sequence analysis of cross-linked L27 is necessary to determine whether L27 actually has reactive residues positioned at the aminoacyl site, the peptidyl site, and also in the region occupied by the nascent protein chain.

This work has unambiguously located, for the first time, structural elements of the ribosome adjacent to a functionally defined site. For example, the reactive 23S rRNA nucleophile appears to be located less than 9 Å from the α-carbon of the C-terminal amino acid in a peptidyl-tRNA bound to the peptidyl site of the peptidyl transferase center. Two different sequences of 23S rRNA have been reported to be adjacent to the peptidyl-transferase center (Yukioka et al. 1977; Eckermann and Symons 1978). It will be interesting to compare these sequences to the one that contains the cross-link to ϵBrAcLys-tRNA, as the last is known, unequivocally, to be located next to the peptidyl site.

The efficiency of covalent reaction between ribosomal components and ϵBrAcLys-tRNA bound in the aminoacyl site is very low. This is not surprising because potent nucleophiles in this region could conceivably attack the aminoacyl ester bond of the aminoacyl-tRNA. The efficiency of cross-linking between ribosomes and recognition-site-bound ϵBrAcLys-tRNA is also very low. Since EF-Tu remains associated with the ribosome under recognition-site binding conditions, ϵBrAcLys-tRNA may react covalently with EF-Tu (as discussed earlier) more efficiently than with the ribosome, thereby reducing the ribosomal cross-linking. In any event, we have been unable to distinguish between the aminoacyl ends of the aminoacyl site and the putative recognition site using ϵBrAcLys-tRNA.

PROSPECTUS

In principle, any type of probe could be covalently attached to the ϵ-amino group of lysyl-tRNA. This includes fluorescent dyes, spin labels, photoreactive moieties, antigenic determinants, trifluoroacetyl ^{19}F nuclear magnetic resonance probes, and chemically reactive groups. One synthetic route requires the appropriate N-hydroxysuccinimide ester of the probe. As noted earlier, fluorescence-labeled lysyl-tRNA was prepared in this way using the N-hydroxysuccinimide ester of N-methylanthranilic acid (A. E. Johnson, unpubl.). A different synthetic pathway utilizes the well-characterized ϵBrAcLys-tRNA as precursor. Any probe reagent containing a sulfhydryl group will react with the ϵBrAcLys-tRNA, thereby placing the probe on the lysine side chain.

The various N^ϵ-modified lysyl-tRNAs provide a unique opportunity to investigate the molecular mechanisms involved in protein synthesis. For example, fluorescence-labeled analogs could be used to measure the kinetics of aminoacyl-tRNA binding to the ribosome and to examine the nature of the catalytic effect of EF-Tu on protein synthesis. Such analogs could also be used to measure, by singlet-singlet energy transfer, the distance from a dye on the lysine side chain to an appropriate dye elsewhere in the ribosomal complex. Any changes in a particular distance as the ϵ-modified lysyl-tRNA moves from the recognition site to the aminoacyl site would be especially interesting, as it would indicate the magnitude of the structural alteration associated with the EF-Tu-mediated GTP hydrolysis. This is extremely pertinent to considerations of the conversion of chemical energy into mechanical work. Using the energy-transfer technique, we have recently obtained results that indicate that the corners of the tRNAs in the aminoacyl and peptidyl sites are immediately adjacent to each other (A. E. Johnson et al., unpubl.). Thus, in the case of tRNA translocation from the aminoacyl to the peptidyl site, GTP hydrolysis is associated with a molecular movement on the order of 20 Å, the thickness of a tRNA molecule (Kim 1976).

REFERENCES

Bodley, J. W., F. J. Zieve, and L. Lin. 1970. Studies on translocation. IV. The hydrolysis of a single round of guanosine triphosphate in the presence of fusidic acid. *J. Biol. Chem.* **245:** 5662.

Cabrer, B., D. Vázquez, and J. Modolell. 1972. Inhibition by elongation factor EF-G of aminoacyl-tRNA binding to ribosomes. *Proc. Natl. Acad. Sci.* **69:** 733.

Chakraburtty, K., A. Steinschneider, R. V. Case, and A. H. Mehler. 1975. Primary structure of tRNA^{Lys} of *Escherichia coli* B. *Nucleic Acids Res.* **2:** 2069.

Eckermann, D. J. and R. H. Symons. 1978. Sequence at the site of attachment of an affinity-label derivative of puromycin on 23S ribosomal RNA of *Escherichia coli* ribosomes. *Eur. J. Biochem.* **82:** 225.

Eilat, D., M. Pellegrini, H. Oen, Y. Lapidot, and C. R. Cantor. 1974a. A chemical mapping technique for exploring the location of proteins along the ribosome-bound peptide chain. *J. Mol. Biol.* **88:** 831.

Eilat, D., M. Pellegrini, H. Oen, N. de Groot, Y. Lapidot, and C. R. Cantor. 1974b. Affinity labelling the acceptor site of the peptidyl transferase center of the *Escherichia coli* ribosome. *Nature* **250:** 14.

Finch, J. T., L. C. Lutter, D. Rhodes, R. S. Brown, B. Rushton, M. Levitt, and A. Klug. 1977. Structure of nucleosome core particles of chromatin. *Nature* **269:** 29.

Friedman, S. 1972. Acylation of transfer ribonucleic acid with the *N*-hydroxysuccinimide ester of phenoxyacetic acid. *Biochemistry* **11:** 3435.

―――. 1973. Patterns of base modification in tRNA. *Nat. New Biol.* **244:** 18.

Friedman, S., H. T. Li, K. Nakanishi, and G. Van Lear. 1974. 3-(3-amino-3-carboxy-*n*-propyl)uridine. The structure of the nucleoside in *Escherichia coli* transfer ribonucleic acid that reacts with phenoxyacetoxysuccinimide. *Biochemistry* **13:** 2932.

Haselkorn, R. and L. B. Rothman-Denes. 1973. Protein synthesis. *Annu. Rev. Biochem.* **42:** 379.

Johnson, A. E. and C. R. Cantor. 1977. Affinity labeling of multicomponent systems. *Methods Enzymol.* **46:** 180.

Johnson, A. E., R. H. Fairclough, and C. R. Cantor. 1977. Some approaches to the study of ribosome-tRNA interactions. In *Nucleic acid-protein recognition* (ed. H. Vogel), p. 469. Academic Press, New York.

Johnson, A. E., D. L. Miller, and C. R. Cantor. 1978. Functional covalent complex between elongation factor Tu and an analog of lysyl-tRNA. *Proc. Natl. Acad. Sci.* **75:** 3075.

Johnson, A. E., W. R. Woodward, E. Herbert, and J. R. Menninger. 1976.

N^ϵ-acetyllysine transfer ribonucleic acid: A biologically active analogue of aminoacyl transfer ribonucleic acids. *Biochemistry* **15**:569.

Kim, S.-H. 1976. Three-dimensional structure of transfer RNA. *Prog. Nucleic Acid Res. Mol. Biol.* **17**:181.

Lake, J. A. 1977. Aminoacyl-tRNA binding at the recognition site is the first step of the elongation cycle of protein synthesis. *Proc. Natl. Acad. Sci.* **74**:1903.

Miller, D. L. 1972. Elongation factors EF-Tu and EF-G interact at related sites on ribosomes. *Proc. Natl. Acad. Sci.* **69**:752.

Miller, D. L. and H. Weissbach. 1977. The interactions of elongation factor Tu. In *Nucleic acid-protein recognition* (ed. H. Vogel), p. 409. Academic Press, New York.

Nauheimer, U. and C. Hedgcoth. 1974. Acylation of several tRNAs of *Escherichia coli* by the phenoxyacetyl derivative of N-hydroxysuccinimide. *Arch. Biochem. Biophys.* **160**:631.

Ohashi, Z., M. Maeda, T. A. McCloskey, and S. Nishimura. 1974. 3-(3-amino-3-carboxypropyl)uridine: A novel modified nucleoside isolated from *Escherichia coli* phenylalanine transfer ribonucleic acid. *Biochemistry* **13**:2620.

Pellegrini, M. and C. R. Cantor. 1977. Affinity labeling of ribosomes. In *Molecular mechanisms of protein biosynthesis* (ed. H. Weissbach and S. Pestka), p. 203. Academic Press, New York.

Pellegrini, M., H. Oen, D. Eilat, and C. R. Cantor. 1974. The mechanism of covalent reaction of bromoacetyl-phenylalanyl-transfer RNA with the peptidyl-transfer RNA binding site of the *Escherichia coli* ribosome. *J. Mol. Biol.* **88**:809.

Richman, N. and J. W. Bodley. 1972. Ribosomes cannot interact simultaneously with elongation factors EF-Tu and EF-G. *Proc. Natl. Acad. Sci.* **69**:686.

Richter, D. 1972. Inability of *Escherichia coli* ribosomes to interact simultaneously with the bacterial elongation factors EF-Tu and EF-G. *Biochem. Biophys. Res. Commun.* **46**:1850.

―――. 1973. Competition between the elongation factors 1 and 2, and phenylalanyl transfer ribonucleic acid for the ribosomal binding sites in a polypeptide-synthesizing system from brain. *J. Biol. Chem.* **248**:2853.

Schiller, P. W. and A. N. Schechter. 1977. Covalent attachment of fluorescent probes to the X-base of *Escherichia coli* phenylalanine transfer ribonucleic acid. *Nucleic Acids Res.* **4**:2161.

Skoultchi, A., Y. Ono, J. Waterson, and P. Lengyel. 1970. Peptide chain elongation: Indications for the binding of an amino acid polymerization factor·guanosine 5'-triphosphate·aminoacyl transfer ribonucleic acid complex to the messenger·ribosome complex. *Biochemistry* **9**:508.

Sopori, M., M. Pellegrini, P. Lengyel, and C. R. Cantor. 1974. Affinity labeling of *Escherichia coli* ribosomal proteins with an analog of the natural initiator tRNA. *Biochemistry* **13**:5432.

Woodward, W. R., J. L. Ivey, and E. Herbert. 1974. Protein synthesis with rabbit reticulocyte preparations. *Methods Enzymol.* **30**:724.

Yukioka, M., T. Hatayama, and K. Omori. 1977. Nucleotide sequence of a region in 23S RNA adjacent to peptidyl transferase catalytic center of *Escherichia coli* ribosomes. *Eur. J. Biochem.* **73**:449.

The Relationship of the Accuracy of Aminoacyl-tRNA Synthesis to That of Translation

Michael Yarus
Department of Molecular, Cellular, and Developmental Biology
University of Colorado
Boulder, Colorado 80309

No real process can be carried out with complete precision. This would require an infinite free-energy difference between a desired reaction and another to be discriminated against. That is, some reactant other than the one desired can always utilize a given pathway at a finite rate, and/or an alternate pathway, and products exist even for an accurately selected reagent. This problem is particularly acute in biology because the evolution of one biochemical from another insures that a variety of structurally similar compounds will confront the cell. In spite of this unfavorable chemical similarity among substrates, large polymeric molecules must be constructed. During their synthesis and accumulation, every one of hundreds or thousands of synthetic steps are usually carried out correctly. Therefore, the accuracy of biosynthesis, e.g., the apparent production of a homogeneous protein, presents an immediate puzzle, and the study of this topic has become a subject in itself. Recent progress in understanding some contributions to the accuracy of translation makes this a good time to try a summary.

Most of what is known applies to the production of a protein with amino acid *a* rather than *b* (e.g., valine rather than isoleucine) at a given position in an otherwise accurately fashioned molecule. We discuss this type of error below, though there are other routes to substitution errors (e.g., incorrect binding of tRNAs on the ribosome), and errors of initiation and termination also occur and may be important.

WHAT IS THE OVERALL ERROR RATE OF TRANSLATION?

At the moment, data bearing on this question are seriously limited in two ways. First, only a small sample of possible measurements exists. There are about 10^6 misreactions possible in aminoacylation and coding, and they may be quite varied in their precision. Second, the measurements we have are of the steady-state substitution rate in completed, stable proteins, whereas we know that the products of induced translational error are degraded (Goldberg 1972). In fact, 8% of pulsed leucine incorporation in exponentially growing *Escherichia coli* K12 is exceedingly unstable and breaks down with a half-life of about 1 minute at 37°C (Brunschede and Bremer 1971). If any substantial fraction of these (1 stable leucine/12 incorporated) represents mistakes of translation, steady-state substitution measurements seriously underestimate the natural error rate. Reliable data on the error rate in nascent and recently released polypeptides are urgently needed.

Most simply, the evident ability of cells to maintain high levels of functional, homogeneous proteins that may contain 10^3 amino acids limits total error rates per amino acid position to less than a few $\times 10^{-3}$. Otherwise, an apparently intolerable fraction of the protein made would have to be degraded to produce the observed homogeneity.

Loftfield (1963) and Loftfield and Vanderjagt (1972) made the first, and still most explicit, measurements of error level. The method required isolation of two tryptic-chymotryptic peptides of rabbit hemoglobin, synthesized in suspensions of reticulocytes incubated with radioactive valine. The critical peptides, four and eight amino acids long, contain isoleucine, but should not contain valine. They were isolated by chromatography and successive electrophoresis in the presence of unlabeled carrier peptides that have valine substituted at the isoleucine position. The result (Loftfield and Vanderjagt 1972) was 2.5×10^{-4} valines per mole for the shorter peptide and 5×10^{-4} for the longer (two determinations each).

To what error can we attribute the incorporation of valine? The experimental design anticipates that isoleucine and valine will be difficult to distinguish in aminoacylation, but later data suggest that this may nevertheless be accomplished (see below). In addition, their codons differ with regard to a purine substitution at the first codon position, and later data also suggest that this type of error may be improbable (Davies et al. 1966). Therefore, we must consider the possibility that valine substitution occurs at other sites; in fact, the isolation procedure would have allowed isolation of a substitution for any neutral amino acid (e.g., it appears from the data to be completely insensitive to the valine-isoleucine difference). The moderately significant doubling in error in going from the shorter peptide (three neutral sites) to the longer (six neutral sites) suggests that more than one site may be active. Thus, an error level of 10×10^{-3} to 30×10^{-3} per amino acid position for 1 of 15 possible neutral substitutions is indicated. This corresponds to about 10×10^{-2} to 30×10^{-2} total substitutions per amino acid position when simply (and hazardously) extrapolated for all possible mistaken amino acids. This is near the presumed upper limit consistent with life, though it already presents a challenge to the enzymes that select isoleucine instead of valine, as Loftfield (1963) emphasized.

Edelmann and Gallant (1977) have measured incorporation of [^{35}S]cysteine into *E. coli* flagellin, a protein that usually contains no cysteine. The extracellular location of the bacterial flagellum and the ability of flagellin to polymerize make it easy to purify, and $^{35}SO_4$-derived radioactivity in cysteine (methionine labeling is suppressed with unlabeled amino acid) can be measured in the polyacrylamide gel band of flagellin. The result is the incorporation of 6×10^{-4} mole of cysteine per flagellin molecule. Flagellin synthesis requires translation of about 450 codons (Silverman and Simon 1977). It is evident that the misincorporation of cysteine into *E. coli* flagellin, calculated per codon, must be lower than that of valine in rabbit hemoglobin, unless some unexpected set of circumstances limits misincorporation into one or two positions in protein. Edelmann and Gallant argue that mistakes probably occur at flagellin's 18 arginine codons because this uniquely requires the type of errors detected during in vitro protein synthesis with defined messages (Davies et al. 1966). In support of this proposal, cysteine incorporation is increased sevenfold by streptomycin, which suggests that ribosomal errors can be made the dominant event in cysteine usage, and by fivefold during arginine (but not isoleucine) starvation, which suggests that cysteine can be inserted as a consequence of inability to translate arginine codons efficiently. Neither experiment, however, really tests the assumption that the basal level of errors is due to ribosomal miscoding rather than, e.g., misaminoacylation of tRNA with cysteine. Therefore, we do not know how many sites are being used. The error per site could be between 1.3×10^{-6} (all sites) and 6×10^{-4} (one site), with values between 10^{-5} and 10^{-4} perhaps being most probable. This latter range would imply that the total number of errors (all possible amino acids) per codon translated would be about 10^{-4} to 10^{-3}.

The low levels of error above are not always maintained. Under conditions of, e.g., amino acid starvation in a *rel*$^-$ strain of *E. coli*, one sees production of inactive and thermolabile β-galactosidase (Hall and Gallant 1972) at a level that suggests >1% erroneous incorporation/ arginine codon or >10% error/histidine codon as judged by the point of inflection of EF-G (O'Farrell 1978) or >20% for the same system in another laboratory (Parker et al. 1978). These large error rates that are dependent on the absence of a given aminoacyl-tRNA are probably due to ribosomal errors in which another aminoacyl-tRNA is used instead of the unacylated one. This mechanism will be assumed below, but there is another plausible contributor to these high error rates. (See the discussion of the accuracy of tRNA selection below.) So, deprived of the proper aminoacyl-tRNA, the translational apparatus' error rate can increase two to three orders of magnitude. The ribosomal complex, even in vivo, is not an intrinsically precise device. Rather, it seems to be very dependent on the presence of normal amounts of the aminoacyl-tRNAs cognate to a given codon.

THE ACCURACY OF AMINOACYL-tRNA SYNTHESIS

There is wide agreement that most, and probably all, amino acids (AA_i) are activated (Eq. 1) and coupled to anticodons (Eq. 2) via the sequential activities of one protein, an aminoacyl-tRNA synthetase (E_j).

$$AA_i + ATP + E_j \longleftrightarrow E_j(AMP\text{-}AA_i) + PP \tag{1}$$
$$E_j(AMP\text{-}AA_i) + tRNA_k \longleftrightarrow E_j + AMP + AA_i - tRNA_k \tag{2}$$

Usually, $i = j = k$, but more easily than one might have expected one of the equalities may fail. Double errors appear too rare to be significant, so we discuss only the cases $i = k \neq j$ and $i = j \neq k$. A priori, one might expect Equation 1 to be the less-precise reaction, as a macromolecule ($tRNA_k$) offers more distinctive features than a micromolecule (AA_i). This appears to be generally true, but the acylation of a tRNA can depend on recognition of a single nucleotide (Yarus et al. 1977) and error rates are easily measurable for both under standard laboratory conditions.

THE ACCURACY OF tRNA SELECTION

Early work showed that when tRNA and enzyme from different species (transspecific reactions; e.g., mixed yeast and *E. coli* reagents) were used, misacylation of tRNAs was readily observed (Holten and Jacobsen 1969; Dudock et al. 1971; Giégé et al. 1971). Such reactions can go almost to completion, especially in the presence of low pH (Roe et al. 1971, 1973) or high pH (Roe et al. 1973), high Mg^{++}/ATP ratios (Giegé et al. 1972), or organic solvents (Ritter et al. 1970). They often have rather low K_m values, within a factor of 10–20 (Giegé et al. 1972), or even approaching the K_m (Roe et al. 1973) of the intraspecific cognate reaction, and a relative V_{max} as high as 0.2 (Roe et al. 1973), though there are many examples 10^{-2} to 10^{-3} that of the cognates (Giegé et al. 1972). The apparent excellent binding of some transspecific combinations has been confirmed by measurement of the equilibrium.

$$E_i + tRNA_k \longleftrightarrow E_i(tRNA_k) \tag{3}$$

Equation 3 uses the quenching of the *E. coli* isoleucyl-tRNA synthetase fluorescence (Lam and Schimmel 1975). The equilibrium dissociation constants range from 3 times to $>10^4$ times that of the cognate. The complexes are also accurately formed, in the sense that the normal part of the noncognate tRNA approaches and may be cross-linked by UV light to synthetases (Budzik et al. 1975; Schoemaker et al. 1975).

On the other hand, there is a distinct tendency for K_m and V_{max} to be directly correlated among the eight *E. coli* tRNAs acylated by yeast phenylalanyl-tRNA synthetase (Roe et al. 1973), which suggests that the best binding of noncognates occurs with the tRNA slightly askew with respect to the site of aminoacyl transfer on the enzyme.

The apparently higher specificity of the velocity of aminoacylation, compared to K_m, which was emphasized by Ebel et al. (1973), may be misleading. The net velocity of misacylation can be small due to a simultaneous editing or proofreading (see below) of the misacylated tRNA before it can escape the enzyme (see Fersht and Kaethner 1976). Thus, the transacylation step may not be so specific, but the steady rate of production of misacylated tRNA is lowered by a subsequent hydrolytic step. In fact, of the above conditions that radically improve the yield in those transspecific misacylations, one, low pH, is now known to inhibit aminoacyl-tRNA synthetase hydrolysis of aminoacyl-tRNA (Schreier and Schimmel 1972), and another, the addition of organic solvent (Yamane and Hopfield 1977), can terminate the excess ATP consumption during misacylation, which suggests a proofreading step.

The values above suggest that in an ordinary in vitro reaction (compiled for the amusement of a biologist) containing equal concentrations of an intraspecific cognate tRNA and an interspecific noncognate, the error rate for aminoacylation would range downward from about 3×10^{-2} to undetectability (independent of total concentration).[1]

The transspecific mischarging reactions are useful for preparation of misacylated tRNAs: for example, almost all *E. coli* tRNAs, under abnormal conditions, will accept phenylalanine from phenylalanyl-tRNA synthetase from brewer's yeast (Kern et al. 1972). They also provide a means of surveying a set of homologs to see if any structures exist in all tRNAs accepted by a single enzyme (Roe et al. 1973). However, it is not certain what they can tell us about normal specificity, so we now consider intraspecific reactions.

When the reactions (Eqs. 1, 2, and 3) occur with tRNA and enzymes from the same species, we are examining the barrier to error that evolution has constructed. In fact, intraspecific complexes (Eq. 3) with some noncognate tRNAs have been detected. They can have binding constants similar to those for cognate tRNAs for *E. coli* glutamyl-tRNA synthetase, measured using fluorescence quenching (Lapointe and Söll 1972), and yeast phenylalanyl-tRNA synthetase, using quenching and polarization of fluorescence (Pachmann et al. 1973). Complexes that are weak have also been studied by the same methods (Lam and Schimmel 1975). There is a definite tendency for the noncognate complexes, even the strong ones, to be abnormal in some respects. They may have bizarre

---

[1] By error rate, I mean the relative rate (v) of production of erroneous product from substrate j in the presence of the correct substrate i:

$$\frac{v_j}{v_i} = \frac{(V_j/K_j)S_j}{(V_i/K_i)S_i}$$

V and K are the V_{max} and K_m, and S the concentration of substrate, which is usually assumed equal for i and j. This equation includes potential competition of the substrates for enzyme. Note that if $S_j = S_i$, the error rate is independent of absolute concentrations.

kinetic properties (Rigler et al. 1970), may not protect from heat inactivation, in contrast to the cognate (Lapointe and Soll 1972; Seno et al. 1978), or may participate in one reaction of the synthetase (tRNA-stimulated pyrophosphate exchange) but not another (aminoacyl transfer) (Seno et al. 1978). Though the binding of noncognate is apparently at the normal tRNA site, as determined by competition with the cognate (Yarus 1972a; Mertes et al. 1972; Pachmann et al. 1973), it is evidently aberrant in some respects.

The most sensitive indicator of interaction by far appears to be the hydrolysis of aminoacyl-tRNA by synthetases; many noncognate combinations can be shown to interact in no other way (Bonnet et al. 1972; Yarus 1973a,b). Once again, as for interspecific reactions, addition of organic solvents stimulates a considerable variety of misacylations (Giegé et al. 1972; Yarus 1973c, 1976). Since both hydrolysis and aminoacyl-tRNA synthesis require a catalytically active complex, some elements of recognition must exist in all, or almost all, tRNAs. For example, isoleucyl-tRNA synthetase of *E. coli* has been shown to interact, in one reaction or the other (Yarus 1973a,b,c), with 18 of 19 possible noncognate tRNAs.

Some intraspecific noncognate complexes aminoacylate (Eq. 2) under normal in vitro conditions at a measurable rate (Yarus 1972a; Mertes et al. 1972; Ebel et al. 1973; Seno et al. 1978). What limits the broad potential for misbinding and misacylation of tRNA revealed in these in vitro experiments?

First in the order of reaction are competitive mechanisms. The cognate tRNA limits, by competition, the access of noncognate tRNA to the cognate's enzyme (Yarus 1972b). There is also a second competition effect, which arises because tRNAs are approximately equimolar or less with the sites that bind them in vivo; and particularly are equimolar with their cognate synthetase. Thus, concentration of a potentially misacylated tRNA is reduced by strong interactions with its own cognate synthetase (and other strong binding sites) and therefore is not available for a weaker interaction with noncognate synthetase (Yarus 1972b). These effects explain why some misacylations will be detectable with a purified synthetase and noncognate tRNA but will disappear with mixed tRNAs (Ebel et al. 1973) or cruder systems. The suppression of misacylation by the cognate tRNA has also been demonstrated in a mixture of only two purified tRNAs; the combination of both effects can make a difference of orders of magnitude in misacylation velocity (Yarus 1972b). Neither of these effects improves the specificity of the enzyme; to put it another way, they are both accounted for when the error rate is calculated as described[1] and equal tRNA concentrations are assumed. But they are potentially germane in vivo. Consider a temperature upshift in a temperature-sensitive aminoacyl-tRNA synthetase mutant for tRNA binding. Because its tRNA would become uncharged, it would presumably not be bound by the translation apparatus and could not be soaked up by the denatured synthetase. The result might be a large increase in its free tRNA concentration and a proportional increase[1] in the aminoacylation error rate due to a failure of the second competition effect. This might be a practical way to test the possibility that a tRNA misrecognition error can limit the overall error rate of protein synthesis. A more limited increase in free tRNA would be expected during amino acid starvation alone, and this could conceivably have a part in the extreme error rates observed in these conditions in rel^- cells (O'Farrell 1978; Parker et al. 1978).

The actual ability of synthetases to discriminate among intraspecific tRNAs can be considered the result of successive binding and aminoacyl-transfer specificities (Ebel et al. 1973), but the previous remarks about the complex nature of the V_{max} again apply. Table 1 lists measured error rates, under normal in vitro conditions, with no variations to favor noncognate acylation. As can be seen in the table, tRNA selections are usually quite precise. Even after multiplication of the largest error rate by about 10 to approximate all possible misacylations of a given tRNA, the error rate just begins to approach the lowest likely error rate in translation as a whole. Therefore, tRNA selection by synthetases should not limit the accuracy of translation as a whole under ordinary conditions.

As for the division of specificity among steps, Yamane and Hopfield (1977) have shown that 40 ATPs are used per isoleucyl-tRNA^{Phe} (*E. coli*) synthesized, and 25 ATPs per isoleucyl-tRNA^{fMet} (*E. coli* synthesized by isoleucyl-tRNA synthetase. Thus, misacylation velocities may be lowered by similar factors (25–40) due to a proofreading step. This, in combination with the data of Table 1, suggests the following generalization: the rather high specificity of tRNA selection is due to large contributions from both K_m and V_{max}. Either may be the larger, though there is a tendency for the contribution from V_{max} to be more important. This latter may itself be due to roughly equal factors resulting from slow transacylation and a proofreading mechanism, whose nature has yet to be proved.

THE ACCURACY OF AMINO ACID SELECTION

The development of this topic may be followed in the history of the valine-isoleucine distinction. After Pauling (1958) had pointed out the physical chemical difficulty, Loftfield (1963) showed that strong discrimination against valine was nevertheless possible. Baldwin and Berg (1966) found that, though valine was indeed activated by the *E. coli* enzyme, the isoleucyl-tRNA synthetase (valyl-AMP) formed broke down in the presence of acceptor tRNA^{Ile}, without any net formation of valyl-tRNA^{Ile}. Since the tRNA^{Ile} was ineffective when its 2'(3') end was blocked or removed, it was suggested that transfer of valine to the tRNA

Table 1 Error rates due to incorrect tRNA selection under normal in vitro conditions

Synthetase	tRNA	Relative K	Relative V	Error rate	Reference
E. coli isoleucyl-tRNA	tRNA^{Phe} (*E. coli*)	$\geq 1.4 \times 10^4$	$\geq 0.8 \times 10^{-3}$	5×10^{-8}	Yarus (1972a)
E. coli isoleucyl-tRNA	tRNA^{fMet} (*E. coli*)	$\geq 2.4 \times 10^3$	$\geq 5.2 \times 10^{-4}$	2.5×10^{-8}	Mertes et al. (1972)
Yeast valyl-tRNA	tRNA^{Phe} (yeast)	360	10^{-4}	2.8×10^{-7}	Ebel et al. (1973)
Yeast valyl-tRNA	tRNA^{Ala} (yeast)	63	2×10^{-4}	3.2×10^{-6}	Ebel et al. (1973)
Yeast arginine-tRNA	tRNA^{Asp} (yeast)	100	4×10^{-4}	4×10^{-6}	Ebel et al. (1973)

Error rates are calculated as described.[1] Relative K and V are the absolute values of those quantities for the noncognate normalized by those for the cognate tRNA.

was a required step. Then Yarus (1972a,c) and Schreier and Schimmel (1972) and Eldred and Schimmel (1972) discovered the hydrolytic activity of the phenylalanyl- and isoleucyl-tRNA synthetases, respectively. Since isoleucyl-tRNA^{Phe} and valyl-tRNA^{Ile} were relatively rapidly hydrolyzed by the enzymes cognate to the tRNA, it was suggested that hydrolysis of a misaminoacylated tRNA would complete the rejection of a mistakenly activated amino acid.

$$(AMP\text{-}AA_i)E_k(tRNA_k) \longleftrightarrow (AA_i\text{-}tRNA_k)E_k + AMP \qquad (4)$$

$$(AA_i\text{-}tRNA_k)E_k \longrightarrow E_k + tRNA_k + AA_i \qquad (5)$$

Eldred and Schimmel (1972) showed that valyl-tRNA^{Ile} hydrolysis was fast enough to account for the rate of ATP usage, and Fersht and Kaethner (1977) showed that the rate of isoleucine transfer to tRNA was the same as the rate of ATP breakdown in the presence of valine at several levels of pH and temperatures covering a 60-fold range of rates. This suggests that valyl-tRNA synthesis (Eq. 4) is a required slow step preceding a faster hydrolytic step (Eq. 5). However, the expected transient accumulation of enzyme-bound valyl-tRNA^{Ile} would not be detected in experiments starting with enzyme (valyl-AMP) (Fersht and Kaethner 1977). This was taken to imply that (1) the enzyme is more active hydrolytically immediately after aminoacyl transfer (the hydrolysis rate was independently estimated by supplying presynthesized valyl-tRNA^{Ile} in the absence of other substrates) or (2) that there is another type of proofreading pathway independent of valyl-tRNA^{Ile} formation. The proofreading reaction very strongly rejects valine. By using elongation factor Tu (EF-Tu) (GTP) to bind and sequester the valyl-tRNA^{Ile} that did escape isoleucyl-tRNA synthetase (to protect it from further hydrolysis), Hopfield (1976) measured the consumption of 270 ATPs for every molecule of valyl-tRNA^{Ile} formed. By contrast, only 1.36 ATPs (recalculated from data of Hopfield 1976) were used per molecule of isoleucyl-tRNA^{Ile} produced. This suggests that $269/270 = 99.6\%$ of the incorrect valyl-AMP is destroyed by proofreading at the cost of $0.36/1.36 = 26\%$ loss of the activated cognate amino acid. In fact, the cost need not be this high, as careful measurements by Mulvey and Fersht (1977) on a variety of synthetases, including *E. coli* isoleucyl-tRNA synthetase, yield 0.90–1.00 cognate aminoacyl-tRNAs per ATP used, with the higher values much more common.

In a similar case, the mechanistic argument is complete: threonine is isosteric with valine and is, in fact, activated by the valyl-tRNA synthetase of *Bacillus stearothermophilus*. However, the enzyme (aminoacyl adenylate) is destroyed in the presence of tRNA^{Val} without net production of threonyl-tRNA^{Val} (Fersht and Kaethner 1976). In this case, threonyl-tRNA^{Val} is detected in the amount predicted by the measured rate constants for its synthesis and destruction. Thus, the misactivated isosteric amino acid is not only destroyed by later editing or proofreading of the misacylated tRNA, as in the case of valyl-tRNA^{Ile}, but the detection of the threonyl-tRNA^{Val} intermediate allows the conclusion that this is the predominant editing pathway.

Early data indicated that the isoleucine on isoleucyl-tRNA^{Ile} did not reside in the normal isoleucine site (Yarus and Berg 1969). Indeed, the hydrolytic activity of the aminoacyl-tRNA synthetases was inhibited only slightly when the normal amino acid site was occupied by formation of aminoacyl adenylate (Eq. 1) (Schreier and Schimmel 1972; Yarus 1972c). This suggested a second hydrolytic amino acid site within reach of an amino acid bound to tRNA. This idea appeared in an entirely new light when von der Haar and Cramer (1975) reported that yeast isoleucyl-tRNA synthetase carried out net synthesis of valyl-tRNA^{Ile} when the terminal A was replaced with 3'-deoxyadenosine. Thus, the enzyme apparently transferred the amino acid to the 2'-OH of tRNA^{Ile} but was unable to edit it without participation of the 3'-OH. Indeed, in a survey of several yeast synthetases (von der Haar and Cramer 1976) it appeared that hydrolysis of cognate aminoacyl-tRNA's enzymes that transfer amino acid to the 2'-OH (isoleucine, phenylalanine, valine) is reduced 10- to 166-fold in rate when the 3'-OH is absent. Seryl-tRNA synthetase, which transacylates to the 3'-OH, is inhibited >20-fold when the 2'-OH is absent, and tyrosyl-tRNA synthetase, which can transacylate to either the 2'- or 3'-OH of tRNA, has little hydrolytic activity at all. The pattern was also confirmed for a misacylated tRNA: threonyl-tRNA^{Ile} was hydrolyzed by valyl-tRNA synthetase at only 1/13 the normal rate when the 3'-OH of the tRNA was absent (Igloi et al. 1977). The requirement for the unused hydroxyl was explained by proposing that it activates a water molecule that participates in the hydrolysis of the adjacent ester bond of valyl-tRNA^{Ile} (von der Haar and Cramer 1976).

Fersht and Kaethner (1976), however, take another point of view: the second hydroxyl group is required so that the aminoacyl group can migrate onto it. From this position, it has access to the hydrolytic site. This notion of two amino acid sites has been elaborated by Fersht (1977) into the double-sieve model. The first sieve is the normal amino acid site; it excludes larger or stereochemically different amino acids very well but allows the activation of isosteric or smaller amino acids. The latter, after transfer to tRNA, are caught by the second sieve, which is the hydrolytic site. It can strongly exclude the larger cognate amino acid or require some feature that it does not have. In fact, a strong negative correlation of rate of hydrolysis of normally acylated tRNAs with amino acid side-chain size has been observed for isoleucyl-tRNA synthetase (Yarus 1973b).

There seems to be little data at the moment that can be used to choose between water-activation and two-site models. However, the two-site model more easily explains the specificity for the various cognate amino acids. Specificity must be added to the other model ad hoc. It also accommodates more easily the relatively small effect (factor of 10–20) of loss of the second hydroxyl. This might usually be compensated by flexibility at the 3' terminal of tRNA, especially in view of the well-confirmed broad specificity of the hydrolytic step (Bonnet et al. 1972; Yarus 1973b).

In fact, this broad specificity has been offered as an argument that the hydrolysis reaction can not contribute to the accuracy of aminoacyl-tRNA synthesis (Bonnet et al. 1972; Bonnet and Ebel 1974; Sourgoutchov et al. 1974). The prototypical observation is that, e.g., valyl-tRNA^{Val} and phenylalanyl-tRNA^{Phe} are deacylated at comparable rates by yeast valyl-tRNA synthetase and phenylalanyl-tRNA synthetase, and therefore no specificity is evident. However, I believe this to be compensation of two specificities. The noncognate aminoacyl-tRNA can be bound very poorly, but its amino acid is usually removed rapidly. The cognate is bound well, but its amino acid, of course, should be hydrolyzed only slowly. This applies to most of the counter examples offered. But there are also times when there is only slow hydrolysis of a misacylation. For example, phenylalanyl-tRNA^{Val} is not rapidly hydrolyzed by valyl-tRNA synthetase (Bonnet et al. 1972; Bonnet and Ebel 1974). However, hydrolysis is not completely general and need not be so to be useful; phenylalanine and the other aromatic amino acids in particular are immune to hydrolysis by isoleucyl-tRNA synthetase of *E. coli* (Eldred and Schimmel 1972; Yarus 1973b). Finally, in one case (Sourgoutchov et al. 1974) the K_m values of isoleucyl- and methionyl-tRNA synthetases of *E. coli* are shown to be the same for methionyl-tRNA^{Met} (*E. coli*), and their hydrolysis V_{max} values are also similar. I cannot explain this, but we have examined the same system (Mertes et al. 1972) and have found the isoleucyl-tRNA synthetase · tRNA^{Met} interaction to be very weak, not as strong as a cognate. Therefore, I believe there is no compelling evidence against a useful role for hydrolysis.

What, then, is the overall precision of amino acid selection? The most difficult distinctions would appear to be those against isosteric and smaller amino acids; e.g., the isoleucyl-tRNA synthetase must accept isoleucine but reject valine that lacks one methylene group. Valyl-tRNA synthetase must accept valine but reject threonine (or α-amino-N-butyrate that lacks one methylene group). The problem is more than theoretical; *E. coli* isoleucyl-tRNA synthetase will exchange PP into ATP (Eq. 1) as rapidly with valine as with isoleucine; valyl-tRNA synthetase will activate α-amino butyrate and exchange PP at 29% the rate stimulated by valine (Loftfield and Eigner 1966). Amino acids with side chains larger by one methylene group are rejected more strongly during activation, and neither larger nor smaller analogs are detectably transferred to tRNA (Berg et al. 1961).

We may probably take the kinetic constants of the pyrophosphate exchange into ATP (Eq. 1) as characteristic of the rate-limiting formation of AMP amino acid (Loftfield and Eigner 1966). In this case, the error rate[1] of *E. coli* isoleucyl-tRNA synthetase for valyl-AMP vs isoleucyl-AMP formation is 8.1×10^{-3} (for the same concentration of amino acid; data of Loftfield and Eigner 1966) or 6.7×10^{-3} (data of Fersht and Kaethner 1977). This is probably entirely due to the fact that the enzyme binds isoleucine better; the binding constants have been measured in the ratio 7.6×10^{-3} (Flossdorf and Kula 1973) and 9.2×10^{-3} (Fersht and Kaethner 1977). But almost every activated isoleucine can be transferred to tRNA^{Ile} ($\geq 95\%$; Mulvey and Fersht 1977). On the other hand, only about 3.7×10^{-3} of the activated valine can be captured by EF-Tu (GTP) as valyl-tRNA^{Ile} (Hopfield et al. 1976), due to subsequent editing, e.g., hydrolysis of the incorrectly acylated tRNA after valyl-tRNA^{Ile} formation (Fersht and Kaethner 1977). Thus, the relative efficiency of conversion of the aminoacyl adenylates to aminoacyl-tRNA is 3.9×10^{-3} ($3.7 \times 10^{-3}/0.95$). Single-step stereochemical criteria and subsequent editing therefore give (7.4×10^{-3}) $(3.9 \times 10^{-3}) = 2.9 \times 10^{-5}$ acting together. This is close to, but below, the measured level of valine replacement per amino acid site in rabbit hemoglobin ($1-3 \times 10^{-4}$; see above). This suggests that mistaken activation of certain amino acids can sometimes make an important contribution to the error rate of protein synthesis. We do not know, however, that the organisms and pathways for error are comparable in the two cases compared.

As for amino acids that may be distinguished more easily by stereochemical criteria (the correct one is larger or more different chemically than valine from isoleucine), A. R. Fersht and C. Dingwall (in prep.) have shown that even smaller error rates than the above can be attained in the activation step alone. Thus, α-amino butyrate is rejected by *E. coli* cysteinyl-tRNA synthetase with an error rate of 2.9×10^{-6}, making it improbable that such distinctions usually limit the accuracy of translation. The situation for erroneous activation of slightly larger similar amino acids (e.g., leucine for isoleucine) needs more examination, as it is unclear what contribution proofreading would make in this case.

ACKNOWLEDGMENTS

It was a pleasure dealing with my editor, Paul Schimmel, and I am grateful for support from the U.S. Public Health Service (research grant GM25627) during the period of preparation.

REFERENCES

Baldwin, A. N. and P. Berg. 1966. Transfer RNA-induced hydrolysis of valyl-adenylate bound to isoleucyl ribonucleic acid synthetase. *J. Biol. Chem.* **241**: 839.

Berg, P., F. Bergmann, E. J. Ofengand, and M. Dieckmann. 1961. Enzymic synthesis of aminoacyl-tRNA derivatives. I. *J. Biol. Chem.* **236**: 17.

Bonnet, J. and J.-P. Ebel. 1974. Correlation of aminoacylation errors: Evidence for a non-significant role of the aminoacyl-tRNA synthetase catalyzed deacylation of aminoacyl-tRNA's. *FEBS Lett.* **39**: 259.

Bonnet, J., R. Giegé, and J.-P. Ebel. 1972. Lack of specificity in the aminoacyl-tRNA synthetase-catalyzed deacylation of aminoacyl-tRNA. *FEBS Lett.* **27**: 139.

Brunschede, H. and H. Bremer. 1971. Synthesis and breakdown of proteins in *E. coli* during amino acid starvation. *J. Mol. Biol.* **57**: 35.

Budzik, G. P., S. S. M. Lam, H. J. P. Schoemaker, and P. R. Schimmel. 1975. Two photo-cross-linked complexes of isoleucine specific tRNA with aminoacyl tRNA synthetases. *J. Biol. Chem.* **250**: 4433.

Davies, J., D. S. Jones, and H. G. Khorana. 1966. A further study of misreading of codons induced by streptomycin and neomycin using RNA containing two nucleotides in alternating sequences as templates. *J. Mol. Biol.* **18**: 48.

Dudock, B., C. DiPeri, K. Scileppi, and R. Reszelback. 1971. The yeast phenylalanyl-tRNA synthetase recognition site: The region adjacent to the dihydrouridine loop. *Proc. Natl. Acad. Sci.* **68**: 681.

Ebel, J.-P., R. Giegé, J. Bonnet, D. Kern, N. Befort, C. Bollack, F. Fasiolo, J. Gangloff, and G. Dirheimer. 1973. Factors determining the specificity of the tRNA aminoacylation reaction. *Biochimie* **55**: 547.

Edelmann, P. and J. Gallant. 1977. Mistranslation in *E. coli*. *Cell* **10**: 131.

Eldred, E. W. and P. R. Schimmel. 1972. Rapid deacylation by isoleucyl RNA synthetase of isoleucine specific tRNA aminoacylated with valine. *J. Biol. Chem.* **247**: 2961.

Fersht, A. R. 1977. *Enzyme structure and mechanism.* A. R. Freeman, San Francisco.

Fersht, A. R. and M. M. Kaethner. 1976. Enzyme hyperspecificity. Rejection of threonine by the valyl-tRNA synthetase by misacylation and hydrolytic editing. *Biochemistry* **15**: 3342.

―――. 1977. Editing mechanisms in protein synthesis. Rejection of valine by the isoleucyl-tRNA synthetase. *Biochemistry* **16**: 1025.

Flossdorf, J. and M.-R. Kula. 1973. Ultracentrifuge studies on binding of aliphatic amino acids to isoleucyl RNA synthetase from *E. coli* MRE 600. *Eur. J. Biochem.* **36**: 534.

Giegé, R., D. Kern, and J.-P. Ebel. 1972. Incorrect aminoacylations catalyzed by *E. coli* valyl-tRNA synthetase. *Biochemie* **54**: 1245.

Giegé, R., D. Kern, J.-P. Ebel, and R. Taglang. 1971. Incorrect heterologous aminoacylation of various yeast tRNA's catalyzed by *E. coli* valyl-tRNA synthetase. *FEBS Lett.* **15**: 281.

Goldberg, A. L. 1972. Degradation of abnormal proteins in *E. coli*. *Proc. Natl. Acad. Sci.* **69**: 422.

Hall, B. and J. Gallant. 1972. Defective translation in RC⁻ cells. *Nat. New Biol.* **237**:131.
Holten, V. Z. and K. B. Jacobson. 1969. Studies on the aminoacylation of valine and alanine-specific-tRNA of *E. coli* by aminoacyl tRNA synthetase from *Neurospora crassa* and *E. coli*. *Arch. Biochem. Biophys.* **129**:283.
Hopfield, J. J., T. Yamane, V. Yue, and S. M. Coutts. 1976. Direct experimental evidence for kinetic proofreading in aminoacylation of tRNA^{Ile}. *Proc. Natl. Acad. Sci.* **73**:1164.
Igloi, G. L., F. von der Haar, and F. Cramer. 1977. Hydrolytic action of aminoacyl-tRNA synthetases from baker's yeast. "Chemical proofreading" of Thr-tRNA by valyl-tRNA synthetase studied with modified tRNA^{Val} and amino acid analogues. *Biochemistry* **16**:1696.
Kern, D., R. Giegé, and J.-P. Ebel. 1972. Incorrect aminoacylations catalyzed by the phenylalanyl- and valyl-tRNA synthetases from yeast. *Eur. J. Biochem.* **31**:148.
Lam, S. S. M. and P. R. Schimmel. 1975. Equilibrium measurements of cognate and non-cognate interactions between aminoacyl-tRNA synthetases and tRNA. *Biochemistry* **14**:2775.
Lapointe, J. and D. Söll. 1972. Glutamyl tRNA synthetase of *E. coli*. *J. Biol. Chem.* **247**:4975.
Loftfield, R. B. 1963. The frequency of errors in protein biosynthesis. *Biochem. J.* **89**:82.
Loftfield, R. B. and E. A. Eigner. 1966. The specificity of enzymatic reactions. Aminoacyl-soluble RNA ligases. *Biochim. Biophys. Acta* **130**:426.
Loftfield, R. B. and D. Vanderjagt. 1972. The frequency of errors in protein biosynthesis. *Biochem. J.* **128**:1353.
Mertes, M., M. A. Peters, W. Mahoney, and M. Yarus. 1972. Isoleucylation of tRNA$_f^{Met}$ (*E. coli*) by isoleucyl-tRNA synthetase from *E. coli*. *J. Mol. Biol.* **71**:671.
Mulvey, S. M. and A. R. Fersht. 1977. Editing mechanisms in aminoacylation of tRNA: ATP consumption and the binding of aminoacyl-tRNA by elongation factor Tu. *Biochemistry* **16**:4731.
O'Farrell, P. H. 1978. The suppression of defective translation by ppGpp and its role in the stringent response. *Cell* **14**:545.
Parker, J., J. W. Pollard, J. D. Friesen, and C. P. Stanners. 1978. Stuttering: High level mistranslation in animal and bacterial cells. *Proc. Natl. Acad. Sci.* **75**:1091.
Pachmann, V., E. Cronvall, R. Rigler, R. Hirsch, W. Wintemeyer, and H. G. Zachau. 1973. On the specificity of interactions between tRNA's and aminoacyl-tRNA synthetases. *Eur. J. Biochem.* **39**:265.
Pauling, L. 1958. The probability of errors in the process of synthesis of protein molecules. In *Festschrift Arthur Stoll*, p. 597. Birkhäuser Verlag, Basel.
Rigler, R., E. Cronvall, R. Hirsch, V. Pachmann, and H. G. Zachau. 1970. Interactions of seryl-tRNA synthetase with serine and phenylalanine specific tRNA. *FEBS Lett.* **11**:320.
Ritter, P. O., F. J. Kull, and K. B. Jacobson. 1970. Aminoacylation of *E. coli* valine tRNA by *N. crassa* phenylalanyl-tRNA synthetase in tris-HCl and K-cacodylate buffers. *J. Biol. Chem.* **245**:2114.
Roe, B., M. Sirover, and B. Dudock. 1973. Kinetics of homologous and heterolo-

gous aminoacylation with yeast phenylalanyl tRNA synthetase. *Biochemistry* **12:** 4146.
Roe, B., M. Sirover, R. Williams, and B. Dudock. 1971. New heterologous mischarging reactions with yeast phenylalanyl tRNA synthetase. *Arch. Biochem. Biophys.* **147:** 176.
Schoemaker, H. J. P., G. P. Budzik, R. Giegé, and P. R. Schimmel. 1975. Three photo-cross-linked complexes of yeast phenylalanine specific tRNA with aminoacyl tRNA synthetases. *J. Biol. Chem.* **250:** 4440.
Schreier, A. A. and P. R. Schimmel. 1972. tRNA synthetase catalyzed deacylation of aminoacyl-tRNA in the absence of AMP and pyrophosphate. *Biochemistry* **11:** 1582.
Seno, T., A. Nakamura, S. Fukahara, and K. Iwata. 1978. Interaction of *E. coli* glutaminyl-tRNA synthetase with noncognate tRNA's. *Nucleic Acids Res.* **5:** 1561.
Silverman, M. and M. I. Simon. 1977. Bacterial flagella. *Annu. Rev. Microbiol.* **31:** 397.
Sourgoutchov, A., S. Blanquet, G. Payat, and J.-P. Waller. 1974. Enzymatic deacylation of Met-tRNA$_f^{Met}$ catalyzed by methionyl, isoleucyl, and phenylalanyl-tRNA synthetases. *Eur. J. Biochem.* **46:** 431.
von der Haar, R. and F. Cramer. 1975. Isoleucyl-tRNA synthetase from baker's yeast: The 3'-hydroxyl group of the 3'-terminal ribose is essential for preventing misacylation of tRNA^{Ile} CCA with misactivated valine. *FEBS Lett.* **56:** 215.
―――. 1976. Hydrolytic action of aminoacyl-tRNA synthetases from baker's yeast. "Chemical proofreading" preventing acylation of tRNA^{Ile} with misactivated valine. *Biochemistry* **15:** 4131.
Yamane, T. and J. J. Hopfield. 1977. Experimental evidence for kinetic proofreading in the aminoacylation of tRNA by synthetase. *Proc. Natl. Acad. Sci.* **74:** 2246.
Yarus, M. 1972a. Solvent and specificity. Binding and isoleucylation of phenylalanine tRNA (*E. coli*) by isoleucyl-tRNA synthetase from *E. coli*. *Biochemistry* **11:** 2352.
―――. 1972b. Intrinsic precision of aminoacyl-tRNA synthesis enhanced through parallel systems of ligands. *Nat. New Biol.* **239:** 106.
―――. 1972c. Phenylalanyl-tRNA synthetase and Ile-tRNA^{Phe}: A possible verification mechanism for aminoacyl-tRNA. *Proc. Natl. Acad. Sci.* **69:** 1915.
―――. 1973a. Pseudoverification. *J. Biol. Chem.* **248:** 6750.
―――. 1973b. Pseudoverification of mixed aminoacyl-tRNA's. *J. Biol. Chem.* **248:** 6755.
―――. 1973c. The variety of intraspecific misacylations carried out by isoleucyl-tRNA synthetase of *E. coli*. *J. Biol. Chem.* **248:** 6744.
―――. 1976. Why do organic solvents enhance mistakes in aminoacyl-tRNA synthesis? *Arch. Biochem. Biophys.* **174:** 350.
Yarus, M. and P. Berg. 1969. Recognition of tRNA by isoleucyl-tRNA synthetase effect of substrates on the dynamics of tRNA-enzyme interaction. *J. Mol. Biol.* **42:** 171.
Yarus, M., R. Knowlton, and L. Soll. 1977. Aminoacylation of the ambivalent Su$^+$7 amber suppressor tRNA. In *Nucleic acid-protein recognition* (ed. H. Vogel), p. 391. Academic Press, New York.

Appendices

At the August 1978 Cold Spring Harbor tRNA meeting, a group met to discuss the problem of a uniform numbering system for tRNA. With the emergence of knowledge regarding the three-dimensional structure of these molecules, it has become apparent that a common numbering system is desirable. There are, unfortunately, several issues that make it difficult to achieve an ideal solution. In particular, recent discoveries demonstrating wider variations in nucleotide sequence, especially among tRNAs from organelles, suggest that at present it may be premature to assemble a wholly rational numbering system. Our present state of knowledge suggests that there are three regions in which there are variable numbers of nucleotides: the region (α) before the two constant GMP residues in the D loop (containing 1-3 nucleotides), the region (β) immediately after these two constant GMP residues (containing 1-3 nucleotides), and the variable loop that generally contains 4-5 nucleotides, but may contain up to 21 nucleotides. The numbering scheme clearly must allow for addition and deletion of nucleotides.

We discussed the relative advantages of adopting a numbering scheme based on our present perception of the minimal tRNA sequence vs modifying the numbering system of yeast tRNA^{Phe}. The consensus from the meeting was that, for the present, we should use a numbering system based on yeast tRNA^{Phe}, and reserve for later a modification based on fuller knowledge.

It is proposed that nucleotides in yeast tRNA^{Phe} be numbered from 1 to 76, and that the corresponding nucleotide positions in other tRNAs have the same numbers as their counterparts in yeast tRNA^{Phe}. The numbers serve as reference points for designating precisely the locations of extra nucleotides. The areas in tRNAs that have extra nucleotides are typically after positions 17, 20, and 47, in the yeast tRNA^{Phe} numbering system. These extra bases are to be designated by the number of the preceding reference nucleotide followed by a colon and a second number. For example, 20:1 and 20:2 mean the first and second nucleotide after position 20. Molecules with a large variable loop would be numbered 47:1, 47:2, etc.

The absence of one of the reference nucleotides would be indicated by the absence of that number. For example, some tRNA sequences would lack residue 17. In the case of a four-numbered variable loop, position 47 would be eliminated, and the numbering in this area would jump from 46 to 48.

A numbering scheme according to these suggestions is displayed in the figure opposite. Heavy circles enclose bases that are common to various tRNAs. In this system all tRNA molecules end with 3'-AMP76, and the total number of nucleotides can be indicated elsewhere. This numbering system may develop difficulties as further discoveries are made, both in sequences and three-dimensional relations among the polynucleotides. However, at present this system will greatly facilitate our discussions of various aspects of tRNA structure and function.

APPENDIX I
Proposed Numbering System of Nucleotides in tRNAs Based on Yeast tRNA[Phe]

A. Compilation of tRNA Sequences

Dieter H. Gauss, Franz Grüter, and Mathias Sprinzl
Abteilung Chemie
Max-Planck-Institut für experimentelle Medizin
D-3400 Göttingen, Federal Republic of Germany

This compilation presents in a small space the tRNA sequences so far published to enable rapid orientation and comparison. The numbering of tRNA^{Phe} from yeast is used as has been done earlier[1] but following the rules proposed by the participants of the 1978 Cold Spring Harbor Meeting on tRNA. This numbering allows comparisons with the three-dimensional structure of tRNA^{Phe}, the only structure known from X-ray analysis. The secondary structure of tRNAs is indicated by specific underlining. In the primary structure, a nucleoside followed by a nucleoside in brackets or a modification in brackets denotes that both types of nucleosides can occupy this position. Part of a sequence in brackets designates a piece of sequence not unambiguously analyzed. Rare nucleosides are named according to the IUPAC-IUB rules (for some, more complicated, rare nucleosides and their identification, see Table 1); those with lengthy names are given with the prefix "x" and specified in the footnotes. Footnotes are numbered according to the coordinates of the corresponding nucleoside and are indicated in the sequence by an asterisk. The references are restricted to the citation of the latest publication in those cases where several papers deal with one sequence. For additional information, refer either to the original literature or to other tRNA sequence compilations.[2-6]

[1]Sprinzl, M., F. Grüter, and D. H. Gauss. 1978. *Nucleic Acids Res.* **5**:r15.
[2]Sodd, M. A. 1976. *Nucleic acids. CRC handbook of biochemistry and molecular biology*, 3rd Ed. (ed. G. D. Fasman), vol. II, p. 423. CRC Press, Cleveland, Ohio.
[3]Dirheimer, G., J. P. Ebel, J. Bonnet, J. Gangloff, G. Keith, B. Krebs, B. Kuntzel, A. Roy, J. Weissenbach, and C. Werner. 1972. *Biochimie* **54**:127.
[4]Sodd, M. A. and B. P. Doctor. 1974. *Methods Enzymol.* **29**:741.
[5]Barrell, B. G. and B. F. C. Clark. 1974. *Handbook of nucleic acid sequences*. Joynson-Bruvvers, Oxford.
[6]Barciszewski, J. and A. J. Rafalski. 1978. *Atlas of transfer ribonucleic acids and modified nucleosides*. Poznan. (In press.)

Table 1 Names of some rare nucleosides and citations regarding their identification

Nucleoside	Identification
o^5U	uridine-5-oxyacetic acid
mo^5U	5-methoxyuridine
mcm^5U	5-methoxycarbonylmethyluridine[1]
mcm^5s^2U	5-methoxycarbonylmethyl-2-thiouridine
mam^5s^2U	5-N-methylaminomethyl-2-thiouridine
i^6A	N-6-(Δ^2-isopentenyl)adenosine
ms^2i^6A	N-6-(Δ^2-isopentenyl)2-methylthioadenosine[3]
t^6A	N-[9-(β-D-ribofuranosyl)purin-6-ylcarbamoyl]threonine
mt^6A	N-[9-(β-D-ribofuranosyl)purin-6-yl-N-methylcarbamoyl]threonine
Q_{34}	7-(4,5-cis dihydroxy-1-cyclopenten-3-ylaminomethyl)-7-deazaguanosine[4]
X	3-N-(3-amino-3-carboxypropyl)uridine[5]
yW	wybutosine[6]
O_2yW	wybutoxosine[7]
N	unknown nucleoside

Compare: Feldman, M. Y. 1978. *Prog. Biophys. Mol. Biol.* **32**:83; Goddard, J. P. 1978. *Prog. Biophys. Mol. Biol.* **32**:233; and McCloskey, J. A. and S. Nishimura. 1977. *Accts. Chem. Res.* **10**:403.

[1]Kuntzel, B., J. Weissenbach, R. E. Wolff, T. D. Tumaitis-Kennedy, B. G. Lane, and G. Dirheimer. 1975. *Biochimie* **57**:61.

[2]H. Ishikura, pers. comm.

[3]Harada, F., H. J. Gross, F. Kimura, S. H. Chang, S. Nishimura, and U. L. RajBhandary. 1968. *Biochem. Biophys. Res. Commun.* **33**:299; Yamada, Y., S. Nishimura, and H. Ishikura. 1971. *Biochim. Biophys. Acta* **247**:170.

[4]Casai, H., Z. Ohashi, F. Harada, S. Nishimura, N. J. Openheimer, P. F. Crain, J. G. Liehr, D. L. von Minden, and J. A. McCloskey. 1975. *Biochemistry* **14**:4198.

[5]Nishimura, S., Y. Taya, Y. Kuchino, and Z. Ohashi. 1974. *Biochem. Biophys. Res. Commun.* **57**:702; Ohashi, Z., M. Maeda, J. A. McCloskey, and S. Nishimura. 1974. *Biochemistry* **13**:2620; Friedman, S., H. J. Li, K. Nakanishi, and G. van Lear. 1974. *Biochemistry* **13**:2932.

[6]Nakanishi, K., N. Furutachi, M. Funamizu, D. Grunberger, and I. B. Weinstein. 1970. *J. Am. Chem. Soc.* **92**:7617.

[7]Blobstein, S. H., D. Grunberger, I. B. Weinstein, and K. Nakanishi. 1973. *Biochemistry* **12**:188; Feinberg, A. M., K. Nakanishi, J. Barciszewski, A. J. Rafalski, H. Augustyniak, M. Wiewiorowski. 1974. *J. Am. Chem. Soc.* **96**:7797.

Table 2 Compilation of known tRNA sequences

		Aminoacyl Stem									D Stem				D Loop										D Stem					Anticodon Stem					Anticodon Loop							Anticodon Stem				
		1	2	3	4	5	6	7	8	9	10	11	12	13	14	15	16	17	17:1	18	19	20	20:1	20:2	21	22	23	24	25	26	27	28	29	30	31	32	33	34	35	36	37	38	39	40	41	42	43
	ALANINE																																														
0010	E.coli 1A	G	G	G	G	G	C	A	(s^4)U	A	G	C	U	C	A	G	C	D		G	G	G			A	G	A	G	C	G	C	C	U	G	C	U	U	o^{5}U	G	C	A	C	G	C	A	G	G
0020	T.utilis 1	G	G	G	C	G	U	G	U	m^{1}G	G	C	G	U	A	G	D	D		G	G	D			A	G	C	G	C	m$^2_2$G	Ψ	U	C	G	C	U	U	I	G	C	m^{1}I	Ψ	G	C	G	A	A
0030*	Yeast 1	G	G	G	C	G	U	G	U	m^{1}G	G	C	G	U	A	G	D	C		G	G	D			A	G	C	G	C	m$^2_2$G	C	U	C	C	C	U	U	I	G	C	m^{1}I	Ψ	G	G	G	A	G
0040	Bombyx mori 1	G	G	G	G	G	C	G	U	A	m^{2}G	C	U	C	A	G	A	D		G	G	U			A	G	A	G	C	m$^2_2$G	C	U	C	G	C	Um	U	I	G	C	m^{1}I	Ψ	Gm	U(Ψ)	G	A	G
0041	Bombyx mori 2	G	G	G	G	G	C	G	U	A	m^{2}G	C	U	C	A	G	A	D		G	G	U			A	G	A	G	C	m$^2_2$G	C	U	C	G	C	Um	U	I	G	C	m^{1}I	Ψ	Gm	C	G	A	G
	ARGININE																																														
0110	E.coli 1	G	C	A	U	C	C	G	s^{4}U	A	G	C	U	C	A	G	C	D		G	G	U(D)			A	G	A	G	U	A	C	U	C	G	G	s^{2}C	U	I	C	G	m^{2}A	A	C	C	G	A	G
0111	E.coli B 2	G	C	A	U	C	C	G	s^{4}U	A	G	C	U	C	A	G	C	D		G	G	A	U(D)		A	G	A	G	U	A	C	U	C	G	G	C	U	I	C	G	m^{2}A	A	C	C	G	A	G
0120	Phage T4	G	U	C	C	C	G	C	U	G	G	U	G	U	A	A	U			Gm	G	A	D		A	G	C	A	U	A	C	G	A	U	C	C	U	N*	C	U	A	A	G	Ψ	U	U	G
0121	Phage T4 psu+ 4 UGA	G	U	C	C	C	G	C	U	G	G	U	G	U	A	A	U			Gm	G	A	D		A	G	C	A	U	A	C	G	A	U	C	C	U	N*	C	A	A	A	G	Ψ	U	U	G
0130	Yeast 2	Ψ	U	C	C	U	C	G	U	m^{1}G	m^{2}G	C	C	C	A	A	D			G	G	D	C		A	C	G	G	C	m$^2_2$G	Ψ	C	U	G	G	C	U	I	C	G	A	A	C	C	A	G	A
0140	Yeast 3a	G	C	G	C	U	C	G	U	m^{1}G	m^{2}G	C	G	U	A	A	D			G	G	C			A	A	C	G	C	m$^2_2$G	Ψ	C	U	G	A	C	U	$_x$U*	C	U	t^{6}A	A	Ψ	C	A	G	A
0141	Yeast 3b	G	C	G	C	U	U	G	U	m^{1}G	m^{2}G	C	G	U	A	A	D			G	G	C			A	A	C	G	C	m$^2_2$G	Ψ	C	U	G	A	C	U	$_x$U*	C	U	t^{6}A	A	Ψ	C	A	G	A
	ASPARAGINE																																														
0210	E.coli	U	C	C	U	C	U	G	s^{4}U	A	G	U	U	C	A	G	D	C		G	G	D			A	G	A	A	C	G	G	C	G	G	A	C	U	Q	U	U	t^{6}A	A	Ψ	C	C	G	U
0260	Mammalian*	G	U	C	U	C	U	G	U	m^{1}G	m^{2}G	C	G	C	A	A	D	C		G	G	D	X		A	G	C	G	C	m$^2_2$G	Ψ	Ψ	C	G	G	C	U	Q	U	U	t^{6}A	A	C	C	G	A	A
	ASPARTIC ACID																																														
0310	E.coli 1	G	G	A	G	C	G	G	s^{4}U	A	G	U	U	C	A	G	D	C		G	G	D	D		A	G	A	A	U	A	C	C	U	G	C	C	U	Q	U	C	m^{2}A	C	G	C	A	G	G
0320	Yeast	U	C	C	G	U	G	A	U	A	G	U	U	Ψ	A	A	D			G	G	D	C		A	G	A	A	U	G	G	G	C	G	C	Ψ	U	G	U	C	m^{1}G	C	G	U	G	C	C

0010 R.J.Williams,W.Nagel,B.Roe,B.Dudock (1974) Biochem.Biophys. Res.Commun. 60, 1215-1221.

0020 S.Takemura,K.Ogawa (1973) J.Biochem.74, 322-333.

0030 J.R.Penswick,R.Martin,G.Dirheimer (1975) FEBS-Lett. 50, 28-31.

0040 + 0041 K.U.Sprague,O.Hagenbüchle,M.C.Zuniga (1977) Cell 11, 561-570.

0110 K.Murao,T.Tanabe,F.Ishii,M.Namiki,S.Nishimura (1972) Biochem. Biophys.Res.Commun. 47, 1332-1337.

0111 K.Chakraburtty (1975) Nucleic Acids Res. 2, 1787-1792.

0120 G.P.Mazzara,J.G.Seidman,W.H.McClain,H.Yesian,J.Abelson, C.Guthrie (1977) J.Biol.Chem. 252,8245-8253.

0121 S.-H.Kao,W.H.McClain (1977) J.Biol.Chem. 252, 8254-8257.

0130 J.Weissenbach,R.Martin,G.Dirheimer (1975) Eur.J.Biochem. 56, 527-532.

0140 + 0141 B.Kuntzel,J.Weissenbach,G.Dirheimer (1974) Biochimie 56, 1069-1087.

0210 K.Ohashi,F.Harada,Z.Ohashi,S.Nishimura,T.S.Stewart,G.Vögeli, T.McCutchan,D.Söll (1976) Nucleic Acids Res. 3, 3369-3376.

0260 E.Y.Chen,B.A.Roe (1978) Biochem.Biophys.Res.Commun. 82, 235-246.

0310 F.Harada,K.Yamaizumi,S.Nishimura (1972) Biochem.Biophys.Res. Commun. 49, 1605-1609.

0320 J.Gangloff,G.Keith,J.P.Ebel,G.Dirheimer (1972) Biochim.Biophys. Acta 259, 210-222.

		Extra Arm													TΨ Stem						TΨ Loop							TΨ Stem					Aminoacyl Stem										
	44	45	46	47	47.1	47.2	47.3	47.4	47.5	47.6	47.7	47.8	47.9	47.10	47.11	47.12	47.13	47.14	47.15	47.16	48	49	50	51	52	53	54	55	56	57	58	59	60	61	62	63	64	65	66	67	68	69	70	71	72	73	74	75	76

ALANINE

- 0010: A G m⁷G U ... C U G C G G T Ψ C G A U U C C G C G C U G C U U C A C C A
- 0020: A G G D ... C U C C G G T Ψ C G A U U C G G C G C U U G U C C A C C A
- 0030: A G G D(U) ... C U G C G G T Ψ C G A U U C G G C A C U C G U C C A C C A
- 0040: A G m⁷G U ... A m⁵C C C G G T Ψ C G m⁷A A C G G G A C C C C U U C A C C A
- 0041: A G m⁷G U ... A m⁵C C C G G T Ψ C G m⁷A U C G G G A C C C U U U C A C C A

ARGININE

- 0110: C G m⁷G X ... C G G A G G T Ψ C G A A A C C U C C C G G A U G C A C C A
- 0111: C G m⁷G X ... C G G A G G T Ψ C G A A A C U U C C C G G A U G C A C C A
- 0120: C G G ... U C C C U G G T Ψ C G A U U C C A G G C G G G A U C A C C A
- 0121: C G G ... U C C C U G G T Ψ C G A U C C A G G C G G G A U C A C C A
- 0130: A G A D ... m⁵C C C A G G T Ψ C A m⁷A G C U U G G C G G A A U C A C C A
- 0140: A G A D ... U A U G G G T Ψ C G m⁷A C C C A U A C G A G U G C A C C A
- 0141: A G A D ... U A U G G G T Ψ C G m⁷A C C C A U C G A G U G C A C C A

ASPARAGINE

- 0210: A U m⁷G U ... C A C U G G T Ψ C G A G U C C A G U C A G A G G A G C C A
- 0260: A G m⁷G D ... U G G U G G N Ψ C G m⁷A G G C C A C C C A G G A C G C C A

ASPARTIC ACID

- 0310: G G m⁷G U ... C G C G G G T Ψ C G A G U C C C G U U C C G U U C G G C C A
- 0320: A G A ... U m⁵C G G G G T Ψ C A A G U C C C C G U G G A G G A G C C A

0030/0 Compare R.W. Holley et al. (1965) Science 147, 1462-1465.
0120/34 N is a not identified derivative of uridine.
0121/34 N is a not identified derivative of uridine.
0140/34 xU is identified as mcm⁵U.
0141/34 xU is identified as mcm⁵U.
0260/0 Isolated from rat liver, human liver and human placenta.

	Aminoacyl Stem								D Stem				D Loop								D Stem				Anticodon Stem					Anticodon Loop							Anticodon Stem				
	1	2	3	4	5	6	7	8	9	10	11	12	13	14	15	16	17	18	19	20	20:1	20:2	21	22	23	24	25	26	27	28	29	30	31	32	33	34	35	36	37	38	39	40	41	42	43

CYSTEINE

0410 E.coli: G G C G C G U s⁴U C A C A A A G C   G G D   D A U G U A G C G G A Ψ U G C A xA* A Ψ C C G U
0440 Yeast: G C U C G U A U G G C C G C A G D   G G D   A G C G C A G C A G A Ψ U G C A iA A Ψ C U G U

GLUTAMINE

0510 E.coli K12 1: U G G G G U A s⁴U C G C C A A G C   Gm G D   A A G G C A C C G G U Um U N* U G m²A Ψ A C C G G U
0520 E.coli K12 2: U G G G G U A s⁴U C G C C A A G D   Gm G D   A A G G C C C G G A Um U U C U G m²A Ψ Ψ G C U A
0530 Phage T4: U G G G A A U s⁴U A G C C A A G D D   G G D   A A G G C U A G C A C C U N* U U G m²A C Ψ G C U A
0531 Phage T4 psu₂⁺ oc: U G G G A A U s⁴U A G C C A A G D D   G G D   A A G G C U A G C A C C U C U A m²A C Ψ G C U A
0532 C34 psu+2 am: U G G G A A U s⁴U A G C C A A G D   G G D   A A G G C U A G C A C C U N* U G m²A C Ψ G C U A
0540 Phage T4 (from precurs.): U G G G A A U s⁴U A G C C C A A G D   G G D A A G G C U A G C A C C U C U G m²A C Ψ G C U A

GLUTAMIC ACID

0610 E.coli B 1: G U C C C C U U C G U C Ψ A G A   G C C C A G G A C C C C C C xU* U C m²A C G G C G G
0620 E.coli 2: G U C C C C U U C G U C Ψ A G A   G C C C A G G A C C C C C C xU* U C m²A C G G C G G
0630 Yeast 3: U C C G A U A U A G U G Ψ A A C   G C C D A U C A C A Ψ C A C G C xU* U C A C C G U G G

0410 G.P.Mazzara,W.H.McClain (1977) J.Mol.Biol. 117,1061-1079.
0440 N.J.Holness,G.Atfield (1976) Biochem.J. 153, 447-454.
0510 + 0520 M.Yaniv,W.R.Folk (1975) J.Biol.Chem. 250, 3243-3253.
0530 + 0531 J.G.Seidman,M.M.Comer,W.H.McClain (1974) J.Mol.Biol. 90, 677-689.
0532 M.M.Comer,K.Foss,W.H.McClain (1975) J.Mol.Biol. 99, 283-293.
0540 C.Guthrie (1975) J.Mol.Biol. 95, 529-548.
0610 M.Uziel,A.J.Weinberg (1975) Nucleic Acids Res. 2, 469-476.
0620 Z.Ohashi,F.Harada,S.Nishimura (1972) FEBS-Lett. 20, 239-241; K.O.Munninger,S.H.Chang (1972) Biochem.Biophys.Res.Commun. 46, 1837-1842.
0630 T.Kobayashi,T.Irie,M.Yoshida,K.Takeishi,T.Ukita (1974) Biochim. Biophys.Acta 366, 168-181.

			Extra Arm			Tψ Stem		Tψ Loop		Tψ Stem		Aminoacyl Stem	
	44	45	46	47(1–16)	48	49 50 51 52 53	54 55	56 57 58 59 60	61 62 63 64 65		66 67 68 69 70 71 72	73 74 75 76
CYSTEINE												
0410	C	U	A		G	U C C G G	T ψ	C G A C U	C C G G A		A C G C G C C	U C C A
0440	U	G	m⁷G	D	m⁵C	C U U A G	T ψ	C G m¹A U C	U G A G...		U G C G A G C	U C C A
GLUTAMINE												
0510	C	A	U	U	C	C C U G G	T ψ	C G A A U	C C A G G		U A C C C C A	G C C A
0520	C	A	U	U	C	C C G A G	T ψ	C G A A U	C C U C G		U A C C C C A	G C C A
0530	G	A	U	U	C	C A A G G	T ψ	C G A G U	C C U U G		A U U C C C A	G C C A
0531	G	A	U	G	C	C A A G G	T ψ	C G A G U	C C U U U		A U U C C C A	G C C A
0532	G	A	U	G	C	C A A G G	T ψ	C G A G U	C C U U U		A U U C C C A	G C C A
0540	G	A	U	G	C	C A A G G	T ψ	C G A G U	C C U U U		A U U C C C A	G C C A
GLUTAMIC ACID												
0610	U	A	A		C	A G G G G	T ψ	C G A A U	C C C C U		G G G G G A C	G C C A
0620	U	A	A		C	A G G G G	T ψ	C G A A A	C C C C U		A G G G G A C	G C C A
0630	A	G	A		m⁵C	G G G G G	T ψ	C G A C U	C C C C G		U A U C G G A	G C C A

0410/37 xA is ms²i⁶A.
0510/34 N is likely a derivative of 2-thiouridine.
0530/34 N is an unknown derivative of uridine.
0531/34 N is an unknown derivative of uridine.

0540/34 N is an unknown derivative of uridine.
0610/34 xU is mam⁵s²U.
0620/34 xU is mam⁵s²U.
0630/34 xU is mcm⁵s²U.

		Aminoacyl Stem									D Stem					D Loop								D Stem					Anticodon Stem					Anticodon Loop							Anticodon Stem			
		1	2	3	4	5	6	7	8	9	10	11	12	13	14	15	16	17	17:1	18	19	20	20:1	20:2	21	22	23	24	25	26	27	28	29	30	31	32	33	34	35	36	37	38	39	40	41	42	43
GLYCINE																																														
0710	E.coli 1	G	C	G	G	G	C	G	s⁴U	A	G	U	U	C	A	A	U	G		G	G				A	G	A	A	C	G	A	G	A	G	C	U	U	C	C	C	A	A	G	C	U	C	U
0711	S.typhimurium	G	C	G	G	G	C	G	U	A	G	U	U	C	A	A	U	G(m)		G					A	G	A	A	C	G	A	G	A	G	C	U	C	C	C	A	A	G	C	U	C	U
0712	S.typhimurium suf D	G	C	G	G	G	C	G	U	A	G	U	U	C	A	A	U	G(m)		G	G				A	G	A	A	C	G	A	G	A	G	C	U	C	C	C	C	A	A	G	C	U	C	U
0720	E.coli 2	G	C	G	G	G	C	A	U	C	G	U	A	U	A	A	U	G		G					A	U	U	A	U	C	A	G	A	G	C	U	C	C	C	C	A	A	G	C	U	G	A
0721	TsuA36	G	C	G	G	G	A	A	U	U	A	G	C	U	A	A	U	G		G	G				A	G	A	G	C	A	C	C	U	G	A	C	U	U(N)*	C	C	U	N*	A	G	C	U	G	A
0730	E.coli 3	G	C	G	G	G	A	A	U	A	G	C	U	C	A	G	D	G		G	G				A	G	A	G	C	A	C	C	U	G	C	U	U(N)*	G*	C	U	N*	A	G	G	U	C	G
0731	E.coli su⁺ A78	G	C	G	G	G	A	A	U	A	G	C	U	C	A	G	D	G		G	G				A	G	A	G	C	A	C	C	U	G	C	U	U	xA*	C	A	A	xA*	A	G	G	U	C	A
0740	S.epidermidis* 1A	G	C	G	G	G	G	G	s⁴U	A	A	U	U	C	A	A	C	G		G	U				A	G	A	A	U	A	U	C	U	U	U	C	C	C	A	A	C	A	G	G	U	U	C
0750	S.epidermidis* 1B	G	C	G	G	G	G	G	s⁴U	A	G	U	U	C	A	A	C	G		G	U				A	G	A	A	U	A	U	C	U	U	U	C	C	C	A	A	C	G	G	A	A	C	G
0760	Phage T4	G	C	G	G	A	U	A	U	C	G	U	A	U	A	A	U	Gm		G					A	U	A	A	U	C	A	C	U	U	A	C	U	xU*	C	C	C	A	A	U	C	U	G	A
0770	Yeast	G	C	G	C	A	A	G	Um¹G	U	G	U	U	A	A	G	D	G		G		C			A	A	A	A	C	C	U	U	C	A	C	G	ψ	C	C	C	A	A	ψ	G	G	U	A	C
0780	Wheat germ 1	G	C	A	Cm	C	A	G	Um¹G	U	G	U	C	ψ	A	G	D	G		G	G				A	G	A	A	U	C	C	U	C	U	C	G	C	C	C	C	A	A m⁷C	A	G	G	U	U	G
0790	Bombyx mori 1	G	C	A	Um	C	G	G	Um¹G	U	U	U	C	ψ	A	G	D	G		G	G				A	A	A	A	C	C	U	U	U	C	G	C	N*	C	C	C	A	A m⁷C	A	G	C	A	G	G
0791	Bombyx mori 2	G	C	A	U	m¹G	U	G	Um¹G	U	U	U	C	ψ	A	A	D	G		G	G	D			A	G	C	A	A	C	C	U	ψ	C	C	U	C	C	C	C	A	A m⁷C	A	G	C	A	G	G
0792	Human Placenta(GCC)	G	C	A	N	U	G	G	U	U	G	m⁵G	U	U	C	A	D	G		G	G				A	A	A	A	C	C	U	A	A	G	A	U	C	G	C	C	C	A	N	G	C	A	G	G
0793	Human Placenta(CCC)	G	C	G	C	C	G	C	U	C	G	G	U	C	A	G	U	G		G	G				A	C	A	U	C	A	G	C	A	A	G	A	N	C	C	C	A	N	C	C	U	U	G
HISTIDINE																																														
0810	E.coli* 1	G	G	U	G	G	C	U	As⁴U	A	G	C	U	C	A	G	D	D		G	G				A	G	A	G	C	C	C	U	G	G	A	U	U	Q	U	G	G	m²A ψ*	ψ*	C	C	A	G

0710 + 0711 C.W.Hill,G.Combriato,W.Steinhart,D.L.Riddle,J.Carbon (1973) J.Biol.Chem. 248, 4252-4262.
0712 D.L.Riddle,J.Carbon (1973) Nature New Biology 242, 230-234.
0720 + 0721 J.W.Roberts,J.Carbon (1975) J.Biol.Chem. 250, 5530-5541.
0730 C.Squires,J.Carbon (1971) Nature New Biology 233, 274-277.
0731 J.Carbon,E.W.Fleck (1974) J.Mol.Biol. 85, 371-391.
0740 + 0750 R.J.Roberts (1974) J.Biol.Chem. 249, 4787-4796.
0760 S.Stahl,G.V.Paddock,J.Abelson (1974) Nucleic Acids Res. 1, 1287-1304;B.G.Barell,A.R.Coulson,W.H.McClain (1973)

0770 FEBS-Lett. 37, 64-69.
0780 M.Yoshida (1973) Biochem.Biophys.Res.Commun. 50, 779-784.
0790 K.B.Marcu,R.E.Mignery,B.S.Dudock (1977) Biochemistry 16, 797-806.
  J.P.Garel,G.Keith (1977) Nature 269,350-352;
  M.C.Zuniga,J.A.Steitz (1977) Nucleic Acids Res. 4, 4175-4196.
0791 M.kawakami,K.Nishio,S.Takemura (1978) FEBS-Lett. 87, 288-290.
0792 + 0793 R.C.Gupta,B.A.Roe,K.Randerath (1978) Cold Spring Harbor Meeting on tRNA,Abstracts,p.5.
0810 C.E.Singer,G.R.Smith (1972) J.Biol.Chem. 247, 2989-3000.

GLYCINE

	44	45	46	47	47.1	47.2	47.3	47.4	47.5	47.6	47.7	47.8	47.9	47.10	47.11	47.12	47.13	47.14	47.15	47.16	48	49	50	51	52	53	54	55	56	57	58	59	60	61	62	63	64	65	66	67	68	69	70	71	72	73	74	75	76
0710	A	U	A																		C	C	G	A	G	G	T	Ψ	C	G	A	U	U	C	C	C	C	U	C	G	C	C	C	C	G	U	C	C	A
0711	A	U	A																		C	C	.G	.A	.G	.G	T	Ψ	C	G	A	U	U	C	.C	.C	.U	.U	C	G	C	C	C	C	G	U	C	C	A
0712	A	U	A																		C	C	.G	.A	.G	.G	T	Ψ	C	G	A	U	U	C	.C	.C	.U	.U	C	G	C	C	C	C	G	U	C	C	A
0720	U	G	A																		U	U	G	.C	.G	.G	T	Ψ	C	G	A	G	U	C	.C	.C	.U	.C	U	G	C	C	C	C	G	U	C	C	A
0721	U	G	A																		U	C	.G	.C	.G	.G	T	Ψ	C	G	A	G	U	C	.C	.C	.U	.C	U	G	C	C	C	C	G	U	C	C	A
0730	G	G	m^{7}G	U																	C	G	.C	.C	.A	.G	T	Ψ	C	G	A	G	U	C	.C	.U	.C	.G	U	U	C	C	C	C	G	U	C	C	A
0731	G	G	m^{7}G	U																	C	G	.C	.C	.A	.G	T	Ψ	C	G	A	G	U	C	.C	.U	.C	.G	U	U	C	C	C	C	G	U	C	C	A
0740	A	G	A																		U	U	A	.U	.A	.G	U	G	C	A	A	A	U	C	.C	.U	.U	.A	C	U	U	C	C	G	G	U	C	C	A
0750	A	G	G																		U	A	.U	.A	.G	.G	U	G	C	A	A	A	U	C	.C	.U	.A	.U	C	U	A	U	C	C	G	U	C	C	A
0760	U	G	A																		U	U	.G	.U	.G	.A	U	U	C	G	A	A	U	C	.C	.U	.C	.A	C	U	G	G	U	C	G	A	C	C	A
0770	G	G	A																		C	m^{1}C	C	.C	.C	.G	U	U	C	G	m^{1}A	A	U	C	.C	.G	.G	.G	C	U	G	G	U	C	G	A	C	C	A
0780	A	G	A																		m^{3}C	m^{5}C	m^{5}C	.m^{5}C	.m^{5}C	.G	U	U	C	G	m^{1}A	A	U	C	.C	.U	.C	.G	C	C	G	A	A	C	G	C	C	C	A
0790	C	G	G																		m^{3}C	m^{5}C	m^{5}C	.m^{5}C	.m^{1}C	.G	T	Ψ	C	G	m^{1}A	A	U	C	.C	.U	.C	.G	C	C	A	A	C	G	C	C	C	A
0791	U	G	A																		U	m^{5}C	m^{5}C	.m^{5}C	.m^{1}C	.G	T	Ψ	C	G	m^{1}A	A	U	C	.C	.U	.C	.G	C	C	A	A	U	G	C	A	C	C	A
0792	A	G	G																		m^{3}C	m^{5}C	m^{5}C	.m^{5}C	.m^{1}C	.G	T	Ψ	C	G	m^{1}A	A	U	C	.C	.U	.C	.G	C	C	A	A	U	G	C	A	C	C	A
0793	C	G	A																		C	m^{5}C	m^{5}C	.m^{5}C	.m^{1}C	.G	T	Ψ	C	G	m^{1}A	A	U	C	.C	.U	.C	.G	G	C	G	G	C	G	C	A	C	C	A

HISTIDINE

	44	45	46	47	48	49	50	51	52	53	54	55	56	57	58	59	60	61	62	63	64	65	66	67	68	69	70	71	72	73	74	75	76
0810	U	U	m^{7}G	U		C	G	U	G	G	T	Ψ	C	G	A	A	U	C	C	C	C	A	U	A	G	C	C	A	C	C	A	C	C	A

0710/35 Mutation C-35→U-35; C.W.Hill, G.Combriato, W.Dolph (1974) J.Bacteriol. 117, 351-359.
0720/34 N is an unidentified derivative of uridine.
0721/34 N is an unidentified derivative of uridine.
0721/37 N is probably a derivative of adenosine.
0730/34 Mutation: E.coli ins has G-34→U-34.
0731/37 xA is ms^{2}i^{6}A.
0740/0   Staphylococcus epidermidis Texas 26.
0750/0   Staphylococcus epidermidis Texas 26.
0760/34 xU is probably related to mam^{5}s^{2}U.
0791/34 N contains 2 unknown modified nucleosides. They are probably derivatives of uridine.
0810/0   Identical with Salmonella typhimurium.
0810/38 + 0810/39 HisT mutation Ψ-38→U-38, Ψ-39→U-39; C.E.Singer, G.R.Smith, R.Cortese, E.N.Ames (1972) Nature New Biology 238, 72-74.

		Aminoacyl Stem								D Stem				D Loop									D Stem				Anticodon Stem					Anticodon Loop							Anticodon Stem				
		1	2	3	4	5	6	7	8	9	10	11	12	13	14	15	16	17	17:1	18	19	20	20:1	20:2	21	22	23	24	25	26	27	28	29	30	31	32	33	34	35	36	37	38	39	40	41	42	43

ISOLEUCINE

		1	2	3	4	5	6	7	8	9	10	11	12	13	14	15	16	17	18	19	20	20:1	20:2	21	22	23	24	25	26	27	28	29	30	31	32	33	34	35	36	37	38	39	40	41	42	43
0910	E.coli 1	A	G	G	C	U	U	G	U	A	G	C	U	C	A	G	G D(U)		G	G	D			A	G	A	G	C	G	C	A	C	C	C	U	G	A	U	t⁶A	A	G	G	G	U	G
0920	T.utilis	G	G	U	C	C	C	U	U	G	m²₆C	C	C	C	A	G	D		G	G	D			A	A	G	G	C	m²₆G	Ψ	G	G	U	G	C	U	A	U	t⁶A	A	C	G	C	C	A

LEUCINE

		1	2	3	4	5	6	7	8	9	10	11	12	13	14	15	16	17	18	19	20	20:1	20:2	21	22	23	24	25	26	27	28	29	30	31	32	33	34	35	36	37	38	39	40	41	42	43
1010	E.coli B/K12* 1	G	C	C	C	G	A	G	G	U	G	G	U	G	A	A	D		G	Gm	G	D		A	A	C	G	C	G	G	C	U	A	G	G	U	U	C	A	G	N*	Ψ	G	G	U	A	G
1011	E.coli K12 2	G	C	C	C	G	G	G	U	G	G	U	G	G	A	A	D		C	Gm	G	D		A	A	C	G	U	C	G	C	U	A	G	G	U	U	C	A	G	N*	Ψ	G	G	U	C	C
1012	E.coli 5	G	C	C	C	G	G	A	U	G	G	U	G	G	A	A	D		D	Gm	G	D		A	G	A	C	A	C	A	A	G	G	A	Ψ	U	U	N	A	A	xA*	Ψ	G	C	C	U	U
1030	Phage T4	G	C	G	A	G	A	A	U	G	G	U	U	C	A	A	D		D	Gm	G	D		A	A	A	G	G	C	A	A	C	C	C	A	Ψ	U	N	A	A	xA*	Ψ	C	A	G	G	G
1040	Yeast 3	G	G	U	U	G	U	U	U	G	m²₆C	C	αt⁴C	G	A	G	C			Gm	G	D		C	A	G	G	C	m²₆G	Ψ	C	C	U	G	A	Ψ	U m⁵C	A	A	m¹⁶G	C	Ψ	C	U	G	A	G
1050	Yeast	G	G	G	A	G	U	U	U	G	m²₆C	C	αt⁴C	G	A	G	D		D	Gm	G	D		A	A	G	G	G	G	A	C	C	A	G	A	Ψ	U U	A	G	m¹⁶G	C	Ψ	C	U	G	A	G
1060	T.utilis	G	G	A	U	C	C	U	U	G	m²₆C	C	αt⁴C	G	A	G	C			Gm	G	D		A	A	G	G	C	m²₆G	Ψ	C	U	C	C	A	Cm	U C m	A	A	m¹⁶G	C	Ψ	C	G	A	G

LYSINE

		1	2	3	4	5	6	7	8	9	10	11	12	13	14	15	16	17	18	19	20	20:1	20:2	21	22	23	24	25	26	27	28	29	30	31	32	33	34	35	36	37	38	39	40	41	42	43
1110	E.coli B	G	G	G	U	C	G	U	U	A	G	C	U	C	A	G	D		D	G	G	D		A	A	G	A	G	C	A	G	U	U	G	A	C	U* xU	U	U	t⁶A	A	Ψ	C	A	A	U
1120	Bacillus subtilis	G	A	G	C	C	A	U	U	A	G	C	U	C	A	A	U		D	G	G	D		A	A	G	A	G	C	A	U	C	U	G	A	C	U U(N)	U	N	A	A	Ψ	C	A	G	A
1130	Yeast(haploid) 1	G	C	C	U	U	G	U	U	G	(m¹)G	C	G	C	A	A	D		C	G	G	D		A	G	C	G	C	m²₆G	Ψ	A	U	G	A	C	U*	U	U	t⁶A	A	Ψ	C	A	U	A
1140	Yeast 2	Ψ	C	C	U	U	G	U	U	A	m²₆C	U	C	A	A	G	D		D	G	G	D		A	A	G	A	G	C	m²₆G	Ψ	C	G	G	C	U U*	U	U	t⁶A	A	C	C	A	G	A

0910 M.Yaris,B.G.Barrell(1971) Biochem.Biophys.Res.Commun. 43, 729-734.
0920 S.Takemura,M.Murakami,M.Miyazaki(1969) J.Biochem. 65, 553-566.
1010 H.U.Blank,D.Söll(1971) Biochem.Biophys.Res.Commun. 43, 1192-1197.
1011 S.K.Dube,K.A.Marcker,A.Yudelevich(1970) FEBS-Lett. 9, 168-170.
1012 H.U.Blank,D.Söll(1971) Biochem.Biophys.Res.Commun. 43, 1192-1197.
1030 Z.Yamaizumi,Y.Kuchino,F.Harada,S.Nishimura,J.A.McCloskey(1978) Cold Spring Harbor Meeting on tRNA,Abstracts,p-4.
1040 T.C.Pinkerton,G.Paddock,J.Abelson(1973) J.Biol.Chem. 248, 6349-6365.
1050 K.Randerath,L.S.Y.Chia,R.C.Gupta,E.Randerath,E.R.Hawkins,C.K.Brum, S.H.Chang(1975) Biochem.Biophys.Res.Commun. 63, 157-163.
1060 A.Murasugi,S.Takemura(1978) J.Biochem. 83, 1029-1038.
1110 K.Chakraburtty,A.Steinschneider,R.V.Case,A.H.Mehler(1975) Nucleic Acids Res. 2, 2069-2075.
1120 Y.Yamada,H.Ishikura(1977) Nucleic Acids Res. 4, 4291-4303.
1130 S.J.Smith,H.S.Teh,A.N.Ley,P.D'Obrenan(1973) J.Biol.Chem. 248, 4475-4485.
1140 J.T.Madison,S.J.Boguslawski(1974) Biochemistry 13, 524-527.
1040 S.H.Chang,S.Kuo,E.Hawkins,N.R.Miller(1973) Biochem.Biophys.Res.Commun. 51, 951-955.

	Extra Arm																	TψStem					TψLoop							TψStem					Aminoacyl Stem										
	44	45	46	47	47	47	47	47	47	47	47	47	47	47	47	47	48	49	50	51	52	53	54	55	56	57	58	59	60	61	62	63	64	65	66	67	68	69	70	71	72	73	74	75	76
				1	2	3	4	5	6	7	8	9	10	11	12	13	14	15	16																										
ISOLEUCINE																																													
0910	A	G	m⁷G	X*													C	C	G	U	G	G	T	ψ	C	A	A	G	U	C	C	A	C	ψ	C	A	G	G	C	C	U	A	C	C	A
0920	A	G	A	D													m⁵C	A	G	C	A	G	T	ψ	C	G	m¹A	U	C	C	U	G	C	U	A	G	G	A	C	C	U	A	C	C	A
LEUCINE																																													
1010	U	G	U	C	C	U	U	A	C	G	G	A	C	G			U	G	G	G	G	G	T	ψ	C	A	A	G	U	C	C	C	C	C	C	C	U	C	G	C	G	A	C	C	A
1011	U	G	C	C	C	A	A	U	A	G	G	G	C	U			U	A	C	G	G	G	T	ψ	C	A	A	G	U	C	C	C	U	C	C	C	U	U	G	G	U	A	C	C	A
1012	C	G	G	C	G	G	C	G	C	G	C	U	G				U	U	G	C	G	G	T	ψ	C	G	A	A	U	C	C	G	U	C	C	U	C	U	C	G	G	A	C	C	A
1030	C	G	G	A	A	U	G	A	U	G	U	C	C				U	U	G	C	G	G	T	ψ	C	G	A	A	U	C	C	G	C	A	C	U	U	C	G	G	G	A	C	C	A
1040	U	U	A	U	C	U	U	C	G	A	U	G					m⁵C	A	A	G	A	G	T	ψ	C	G	m¹A	A	U	C	C	U	C	G	U	A	G	C	A	A	C	C	A	C	C	A
1050	U	C	C	U	U	C	G	G	A	G	G						m⁵C	A	U	G	A	G	T	ψ	C	G	m¹A	A	U	C	U	C	A	U	C	A	G	C	U	C	C	A	C	C	A
1060	U	A	U	C	G	U	A	A	G	A	U	G					m⁵C	A	U	G	A	G	T	ψ	C	G	m¹A	A	U	C	U	C	A	U	C	A	G	G	A	U	C	C	A	C	C	A
LYSINE																																													
1110	U	G	m⁷G	X													C	C	G	A	G	G	T	ψ	C	G	A	A	U	C	C	U	U	G	A	C	G	A	C	C	A	C	C	A
1120	G	G	m⁷G	U													C	C	G	A	A	G	T	ψ	C	G	A	A	U	C	U	U	C	C	A	U	G	G	C	U	C	U	C	C	A
1130	A	G	m⁷G	U													U	U	A	G	G	G	T	ψ	C	G	m¹A	U	C	C	C	C	C	C	A	C	A	G	G	C	U	C	U	C	C	A
1140	A	U	m⁷G	DI(U)													m⁵C	A	G	G	G	G	T	ψ	C	G	m¹A	A	U	C	C	C	C	U	A	ψ	G	A	G	G	A	G	C	C	A

0910/47 Probably X, 3N-(3-amino-3-carboxypropyl)uridine,S.Friedman,H.J.Li, K.Nakanishi,G.van Lear(1974) Biochemistry 13, 2932-2937.

1010/0 Identical with Salmonella typhimurium LT2 tRNA$^{Leu}_1$

1010/38 His T mutant of Salmonella typhimurium tRNA$^{Leu}_1$ has ψ-38→U-38 and
1010/40 ψ-40→U-40, H.S.Allaudeen,S.K.Yang,D.Söll(1972) FEBS-Lett. 28, 205-208.

1010/0 For numbering of E.coli leucine tRNAs see R.E.Hurd,G.T.Robillard,
1011/0 B.A.Reid(1977) Biochemistry 16, 2095-2100.

1010/37 N is an unknown derivative of guanosine.

1011/37 N is an unknown derivative of guanosine.

1012/37 xA is ms^{2}i^{6}A.

1030/34 N is an unknown derivative of uridine.

1030/37 xA is ms^{2}i^{6}A.

1110/34 xU is mam^{5}s^{2}U.

1120/37 U is partially replaced by N,which is probably a derivative of 2-thiouridine.

1120/0 N is an unknown derivative of guanosine.

1140/0 Is identical with Saccharomyces cerivisiae haploid 2, C.J.Smith, H.-S.Teh,A.N.Ley,P.D'Obrenan(1973) J.Biol.Chem. 248, 4475.4485.

1140/34 xU is mcm^{5}s^{2}U.

		Aminoacyl Stem							D Stem				D Loop									D Stem					Anticodon Stem					Anticodon Loop							Anticodon Stem			
		1	2	3	4	5	6	7	8	9	10	11	12	13	14	15	16	17	18	19	20	20 1	20 2	21	22	23	24	25	26	27	28	29	30	31	32	33	34	35	36	37	38	39	40	41	42	43
METHIONINE																																													
1210	E.coli CA 265 1	G	G	C	U	A	C	G	s⁴U	A	G	C	U	C	A	G	D(U)	D	G(m)	G	D			A	G	A	G	C	A	C	A	U	C	A	C	U	ac⁴C	A	U	t⁶A	A	ψ	G	A	U	G
1240	Yeast 3	G	C	U	U	C	A	G	U	A	m²G	C	U	C	A	G	D	A	G	G	D			A	G	A	G	C	m²₂G,ψ	ψ(U)	C	A	G	ψ	C	U	C	A	U	t⁶A	A	ψ	C	U	G	A
1250	Mammalian*	G	C	C	U	Cm²⁶G	U	A	U	C	G	C	G	C	A	G	D	A	G	G	D			A	G	C	G	C	m²₂G	ψ	C	A	G	ψ	C	U	Cm	A	U	t⁶A	A	ψ	C	U	G	A
METHIONINE-INITIATOR																																														
1310	E.coli CA 265	C	G	C	G	G	G	G	s⁴U	G	G	A	G	C	A	G	C	C	U	G	G			A	G	C	U	C	G	U	C	G	G	G	C	U	C	A	U	A	A	C	C	C	G	A
1320	Thermus thermophilus C	G	C	G	G	G	G	s⁴U	G	G	A	G	C	A	G	C	C	U	G	G			A	G	C	U	C	G	U	C	G	G	G	C	U	C	A	U	t⁶A	A	C	C	C	G	A
1330	Bacillus subtilis	C	G	C	G	G	G	G	U	G	G	A	G	C	A	G	C	C	U	G	G			A	G	C	U	C	G	U	C	G	G	G	Cm	U	C	A	U	A	A	C	C	C	G	A
1340	Anacystis nidulans	C	G	C	G	G	G	G	U	A	G	A	G	C	A	G	U	C	G	G	D			A	G	C	U	C	G	U	C	G	G	G	C	U	C	A	U	A	A	C	C	C	G	A
1350	Mycoplasma	C	G	C	G	G	G	G	s⁴U	A	G	A	G	C	A	G	U	C	U	G	G			A	G	C	U	C	G	U	C	G	G	G	Cm	U	C	A	U	A	A	C	C	C	G	G
1360	Neurospora crassa (mito)	U	G	C	G	G	G	A	U	U	G	A	U	G	U	A	A	D	A	G	D			A	A	C	A	U	A	U	U	U	G	ψ	C	U	C	A	U	m²G	N*	C	C	G	A	A
1370	Neurospora crassa (cyto)	A	G	C	U	U	G	C	A	U	G	G	C	G	C	A	G	U	C	G	A			A	G	C	G	C	m²₆G	C	A	G	G	G	C	U	C	A	U	t⁶A	A	C	C	C	G	A
1375	Wheat germ	A	U	C	A	G	A	G	Um¹G	Um⁶m²G	C	G	C	A	G	C	C	G	G	A			A	G	C	U	G	m²₆G	C	ψ	G	ψ	C	A	G	C	A	U	t⁶A	A	C	C	C	A	C
1380	Yeast	A	G	C	(C	G)	C	G	Um⁶m²G	Um⁶m²G	C	G	C	A	G	D	C	G	G	A			A	G	C	G	C	m²₆G	C	A	G	G	G	C	U	C	A	U	t⁶A	A	C	C	C	U	G
1390	Mammalian*	A	G	C	A	G	A	G	U	G	G	C	G	C	A	G	C	G	G	A	A			A	G	C	G	C	m²₆G	C	A	G	G	G	C	U	C	A	U	t⁶A	A	C	C	C	A	G

1210 S.Cory,K.A.Marcker (1970) Eur.J.Biochem. 12, 177-194.
1240 H.Gruhl,H.Feldmann (1976) Eur.J.Biochem. 68, 209-217;
       O.Koiwai,M.Miyazaki (1976) J.Biochem. 80, 951-959.
1250 P.W.Piper (1975) Eur.J.Biochem. 51, 283-293;
       G.Petrissant,M.Boisnard (1974) Biochimie 56, 787-789.
1310 S.K.Dube,K.A.Marcker (1969) Eur.J.Biochem. 8, 256-262.
1320 K.Watanabe,T.Oshima,S.Nishimura (1976) Nucleic Acids Res. 3, 1703-1713.
1330 Y.Yamada,H.Ishikura (1975) FEBS-Lett. 54, 155-158.
1340 B.Ecarot-Charrier,R.J.Cedergren (1976) FEBS-Lett. 63, 287-290.
1350 R.T.Walker,U.L.RajBhandary (1978) Nucleic Acids Res. 5, 57-70.

1360 J.E.Heckman,L.I.Hecker, S.D.Schwartzbach,W.E.Barnett,B.Baumstark,
       U.L.RajBhandary (1978) Cell 13, 83-95.
1370 A.M.Gillum,L.I.Hecker,M.Silberklang,S.D.Schwartzbach,U.L.RajBhandary,
       W.E.Barnett (1977) Nucleic Acids Res. 4, 4109-4131.
1375 H.P.Ghosh,K.Ghosh,M.Simsek,U.L.RajBhandary (1978) Cold Spring Harbor
       Meeting on tRNA, Abstracts,p.6.
1380 M.Simsek,U.L.RajBhandary (1972) Biochem.Biophys.Res.Commun. 49, 508-515.
1385 M.Simsek,U.L.RajBhandary,M.Boisnard,G.Petrissant (1974) Nature 247, 518-520;
       A.M.Gillum,N.Urquhart,M.Smith,U.L.RajBhandary (1975) Cell 6, 395-405;
       P.W.Piper,B.F.C.Clark (1974) Eur.J.Biochem. 45, 589-600;
       M.Wegnez,A.Mazabraud,H.Denis,G.Petrissant,M.Boisnard (1975) Eur.J.Biochem.
       60, 295-302.

	Extra Arm																TψStem					TψLoop							TψStem					Aminoacyl Stem											
	44	45	46	47	47:1	47:2	47:3	47:4	47:5	47:6	47:7	47:8	47:9	47:10	47:11	47:12	47:13	47:14	47:15	47:16	48	49	50	51	52	53	54	55	56	57	58	59	60	61	62	63	64	65	66	67	68	69	70	71	72	73	74	75	76
METHIONINE																																																	
1210	G	G	m⁷G	X																	C	C	A	C	A	G	G	T	ψ	C	G	A	A	U	C	C	C	G	U	A	G	C	C	A	C	C	A
1240	A	G	m⁷G	D(U)																	m⁵C	G	A	G	A	G	T	ψ	C	G	m¹A	A	C	C	U	C	U	C	U	G	G	A	G	C	A	C	C	A
1250	A	G	m⁷G	D																	m⁵C	G	U	G	A	G	T	ψ	C	G	m¹A	U	C	C	U	C	A	C	A	C	G	G	G	C	A	C	C	A
METHIONINE-INITIATOR																																																	
1310	A	G	m⁷G*	U																	C	G	U	C	G	G	T	ψ	C	A	A	A	U	C	C	G	C	C	C	C	G	C	A	A	C	C	A
1320	A	G	m⁷G	U																	C	G	C	G	G	G	T	ψ	C	A	m¹A	A	U	C	C	U	C	C	C	C	C	G	C	A	A	C	C	A
1330	A	G	G	U																	C	G	C	A	G	G	s²T	ψ	C	A	A	A	U	C	C	U	C	C	C	C	C	G	C	A	A	C	C	A
1340	A	G	m⁷G	U																	C	A	G	C	A	G	T	ψ	C	A	A	A	U	C	C	U	C	C	C	C	C	G	C	A	A	C	C	A
1350	A	G	G	C																	C	G	C	A	U	G	T	ψ	C	G	A	G	U	C	C	U	C	C	C	C	G	C	A	A	C	C	A
1360	U	G	A																		C	A	U	A	G	G	U	ψ	C	G	A	A	A	U	C	C	U	A	U	C	C	G	C	A	U	C	C	A
1370	A	G	m⁷G	U(D)																	C	A	C	U	C	G	A	ψ	C	G	m³A	A	A	C	G	A	N*	U	U	G	C	A	G	C	U	A	C	C	A
1375	A	G	m⁷G	D																	m³C	C	C	A	G	G	A	ψ	C	G	m³A	A	A	C	C	U	G	*	C	U	C	U	G	A	U	A	C	C	A
1380	A	U	m⁷G	D																	m⁵C	m⁵C	U	C	G	G	A	U	C	G	m³A	A	A	C	C	G	N*	N*	C	G(C G)	G	C	U	A	C	C	A
1390	A	G	m⁷G	D																	m⁵C	G	A	U	G	G	A	ψ	C	G	m³A	A	A	C	C	A	U	C	C	U	C	U	G	C	U	A	C	C	A

1250/0. Mouse myeloma and rabbit liver.

1310/46 m⁷G-46→A-46 in the minor species of tRNA^fMet from E.coli, S.K.Dube, K.A.Marcker,B.F.C.Clark,S.Cory.(1968) Nature 218, 231-233; B.Z.Egan,J.F.Weiss,A.D.Kelmers' (1973) Biochem.Biophys.Res.Commun. 55, 320-327.

1360/33 N is most probably pseudouridine.

1370/28 N is an unidentified derivative of pyrimidine.

1370/64 N is an unidentified derivative of guanosine.

1375/65 Is probably a modified derivative of guanosine.

1380/64 N is an unidentified derivative of adenosine.

1380/65 N is an unidentified derivative of guanosine.

1390/0 Rabbit liver,sheep mammary glands,salmon testes,salmon liver,human placenta,mouse myeloma cells,oocytes and somatic cells of Xenopus laevis.

531

		Aminoacyl Stem								D Stem					D Loop								D Stem				Anticodon Stem					Anticodon Loop							Anticodon Stem				
		1	2	3	4	5	6	7	8	9	10	11	12	13	14	15	16	17	17 1	18	19	20	20 1	20 2	21	22	23	24	25	26	27	28	29	30	31	32	33	34	35	36	37	38	39	40	41	42	43

PHENYLALANINE

		1	2	3	4	5	6	7	8	9	10	11	12	13	14	15	16	17	17₁	18	19	20	20₁	20₂	21	22	23	24	25	26	27	28	29	30	31	32	33	34	35	36	37	38	39	40	41	42	43
1410	E.coli	G	C	C	C	G	G	A	s⁴U	A	G	C	U	C	A	G	D	C		G	G				A	G	A	G	C	A	G	G	G	A	Ψ	U	G	A	A	xA*	A	Ψ	C	C	C	C
1420	B.stearothermophilus	G	G	C	U	C	G	G	s⁴U	A	G	C	U	C	A	G	U	C		G	G				A	G	A	G	C	A	A	A	G	A	C	U	Gm	A	A	xA*	A	Ψ	C	C	C	U
1430	Bacillus subtilis	G	G	C	U	C	G	G	U	A	G	C	U	C	A	G	U	D		G	G				A	G	A	G	C	A	C	G	G	A	C	U	Gm	A	A	xA*	A	Ψ	C	C	C	U
1440	Mycoplasma	G	G	U	C	G	U	G	U	A	G	C	U	C	A	G	U	C		G	G				A	G	A	G	C	A	A	C	A	G	A	C	U	G	A	A	mG	A	Ψ	C	C	U	C
1450	Bean chloroplast	G	U	C	G	G	G	A	U	A	G	C	U	C	A	G	U	D		Gm	G				A	G	A	G	C	A	G	C	A	G	A	C	U	Gm	A	A	xA*	A	Ψ	C	C	U	C
1460	Euglena grac.chloro.	G	C	U	G	G	G	A	U	A	G	C	U	C	(mb)A	G	U	D		G	Gm	U(D)			A	G	A	G	C	A	C	U	A	G	A	C	U	Gm	A	A	xA*	A	Ψ	C	C	U	U
1461	Euglena grac. cyto.	G	C	C	A	G	G	A	U	A	G	C	U	C	A	G	U	U		G	Gm				A	G	A	G	C	m²G	Ψ	A	G	A	C	U	Gm	A	A	N	A	Ψ	C	U	A	A
1462	Blue green algae	G	C	C	A	G	G	A	U	A	m²⁶C	C	U	C	A	G	D	D		G	G				A	G	A	G	C	m²⁶G	C	C	A	G	A	Cm	U	Gm	A	A	yW	A	Ψ	mC	C	U	C
1470	Yeast	G	U	C	G	C	A	A	U	A	m²⁶C	C	U	C	A	G	D	D		G	Gm				A	G	C	A	Y	A	m²⁶G	A	C	A	C	U	Gm	A	A	yW	A	Ψ	mC	U	G	U
1471	S. pombe	G	C	G	G	A	U	U	U	A	m²⁶C	C	U	C	A	G	D	D		G	G				A	G	A	G	C	m²⁶G	A	C	C	A	Cm	U	Gm	A	A	O₂yW	A	Ψ	C	U	G	U
1480	Wheat,pea,lupin	G	C	G	G	G	A	A	U	A	m²⁶C	C	U	C	A	G	D	D		G	G				A	G	A	G	C	m²⁶G	Ψ	Ψ	A	G	A	Cm	U	Gm	A	A	O₂yW	A	Ψ	C	U	G	A
1490	Mammalian*	G	C	C	G	A	A	A	U	A	m¹A	C	U	C	m¹A	G	D	D		G	G				A	G	A	G	C	m²⁶G	U	U	A	G	A	Cm	U	Gm	A	A	O₂yW	A	Ψ	C	U	A	A

PROLINE

| 1510 | Phage T4 | C | U | C | C | G | U | G | (s⁴)U | A | G | C | U | C | A | G | U | U | | G | G | D | | | A | G | A | G | C | G | C | C | C | U | G | A | Um | U | N* | G | Gm¹⁶ | A | Ψ | C | A | G | G |

1410 B.G.Barrell,F.Sanger(1969) FEBS-Lett. 3, 275-278.
1420 G.Keith,C.Guerrier-Takada,H.Grossjean,G.Dirheimer(1977)FEBS-Lett. 84, 241-244.
1430 H.Arnold,G.Keith(1977) Nucleic Acids Res. 4, 2821-2829.
1440 M.E.Kimball,K.S.Szeto,D.Söll(1974) Nucleic Acids Res. 1, 1721-1732.
1450 P.Guillemaut,G.Keith(1977) FEBS-Lett. 84, 351-356.
1460 S.H.Chang,L.Hecker,M.Siberklang,C.K.Brum,W.E.Barnett,U.L.RajBhandary (1976) Cell 9, 717-724.
1461 S.H.Chang,C.K.Brum,J.J.Schnabel,J.E.Heckman,U.L.RajBhandary,W.E.Barnett (1978) Fed.Proc. 37, 1768-1768.
1462 S.H.Chang,F.K.Lin,L.I.Hecker,J.E.Heckman,U.L.RajBhandary,W.E.Barnett (1978) Cold Spring Harbor Meeting on tRNA,Abstracts,p.45.

1470 U.L.RajBhandary,S.H.Chang(1968) J.Biol.Chem. 243, 598-608.
1471 T.McCutchan,S.Silverman,J.Kohli,D.Söll(1978)Biochemistry 17,1622-1628.
1480 B.S.Dudock,G.Katz(1969) J.Biol.Chem. 244, 3069-3074;
G.A.Everett,J.T.Madison(1976) Biochemistry 15, 1016-1021;
A.J.Rafalski,J.Barciszewski,K.Gulewicz,T.Twardowski,G.Keith(1977) Acta Bioch.Polonica 24, 301-318.
1490 G.Keith,G.Dirheimer(1978) Biochim.Biophys.Acta 517, 133-149;
B.A.Roe,M.P.J.S.Anandaraj,L.S.Y.Chia,E.Randerath,R.C.Gupta,K.Randerath (1975) Biochem.Biophys.Res.Commun. 66, 1097-1105.
1510 J.G.Seidman,B.G.Barrell,W.H.McClain(1975) J.Mol.Biol. 99, 733-760.

	Extra Arm		TΨ Stem	TΨ Loop	TΨ Stem	Aminoacyl Stem	
	44 45 46	47(1-16) 48	49 50 51 52 53 54	55 56 57 58 59 60	61 62 63 64 65	66 67 68 69 70 71 72	73 74 75 76
PHENYLALANINE							
1410	G U m⁷G	X	C C U C G G	T ψ C G A A U	C C G A G	U C C G G G C	A C C A
1420	G U m⁷G	U	C C G C G G	T ψ C G A A U	C C G C G	U C C G A G C C	A C C A
1430	G U m⁷G	U	C C G C C G	T ψ C G A A U	C G G U G	C C G A G C C	A C C A
1440	G U m⁷G	U	C A C C A G	T ψ C A A A U	C U G G U	U C C U G G C	A C C A
1450	G U m⁷G	X	C C A C A G	T ψ C A A A U	C U G U G	U C C U A G C	A C C A
1460	G U m⁷G	X	C C C U G G	T ψ C G m¹A U	C C A G G	A G ψ C G G C	A C C A
1461	A G m⁷G	xU*	m⁵C C G C C G G	T ψ C A A A U	C C G G C	U C C G G G C	A C C A
1462	G U m⁷G	U	C C U G U G	T ψ C G m¹A U	C A C A G	A A U G U A C	A C C A
1470	A G m⁷G	U	A U C C G	T ψ C G A U	C C A C G	U C G G A C	A C C A
1471	U G m⁷G	xU*	C G* C U G	T ψ C G m¹A C	C A G C G	U C A C G G C	A C C A
1480	A G m⁷G	D	C C G C G G	T ψ C G A U	C C G C	U U U G G C	A C C A
1490	A G m⁷G	U (D)	m⁵C C C G G	T(U)*ψ C G m¹A C	C G G G	U U G G A C	A C C A
PROLINE							
1510	A G m⁷G	U	C C A A G G	T ψ C A A A U	C C U U G	U A U G G A G	A C C A

1410/37 xA is ms²i⁶A.
1420/37 xA is ms²i⁶A.
1430/37 xA is ms²i⁶A.
1450/37 xA is ms²i⁶A.
1460/37 xA is ms²i⁶A.
1461/47 xU is probably a derivative of uridine.
1462/39 xU is probably a derivative of uridine.
1471/9 N is an unidentified derivative of guanosine.
1471/10 Is probably m²G.
1471/26 Is probably m²G.
1471/47 xU is probably a derivative of uridine.
1480/49 The Lupinus luteus sequence has mainly adenosine.
1480/65 The Lupinus luteus sequence has mainly uridine.
1490/0 Rabbit liver, calf liver, bovine liver and human placenta.
1490/54 Content of T is different for different species.
1510/34 N is an unidentified derivative of uridine.

			Aminoacyl Stem								D Stem					D Loop								D Stem				Anticodon Stem					Anticodon Loop					Anticodon Stem			
			1	2	3	4	5	6	7	8	9	10	11	12	13	14	15	16	17	18	19	20	20:1	20:2	21	22	23	24	25	26	27	28	29	30	31	32	33	34	35	36	37	38	39	40	41	42	43
SERINE																																													
1610	E.coli	1	G	G	A	A	G	U	G	s⁴U	G	G	C	C	A	G	C	Gm	G	D			A	A	G	G	C	A	C	C	G	G	Cm	U	G⁵U	G	AxA*	A	G	A	C	C	G	G
1620	E.coli	3	G	G	U	G	A	G	G	s⁴U	G	G	C	C	G	A	G	G	D	D			A	A	G	G	C	G	C	U	C	C	s³t*	U	G	C	U	t⁶A	A	G	G	A	G	G
1630	Phage T4		G	G	A	G	G	C	G	s⁴U	G	G	C	A	A	G	A	Gm	G	D			A	A	G	G	C	A	C	C	U	G	Cm	U	G	C	A	xA*	A	A	C	C	G	G
1631	Phage T4 psu⁺am		G	G	A	G	A	A	C	U	U	G	Cα⁴c⁴G	G	D	D			A	A	G	G	C	C	C	U	G	U	N	C	A	i⁶A	A	ψ	C	U	G	G
1640	Yeast	1	G	G	C	A	A	C	U	U	G	G	Cα⁴c⁴G	G	A	G	D			A	A	G	G	C	A	A	A	G	A	U	I	G	A	i⁶A	A	ψ	C	U	U	G
1650	Yeast	2	G	G	C	U	A	C	A	U	G	G	Cα⁴c⁴G	G	A	G	D			A	A	G	G	C	C	A	G	A	N	C	m¹C	G	A	i⁶A	A	ψ	C	U	G	G
1651	Yeast(UCG)*		G	U	A	G	U	C	A	U	G	G	Cα⁴c⁴G	G	A	G	D			A	A	G	G	C	A	ψ	G	G	A	U	U	G	A	i⁶A	A	ψ	C	C	A	U
1660	Rat liver	1	G	U	C	G	A	G	G	U	G	G	Cα⁴c⁴G	G	A	G	D			A	A	G	G	C	m⁵C	A	G	A	G	A	C	m¹C	U	m⁵A	A	ψ	C	C	A	U
1670	Rat liver	3	G	A	C	G	A	G	G	U	G	G	Cα⁴c⁴G	G	A	U	D			A	A	G	G	C	m⁵C	A	G	A	G	A	C	m¹C	U	m⁵A	A	U	C	C	A	U
THREONINE																																											
1710	E.coli		G	C	U	G	A	U	A	U	A	G	C	U	C	A	G	D	D			A	G	A	G	C	A	C	C	C	U	G	Cm	U	G	G	Um⁵U	A	ψ	G	G	U	G
1720	Bacillus subtilis		G	C	C	G	A	U	A	G	A	A	C	U	C	A	G	D	D			A	G	A	G	C	A	C	U	G	A	C	U	U	m⁵U	G	U	A	ψ	C	A	G	U
1730	Phage T4		G	C	U	G	A	U	A	U	A	G	C	U	C	A	G	D	D			A	G	A	G	C	A	C	C	U	G	A	C	U	G	G	U	A	ψ	G	A	G	G
1760	Yeast 1a, 1b		G	C	U	U	C	U	A	U	G	m²G	C	C	A	A	G	U(D)	G				G	A	G	C	A	C	A	C	A	C	U	I	G	U	A	U	C	G	U	G	G
TRYPTOPHAN																																											
1810	E.coli CA244		A	G	G	G	G	C	G	U	A	G	U	U	C	A	A	D	D			A	G	A	G	C	A	C	C	C	G	A	Cm	U	C	C	A	xA*	A	G	A	C	C	G	G
1811	E.coli psu⁺ UGA		A	G	G	G	G	C	G	s⁴U	A	G	U	U	C	A	A	D	D			A	G	A	G	C	A	C	C	U	G	A	Cm	U	C	C	A	xA*	A	G	A	C	C	G	G
1812	psu⁺ 7am		A	G	G	G	G	C	G	s⁴U	A	G	U	U	C	A	A	D	D			A	G	A	G	C	A	C	C	U	G	A	Cm	U	U	A	A	xA*	A	G	A	C	C	G	G
1813	psu⁺ 7oc		A	G	G	G	G	C	G	s⁴U	A	G	U	U	C	A	A	D	D			A	G	A	G	C	A	C	C	U	A	A	Cm	U	U	A	A	xA*	A	G	A	C	C	G	G
1840	Yeast		A	A	G	G	C	C	U	Um⁶G	G	m²G	C	U	C	A	C	D				A	G	C	G	C	ψ	C	U	G	A	Cm	U	C	C	A	A	A	ψm	C	A	G	A
1850	Chicken cells*		G	A	C	C	U	C	G	Um⁶G	U	m²G	C	G	C	A	G	C	D			A	G	C	G	C	A	C	U	C	G	A	C	(m)	U	C	C	A	m⁵C	A	ψm	C	A	G	A
1860	Bovine liver		G	A	C	C	U	C	G	U	G	m²G	C	G	C	A	D(C)					A	G	C	G	C	A	C	U	G	A	Cm	u(m)	C	C	A	m⁵C	A	ψ	C	A	G	A

1610 H.Ishikura,Y.Yamada,S.Nishimura(1971) FEBS-Lett. 16, 68-70;
     Y.Yamada,H.Ishikura(1975) Biochim.Biophys.Acta 402, 285-287.
1620 Y.Yamada,H.Ishikura(1973) FEBS-Lett. 29, 231-234;
     D.Ish-Horowicz,B.F.C.Clark(1973) J.Biol.Chem. 248, 6663-6673.
1630 W.H.McClain,B.G.Barrell,I.G.Seidman(1975) J.Mol.Biol. 99, 717-732.
1631 W.H.McClain,C.Guthrie,B.G.Barrell(1973) J.Mol.Biol. 81, 157-171.
1640 + 1650 H.G.Zachau,D.Dütting,H.Feldmann(1966) Hoppe-Selyer's Z.
     Physiol.Chem. 347, 212-235.
1651 P.W.Piper(1978) J.Mol.Biol. 122, 217-235.
1660 T.Ginsberg,H.Rogg,M.Staehelin(1971) Eur.J.Biochem. 21, 249-257.
1670 H.Rogg,P.Müller,M.Staehelin(1975) Eur.J.Biochem. 53, 115-127.

1710 L.Clarke,J.Carbon(1974) J.Biol.Chem. 249, 6874-6885.
1720 T.Hasegawa,H.Ishikura(1978) Nucleic Acids Res. 5, 537-548.
1730 C.Guthrie,C.A.Scholla,H.Yesian,J.Abelson(1978) Nucleic Acids
     Res. 5, 1833-1844.
1760 J.Weissenbach,I.Kiraly,G.Dirheimer(1977) Biochimie 59, 381-391.
1810 + 1811 D.Hirsh(1971) J.Mol.Biol. 58, 439-458.
1812 + 1813 M.Yaniv,W.R.Folk,P.Berg,L.Soll(1974) J.Mol.Biol. 86, 245-260.
1840 G.Keith,A.Roy,J.P.Ebel,G.Dirheimer(1972) Biochimie 54, 1405-1426.
1850 F.Harada,R.C.Sawyer,J.E.Dahlberg(1975) J.Biol.Chem. 250, 3487-3497.
1860 M.Fournier,J.Labouesse,G.Dirheimer,C.Fix,G.Keith(1978) Biochim.
     Biophys.Acta 521, 198-208.

		Extra Arm																			TΨ Stem					TΨ Loop							TΨ Stem					Aminoacyl Stem										
	44	45	46	47	47	47	47	47	47	47	47	47	47	47	47	47	47	47	47	48	49	50	51	52	53	54	55	56	57	58	59	60	61	62	63	64	65	66	67	68	69	70	71	72	73	74	75	76
					1	2	3	4	5	6	7	8	9	10	11	12	13	14	15	16																												
SERINE																																																
1610	C	G	A	C	C	C	G	A	A	A	G	G	U	U							C	C	A	G	A	G	T	Ψ	C	G	A	A	U	C	U	C	U	G	C	C	U	U	C	C	G	C	C	A
1620	U	A	U	G	U	C	G	G	U	A	A	A	A	G	C	U	G		C	A	U	C	C	A	U	A	G	G	T	Ψ	C	G	A	U	U	C	C	U	G	C	C	A	C	C	G	C	C	A
1630	C	A	G	U	C	G	C	C	G	U	U	C	C	G							C	A	U	A	G	C	T	Ψ	C	C	A	U	C	U	G	U	U	C	C	U	C	C	U	G	C	C	A
1631	C	A	G	U	C	G	C	C	G	C	C	C	G	C							C	A	U	A	G	G	T	Ψ	C	C	A	U	C	U	G	U	A	U	G	C	C	U	C	C	G	C	C	A
1640	Um	G	G	G	C	U	U	U	C	C	C	C	G	G							m⁵C	G	C	A	G	G	T	Ψ	C	A	A	U	C	U	C	U	G	C	A	G	U	U	C	G	G	C	C	A
1650	Um	G	G	G	C	C	U	U	U	C	C	C	C	G							m⁵C	G	C	A	G	G	T	Ψ	C	A	A	U	C	U	G	U	G	A	G	U	U	C	G	A	C	C	A
1651	Um	G	G	G	C	C	U	U	U	C	C	C	G	C							m⁵C	G	C	A	G	G	T	Ψ	C	A	A	U	C	U	G	U	G	A	C	U	A	C	G	A	C	C	A
1660	Um	G	G	U	C	U	U	C	C	C											m⁵C	G	C	U	G	C	T	Ψ	C	G	m¹A	A	U	C	U	G	U	C	C	G	U	C	G	G	C	C	A
1670	Um	G	Ψ	C	C	C	C	A	C	G											m⁵C	G	U	G	G	G	T	Ψ	C	G	m¹A	A	U	C	C	C	C	C	C	G	A	C	G	G	C	C	A
THREONINE																																																
1710	A	G	m⁷G	U																	C	G	G	C	A	G	T	Ψ	C	G	A	A	U	C	U	G	C	C	U	A	U	C	A	G	A	C	C	A
1720	A	G	m⁷G	U																	U	G	G	C	A	G	T	Ψ	C	A	A	U	C	U	G	C	C	U	U	G	C	C	G	A	C	C	A
1730	A	U	m⁷G	U																	C	G	G	C	G	G	T	Ψ	C	G	A	U	C	C	C	G	C	C	A	A	U	C	G	A	C	C	A
1760	A	G	A	D																	(m⁵)C	A(G)	U	C	G	G	T	Ψ	C	A	m¹A	A	C	C	C	G	A U(C)	U	G	G	A	A	G	C	A	C	C	A
TRYPTOPHAN																																																
1810	G	U	m⁷G	U																	U	G	G	G	A	G	T	Ψ	C	G	A	U	C	U	C	C	C	C	C	G	C	C	C	G	C	C	A
1811	G	U	m⁷G	U																	U	G	G	G	A	G	T	Ψ	C	G	A	U	C	U	U	C	C	C	C	C	U	C	C	G	C	C	A
1812	G	U	m⁷G	U																	U	G	G	G	A	G	T	Ψ	C	G	A	U	C	U	C	C	C	C	G	C	U	C	C	G	C	C	A
1813	G	U	m⁷G	U																	U	G	G	C	A	G	T	Ψ	C	G	A	U	C	U	G	C	C	G	C	U	U	C	C	G	C	C	A
1840	G	G	m⁷G	D																	U	G	C	A	G	G	T	Ψ	C	C	m¹A	A	C	C	C	C	U	C	C	G	U	U	U	G	C	C	A
1850	A	G	m⁷G	C																	U	G	C	U	G	U	T	Ψ	C	G	m¹A	A	C	A	C	C	G	G	G	U	U	C	G	A	C	C	A
1860	A	G	(m⁷)G	D(C)																	U	G	C	G	U	G	Ψ	Ψ	C	G(A)	m¹A	A	U	C	A	C	C	G	G	G	U	U	C	G	C	C	A

1610/37 xA is ms²i⁶A.
1620/32 In the position 32 is most probably 2-thiocytidine.
1630/37 xA is ms²i⁶A.
1631/37 xA is ms²i⁶A.
1651/0 A minor species has G-28,C-42 and U(m)-44.

1651/35 Suppressor sup-R11 has U-35.
1730/34 N is an unknown derivative of uridine.
1730/37 N is an unknown derivative of adenosine.
1810/8 The s⁴U-8——C-13 cross link was identified.
1810/37 xA is ms²i⁶A.
1811/37 xA is ms²i⁶A.

1812/37 xA is ms²i⁶A.
1813/37 xA is ms²i⁶A.
1850/0 The sequence was determined on primer RNA for initiation of in vitro Rous-Sarcoma virus DNA synthesis; tRNA-Trp from chicken cells has an identical composition; L.C. Waters,W.-K.Yang(1975)J.Biol.Chem.250, 6627-6629.

TYROSINE

		Aminoacyl Stem							D Stem					D Loop								D Stem				Anticodon Stem				Anticodon Loop							Anticodon Stem			
		1	2	3	4	5	6	7	8	9	10	11	12	13	14	15	16	17	18	19	20	20:1	20:2	21	22	23	24	25	26	27	28	29	30	31	32	33	34	35	36	37	38	39	40	41	42	43
1910	E.coli	G	G	U	G	G	G	G	s⁴U s⁴U		C	C	C	A	G	C			Gm	G	C			A	A	G	G	G	A	G	C	A	G	A	C	U	Q	U	A	A*	A	ψ	C	U	G	C
1911	E.coli psu+ 3am	G	G	U	G	G	G	G	s⁴U s⁴U		C	C	C	A	G	C			Gm	G	C			A	A	G	G	G	A	G	C	A	G	A	C	U	C	U	A	xA*	A	ψ	C	U	A	C
1912	A2 psu+ 3oc	G	G	U	G	G	G	G	s⁴U s⁴U		C	C	C	A	G	C			Gm	G	C			A	A	G	G	G	A	G	C	A	G	A	C	U	U	U	A	xA*	A	ψ	C	C	G	C
1920	B.stearothermophilus	G	G	A	G	G	G	G	U	A	C	U	C	G	A	A			Gm	G	C			A	m²A	C	G	C	A	A	G	C	A	G	A	C	U	G	U	A	xA*	A	ψ	C	U	U	G
1930	Yeast	C	U	C	U	C	G	G	U	A	m²G	C	C	A	A	G	D	D	Gm	G	G			A	A	G	G	C		m²₂G	C	A	A	G	A	C	U	Q	U	A	xA*	A	ψ	C	U	G	C
1931	Yeast sup-Sam	C	U	C	U	C	G	G	U	A	m²G	C	C	A	G	G	D	D	Gm	G	G			A	A	G	G	C		m²₂G	C	A	A	G	A	C	U	C	U	A	i⁶A	A	ψ	C	C	G	C
1940	T.utilis	C	U	C	U	C	G	G	Um⁶G		m²G	C	C	A	G	D	D		Gm	G	G			A	A	G	G	C		m²₂G	C	A	A	G	A	C	U	G	ψ	A	i⁶A	A	ψ	C	U	G	C

VALINE

		1	2	3	4	5	6	7	8	9	10	11	12	13	14	15	16	17	18	19	20	20:1	20:2	21	22	23	24	25	26	27	28	29	30	31	32	33	34	35	36	37	38	39	40	41	42	43
2010	E.coli K12,B 1	G	G	U	G	G	A	U	s⁴U	A	G	C	U	C	A	G	C	D	G	G	D			A	G	A	G	C	A	C	C	U	C	C	U⁵U	U	A	A	C	m⁵A	A	G	G	A	G	G
2020	E.coli 2a	G	C	G	U	U	C	G	s⁴U	A	G	C	U	C	A	G	D	D	G	G	D			A	G	A	G	C	A	A	C	C	A	C	C	U	G	A	C	A	U	G	G	U	G	G
2021	E.coli 2b	G	C	G	U	U	C	A	s⁴U	A	G	C	U	C	A	G	D	D	G	G	D			A	G	A	G	C	A	A	C	C	A	C	C	U	U	A	C	A	U	G	G	U	G	G
2030	B.stearothermophilus	G	A	U	U	C	C	G	U	A	G	C	U	C	A	G	D		G	G	D			A	U	G	A	G	C	A	C	U	C	G	G	U	U	A	C	m⁶A	A	G	C	A	G	A
2040	Yeast 1	G	G	U	U	U	C	A	Um⁵C		G	U	C	U	G	G	D	D	G	G	D			A	U	A	G	G		m²₂G	C	U	C	G	U	U	U	A	C	A	C	A	C	G	A	G	A
2050	Yeast 2a	G	U	U	U	C	C	A	U	A	m⁵G	U	C	U	C	C	D	D	G	G	D			A	U	C	A	G		m²₂G	C	ψ	C	G	G	C	ψ	I	A	C	A	C	C	G	A	G	A
2051	Yeast 2b	G	U	U	U	C	C	A	U	A	m⁵G	U	C	U	C	C	D	D	G	G	D	C		A	U	C	A	G		m²₂G	C	ψ	C	G	G	C	ψ	N*	A	C	A	C	C	G	A	G	A
2060	T.utilis	G	G	U	U	C	C	G	U	A	G	U	G	U	A	G	D		G	G	D			A	U	C	A	C		m²₂G	C	ψ	U	C	G	U	U	I	A	C	A	C	C	G	A	A	A
2070	Mammalian*	G	U	U	U	C	C	G	U	A	G	U	G	U	A	G	D		G	G	D	D		A	U	C	A	C		m²₂G	C	Cm	U	C	Cm	U	I	A	C	A	m⁵C	A	G	C	G	A	A
2071	Human placenta 1b	G	U	U	U	C	C	G	U	A	G	U	G	U	A	G	D		G	G	D			A	U	C	A	C		m²₂G	C	U	C	G	C	U	U	I	A	C	A	A	C	G	C	G	A

1910 + 1911 H.M.Goodman,J.Abelson,A.Landy,S.Brenner,J.D.Smith(1968) Nature 217, 1019-1024.
1912 S.Altman,S.Brenner,J.D.Smith(1971) J.Mol.Biol. 56, 195-197.
1920 R.S.Brown,J.R.Rubin,D.Rhodes,H.Guilley,A.Simoncsits,G.G.Brownlee (1978) Nucleic Acids Res. 5, 23-36.
1930 J.T.Madison,H.-K.Kung(1967) J.Biol.Chem. 242, 1324-1330.
1931 P.W.Piper,M.Wasserstein,F.Engbaek,K.Kaltoft,J.E.Celis,J.Zeuthen, S.Liebman,F.Sherman(1976) Nature 262, 757-761.
1940 S.Hashimoto,S.Takemura,M.Miyazaki(1972) J.Biochem. 72, 123-134.
2010 M.Yaniv,B.G.Barrel(1969) Nature 222, 278-279;
F.Kimura,F.Harada,S.Nishimura(1971) Biochemistry 10, 3277-3283.
2020 + 2021 M.Yaniv,B.G.Barrel(1971) Nature New Biol. 233, 113-114.
2030 C.Takada-Guerrier,H.Grosjean,G.Dirheimer,G.Keith (1976) FEBS-Lett. 62, 1-3.
2040 J.Bonnet,J.P.Ebel,G.Dirheimer,L.P.Shershneva,A.I.Krutilina, T.V.Venkstern,A.A.Bayev(1974) Biochimie 56, 1211-1213.
2050 V.D.Axelrod,V.M.Kryukov,S.N.Isaenko,A.A.Bayev(1974) FEBS-Lett. 45, 333-336.
2051 V.G.Gorbulev,V.D.Axelrod,A.A.Bayev(1977) Nucleic Acids Res. 4, 3239-3258.
2060 T.Mizutani,M.Miyazaki,S.Takemura(1968) J.Biochem. 64, 839-848.
2070 P.W.Piper(1975) Eur.J.Biochem. 51, 295-304;
P.Jank,N.Shinda-Okada,S.Nishimura,H.J.Gross(1977) Nucleic Acids Res. 4, 1999-2008.
2071 E.Y.Chen,B.A.Roe(1977) Biochem.Biophys.Res.Commun. 78, 631-640.

	Extra Arm													TΨStem					TΨLoop						TΨStem					Aminoacyl Stem										
	44	45	46	47	47 1	47 2	47 3	47 4	47 5	47 6	47 7	47 8	47 9	47 10	47 11	47 12	47 13	47 14	47 15	47 16	48	49	50	51	52	53	54	55	56	57	58	59	60	61	62	63	64	65	66	67	68	69	70	71	72	73	74	75	76

TYROSINE

	44	45	46	47	1	2	3	4	5	6	7	8	9	10	11	12	13	14	15	16	48	49	50	51	52	53	54	55	56	57	58	59	61	62	63	64	65	66	67	68	69	70	71	72	73	74	75	76
1910	C	G	U	C	A	U(C)	C(A)	G	A	C	U	U									C	C	G	A	A	G	T	Ψ	C	G	A	U	C	C	U	U	C	C	C	C	A	C	C	A	C	C	A
1911	C	G	U	C	A	U	C	G	A	C	U	U									C	C	G	A	A	G	T	Ψ	C	G	A	U	C	C	U	U	C	C	C	C	A	C	G	C	A	C	C	A
1912	C	G	U	C	A	U	C	G	A	C	U	U									C	C	G	A	A	G	T	Ψ	C	G	A	U	C	G	U	C	C	C	C	C	A	C	C	A	C	C	A
1920	U	C	C	C	U	U	U	G	G	G	U	U									C	C	G	G	C	C	T	Ψ	C	G	A	U	C	G	C	C	C	C	C	U	C	G	C	A	C	C	A
1930	A	G	A	D																	m⁵C	C	G	G	G	C	T	Ψ	C	Gm⁵A	A	U	C	G	C	C	C	C	C	G	G	A	G	A	C	C	A
1931	A	G	A	D																	m⁵C	C	G	G	C	C	T	Ψ	C	Gm⁵A	A	U	C	G	C	C	C	C	C	G	G	A	G	A	C	C	A
1940	A	C	A	D																	m⁵C	C	G	G	C	G	T	Ψ	C	Gm¹A	A	J	C	G	C	C	C	C	G	A	G	A	G	A	C	C	A

VALINE

2010	G	Gm⁷G	U																		C	C	G	G	C	G	T	Ψ	C	G	A	U	C	C	C	G	U	C	A	U	C	A	C	C	A	C	C	A
2020	G	Gm⁷G	X																		C	C	G	U	U	G	T	Ψ	C	G	A	U	C	C	C	A	G	C	G	G	A	C	G	C	A	C	C	A
2021	G	Gm⁷G	X																		C	C	G	C	U	G	T	Ψ	C	G	A	G	C	U	C	C	A	U	G	A	A	U	C	A	C	C	A
2030	A	Gm⁷G	U																		C	C	G	m⁵C	C	A	T	Ψ	C	Gm¹A	U	C	C	U	C	A	G	C	G	A	A	A	U	A	C	C	A
2040	A	Cm⁷G	D																		C	m⁵C	m⁵C	C	A	G	T	Ψ	C	Gm¹A	A	C	C	U	U	G	G	C	G	A	A	A	C	A	C	C	A
2050	A	G	A	D																	m⁵C	m⁵C	C	C	A	G	T	Ψ	C	Gm¹A	A	U	C	U	U	G	G	U	U	G	A	A	U	A	C	C	A
2051	A	Gm⁷G	D																		m⁵C	m⁵C	C	C	A	G	T	Ψ	C	Gm¹A	A	C	C	U	U	G	G	C	G	A	A	A	U	A	C	C	A
2060	A	C																			m⁵C	C	C	C	G	G	T	Ψ	C	Gm¹A	A*	C	C	A	G	G	U	U	G	A	A	A	C	A	C	C	A
2070	A	Gm⁷G	D																		m⁵C	m⁵C	C	C	G	G	U*	Ψ	C	Gm¹A	A	A	C	C	G	G	G	C	G	G	A	A	U	A	C	C	A
2071	A	Gm⁷G	D																		m⁵C	C	C	C	G	G	U	Ψ	C	Gm¹A	A	C	C	C	G	G	G	C	G	G	A	A	A	C	A	C	C	A

1910/37 xA is ms²i⁶A
1911/37 xA is ms²i⁶A
1912/34 Uridine may be modified; S.Altman(1976) Nucleic Acids Res. 3, 441-448.
1912/37 xA is ms²i⁶A.
1920/37 xA is ms²i⁶A.

2050/34 N is an unknown derivative of uridine.
2070/0   Mouse myeloma,rabbit liver and human placenta 1a, in the latter case C-32 and C-38 are unmodified.
2070/54 The U-54 A-60 base pair was detected by P.Jank, D.Riesner,H.J.Gross (1977) Nucleic Acids Res. 4, 2009-2020.
2070/60

B. Collection of Mutant tRNA Sequences

Julio E. Celis
Biostructural Chemistry, Department of Chemistry
Aarhus University
8000 Aarhus C, Denmark

As a supplement to the compilation of tRNA sequences, this section presents the nucleotide sequence of published mutant tRNAs, together with a brief description of properties of these mutants.

Mutant tRNAs have been derived by mutation of suppressor tRNA genes present in *Escherichia coli*, $\phi 80$, or bacteriophage T4. In all cases, mutants having different levels of suppressor activity have been isolated. These mutants present single-base substitutions and in many cases (those mutants having no suppressor activity) it has been possible to isolate double mutants among the revertants having suppressor activity. Double mutants have also been constructed by genetic recombination.

To designate a particular nucleotide substitution one refers to the identity of the new nucleotide followed by a number indicating its position in the tRNA sequence starting from the 5' end. For example, the mutant A31 in suppressor tRNA^{Tyr} refers to a tRNA that contains an A at position 31 instead of the usual base (G). To conform to the rules for numbering tRNAs used by Gauss et al. (see Appendix IA), the new numbering system for these mutants has been added. Thus, mutant A31 will correspond to A30 according to the new numbering system.

In the case of the mutant T4 tRNAs, some of these mutants have been sequenced in the precursor, since they do not produce any detectable amount of mature tRNA. These mutants have been listed in a separate table and that part of the tRNA precursor carrying the mutation is indicated. The position of the nucleotide substitution in this case has been numbered from the 5' terminal of the altered tRNA in the precursor sequence.

Table 3 Mutant suppressor tRNAs^{Tyr}

Mutant[a]	Base change position[b]	from	to	Interesting properties[c]	References[d]
Mutant Suppressor tRNAs^{Tyr} (su$^+$3, 1191)					
A1	1	G	A	mischarger; inserts glutamine in vivo; strong suppressor	1, 2
A1-U81	1	G	A	mischarger; inserts tyrosine and glutamine in vivo; strong suppressor	2, 3
	72	C	U		
A1-G82	1	G	A	mischarger; inserts glutamine in vivo; strong suppressor	4
	73	A	G		
A2	2	G	A	temperature sensitive; mischarger; inserts tyrosine and glutamine in vivo; weak suppressor.	5, 2
A2-U80	2	G	A	temperature resistant; indistinguishable from original suppressor; strong suppressor	5, 2
	71	C	U		
A15	15	G	A	defective in a step subsequent to its binding to the ribosome; accumulates tRNA precursor; weak suppressor	6
A15-D19	15	G	A	strong suppressor	5
	20	C	D		
A15-D20	15	G	A	strong suppressor	5
	20:1	C	D		
A17	18	Gm	A	accumulates tRNA precursor; weak suppressor	6
A25	24	G	A	accumulates tRNA precursor; no detectable suppressor activity	5

Mutation	Position			Description	Ref.
A25-U11	24 / 11	G / C	A / U	indistinguishable from original suppressor; strong suppressor	5
A25-U19	24 / 20	G / C	A / U	strong suppressor	5
A31	30	G	A	altered K_m of aminoacylation; accumulates precursor; weak suppressor	6, 7
A31-U16	30 / 16	G / C	A / U	strong suppressor	7
A31-U41	30 / 40	G / C	A / U	strong suppressor	7
A31-U45	30 / 44	G / C	A / U	strong suppressor	7
U31	30	G	U	accumulates tRNA precursor; weak suppressor	7
U31-U16	30 / 16	G / C	U / U	strong suppressor	7
U31-A41	30 / 40	G / C	U / A	strong suppressor	7
U31-U45	30 / 44	G / C	U / U	strong suppressor	7
A46	45	G	A	altered K_m of aminoacylation; accumulates tRNA precursor; weak suppressor	8
A46-U54	45 / 47:6	G / C	A / U	strong suppressor	8

Mutant Suppressor tRNAs^{Tyr} (su$^+$3, 1192)

Mutant[a]	Base change position[b]	from	to	Interesting properties[c]	References[d]
A62	53	G	A	accumulates tRNA precursor; no detectable suppressor activity	9
+C(73–78)	64–69	C insertion		cannot be aminoacylated under normal conditions; no detectable suppressor activity	10
U80	71	C	U	mischarger; inserts tyrosine and a neutral amino acid in vivo; strong suppressor	5, 2
A81	72	C	A	mischarger; inserts tyrosine and a neutral amino acid in vivo; strong suppressor	5, 2
U81	72	C	U	mischarger; inserts tyrosine and glutamine in vivo; strong suppressor	11, 12, 2
G82	73	A	G	mischarger; inserts glutamine in vivo	11, 12, 13, 2

[a]Mutant denomination in the original literature.
[b]Numbering according to tRNA compilation.
[c]Strong suppressors are defined as those having >50% of the wild-type suppressor activity. Weak suppressors present <20% of the wild-type suppressor activity.
[d]References: [1]Smith, J. D. and J. E. Celis. 1973. *Nature* 243:66; [2]Ghysen, A. and J. E. Celis. 1974. *J. Mol. Biol.* 83:333; [3]Celis, J. E., et al. 1977. *Nucleic Acids Res.* 4:2799; [4]Inokuchi, H., et al. 1974. *J. Mol. Biol.* 85:187; [5]Smith, J. D., et al. 1970. *J. Mol. Biol.* 54:1; [6]Abelson, J. N., et al. 1969. *FEBS Lett.* 3:1; [7]Anderson, K. W. and J. D. Smith. 1972. *J. Mol. Biol.* 69:349; [8]Hooper, M. L. 1972. Ph.D. thesis, University of Cambridge; [9]J. D. Smith (unpubl.); [10]D. Riddle et al. (in prep.); [11]Hooper, M. L., et al. 1972. *FEBS Lett.* 22:149; [12]Celis, J. E., et al. 1973. *Nat. New Biol.* 224:261; and [13]Shimura, Y., et al. 1972. *FEBS Lett.* 22:144.

Table 4 Mutant T4 tRNA^{Gln} (psu^+2, 0531) and tRNA^{Ser} (psu^+1, 1631)

tRNA	Mutant[a]	Base change position[b]	Base change from	Base change to	Interesting properties	References[c]
tRNA^{Gln}	U11	11	C	U	synthesizes reduced amount of tRNA^{Gln} (49% of psu^+2); no detectable suppressor activity; it accumulates small amount of tRNA^{Gln}-tRNA^{Leu} precursor	1
tRNA^{Gln}	A40	40	G	A	synthesizes reduced amount of tRNA^{Gln} (41% of psu^+2); no detectable suppressor activity; it accumulates small amount of tRNA^{Gln}-tRNA^{Leu} precursor	1
tRNA^{Gln}	A62	62	C	U	synthesizes reduced amount of tRNA^{Gln} (17% of psu^+2); no detectable suppressor activity; it accumulates small amount of tRNA^{Gln}-tRNA^{Leu} precursor	1
tRNA^{Ser}	U26	25	C	U	synthesizes normal amount of tRNA^{Ser}; no suppressor activity	2
tRNA^{Ser}	G27	26	A	G	synthesizes normal amount of tRNA^{Ser}; no suppressor activity	2
tRNA^{Ser}	C48	47	U	C	synthesizes reduced amount of tRNA^{Ser} (25% of psu^+1); no suppressor activity	2

[a]Mutant denomination in the original literature.
[b]Numbering according to tRNA compilation.
[c]References: [1]Seidman, J. D., M. M. Comer, and W. H. McClain. 1974. *J. Mol. Biol.* **90**:677; [2]McClain, W. H. 1977. *Accts. Chem. Res.* **10**:418.

Table 5 Mutant T4 tRNA^{Ser}, tRNA^{Gln}, and tRNA^{Leu} sequenced at the precursor level

tRNA precursor[a]	Mutant[b]	Base change position[c]	from	to	Interesting properties	References[d]
Pro-<u>Ser</u>	U1	1	G	U	accumulates tRNA^{Pro}-tRNA^{Ser} precursor lacking many modified nucleotides; blocks RNase P reaction; precursor RNA terminates in CCA$_{OH}$	1
Pro-<u>Ser</u>	-3	3	A	–[e]	same as U1 (G → U)	1
Pro-<u>Ser</u>	C8	8	U	C	same as U1 (G → U)	1
Pro-<u>Ser</u>	G12	12	A	G	accumulates tRNA^{Pro}-tRNA^{Ser} precursor lacking many modified nucleotides; defective in the first step of the biosynthetic pathway, the removal of UAA$_{OH}$ residues at the serine terminal	1
Pro-<u>Ser</u>	U17	18	G	U	same as G12 (A → G)	1
Pro-<u>Ser</u>	A29	28	C	A	same as G12 (A → G)	1
Pro-<u>Ser</u>	A30	29	G	A	same as G12 (A → G)	1
Pro-<u>Ser</u>	U58	47:10	G	U	same as G12 (A → G)	1
Pro-<u>Ser</u>	A67	53	G	A	same as G12 (A → G)	2
Pro-<u>Ser</u>	C68	54	T	C	same as G12 (A → G)	1

Pro-Ser	U70	56	C	U	same as G12 (A→G)	2
Pro-Ser	U75	61	C	U	same as G12 (A→G)	2
Pro-Ser	C77	63	U	C	same as G12 (A→G)	1
<u>Gln-Leu</u>	U11	11	C	U	accumulates tRNA^{Gln}-tRNA^{Leu} precursor lacking many modified nucleotides; partial reduction of RNase P and CCA repair at glutamine moiety	3
<u>Gln-Leu</u>	A40	40	G	A	same as U11 (C→U)	3
<u>Gln-Leu</u>	U62	62	C	U	same as U11 (C→U)	3
<u>Gln-Leu</u>	U72	61	C	U	accumulates tRNA^{Gln}-tRNA^{Leu} precursor lacking many nucleotide modifications; RNase P cleavage reduced at both moieties	3

[a] In all cases nucleotide substitutions have been localized in the precursor. The underlined tRNA contains the mutation. The tRNA^{Ser} in the tRNA^{Pro}-tRNA^{Ser} precursor has the anticodon N$_2$UA (psu$^+$1, 1631). The tRNA^{Gln} in the tRNA^{Gln}-tRNA^{Leu} precursor has the anticodon NUA (psu$^+$2, 0531) and the tRNA^{Leu} has the normal anticodon NAA (su$^-$, 1030).
[b] Mutant denomination in the original literature.
[c] Numbering according to tRNA compilation.
[d] References: [1]McClain, W. H. 1977. Accts. Chem. Res. 10:418; [2]McClain, W. H., B. G. Barrell, and J. G. Seidmann. 1975. J. Mol. Biol. 99:717; [3]McClain, W. H. and J. G. Seidmann. 1975. Nature 257:106.
[e] Deletion.

APPENDIX II
Structures of Modified Nucleosides Found in tRNA

Susumu Nishimura
Biology Division
National Cancer Center Research Institute
Chuo-ku, Tokyo 104, Japan

The structures of modified nucleosides of tRNA found to date are displayed.

- *a* Pseudouridine (ψ)
- *b* Dihydrouridine (D)
- *c* Ribothymidine (T)
- *d* 5-Methoxyuridine (mo^{5}U)
- *e* Uridin-5-oxyacetic acid (V)
- *f* Uridin-5-oxyacetic acid methyl ester (mV)[a]
- *g* 5-(Methoxycarbonylmethyl)uridine (mcm^{5}U)
- *h* 5-Carboxymethylaminomethyluridine (cmnm^{5}U)[b]
- *i* 4-Thiouridine (s^{4}U)
- *j* 5-Methyl-2-thiouridine (m^{5}s^{2}U)
- *k* 5-(Methoxycarbonylmethyl)-2-thiouridine (mcm^{5}s^{2}U)
- *l* 5-Methylaminomethyl-2-thiouridine (mnm^{5}s^{2}U)
- *m* 5-Carboxymethylaminomethyl-2-thiouridine (cmnm^{5}s^{2}U)[c]
- *n* 3-(3-Amino-3-carboxypropyl)uridine (acp^{3}U)
- *o* 3-Methylcytidine (m^{3}C)
- *p* 5-Methylcytidine (m^{5}C)
- *q* N^4-Acetylcytidine (ac^{4}C)
- *r* 2-Thiocytidine (s^{2}C)
- *s* 1-Methylinosine (m^{1}I)
- *t* Inosine (I)
- *u* N^6-Methyladenosine (m^{6}A)
- *v* 1-Methyladenosine (m^{1}A)
- *w* N^6-Isopentenyladenosine (i^{6}A)
- *x* 2-Methylthio-N^6-isopentenyladenosine (ms^{2}i^{6}A)
- *y* N-[(9-β-D-Ribofuranosylpurin-6-yl)carbamoyl]threonine (t^{6}A)
- *z* N-[(9-β-D-Ribofuranosylpurin-6-yl)N-methylcarbamoyl]threonine (mt^{6}A)
- *aa* N-[N-[(9-β-D-Ribofuranosylpurin-6-yl)carbamoyl]threonyl]2-amido-2-hydroxymethylpropane-1,3-diol[d]
- *bb* 1-Methylguanosine (m^{1}G)
- *cc* N^2-Methylguanosine (m^{2}G)
- *dd* N^2,N^2-Dimethylguanosine (m$_2^2$G)
- *ee* 7-Methylguanosine (m^{7}G)
- *ff* Base "Y" (yW)
- *gg* Base "peroxy Y" (oyW)
- *hh* Base "Yt" (W)
- *ii* R = H Q (queuosine or Quo),[e] R = β-D-mannosyl manQ,[f] R = β-D-galactosyl galQ[f]

References to characterization of the modified nucleosides are cited in: S. Nishimura. 1972. *Prog. Nucleic Acid Res. Mol. Biol.* **12**:49; and D. B. Dunn and R. H. Hall. 1975. In *Handbook of biochemistry and molecular biology*, 3rd ed. (ed. G. D. Fasman), *Nucleic Acids*, vol. 1, p. 216. CRC Press, Cleveland, Ohio. [a]Lesiewicz, J. and B. Dudock. 1977. *Fed. Proc.* **36**:705; [b]Murao, K. and H. Ishikura. 1978. *Nucleic Acids Res.* (special publication) **5**:s333; [c]H. Ishikura, pers. comm.; [d]Kasai, H., K. Murao, S. Nishimura, J. G. Liehr, P. F. Crain, and J. A. McCloskey. 1976. *Eur. J. Biochem.* **69**:435; [e]Kasai, H., Z. Ohashi, F. Harada, S. Nishimura, N. J. Oppenheimer, P. F. Crain, J. G Liehr, D. L. von Minden, and J. A. McCloskey. 1975. *Biochemistry* **14**:4198; [f]Kasai, H., K. Nakanishi, R. D. Macfarlane, D. F. Torgerson, Z. Ohashi, J. A. McCloskey, H. J. Gross, and S. Nishimura. 1976. *J. Am. Chem. Soc.* **98**:5044.

548

549

APPENDIX III
Chromatographic Mobilities of Modified Nucleotides

Susumu Nishimura
Biology Division
National Cancer Center Research Institute
Chuo-ku, Tokyo 104, Japan

These figures illustrate chromatographic properties of modified nucleotides found in tRNA. The information has been compiled from the most widely used separation systems. (See Appendix IA and M. Sprinzl, F. Grüter, and D. H. Gauss. 1979. *Nucleic Acids Res.* **6:**r1.)

Two-dimensional thin-layer chromatography was carried out on Avicel-SF cellulose thin-layer plates at 20°C. Solvent systems used were: first dimension, isobutyric acid:0.5 M ammonium hydroxide, 5:3 (v/v); second dimension, isopropanol:concentrated HCl:water, 70:15:15 (v/v/v). The relative positions of the modified nucleotides vary slightly depending on the time of development and the solvents used (freshly prepared or after repeated use). The corresponding nucleosides move faster in the first dimension and slower in the second dimension.

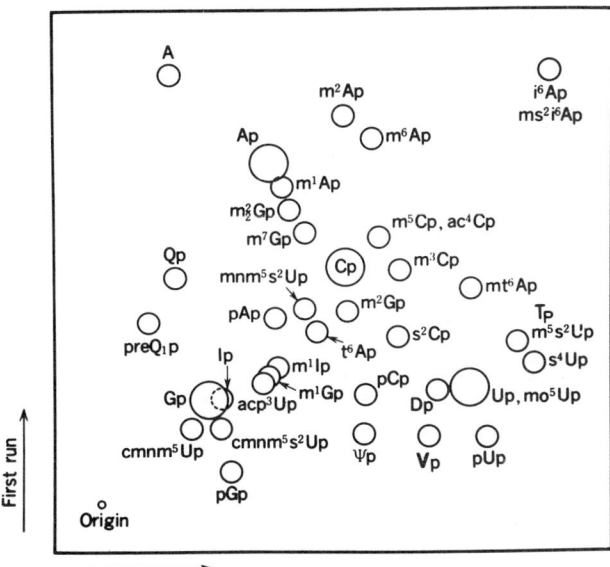

Figure 1 Positions of modified nucleoside 3′-phosphates. Nucleoside 5′-monophosphates move slightly slower than corresponding nucleoside 3′-monophosphates in both dimensions.

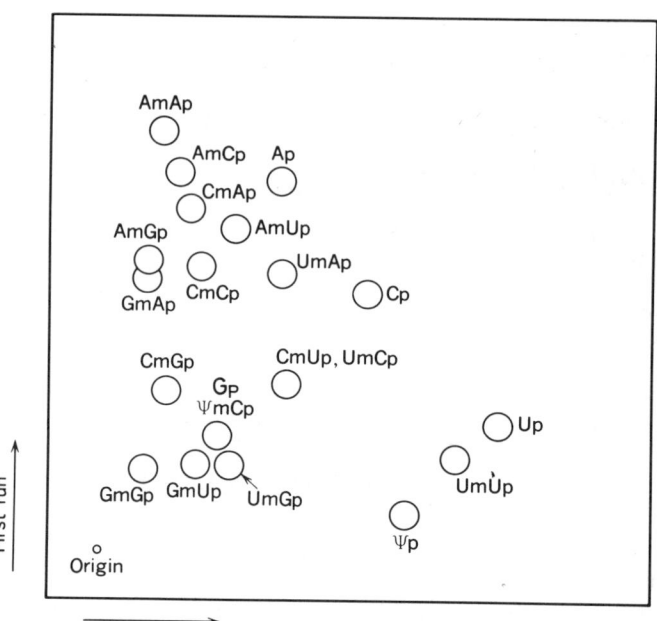

Figure 2 Positions of modified 2'-O-methylated dinucleotides with 3'-phosphate (Hashimoto, S., M. Sakai, and M. Muramatsu. 1975. *Biochemistry* **14:**1956).

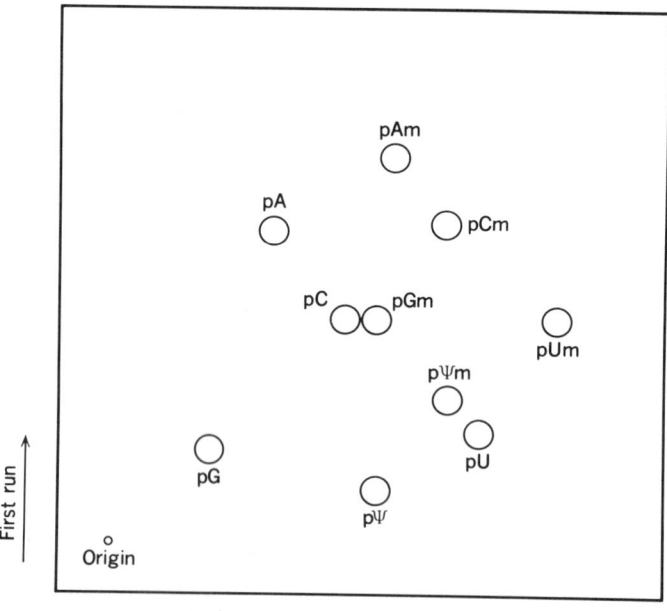

Figure 3 Positions of modified 2'-O-methyl nucleoside 5'-monophosphates (Harada, F., R. C. Sawyer, and J. E. Dahlberg. 1975. *J. Biol. Chem.* **250:**3487).

APPENDIX IV
Characteristics
of Aminoacyl-tRNA Synthetases

Dieter Söll
Department of Molecular Biophysics and Biochemistry
Yale University
New Haven, Connecticut 06520

Paul R. Schimmel
Biology Department
Massachusetts Institute of Technology
Cambridge, Massachusetts 02139

Table 1 gives molecular weights and subunit compositions, and information on amino acid compositions and binding sites, of aminoacyl-tRNA synthetases. Only enzymes that have been highly purified are included in the table. Some of the molecular weights may be adjusted upward in the future because proteolysis during isolation can occur. The table shows that synthetases have several subunit types. These subunit types (α, α_2, $\alpha_2\beta_2$, α_4) represent the predominant aggregation states of the enzymes as isolated in vitro and it is conceivable that, in some cases, further studies will suggest alternate aggregation states (with full activity) can also exist. (Reprinted, with permission, from Schimmel, P. R. and D. Söll. 1979. *Annu. Rev. Biochem.* **48**:601.)

Table 2 summarizes current information on aminoacyl-tRNA synthetase mutants. (Reprinted, with permission, from Schimmel, P. R. and D. Söll. 1979. *Annu. Rev. Biochem.* **48**:601.)

Table 1 Molecular weight, quaternary structure, and substrate binding sites of highly purified aminoacyl-tRNA synthetases

Synthetase	Source	Structural parameters				Binding sites	
		m.w.	subunits	type	AA comp.[a]	type	no.[b]
Alanyl-	yeast	128,000 (1)	none (1)	α	unknown	amino acid	1 (2)
Arginyl-	Bacillus stearothermophilus	78,000 (2)	none (2)	α	unknown	ATP	1 (2)
						aminoacyl adenylate	none (2)
Aspartyl-	Escherichia coli	63,000 (3) 74,000 (4, 5)	none (3, 4)	α	known (5)		
Cysteinyl-	Neurospora crassa	85,000 (6)	none (6)	α	unknown		
	yeast	106,000 (7)	none (7)	α	unknown		
	yeast	160,000 (8)			unknown		
	rat liver	240,000 (9)	2 (9)	α_2	unknown	aminoacyl adenylate	2 (9)
Glutaminyl-	E. coli	69,000 (10)	none (10)	α	known (10)		
Glutamyl-	E. coli	102,000 (11)	2: $\alpha = 56,000$; $\beta = 46,000$ (11)	$\alpha\beta$	known (11)	tRNA	1 (12)
Glycyl-	E. coli	59,000 (13)	none (13)	α			
		227,000 (14, 15)	4: $\alpha = 33,000$; $\beta = 80,000$ (14, 15)	$\alpha_2\beta_2$	known (15)	glycyl-AMP	2 (15)
Histidyl-	E. coli	84,000 (16)	2 (16)	α_2	known (16)	amino acid	2 (16)
						ATP	2 (16)
	Salmonella typhimurium	80,000 (17)	2 (17)	α_2	known (17)		
Isoleucyl-	B. stearothermophilus	115,000 (18)			unknown	aminoacyl adenylate	1 (18)
						tRNA	1 (18)
	E. coli	112,000 (19, 20)	none (19)	α	known (20)	amino acid	1 (21)
						ATP	1 (22)
						aminoacyl adenylate	1 (23)
	yeast	124,000 (1)	none (1)	α	unknown	tRNA	1 (24, 25)

Enzyme	Source	MW	Subunits	Structure	Intermediate	Substrate	n (ref)
Leucyl-	B. stearothermophilus	110,000 (26)	none (26)	α	known (26)		
	E. coli	105,000 (27, 28)	none (27, 28)	α	known (27)	ATP	1 (28)
						aminoacyl adenylate	1 (28)
						tRNA	1 (28)
Lysyl-	Saccharomyces cerevisiae	120,000 (92)	2 (92)	α₂	unknown		
	Candida utilis	128,000 (93)	none (93)	α	unknown		
	E. coli	104,000 (29)	2 (29)	α₂	known (29)	amino acid	2 (29)
						ATP	2 (29)
	yeast	138,000 (29, 30)	2 (29, 30)	α₂	known (29)	amino acid	2 (29)
						ATP	2 (29)
						aminoacyl adenylate	1 (31)
						tRNA	1 (31)
	rabbit reticulocytes	122,000 (32)	2 (32)	α₂	unknown		
Methionyl-	B. stearothermophilus	135,000 (26)	2 (26)	α₂	known (26)	aminoacyl adenylate	2 (32)
	E. coli	170,000 (33, 34)	2 (33)	α₂	known (35)	amino acid	2 (36)
						ATP	4 (36)
						aminoacyl adenylate	2 (32)
						methionyl-AMP	2 (36)
						tRNA	2 (37)
Phenylalanyl-	wheat germ A	105,000 (38)	none (38)	α	known (38)		
	wheat germ B	70,000 (38)	none (38)	α	known (38)		
	Lupinus luteus seeds	170,000 (39)	2 (39)	α₂	unknown		
	rat liver	287,000 (40)	4: α = 69,000; β = 75,000 (40)		unknown		
	Drosophila melanogaster	180,000 (41)			unknown		
	E. coli	267,000 (42, 43)	4: α = 39,000; β = 94,000 (42, 43)	α₂β₂	known (44)	aminoacyl adenylate	2 (45)
						tRNA	2 (45)

		Structural parameters				Binding sites	
Synthetase	Source	m.w.	subunits	type	AA comp.[a]	type	no.[b]
Phenylalanyl-	yeast	220,000 (46)	4: α = 50,000; β = 60,000 (46)	$\alpha_2\beta_2$	known (46)	amino acid aminoacyl adenylate tRNA	2 (47) 2 (48) 2 (47)
		262,000 (49)	4: α = 61,000; β = 70,000 (49)	$\alpha_2\beta_2$			
		286,000 (50)	4: α = 66,000; β = 77,000 (50)	$\alpha_2\beta_2$			
Prolyl-	E. coli	94,000 (51)	2 (51)	α_2	unknown		
	hen liver	120,000 (52)	2 (52)	α_2	unknown		
Seryl-	E. coli B	103,000 (53)	2 (53)	α_2	unknown	ATP	2 (54)
	E. coli K12	95,000 (55)	2 (55)	α_2	known (55)	aminoacyl adenylate amino acid ATP tRNA	2 (54) 2 (56) 2 (56) 1 (57), 2 (56)
	yeast	95,000 (58)	2 (58)	α_2	unknown	tRNA	1 (58, 59)
		120,000 (60)	2 (60)	α_2	unknown	tRNA amino acid ATP	2 (61–64) 2 (65) 2 (65)
Threonyl-	L. luteus seeds	110,000 (66)	2 (66)	α_2	unknown		
	E. coli	152,000 (67)	none (67)	α_2	unknown		
Tryptophanyl-	human placenta	116,000 (68)	2 (68)	α_2	unknown		
	bovine pancreas	108,000 (69)	2 (69)	α_2	known (70)	amino acid aminoacyl adenylate tRNA	2 (69) 2 (71) 2 (72)

Enzyme	Source	MW	Subunits	Structure	Mechanism	Substrate	
	buffalo brain	155,000 (73)	3 (73)	α_3	unknown		
	B. stearothermophilus	70,000 (26)	2 (26)	α_2	known (74)	aminoacyl adenylate	2 (32)
	E. coli	74,000 (75)	2 (75)	α_2	known (76)	amino acid	2 (77)
						tryptophanyl-ATP	2 (76)
						tRNA	2 (77)
Tryptophanyl-	yeast	110,000 (78)	2 (78)	α_4	unknown		
Tyrosyl-	L. luteus seeds	200,000 (66)	4 (66)		unknown		
	B. stearothermophilus	88,000 (26)	2 (26)	α_2	known (26)	amino acid	2 (79)
						ATP	2 (79)
						aminoacyl adenylate	2 (79)
	B. subtilis	88,000 (80)			known (80)		
	E. coli	95,000 (80, 81)	2 (81)	α_2	known (80, 81)	amino acid	2 (82)
						aminoacyl adenylate	2 (82)
						tRNA	1 (83, 84)
							2 (81)
	yeast	80,000 (85)	2 (85)	α_2	unknown		
	rat liver	116,000 (1)	4 (1)	α_4	unknown		
		124,000 (86)	2 (86)	$\alpha\beta$	unknown		
		$\alpha = 62,000$					
		$\beta = 61,000$					
	soybean cytoplasm	122,000 (87)	2 (87)	α_2	unknown		
	soybean chloroplast	86,000 (87)	2 (87)	α_2	unknown		
Valyl-	B. stearothermophilus	110,000 (26)	none (26)	α	known (26)	amino acid	2 (82)
						ATP	2 (82)
						aminoacyl adenylate	2 (82)
	E. coli	110,000 (88)	none (21)	α	known (88)	amino acid	1 (21)
						aminoacyl adenylate	1 (21)
						tRNA	1 (89, 90)
	yeast	122,000 (29)	none (29)	α	known (29)	amino acid	1 (29)
						ATP	1 (29)
						tRNA	1 (91)
	L. luteus seeds	125,000 (66)	none (66)	α	unknown		

Criteria for including data for an enzyme were: (1) a check of the purity of the enzyme preparation by several methods; and (2) a determination of the m.w., preferably by polyacrylamide gel electrophoresis in the presence of SDS. Since the accuracy of these methods is limited and since proteolytic cleavage is sometimes a problem during enzyme purification (30), values for m.w. must be regarded as approximate.

[a]AA comp. = amino acid composition.
[b]Figures in parentheses are reference numbers.

REFERENCES

1. Bhanot, O. S., Z. Kucan, S. Aoyagi, F. C. Lee, and R. W. Chambers. 1974. *Methods Enzymol.* **29**:547.
2. Parfait, R. and H. Grosjean. 1972. *Eur. J. Biochem.* **30**:242.
3. Craine, J. and A. Peterkofsky. 1974. *Arch. Biochem. Biophys.* **168**:343.
4. Hirshfield, I. N. and H. P. J. Bloemers. 1969. *J. Biol. Chem.* **244**:2911.
5. Marshall, R. D. and P. C. Zamecnik. 1969. *Biochim. Biophys. Acta* **181**:454.
6. Nazario, M. and J. A. Evans. 1974. *J. Biol. Chem.* **249**:4934.
7. Gangloff, J. and G. Dirheimer. 1973. *Biochim. Biophys. Acta* **294**:263.
8. James, H. L. and E. T. Bucovaz. 1969. *J. Biol. Chem.* **244**:3210.
9. Pan, F., H. H. Lee, S. H. Pai, T. C. Yu, J. Y. Guoo, and G. M. Duh. 1976. *Biochim. Biophys. Acta* **452**:271.
10. Folk, W. R. 1971. *Biochemistry* **10**:1728.
11. Lapointe, J. and D. Söll. 1972. *J. Biol. Chem.* **247**:4966.
12. Lapointe, J. and D. Söll. 1972. *J. Biol. Chem.* **247**:4975.
13. Willick, G. E. and C. M. Kay. 1976. *Biochemistry* **15**:4347.
14. Ostrem, D. L. and P. Berg. 1970. *Proc. Natl. Acad. Sci.* **67**:1967.
15. Ostrem, D. L. and P. Berg. 1974. *Biochemistry* **13**:1338.
16. Kalousek, F. and W. H. Konigsberg. 1974. *Biochemistry* **13**:999.
17. De Lorenzo, F., F. P. Di Natale, and A. N. Schechter. 1974. *J. Biol. Chem.* **249**:908.
18. Charlier, J. and H. Grosjean. 1972. *Eur. J. Biochem.* **25**:163.
19. Arndt, D. J. and P. Berg. 1970. *J. Biol. Chem.* **245**:665.
20. Baldwin, A. N. and P. Berg. 1966. *J. Biol. Chem.* **241**:831.
21. Berthelot, F. and M. Yaniv. 1970. *Eur. J. Biochem.* **16**:123.
22. Holler, E., E. L. Bennett, and M. Calvin. 1971. *Biochem. Biophys. Res. Commun.* **45**:409.
47. Fasiolo, F., P. Remy, J. Pouyet, and J.-P. Ebel. 1974. *Eur. J. Biochem.* **50**:227.
48. Fasiolo, F. and J.-P. Ebel. 1974. *Eur. J. Biochem.* **49**:257.
49. Schmidt, J., R. Wang, S. Stanfield, and B. R. Reid. 1971. *Biochemistry* **10**:3264.
50. Lanks, K. W., R. K. Eng, and L. Baltusis. 1973. *Fed. Proc.* **32**:460 (Abst. 1336).
51. Lee, M.-L. and K. H. Muench. 1969. *J. Biol. Chem.* **244**:223.
52. Lemeur, M. A., P. Gerlinger, J. Clavert, and J.-P. Ebel. 1972. *Biochimie* **34**:1391.
53. Boeker, E. A., A. P. Hays, and G. L. Cantoni. 1973. *Biochemistry* **12**:2379.
54. Boeker, E. A. and G. L. Cantoni. 1973. *Biochemistry* **12**:2384.
55. Katze, J. R. and W. Konigsberg. 1970. *J. Biol. Chem.* **245**:923.
56. Waterson, R. M., S. J. Clarke, F. Kalousek, and W. H. Konigsberg. 1973. *J. Biol. Chem.* **248**:4181.
57. Knowles, J. R., J. R. Katze, W. Konigsberg, and D. Söll. 1970. *J. Biol. Chem.* **245**:1407.
58. Pingoud, A., D. Riesner, D. Boehme, and G. Maass. 1973. *FEBS Lett.* **30**:1.
59. Engel, G., H. Heider, A. Maelicke, F. von der Haar, and F. Cramer. *Eur. J. Biochem.* **29**:257.
60. Heider, H., E. Gottschalk, and F. Cramer. 1971. *Eur. J. Biochem.* **20**:144.
61. Hörz, W. and H. G. Zachau. 1973. *Eur. J. Biochem.* **32**:1.
62. Pachmann, U., E. Cronvall, R. Rigler, R. Hirsh, W. Wintermeyer, and H. G. Zachau. 1973. *Eur. J. Biochem.* **39**:265.
63. Rigler, R., E. Cronvall, R. Hirsch, U. Pachmann, H. G. Zachau. 1970. *FEBS Lett.* **11**:320.

23. Norris, A. T. and P. Berg. 1964. *Proc. Natl. Acad. Sci.* **52**:330.
24. Yarus, M. and M. Berg. 1967. *J. Mol. Biol.* **28**:479.
25. Yarus, M. and P. Berg. 1969. *J. Mol. Biol.* **42**:171.
26. Koch, G. L. E., Y. Boulanger, and B. S. Hartley. 1974. *Nature* **249**:316.
27. Hayashi, H., J. R. Knowles, J. R. Katze, J. Lapointe, and D. Söll. 1970. *J. Biol. Chem.* **245**:1401.
28. Murasugi, A. and H. Hayashi. 1975. *Eur. J. Biochem.* **57**:169.
29. Rymo, L., L. Lundvik, and U. Lagerkvist. 1972. *J. Biol. Chem.* **247**:3888.
30. Dimitrijevic, L. 1972. *FEBS Lett.* **25**:170.
31. Rymo, L., U. Lagerkvist, and A. Wonacott. 1970. *J. Biol. Chem.* **245**:4308.
32. Fersht, A. R., J. S. Ashford, C. J. Bruton, R. Jakes, G. L. E. Koch, and B. S. Hartley. 1975. *Biochemistry* **14**:1.
33. Koch, G. L. E. and C. J. Bruton. 1974. *FEBS Lett.* **40**:180.
34. Lemoine, F., J. P. Waller, and R. van Rapenbusch. 1968. *Eur. J. Biochem.* **4**:213.
35. Cassio, D. and J. P. Waller. 1971. *Eur. J. Biochem.* **20**:283.
36. Blanquet, S., G. Fayat, J. P. Waller, and M. Iwatsubo. 1972. *Eur. J. Biochem.* **24**:461.
37. Blanquet, S., M. Iwatsubo, and J. P. Waller. 1973. *Eur. J. Biochem.* **36**:213.
38. Rosa, M. D. and P. B. Sigler. 1977. *Eur. J. Biochem.* **78**:141.
39. Joachimiak, A., J. Barciszewski, T. Twardowski, M. Barciszewska, and M. Wiewiorowski. 1978. *FEBS Lett.* **93**:51.
40. Tscherne, J. E., K. W. Lanks, P. D. Salim, D. Grunberger, C. R. Cantor, and I. B. Weinstein. 1973. *J. Biol. Chem.* **248**:4052.
41. Christopher, C. W., D. B. Sittman, and D. W. Stafford. 1975. *Arch. Biochem. Biophys.* **166**:94.
42. Fayat, G., S. Blanquet, P. Dessen, G. Batelier, and J. P. Waller. 1974. *Biochimie* **56**:35.
43. Hanke, T., P. Bartmann, H. Hennecke, H. M. Kosakowski, R. Jaenicke, E. Holler, and A. Böck. 1974. *Eur. J. Biochem.* **43**:601.
44. Kosakowski, M. H. J. E. and A. Böck. 1970. *Eur. J. Biochem.* **12**:67.
45. Bartmann, P., T. Hanke, and E. Holler. 1975. *J. Biol. Chem.* **250**:7668.
46. Fasiolo, F., N. Befort, Y. Boulanger, and J.-P. Ebel. 1970. *Biochim. Biophys. Acta* **217**:305.
64. Rigler, R., E. Cronvall, M. Ehrenberg, U. Pachmann, R. Hirsch, and H. G. Zachau. 1971. *FEBS Lett.* **18**:193.
65. Pachmann, U. and H. G. Zachau. 1978. *Nucleic Acids Res.* **5**:961.
66. Jakubowski, H. and J. Pawelkiewicz. 1975. *Eur. J. Biochem.* **52**:301.
67. Hennecke, H., A. Böck, J. Thomale, and G. Nass. 1977. *J. Bacteriol.* **131**:943.
68. Penneys, N. S. and K. H. Muench. 1974. *Biochemistry* **13**:560.
69. Gros, C., G. Lemaire, R. van Rapenbusch, and B. Labouesse. 1972. *J. Biol. Chem.* **247**:2931.
70. Lemaire, G., R. van Rapenbusch, C. Gros, and B. Labouesse. 1969. *Eur. J. Biochem.* **10**:336.
71. Dorizzi, M., B. Labouesse, and J. Labouesse. 1971. *Eur. J. Biochem.* **19**:563.
72. Dorizzi, M., G. Merault, M. Fournier, J. Labouesse, G. Keith, G. Dirheimer, and R. H. Buckingham. 1977. *Nucleic Acids Res.* **4**:31.
73. Liu, C.-C., C.-H. Chung, and M.-L. Lee. 1973. *Biochem. J.* **135**:367.
74. Winter, G. P. and B. S. Hartley. 1977. *FEBS Lett.* **80**:340.
75. Joseph, D. R. and K. H. Muench. 1971. *J. Biol. Chem.* **246**:7610.
76. Joseph, D. R. and K. H. Muench. 1971. *J. Biol. Chem.* **246**:7602.
77. Muench, K. H. 1976. *J. Biol. Chem.* **251**:5195.
78. Hossain, A. and N. R. Kallenbach. 1974. *FEBS Lett.* **45**:202.
79. Fersht, A. R., R. S. Mulvey, and G. L. E. Koch. 1975. *Biochemistry* **14**:13.
80. Calender, R. and P. Berg. 1966. *Biochemistry* **5**:1681.
81. Chousterman, S. and F. Chapeville. 1973. *Eur. J. Biochem.* **35**:51.
82. Fersht, A. R. 1975. *Biochemistry* **14**:5.
83. Buonocore, V. and S. Schlesinger. 1972. *J. Biol. Chem.* **247**:1343.
84. Jakes, R. and A. R. Fersht. 1975. *Biochemistry* **14**:3344.
85. Faulhammer, and F. Cramer. 1977. *Eur. J. Biochem.* **75**:561.
86. Rao, Y. S. P. and P. R. Srinivasan. 1978. *Nucleic Acids Res.* **4**:3887.
87. Locy, R. D. and J. H. Cherry. 1978. *Phytochemistry* (Oxf) **17**:19.
88. Yaniv, M. and F. Gros. 1969. *J. Mol. Biol.* **44**:1.
89. Hélène, C., F. Brun, and M. Yaniv. 1971. *J. Mol. Biol.* **58**:349.
90. Yaniv, M. and F. Gros. 1969. *J. Mol. Biol.* **44**:17.
91. Lagerkvist, U. and L. Rymo. 1969. *J. Biol. Chem.* **244**:2476.
92. Rouget, P. and F. Chapeville. 1970. *Eur. J. Biochem.* **14**:498.
93. Chirikjian, J. G., H. T. Wright, and J. R. Fresco. 1972. *Proc. Natl. Acad. Sci.* **69**:1638.

Table 2 Aminoacyl-tRNA synthetase mutants

Synthetase	Organism	Type of mutant	Enzyme defect[a]	Genetic location	Reference
Alanyl-	Chinese hamster	auxotrophic			1
	Escherichia coli	temperature sensitive	tRNA	near recA	2–4
	mouse cells	temperature sensitive			5
Arginyl-	Chinese hamster	temperature sensitive			6
	E. coli	analog resistant	ATP, tRNA	between pheS and his	7, 8
					9
Asparaginyl-	Chinese hamster	temperature sensitive	amino acid		10, 11
	E. coli	temperature sensitive		at 21 min	12
	hamster	temperature sensitive			13
Glutaminyl-	Chinese hamster	temperature sensitive			6
	E. coli	temperature sensitive		near lip	14
Glutamyl-	E. coli	temperature sensitive		near dsdA	15
		streptomycin dependent	amino acid	near xyl, near his	16, 17
Glycyl-	E. coli	auxotrophic	amino acid	near xyl	18, 19
		temperature sensitive		near xyl	20
Histidyl-	Chinese hamster	temperature sensitive			6
	E. coli	bradytrophic	amino acid		21
	Salmonella typhimurium	analog resistant	amino acid	near strB	22
Isoleucyl-	E. coli	auxotrophic	ATP, amino acid	between thr and pyrA	23, 24
		analog resistant			25
		temperature sensitive			26
	S. typhimurium	auxotrophic		near pyrA	27
	Saccharomyces cerevisiae	temperature sensitive		on chromosome 2	28, 29

Leucyl-	Chinese hamster	temperature sensitive	amino acid		30–32
	E. coli	temperature sensitive	amino acid	near lip	33
	Salmonella typhimurium	analog resistant		near gal	34, 35
	Neurospora crassa	temperature sensitive, auxotrophic			36, 37
Lysyl-	Bacillus subtilis	temperature sensitive	amino acid		38, 39
	E. coli	analog resistant	amino acid, ATP	between purA and sul	40
Methionyl-	Chinese hamster	temperature sensitive			6
	E. coli	auxotrophic	amino acid	at 47 min	41
		analog resistant	tRNA		42
	S. typhimurium	auxotrophic	amino acid	at 67 min	43, 44
	Saccharomyces cerevisiae	temperature sensitive	amino acid		29
Phenylalanyl-	E. coli	analog resistant	amino acid	near pps	45, 46
		temperature sensitive		near pps	15
Seryl-	E. coli	temperature sensitive		near serC	33, 47
		serine hydroxamate resistant			48
Threonyl-	E. coli	borrelidin resistant	amino acid, ATP[b]	at 37.7 min	49, 50
		auxotrophic	amino acid	near trp	51
Tryptophanyl-	B. subtilis	temperature sensitive, analog resistant		between argG and metA	52
	E. coli	auxotrophic, temperature sensitive	amino acid	between purA and aroB	53, 54
					15
	N. crassa	auxotrophic			55
Tyrosyl-	E. coli	analog resistant	amino acid	at 35 min	56, 57
		auxotrophic			15
	Salmonella typhimurium				58
Valyl-	E. coli	temperature sensitive	ATP, amino acid	near pyrB	59, 60
					61

[a]The affinity of the enzyme for the specified substance is lowered unless otherwise indicated.
[b]Affinity increased.

REFERENCES

1. Hankinson, O. 1976. *Somatic Cell Genet.* **2**:497.
2. Buckel, P., W. Lubitz, and A. Böck. 1971. *J. Bacteriol.* **108**:1008.
3. Yaniv, M. and F. Gros. 1966. Genet. Elem. Prop. Funct. Symp., 3rd FEBS Proc. Meet. P. 157.
4. Theall, G., K. B. Low, and D. Söll. 1977. *Mol. Gen. Genet.* **156**:221.
5. Sato, K. 1975. *Nature* **257**:813.
6. Thompson, L. H., D. J. Lofgren, and G. M. Adair. 1977. *Cell* **11**:157.
7. Hirshfield, I. N. and H. P. J. Bloemers. 1969. *J. Biol. Chem.* **244**:2911.
8. Cooper, P. H., I. N. Hirshfield, and W. K. Maass. 1969. *Mol. Gen. Genet.* **104**:383.
9. Williams, L. S. 1973. *J. Bacteriol.* **113**:1419.
10. Thompson, L. H., C. P. Stanners, and L. Siminovitch. 1975. *Somatic Cell Genet.* **1**:187.
11. Andrulis, I. L., C. S. Chiang, S. M. Arkin, T. A. Miner, and G. W. Hatfield. 1978. *J. Biol. Chem.* **253**:58.
12. Yamamoto, M., M. Nomura, H. Ohsawa, and B. Maruo. 1977. *J. Bacteriol.* **132**:127.
13. Wasmuth, J. J. and C. T. Caskey. 1976. *Cell* **9**:655.
14. Körner, A., B. B. Magee, B. Liska, K. B. Low, E. A. Adelberg, and D. Söll. 1974. *J. Bacteriol.* **120**:154.
15. Russel, R. R. B. and A. J. Pittard. 1971. *J. Bacteriol.* **108**:790.
16. Murgola, E. J. and E. A. Adelberg. 1970. *J. Bacteriol.* **103**:20.
17. Murgola, E. J. and E. A. Adelberg. 1970. *J. Bacteriol.* **103**:179.
18. Böck, A. and F. C. Neidhardt. 1966. *Z. Vererbungsl.* **98**:187.
19. Folk, W. R. and P. Berg. 1970. *J. Bacteriol.* **102**:204.
20. Roback, E. R., J. D. Friesen, and N. P. Fiil. 1973. *Can. J. Microbiol.* **19**:425.
21. Nass, G. 1967. *Mol. Gen. Genet.* **100**:216.
22. Roth, J. R. and B. N. Ames. 1966. *J. Mol. Biol.* **22**:325.
23. Iaccarino, M. and P. Berg. 1971. *J. Bacteriol.* **105**:527.
24. Treiber, G. and M. Iaccarino. 1971. *J. Bacteriol.* **107**:828.
25. Coker, M. and H. E. Umbarger. 1970. *Bacteriol. Proc.* P. 135 (Abstr.).
32. Farber, R. A. and M. P. Deutscher. 1976. *Somatic Cell Genet.* **2**:509.
33. Low, B., F. Gates, T. Goldstein, and D. Söll. 1971. *J. Bacteriol.* **108**:742.
34. Alexander, R. R., J. M. Calvo, and M. Freundlich. 1971. *J. Bacteriol.* **106**:213.
35. Mikulka, T. W., B. I. Stieglitz, and J. M. Calvo. 1972. *J. Bacteriol.* **109**:584.
36. Beauchamp, P. M., E. W. Horn, and S. R. Gross. 1977. *Proc. Natl. Acad. Sci.* **74**:1172.
37. Gross, S. R., M. T. McCoy, and E. B. Gilmore. 1968. *Proc. Natl. Acad. Sci.* **61**:253.
38. Racine, F. and W. Steinberg. 1974. *J. Bacteriol.* **120**:372.
39. Racine, F. and W. Steinberg. 1974. *J. Bacteriol.* **120**:384.
40. Hirshfield, I. N., J. W. Tomford, and P. Zamecnik. 1972. *Biochim. Biophys. Acta* **259**:344.
41. Somerville, C. R. and A. Ahmed. 1977. *J. Mol. Biol.* **111**:77.
42. Archibald, E. R. and L. S. Williams. 1973. *J. Bacteriol.* **114**:1007.
43. Sanderson, K. E. and M. Demerec. 1965. *Genetics* **51**:897.
44. Gross, T. S. and R. J. Rowbury. 1969. *Biochim. Biophys. Acta* **184**:233.
45. Fangman, W. L. and F. C. Neidhardt. 1964. *J. Biol. Chem.* **239**:1839.
46. Comer, M. M. and A. Böck. 1976. *J. Bacteriol.* **127**:923.
47. Clarke, S. J., B. Low, and W. Konigsberg. 1973. *J. Bacteriol.* **113**:1096.
48. Tosa, T. and L. I. Pizer. 1971. *J. Bacteriol.* **106**:972.
49. Hennecke, H., A. Böck, J. Thomale, and G. Nass. 1977. *J. Bacteriol.* **131**:943.
50. Nass, G. and J. Thomale. 1974. *FEBS Lett.* **39**:182.
51. Johnson, E. J., G. N. Cohen, and I. Saint-Girons. 1977. *J. Bacteriol.* **129**:66.
52. Steinberg, W. and C. Anagnostopoulos. 1971. *J. Bacteriol.* **105**:6.
53. Doolittle, W. F. and C. Yanofsky. 1968. *J. Bacteriol.* **95**:1283.

26. McGinnis, E. and L. S. Williams. 1974. *Annual meeting of the American Society of Microbiology.* P. 172 (Abstr.).
27. Blatt, J. M. and H. E. Umbarger. 1972. *Biochem. Genet.* **6**:99.
28. Hartwell, L. L. and C. S. McLaughlin. 1968. *Proc. Natl. Acad. Sci.* **59**:422.
29. McLaughlin, C. S. and L. H. Hartwell. 1969. *Genetics* **61**:557.
30. Thompson, L. H., J. L. Harkins, and C. P. Stanners. 1973. *Proc. Natl. Acad. Sci.* **70**:3094.
31. Haars, L., A. Hampel, and L. Thompson. 1976. *Biochim. Biophys. Acta* **454**:493.
54. Kano, Y., A. Matsushiro, and Y. Shimura. 1968. *Mol. Gen. Genet.* **102**:15.
55. Nazario, M., J. A. Kinsey, and M. Admad. 1971. *J. Bacteriol.* **105**:121.
56. Schlessinger, S. and E. W. Nester. 1969. *J. Bacteriol.* **100**:167.
57. Buonocore, V., M. H. Harris, and S. Schlessinger. 1973. *J. Biol. Chem.* **247**:4843.
58. Heinonen, J., S. W. Artz, and H. Zalkin. 1972. *J. Bacteriol.* **112**:1254.
59. Eidlic, L. and F. C. Neidhardt. 1965. *J. Bacteriol.* **89**:706.
60. Tingle, M. A. and F. C. Neidhardt. 1969. *J. Bacteriol.* **98**:837.
61. Yaniv, M. and F. Gros. 1969. *J. Mol. Biol.* **44**:31.

Author Index

Ackerman, E., 3
Alzner-DeWeerd, B., 3

Baltzinger, M., 325
Bertram, S., 459
Beurling, K., 145
Blanquet, S., 281
Bonnet, J., 325

Cantor, C.R., 363
Celis, J.E., 539
Cramer, F., 267
Crothers, D.M., 163

Dell, A., 255
Dessen, P., 281
Dietrich, A., 325
Dirheimer, G., 19
Ducruix, A., 101

Ebel, J.-P., 325
Ehrlich, R., 325

Fasiolo, F., 325
Favorova, O.O., 235, 325
Fayat, G., 281
Fersht, A.R., 247
Ford, N.C., Jr., 207
Fournier, M.J., 207
Fresco, J.R., 145
Fritzinger, D.C., 207

Gassen, H.G., 459
Gauss, D.H., 521
Giegé, R., 325
Grüter, F., 521
Gupta, R.C., 43

Hartley, B.S., 223, 255
Hecht, S.M., 345
Heckman, J.E., 3
Hughes, J.J., 133
Hurd, R.E., 177

Igloi, G.L., 267

Johnson, A.E., 487
Johnston, P.D., 191

Karpel, R.L., 145
Keith, G., 19, 325
Kim, S.-H., 83
Kisselev, L.L., 235
Koch, G.L.E., 255
Kovaleva, G.K., 235
Kuechler, E., 413

Lake, J.A., 393
Lipecky, R., 459

Manderschied, U., 459
Manor, P.C., 145
Martin, R.P., 19
Möller, A., 459

Nishimura, S., 59, 547, 551

Ofengand, J., 413

Podjarny, A.D., 133
Potts, R.O., 207

Quigley, G.J., 101

RajBhandary, U.L., 3
Randerath, E., 43
Randerath, K., 43
Redfield, A.G., 191
Reid, B.R., 177
Remy, P., 325
Renaud, M., 325
Rich, A., 101
Robertson, J.M., 445

Schevitz, R.W., 133
Schimmel, P.R., 297, 553
Schmitt, M., 459
Schulman, L.H., 311
Sibler, A.-P., 19
Sigler, P.B., 133
Söll, D., 553
Sprinzl, M., 473, 521
Sundaralingam, M., 115

Teeter, M.M., 101

Vassilenko, S., 325
von der Haar, F., 267

Wagner, T., 473
Wang, C.-C., 207
Weidner, H., 445
Winter, G., 255
Wintermeyer, W., 445
Woo, N., 101
Wright, H.T., 145

Yarus, M., 501
Yin, S., 3

Zachau, H.G., 445

Subject Index

Acceptor stem, 30
  specificity, role in, 155
  in three-dimensional structure, 299–300
  yeast, tRNA^{Gly}, 148–151
N-acetyl-phenylalanyl-tRNA^{Phe}, 105–106
acp^{3}U
  affinity labeling probe, 414, 418, 423
  chemical modification of, 488, 495
Affinity labeling
  aminoacyl moiety of tRNA, 414–422
  aminoacyl-tRNA affinity analogs, 399–401, 415–417, 420–421
  aminoacyl-tRNA probe, eukaryotic ribosomes, 422
  of EF-Tu, 490–491
  mRNA-ribosomal contacts, 397, 424–431
  tRNA-ribosomal contacts, 368–391, 399–401, 413–424
  tRNA affinity analogs, 368–371, 413–415
  tRNA bases as probes
    acp^{3}U47, 414, 418, 423
    s^{4}U8, 414, 418, 422
  tRNA binding sites, 413–415
  of tryptophanyl-tRNA synthetase, 241–244
Affinity-labeling reagents, 241–244, 415–423
Allosteric effects
  of IF-2 on initiation complex, 464–466, 468
  of Mg^{++} binding to tRNA, 163–168
Amino acids
  activation, 250, 271
  methionine activation, 286–287
  misactivation, 249
  recognition, 271–277
Aminoacylation, 243, 268, 270–271, 304–306
  editing mechanism, 247–253
    cysteinyl-tRNA synthetase, 251
    isoleucyl-tRNA synthetase, 253
    tyrosyl-tRNA synthetase, 252

    valyl-tRNA synthetase, 248–251
  fidelity, 145–146, 154–155, 501–512
  heterologous, 300
  initial position of, (2′ vs 3′), 348–353, 474–476
  of modified $E.$ $coli$, tRNA^{Met}, 314–318
  proofreading of tRNA, 252–253
  of tRNA by methionyl-tRNA synthetase, 287–290
  tRNA structure, effect on, 207–219, 338–339
    conformational changes, 104, 106
  of tRNA^{Trp} from beef pancreas, 235–237, 241–244
  2′-OH vs 3′-OH specificity, 345, 348–357, 474–484, 510–511
Aminoacyl binding site
  affinity labeling of, 413–415, 424, 429–430
    by aminoacyl-tRNA analogs, 414
    by N^ϵ-modified lysyl-tRNA, 494–498
  binding reversibility, 480–481
  localization on ribosome, 496–498
  nature and location, 368–373, 400–403
  probed with fluorescent tRNA derivatives, 452–454
  protein biosynthesis, role in, 364–368, 403–407
  tRNA–mRNA interactions, 372–373, 381–382, 384
  of tRNA to ribosome, 364–383, 400–409, 413–415
Aminoacyl residue migration. See Transaminoacylation
Aminoacyl-tRNA
  accuracy of synthesis, 501–512
  binding by EF-Tu, 354–357, 477–484
  EF-Tu-dependent ribosome binding, 478–481
  nonisomerizable analogs, 473

Aminoacyl-tRNA *(continued)*
  positional ($2' \rightarrow 3'$) isomerization, 345–346, 349–357, 474–478
Aminoacyl-tRNA analogs, side-chain modification, 487–498
Aminoacyl-tRNA · EF-Tu · GTP ternary complex, 356–357
  aminoacyl-tRNA isomerization, 477–478, 481–484
  complementary oligonucleotide binding, 460–461, 463–466
  intermediate in protein synthesis, 365, 374, 378
  modified aminoacyl-tRNA studies, 488–495
Aminoacyl-tRNA synthetase, 223–232. *See also individual synthetases*
  accuracy. *See also* proofreading
    amino acid selection, 507–512
    tRNA selection, 504–507
  acylation
    effect of dioxane, 305–306
    effect of DMSO, 305–306
  affinity labeling of, 287
  amino acid binding sites, 304
  amino-acid-dependent ATP pyrophosphatase, 248
  amino-acid-sequencing strategy, 256–258
  ATP · PPi exchange, 334
  catalytic cycle, 273, *See also* Aminoacylation
  characteristics of, 553–559
  chemical labeling of active site, 229–230
  chemical modifications, 224–225, 229–230
  chemical proofreading mechanism, 276–277
  conformational changes
    induced by cognate tRNA, 336–337
    induced by cognate tRNA 3' terminal, 270
  deacylation at hydrolytic site, 247–249
  editing mechanisms
    double-sieve editing model, 275
    steric expulsion principle, 250
  effector sites of, 237, 240–244
  error frequency, 277, 501–512
  evolution, 255–256, 259–260, 262
  fidelity, 145–146, 154–156
  internal repeats. *See* Sequence duplication of aminoacyl-tRNA synthetases
  misacylation, 248–249
    effect of dioxane on, 305–306
    effect of DMSO on, 305–306
    induced by special reaction conditions, 298
  molecular weights, 225–231
  mutants, 560–563
  mutations, 306–308
  nucleotide binding sites, 238–240, 285–287, 291
  photoaffinity labeling, 229
  primary structure, 225–231
    homology, 230–231
  proofreading, 267–277
  recognition of amino acid, 267, 271
    activation, 250
    misactivation, 249, 251
  recognition of substrates, 267–277
  recognition of tRNA, 267–271
    all-or-none, 298
    anticodon, 301
    cognate, 305
    D stem, 300
    Dudock hypothesis, 300
    mechanism, 304–306
    noncognate, 305, 348–357
    3' terminal, 298
  recognition complex of two tRNA molecules, 155
  recognition properties, dual discrimination, 298
  sequence duplication, 225, 255–256, 281–282
    valyl-tRNA synthetase, 337
  sequence heterogeneity, 224–225
  sequence homology, 230–231, 259–263
  specificity, 145–146, 154–155, 345–357, 474–477
  structural studies, amino group reactivity, 230–232
  structural twofoldness, 154
  subunit structure, 225–231, 298
  tertiary structure
    domain homology, 230
    structural domains, 225, 230–231
  tRNA binding, salt dependence, 262
  tRNA binding sites, 270, 305
  tyrosyl-tRNA synthetase, 262
Aminoacyl-tRNA synthetase · tRNA complex
  competitive labeling, 230
  conformation transitions, 271–272
  noncognate tRNA, 298
  nuclease digestions, 330–331
  structural organizations, 297–308
Aminoacyl-tRNA synthetase–tRNA interaction, 325–340
  dissection of tRNA, 331–333
  important regions, 298–304
    in three-dimensional configuration 335–336
  induction of misacylation, 304
  methods of study
    competitive labeling, 230
    isotopic labeling, 301
    m⁷G, 331–332
    minor bases excision, 331–333
    oligonucleotide hybridization, 301
    photochemical cross-linking, 301, 304, 326–329
Amino group reactivity, 230–232
7-(aminomethyl)-7-deazaguanosine, 66–68

Subject Index

Anticodon
  nucleotide modification, 35, 37
  recognition by *E. coli*, methionyl-tRNA synthetase, 315–316
  wobble position, 34
Anticodon loop
  base stacking in, 384, 406–407, 429–431
  base stacking probed with Y-base fluorescence, 459
  conformation, 89, 406–407
  conformational change
    binding of complementary oligonucleotides, 459–463
    ribosome binding, 130
    tRNA-tRNA fluorescent studies, 448–454
  initiator tRNA, 111–112
  in intermolecular interactions, 152
  three-dimensional structure, 111–112
  yeast, tRNA^{fMet}, 140–141
Anticooperative binding
  of tRNA^{Met}, 285–286, 288–289
  of tRNA^{Trp}, 242
A-RNA double-helical structure, 84, 86
ATP analogs, probe of nucleotide binding sites, 238–241
ATP pyrophosphatase, 248
Atypical nucleosides. *See* modified nucleosides

Bacillus stearothermophilus. *See individual tRNA synthetases*
Backbone-backbone interaction of tRNA, 89
Backbone structure of yeast, tRNA^{Phe}, 83–91, 115–116, 124–130
Base analogs, inhibition of tRNA modification, 46
Base-backbone interaction in yeast tRNA^{Phe}, 89
Base-base interaction
  base stacking, 86
  determination by ^{3}H derivative methods, 43–46
  pairing by hydrogen bonds, 86–89, 95–97
Base composition analysis of tRNA by ^{3}H derivative methods, 43–46
Base conservation in tRNA, 84, 97–98
Base interdigitation, 121–122
Base pairing, 116, 118–121, 128–129
  GU wobble pair, 195–197, 203
  NMR analysis, 178–188
  in yeast tRNA^{Phe}, 116–119, 129
Base stacking, 118–122, 128, 130
  in anticodon loop, 384, 406–407, 429–431
Binding sites on aminoacyl-tRNA synthetases, 270, 305
  on ribosome, 364–386, 395–397, 400–408, 413–439
Bisulfite modification of *E. coli* tRNA^{fMet}, 311–313, 320, 321

Borotritide. *See* [^{3}H] Borohydride

Calorimetry, 167
Cationic molecules, effect on tRNA three-dimensional structure, 102–105
Cell-wall peptidoglycan synthesis, role of tRNA in, 28–29
Chemical aminoacylation, 311–319, 354–356
Chemical modification
  of aminoacyl-tRNA side chain, 488–489
  of aminoacyl-tRNA synthetase, 224–225, 229–230
  of tRNA, 311–319. *See also* tRNA modification
  of yeast tRNA^{Phe} by kethoxal, 214–215, 218
Chloroacetaldehyde modification of tRNA^{fMet}, 313, 315
Chloroplast tRNA, 9
  modification methyl transferases, 35
Chromatographic mobilities of modified nucleotides, 551–552
cmo^{5}U, affinity-labeling probe, 415, 431–432
Codon–anticodon interaction
  codon recognition, 384–386
  codon recognition, role in proofreading, 450–454
  model tRNA–mRNA interaction, 373, 381–384
  modified nucleosides, role of, 60–65
  probed
    with mRNA affinity analogs, 424, 429–431
    with tRNA with proflavine in place of Y base, 373
    with Y-base fluorescence, 373–374, 381
  ribosome site, 397
  stabilization of TψCG complementary oligonucleotides, 461–462
Codon recognition in mitochondrial tRNA, 24–25
Cognate tRNA, recognition by aminoacyl-tRNA synthetase of, 305
Complementary oligonucleotide binding to tRNA
  in the anticodon, 459, 462–468
  in the TψC loop, 460–462, 466–468
Conformational changes of aminoacyl-tRNA synthetase, 267–272
  in enzyme · tRNA complex, 271–272
  induced by cognate tRNA, 336–337
  induced by 3′-terminal A, 268
Conformational changes of tRNA, 163
  effect of cations, 105
  induced by aminoacylation, 106
  structural transitions in solution, 207–219
Conformational wheels of bond torsion angles of yeast, tRNA^{Phe}, 124–128

Cooperative binding of Mg^{++} by yeast, tRNA^{Phe}, 212–219
Crystal structure. *See also* X-ray crystallography
  E. coli, tRNA^{fMet}, 107–112
  E. coli, tRNA^{Ser}, 119
  yeast, tRNA^{fMet}, 133–141
  yeast, tRNA^{Gly}, 145–152
  yeast, tRNA^{Phe}, 83–98, 102–106
7 (cyano)-7-deazaguanosine, 66–68
Cyclobutane dimers of pyrimidines, 431
Cysteinyl-tRNA synthetase, 225
  editing mechanism, 251

D loop, three-dimensional structure of, 114
Deacylation, 247–249
Decoding site, 434–438. *See also* tRNA binding sites
Deoxyadenosine analogs, 345–357, 473–481
Dioxane
  effect of acylation and misacylation, 305–306
  effect on crystal structure, 152–154
  effect on tRNA melting, 154
  effect on tRNA structure, 152–154
Divalent cations
  binding, 165, 167–169, 375
  Mg^{++} effect on tRNA conformation, 446–448
  tRNA crystallographic binding sites, 183
Divalent ion binding to tRNA, 167–169
  anticooperative binding, 169
  cooperative binding, 168
  Mg^{++} binding, 165, 167–168
DMSO, effect on acylation and misacylation, 305–306
Double-resonance solvent-exchange rates, 192–204
Double-sieve editing model, 275
Dudock hypothesis, 300
Dye binding to tRNA, 122–123

Editing mechanisms, 249–252. *See also individual synthetases*; Aminoacylation
  chemical proofreading, 249
  double-sieve editing mechanism, 249–252
  hydrolytic editing, 249
  kinetic proofreading, 249
Electrostatic changes, 211, 213–215
Electrostatic charge of yeast, tRNA^{Phe}, 208–214
Elongation factor G (EF-G)
  ribosome binding site, 399
  translocation, role in, 365–367, 379, 493
Elongation factor Tu (EF-Tu)
  affinity labeling of, 490–491
  binding of aminoacyl-tRNA isomers, 354–357, 477–484
  component of elongation cycle, 365
  codon-anticodon complex stability, influence on, 463–464
  function probed with aminoacyl-tRNA analogs, 487–495
  GTP hydrolysis mediated by, 492, 498
  recognition of aminoacyl-tRNAs, 320–321
  recognition of tRNA, 297
  ribosome binding site, 399
  ternary complex with tRNA and GTP, 314, 320–321
  tRNA binding, 374–378
  tRNA structure, effect on, 460–464, 468
Escherichia coli. See individual tRNA synthetases; Initiator tRNA; Lysyl-tRNA^{Lys}; Methionyl-tRNA synthetase; Peptidyl-tRNA hydrolase
Ethionine, misactivation by methionyl-tRNA synthetase, 251

Fidelity
  aminoacylation, 145–146, 154–155
  translation, 273, 501–512
5S interaction with TψC loop, 460
Fluorescence spectroscopy
  tRNA conformation studies, 446–455
  tRNA modifications
    ethidium replacement of D, 446–447, 449–451
    ethidium replacement of Y base, 446–453
    proflavine replacement of D, 452–453
    proflavine replacement of Y base, 373, 446, 452–454
  of s^4U, 414, 418, 422, 438
  Y base of yeast tRNA^{Phe}
    codon-anticodon interaction, 373–374
    multiple anticodon environments, 381, 446
Formylation of modified tRNA, 319
Formylmethionyl-tRNA synthetase. *See* Methionyl-tRNA synthetase
Fourier transform NMR (FT NMR). *See* Nuclear magnetic resonance

G + C content of yeast tRNA^{Phe}, 96–97
Glutamyl-tRNA synthetase, *E. coli*, 225
GTP hydrolysis, 375, 379, 403
  EF-Tu-mediated, 492, 498

Half-of-the-sites reactivity of beef pancreas tryptophanyl-tRNA synthetase, 237, 241–243
Helix conformations of yeast tRNA^{Phe}, 123–130

Heterologous aminoacylation, 300
Hydroxylamine
  modification of tryptophan, 235–237
  modification of beef pancreas tryptophanyl-tRNA synthetase, 235–237
Hypermodified nucleosides, 30–32. See Modified nucleosides
  location in tRNA, 59
  in mitochondrial tRNA, 37
  queuosine and derivatives, 30, 32, 60, 65
  wyosine, 30, 32, 69

Immuno-electron microscopy
  localization of ribosomal proteins, 433
  ribosome structure, 393–400
Initiation factor 2 (IF-2)
  binding of 30S ribosomal subunit to formylmethionyl-tRNA^{fMet}, 365, 398
  catalysis of GTP, 398
  initiation-complex formation, role in, 464–466, 468
  recognition of initiator tRNA, 318–319
  ribosome binding site, 399
Initiation factor 3 (IF-3), 397–398
Initiator tRNA. See tRNA^{fMet}
Invariant nucleotides, 26–32
  in three-dimensional configuration, 101
Iodination of tRNA^{fMet} crystals, 137–138
Ion binding properties of tRNA, 163
Isoaccepting tRNA
  regulation, role in, 69
  in tumor cells, 69–72
Isoleucyl-tRNA synthetase, E. coli, 223
  editing mechanism, 247–249, 252
  error rate of selection of amino acids, 4, 277
  substrate specificity, 351

Kinetics of tRNA unfolding, 194, 198–202

Laser light scattering, 207–219
  electrostatic-charge measurements, 211–213
  translational diffusion coefficient, 208–211
Loop folding in tRNA, 123–128, 130
Low-angle neutron-scattering analysis
  methionyl-tRNA synthetase, 284, 289, 291
  trypsin-modified methionyl-tRNA synthetase, 284
  valyl-tRNA synthetase, 333–336
Lysyl-tRNA^{Lys}, N^{ϵ}-modified E. coli, 489–498

Magnesium ion
  effect on tRNA three-dimensional structure, 103–105

interaction with tRNA, allosteric effector, 163–168
Metal–tRNA interaction, 89–91
Methionyl-tRNA synthetase
  B. stearothermophilis, 290–291
  E. coli, 223–224, 229–230, 281–291
    adenylation site, 284–285
    aminoacylation, 287–290
    crystal structure of tryptic fragment, 283–284, 291
    methionine activation, 286–287
    methionyl-tRNA synthetase · tRNA^{fMet} complex, 231–233
    methionyl-tRNA synthetase · tRNA^{mMet} complex, 231–233
    neutron-scattering analysis of, 284, 289, 291
    nucleotide binding sites, 285–287, 291
    recognition by tRNAs^{Met}, 285–286, 314–318
    trypsin modification, 283–291
    X-ray-scattering analysis, 284
  misactivation of ethionine by, 251
Methionyl-tRNA transformylase, E. coli
  recognition of initiator tRNA, 318–320
  recognition of methionyl-tRNA, 318–319
Methylation, 45–48, 59
  m^{7}G in yeast tRNA^{Phe}, role in synthetase recognition, 331–332
O-Methylthreonine, misactivation by valyl-tRNA synthetase, 251
Methyl transferase. See tRNA methyl transferase
Minor bases, role in aminoacyl-tRNA synthetase recognition, 331–333
Misacylation, 248–249, 507–512
  yeast, phenylalanyl-tRNA synthetase, 270
  yeast, valyl-tRNA synthetase, 270
Mischarging mutants, 300, 302–303
Mitochondria
  evolution, 26
  methylases, 36
  N. crassa, tRNAs, 9–14
    coding origin, 23–25
    codon recognition, 24–25
    gene mapping and cloning, 14
    initiator, 8, 13
    methylation of, 37
    number of species, 13
    structure of, 25
  transamidase, 22
  tRNA, hypermodified nucleosides in, 37
  yeast, tRNA, 13, 20–26
Modification of tRNA, 65–69
  tRNA transglycosylase, 65–68
Modified nucleosides, 30–37, 59–73, 522, See also Hypermodified nucleosides
  in anticodon, 35

Modified nucleosides *(continued)*
  detection methods, 59
  effect on tRNA-tRNA base pairing, 168–169
  in eukaryotes, 35
  in fidelity of translation, 62–65
  hydrophilic, 60
  hydrophobic, 60
  location in tRNA, 33–35, 61
  location-function relationship, 60–65
  methyl transferases, 35
  in mitochondrial tRNA, 36–37
  positions, 33–35, 59, 61
  possible functions, 60, 63–65
  in prokaryotes, 35
  queuosine, 30, 32, 60, 65
    tRNA transglycosylase, 65–68
  $2'$-O-ribose-methylated nucleosides, 30–31
  structures of, 547–549
  in tumor tRNA, 69–72
  wyosine, 30, 32, 69
Modified nucleotides, chromatographic mobilities of, 551–552
mo^5U, photoaffinity-labeling probe, 430–432
mRNA
  affinity-labeling analogs, 424–426, 428
  photoaffinity-labeling analogs, 424, 427–431
  photolabeled by yeast phenylalanyl-tRNA^{Phe} Y base, 428–430
Neurospora crassa. See Mitochondria
Neutron-scattering analysis
  of methionyl-tRNA synthetase, 284, 289, 291
  of valyl-tRNA synthetase–tRNA^{Val} interaction, 333, 336
Mutant tRNA sequences, 539–545

Noncognate tRNA, recognition by aminoacyl-tRNA synthetases, 305
Nuclear magnetic resonance (NMR)
  of *E. coli* tRNA^{fMet}, 185–186, 312
  FT NMR
    of *E. coli* tRNA^{fMet}, 195–196
  saturation recovery, 185–189, 192–196
  tRNA dynamics studies, 191–204
  of yeast, tRNA^{Phe}, 195–204
  NOE, 193–197, 203
  paramagnetic ion studies
    E. coli tRNA^{Lys}, 181, 185
    E. coli tRNA$_1^{Val}$, 178, 180–182, 184–185
    yeast, tRNA^{Phe}, 178, 184–188, 197–204
  saturation recovery, 186–189, 192–196
  solvent-exchange measurements, 192–194
  tRNA solution structure, 177–189
  tRNA thermal unfolding, 166–167, 197–204
Nucleotide sequences, 519–545
Nuclear Overhauser effect (NOE), 191–197, 203

Nuclease digestion as probe of aminoacyl-tRNA synthetase in tRNA complexes, 330–331
Nucleoside modification
  chloroplast methyl transferases, 35
  mitochondrial
    methylases, 36
    transamidase, 22
Nucleoside transdeoxyribosidase, 68
Nucleotide analogs of ATP, 238–240
Nucleotide binding site
  of methionyl-tRNA synthetase, 285–287, 291
  of tryptophanyl-tRNA synthetase, 238–240
Nucleotidyl transferase modification of tRNAs, 137, 346–347, 349–350, 474

Oligonucleotide binding to tRNA, 459–469
One-site enzyme
  tryptophanyl-tRNA synthetase, 242–244
  valyl-tRNA synthetase, 244
Organelle tRNA, 19–20

Paramagnetic ion effects, NMR binding of metals by tRNA, 180–182
Peptide-bond formation, aminoacyl-tRNA selection, 473–484
Peptidyl binding site
  affinity labeling of, 424, 429–431, 494–498
  fluorescent tRNA derivatives, probed with, 452–454
  localization on ribosome, 496–497
  mRNA-tRNA interactions, 373, 381–384
  nature and location of, 368–372, 400–403
  protein synthesis, role in, 364–368, 403–407
  of tRNA to ribosome, 364–386
Peptidyl transferase
  affinity probes of, 434
  location on ribosome, 399–402, 434, 438
  substrate specificity, 364, 378
Peptidyl-tRNA, tertiary structure of, 207
Peptidyl-tRNA hydrolase, *E. coli*, 30, 314, 321
Periodate oxidation of tRNA^{Phe} · phenylalanyl-tRNA synthetase, 334
Phenoxyacetylation in isolation of charged tRNAs, 314
Phenylalanyl-tRNA synthetase, yeast, 227
  aminoacylation properties, 300
  ATP-PPi exchange, 334
  interaction with tRNA^{Phe}, 334
  misacylation, 270
  recognition of tRNA^{Phe}, 268–269, 272
  substrate specificity, 348, 354, 356
  UV cross-linking, 333–334

Subject Index

Photoaffinity labeling
  by aminoacyl-tRNA analogs of 23S RNA, 418–419, 421
  of aminoacyl-tRNA synthetases, 229–230
  of aminoacyl-tRNA synthetase · tRNA complexes, 304, 326–329
  by mo^{5}U of *E. coli* tRNAs, of 16S RNA, 430–432
  mRNA photoaffinity analogs, 424, 427–431
  natural tRNA bases as probes
    cmo^{5}U, 415, 431
    mo^{5}U, 415, 430–431
    Y base, 415
  reagents, 416–419, 421–423
  tRNA photoaffinity analogs, 415, 418–423
Photooxidation of tRNA, 313, 315
Platform (30S). *See* Ribosome, small subunit
Polynucleotide phosphorylase in preparation of modified tRNA, 347
Primer tRNA, 29, 156
Proofreading
  by aminoacyl-tRNA synthetases, 351–353, 507
  codon-anticodon recognition, 450, 454
  conservation of initial positions of tRNA activation, 352–353
  deacylation by aminoacyl-tRNA synthetases, 273
  mechanisms, 252–253, 267–277
  3'-OH on 3'-terminal base, role of, 276–277
Protein synthesis
  elongation, 133, 365, 378–379, 403–408
  fidelity of, 145–146, 154–156, 273
  initiation of, 133, 365, 378, 403–405
  termination of, 366
  transformylase, 297
  translocation, photoaffinity-labeling studies, 428–430
  tRNA binding to ribosome, 364–368, 401–408
Protein-tRNA recognition, 311, 314–321
Proton exchange with solvents, 194–204
Proton NMR. *See* Nuclear magnetic resonance
Pyrimidines, cyclobutane dimers of, 431

Queuosine (Q base), 30, 32, 60, 65–68

Rare nucleosides. *See* Modified nucleosides
Regulation, tRNA role in, 30, 65
Release factors (RF-1, RF-2, and RF-3), 366
Ribosome (*E. coli*). *See also* Aminoacyl binding site; Peptidyl binding site
  binding sites, 413–438
  interactions, 363–386
    with aminoacyl-tRNA, affinity labeling, 493–497
  with EF-Tu, 492–495
  equilibrium studies by fluorescence, 451–453
  kinetic studies, 453–454
  with tRNA, 363–386, 393–408, 450–455
  large subunit
    affinity labeling by aminoacyl-tRNA analogs, 415–420
    factor binding sites, 398–399
    L7/L12 proteins, role in GTP hydrolysis, 398
    L7/L12 stalk, 393, 398–399
    peptidyl transferase site, 399–401, 434
    surface protein distribution, 434
  small subunit
    antibody attachment sites, 435–438
    anticodon binding site, 397
    exterior proteins in misreading, role of, 395–396
    initiation-factor binding sites, 397
    mRNA analog affinity labeling of, 424–431
    platform region, 396–397
    surface protein distribution, 395–397, 435
    tRNA binding by exterior proteins, 395–396
  structure-function relationships
    antibody mapping studies, 393–400, 433–439
    large subunit, 398–400
    small subunit, 393–398
    70S unit, 394–395, 400–408, 433
    tRNA binding sites, 363–386, 393–408, 450–455
    affinity labeling, 413–415, 428–431
  tRNA recognition site
    affinity labeling of, 497
    location on ribosome, 405
    protein synthesis, role in, 403–408
RNA ligase in tRNA modification, 347
RNA sequencing by ^{3}H labeling, 46–54
RNA unwinding proteins, 146

Secondary structure
  fidelity in aminoacylation, 154
  solvent effects, 152–154
Second conformational state of tRNA, 145–157
Semiinvariant nucleotides, 26–32
Sequence analysis of tRNA
  in vitro postlabeling 3–14
  ^{3}H derivative methods, 46–54
Sequence duplication of aminoacyl-tRNA synthetases, 225, 255–256, 281–284, 286–291
Solution dynamics of tRNA, 185–188
Solution structure of tRNA, 177–189. *See also* tRNA confirmation

Solvent accessibility of yeast, tRNA^{Phe}, 91–92
Solvent effects
  on tRNA melting, 154
  on tRNA structure, 152–154
Solvent interaction with tRNA, 91–92
Space-filling model, X-ray crystallography, yeast tRNA^{Phe}, 85
Specificity
  acceptor stem, role of, 155
  of aminoacyl-tRNA synthetases, 145–146, 154–155
Spermine
  binding to yeast tRNA^{Phe}, 90–91
  interaction with tRNA, 90–91, 375
  translocation, role in, 103
  tRNA folding, effect on, 103
Steric expulsion principle as editing mechanism, 250
Stringent response, 379
Structure of tRNA. *See also* Three-dimensional structure of tRNA
  modified nucleosides in, 547–549
Structure-function relationships of tRNA, 128–130
s^{4}U, site of affinity probe substitution, 414, 418, 422, 438

T4 RNA ligase, modification of tRNA, 347, 354. *See also* RNA ligase
Temperature, effect on tRNA melting, 154
Temperature-jump relaxation kinetics of, tRNA unfolding, 166–167
Tertiary base interactions
  conservation in all tRNAs, 95–98, 116, 129
  NMR resonances 178–187, 197–201, 203–204
  TψC-D loops, 118, 382–383
  structure-function relationships of, 123, 128–130
  yeast, tRNA^{fMet}, 140
  yeast, tRNA^{Phe}, 86–89, 95–98
Tertiary structure, domain homology of aminoacyl-tRNA synthetases, 230
Thermal unfolding of tRNA, 194–195, 197–204
Thermal vibrational motion of yeast, tRNA^{Phe}, 92–93
Three-dimensional structure of tRNA, 83–98, 115–130
  N-acetyl-phenylalanyl-tRNA^{Phe}, 106
  cationic molecules, effect of, 102–105
    magnesium ions, 103–105
    spermine, 103
  initiator tRNA (tRNA^{fMet})
    E. coli, 107–112
    yeast, 107–112, 133–141
  yeast tRNA^{Gly}, 145-152

yeast tRNA^{Phe}, 83–89, 101–106
  N-acetyl-phenylalanyl-tRNA^{Phe}, 105–106
[^{3}H]Borohydride for tRNA analysis, 43–54
^{3}H Derivative methods, 43–46
^{3}H-labeling procedure, 43–54. *See also* tRNA sequence analysis
3′-Terminal nucleotides, 268–270
  aminoacylation, role in, 338–339
Threonyl-tRNA synthetase · tRNA^{Val} complex, enzymatic destruction of, 275
Transaminoacylation (2′, 3′-isomerization), 483, 484
Transformylase, recognition of tRNA, 297
Translation
  fidelity of. *See also* Aminoacyl-tRNA synthetase; Editing mechanisms
    editing mechanisms, 506, 507, 509–512
    error frequency, 501–512
    modified nucleosides, role of, 60, 63–64
  studied by photoaffinity labeling, 428–430
Translational diffusion coefficient ($D_{20,w}$), 208–218
  effect of magnesium, 209–217
tRNA, general features of, 19, 20–21, 26–37, 95–98
tRNA^{Asn}, mammalian, sequence of, 52
tRNA conformation
  organic solvent, effect of, 306
  in solution, 207–219
    changes with ligand binding, 459–468
    codon-dependent rearrangement, 450–455
    fluorescent tRNA derivation, 445–450
    magnesium effect, 163–168, 446–448
  on ribosome, 376–377, 380–386
tRNA conformational changes
  cations, effect of, 105
  codon binding, induced by, 14, 459–469
  in solution, 163–169
tRNA conformational mobility, 145–146
tRNA conformational states, 146, 207–219
tRNA fluorescent derivatives, 445–455
tRNA^{fMet}
  E. coli, 283, 285–289, 291
    anticodon loop, 111–112
    bisulfite modification of, 311–313, 320, 321
    chemical modification of anticodon, 301
    conformation of, 215–217, 219
    contact with IF-2, 374, 378, 398
    crystal structure of, 107–112
    D-loop structure of, 112
    general features of, 107–112, 283–289, 291
    initiation-complex formation, tetranucleotide codon, 464–466, 468
    melting pathways, 166–167
    modification of, 311–313
    NMR resonances, 185–186, 195–196

Subject Index

recognition by EF-Tu, 320–321
  methionyl-tRNA synthetase, 314–318
  methionyl-tRNA transformylase, 318–320
  peptidyl-tRNA hydrolase, 321
  secondary structure of, 179
  solution structure of, 215–219
  X-ray crystallography, 107–112
eukaryotic
  cytoplasmic, 12
  N. crassa mitochondrial, 8–9, 13, 28
  TψC loop, 12
prokaryotic, 12
structure-function relationship, 133
yeast
  iodination of tRNA^{fMet} crystals, 137–138
  sequence of, 137
  tertiary-base interactions, 140
  three-dimensional structure of, 101–106, 138–141
  X-ray crystallography, 133–141
    crystallization conditions, 134
    difference maps, 135, 137
    heavy atom markers, 134, 135–138
    molecular packing, 140–141
    phase improvement, 133–141
tRNA^{Gln}, mutant T4, 543, 544–545
tRNA$_2^{Glu}$, *E. coli*, conformation of, 215–216
tRNA^{Gly}, *E. coli*, cell-wall peptidoglycan synthesis, 28
tRNA$_{CCC}^{Gly}$, human, sequence of, 52–54
tRNA$_{GCC}^{Gly}$, human, sequence of, 52–54
tRNA^{Gly}, yeast
  acceptor stem, 148–149, 151
  crystal forms, 147–148
  melting behavior of, 153
  sequence of, 147
  tertiary interactions, 151–152, 156
  tertiary structure 147–151
  X-ray crystallography, 145–152
tRNA^{Ile}, yeast, 3'-terminal A in aminoacylation, 269–270
tRNA^{Leu}, mutant T4, 544–545
tRNA$_1^{Leu}$, *E. coli*, conformation of, 215–216, 218
tRNA$_{UAG}^{Leu}$, yeast, sequence of, 51–52
tRNA^{Lys}, *E. coli*
  NMR resonances, 181, 185
  secondary structure of, 179
tRNA melting, 153, 164–167
  close variant, 164–165
  coil form, 165
  conformational equilibrium, 164–165
  conformational phase diagram, 164–166
  denaturable tRNA, 165
  dioxane effect on, 154
  elevated temperature, effect of, 154
  extended forms, 165
  kinetics, 194, 197–202
  magnesium, effect of, 167
  NMR studies, 166, 194–195, 197–203
  salt concentration, effect of, 164–166
  solvent effect, 154
  temperature-jump studies, 166
  yeast tRNA^{Phe} melting curve, 332
tRNA^{Met}
  anticooperative binding of, 285–286, 288–289
  E. coli, chemical modification of, 311–313
tRNA modification, 30–37, 59–73, 547–549
  base analog inhibition, 46
  CTP(ATP):tRNA nucleotidyl transferase, 137, 346–347, 349–350, 474
  methylation, 59
  methyl transferases, 35, 46
  nucleoside transdeoxyribosidease, 68
  periodate oxidation, 334
  polynucleotide phosphorylase, 347
  RNA ligase, 347
  tRNA transglycosylase, 69, 72–73
tRNA–mRNA interaction on ribosomes
  codon–anticodon interaction, 372–373, 381–384
  30S platform, 396–397
tRNA^{Phe}, yeast
  affinity labeling of, 416–423
  aminoacylation of, 348–349, 353, 356
  anticodon-loop conformation, 89, 130
  backbone structure of, 83–91, 115–116, 124–130
  base–backbone interaction in, 89
  base interdigitation, 121–122
  calorimetry data, 167
  chemical modification of, by kethoxal, 214–215, 218
  circular dichroism of, 152
  conformation and electrostatic charge, 208–210
  conformational analysis
    conformational wheels, 123–128
    helices, 86, 123
    loop folding, 123–124, 129
  G + C content of, 96–97
  helix conformations, 123–130
  interaction with synthetase, 331–333
  melting curve of, 332
  NMR resonances of, 178, 184–188, 195–204
  nucleotide numbering system, 518–519
  secondary structure of, 84
  sequence of, 299
  solution conformation of
    fluorescence studies, 446–454
    light scattering, 208–219
  solvent accessibility of, 91–92

tRNA^{Phe}, yeast *(continued)*
  structure-function relationships, 128–130
  tertiary base pairing, 86–89, 95–98, 116, 128–130
  thermal transitions, 197–204
  thermal unfolding, 166–167
  thermal vibrational motion of, 92–93
  X-ray crystallography of, 83–98, 101–112, 133–141, 145–152, 182–183, 185
    of N-acetyl-phenylalanyl-tRNA^{Phe}, 105–106
    space-filling model, 85
tRNA^{Phe}–phenylalanyl-tRNA-synthetase interaction, 326–334
tRNA primer, 29, 156
tRNA recognition
  by aminoacyl-tRNA synthetase, 155, 298–304, 311–322, 353
    acceptor end, 299–300
    anticodon, 267–271, 301
    D loop, 300
    yeast phenylalanyl-tRNA synthetase, 268–269
  by EF-Tu, 297
  by transformylase, 297
tRNA sequence, 521–538. *See also individual tRNAs*
tRNA sequence analysis
  postlabeling method, 3–14
    mobility shifts, 6–7
    modified nucleoside analysis, 5, 8
  ^{3}H derivative methods, 43, 46–84
tRNA sequence conservation, 26–32, 84, 97–98, 101
  at sites of ribosome contact, 376–377
tRNA^{Ser}
  E. coli, crystal structure of, 119
  mutant T4, 543, 544–545
tRNA–16S-RNA interaction, photoaffinity labeling, 430–431
tRNA tertiary structure, 149–151, 163. *See also individual tRNAs*
  aminoacylation, effect of, 207–219
  tertiary interactions, 95–98, 101
tRNA three-dimensional structure. *See individual tRNAs*
tRNA transglycosylase, 69, 72–73
tRNA–tRNA interactions, 168–169
  α_2 dimers, 169
  $\alpha\beta$ complex, 168
  anticodon-anticodon, 168–169
  fluorescence studies, 448–450
  wobble base pairing, 169
tRNA^{Trp}
  anticooperative binding, 242
  beef pancreas, 236–237, 241–243
tRNAs^{Tyr}, mutant suppressor, 540–542

tRNA unfolding. *See* tRNA melting
tRNA$^{Val}_1$, *E. coli*, NMR of, 178, 180–182, 184–185, 195–196
tRNA–water interaction, 91–92
tRNA X-ray crystallography. *See individual tRNAs;* Three-dimensional structure of tRNA
Tryptophan fluorescence, methionyl-tRNA synthetase, ligand binding, 285–286, 288
Tryptophanyl-tRNA synthetase
  B. stearothermophilis, 255–263
  beef pancreas, 235–244
    chemical labeling of active site, 229–230
    half-of-the-sites reactivity, 237, 241–243
    inhibition by nucleotide analogs, 238–241
tRNAs^{Tyr}, mutant suppressor, 540–542
  modification by hydroxylamine, 235–237
  nucleotide binding sites, 238–240
Tryptophanyl-tRNA^{Trp}, *B. stearothermophilis*, 236, 241–243
  amino acid sequence of, 262
  analogs of, 241, 243
  cysteine residues, 260–262
  sequence homologies, 259–263
Tumor cells
  isoaccepting tRNA, 69–72
  primer tRNA, 29
  tRNA base composition, 45–46
  tRNA modification in
    modified nucleosides, 69–72
    tRNA transglycosylase, 69–73
    tRNA methyl transferases, 46
Turnip yellow mosaic virus, 334, 338
Tyrosyl-tRNA synthetase, *B. stearothermophilis*, 225–230, 255–263
  amino acid sequence, 262
  amino group reactivity, 230
  editing mechanism, 252
  sequence homologies, 259–263
  tRNA binding site, 262
  tyrosine cleft, 263

Unique sequences, in yeast, tRNA^{Leu}, 52
Unnatural amino acid activation, 251
Unwinding of tRNA, 156
UV cross-linking of yeast phenylalanyl-tRNA synthetase and tRNA^{Phe}, 333–334

Valyl-tRNA synthetase
  editing mechanism of, 248–251
  interaction with tRNA^{Val}, 326–327, 330–333
  neutron scattering of, 333–336
  one-site enzyme model, 244
  sequence duplications in, 337

Wobble hypothesis
  base pairing, 116, 118–119, 129
  NMR resonances, 195–196
  wobble bases, 62
Wybutine (WY). *See* Y base in yeast tRNA^{Phe}
Wyosine, 30, 32, 69

X base. *See* acp^{3}U
X-ray crystallography. *See also individual tRNA species*
  initiator *E. coli*, tRNA, 107–112
  initiator yeast, tRNA, 133–141
  trypsin-modified methionyl-tRNA synthetase, 283–284, 291
  yeast, tRNA^{Gly}, 145–152
  yeast, tRNA^{Phe}, 83–98, 101–106
X-ray-scattering analysis of *E. coli* methionyl-tRNA synthetase, 284

Y base in yeast tRNA^{Phe}. *See also* Wyosine
  fluorescent codon-anticodon probe, 373–374, 459
  photoaffinity-labeling probe, 415, 428–430, 432
Yeast. *See individual tRNAs; N*-acetyl-phenylalanyl-tRNA^{Phe}; isoleucyl-tRNA synthetase; mitochondrial tRNA; phenylalanyl-tRNA synthetase; valyl-tRNA synthetase

THE LIBRARY
ST. MARY'S COLLEGE OF MARYLAND
ST. MARY'S CITY, MARYLAND 20686